H.W. Borns
1985

Memoir 10

SEDIMENTOLOGY OF GRAVELS AND CONGLOMERATES

Edited by

Emlyn H. Koster
Alberta Geological Survey
Alberta Research Council
4445 Calgary Trail South
Edmonton, Alberta T6H 5R7, Canada

and

Ron J. Steel
Norsk Hydro Research Centre
P.O. Box 4313
N-5013 Bergen, Norway

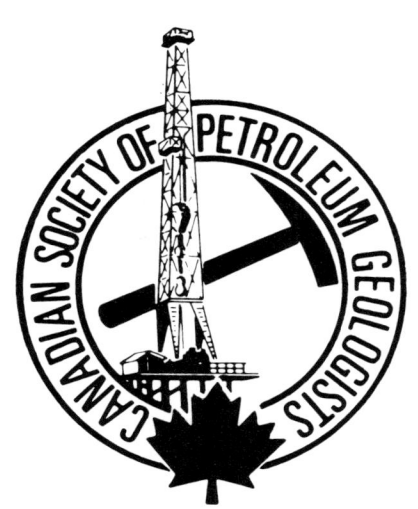

CANADIAN SOCIETY OF PETROLEUM GEOLOGISTS

Calgary, Alberta, Canada   November 1984

© Canadian Society of Petroleum Geologists
ISBN 0-920230-27-X
Calgary, Alberta, Canada   November, 1984

Publications of the Society may be obtained from:—
Canadian Society of Petroleum Geologists,
#505, 206 - 7th Avenue S.W.,
Calgary, Alberta T2P 0W7, Canada
Phone (403) 264-5610

McAra Printing Limited
Calgary, Alberta, Canada

FOREWORD

The International Association of Sedimentologists held its first Congress in North America during August 22-27, 1982 at McMaster University in Hamilton, Ontario. This 11th Congress was jointly sponsored by the Canadian Society of Petroleum Geologists and the Geological Association of Canada. Approximately 1,200 delegates attended five days of technical sessions. Authors from nearly 50 countries presented papers in 13 general themes and over 40 symposia on more specialized topics.

Amongst all these sessions, the 1983 Executive of the C.S.P.G. chose to publish a collection of papers on conglomerates that arose from a symposium organized by E.H. Koster and R.J. Steel. Our purpose in this decision was to initiate a series on clastic sedimentology, following our highly successful Memoir 5 on Fluvial Sedimentology, and to precede an anticipated publication on Shelf Sands and Sandstones resulting from a research symposium recently held in Calgary.

Our editors of this Memoir, E.H. Koster and R.J. Steel, are to be highly commended for bringing together in this publication a large international collection of papers on the basinal controls, sedimentary processes, facies sequences and economic aspects of gravel-sized sediment ranging in age from Proterozoic to Recent. With a Preface by R.G. Walker, this volume represents the first comprehensive collection of research to deal exclusively with coarse clastic sedimentation in environments stretching from foreland molasse basins to the continental rise.

Ian A. McIlreath
1983 President, Canadian Society
of Petroleum Geologists

ACKNOWLEDGEMENTS

This Memoir stems from a symposium entitled 'Rudites Formed By Unidirectional Flow' during the 11th Congress of the International Association of Sedimentologists held at McMaster University, Hamilton, Ontario, Canada, during August 1982. The co-editors, who were also the co-chairmen for this symposium, wish to gratefully acknowledge the Congress Organising Committee (Chairman, G.V. Middleton) for their endorsement of the original proposal to gather researchers working on modern and ancient gravel-sized sediment.

Following the Congress, the enthusiastic approval by the Canadian Society of Petroleum Geologists to publish an enlarged collection of papers as a Memoir, and the continuing guidance and advice of their 1983/84 Executive Committees, in particular Publications Editor James Dixon, are recognised here with our appreciation. The Alberta Geological Survey (EHK) as well as Norsk Hydro and Stanford University (RJS) are thanked for their provision of clerical support and expenses for communication with contributing authors.

In the editorial process, each paper was reviewed externally: the co-editors wish to take this opportunity to thank the following international group of sedimentologists who assisted them by consenting to provide thorough, constructive reviews:

J.R.L. Allen	A.B. Hayward	A.D. Miall
G.M. Ashley	F.J. Hein	L.T. Middleton
J. Bourgeois	H.E. Hendry	W. Nemec
A.C. Brayshaw	C.S. Kingsley	A. Nøttvedt
L.K. Burns	K.L. Kleinspehn	G. Parker
D.J. Cant	R.C. Kochel	G. Postma
M. Church	P.D. Komar	I. Reid
H.E. Clifton	M.J. Kraus	B.R. Rust
J.E. Costa	R. Kreisa	J. Shaw
T. Elliott	D.K. Larue	N.D. Smith
F.G. Ethridge	M.R. Leeder	F. Surlyk
M.G. Foley	I.P. Martini	R.J. Wasson
M.R. Gibling	P.J. McCabe	N.A. Wells
A.M. Harvey	J.R. McLean	

The particular assistance to the co-editors of Linda Jones and Charlotte Mougeot, Alberta Geological Survey, and Karen Kleinspehn, Wojtek Nemec and George Postma during their terms at the University of Bergen, is especially acknowledged. The excellent cooperation of McAra Printing Limited of Calgary, in particular Diane Baldridge, is recognized with our appreciation.

Translation of abstracts by Dr. Robert Ledoux, Départment de Géologie, Université Laval, Québec was financed by the Alberta Geological Survey, Alberta Research Council. Other funds to help offset the preparation and publication costs of this memoir have been gratefully received from the Geological Survey of Canada.

Emlyn H. Koster
Ron J. Steel

INTRODUCTION

The theme of this Memoir originated with the 11th Congress of the International Association of Sedimentologists held at McMaster University, Hamilton, Canada during August 1982. It was in March 1979 that a University of Saskatchewan colleague, Hugh E. Hendry and myself proposed to the Congress Organizing Committee that a comprehensive synthesis of current knowledge on conglomeratic sediments was timely. This proposal materialised into a 1½-day symposium, chaired by the present co-editors, of 29 oral papers with published abstracts — 14 of which became material for this Memoir.

Immediately following the Congress and with a view to publication, additional papers aimed at improving the coverage of processes and environments as well as the spread of field areas and geologic age were solicited. The C.S.P.G. Memoir series was selected for publication because of the Canadian site for the 11th Congress, the prior international success of Memoir 5 on Fluvial Sedimentology edited by A.D. Miall, and the now popular large-format page size. After this decision in March 1983, papers were refereed, revised and edited: the contract with McAra Printing Limited commenced in April 1984.

The result is 29 original papers by authors from 13 countries on the sedimentology of modern and ancient gravelly deposits (their basinal controls, depositional processes, facies sequences and economic aspects) tackled using outcrop and sub-surface data ranging in age from Proterozoic to Recent from five continents, as well as experimental work and hydrodynamic principles. The authors are an international group of researchers from resource companies, government surveys and universities — the papers representing both recent thesis work and grant- or company-supported projects of established sedimentologists.

The Memoir is organized into six sections, following prefatorial comments by Roger G. Walker who has published extensively on gravelly deposits in fluvial and marine settings. In 'Background Considerations', two papers deal with the important descriptive parameters and recognition of fluvial, coastal and deep-sea occurrences of conglomerates. The six papers that follow in 'Fluvial Processes' are mostly investigations of the hydraulic controls on gravel movement and deposition at scales ranging from bed microtopography to bars to deep-flood features.

The remaining four sections, which concentrate on the recognition and interpretation of facies, are arranged according to environment beginning with Quaternary landforms within orogenic belts and Cenozoic molasse basins, and ending with ancient submarine accumulations along active continental margins. In 'Modern and Ancient Alluvial Systems', eight papers are ordered with respect to the relative proximal to distal location of environment. Using examples from North America, Europe and Antarctica, they investigate the basin controls, geomorphic processes, facies assemblage and reservoir properties of fan, braidplain and braided river deposits. In 'Ancient Fan-Delta Systems', six papers beginning with a general review and assessment of reservoir potential are then arranged according to geologic age in areas stretching from Spitsbergen to the South African goldfields. They deal with the tectonic setting, sedimentary processes and facies sequences in the deposits of coastal fans. This is followed by two papers under 'Wave/Tide-Dominated Systems' which examine structure and process in the reworked gravel deposits of littoral and shelf zones. In 'Ancient Submarine Slope-Fan Systems', five papers are ordered in terms of relative depositional site and extend from the Californian Paleocene to the Jurassic under the British North Sea. These investigate the basin controls, depositional mechanics and facies modelling of resedimented conglomerates.

The co-editors, who have PhD degrees dealing with coarse clastic deposits under the supervision of Brian J. Bluck (R.J.S.) and Brian R. Rust (E.H.K.), have found the assembling of the contents of this Memoir to be a rewarding and educational experience. Our overall aim is to have produced a broad, international compilation of new research findings to help meet the modern reference needs of the explorationist, academic and new student alike. As well, it is hoped that the Memoir will stimulate further enhancement in the global research effort on a category of clastic sediments which holds valuable clues in deciphering basin development and contains significant quantities of various energy resources.

Emlyn H. Koster

PREFACE

I have never really been sure why books have prefaces. I am even less sure about who reads them. Part of my audience consists of those of you who perhaps come to the study of conglomerates for the first time, and wonder what the background of the subject is. Perhaps you share my feelings of fifteen years ago, faced with thick successions of conglomerates — what on earth do you do with these rocks; what is useful to measure in order to distinguish facies, to interpret environments, and to deduce transport processes? Others of you may be well versed in conglomerate research, and are reading this because it is much less of an intellectual challenge than the rest of the book!

At Emlyn Koster's request, I will try to give my reactions to the research embodied in this volume, and will try to assess what has happened in the field of conglomerates since my paper (Walker, 1975), which first attempted to set up some models for deep water conglomerates.

My first impression is that ways of studying conglomerates in the field have changed little over the last ten years. I rather hoped that someone would have discovered or defined a new structure or texture that would give conglomerate research a boost, comparable to the surge of interest in shallow marine sands following the definition of hummocky cross stratification. But I do not believe this to be the case. What is new is the amount of information being extracted from tried and true techniques, for example plots of maximum pebble size (MPS) versus bed thickness (BTh). These have been used in studies of conglomerates from several environments, with suggestions that different flow processes might be deduced from different MPS/BTh relationships. It will become increasingly important to relate these relationships to textural patterns, particularly the clast fabrics and the overall sorting of the conglomerates, paying particular attention to the range of grain sizes present (bimodal, polymodal, or a continuous distribution from the coarsest to finest fractions). In this way, interpretations of processes can be built on as many different lines of evidence as possible.

It should perhaps be pointed out here that the study of conglomerates and gravels is severely hampered by the lack of experimental work. Imagine what our understanding of sandstones would be like without a depth-velocity-grain size phase diagram for sedimentary structures, or without any experimental work on the initiation and rate of bed load transport. With conglomerates, it is bad enough not to have the experimental work, but the problem is compounded by the immense range of grain sizes present in many mud-sand-gravel flows, and by the possibility that the framework of the bed was deposited by one event, and the pore-filling matrix by another. This latter possibility has not been sufficiently studied, but is clearly of great importance in the study of the distribution and reservoir characteristics of many conglomerates in Alberta (Fahler, Cardium and Viking to name a few). There are two ways around the difficulty of doing experimental work — one is to use rivers as flumes, as has been done by several of the authors in this volume. The other is to isolate particular aspects of conglomerates, and design relatively small-scale experiments to attack specific problems. An example is John Southard's study of conglomerate fabrics deposited from single surges of gravel down a pipe — this work is currently in progress.

I cannot help comparing some aspects of conglomerate research with parallel aspects of turbidite research in the late 1950's, before the Bouma sequence. Many authors are faced with a rather wide spectrum of deposits that range from clean, clast-supported gravels, through various mud-matrix clast-supported gravels, into purely matrix-supported types. The wide range of facies leads to the suggestion of a wide range of transportational and depositional processes, just as the range of sandstone types in the 1950's led to "fluxoturbidites" and vague processes "transitional between watery slides and slumps". It might be rewarding (but difficult) to do an analysis of matrix-supported conglomerates along the lines of my 1975 analysis of clast-supported types, just to see if there are distinct recurrent types, or whether each deposit is different. The documentation of a few general types would lead inevitably to fewer postulated processes. This volume would be an ideal place to start such an analysis, because many new data are presented, and the papers that discuss matrix-supported conglomerates afford an unparalleled entry into the literature through their bibliographies.

Readers familiar with work on conglomerates over the last ten years will notice that the papers in the volume do not completely mirror the work on conglomerates at large. The emphasis on fluvial gravels and conglomerates is not surprising; they are the most easily studied, and ancient waterlain fluvial conglomerates are among the easier ones to interpret. Studies of modern rivers also substitute for experimental work to a certain extent. The emphasis on fan-deltas is exciting, economically important, and intellectually challenging. Careful reading of these contributions leaves the impression that fan-deltas commonly grade rather imperceptibly into submarine fans. Similar processes are hypothesized in both environments, making their separation often extremely difficult. This has led to interpretive disagreements for some North Sea oil and gas bearing conglomerates. Solutions to the problems are clearly important; it helps in reservoir prediction and development to know whether associated sandbodies might be of deep water turbidite character, or whether they may have geometries associated with shoreline or fluvial environments. More thought is clearly needed on the processes envisaged to occur on the fan-deltas and associated deeper marine submarine fans. Some authors postulate similar processes on both, but this raises the question of where and how the large gravel mass movements originate, and whether subaerial flows can move across the fan-deltas and continue to flow essentially uninterrupted into deep water. One triggering mechanism for the large flows would be tectonic, but the relationship of tectonics and sedimentation is scarcely considered in this volume.

Readers perusing the table of contents might get the impression that conglomerates are only found in subaerial and deep marine settings — and here I use the term subaerial loosely to include alluvial fans, braided rivers, fan-deltas, and beaches. However, some of the most fascinating occur in the Alberta basin, in shallow marine units such as the Cretaceous Viking and Cardium Formations. In one interpretation, Cardium conglomerates may have been deposited tens to a couple of hundred kilometers offshore (Walker, 1983), posing severe problems concerning transport processes and flow initiation. Conglomerates accumulate in thicknesses up to about 20 m in the Carrot Creek area, and some wells produce about 500 barrels of oil per day from these conglomerates. In an alternative interpretation, various workers have discussed with me the possibility that the conglomerates are transgressive lags. I think this is unlikely, especially when thicknesses up to 20 m are involved, and where textural changes clearly indicate that the conglomerate has been built up from a series of distinct depositional events. I use this particular controversy only to highlight the fact that there is no agreement about how to distinguish a transgressive lag conglomerate from one originally emplaced tens of kilometers offshore. Again, the problem is not trivial. These conglomerates are currently attractive exploration targets — not only do we need to know their depositional processes and settings for exploration purposes; we also need to know whether to pursue a transgressive lag interpretation, with major implications concerning basin-wide changes of sea level.

I have mentioned omissions only to highlight some of the aspects of the Sedimentology of Gravels and Conglomerates that I think will be important in the coming years. It is easy to call for more experimental work, but much more difficult to devise ways of doing it. It is also easy to call for more observations of modern fan-deltas, but it will be very difficult to monitor the fluid mechanics of debris flows whilst they are flowing and transporting boulders. I feel, therefore, that the sharp-eyed geologist still has a lot to contribute to the understanding of gravels and conglomerates, even if the results are ambiguous. This brings me back to the whole idea of this preface — what has happened in conglomerate studies since the four models I proposed in 1975. Well, Finn Surlyk presents convincing evidence that from the fault scarp to the basin axis, there is no downslope change in conglomerate facies types, or in the relative proportion. "These observations are incompatible with the predicted downslope changes in published conglomerate models [mine, amongst others!] . . . the use of the published models in basin analysis should be discontinued". However, in a subsequent paper, Szczepan Porębski carefully documents proximal to distal changes which involve a decrease in thickness and proportion of ungraded, inversely graded, and inverse-to-normally graded facies, and an increase in thickness and abundance of normally graded and graded-stratified facies. Porębski notes that "the facies sequence suggested here . . . closely resembles the downcurrent grading pattern in deep-sea fan resedimented conglomerates, predicted by Walker". The reader may therefore wonder whether *any* progress has been made in the last ten years with respect to facies models and lateral facies sequences in deep marine resedimented conglomerates!

A careful reading of the case histories and their interpretations presented in this volume will bring many other ideas to mind. It will serve as a mine of information and ideas, and will certainly help to focus future research on many outstanding problems.

References

Walker, R.G. 1975. Generalized facies models for resedimented conglomerates of turbidite association. Geological Society of America Bulletin, v. 86, p. 737-748.

Walker, R.G. 1983. Cardium Formation 3. Sedimentology and stratigraphy in the Garrington-Caroline area, Alberta. Bulletin of Canadian Petroleum Geology, v. 31, p. 213-230.

Roger G. Walker
Department of Geology
McMaster University
Hamilton, Ontario L8S 4M1

CONTENTS

FOREWORD .... *I.A. McIlreath*   iii

ACKNOWLEDGEMENTS .... *E.H. Koster and R.J. Steel*   v

INTRODUCTION .... *E.H. Koster*   vii

PREFACE .... *R.G. Walker*   ix

BACKGROUND CONSIDERATIONS

ALLUVIAL AND COASTAL CONGLOMERATES: THEIR SIGNIFICANT FEATURES AND SOME COMMENTS ON GRAVELLY MASS-FLOW DEPOSITS .... *W. Nemec and R.J. Steel*   1

DEEP-SEA AND FLUVIAL BRAIDED CHANNEL CONGLOMERATES: A COMPARISON OF TWO CASE STUDIES .... *F.J. Hein*   33

FLUVIAL PROCESSES

CHUTES AND LOBES: NEWLY IDENTIFIED ELEMENTS OF BRAIDING IN SHALLOW GRAVELLY STREAMS .... *J.B. Southard, N.D. Smith and R.A. Kuhnle*   51

PARTICLE INTERACTION AND ITS EFFECT ON THE THRESHOLDS OF INITIAL AND FINAL BEDLOAD MOTION IN COARSE ALLUVIAL CHANNELS .... *I. Reid and L.E. Frostick*   61

RELATIONSHIP BETWEEN FLOWS AND SEDIMENT SIZE IN SOME GRAVEL STREAMS OF THE ARID NEGEV, ISRAEL .... *Z.B. Begin and M. Inbar*   69

CHARACTERISTICS AND ORIGIN OF CLUSTER BEDFORMS IN COARSE-GRAINED ALLUVIAL CHANNELS .... *A.C. Brayshaw*   77

FLOOD SEDIMENTATION IN BEDROCK FLUVIAL SYSTEMS .... *V.R. Baker*   87

PALEOHYDROLOGIC TECHNIQUES WITH ENVIRONMENTAL APPLICATIONS FOR SITING HAZARDOUS WASTE FACILITIES .... *M.G. Foley, J.M. Doesburg and D.A. Zimmerman*   99

MODERN AND ANCIENT ALLUVIAL SYSTEMS

GEOMORPHOLOGY AND SEDIMENTOLOGY OF HUMID-TEMPERATE ALLUVIAL FANS, CENTRAL VIRGINIA .... *R.C. Kochel and R.A. Johnson*   109

DEBRIS FLOWS AND FLUVIAL DEPOSITS IN SPANISH QUATERNARY ALLUVIAL FANS: IMPLICATIONS FOR FAN MORPHOLOGY .... *A.M. Harvey*   123

SHEET DEBRIS FLOW AND SHEETFLOOD CONGLOMERATES IN CRETACEOUS COOL-MARITIME ALLUVIAL FANS, SOUTH ORKNEY ISLANDS, ANTARCTICA .... *N.A. Wells*   133

A UNIQUE MASS FLOW MARKER BED IN A MIOCENE STREAMFLOW MOLASSE SEQUENCE, SWITZERLAND .... *H.M. Bürgisser*   147

SEDIMENTOLOGY OF A PRECAMBRIAN QUARTZ-PEBBLE CONGLOMERATE, SOUTHWEST COLORADO .... *F.G. Ethridge, N. Tyler and L.K. Burns*   165

SEDIMENTOLOGY AND HYDROCARBON DISTRIBUTION OF THE LOWER CRETACEOUS CADOMIN FORMATION, NORTHWEST ALBERTA .... *C.J. Varley*   175

SEDIMENTOLOGY AND DEPOSITIONAL SETTING OF THE UPPER PROTEROZOIC SCANLAN CONGLOMERATE, CENTRAL ARIZONA .... *L.T. Middleton and A.P. Trujillo*   189

SEDIMENTOLOGY AND TECTONIC SETTING OF EARLY TERTIARY QUARTZITE CONGLOMERATES, NORTHWEST WYOMING .... *M.J. Kraus*   203

ANCIENT FAN-DELTA SYSTEMS

TECTONIC SETTING, RECOGNITION AND HYDROCARBON RESERVOIR POTENTIAL OF FAN-DELTA DEPOSITS .... *F.G. Ethridge and W.A. Wescott* 217

MASS-FLOW CONGLOMERATES IN A SUBMARINE CANYON: ABRIOJA FAN-DELTA, PLIOCENE, SOUTHEAST SPAIN .... *G. Postma* 237

RESEDIMENTED CONGLOMERATES OF A MIOCENE FAN-DELTA COMPLEX, SOUTHERN ALPS, ITALY .... *F. Massari* 259

CONGLOMERATIC FAN-DELTA SEQUENCES, LATE CARBONIFEROUS — EARLY PERMIAN, WESTERN SPITSBERGEN .... *K.L. Kleinspehn, R.J. Steel, E. Johannessen and A. Netland* 279

DOMBA CONGLOMERATE, DEVONIAN, NORWAY: PROCESS AND LATERAL VARIABILITY IN A MASS FLOW-DOMINATED LACUSTRINE FAN-DELTA .... *W. Nemec, R.J. Steel, S.J. Porębski and A. Spinnangr* 295

DAGBREEK FAN-DELTA: AN ALLUVIAL PLACER TO PRODELTA SEQUENCE IN THE PROTEROZOIC WELKOM GOLDFIELD, WITWATERSRAND, SOUTH AFRICA .... *C.S. Kingsley* 321

WAVE/TIDE-DOMINATED SYSTEMS

WAVE-WORKED CONGLOMERATES — DEPOSITIONAL PROCESSES AND CRITERIA FOR RECOGNITION .... *J. Bourgeois and E.L. Leithold* 331

DEPOSITIONAL FEATURES OF LATE MIOCENE, MARINE CROSS-BEDDED CONGLOMERATES, CALIFORNIA .... *R.L. Phillips* 345

ANCIENT SUBMARINE SLOPE-FAN SYSTEMS

FAN-DELTA TO SUBMARINE FAN CONGLOMERATES OF THE VOLGIAN-VALANGINIAN WOLLASTON FORLAND GROUP, EAST GREENLAND .... *F. Surlyk* 359

DEPOSITIONAL PROCESSES AND FLUID MECHANICS OF UPPER JURASSIC CONGLOMERATE ACCUMULATIONS, BRITISH NORTH SEA .... *L.G. Kessler II and K. Moorhouse* 383

CLAST SIZE AND BED THICKNESS TRENDS IN RESEDIMENTED CONGLOMERATES: EXAMPLE FROM A DEVONIAN FAN-DELTA SUCCESSION, SOUTHWEST POLAND .... *S.J. Porębski* 399

RESEDIMENTED CONGLOMERATES IN A MIOCENE COLLISION SUTURE, HOKKAIDO, JAPAN .... *H. Okada and S.K. Tandon* 413

SEDIMENTATION UNITS IN STRATIFIED RESEDIMENTED CONGLOMERATE, PALEOCENE SUBMARINE CANYON FILL, POINT LOBOS, CALIFORNIA .... *H.E. Clifton* 429

BACKGROUND CONSIDERATIONS

… ALLUVIAL AND COASTAL CONGLOMERATES: THEIR SIGNIFICANT FEATURES AND SOME COMMENTS ON GRAVELLY MASS-FLOW DEPOSITS

W. Nemec[1] and R. J. Steel[2]

Abstract

Conglomerates originating in coastal environments represent mainly beachface, shoreface, fan-deltaic or deltaic mouth bar, and Gilbert-type delta sequences. They show structures, textures and other features created mainly by the varied influence of waves and fluvial output in the shallow marine setting. Transitional, alluvial/marine systems show a broad range of facies characteristics and sequences, and these are discussed in detail.

Conglomerates originating in alluvial environments comprise mainly braided stream and mass flow sequences. The former include regular braided river and fan (distributary) channel deposits, and show textures and structures which vary greatly with source and climatic setting. Braided stream sequences commonly show an upward fining motif, due to falling flood stage or to gradual abandonment of alluvial tracts. Mass flow conglomerates originate from a variety of debris flows in subaerial settings, but fluidal gravelly flows (like many 'sheetfloods' or 'streamfloods') may also be important, and they often become prominent subaqueously (high-density gravelly turbidites). In both instances, the deposits show remarkably varied texture, structure, and fabric. Subaerial flows are often considerably transformed when passing into water. A review of diagnostic features and facies sequences is presented.

When interpreting the emplacement mechanics of mass flow conglomerates, particular effort must be made to extract maximum information from the individual bed characteristics. We illustrate with examples that even such basic data as bed thickness and maximum clast size may serve as a valuable source for some genetic inferences.

Résumé

Les conglomérats formés dans des milieux côtiers représentent principalement des séquences de zone infralittorale de grande énergie, de zone infratidale, de cône de déjection ou de flèche barrante deltaïque dans une embouchure, et de delta du type de Gilbert. Ils exhibent des structures, des textures et autres particularités créées principalement par les effets variés des vagues et des régimes fluviatiles dans un contexte marin de faible profondeur.

Les conglomérats développés dans des milieux alluvionnaires comprennent principalement des séquences de cours d'eau anastomosés et d'écoulements en masse. Le premier inclut des dépôts de rivière anastomosée régulière et de chenaux de cône de déjection (de distribution); on y observe des textures et des structures qui varient grandement selon la région source et le contexte climatique. Les séquences des cours d'eau anastomosés présentent fréquemment un motif de granulométrie décroissante vers le sommet des couches qui résulte d'un retranchement du stade de crue ou de l'abandon graduel des zones alluvionnaires. Les conglomérats d'écoulement en masse dérivent d'un grand nombre de coulées de débris dans des contextes subaériens, mais les coulées graveleuses fluidales commes beaucoup d'innondations en nappe ou de cours d'eau peuvent aussi avoir contribué largement, et souvent elles dominent en milieu sous-aquatique (turbidites graveleuses de densité élevée). Dans les deux cas, les dépôts exhibent remarquablement des diversités dans la texture, la structure et la fabrique. Les coulées subaériennes sont souvent considérablement transformées par le passage en milieu aqueux.

Lors de l'interprétation du mécanisme de mise en place des conglomérats d'écoulement en masse, il faut fournir un effort particulier pour acquérir le maximum de renseignements dévoilés par les caractéristiques de chaque lit. Nous illustrons au moyen d'exemples qui montrent que même des données de base comme l'épaisseur des lits et la grosseur maximale des fragments peuvent servir d'enseignement précieux pour les déductions génétiques.

Introduction

Conglomerates have long attracted attention as possible indicators of tectonic activity and from the viewpoint of provenance studies and local palaeotransport directions. Only more recently has specific effort gone into the study of gravelly depositional environments and the interpretation of conglomerate facies. Although quite a few modern gravelly environments have been well-investigated, especially gravelly alluvium, a great number of ancient depositional systems still have no available or well-studied modern analogue. Synthesis of conglomerate facies characteristics is therefore approached mainly or entirely from the ancient record, and one may find it surprising, indeed, that models have been attempted only for certain ancient submarine fans (e.g., Walker, 1975; Krause and Oldershaw, 1979; Porębski, 1981; Hein, 1982). In terms of facies analysis, perhaps the main recent efforts have gone to distinguish specific sedimentary sequences which may be typical for particular processes or depositional settings, and to establish criteria which may serve for the recognition of such settings in other rock successions.

A research direction that has been prominent recently is the analysis of fluvially derived gravels in ancient and modern shoreline/nearshore environments, notably on fan-deltas. Considerable attention is now being paid to lacustrine fan-delta settings where the deposition of fan-derived gravels is influenced by a passive body of water, and also involves mixing with lake muds, and to marine fan-delta settings where fan gravels are variously reworked, and

[1]Geological Institute (A), University of Bergen, Allégt. 41, 5014 Bergen, Norway.
[2]Norsk Hydro Research Centre, P.O. Box 4313, 5013 Bergen, Norway.
Copyright © 1984, Canadian Society of Petroleum Geologists

mix with marine sediments under the action of wave and tidal processes. In the first part of this contribution we briefly review those aspects of shallow marine and alluvial conglomerates which we have found to be important to their interpretation, and we put some emphasis on transitional subaqueous environments.

During the last decade considerable progress has also been made on understanding the sedimentary features of conglomerates deposited from sediment gravity flows, particularly through the 'benchmark' work of Middleton and Hampton (1973, 1976) and Lowe (1979, 1982). This progress pertains mainly to the rheology of flow and the mechanics of gravel transport, but includes also some suggested depositional models for conglomerate beds. In addition, the classification of sediment gravity flows and their deposits has been considerably improved.

However, there are still many difficulties in the field application of this theoretically-based work. In the second part of this contribution we attempt to review some important aspects of the classification of sediment gravity flows, emphasising application to subaerial deposits and commenting on other points of potential practical importance. We suggest, in particular, that *MPS/BTh* plots, when based on carefully collected data, can serve to make some important inferences about the depositing flows.

Conglomerates Originating in Coastal Environments

The existing literature provides a fairly limited documentation of conglomerates that are thought to have originated in coastal or nearshore environments. This is due partly to a real scarcity of gravels in these environments, but also partly to research inadequacy. In addition, there is a complicating tendency for gravel to be trapped in or near river mouths, and it is difficult later to distinguish such marine deposits from the closely associated alluvial conglomerates. However, we feel that greater efforts should be made in studying this transitional zone, not least because of the importance of distinguishing between the products of fluvial output and wave processes. In cases where the depositional systems are known to be dominated by the input of subaerial, gravelly mass flows, it is important to evaluate the likelihood or nature of flow transformations between the subaerial and subaqueous realms. The growing use of the fan-delta concept, for example, is clearly a step towards a more critical examination and understanding of this zone, because it focuses attention on the subaerial-subaqueous transition.

Lateral redistribution of gravel beyond river mouths along beaches and shorefaces is also common, but gravels here are usually thin and dominated by associated sand. Modern beaches with impressive amounts of gravel are usually associated with gravelly river mouths or derive their gravel from adjacent outcrops of older rudites.

Gravel can also be transported beyond the nearshore zone, but this requires rather unusual fluvial output or storm conditions, and happens particularly in regions with a narrow or non-existent shelf, where shoreline transport has easy access to slopes beyond.

WAVE-DOMINATED SETTING

In a wave-dominated setting, the areas of significant gravel accumulation are the beachface and shoreface zones. Within the latter, thick gravel sequences are rare, but significant amounts of thin gravelly beds or pebbly sandstones, associated with thicker sandstone sequences, are common.

Beachface conglomerates have only rarely been recognised in ancient successions, perhaps partly because of poor preservation and partly because they tend to be thin and can be overlooked easily. However, they can be recognised by the following features:

(1) Depositional sequences are well-stratified, usually because of sandy interbeds, and the individual conglomerate beds are laterally more persistent (Clifton, 1973) and texturally uniform than in equally coarse-grained alluvial sequences (Figs. 1A, B, C). Stratification on the beachface generally dips gently seawards (Fig. 1A) but landward dips may occur in backshore sequences (Maejima, 1982).

(2) The vertical variation in shape-sorting between adjacent beds is often marked, for example by interbedding of the assemblages dominated by disc/blade and sphere/rod shapes respectively (Fig. 2A). This results from the prominent shape-sorting zonation of gravel beaches, as documented by Bluck (1967a) and Dobkins and Folk (1970).

(3) Seaward dipping imbrication is often prominent.

(4) The framework component is commonly well-sorted in individual conglomerate beds (Fig. 2A), though successive beds may vary greatly in coarseness. Sand-infill or small pebble-infill can commonly cause bimodal or polymodal textures (Figs. 2B and 2A, uppermost bed). This textural inversion appears to occur particularly in beds where the primary framework was dominated by spherical or rod-shaped clasts (see also Bluck, 1967a, Fig. 27). The process of shape-filtering on gravel beaches (Bluck, 1967a) may also cause pebble entrapment on sand beds, producing pebbly sandstones (*e.g.*, Fig. 1B, lowermost pebbly layers).

(5) Normal grading in individual conglomerate beds is a less common product in this setting than in the braided stream setting.

(6) Progradational beach sequences tend to coarsen upwards through the beachface, but may become finer grained in the backshore beds (Maejima, 1982). Figure 3 shows two local models for progradational beach sequences, emphasising: vertical change in shape-sorting (A), and vertical change in grain size and stratification dip (B).

Shoreface conglomerates are usually only minor components in sandy shoreface sequences and may result from either fluvial output (mouth bar) or, as more likely in the wave-dominated setting, may be the product of reworking of this output by waves and marine currents. In their discussion of shoreface conglomerates, Bourgeois and Leithold (this volume) emphasise the following characteristics of the lower and upper shoreface:
(1) Lower shoreface conglomerates reflect mainly onshore transport and consist of lenticular or sheet-like units which are either lag deposits (in scours) or are dominated by low-angle cross-strata. The latter, produced by the shoreward movement of low bars or megaripples, commonly become sandy in the upper half of individual sets, *i.e.*, only the bottom sets and lower foresets are conglomeratic. The former, the lag deposits, are sometimes moulded into megaripple forms, the product of oscillatory current reworking. Conglomerates of the lower shoreface tend to associate with hummocky cross-strata.

Fig. 1. Stratification in beachface gravels and conglomerates. Strata tend to be well-marked, thinly developed, laterally persistent and texturally uniform. Examples from: **(A)** Recent beachface from Van Keulenfjorden, Spitsbergen (note the low seaward dip of strata); **(B)** Pleistocene beachface, Cape Blanco, southwestern Oregon; and **(C)** Late Carboniferous beachface, Hornsund, Spitsbergen.

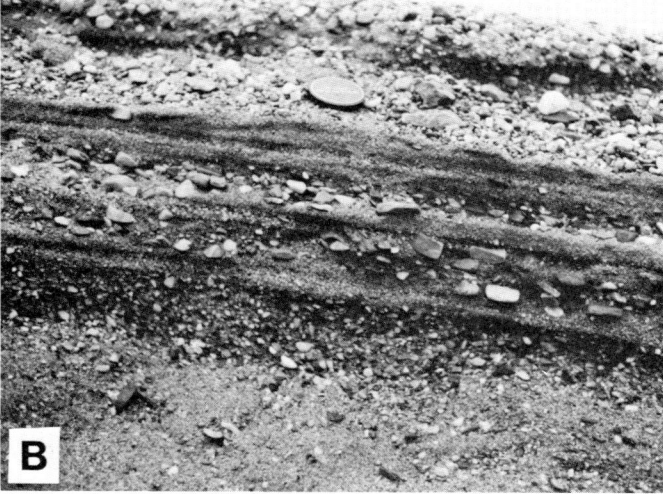

Fig. 2. Texture in beachface gravels. Example **A**: Sub-Recent beachface, southern England. The spherical clasts (top of sequence) are sorted from blade/disc clasts. The former assemblage is sand-rich (bimodal), the latter is imbricate. Example **B**: Laterally persistent sandy gravel (polymodal), Van Keulenfjorden, Spitsbergen.

(2) Upper shoreface conglomerate sequences may reflect onshore, offshore or longshore transport and consist of (a) clast-supported conglomerate sheets, often normally graded, which are thought to be the product of variably reworked river mouth gravels; (b) trough cross-stratified pebbly sandstones; and (c) high angle scours, partly conglomerate-filled. The trough cross-strata originate from gravelly megaripples which migrate shorewards or occur in longshore troughs or rip-current channels. The clast-supported conglomerates are also sometimes reworked, by oscillatory currents, into megaripple forms.

The key attributes of conglomerate beds in a shoreface succession are illustrated in Figure 4.

SETTING DOMINATED BY FLUVIAL OUTPUT

Thick sequences of conglomerate in shallow marine settings are likely to be the product of a pronounced fluvial flooding out from fan-deltaic or deltaic systems. Such sequences represent the innermost part of the deltaic front, where they are deposited as variably reworked, gravelly mouth bars. In the systems of fluvial-wave interaction, gravels may accumulate laterally from the channel mouth in spits and barrier bars, and the conglomerates of this origin will have the essential characteristics of wave-dominated settings.

Facies Sequences

One of the most prominent features of conglomeratic mouth bar sequences are low-angle, curved erosion surfaces (Figs. 5A, B). These are usually infilled by broadly

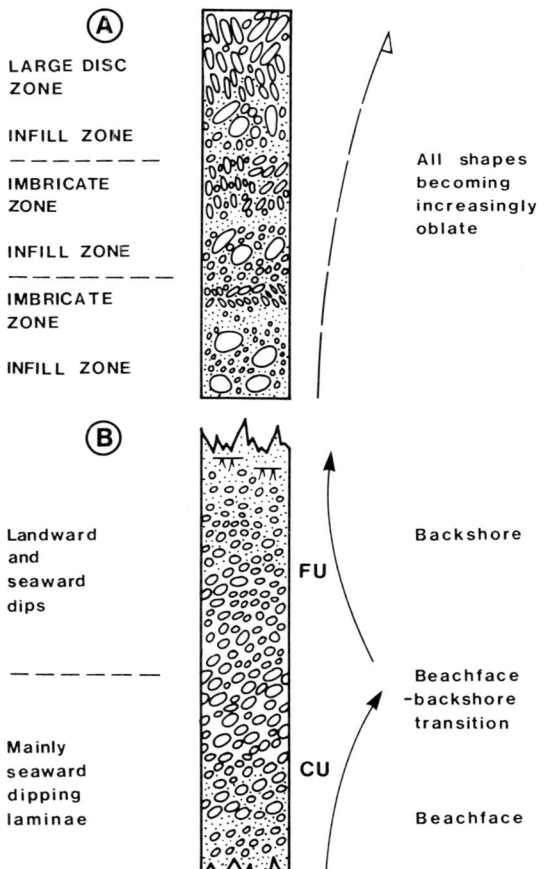

Fig. 3. Local models of progradational beachface gravel showing vertical changes in texture and lamination dip. Examples from: **(A)** Sker Point, S. Wales (from Bluck, 1967); and **(B)** Enju Beach (from Maejima, 1982).

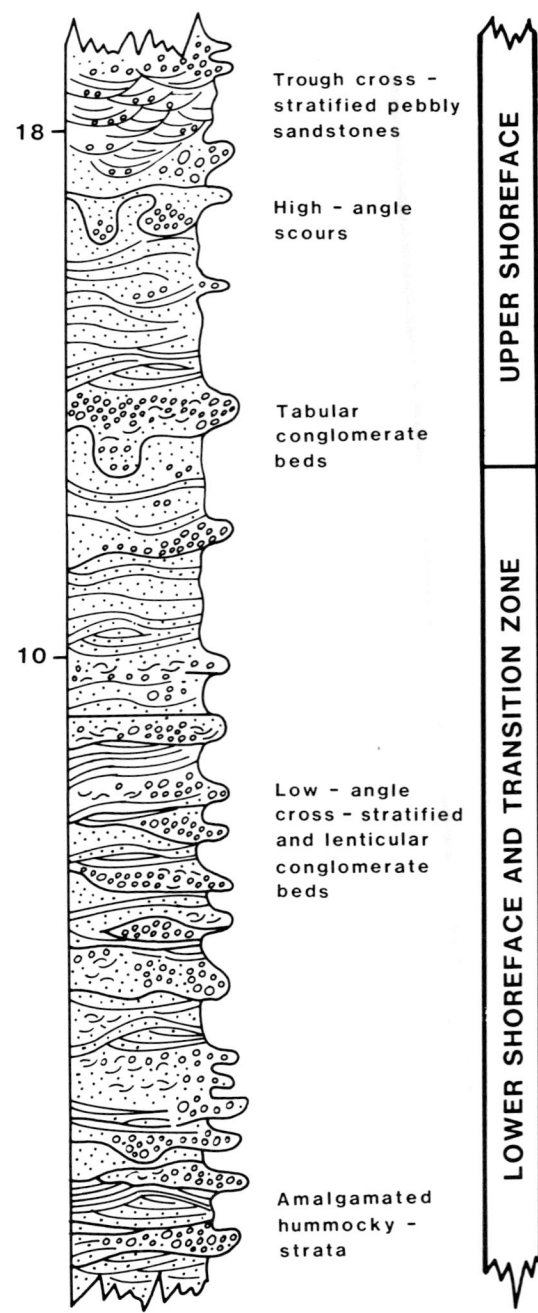

Fig. 4. Features of marine conglomerates in a wave-dominated setting, in the Sandstone of Flores Lake, SW. Oregon (redrawn from Fig. 3 in Leithold and Bourgeois, in press).

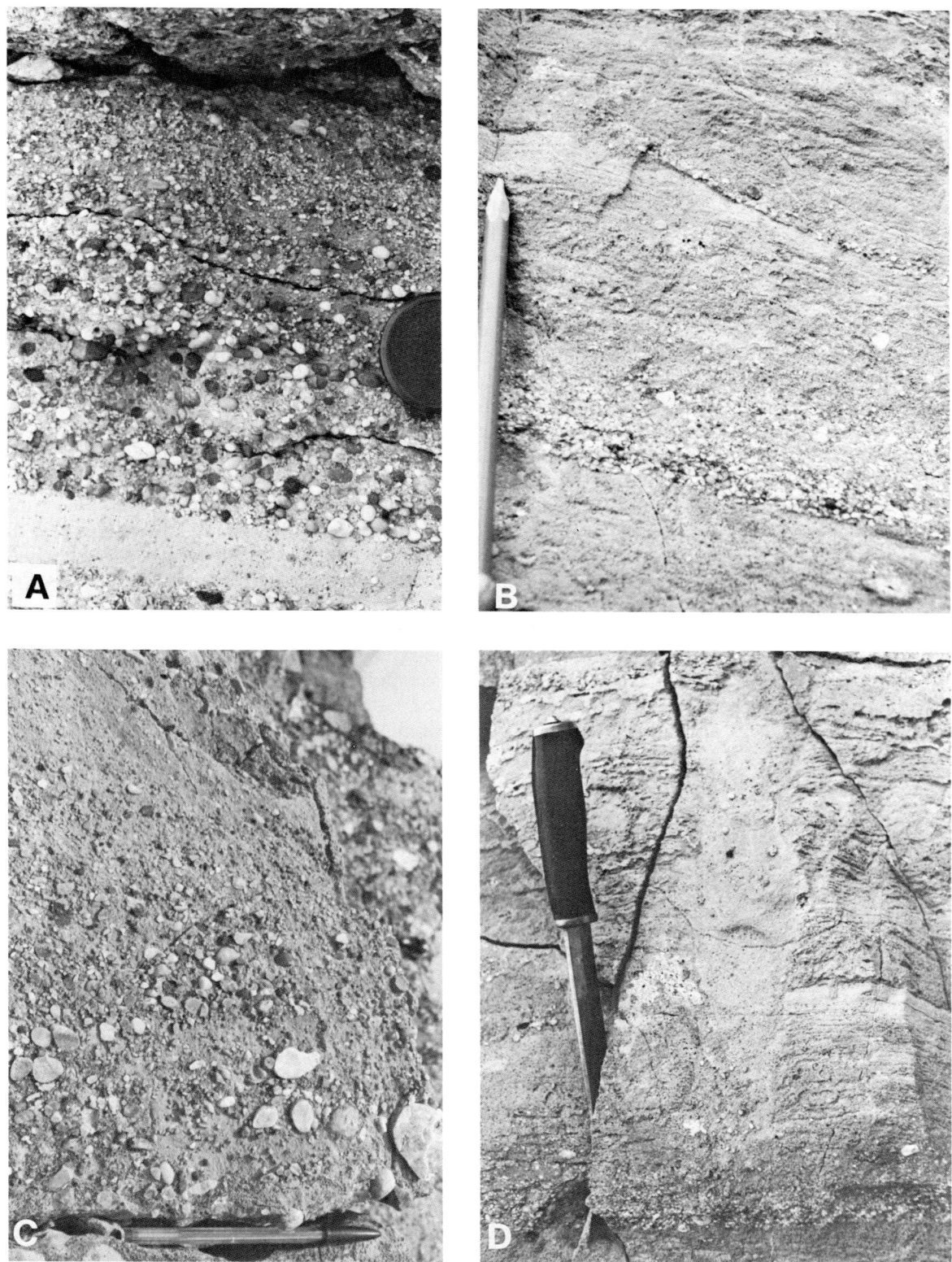

Fig. 5. Features of Carboniferous marine conglomerates from a setting dominated by fluvial output (Aldegondaberget, W. Spitsbergen). Examples show: **(A)** Sharply-based conglomerate beds, crudely normally graded; **(B)** low-angle curved erosion surfaces overlain by crudely graded, high-density turbidite; **(C)** conglomerate bed disturbed by bioturbation, with resultant matrix-supported conglomerate 'patches'; and **(D)** probable arthropod burrow, prominently cutting a graded-laminated mouth bar unit.

lenticular units of normally graded conglomerate (Fig. 5C), or conglomerate and sandstone in alternation. Vertical burrowing of beds (Fig. 5D) is often profound, being a symptom of the intermittent fluvial flooding out onto the mouth bar. The actual type of facies sequence present above the erosion surface is dependent on: (1) the varying discharge of sediment and fresh water during and between the individual flood events in the parent distributary channel; (2) the mode of sediment dispersal out onto, and beyond the channel mouth slope; and (3) the intensity of independent wave, tide and current processes across the mouth bar zone. It is the characteristic heterogeneity of facies sequences, within individual successions of this type, which reflects the environmental interplay of these processes. Details of illustrative facies sequences, encompassing the depositional products of marine and fluvial traction currents, high-density turbidity currents, and cohesive debris flows, are given by Kleinspehn *et al.* (this volume). This documentation of the importance of sediment gravity flows in the fan-delta front setting, emphasised also by Gnaccolini (1981), Massari (1981, this volume), Wiley and Moore (1983) and Postma (this volume) is perhaps surprising, but presumably reflects both slope instability and the heavily sediment-laden nature of the fresh-water currents emerging from the distributary channels. The contribution of flood-generated turbidity currents to marine delta fronts (Elliott, 1978), and even of various debris flows on marine and lacustrine fan-delta fronts (Larsen and Steel, 1978; Gloppen and Steel, 1981; Porębski, 1981) is well documented, but knowledge of the depositional role of high sediment-concentration underflow on both marine delta and fan-delta fronts is still emerging.

Vertical Organisation

Upward-coarsening sequences are common in most progradational nearshore/shoreline successions, but are particularly characteristic of flood-dominated settings. This results from the natural tendency of a distributary channel mouth to produce depositional outbuilding. Two examples of such upward coarsening conglomeratic sequences, representing deltaic setting (Brent Group, Northern North Sea) and fan-deltaic setting (sequence 6 in Kleinspehn *et al.*, this volume, Fig. 1), are shown in Figures 6A and B.

OTHER SHORELINE CONGLOMERATES

Worthy of special mention are the occasionally described giant conglomeratic foresets, which developed in shallow standing water of either lacustrine or marine coastal settings. The conglomerates here represent Gilbert-type deltas, formed by rapid flow deceleration and expansion, as gravel-laden braided streams discharged across the shoreline (see also Van der Meulen, 1983). Presumably the density of the sediment-charged outflow from the river mouth about equalled the density of the receiving water (homopycnal flow model of Bates, 1953), causing thorough mixing of the two systems and immediate deposition of the coarse-grained delta (see also Farquarson, 1982; Wright, 1977). The height of the giant foresets simply reflects the water depth in front of the river mouth. Examples from a lacustrine setting are given by McLaughlin and Nilsen (1982; 10 m high foresets), and from a brackish to marine embayment setting by Casey and Scott (1979; up to 4 m high foresets). Netland (1981) documented cases (19 m high foresets) from marine carbonate platforms, whereas Gradstein and Van Gelder's (1971) examples (up to 30 m high foresets) are from a marine clastic setting.

The conglomeratic foresets can reach 100 cm in thickness, are usually clast-supported internally, often show crude to well-developed normal grading, usually dip at angles less than 22°, and can be either sharply or tangentially based. It is probably the abrupt settling of coarse river load and the slope instability (slumping) that result in viscous, cohesionless debris flows which then rapidly 'freeze' when spreading out and thinning through their downslope movement (frictional freezing due to normal-stress reduction). When incorporating water, such flows reduce also their viscosity and become more mobile, thus giving rise to long tangential foresets. The conglomerate bodies usually develope in settings of insignificant wave energy, but at least one case has been documented where clast shape-sorting, produced by wave action, was superimposed on low angle (10°) foresets (Rainone *et al.*, 1981).

The Gilbert-delta type of conglomerate body contrasts markedly with the lower relief, conglomeratic mouth bar accumulations described by Kleinspehn *et al.* (this volume). The latter, at least where sediment gravity flows figure prominently, presumably reflect situations of greater density contrast (hyperpycnal flow) between fluvial outflow and receiving basin waters, and higher impact of these latter on both mass-flow mobility and the subaqueous slope morphology (reworking).

Conglomerates Originating in Alluvial and Related Environments

The stratigraphic context of alluvial conglomerates, in particular their association with sandstone sequences showing evidence of subaerial flood or soil-forming processes, or containing a non-marine fauna or ichnofauna, may suffice to indicate their depositional setting. Only a brief summary of those features of alluvial conglomerates which we believe are useful for field recognition, and for distinction from marine conglomerates, is presented below.

BRAIDED STREAM CONGLOMERATES

Conglomerates originating from the braided stream environment are, by far, the largest group within the alluvial conglomerate association. The group includes both basin axis and alluvial-fan braided stream deposits, and, within the latter, encompasses primary fan channel, distributary braided stream, and sheetflood/streamflood deposits. Much

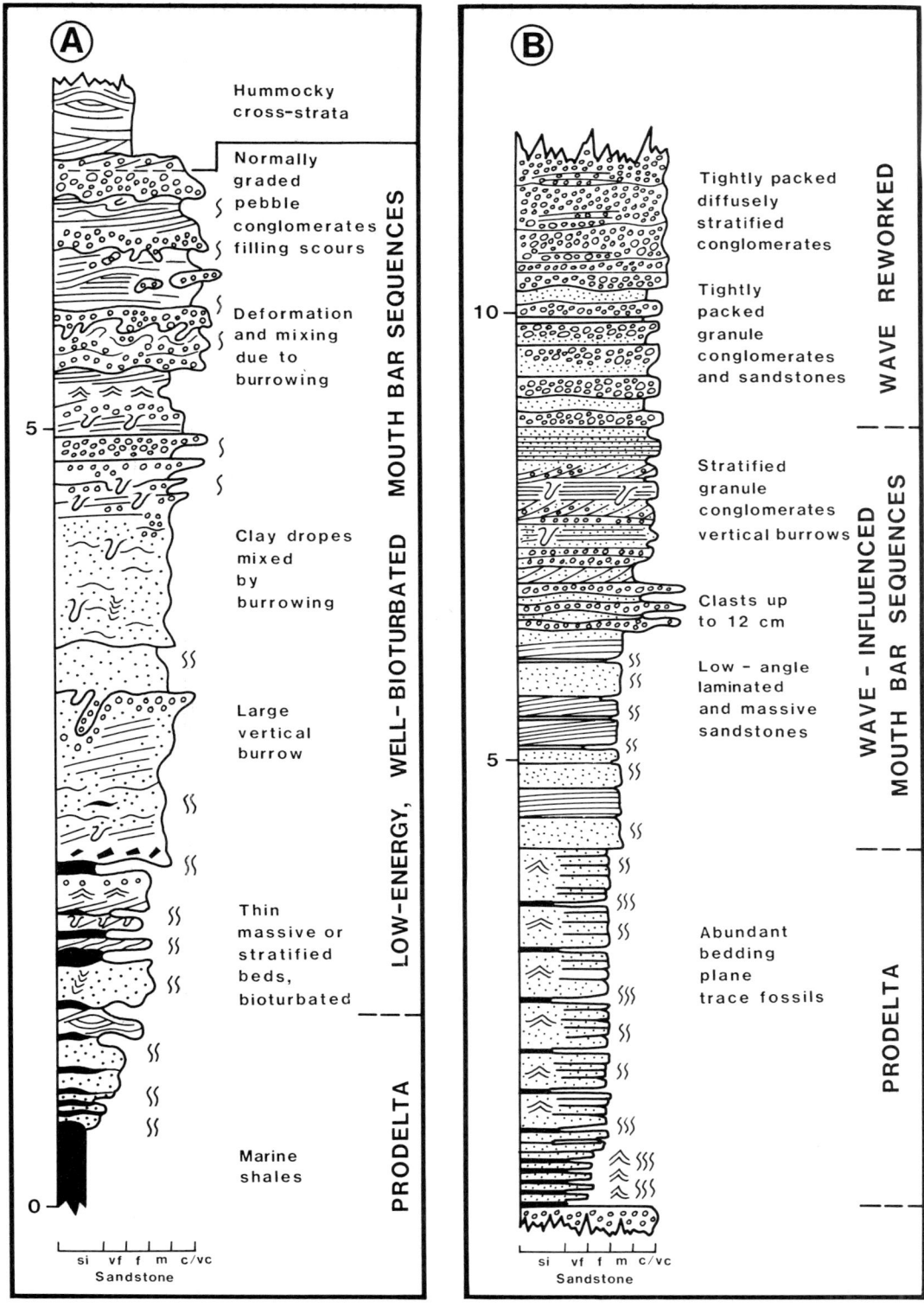

Fig. 6. Two sequences of marine conglomerates from setting dominated by fluvial output. The sequences show a prominent upward coarsening, due to mouth bar progradation. Examples from: **(A)** Middle Jurassic Brent Group (Hild Field, Norwegian sector of North Sea); and **(B)** Upper Carboniferous/Lower Permian Reinodden Formation, Aldegondaberget, Spitsbergen (see also Kleinspehn et al., this volume, Fig. 1, sequence 6).

Fig. 7. Texture and fabric in braided stream conglomerates. Examples show: (A) tightly packed, clast-supported Devonian conglomerates (Hovden Fm., Hornelen Basin, Norway); and (B) Recent gravel from 'flashy' braided streams on an alluvial fan, S. California.

of our knowledge of gravelly braided alluvium comes from studies of paraglacial braided rivers (Rust, 1972, 1975; Bluck, 1974; Boothroyd and Ashley, 1975; Church and Gilbert, 1975). Rust (1978) classified braided-stream gravels into proximal and distal assemblages. Generalised, vertical-profile facies models were presented by Miall (1977, 1978).

Texture alone cannot be used to distinguish stream-deposited conglomerate from shoreline or nearshore conglomerate, but the clast-supported nature of most braided stream conglomerates is characteristic (Figs. 7 and 9) (Rust 1978). However, textural maturity can vary considerably, as judged from (a) amount of matrix retained in the framework, (b) degree of sorting of the framework components, and (c) roundness of the clasts. This reflects mainly the mode of discharge and degree of contemporaneous reworking. Gravels deposited from ephemeral (flashy) flooding, particularly alluvial fan stream deposits, tend to show greatest textural immaturity. Such deposits retain most of the textural characteristics of the primary weathered debris, and only the evidence of scouring or channelised flow conditions, widespread clast imbrication, vague to distinct stratification, and lack of well-defined beds allow this type of braided alluvium (Fig. 7B) to be distinguished from typical clast-supported debris-flow deposits. However, it may be sometimes difficult to differentiate between such 'flashy' alluvium and the deposits of fluidal sediment-gravity flows or even those of turbulent and/or surging debris-flows (*e.g.*, compare Figs. 7B and 14B; for further discussion see next section). At the other end of the spectrum, many braided stream conglomerates are texturally mature (Fig. 9C), probably resultant from stronger channelised transport, more continuous runoff, and effective contemporaneous reworking. The latter conglomerates are matrix-free (openwork) in places (Fig. 9C, middle zone), but may also exhibit textural inversions (bimodality) (Fig. 9C, lower zone) due to the entrainment of loose sand (from the stream bed) during transport of mature gravel.

Stratification is an important feature of braided stream conglomerates and may be crudely (Fig. 7B) or well-developed (Fig. 8) and either planar-horizontal (Figs. 8A, D) or inclined (Figs. 8B, C). The stratification usually records changes in grain size or sorting, reflecting varying discharge and discontinuous accretion, and is often highlighted by an intimate association with well-laminated sandstone. The thickness of conglomerate strata, or of cross-strata sets, may provide a rough indication of water depth. Thin strata (few cm) (Fig. 8A, upper part; 8D) typically result from rapidly shifting, shallow braided channels, from shallow flow on the tops of braid bars or gravel sheets, or from unchannelised flooding on the lower reaches of alluvial fans. Thicker strata (dcm) and thicker cross-sets (few dcm to metres) (Fig. 8B) imply deeper, channelised flow, as in primary fan channels, or strong, channelised flood events closely related to the latter. A frequently described feature of braided, gravelly alluvium is a lensoid geometry of lithosomes and the abundance of curved erosion surfaces (Fig. 8C) (Rust, 1979, Figs. 14-16), reflecting the characteristic, combined effect of both entrenching and avulsive behaviour of the braided channel

Fig. 8. Stratification in braided stream conglomerates. Examples show: **(A)** dominantly flat-lying strata (Triassic, NW. Scotland); **(B)** planar cross-stratified set (Triassic, Staffordshire, England); **(C)** curved erosion surfaces and lensoid conglomerate sequences (Devonian, Hornelen Basin, Norway); and **(D)** thinly bedded, fine-grained conglomerates with some outsized clasts (Devonian, Hornelen Basin, Norway).

system, and the rapid, lateral shifting and abandonment of the individual channels.

Some unusual examples of giant-scale (up to 25 m thick) cross-strata sets have been documented in alluvial conglomerates. Foresets are inclined at 10 - 20° (Ori and Ricci Lucchi, 1981), occasionally being much steeper (Massari, 1983), and are often normally graded (Steel and Thompson, 1983; Massari, 1983). Such giant sets are generally viewed as the result of deep flood water, most probably enhanced by bank or valley wall confinement, but have been interpreted in various terms: as point bars in sinuous channels (Collela, 1980), as transverse bars in low-sinuosity distributary channels (Massari, 1983), and as medial or longitudinal bars in deep braided streams (Steel and Thompson, 1983). These bed forms should be carefully distinguished from the marine or lacustrine cross-strata of Gilbert-delta origin (discussed above); in their depositional aspect, the former more closely resemble the giant (up to 45 m high) gravel bars produced by Pleistocene catastrophic flooding (Malde, 1968; Baker, 1973; Baker and Nummedal, 1978).

Important attempts have been made to generalise about alluvial morphology or bar type from stratification characteristics. Poorly defined, flat-lying strata have been commonly related to longitudinal bar development, and high-angle cross-strata to transverse bar or bar margin development (Rust, 1972; Smith, 1974). Hein and Walker (1977) investigated the factors which influence bar morphology and, in particular, the reasons why some braid bars develope foresets whereas others do not. They emphasise the role of rapid decrease in fluid and sediment discharge, across an incipient bar core, in the creation of cross-strata. Bluck (1976, 1979) emphasised the internal complexity of braid bars in some rivers, implying that much effort may be necessary in order to reconstruct bar type in ancient alluvium.

Facies sequence that typifies the deposits of gravelly braided streams, whether comprising (sub-)horizontal strata or foreset cross-strata, results from waning flood action or declining discharge. The sequence consists of a composite unit of conglomerate, usually a few decimetres to few

Fig. 9. Facies sequences in braided stream conglomerates. Examples show: **(A)** horizontally stratified conglomerate and sandstone in alternation; **(B)** crudely graded conglomerate with laminated sandstone capping; and **(C)** texturally mature, well-graded conglomerate unit. Examples A and B are from Devonian Hornelen Basin, Norway; example C is from Traissic succession of Staffordshire, England.

metres in thickness, showing a crude to well-developed (Figs. 9B, C) upward fining signature, and often capped by sandstone. Matrix-winnowing or by-passing during falling stage commonly creates an openwork lower or middle part to the unit (Smith, 1974; Bluck, 1976), whereas the uppermost part is matrix-filled due to low-stage infiltration. The variability of this type of facies sequence has been documented by Steel and Thompson (1983) from Triassic alluvium. Another common signature, especially in sand-dominated sequences, is a simple alternation of broadly lenticular, gravelly and sandy strata (Fig. 9A). As pointed out by Walker (in Harms *et al.*, 1975), such sequences may simply represent deposition due to *slight* fluctuations in stream velocity.

The upward fining facies motif is commonly prominent in conglomerate successions originating from braided streams (see the Donjek and Scott models of Miall, 1977), but it is not easy to distinguish between a flood-generated facies sequence and an upward fining sequence, often on a scale of several to tens of metres, generated by large-scale abandonment of braided alluvial tract. The two successions sketched in Figure 10 illustrate this difficulty. The Cannes de Roches succession (Fig. 10B) shows fining-up sequences which are probably resultant from the gradual abandonment of segments of a braided alluvial system (Rust, 1978). However, in the Hovden Formation example (Figs. 10 A, C and D), taken from Spinnangr (1975) and Steel and Aasheim (1978), it is much less clear whether the fining-up sequences result from such longer-term abandonment process, or simply from waning of very large floods. The latter possi-

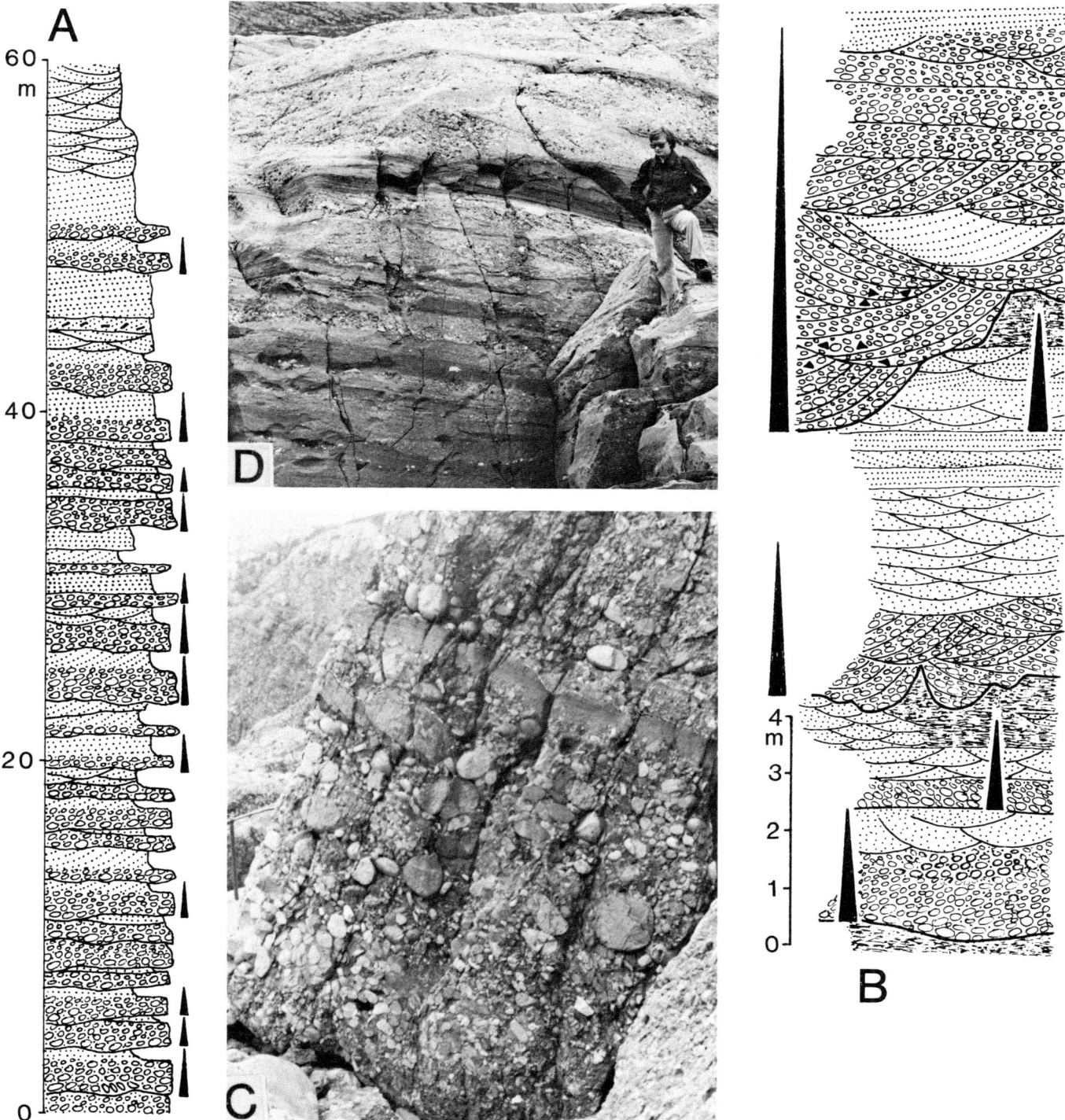

Fig. 10. Successions of braided stream conglomerates. Examples show: **(A)** Repeated fining upward (*FU*) sequences (usually less than 2m thick); **(B)** Repeated *FU* sequences which are structurally and lithologically more complex than in A, and commonly thicker than 2m; **(C)** Details of a thick conglomerate-sandstone sequence from near the base of example A; and **(D)** Repeated conglomerate-sandstone sequences in a local coarsening upward succession. Examples A, C and D are from Hovden Formation, Devonian Hornelen Basin, Norway; example B is fragment of Cannes de Roches Formation, Gaspé, Canada (slightly modified from Rust, 1978).

bility may be especially likely in Devonian and pre-Devonian ('pre-vegetation') settings.

MASS-FLOW CONGLOMERATES

Sediment gravity flows are known to be an important, often predominant mechanism for the transport and deposition of gravels in many alluvial and coastal settings. A variety of gravelly mass flows occurs in the alluvial environment, particularly on alluvial fans which have a relatively steep slope, small radius, and scarce vegetation (Hooke, 1967; Bull, 1977). Conglomerates and breccias of mass flow origin are well-documented from ancient alluvial fans (for useful review see Heward, 1978b), and Hooke's (1967) contention that mass flows preferentially occur on the upper and middle reaches of alluvial fans has been confirmed from a number of ancient fan bodies (Steel, 1974; Steel et al., 1977). Even some glaciomarine, seemingly water-rich settings may be dominated by mass flows (Miall, 1983).

The flows may range from fairly slow-moving, high-strength/high-viscosity debris flows, to swift, fully-turbulent fluidal flows (transitional to heavily sediment-laden stream flow) (e.g., Bull, 1977; Pierson, 1981; Lawson, 1982; Innes, 1983). This whole range of flow behaviour is accompanied by a variety of deposits, though only recently this continuum (cf. Beverage and Culbertson, 1964) become subject to more detailed investigation (e.g., Lawson, 1982).

The variety of mass flow deposits is perhaps greatest on alluvial fans whose distal reaches are subaqueous, either submarine or sublacustrine, and whose depositional surface and sediments are thus actively agitated (reworking), or only passively influenced (saturation) by the adjacent water reservoir. Much recent interest has been drawn to such depositional settings, giving rise to the *fan-delta* concept. The gravelly mass flows are subject to substantial physical transformation when coming into water (see next section), and the resulting deposits are often subject to reworking (as by waves or tides on an active coast) and/or subject to further resedimentation (when emplaced unstably under water). The fan-derived mass flow gravels may also mix with the fine-grained marine or lacustrine sediment, making the variety of depositional products even wider. Some illustrative examples from ancient sublacustrine settings are presented by Larsen and Steel (1978), Gloppen and Steel (1981), and Nemec et al. (this volume); ancient submarine examples are described by Kleinspehn et al. (this volume), Postma (this volume), and Massari (this volume) among many others.

Figure 11 shows an example of mass flow-dominated, ancient lacustrine fan-delta, namely the Violin Breccia succession of the Ridge Basin, southern California. Subaerial debris-flow deposits and associated fluvial conglomerates intimately interfinger here with subaqueous debris-flow conglomerates, high-density turbidites, and lacustrine mudstones and siltstones. This cross-section well illustrates the active depositional interplay between sub-

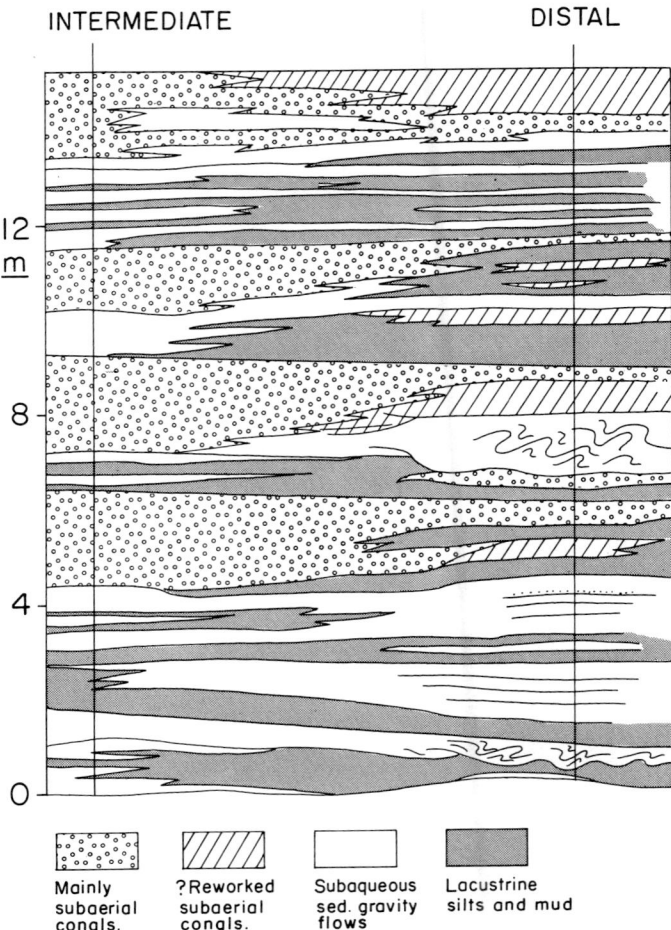

Fig. 11. Intricate interfingering of subaerial debris flow deposits (and associated, fluvially-reworked deposits) with subaqueous deposits of gravity flows (debris flow deposits and high density turbidites) and lacustrine mudstones; this section shows segment of the Violin Breccia, Ridge Basin, S. California. Distance between profiles approx. 150 m.

aerial and subaqueous processes along an ancient lacustrine shoreline.

Because of the potential variability of mass flows, characteristics of their deposits such as the type of clast fabric, type of clast-size vertical grading (if any), and matrix percentage are not diagnostic. However, features that are generally characteristic of debris flow deposits include at least the following few attributes:

(1) beds are usually sheetlike, with limited or insignificant basal erosion (Figs. 12A, 13 and 14) though often with a highly lenticular *overall* geometry;

(2) beds are ungraded (Fig. 12) to well-graded (Figs. 13 A and 14 A), dependent on the internal regime (clast-support mode) of the depositing flow, and the grading type often changes downslope;

Fig. 12. Examples of ungraded, matrix-supported (A) and clast-supported (B) beds deposited from debris flows. Examples from: **(A)** Tertiary of Split Mountain Canyon, S. California: and **(B)** Devonian Hornelen Basin, SW. Norway.

(3) beds show no obvious stratification (Figs. 12 and 13), though some may be crudely layered (Fig. 14) when deposited from surging flows, and a succession of deposits may be well-bedded due to distinct bed boundaries (Fig. 13C);

(4) beds may range from texturally polymodal to bimodal, and from clast-supported to matrix-supported (see spectrum of examples in Figs. 12-14), quite often containing some large, 'outsized' cobbles or boulders;

(5) beds commonly display a significant positive correlation, in statistical terms, between their thicknesses and maximum clast sizes (see next section).

Some debris flows may be turbulent, and a turbulent flow regime is usually inferred from such depositional features as normal grading (Fig. 14A) (cf. Lowe, 1982, Fig. 12) and marked signs of basal scour. A tendency for channelling, some faint stratification, and 'tractive' clast-fabric may develop from relatively dilute (water-rich) flows.

Disorganised clast fabric may simply reflect short travel distance, but often suggests non-sheared (high strength) 'plug' flow, or only weakly sheared (high viscosity) flow. Preferred clast orientation, often subhorizontal and parallel-to-flow (e.g., Fig. 13B), may originate from strongly sheared, laminar flow, most likely due to clast interactions and dispersive pressure (cf. Enos, 1977; Lewis et al., 1980). If experiments on sheared sand-size dispersions are relevant also to the gravelly ones, then the alignment of clasts (whether parallel or transverse to flow direction) can be expected to vary with clast concentration (cf. Rees, 1983). Vertical variability in the shear rate and in clast concentration within the flow may thus produce vertical variation in clast fabric, or in imbrication (Hiscott and Middleton, 1979; Massari, this volume). Clast upflow imbrication is also common, and may be of either a-axis or b-axis type. Transverse alignment of clasts, frozen from their movement by rolling along the bed, can be expected to develop from more fluidal, turbulent flow with a significant tractional component (cf. Hand, 1961); for instance, from some gravelly turbidity currents (as discussed by Lowe, 1982) or from markedly fluidal subaerial flows (as represented by part of the spectrum studied by Pierson, 1981; and Lawson, 1982).

Subaerial Mass-Flow Deposits

Some of their significant features are reviewed in Figure 15. Beds may range from mud-rich, matrix-supported (Figs.

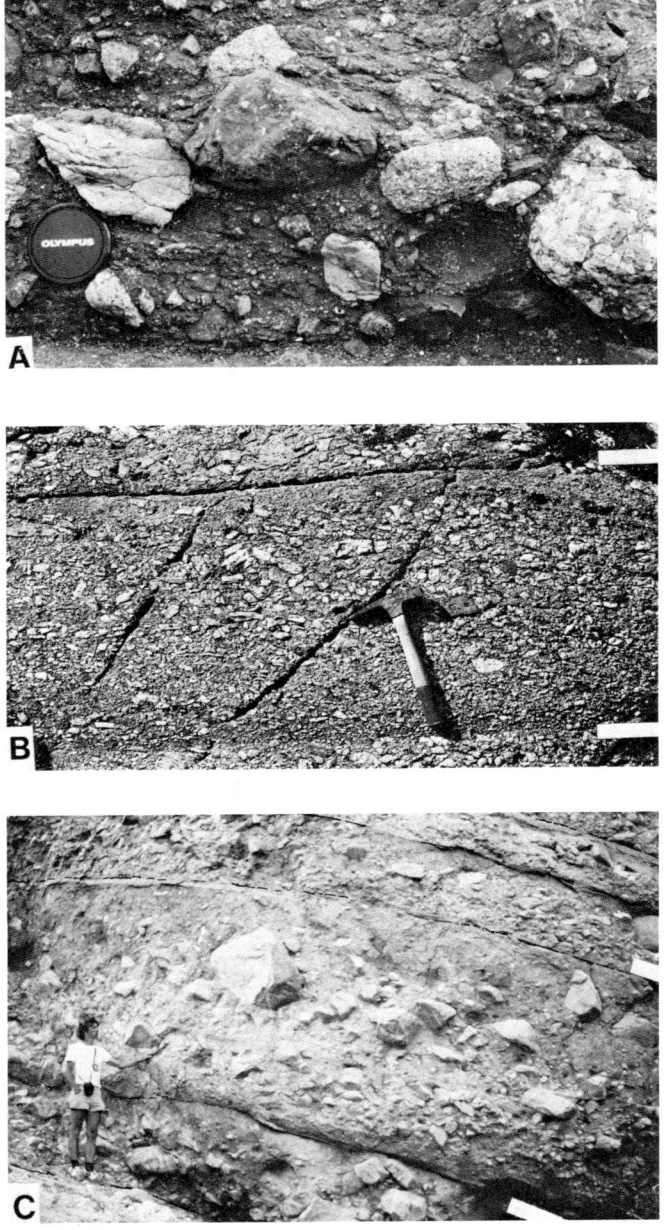

Fig. 13. Examples of ancient debris-flow deposits: **(A)** lower part of an ungraded, subaerial debris flow bed (Permian, Oslofjord, Norway); **(B)** inversely graded, clast-supported bed deposited from subaqueous debris flow (Devonian Hornelen Basin, SW Norway); and **(C)** thick, clast- to matrix-supported, inversely graded bed deposited from subaqueous debris flow (Tertiary of Split Mountain Canyon, S. California).

12A and 13A) to clast-supported (Figs. 12B and 14B). They are usually ungraded, and most often represent 'plug' flow deposition, as described and modelled by Johnson (1970); signs of inverse grading are restricted to basal few cm (*e.g.*, Gloppen and Steel, 1981, Fig. 2). However, some beds tend to show better internal organisation (grading, fabric), usually implying either a considerable role of dispersive pressure (inverse grading) or deposition from viscous,

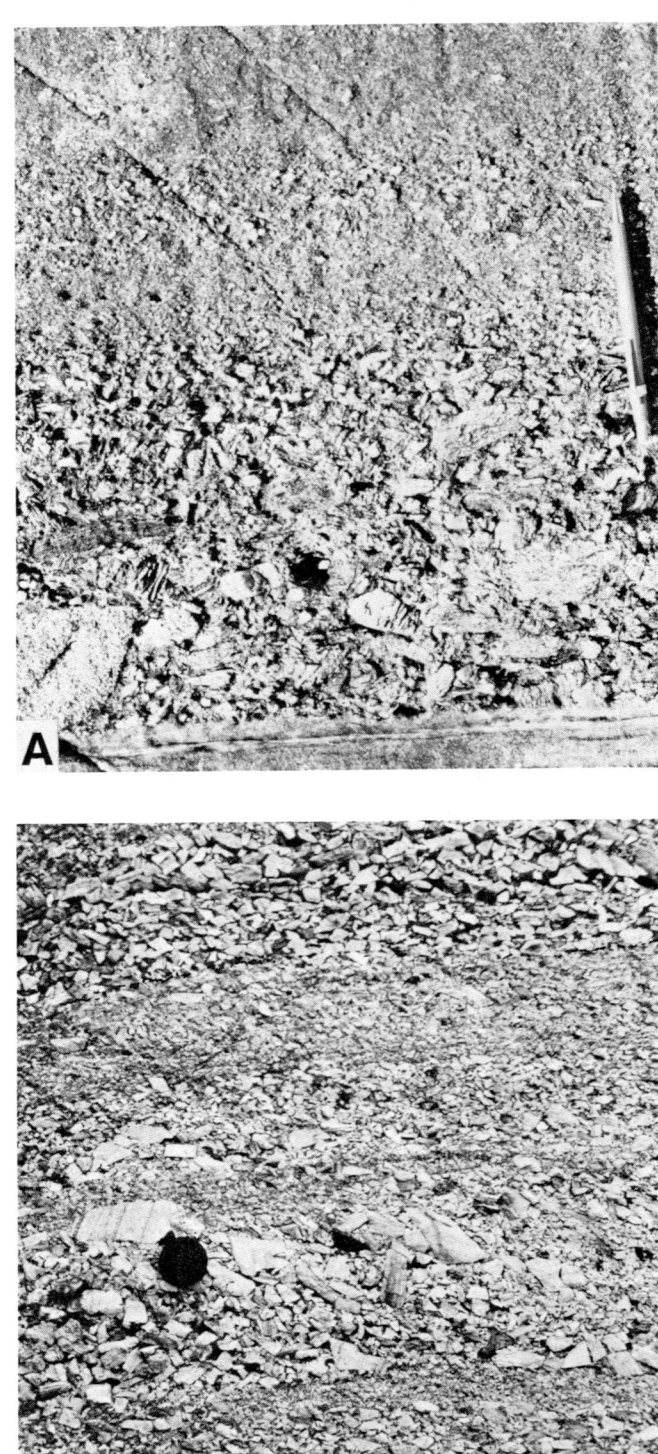

Fig. 14. Examples of ancient debris-flow deposits: **(A)** clast-supported conglomerate bed with marked normal grading, emplaced by subaqueous, turbulent debris flow which probably transformed to high-density turbidity current; and **(B)** composite sequence of clast-supported conglomerates deposited from surging debris flows. Both examples from Devonian Hornelen Basin, Norway.

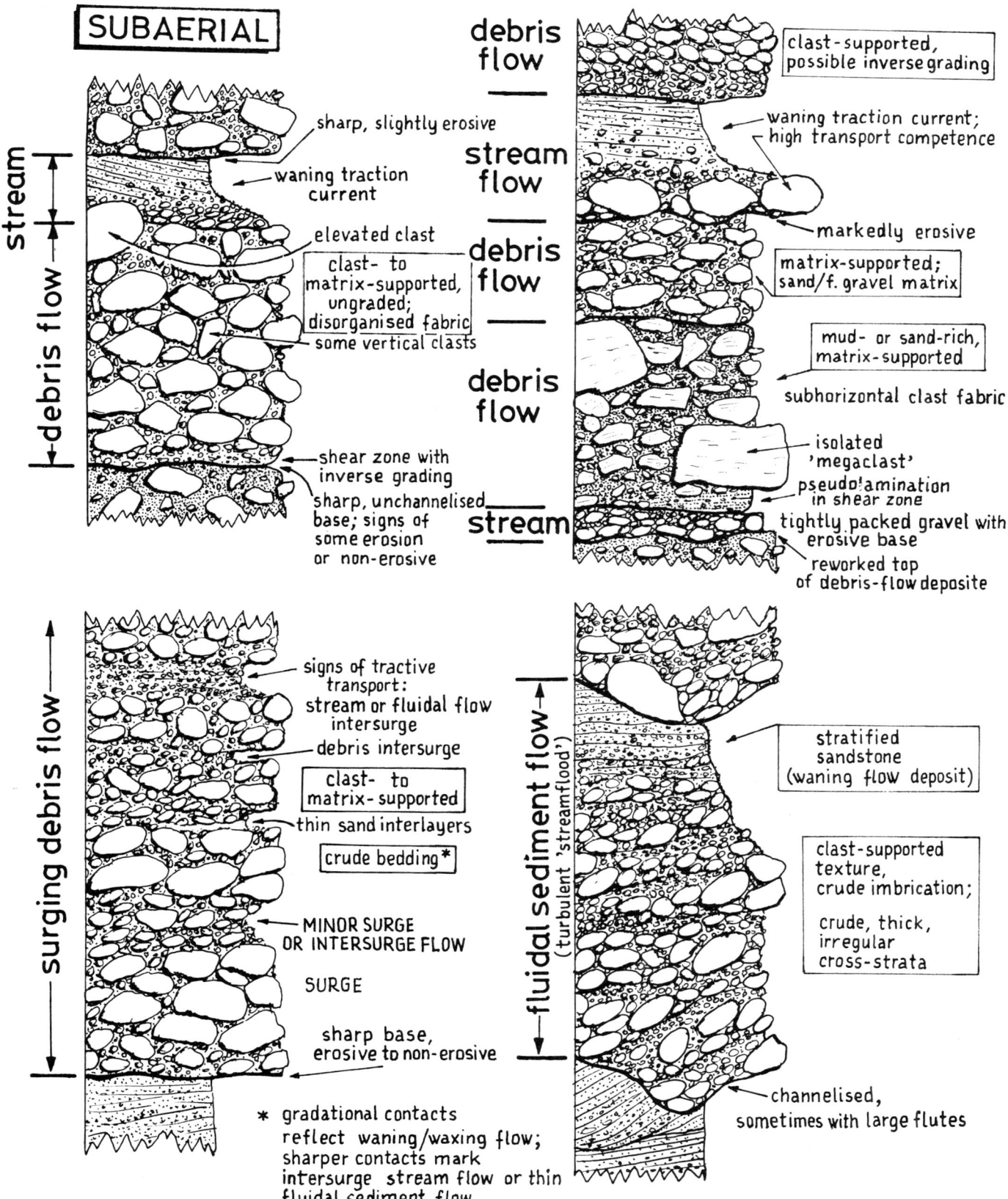

Fig. 15. Some typical features of subaerial mass-flow deposits as discussed in text (range of bed thickness from few decimetres to metres, as illustrated in Figs. 12-14).

more watery and possibly turbulent flows (normal grading). Some debris flows are turbulent, but signs of only transient turbulence are more common. During the deposition of a fully-turbulent debris flow, the coarser and denser material settles directly from suspension and forms a coarse and normally-graded basal part of the resulting deposit, while the remaining sediment 'freezes' as a fairly uniform dispersion which may still contain some better-buoyed (intraformational) large clasts. Examples of such bipartite, 'layered' beds are described by Lowe (1982, Fig. 13C).

Presence of an upward-fining sandy capping, sometimes with an erosive base and signs of stratification (Fig. 15), may result from turbulent fluidal flow or heavily sediment-laden stream flow following the debris flow (or its main surge). In cases where this gravelly-sand layer represents intersurge deposition, its contact with the overlying debris-flow still tends to be sharp, but is non-erosive (Fig. 14B, lower part). Especially in proximal depositional reaches, the interflow (not intersurge) capings may be represented by tightly packed, fine pebble conglomerate, also the product of stream flow. In some instances, the transport (tractive) competence of the interflow medium may actually be greater than the suspension competence of the preceding debris flow, causing a markedly coarser grained capping composed of a gravelly 'lag' (large, scattered or packed cobbles) that goes abruptly upwards into sandstone or pebbly sandstone (Fig. 15).

Composite units with thick, crude to distinct internal layering (Fig. 14B) or with the presence of thin, discontinuous sandy zones may result from rapidly surging flows (Fig. 15). Other units may be variably channelised and show a clast-supported texture, crude (subhorizontal to inclined) strata, some imbrication and well-stratified sandstone cappings (Fig. 15). An abundance of such units strongly suggests depositional systems conducive to more watery flows (*e.g.*, 'wet' fans of paraglacial settings).

Subaqueous Mass-Flow Deposits

Some of their significant features are summarised in Figure 16, and are discussed below. In a succession of gravelly mass-flow deposits, their subaqueous emplacement can often be inferred directly from such features as: (1) association with turbidites; (2) presence of fossiliferous or biomicritic interlayers; (3) bioturbation; (4) presence of wave-generated structures in the sandy interlayers; or at least (5) common presence of mud/silt interbeds or partings. Association with pebbly mudstones and/or slump-contorted muddy units, and possibly a common presence of sand sills/dikes also may be considered diagnostic. Otherwise, as in the absence of such self-evident features, the subaqueous emplacement of conglomeratic deposits may not be obvious or easy to infer. Below, we emphasise some other features of potential importance.

In general, main characteristics of the subaqueous gravelly debris-flow deposits (*e.g.*, their textural variety, types of fabric and grading) may be similar or transitional to those of the subaerial deposits. However, there are several features, or tendencies, which we find particularly typical of the subaqueous debris-flow conglomerate beds.

They tend to be better organised internally (clast alignment, imbrication, grading), with the well-developed grading (inverse, inverse-to-normal, and normal) and sandy cappings being here far more common (*eg.*, Figs. 13B, C and 14A). This reflects the tendency of the subaqueous debris flows to evolve towards high-density turbidity currents (Lowe, 1982; see also ancient example at the end of next section). An increase in water content or shearing induced by some initial, transient turbulence (enhanced by water admixed into the flow) causes the debris to loose its strength and the flow, finally, to become fully turbulent. Such flow will considerably decrease its viscosity, hence the deposition occurs through settling from suspension and possibly even with some traction (high-density turbidite). But still at any stage some surges may occur, producing vertical irregularities in the resulting graded deposit (*eg.*, Fig. 14A).

Irrespective of the type of grading, many subaqueous debris-flow beds show a marked upward increase in their matrix content, particularly near the top (Figs. 13B, C) (Nemec *et al.*, 1980; Kelling and Holroyd, 1979). We find this feature virtually atypical to subaerial debris-flow deposits, where, in contrast, beds rather tend to terminate with a finer-grained, tightly-packed conglomerate overlain by stratified sandstone (*i.e.*, with the product of stream reworking and deposition).

When passing into water, the debris flow may considerably reduce its thickness and hence distally split into small lobes. This leaves thin, essentially scourless (though possibly loaded) gravel lenses which then become enveloped in marine or lacustrine mud (Fig. 16). Lower-viscosity debris flows are particularly likely to terminate with such finger-like lobes.

On the lower reaches of certain fan-delta slopes, the gravelly debris-flow deposits may associate with fine-grained (sand-granule), crudely to evenly layered sediments which texturally resemble closely the matrix material of the debris flows themselves. As interpreted by Postma (this volume) and Nemec *et al.* (this volume, facies C), such sediments (example see Fig. 17) represent the subaqueously resedimented ('gravity-winnowed') material derived from some unstable portions of the gravelly fan-delta slope; active slump scars or steep, unstable 'noses' of the debris flows that have become frozen farther upslope are the likely sources. This process of resedimentation probably begins with partial, selective liquefaction, and then continues as a series of thin, surging flows that range from density-modified grain flows to high-density turbidity currents.

This distinct depositional facies is likely to be preserved in a relatively passive or protected water reservoir, because otherwise it must probably be reworked by waves or tides. However, fairly similar deposits, though much thicker-bedded and coarser grained, are described by Lewis *et al.* (1980) from large, unprotected ancient slope and are inter-

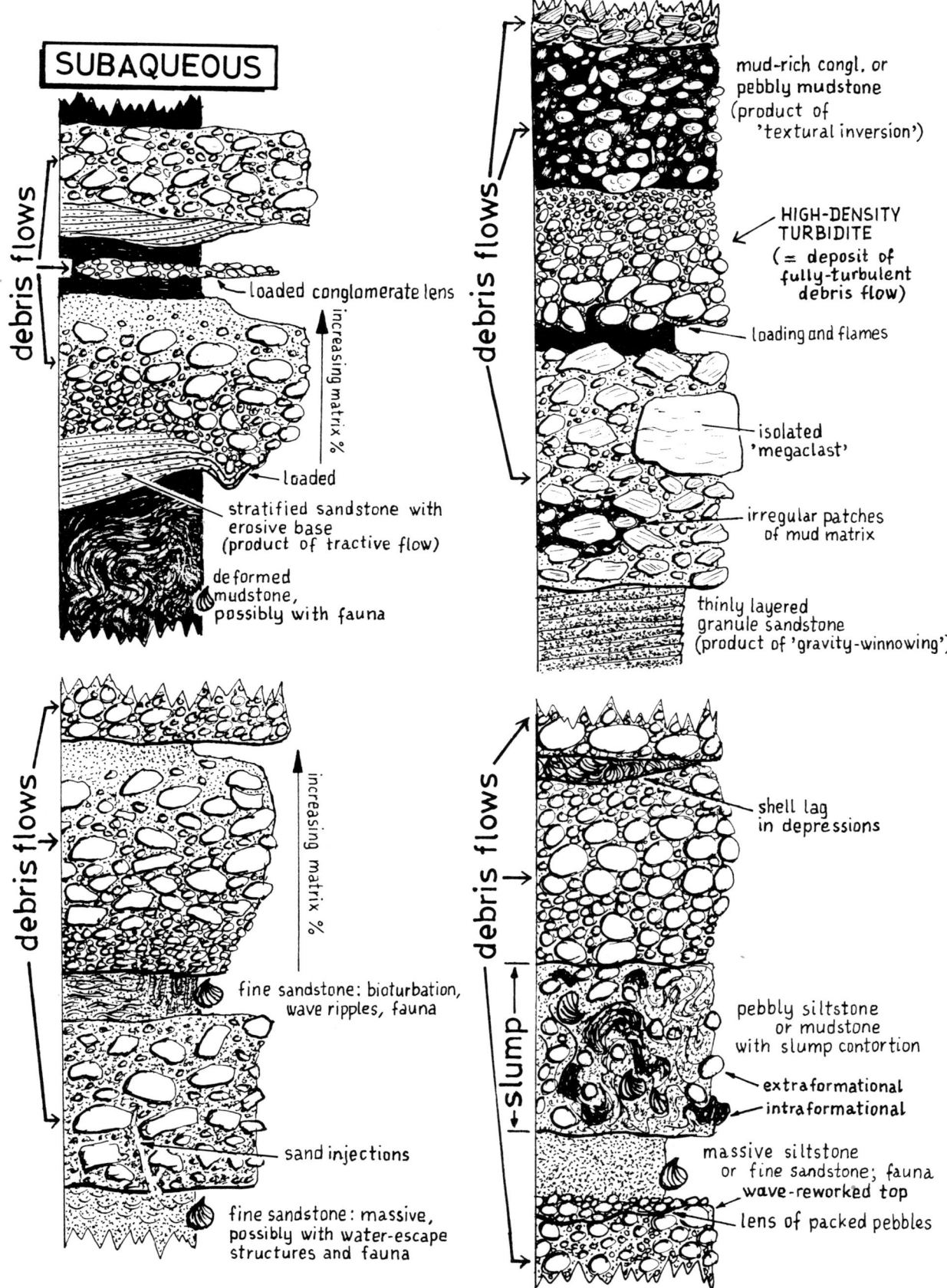

Fig. 16. Some typical features of subaqueous mass-flow deposits as discussed in text (range of bed thickness illustrated in Figs. 12-14 and 17).

preted in terms of a slow-moving, turbulent flow analogous to "slump creep" (*sensu* Carter, 1975).

We have also noticed another type of sandy sediment often associated with the gravelly deposits of subaqueous debris flows. This sediment invariably occurs at the base of a debris flow bed (Fig. 16), and seems to be quite intimately associated with the latter. Such basal sandy portions are erosively based and show distinctive tractional structures (crude to well-developed, subhorizontal or inclined stratification); some cross-lamination and occasional small pebbles may also be present. This sort of sandy deposit probably originates from a turbulent, tractive flow of sediment-laden water (again see Lewis *et al.*, 1980) which, after being entrained above the moving debris, eventually bypasses the deccelerating debris flow and immediately starts to drop its tractive load; the deposited sand is then overridden by the slower-moving debris flow from behind.

The fan-derived, mass flow gravels may also mix with the marine or lacustrine muds, hence giving rise to bimodal, mud-rich conglomerates ('textural inversion'). Some possible modes of such mixing are suggested by Larsen and Steel (1978), and Nemec *et al.* (this volume).

Fig. 17. Thinly layered granule sandstone deposited on lower reaches of the subaqueous slope of a gravelly fan-delta. The layers tend to be inversely or inverse-to-normally graded, and locally display water-escape disruption. Example from the Paleogene of Ridge Basin, S. California.

CONGLOMERATES DEPOSITED FROM SEDIMENT GRAVITY FLOWS: SOME ASPECTS OF THEIR CLASSIFICATION AND FIELD ANALYSIS

CLASSIFICATION OF GRAVELLY MASS-FLOWS

In an attempt to refine some of the nomenclature problems associated with *sediment gravity flows*, Lowe (1979) recently revised the general classification scheme for these processes. The genetic terminology proposed by Lowe (1979, 1982) is perhaps the most straightforward and useful classification available at the present stage of our knowledge of this category of processes. He classified sediment gravity flows by their rheological behaviour, and distinguished *fluidal flows* (with viscous fluid behaviour) and *debris flows* (with plastic behaviour). He further divided these flow types on the basis of a modification of Middleton and Hampton's (1973, 1976) classification of grain-support mechanisms. The fluidal flows include turbidity currents (with support from fluid turbulence), fluidised flows (with full support from escaping pore fluid), and some liquefied flows (with only partial support from escaping pore fluid). In this group, perhaps only the high-density type of turbidity currents represents a sole-operating mechanism that can be important in the transport and deposition of gravels (Lowe, 1982; see also Nardin *et al.*, 1979, p. 65); illustrative examples are described by Walker (1975), Aalto (1976), Winn and Dott (1979), Hein (1982), Hein and Walker (1982), and Massari (this volume). However, it has been suggested in the literature, and is also partly considered in the next section, that any of these three mechanisms can become active in some gravelly debris flows, to considerably modify the flow's behaviour, though not necessarily to offer a prime clast-supporting force (*e.g.*, Pierson, 1981; Lowe, 1982).

Some markedly fluidal, fully turbulent subaerial gravity flows, as for example some of the flows studied by Lawson (1982) and Pierson (1981) and perhaps the 1982 Mount St. Helens sediment flows (Harrison and Fritz, 1982), should also be classified into the fluidal-flow category.

Such flows tend to be markedly channelised and characteristically display high water-to-sediment ratios (roughly 20% water by weight) (*e.g.*, Lawson, 1982). In Beverage and Culbertson's (1964) terminology, these are 'hyperconcentrated flows', defined as sediment-water mixtures having between 40 and 80% sediment by weight. In terms of the theoretical flow continuum predicted by these latter authors, the hyperconcentrated flows are intermediate between ordinary stream-flow, in which there is no more than 40% sediment by weight, and debris-flow (or mudflow), which has more than 80% sediment by weight.

The deposits of such intermediate-type flows still tend to form sheet-like beds, but these may probably range from only slightly channelised and weakly graded, to strongly channelised, normally graded, and even thoroughly stratified. Various examples representing this depositional spectrum were described under the terms 'intermediate deposits' (Bull, 1963), 'streamflood deposits' (Bluck, 1967b; Steel,

1974), and 'sheetflood deposits' (Wasson, 1977, 1979; Heward, 1978a; Nemec and Muszyński, 1982). However, this first term does not seem particularly meaningful or useful, and the usage of the other two terms does not follow closely their original definitions (introduced strictly in the context of intense sheetwash erosion processes; see Rich, 1935; Davies, 1938). Accordingly, we rather suggest that the general term *fluidal sediment flow* is used at the present stage, until further research is done and more specific terminology is developed (*cf.* Lawson, 1982, Table 1).

The fluidal subaerial sediment flows, with pronounced turbulence, are likely to be an important agent for the transport and deposition of gravels in relatively water-rich alluvial-fan systems, but are expected to be even more significant in glacial (meltwater) settings.

We will concentrate our present discussion on the rheological *debris-flow* category. The (rheological) debris flows distinguished by Lowe (1979) correspond to 'mass flows' as defined by some other authors (*e.g.*, Dott, 1963; Nardin *et al.*, 1979). These flows are postulated by Lowe to be collectively described by the Coulomb-viscous rheological model (Fig. 18D), which was developed by Johnson (1970, p. 518-519) and has already been adopted for similar considerations by several other authors (*e.g.*, Hampton, 1972; Middleton and Hampton, 1973, 1976; Carter, 1975; Middleton and Southard, 1978, chapter 8; Enos, 1977). In terms of this one-dimensional model, the shear strength of debris is described as:

$$S = \underbrace{C + \sigma'_n \cdot \tan \varphi'}_{\text{yield strength } (k)} + \mu_c \cdot \frac{dU}{dy}$$

and the shear stress (flow condition) as:

$$\left. \begin{array}{l} \sigma_s = S \\ \sigma_s > k \end{array} \right\} \text{ (at least along the base),}$$

where:
- S = shear strength;
- σ_s = applied internal shear stress;
- C = cohesion;
- σ'_n = effective internal normal stress;
- φ' = angle of effective internal friction;
- μ_c = coefficient of flow's viscosity; and
- $\frac{dU}{dy}$ = vertical velocity gradient (expression for the rate of shear strain, ϵ_s, as used below).

Lowe (1979, p. 77-78) proposed that on the basis of this rheological model, the debris flows are classified into *cohesive debris flows* (or mudflows) and *grain flows*, depending on whether the cohesive component or frictional component dominates in determining the debris yield-strength.

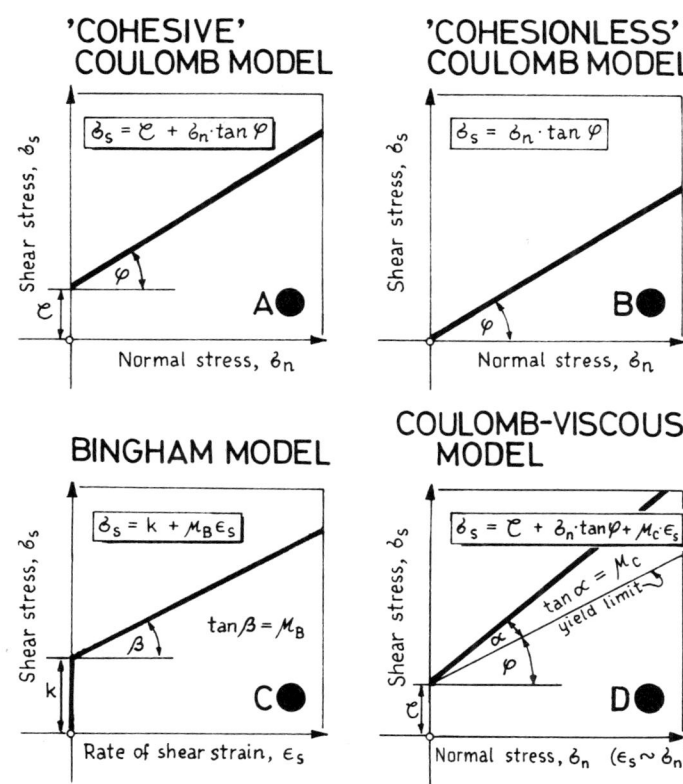

Fig. 18. Diagrams showing three rheological models pertinent to the debris flow mechanics: Coulomb model (**A, B**), Bingham model (**C**), and Coulomb-viscous model (**D**). Note that this latter model (considered in text) is a combination of the other two.

Although we slightly modify this terminology in our reasoning below, we entirely agree with Lowe (1979, p. 77) that his classification is physically substantiated and it offers a clear theoretical distinction between two important categories of sediment gravity flows, a distinction that has generally been lacking in previous discussions. The importance of the presence or absence of cohesion to the mode of the flow's behaviour, particularly to the clast-support mechanisms, has already been pointed out by numerous authors, and thus does not need to be repeated here.

In following Lowe's (1979) nomenclature, we prefer to simplify it slightly, by focusing on the distinction between *cohesive* debris flows (or mudflows) and *cohesionless* debris flows (or grainflows). Although the term 'cohesionless' does not necessarily imply any particular grain-support mechanism, we deliberately introduce it here to emphasise the unimportance of cohesion in a given flow type. In other words, we still prefer to accentuate the rheological aspect of the two debris-flow categories, rather than a single (though possibly dominant) grain-support mechanism. As stressed by Lowe (1982), the two types of debris flow differ both in the mode of clast support, and in the mode of their *en masse* deposition (cohesive freezing and frictional freezing, respectively); apparently, both differences derive directly from the rheological properties of flowing debris.

When considering natural, gravelly debris flows, we find it also more realistic to introduce slight modification to Lowe's (1979, p. 77-78) rheological 'end-member' definitions for the two flow types. This is primarily because, in terms of a natural depositional system, one can easily envisage a fully cohesionless flow, but it is rather difficult to envisage a virtual lack of any frictional resistance in even a highly cohesive gravelly flow (though we are obviously aware of the fact that the frictional strength-component may be considerably reduced in a cohesive, particularly water-rich, debris flow; cf. Rodine and Johnson, 1976; Pierson, 1981). Accordingly, we suggest that the two types of *gravelly* debris flows are probably better defined by the following end-members (symbols as used above):

(1) cohesive debris flow (or mudflow)
$$\sigma_s = C + \sigma'_n \cdot \tan\varphi' + \mu_c \cdot \epsilon_s$$
with $C >> \sigma'_n \cdot \tan\varphi'$; and

(2) cohesionless debris flow (or grainflow)
$$\sigma_s = \sigma'_n \cdot \tan\varphi' + \mu_c \cdot \epsilon_s.$$

It is worthy to notice that the main difference between the two is essentially in the presence or absence of the cohesive strength-component (see also Figs. 18A, B). Therefore, we will especially concentrate on this point in the following discussion.

Although the theoretical distinction between the two debris-flow categories is clear, both Lowe (1979, 1982) in his proposal, and we here are well aware that this distinction may often be difficult to clearly detect in the field, especially when studying ancient deposits. The presence of abundant mud matrix, as in pebbly mudstones and muddy, matrix-supported conglomerates and breccias, should probably leave little doubt that they were deposited from cohesive (though possibly turbulent) debris flows. In general, however, both cohesive and cohesionless debris flows may deposit gravels that range from fully matrix-supported to fully clast-supported, and thus the abundance or scarcity of fine-grained matrix does not seem to be a reliable, discriminative criterion (Lowe, 1979, 1982). The clay content of a debris flow may be a critical feature, but this is difficult to estimate in ancient deposits, and it has also been argued in the literature that the clay content in a debris flow with cohesive appearance may not necessarily be high (and is often reported to be very low).

In the subsequent two sections, we attempt to partly relieve the problem of the distinction discussed above. We develop and evaluate a simple criterion which, in terms of a semi-quantitative statistical approach, appears to be useful to discriminate between the two debris-flow categories. Our reasoning below is essentially based on a simplified, conceptual consideration of the likely relationship between debris-flow's competence and thickness.

FLOW COMPETENCE VS. FLOW THICKNESS

Competence (*sensu* Hampton, 1975, 1979), or *suspension competence* (*sensu* Pierson, 1981), is the size of the largest particle that can be completely supported above the bed (*i.e.*, suspended in the debris mixture) over the life of the flow. The problem of competence of a debris flow has already been analytically considered by numerous authors, both on empirical grounds (Kuenen, 1951; Johnson, 1970, chpt. 13; Hampton, 1975, 1979; Rodine and Johnson, 1976; Pierson, 1981), and through theoretical reasoning (*e.g.*, Dott, 1963; Fisher, 1971; Hampton, 1972; Carter, 1975; Middleton and Hampton, 1973, 1976; Lowe, 1976, 1979, 1982; Middleton and Southard, 1978; Hiscott and Middleton, 1979; Nardin et al., 1979; Lewis et al., 1980; Innes, 1983). Based on these previous studies, we attempt here a simplified, quasi-quantitative reconsideration of the problem, solely from the point of view of a potential relationship between the debris-flow competence and thickness. It has commonly been emphasised in the literature that the high density of sediment gravity flows makes their thickness particularly important. However, the purpose and practical aspects of our approach will become clear in the next section.

By adopting a two-phase model (Fisher, 1971, p. 918), the competence of a debris flow is often conceptually considered in terms of a force-equilibrium equation (Hampton, 1975, 1979; Rodine and Johnson, 1976). A simplified relationship equating the downward force of clast weight to the upward force of buoyancy, resistance force of debris shear-strength, and other possible lift forces may be written as follows (modified from Hampton, 1979, p. 754, eq. 1):

$$\underbrace{\rho_c \cdot g \cdot \tfrac{1}{6} \cdot \pi \cdot D^3}_{\text{clast weight}} =$$

$$\underbrace{\rho_m \cdot g \cdot \tfrac{1}{6} \cdot \pi \cdot D^3 + (\rho_c - \rho_m) \cdot c_s \cdot g \cdot \tfrac{1}{6} \cdot \pi \cdot D^3}_{\text{buoyancy}}$$

$$+ \underbrace{a_1 \cdot S \cdot \tfrac{1}{2} \cdot \pi \cdot D^2}_{\text{strength}} + \underbrace{a_2 \cdot \Sigma_i f_i \cdot \tfrac{1}{2} \cdot \pi \cdot D^2}_{\text{other lift forces}}$$

where: D = clast diameter (note that the equation above assumes spherical clast);
ρ_c = clast density;
ρ_m = matrix-phase density;
g = acceleration due to gravity;
c_s = volume concentration of the clasts which are supported by the matrix (clasts whose weight is effectively borne by the fluidal matrix phase);
S = shear strength;
f_i = upward component of any other possible factor which offers lift force to the clast (see below);
a_1, a_2 = dimensionless proportionality coefficients (constants).

In this equation, the buoyancy term includes two components (Hampton, 1979, p. 754): (1) component for the density of matrix ('pure' buoyancy, as defined by Archimedes' principle); and (2) component for the loading of the matrix by the dispersed clasts (*i.e.*, excess pore pressure, as discussed by Pierson, 1981). This second term takes also account of static clast-to-clast contacts, which may become significant at the bulk clast concentrations, c_b, greater than 50% (Hampton, 1979, p. 756; Rodine and Johnson, 1976; Pierson, 1981); thus, in some flows the c_s value may actually be lower than c_b (but we follow Hampton here, and assume $c_s \simeq c_b$ for simplicity).

It is also important to note that most forces considered under the last ('flow-modification') term in our equation, such as dispersive pressure, turbulence, and pore-fluid expulsion, can only become effective if the debris strength is at least considerably reduced in a given part of the flow. This condition, however, is formally accounted for by the two proportionality coefficients in the equation.

After rearrangement (see Hampton, 1979, p. 755, eq. 3), and following the Coulomb-viscous strength model (see previous section), a generalised expression for competence can be written as:

$$D = \underbrace{\frac{b_1 \cdot C}{\Delta\rho' \cdot g}}_{\substack{\text{cohesive} \\ \text{strength} \\ \text{factor}, F_C^*}} + \underbrace{\frac{b_2 \cdot \sigma'_n \cdot \tan\varphi'}{\Delta\rho' \cdot g}}_{\substack{\text{frictional} \\ \text{strength} \\ \text{factor}, F_F^*}} + \underbrace{\frac{b_3 \cdot \mu_c \cdot \epsilon_s}{\Delta\rho' \cdot g}}_{\substack{\text{viscous} \\ \text{resistance} \\ \text{factor}, F_V^*}} + \underbrace{\frac{b_4 \cdot f_i}{\Delta\rho' \cdot g}}_{\substack{\text{other} \\ \text{supportive} \\ \text{factor(-s)}, \\ F_M^*}},$$

where:
- D = clast longest dimension;
- $\Delta\rho'$ = density contrast between dispersed clast and the total debris mixture ($\rho_c - \rho_d$);
- b_{1-4} = constants (which appear here as proportionality coefficients, but also take account of the clast's shape);
- F^* = general symbols for the individual 'factors', as further used in text;

and remaining symbols as already used above.

It should be noted that these various factors, as distinguished in the above equation, are obviously interdependent. Some of them can co-operate together, while others are mutually exclusive or nearly so. For example, only when the debris loses its yield strength (accounted for by the first two factors), does the viscous-resistance factor become important (Fig. 18C).

In other words, the competence is not a simple sum of all the possible factors, but rather a resultant of the factors' interaction in a given flow (as accounted for by the proportionality coefficients in the equation above). Here, however, we leave this problem, and we consider the individual 'factors' only from the point of view of their potential dependence on the flow's thickness (or applied normal stress).

Cohesion, as the resistance of a sediment against shear along a surface which is under *no* pressure, obviously represents a 'material constant' and by definition does not depend on the flow's thickness. This same statement pertains to the entire cohesive-strength factor (F_C^*) here.

As predicted by Coulomb's model (Figs. 18A, B), the frictional strength factor (F_F^*) is directly proportional to the applied normal stress, and hence to the flow's thickness. Also the rate of shear strain (ϵ_s) in the viscous-resistance factor (F_V^*) is directly proportional to the force applied (Fig. 18C), thus making the magnitude of viscous grain interactions proportional to the flow thickness; this is considered below in terms of viscous or inertial dispersive pressure (*cf.* Lowe, 1979, p. 78).

Since the mechanism of a plastico-viscous flow has commonly been inferred to be often *modified* by some other factors (*e.g.*, dispersive pressure, turbulence, and pore-fluid expulsion), these 'modification factors' (F_M^*) are also to be considered here.

As predicted by Bagnold's (1954) theory, the dispersive pressure arising from clast collisions is strongly dependent on, and directly proportional to the rate of shear strain. Thus, in our reasoning, the dispersive-pressure lift factor is to be considered directly proportional to the flow's thickness (Lowe, 1976, 1982, p. 286).

When a debris flow becomes turbulent, its clasts can be at least partly supported by the turbulence factor. When considering such a case, we can make use of the criterion that suspension of a clast requires shear velocities at least as large as the clast's settling velocity; however, it needs also to be mentioned that the effective settling velocity of the large clasts is expected to be considerably reduced by the concentration of the smaller, particularly clay-size particles (Hiscott and Middleton, 1979, p. 322). Anyway, for a given flow and slope the shear velocity is proportional to the flow's thickness (Middleton and Southard, 1978, chpt. 5; Hiscott and Middleton, *op. cit.*), and hence also the turbulence-lift factor must be considered proportional to the latter.

In a debris flow undergoing fluidisation, the drag force exerted on clasts by upward escaping fluid can be considered proportional to the fluid density and its velocity relative to the clasts (Middleton and Southard, 1978, chpt. 1, p. 15). Here again it can be argued that the pore-fluid expulsion is proportional to the flow's thickness, primarily because its magnitude (velocity) depends directly on the magnitude of the pore pressure (for appropriate formula see Pierson, 1981, p. 54, eq. 5).

A general implication of our simplified reasoning above is that the clast-support mechanism generated by a debris flow (see last equation) may be considered conceptually in terms of two components (Fig. 19): (1) *thickness-independent* component represented by the cohesive-strength factor;

and (2) *thickness-dependent* component represented by one or more of the other discussed factors. By referring to the previous section, this highly-idealised conceptual model can now be written as (Fig. 19):

$$D = F_c^* + \Sigma_i F_i^* (Y,...)$$ Cohesive debris flow

and

$$D = \Sigma_i F_i^* (Y,...)$$ Cohesionless debris flow

where: D = maximum clast size;
F_c^* = cohesive strength factor (as defined earlier in text);
$F_i^* (Y,...)$ = other physically *possible* factor (from among those reviewed above), considered only in terms of its dependence on the flow's thickness (Y).

Although these two equations above do not need to be necessarily first-order linear ones (and in theory they are not always so in fact), we simply find it more convenient to consider them as such here. In other words, we do not attempt to develop any strict mathematical model, but rather take its first-order linear approximation, which we find close enough to reality and easiest to evaluate.

In the following section, we attempt to evaluate this simple model by confronting it with available field data.

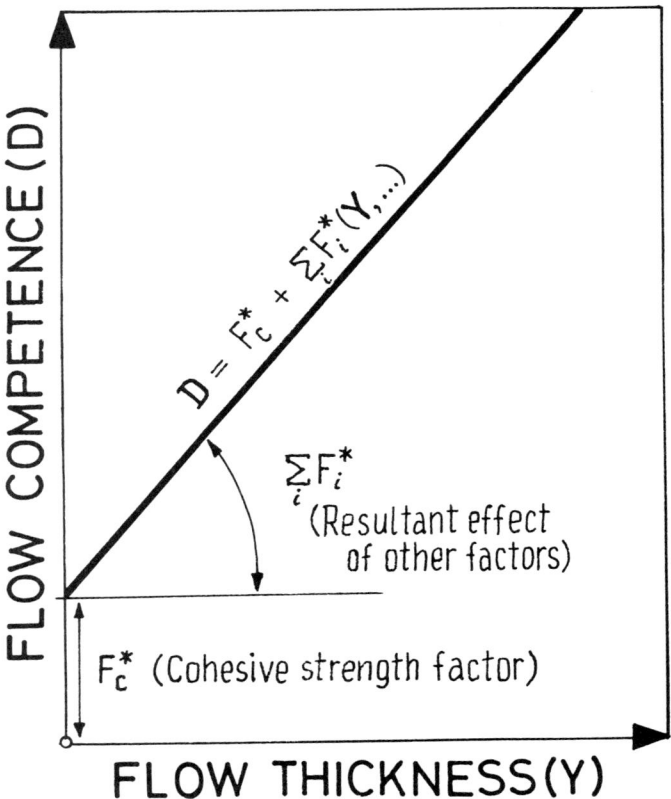

Fig. 19. Idealised, conceptual model relating the competence and thickness of debris-flows (for explanation and discussion see text).

APPLICATION OF *MPS/BTh* DIAGRAMS

Following Bluck's (1967b, p. 142) original suggestion, the thickness of a conglomerate bed (*BTh*) and the bed's maximum particle size (*MPS*, defined as an arithmetic mean from ten largest clasts) have often been used as an approximation of the thickness and competence of depositing flow at the point of measurement. Evidence from various sedimentary successions clearly indicates that the deposits of sediment gravity flows display a significant, positive *linear correlation* between their bed thicknesses and maximum particle sizes (Bluck, 1967b; Steel 1974; Larsen, 1977; Larsen and Steel, 1978; Kelling and Holroyd, 1978; Nemec *et al.*, 1980; Gloppen and Steel, 1981; Porębski, 1981, this volume; Rye Larsen, 1982; Nemec and Muszyński, 1982; Lajoie and Saint-Onge, 1984; Nemec *et al.*, this volume; Kessler and Moorhouse, this volume).

In contrast to the mass flow deposits, no such correlation has been identified in the conglomerates of fluvial origin, most likely because of their 'grain-by-grain' mode of deposition from stream bedload, and because of the considerable amounts of contemporaneous erosion invariably involved in fluvial sedimentation (Bluck, 1967b; see also Steel, 1974; Nemec and Muszyński, 1982). Based on this striking difference, the evidence of a significant *MPS/BTh* correlation has often been used as a supportive argument in inferring a mass-flow mode of deposition for the conglomerates (see papers cited above).

We strongly suggest that this simple criterion is useful for such inferences, but we also caution the reader that: (1) an appropriate statistical test is always necessary before a significant linear correlation is inferred from the data set; and (2) insignificant correlation does not necessarily negate the mass-flow mode of conglomerate emplacement. The lack of *MPS/BTh* correlation in mass flow conglomerates may result from at least two sources: misidentification of bed boundaries (*e.g.*, overestimated *BTh* in the cases of amalgamated beds) and/or considerable effects of contemporaneous interstratal erosion (underestimated *BTh*). However, although any data set must be considered as potentially biased with at least this sort of error, the entire problem is usually relieved by the statistical approach.

In this section, we use some available *MPS/BTh* data from various mass-flow conglomerates to evaluate the conceptual, quasi-quantitative model we inferred earlier for the relationship between debris-flow's competence and thickness (previous section). Obviously, the *MPS/BTh* diagram must be considered as a rough and purely statistical approximation of the latter relationship, and the entire estimation necessarily pertains to an assemblage of debris-flow beds, rather than to a single bed; hence, it reveals tendencies rather than any exact relationship. Nevertheless, we suggest that the *MPS/BTh* data, when carefully collected and statistically tested, may serve as the basis for some important inferences. Segregation of field data according to facies is advised.

Cohesive vs. Cohesionless Debris-Flows

For carefully collected data, the value of the correlation coefficient can serve as an approximate measure of consistency (or similarity) in the physical behaviour of individual flows; additional information can be obtained from the *MPS* and *BTh* frequency distributions. With the use of the *MPS/BTh* diagram, it seems possible to infer whether the debris-flow conglomerate beds have been deposited from dominantly cohesive or dominantly cohensionless flows. Providing that the data reveal high positive correlation, the *MPS/BTh* diagram and its 'best-fit' regression line can probably be considered in terms of the model presented in Figure 19. If this assumption is true, then the presence or absence of the "cohesive strength factor" (as graphically indicated in Fig. 19) may serve as a simple criterion to discriminate between the deposits emplaced by the two respective types of debris-flow.

Apparently, this suggestion is well supported by the published data on various debris-flow deposits. Figure 20 shows examples from conglomerates interpreted in terms of cohesive debris-flows, and Figure 21 shows examples interpreted as the deposits of cohesionless debris-flows.

The diagrams shown in Figure 22, in turn, not only serve as another example of the deposits of cohesionless debris-flows, but also illustrate our earlier contention that the amount of matrix in conglomerates cannot be used as a reliable criterion for discrimination between cohesive and cohesionless debris flows.

In terms of the *MPS/BTh* diagram, one may further attempt to speculate on the magnitudes of the cohesive strength (see previous section); Johnson's (1970, p. 487) criterion may also be adopted here, if the debris yield strength can be assumed to be mainly the cohesive strength. However, we emphasise the necessity of using the size of the *largest* clast, rather than *MPS*, because this former provides a better estimate of the flow's competence; perhaps the same caution pertains also to the *MPS/BTh* diagram itself.

It should be noted that our entire approach concerns the identification of the *active role* of cohesive strength in the clast-support mechanism, not merely the presence or absence of a cohesive *material*. Even a mud-rich debris-flow may sometimes be fully turbulent and behave like a viscous fluid, with no role being played by the cohesive strength. However, such a flow often must be classified as a fluidal flow, rather than as a cohesive debris-flow (see previous sections).

It is also important to note that we expect to identify here only those cohesive debris-flows in which the role of cohesive strength (or at least its clast-support effects) have not been cancelled over the life of the flow. If, for example, a cohesive debris-flow looses its strength and its bulk competence happens to decrease below the original value, a new ratio between the flow's competence and thickness arises, and this new relationship will probably contain no effect of the cohesive strength factor. Thus, the possibility of flow transformations during movement makes the inferences more intricate. However, some important transformations may be detected with the aid of the MPS/BTh diagrams (see examples below).

Flows of Higher and Lower Competence

From the presented examples (Figs. 20-22) it is clear that, irrespective of the presence or absence of the cohesive-strength effect, the slopes of the *MPS/BTh* regression lines may vary greatly. As already argued in the previous section, the slope of the line here reflects the role played by factors *other* than cohesive strength (Fig. 19). We therefore suggest that the regression line gradient can be used to evaluate the combined magnitude of these other factors, and (together with the magnitude of the cohesive strength factor, if any) to compare various groups of debris flow deposits. By this means, it is possible to distinguish between groups of flows with lower and higher competence. Such comparison is reasonable, primarily because it considers flow competence *relative* to flow thickness. The *MPS/BTh* ratio may possibly be adopted for this purpose (see Gloppen and Steel, 1981; Porębski, this volume), especially for cohesionless debris-flow data.

The nature of the other factors mentioned above (see also previous section) can often be inferred from the conglomerate beds, particularly from the type of clast grading. For instance, a well-developed inverse grading throughout the bed most likely indicates the considerable role of dispersive pressure. Beds with inverse grading in the lower part and lack of grading in the upper part are likely to indicate the existence of a non-sheared 'plug', and hence the considerable role of frictional strength. Ungraded beds usually indicate high shear-strength, or high viscosity (which prevents both turbulence and effective grain-interactions). Normal grading usually suggests some role of turbulence, though this type of vertical clast-segregation does not necessarily imply reduced competence of the flow (see example below); a fully turbulent debris-flow may suspend clasts even larger than those which could be supported in it by the shear strength and buoyancy alone (Lowe, 1982, p. 294).

Subaerial vs. Subaqueous Emplacement

Available data from ancient fan-deltas indicate that there is always a significant change in the competence of debris-flows when these latter come into water (Fig. 23; see also Larsen and Steel, 1978, Fig. 17). Water, when suddenly admixed into the flow, reduces density and viscosity of the matrix phase and concentration of the dispersed phase; hence also considerably aids the transformation of the flow's internal regime. We thus suggest that, for the products of one depositional system, some reliable *MPS/BTh* data can help to distinguish between subaerially deposited and subaqueously deposited debris-flow conglomerates; this particular problem is important in many ancient fan-deltaic successions. In some instances, such data may aid

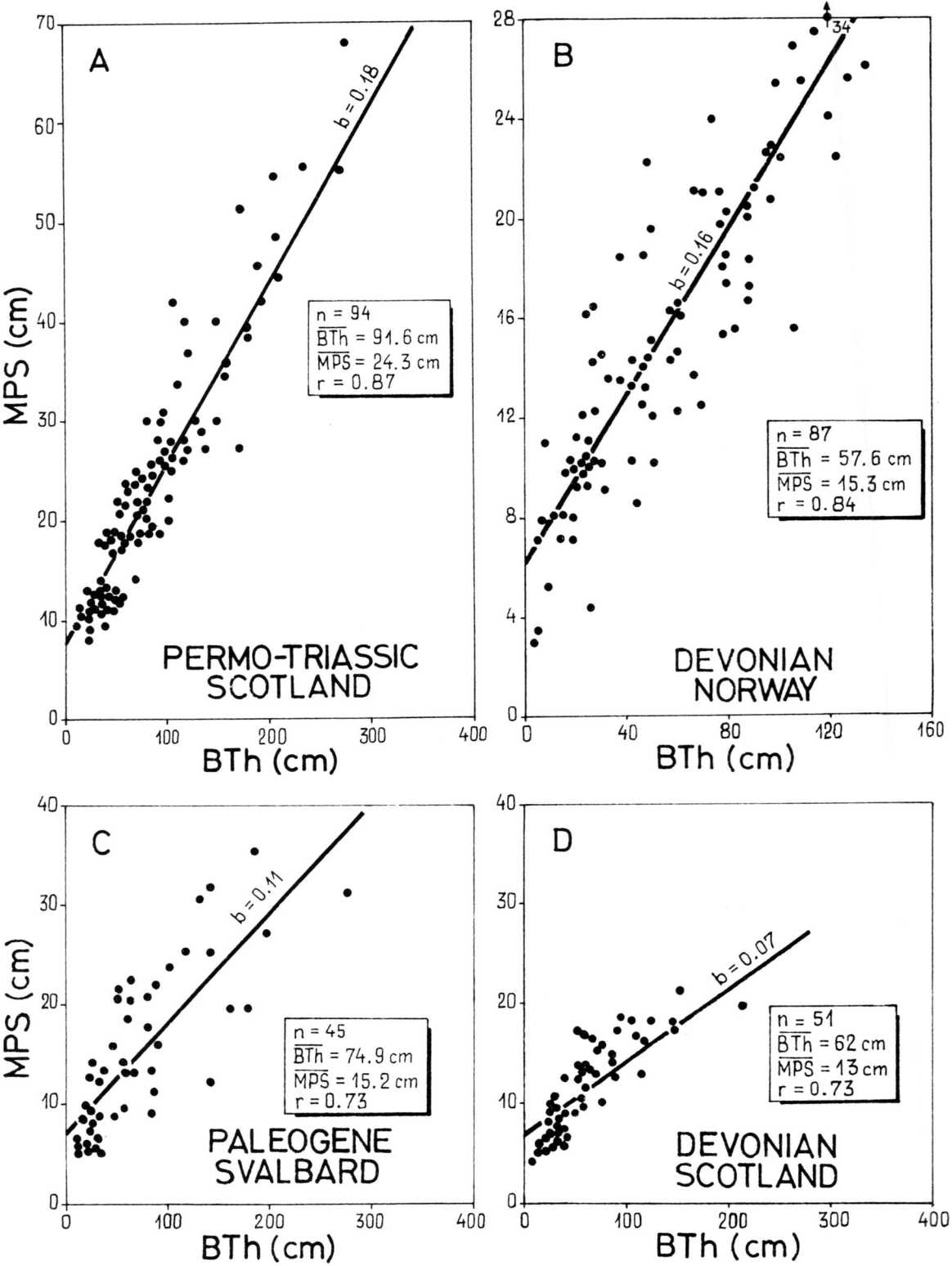

Fig. 20. Examples of *MPS/BTh* data from subaerial deposits of *cohesive* debris flows. Data from: **(A)** New Red Sandstone fanglomerates of Stornoway and Inch Kenneth, W. Scotland (modified from Steel, 1974, Fig. 4); **(B)** Devonian fanglomerates of the northern part of Hornelen Basin, SW. Norway (modified from Larsen, 1977, p. 230-246, facies D); **(C)** Paleogene fanglomerates of Prins Karls Foreland, Spitsbergen (modified from Rye Larsen, 1982, Fig. 2.10c); and **(D)** Old Red Sandstone fanglomerates of Clyde area, Scotland (modified from Bluck, 1967, Fig. 3). The slant lines are least-square regression lines fitted to the data. Explanation of symbols: b = regression coefficient (line gradient); n = number of data; $\overline{BTh}$ = mean bed thickness; $\overline{MPS}$ = mean maximum particle size; r = Pearsonian correlation coefficient.

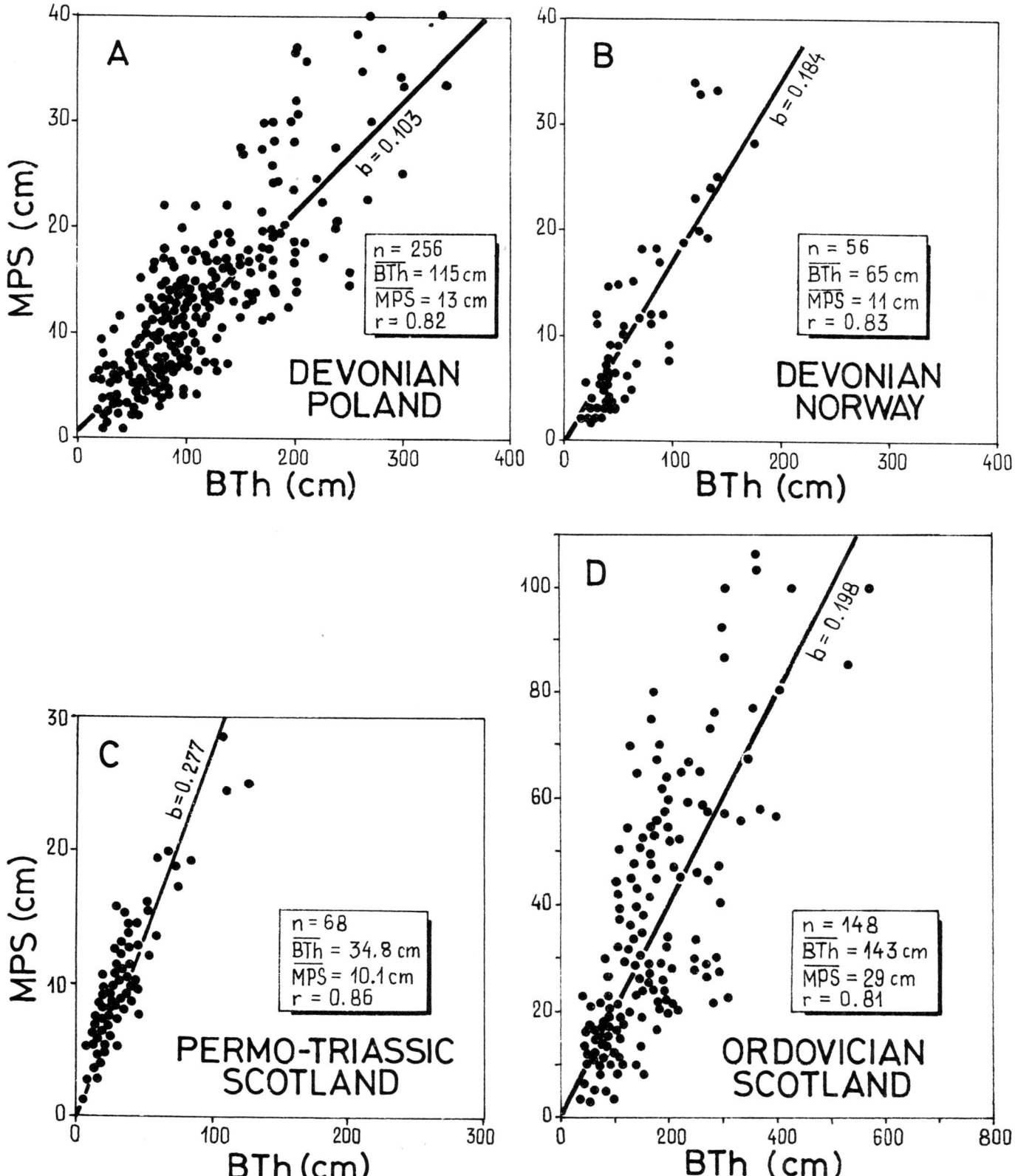

Fig. 21. Examples of *MPS/BTh* data from subaerial and subaqueous deposits of *cohesionless* debris flows. Data from: **(A)** Chwaliszów Formation (submarine fan-delta complex, Famennian-?lowest Tournaisian) of Świebodzice Depression, SW. Poland (modified from Porębski, this volume, Fig. 9); **(B)** Domba Conglomerate Mb. (data from subaerial, 'proximal' part of this lacustrine fan-delta from Devonian Hornelen Basin, SW. Norway; modified from Nemec et al., this volume, Fig. 25); **(C)** New Red Sandstone fanglomerates of Rhum and Scalpay islands, W. Scotland; modified from Steel, 1974, Fig. 4); and **(D)** Ordovician conglomerates (Caradoc submarine canyon-fan complexes) of S. Scotland (modified from Kelling and Holroyd, 1979, Figs. 11-4 A, C and E; rudite complexes S1, S5 and S6). Explanations as for Figure 20.

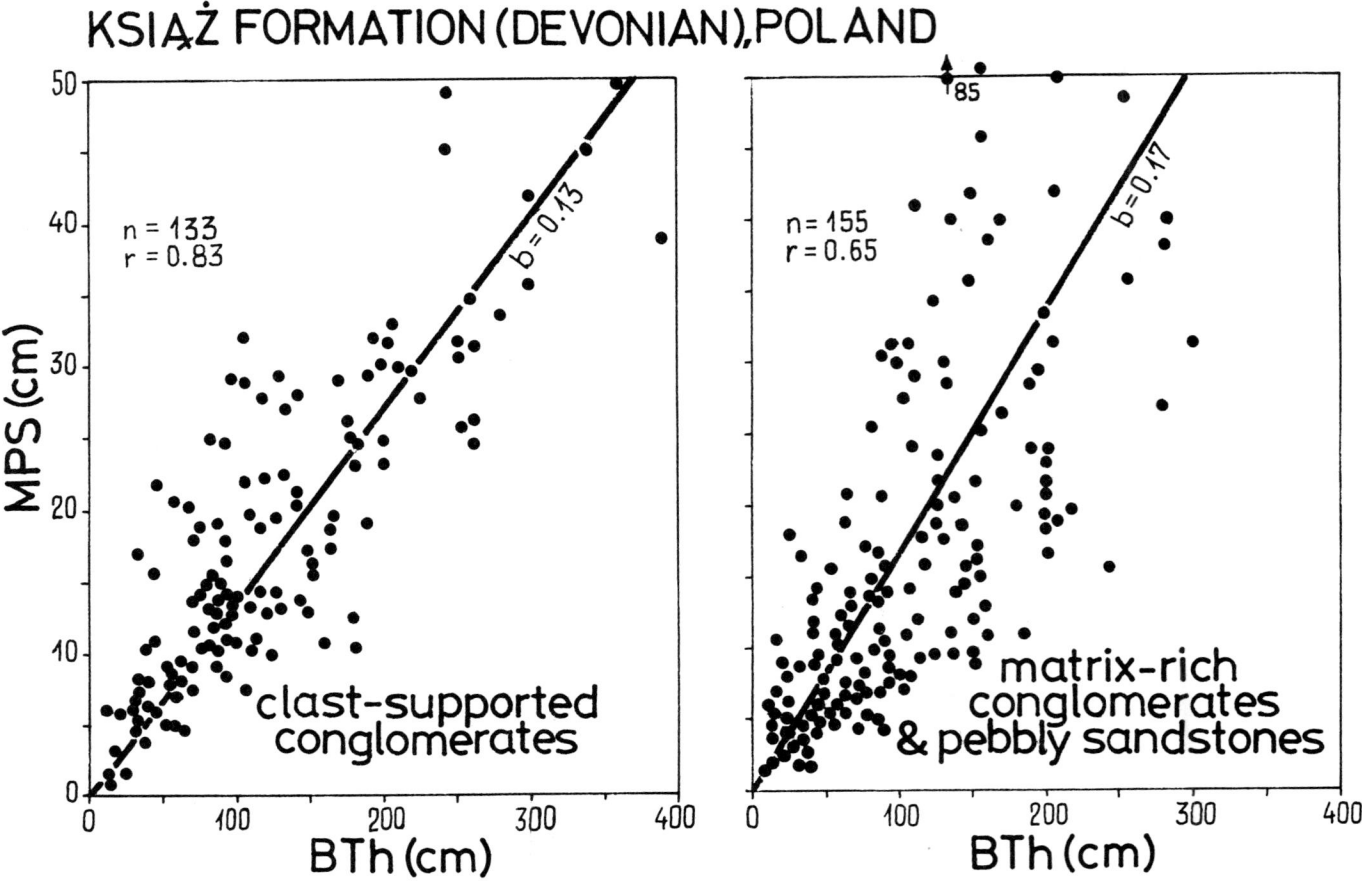

Fig. 22. Examples of *MPS/BTh* data from a submarine fan-delta complex (Książ Formation, Devonian, SW. Poland) comprising clast-supported conglomerates and sand-rich, matrix-supported conglomerates and pebbly sandstones (compiled and modified from Nemec et al., 1980, Figs. 11 and 12).

the detection of a subaqueous setting, and hence possibly also the identification of a fan-delta segment in the alluvial succession.

However, it should be noted that we cannot expect to detect such differences between subaerial and subaqueous conglomerates from two entirely different depositional systems (see examples in Figs. 20 and 21).

Other Inferences

The *MPS/BTh* diagrams may also aid analysis of the gravelly depositional systems. We illustrate this with the following example, based on Porębski's (1981, this volume) study of a Devonian, submarine fan-delta succession composed of clast-supported, subaqueously deposited, debris flow conglomerates. The four conglomerate facies shown in Figure 24 belong, both in terms of our *MPS/BTh* criterion and in the context of Porębski's interpretation, to the cohesionless category of debris flows. Porębski deduced that these facies represent a progressive downfan development of sediment gravity flows, in the order: $IG \rightarrow UG \rightarrow ING \rightarrow NG$ (facies symbols as in Fig. 24).

The inversely graded conglomerates suggest a considerable role of dispersive pressure in clast support, and they were inferred to have been deposited on a fairly steep, proximal slope of the fan delta system. A very high correlation between *MPS* and *BTh* (Fig. 24) suggests a well-established equilibrium between the competence and thickness of the depositing flows. From our experience on ancient deposits, we find this high correlation typical for the debris flows with a high component of dispersive pressure.

The ungraded conglomerates were interpreted to have been deposited in 'hydraulic jump' conditions related to submarine slope break, where sudden dissipation of the flow's kinetic energy caused rapid deposition; the latter was accompanied by partial settling (redistribution) of the larger clasts, an effect of the dispersive pressure cessation. The data (Fig. 24) show relatively low correlation between *MPS* and *BTh*, hence suggesting poor overall equilibrium between the competence and thickness of the depositing flows. The competence of the flows has been considerably reduced here in comparison to the inversely graded facies

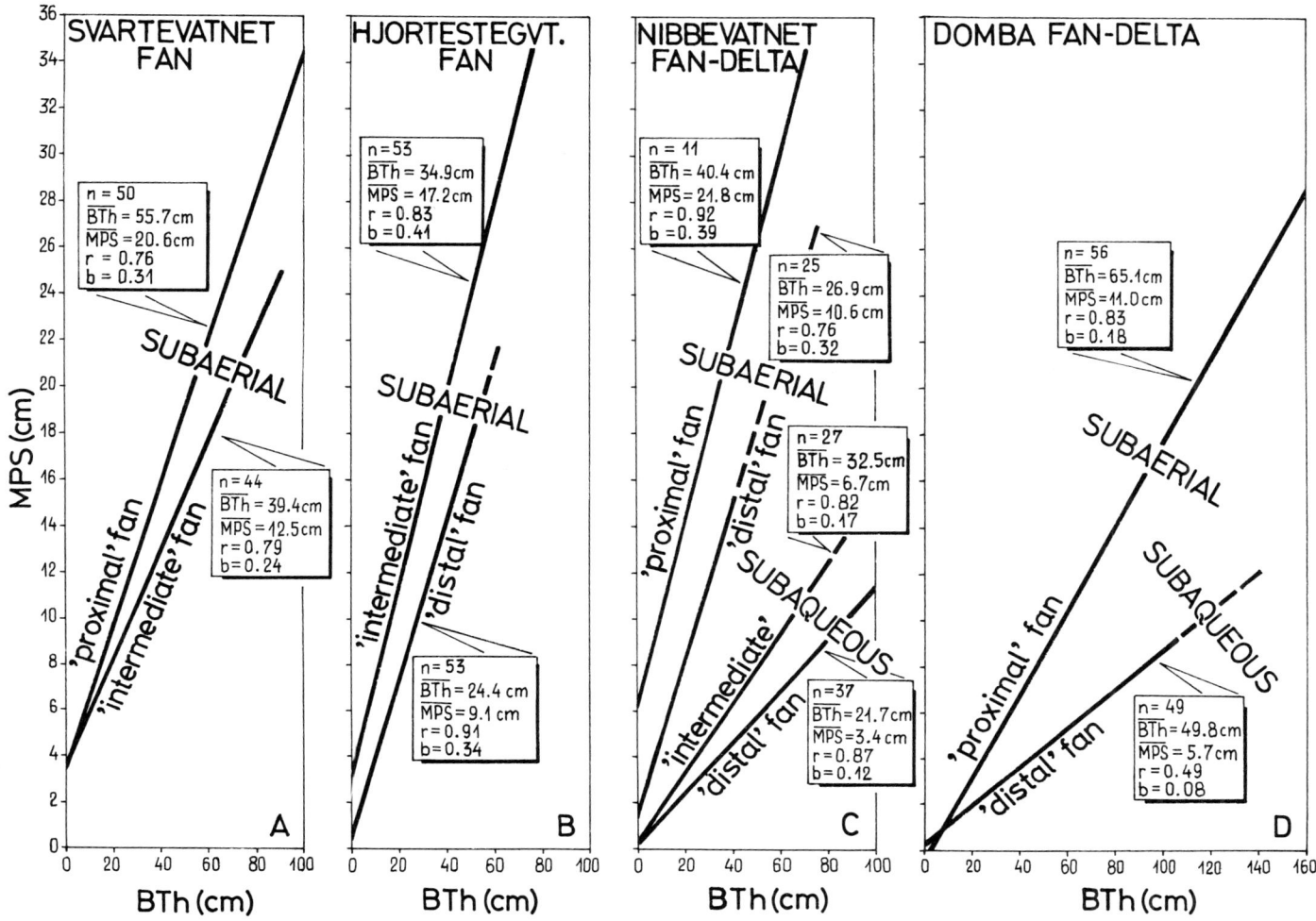

Fig. 23. Examples of *MPS/BTh* relationship in alluvial-fan and fan-delta deposits of the Devonian Hornelen Basin, SW. Norway; for simplicity and easy comparison, only the calculated 'best-fit' regression lines and the statistical characteristics of data sets are given. (Diagrams **A-C** are modified versions from Gloppen and Steel, 1981, Figs. 3 and 6; diagram **D** is modified from Nemec *et al.*, this volume, Fig. 25).

(compare regression line gradients, Fig. 24). However, many beds plot far above the regression line, and these seem to have been deposited from some 'over-competent' flows. These were probably flows that thinned drastically and rapidly froze before the inherited large clasts adjusted to the new, reduced competence. Thus, the thinning of flows and their rapid deposition, with varying amount of time left for the clast sizes to adjust to the reduced thicknesses, are thought to be the circumstances reflected by the scatter plot for this facies.

The inverse-to-normally graded conglomerates were inferred to have been deposited mainly from those higher-energy flows which passed the slope-break zone and probably continued for some distances beyond. At the slope break, however, water was probably induced into the flow and reduced its density and frictional/viscous resistance; hence, turbulence was allowed to develop in the upper part of the flow, while dispersive pressure still dominated in the lower part (as reflected by clast segregation). High correlation between *MPS* and *BTh* (Fig. 24) reflects undisturbed equilibrium between the competence and thickness of the depositing flows, but the regression line gradient indicates a relative increase of the flows' competence. Hence, although the flows became both thinner and finer (see Fig. 24, bottom), their actual competence, relative to thickness, apparently increased. Accordingly, we infer that the magnitude of dispersive pressure in these flows was probably enhanced by the turbulent shear stresses exerted from the top (*cf.* Lowe, 1976, 1982), thus making the flows more competent.

The normally graded conglomerates were deposited from those flows which, after passing the slope break, soon started to evolve towards high-density turbidity currents; more distally, these beds give way to the graded-stratified ones, with sandstone cappings (Porębski, 1981, this volume). Their clast segregation suggests rapid reduction of the

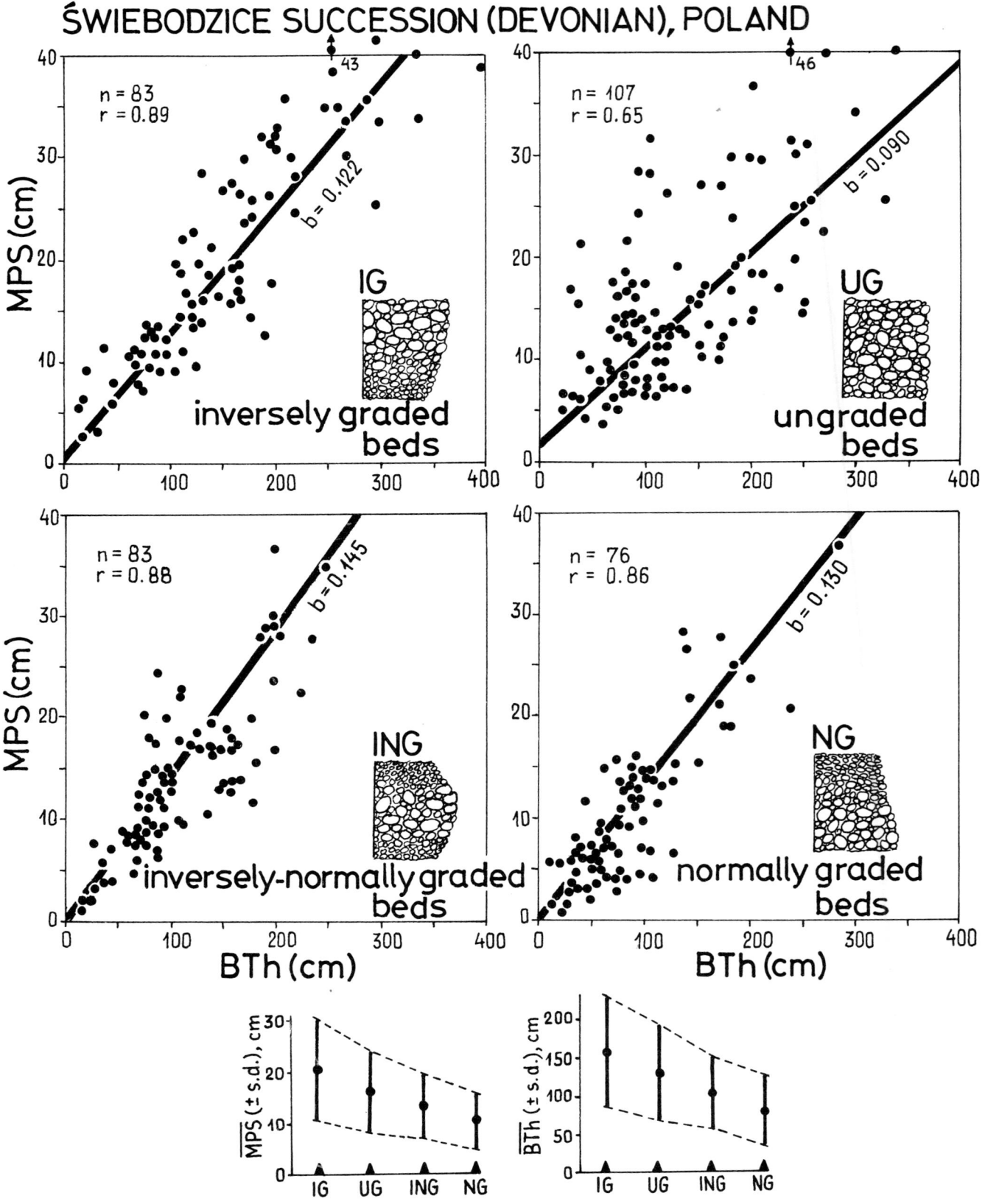

Fig. 24. *MPS/BTh* data on clast-supported conglomerates from the upper part of Świebodzice submarine fan-delta succession (Fammenian-?lower Tournaisian, SW. Poland); compiled from Porębski (this volume, Fig. 9) and Nemec *et al.* (1980, Fig. 11). The two diagrams at the bottom show the mean values of *MPS* and *BTh* (± standard deviation) for the respective bed types (from Porębski, 1981, Fig. 42B).

dispersive pressure and its replacement by turbulence (though with sediment concentration still high enough to prevent the development of any traction structures). An important suggestion from the respective diagram (Fig. 24) is that, despite this latter change, the competence of the flows essentially remained unchanged. Thus, for the flow of a given thickness, the turbulence factor was able to support clasts as large as those that were supported by the dispersive-pressure factor.

This brief example above clearly shows that the *MPS/BTh* diagrams, when based on reliable data, can considerably aid the understanding of a depositional system. Many of these inferences would otherwise be difficult or impossible to arrive at.

Final Remarks

Both the alluvial environment and the marine shoreline environment allow the transport and deposition of gravels, and in both these environments a variety of processes is later reflected by the resultant depositional sequences. Depositional facies commonly comprise gravels intimately intercalated with finer clastics, and it is only the analysis of all types of sediment that provides the critical information about the individual processes and environment.

Until recently, emphasis on the analysis of gravels and conglomerates has been directed to these two environments individually. Now, researchers have stressed that when these two environments occur adjacent to each other the depositional settings of gravel become particularly intricate and give rise to a new spectrum of facies. Within the concept of gravelly fan-delta, particularly when referring to ancient submarine cases, a much broader range of depositional settings must then be accounted for, and hence an even broader view of diagnostic features is required.

In our contribution to this volume, we have attempted to present a brief guide for recognising depositional facies and facies sequences which we have found particularly useful in our own research on alluvial and coastal gravelly environments. Not all of the reviewed depositional settings have been equally well studied, thus it is not always obvious which facies characteristics are most diagnostic of a particular setting. We have used mainly our knowledge from ancient sedimentary successions. Most sedimentologists perhaps agree that the ancient record provides, indeed, a much better view of depositional features, and hence offers also a better choice of features which are potentially diagnostic.

We have presented a review of those features and facies criteria which we believe are common, or potentially common, to most ancient situations. Our primary intention here is to stimulate additional interest in gravelly depositional settings, especially those in the important transition from subaerial to subaqueous sedimentation.

References

Aalto, K.R. 1976. Sedimentology of a mélange: Franciscan of Trinidad, California. Journal of Sedimentary Petrology, v. 46, p. 913-929.

Bagnold, R.A. 1954. Experiments on a gravity-free dispersion of large solid spheres in a Newtonian fluid under shear. Proceedings of the Royal Society of London, Series A, v. 225, p. 49-63.

Baker, V.R. 1973. Paleohydrology and sedimentology of Lake Missoula flooding of eastern Washington. Geological Society of America, Special Paper 144, 79 p.

Baker, V.R. and Nummedal, D. 1978. The Channeled Scabland. Washington: National Aeronautics and Space Administration, Comparative Planetary Geology Field Conference Guide to Columbia Basin, 186p.

Bates, C.C. 1953. Rational theory of delta formation. American Association of Petroleum Geologists Bulletin, v. 39, p. 2119-2162.

Beverage, J.P. and Culbertson, J.K. 1964. Hyperconcentration of suspended sediment. Proceedings of the American Society of Civil Engineers, Journal of Hydraulics Division, v. 90, p. 117-128.

Bluck, B.J. 1967a. Sedimentation of beach gravels: examples from South Wales. Journal of Sedimentary Petrology, v. 37, p. 128-156.

Bluck, B.J. 1967b. Deposition of some Upper Old Red Sandstone conglomerates in the Clyde area: a study in the significance of bedding. Scottish Journal of Geology, v. 3, p. 139-167.

Bluck, B.J. 1974. Structure and directional properties of some valley sandur deposits in southern Iceland. Sedimentology, v. 21, p. 533-554.

Bluck, B.J. 1976. Sedimentation in some Scottish rivers of low sinuosity. Transactions of the Royal Society of Edinburgh, v. 69, p. 425-456.

Bluck, B.J. 1979. Structure of coarse-grained braided alluvium. Transactions of the Royal Society of Edinburgh, v. 70, p. 181-221.

Boothroyd, J.C. and Ashley, G.H. 1975. Process, bar morphology, and sedimentary structures on braided outwash fans, northeastern Gulf of Alaska. *In*: Jopling, A.V. and McDonald, B.C. (Eds.), Glaciofluvial and Glaciolacustrine Sedimentation. Society of Economic Paleontologists and Mineralogists, Special Publication 23, p. 193-222.

Bull, W.B. 1963. Alluvial fan deposits in western Fresno County, California. Journal of Geology, v. 71, p. 243-251.

Bull, W.B. 1977. The alluvial-fan environment. Progress in Physical Geography, v. 1, p. 222-270.

Carter, R.M. 1975. A discussion and classification of subaqueous mass-transport with particular application to grain-flow, slurry-flow and fluxoturbidites. Earth Science Reviews, v. 11, p. 145-177.

Casey, J.M. and Scott, A.J. 1979. Pennsylvanian coarse-grained fan deltas associated with the Uncompahgre Uplift, Talpa, New Mexico. New Mexico Geological Society Guidebook, 30th Field Conference, Santa Fe Country, p. 211-218.

Church, M. and Gilbert, R. 1975. Proglacial fluvial and lacustrine environments. *In*: Jopling, A.V. and McDonald, B.C. (Eds.), Glaciofluvial and Glaciolacustrine Sedimentation. Society of Economic Paleontologists and Mineralogists, Special Publication 23, p. 22-100.

Clifton, H.E. 1973. Pebble segregation and bed lenticularity in wave-worked versus alluvial gravel. Sedimentology, v. 20, p. 173-187.

Collela, A. 1980. Coarse-grained point bars and channel mouth-bars along Zena Valley (Pliocene, Intra-Appenninic Basin, Bologna, Italy). Bochum: International Association of Sedimentologists, 1st European Regional Meeting, Abstracts, p. 83-85.

Davies, W.M. 1938. Sheetfloods and stream floods. Geological Society of America Bulletin, v. 49, p. 1337-1406.

Dobkins, J.E. and Folk, R.L. 1970. Shape development on Tahiti-Nui. Journal of Sedimentary Petrology, v. 40, p. 1161-1203.

Dott, R.J., Jr. 1963. Dynamics of subaqueous gravity depositional processes. American Association of Petroleum Geologists Bulletin, v. 47, p. 104-128.

Elliot, T.E. 1978. Deltas. *In*: Reading, H.G. (Ed.), Sedimentary Environments and Facies. Oxford: Blackwell, p. 97-142.

Enos, P. 1977. Flow regimes in debris flow. Sedimentology, v. 24, p. 133-142.

Farquarson, G.W. 1982. Lacustrine fan deltas in a Mesozoic alluvial sequence from Camp Hill, Antarctica. Sedimentology, v. 29, p. 717-726.

Fisher, R.V. 1971. Features of coarse-grained, high-concentration fluids and their deposits. Journal of Sedimentary Petrology, v. 41, p. 916-927.

Gloppen, T.G. and Steel, R.J. 1981. The deposits, internal structure and geometry in six alluvial fan-fan delta bodies (Devonian, Norway) — a study in the significance of bedding sequences in conglomerates. In: Ethridge, F.G. and Flores, R.M. (Eds.), Recent and Ancient Non-Marine Depositional Environments: Models for Exploration. Society of Economic Paleontologists and Mineralogists, Special Publication 31, p. 49-69.

Gnaccolini, M. 1981. A fan-delta depositional model from the Oligocene of southern Piedmont. Bologna: International Association of Sedimentologists, 2nd European Regional Meeting, Abstracts, p. 75-78.

Gradstein, F.M. and Van Gelder, A. 1971. Prograding clastic fans and transition from a fluviatile to a marine environment in Neogene deposits of eastern Crete. Geologie en Mijnbouw, v. 50, p. 383-392.

Hampton, M.A. 1972. The role of subaqueous debris flow in generating turbidity currents. Journal of Sedimentary Petrology, v. 42, p. 775-793.

Hampton, M.A. 1975. Competence of fine-grained debris flows. Journal of Sedimentary Petrology, v. 45, p. 834-844.

Hampton, M.A. 1979. Buoyancy in debris flows. Journal of Sedimentary Petrology, v. 49, p. 753-758.

Hand, B.M. 1961. Grain orientation in turbidites. Compass Sigma Gamma Epsilon, v. 38, p. 133-144.

Harms, J.C., Southard, J., Spearing, D.R. and Walker, R.G. 1975. Depositional Environments as Interpreted from Primary Sedimentary Structures and Stratification Sequences. Dallas, Texas: Society of Economic Paleontologists and Mineralogists, Short Course No. 2, 161 p.

Harrison, S. and Fritz, W.J. 1982. Depositional features of March 1982 Mount St. Helens sediment flows. Nature, v. 299, p. 720-722.

Hein, F.J. 1982. Depositional mechanisms of deep-sea coarse clastic sediments, Cap Enragé Formation, Québec. Canadian Journal of Earth Sciences, v. 19, p. 267-287.

Hein, F.J. and Walker, R.G. 1977. Bar evolution and development of stratification in the gravelly, braided Kicking Horse River, British Columbia. Canadian Journal of Earth Sciences, v. 14, p. 562-570.

Hein, F.J. and Walker, R.G. 1982. The Cambro-Ordovician Cap Enragé Formation, Québec, Canada: conglomeratic deposits of a braided submarine channel with terraces. Sedimentology, v. 29, p. 309-329.

Heward, A.P. 1978a. Alluvial fan and lacustrine sediments from the Stephanian A and B (La Magdalena, Cinera-Matallana and Sabero) coalfields, northern Spain. Sedimentology, v. 25, p. 451-488.

Heward, A.P. 1978b. Alluvial fan sequence and megasequence models: with examples from Westphalian D — Stephanian B coalfields, N. Spain. In: Miall, A.D. (Ed.), Fluvial Sedimentology. Canadian Society of Petroleum Geologists, Memoir 5, p. 669-702.

Hiscott, R.N. and Middleton, G.V. 1979. Depositional mechanics of thick-bedded sandstones at the base of submarine slope, Tourelle Formation (Lower Ordovician), Québec, Canada. In: Doyle, L.J. and Pilkey, O.H. (Eds.), Geology of Continental Slopes. Society of Economic Paleontologists and Mineralogists, Special Publication 27, p. 307-326.

Hooke, R.LeB. 1967. Processes on arid region alluvial fans. Journal of Geology, v. 75, p. 438-460.

Innes, J.L. 1983. Debris flows. Progress in Physical Geography, v. 7, p. 469-501.

Johnson, A.M. 1970. Physical Processes in Geology. San Francisco: Freeman, Cooper and Company, 577 p.

Kelling, G. and Holroyd, J. 1978. Clast size, shape, and composition in some ancient and modern fan gravels. In: Stanley, D.J. and Kelling, G. (Eds.), Sedimentation in Submarine Canyons, Fans, and Trenches. Stroudsburg, Pennsylvania: Dowden, Hutchinson and Ross, p. 136-159.

Kuenen, Ph.H. 1951. Properties of turbidity currents of high density. In: Hough, J.L. (Ed.), Turbidity Currents and the Transportation of Coarse Sediments to Deep Water. Society of Economic Paleontologists and Mineralogists, Special Publication 2, p. 14-33.

Krause, F.F. and Oldershaw, A.E. 1979. Submarine carbonate breccia beds — a depositional model for two-layer, sediment gravity flows from the Sekwi Formation (Lower Cambrian), Mackenzie Mts, Northern Territories, Canada. Canadian Journal of Earth Sciences, v. 16, p. 189-199.

Lajoie, J. and Saint-Onge, D.A. 1984. Characteristics of two Pleistocene channel-fill deposits and their implications for the interpretation of megasequences in ancient sediments. Marseille: International Association of Sedimentologists, 5th European Regional Meeting, Abstracts, p. 248.

Larsen, V. 1977. Aspects of the sedimentology and palaeogeography along a segment of the northern margin of Hornelen Basin (Devonian), western Norway, with emphasis on alluvial fan bodies and their deposits between Mykebustsætra and Karlskaret. Can. Real. Thesis, University of Bergen, Bergen, 246 p.

Larsen, V. and Steel, R.J. 1978. The sedimentary history of a debris flow-dominated alluvial fan: a study of textural inversion. Sedimentology, v. 25, p. 37-59.

Lawson, D.E. 1982. Mobilization, movement and deposition of active subaerial sediment flows, Matanuska Glacier, Alaska. Journal of Geology, v. 90, p. 279-300.

Leithold, E.L. and Bourgeois, J. (in press). Characteristics of coarse-grained sequences deposited in nearshore, wave-dominated environments: examples from the Miocene of southwest Oregon. Sedimentology.

Lewis, D.W., Laird, M.G. and Powell, R.D. 1980. Debris flow deposits of early Miocene age, Deadman Stream, Marlborough, New Zealand. Sedimentary Geology, v. 27, p. 83-118.

Lowe, D.R. 1976. Grain flow and grain flow deposits. Journal of Sedimentary Petrology, v. 46, p. 188-199.

Lowe, D.R. 1979. Sediment gravity flows: their classification and some problems of application to natural flows and deposits. In: Doyle, L.J. and Pilkey, O.H. (Eds.), Geology of Continental Slopes. Society of Economic Paleontologists and Mineralogists, Special Publication 27, p. 75-82.

Lowe, D.R. 1982. Sediment gravity flows: II. Depositional models with special reference to the deposits of high-density turbidity currents. Journal of Sedimentary Petrology, v. 52, p. 279-297.

Malde, H.E. 1968. The catastrophic late Pleistocene Bonneville Flood in the Snake River Plain, Idaho. United States Geological Survey Professional Paper 596, 52 p.

Maejima, W. 1982. Texture and stratification of gravelly beach sediments, Enju Beach, Kii Peninsula, Japan. Osaka City University, Journal of Geosciences, v. 25, p. 35-51.

Massari, F. 1981. Giant bedforms in Messinian distributary channels. Bologna: International Association of Sedimentologists, 2nd European Regional Meeting, Abstracts, p. 104-106.

Massari, F. 1983. Tabular cross-bedding in Messinian fluvial channel conglomerates, southern Alps, Italy. International Association of Sedimentologists, Special Publication 6, p. 287-300.

McLaughlin, R.J. and Nilsen, T.H. 1982. Neogene non-marine sedimentation and tectonics in small pull-apart basins of the San Andreas fault system, Sonoma County, California. Sedimentology, v. 29, p. 865-876.

Miall, A.D. 1977. A review of the braided river depositional environment. Earth Science Reviews, v. 13, p. 1-62.

Miall, A.D. 1978. Lithofacies types and vertical profile models in braided river deposits: a summary. In: Miall, A.D. (Ed.), Fluvial Sedimentology. Canadian Society of Petroleum Geologists, Memoir 5, p. 597-604.

Miall, A.D. 1983. Glaciomarine sedimentation in the Gowganda Formation (Huronian), northern Ontario. Journal of Sedimentary Petrology, v. 53, p. 477-491.

Middleton, G.V. and Hampton, M.A. 1973. Sediment gravity flows: mechanics of flow and deposition. In: Middleton, G.V. and Bouma, A.H. (Co-chairmen), Turbidites and Deep-Water Sedimentation. Los Angeles: Society of Economic Paleontologists and Mineralogists, Pacific Section, Short Course, p. 1-38.

Middleton, G.V. and Hampton, M.A. 1976. Subaqueous sediment transport and deposition by sediment gravity flows. In: Stanley, D.J. and Swift, D.J.P. (Eds.), Marine Sediment Transport and Environmental Management. New York: Wiley, p. 197-298.

Middleton, G.V. and Southard, J.B. 1978. Mechanics of Sediment Movement (2nd printing). Binghamton, New York: Society of Economic Paleontologists and Mineralogists, Short Course No. 3, 242 p.

Nardin, T.R., Hein, F.J., Gorsline, D.S. and Edwards, B.D. 1979. A review of mass movement processes, sediment and acoustic characteristics, and contrasts in slope and base-of-slope systems versus canyon-fan-basin floor systems. *In*: Doyle, L.J. and Pilkey, O.H. (Eds.), Geology of Continental Slopes. Society of Economic Paleontologists and Mineralogists, Special Publication 27, p. 61-73.

Nemec, W., Porębski, S.J. and Steel, R.J. 1980. Texture and structure of resedimented conglomerates — examples from Książ Formation (Famennian-Tournaisian), southwestern Poland. Sedimentology, v. 27, p. 519-538.

Nemec, W. and Muszyński, A. 1982. Volcaniclastic alluvial aprons in the Tertiary of Sofia district (Bulgaria). Annales Societatis Geologorum Poloniae, v. 52, p. 239-303.

Netland, A. 1981. Facies analyse av Drevbreen Beds og Nordenskioldbreen Formasjonen, Øvre Karbon til Undre Perm, Bellsund område, Svalbard. Cand. Real. Thesis, University of Bergen, Bergen, 212 p.

Ori, G.G. and Ricci Lucchi, F. 1981. Giant epsilon bedding in coarse-grained point bars of late Pliocene fan-delta systems. Bologna: International Association of Sedimentologists, 2nd European Regional Meeting, Abstracts, p. 137-141.

Pierson, T.C. 1981. Dominant particle support mechanisms in debris flows at Mt Thomas, New Zealand, and implications for flow mobility. Sedimentology, v. 28, p. 49-60.

Porębski, S.J. 1982. Świebodzice succession (Upper Devonian-lowest Carboniferous, Western Sudetes): a prograding, mass-flow dominated fan-delta complex. Geologia Sudetica, v. 16, p. 101-192.

Rainone, M., Nanni T., Ori, G.G. and Ricci Lucchi, F. 1981. A prograding gravel beach in Pleistocene fan delta deposits, south of Ancona, Italy. Bologna: International Association of Sedimentologists, 2nd European Regional Meeting, Abstracts, p. 155-156.

Rees, A.J. 1983. Experiments on the production of transverse grain alignment in a sheared dispersion. Sedimentology, v. 30, p. 437-448.

Rich, J.L. 1935. Origin and evolution of rock fans and pediments. Geological Society of America Bulletin, v. 46, p. 999-1024.

Rodine, J.D. and Johnson, A.M. 1976. The ability of debris, heavily freighted with coarse clastic materials, to flow on gentle slopes. Sedimentology, v. 23, p. 213-234.

Rust, B.R. 1972. Structure and process in a braided river. Sedimentology, v. 18, p. 221-245.

Rust, B.R. 1975. Fabric and structure in glaciofluvial gravels. *In*: Jopling, A.V. and McDonald, B.C. (Eds.), Glaciofluvial and Glaciolacustrine Sedimentation. Society of Economic Paleontologists and Mineralogists, Special Publication 23, p. 238-248.

Rust, B.R. 1978. Depositional models for braided alluvium. *In*: Miall, A.D. (Ed.), Fluvial Sedimentology. Canadian Society of Petroleum Geologists, Memoir 5, p. 605-625.

Rust, B.R. 1979. Facies models 2: Coarse alluvial deposits. *In*: Walker, R.G. (Ed.), Facies Models. Geoscience Canada Reprint Series 1, p. 9-23.

Rye Larsen, M. 1982. Sedimentasjon og tektonisk utvikling av et basseng ved en transform plategrense. Cand. Real. Thesis, University of Bergen, Bergen, 380 p.

Smith, N.D. 1974. Sedimentology and bar formation in the Upper Kicking Horse River, a braided outwash stream. Journal of Geology, v. 82, p. 205-224.

Spinnangr, Å. 1975. Some sedimentary and stratigraphic studies of the Devonian strata across the western part of Hornelen Basin, western Norway. Cand. Real. Thesis, University of Bergen, Bergen, 247 p.

Steel, R.J. 1974. New Red Sandstone floodplain and piedmont sedimentation in the Hebridean province, Scotland. Journal of Sedimentary Petrology, v. 44, p. 336-357.

Steel, R.J., Mæhle, S., Nilsen, H., Røe, S.L. and Spinnangr, Å. 1977. Coarsening-upward cycles in the alluvium of Hornelen Basin (Devonian), Norway: Sedimentary response to tectonic events. Geological Society of America Bulletin, v. 88, p. 1124-1134.

Steel, R.J. and Aasheim, S. 1978. Alluvial sand deposition in a rapidly subsiding basin (Devonian, Norway). *In*: Miall, A.D. (Ed), Fluvial Sedimentology. Canadian Society of Petroleum Geologists, Memoir 5, p. 385-412.

Steel, R.J. and Thompson, D.B. 1983. Structures and textures in Triassic braided stream conglomerates (Bunter Pebble Beds) in the Sherwood Sandstone Group, North Staffordshire, England. Sedimentology, v. 30, p. 341-367.

Van der Meulen, S. 1983. Internal structure and environmental reconstruction of Eocene transitional fan-delta deposits, Monllobat-Castigalen Formations, southern Pyrenees, Spain. Sedimentary Geology, v. 37, p. 85-112.

Walker, R.G. 1975. Generalized facies models for resedimented conglomerates of turbidite association. Geological Society of America Bulletin, v. 86, p. 737-748.

Wasson, R.J. 1977. Last-glacial alluvial fan sedimentation in the Lower Derwent Valley, Tasmania. Sedimentology, v. 24, p. 781-799.

Wasson, R.J. 1979. Sedimentation history of the Mundi-Mundi alluvial fans, western New South Wales. Sedimentary Geology, v. 22, p. 21-51.

Wiley, T.J. and Moore, E.L. 1983. Pliocene shallow-water sediment gravity flows at Moss Beach, San Mateo County, California. *In*: Larue, D.K. and Steel, R.J. (Eds.), Cenozoic Marine Sedimentation, Pacific Margin, USA. Society of Economic Paleontologists and Mineralogists, Pacific Section, p. 29-43.

Winn, R.D., Jr. and Dott, R.H., Jr. 1979. Deep-water fan-channel conglomerates of Late Cretaceous age, southern Chile. Sedimentology, v. 26, p. 203-228.

Wright, L.D. 1977. Sediment transport and deposition at river mouth: a synthesis. Geological Society of America Bulletin, v. 88, p. 857-868.

DEEP-SEA AND FLUVIAL BRAIDED CHANNEL CONGLOMERATES: A COMPARISON OF TWO CASE STUDIES

FRANCES J. HEIN[1]

ABSTRACT

Detailed comparisons of modern gravelly fluvial deposits from rivers in the Canadian Rocky Mountains and Cambro-Ordovician deep-sea valley-fill conglomerates (Cap Enragé Formation, Québec) reveal some similarities and differences.

Braided channel deposits in both deep-sea and fluvial settings occur in laterally fining/thinning units within concave-up lenses bounded by major scour surfaces at the base. Vertical sequences are mainly fining-upward, less commonly coarsening, 5-10 m thick for deep-sea and 2-3 m thick for fluvial channel fills. Main channel deposits in both settings are dominantly structureless.

Braid bar deposits in deep-sea and fluvial sediments occur in laterally coarsening/thickening units with flat bases and convex-up top surfaces. Vertical sequences vary depending upon locations within the bar complex and the type of braid bar. Generally, vertical sequences are fining-upward, 1-2 m thick, horizontally stratified or cross-bedded units. Planar tabular cross-stratification is more common in fluvial bars than in deep-sea bar deposits. Graded trough cross-stratification, graded horizontal stratification and irregular inclined cross-stratification are very common in deep-sea bar deposits. These features were not recognized in the fluvial conglomerates. Horizontal stratification consists of layers with different clast sizes or layers with alternating matrix-filled and open-work texture in fluvial conglomerates. Horizontal stratification consists of layers with different clast sizes or layers with alternating clast-supported and clast-dispersed texture in deep-sea conglomerates. Open-work texture was not observed in the deep-sea conglomerates.

Grading types and gravel fabric patterns are perhaps the most useful criteria in the distinction of fluvial from deep-sea conglomerates. Fluvial conglomerates are mainly ungraded, with less common normal or inversely-graded beds. Deep-sea conglomerates are mainly normally graded, with less common ungraded conglomerates and rare inversely or complexly graded beds. Deep-sea channel conglomerates have *a*-axis flow-parallel, *a*-axis upstream imbricate fabrics. In fluvial channels the smaller clasts may also assume an *a*-axis flow-parallel, *a*-axis imbricate upstream pattern. However, the coarser clasts are generally aligned in fluvial channels with *a*-axis flow-transverse, *b*-axis imbricate upstream. Braid bar deposits are distinguished on the basis of imbrication: in fluvial deposits imbrications are *b*-axis upstream; in deep-sea deposits imbrications are either *a*-axis upstream or *a*-axis upstream and downstream (bimodal). A-axis orientations in bedding are quite variable in both fluvial and deep-sea bar deposits and are not very reliable.

RÉSUMÉ

Des comparaisons détaillées de dépôts fluviatiles graveleux actuels de rivières dans le Montagnes Rocheuses du Canada avec des conglomérats cambro-ordoviciens de remplissage de canyons sous-marins profonds (formation du Cap Enragé, Québec) indiquent certaines similitudes et différences.

Les dépôts des chenaux anastomosés, tant dans un contexte de mer profonde que fluviatile, apparaissent en unités qui latéralement s'amincissent et déploient une texture plus fine avec des lentilles concaves vers le haut limitées à la base par d'importantes surfaces affouillées. En général, dans les séquences verticales, la finesse de la texture croît vers le haut, moins souvent vers le bas, l'épaisseur varie de 5-10 m pour les dépôts de mer profonde et de 2-3 m pour les dépôts de remplissage de chenaux fluviaux. Les dépôts majeurs des chenaux dans les deux contextes sont généralement non-structurés.

Les dépôts de barres de courants anastomosés en mer profonde et les sédiments fluviatiles forment des unités qui déploient latéralement une granulométrie et une épaisseur croissantes sur un bas-fond plat et avec les surfaces supérieurs convexes vers le haut. Les séquences verticales varient selon les endroits au sein du complexe de barres et le type de barre de courants anastomosés. Généralement, dans les séquences verticales la finesse de la texture est ascendante, l'épaisser est de 1-2 m, et les unités sont à stratification horizontale ou oblique. La stratification oblique de corps tabulaires planaires est plus fréquente dans les dépôts de barres fluviatiles que dans ceux de mer profonde. Dans les dépôts de barres de mer profonde on observe abondamment des couches granoclassées à stratification oblique occupant des creux, des couches granoclassées à stratification horizontale et des couches irrégulières inclinées à stratification oblique. Ces particularités n'ont pas été reconnues dans les conglomérats fluviatiles. La stratification horizontale est formée de couches incluant des fragments de différentes grosseurs ou des couches exposant une alternance d'une texture de matrice réaménagée par colmatage ou ouverture dans les conglomérats fluviatiles. La stratification horizontale renferme des couches de fragments de différentes grosseurs ou des couches avec une texture alternante de fragments jointifs et de fragments dispersés dans les conglomérats de mer profonde. La texture réaménagée ouverte n'est pas observée dans les conglomérats de mer profonde.

Les types de granoclassement et les motifs des fabriques dans les graviers constituent probablement les meilleurs critères de différenciation entre les conglomérats fluviatiles et ceux de mer profonde. Les conglomérats fluviatiles sont en général non-granoclassés, avec une plus faible proportion de couches à granoclassement normal ou inverse. Les conglomérats de mer profonde sont majoritairement à granoclassement normal, avec une quantité inférieur de couches de conglomérats non-granoclassés et rarement avec des couches à granoclassement inverse ou complexe. Les conglomérats des chenaux de mer profonde ont un axe -*a* parallèle à la direction de l'écoulement et dans les fabriques imbriquées l'axe -*a* est orienté vers l'amont. Les plus petits fragments déposés dans les chenaux fluviaux peuvent également présenter l'axe -*a* parallèle à la direction de l'écoulement et pur les imbrications l'axe -*a* est orienté vers l'amont. Cependant, les

[continued]

---

[1]Department of Geology, University of Alberta, Edmonton, Alberta T6G 2E3, Canada.

This paper is based upon two theses supervised by Roger G. Walker at McMaster University. Funds for the modern river studies were from a Geological Society of America Penrose Research Grant, 1973 to F.J. Hein. The Cap Enragé study was funded by a National Research Council of Canada grant to R.G. Walker. Funds for further data analysis and preparation of this manuscript were from a National Sciences and Engineering Research Council (NSERC) grant to F.J. Hein. Susan Meyer typed preliminary versions of this manuscript and Frank Dimitriov assisted in the preparation of some of the diagrams. Douglas J. Cant, Ron J. Steel and Finn Surlyk reviewed an earlier version of this manuscript.

Copyright © 1984, Canadian Society of Petroleum Geologists

plus gros fragments sont généralement alignés dans les cheanus fluviaux avec leur axe -*a* oblique à la direction de l'écoulement et dans les imbrications leur axe -*b* est orienté vers l'amont. La différenciation des dépôts de barres de courants anastomosés est fondée sur le type d'imbrication: dans les dépôts fluviatiles, les imbrications présentent un axe -*b* orienté vers l'amont; dans les dépôts de mer profonde, les imbrications sont soit avec l'axe -*a* vers L'amont ou l'axe -*a* vers l'amont et vers l'aval (distribution bimondale). Les orientations des axes -*a* sont passablement aléatoires dans les dépôts de barres fluviatiles et de mer profonde et elles ne constituent pas un critère fiable.

Introduction

Channelled conglomerates occur in both deep-sea and fluvial settings. In both, braided channel processes are important. Some possibilities for confusion exist in the distinction of ancient deep-water from fluvial braided channel conglomerates, especially since trough cross-bedding and multiple cross-cutting scour fills may occur in both types of conglomerates (Figs. 1 and 2). Problems arise in the deposits of fault-bounded basins in which there are only narrow shoreline or shelf zones. A recent case is from the Brae Oilfield area, North Sea, where Stow *et al.* (1982) interpret the Upper Jurassic reservoir section as submarine fans deposits; whereas, Harms *et al.* (1981) and Harms and McMichael (1983) interpret the same data as a series of stacked prograding fan-deltas. The deep-water submarine fans and subaerial to shallow-water fan-deltas do co-exist in modern fjords of British Columbia (Syvitski and Farrow, in press) and Baffin Island (Syvitski and Blakeney, 1983) as well along the southern coast of Jamaica (Wescott and Ethridge, 1980).

Distinction of channelled conglomerates of deep-water affinity from those dominated by subaerial, fluvial association is critical in the development of accurate facies models in these complex settings. Nemec *et al.* (1980) have discussed the characteristics of subaerial versus subaqueous mass-flow conglomerates in such basins. The purpose of this paper is to discuss those sedimentary features which can be used to distinguish clast-supported conglomerates deposited in fluvial braided channel systems from those deposited in deep-water braided channel systems. These criteria are based upon detailed sedimentological studies of modern fluvial conglomerates in rivers of the Canadian Rockies and the deep-sea channel deposits of the Cambro-Ordovician Cap Enragé Formation of Québec (Fig. 3).

Fluvial and Deep Marine Conglomerate Studies

CAP ENRAGÉ FORMATION

This unit was studied regionally by Hein (1979) with specific conglomerate localities examined in detail by

Fig. 1. Medium scale trough cross-bedding in cross-section view Cap Enragé deep-water braided channel deposits. Cap à la Carre Ouest, near St. Simon de Rimouski, Québec. Clipboard is 35 cm long. Paleoflow is to the right and into the page. Three sets of cross-bedding occur, with a reactivation structure (arrow) separating the top two sets.

Fig. 2. Small multiple scour fills in cross-section view. Cap Enragé deep-water braided channel deposits. St. Simon sur Mer (Two Cottages section) Rectangles on notebook are 5 cm long.

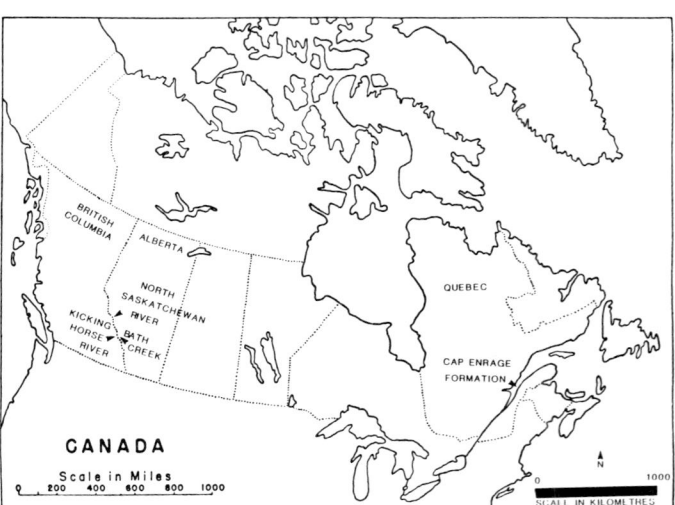

Fig. 3. Location of study areas. Rivers studied include the upper North Saskatchewan River and Bath Creek in Alberta and the Kicking Horse River in British Columbia. The ancient deep-water Cambro-Ordovician Cap Enragé braided channel fills were studied in a 70 km long belt near St. Simon de Rimouski, Québec.

Davies (1972); Davies and Walker (1974); Johnson (1974); Johnson and Walker (1979). This sequence is known to be of a deep-sea origin because:

(1) The local stratigraphic sequence begins with thick, abyssal plain claystones and siltstones of the Orignal Formation (Hubert et al., 1970);

(2) The Cap Enragé Formation conformably overlies the Orignal Formation and is, in turn, overlain conformably by the Ladriére Formation, which consists of turbidite sequences (Hubert et al., 1970);

(3) Local and regional dispersal trends and tectonic reconstructions indicate that gravel was transported down a south-eastward dipping continental slope (Hubert et al., 1970) and swung southwestward, being constrained by a base-of-rise channel, about 300 m deep, 70 km long (minimum) and 10 km wide (Hein and Walker, 1982).

Three major facies associations occur: (1) Coarse Channelled Association, with boulders and graded-stratified conglomerates, characterized by abundant channels 1-10 m deep and up to 250 m wide; (2) Multiple Scoured Coarse Sandstones with multiple channelling on the scale of 0.5-1 m; (3) Unchannelled Sandstones, with scouring on the scale of a few tens of centimetres. These sediments are interpreted as deposits of a braided bar and channel system (Association 1), flanked by higher terrace deposits (Associations 2 and 3). Full data and interpretations of the paleotopographic reconstructions and depositional mechanisms are given in Hein and Walker (1982) and Hein (1979, 1982). Field techniques are elaborated in Hein and Walker (1982) and Hein (1979).

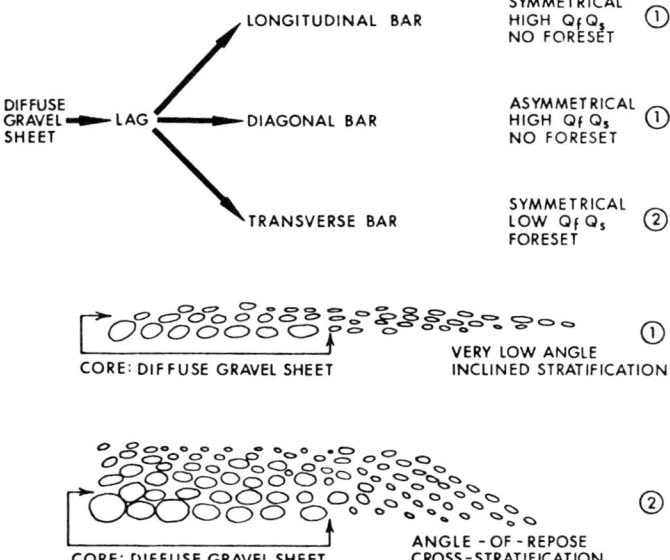

Fig. 4. Model of Hein and Walker (1977) relating growth of a diffuse gravel sheet into one of three bar types, depending on flow symmetry, and fluid discharge (Q_f) and sediment discharge (Q_s). The development of planar tabular or low angle cross-stratification depends upon whether the bars have an angle-of-repose foreset margin or a riffle margin without an angle-of-repose foreset.

MODERN RIVERS

In the Canadian Rockes several rivers were studied: the Kicking Horse River, near Field, B.C.; the North Saskatchewan River, in Banff National Park; and, Bath Creek, near Lake Louise, Alberta. Main results from the Kicking Horse River studies are published in Hein and Walker (1977) and from the North Saskatchewan River are presented in Hein (1974). Some additional data from these studies will be given here.

The rivers studied are all glacial meltwater, gravelly braided systems characterized by very low winter discharges, peak spring floods and summer diurnal discharge fluctuations (Hein, 1974; Smith, 1974; Smith, 1973a, b). The Kicking Horse River flats were studied by several workers (Smith, 1974; Hein and Walker, 1977). Gravel flats and bars have complex origins due to continual reworking during summer diurnal discharge periods. Four types of simple 'unit bars' were recognized by Smith (1974), including transverse, diagonal, longitudinal and point bars. Hein and Walker (1977) found a major difference in the distributions of active gravel bars in upstream and downstream reaches. Gravel was initially transported as diffuse gravel sheets during flood conditions. During lower flows these gravel sheets modified the flow around them and served as nuclei for simple unit bars to grow in protected areas in upstream reaches and in most channels of downstream reaches.

A possible model for the origins of stratification and cross-bedding in fluvial gravels was proposed by Hein and Walker (1977), based upon hydraulic and bed load measurements, geomorphological development of bars, and types of stratification and cross-bedding as seen in trenched bar sections (Fig. 4).

Hydrology, geomorphology and aggradational rates of the North Saskatchewan River were studied by Smith (1973a, b). Preservability of stratification and cross-bedding in the North Saskatchewan River was studied by Hein (1974). These results along with gravel fabric patterns from the Kicking Horse River, North Saskatchewan River and Bath Creek will be discussed in following sections. Details of field techniques of the modern river studies are given by Hein (1974) and Hein and Walker (1977).

COMPARISONS OF CONGLOMERATES — BRAIDED FLUVIAL AND DEEP-SEA CHANNELS

In this paper distinction of dominantly clast-supported braided channel conglomerates is based upon detailed comparisons of the internal geometry of channel fills, sorting and grain-size distributions of bar and channel sediments, grading patterns, fabric, stratification and cross-bedding. Larger scale relationships, cycles and surrounding lithologies are also discussed.

INTERNAL GEOMETRY OF CHANNEL FILLS

In the rivers studied relief from channel bases to channel sides was 1-2 m and bars were generally less than 1 m in height. The resultant maximum topographic relief in the systems was 3 m. Similar scales of relief were noted by Smith (1973a, b) for pool-and-bar couplets in the North Saskatchewan River and by Fahnestock and Bradley (1973) for the Knik River in Alaska. This relief is less than that reported by Baker (1973) for the fluvial bedforms in the scablands, formed by glacial dam break of Lake Missoula. In the catastrophic Lake Missoula flooding, bedforms up to 12 m high were deposited in the bottoms of the fluvial break-out channels.

In the deep-sea Cap Enragé deposits, sediments of the Coarse Channelled Association (main channel and bar deposits) occur in lenses developed on basal scour surfaces with 1-5 m of relief. In addition, preserved topographic highs reach 1-2 m above the scour margins. This results in a maximum relief of 7 m for the deep-sea bar-and-channel topographies, about twice the scale of that surveyed in modern braided river systems. The scale of the Cap Enragé bars and channels compares with the relief of other deep-sea channel deposits: 1-12 m for the Lago Sofio conglomerates (Winn and Dott, 1979); 0.5-2.5 m for possible bar deposits in the St. Roch Formation, L'Islet Wharf, Québec (Kessler, 1979; Walker, 1979); 0.3-1 + m of hummocky topography on modern deep-sea channels (Hess and Normark, 1976); and, 4-10 m from channels to point bar tops in the Northwest Atlantic Mid-Ocean Channel, Labrador Sea (Chough and Hesse, 1976).

GRADING

Most models of shallow braided pebbly river facies have been based upon sediment distributions of modern outwash surfaces with little attention given to the potential of such facies to be incorporated in the ancient record. The base map of the modern upper North Saskatchewan River terrace that was trenched by Hein (1974) is shown in Figure 5. This terrace has bar-and-channel features which are morphologically similar to bar-and-channel complexes mapped and hydraulically monitored in the Kicking Horse River. Consequently, the stratification and textures preserved in this terrace were thought to be representative of an outwash system subject to extensive bar-and-channel development.

Areal percentages show that on this terrace, reworked gravel flats of unknown origin account for over 99% of the exposed surface area. Bars comprise 0.09%, convergence channels or pools account for 0.01% and major channels comprise 0.09% of the surface area. Individual stratigraphic sections from the different trenches (Fig. 5) are given by Hein (1974, Appendix 5). Percentages (on the basis of number of beds) from 35 m of cumulative stratigraphic section suggest that in modern outwash gravel flats, ungraded gravels are the dominant sediment type (74%). The remaining fraction is almost equally composed of inversely-graded (16%), and normally-graded gravels (10%) (Fig. 6).

In the deep-sea Cap Enragé system the Coarse Channelled Association consists of coarse conglomerates (Facies 1), graded-stratified fine conglomerates and pebbly sandstones (Facies 2) (Table 1). Less commonly, structureless pebbly sandstones (Facies 6) and rare trough cross-bedded sediments (Facies 5) occur (Table 1). Of these facies, most are normally graded, with the exception of Facies 5 sets which are ungraded. Facies 1 beds are normally graded (41%, on basis of number of beds), ungraded (37%),

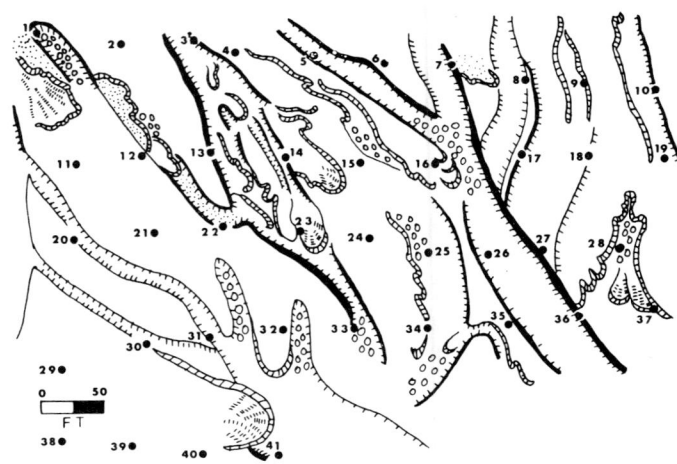

Fig. 5. Surficial plane table map of unmodified bar-and-channel complex, upper North Saskatchewan River terrace, where the preservabillity of stratification was determined by measurment of stratigraphic sections in shallow trenches (numbered dots). 35 m of cumulative stratigraphic section were measured.

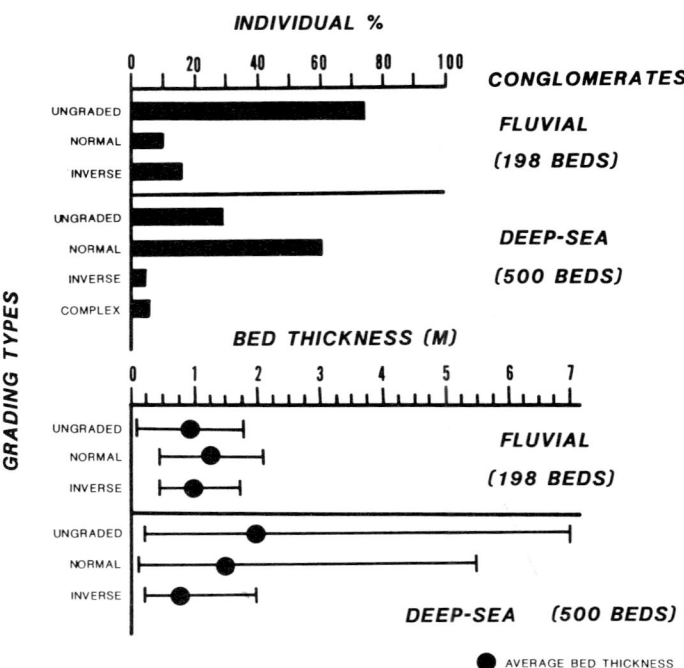

Fig. 6. Percentages of grading types in conglomerates of the North Saskatchewan River terrace and the deep-sea Cap Enragé deposits. Average bed thicknesses and ranges are also given for the different types of conglomerates.

inversely-graded (10%) or complexly graded (11%). Facies 2 beds are usually normally graded (72%), less commonly ungraded (23%) and rarely inversely-graded (1%) or complexly graded (4%). Structureless pebbly sandstones are usually normally graded (71%), less commonly ungraded (26%) and rarely inversely-graded (1%) or complexly graded (2%). Compared with the fluvial conglomerates there is a much greater proportion of normally-graded beds (average 61%); a lower proportion of ungraded beds (average 29%) and inversely-graded beds (average 4%). Rare occurrences of complex grading occur in the deep-sea channel fills, which were not recognized in the fluvial conglomerates.

These percentages of graded beds observed in the North Saskatchewan terrace compare well with the fluvial Bunter Pebble Beds (Steel and Thompson, 1983). Those from the Cap Enragé beds were similar to those of the deep-sea St. Roch channel fills (Walker, 1979; Rocheleau and Lajoie, 1974), but contrast with the percentage of grading found in the deep-sea Lago Sofio conglomerates, which were mainly ungraded (Winn and Dott, 1979).

STRATIFICATION AND CROSS-BEDDING

In the North Saskatchewan River terrace over 40% of the gravel was structureless, with 6% consisting of horizontally stratified beds and 22% comprising planar tabular cross-bedded gravels (Fig. 7). Stratification and cross-bedding is usually defined by alternating layers of clast-supported, matrix-filled conglomerate of differing grain sizes. Less commonly, stratification or cross-bedding consists of alternating layers of clast-supported, matrix-filled conglomerate and clast-supported, open-work conglomer-

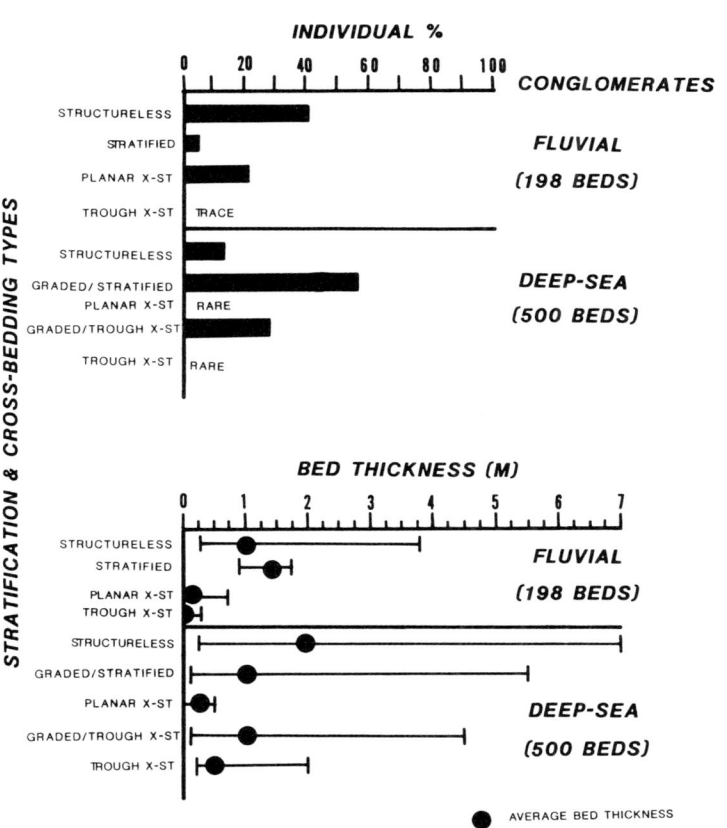

Fig. 7. Percentages of stratification and cross-bedding types in conglomerates of the North Saskatchewan River terrace and the deep-sea Cap Enragé deposits. Average bed thicknesses and ranges are also given for the different types of stratified or cross-bedded conglomerates.

Facies*	Lithology	No. of beds measured	Bed thickness (m) Average	Range	Bases†	Grading†	Characteristic structures
1	Coarse conglomerate	160	1.9	0.25- 7.0	Flat, scoured	Normal, ungraded, inverse	Structureless or horizontal stratification
2	Fine conglomerate, pebbly sandstone	300	1.0	0.15- 5.5	Flat, scoured	Normal, abrupt normal	Horizontal stratification and (or) cross-bedding (trough or low-angle oblique)
3	Fine conglomerate, pebbly sandstone	265	1.3	0.20-11.6	Flat	Normal, abrupt normal, ungraded	Dispersed texture
4	Fine conglomerate, pebbly sandstone	166	1.6	0.25-16.0	Flat, scoured	Normal, abrupt normal, ungraded	Fluid escape tubes and (or) dish structures
5	Fine conglomerate, pebbly sandstone, sandstone	53	0.5	0.02- 2.0	Loaded, scoured	Ungraded	Trough cross-bedding
6	Pebbly sandstone, sandstone	200	1.5	0.15-13.2	Flat, scoured	Abrupt normal, normal, ungraded	Structureless

*Hein (1979).
†Arranged in order of decreasing abundance.

Table 1. Characteristics of facies in the Cap Enragé braided channel fill sequence (from Hein, 1982).

ate (Fig. 8A). Trough cross-bedding is rare (0.25%) and only occurs in pebbly sands, not in the dominant conglomerate lithology.

In the Cap Enragé coarse-grained channel fills, over 50% of the conglomerate is graded-stratified, with about 30% graded-trough cross-bedded and, less commonly, structureless conglomerates. Planar tabular and trough cross-bedding rarely occurred (Fig. 7). As discussed by Hein (1982), stratification in the lower parts of graded-stratified conglomerates is defined by alternating clast-supported and clast-dispersed bands (Fig. 8B). The clast-dispersed bands are characterized by an unusual dispersed texture, in which the coarser clasts do not touch one another, but are scattered throughout a finer grained, sandy matrix (Hein, 1982; Hein and Walker, 1982). Upper parts of graded-stratified conglomerates show stratification clearly defined by alternating bands of clast-supported conglomerate of differing grain sizes, similar to horizontal stratification in fluvial gravels. The hydrodynamic implications of the unusual clast-dispersed/clast-supported stratification at the base of the graded-stratified beds is discussed by Hein (1982).

A final distinction between the stratification of deep-sea and fluvial conglomerates is the occurrence of an irregular inclined stratification in the deep-sea conglomerates and pebbly sandstones (Hein, 1979, p.348-349). This type of stratification is not commonly found in the bases of main channel fills, but rather in sediments interpreted as channel margin, braid bar or point bar sites.

This irregular inclined stratification is defined as stratification that may undulate up and down in cross-section, but generally follows the basal margin of the bed. In cross-section view it has a sweeping or wavy appearance (Fig. 9). No plan (bedding plane) views were seen of this feature. Laminations are 0.5-1 cm thick and are commonly marked by quartz pebble or granule bands. This stratification is most common in deposits that have many scours along the base of the bed, giving the basal margin a scalloped appearance. Irregular inclined stratified units vary from a few centimetres to 0.4 m thick. Successive laminae tend to follow the form of the previous laminae, although generally the lamination becomes nearly horizontal up-section. Laminae are well-defined and traceable for many metres along strike. This type of stratification was observed in 16% of the graded-stratified Facies 2 and Facies 3 (Table 1) conglomerates and pebbly sandstones. A similar type of wavy stratification was noted by Hein and Arnott (1983) in pebbly sandstones and fine conglomerates of the Precambrian Hector Formation, interpreted as part of a deep-sea channel-fill sequence.

The scale of bedding also differs between fluvial and deep-sea stratified and cross-bedded conglomerates. Structureless and stratified beds have about the same bed thickness for both types of conglomerates, but the range of bed thicknesses is much higher for the deep-sea conglomerates, with a maximum bed thickness of 5.5 m for the graded-stratified types (Fig. 7). Planar tabular cross-beds were

Fig. 8. Horizontally stratified conglomerates. **(A)** North Saskatchewan River gravels showing alternating clast-supported, matrix-filled conglomerates and clast-supported, open-work conglomerates. **(B)** Cap Enragé deep-sea fine conglomerates and pebbly sandstones, showing alternating clast-supported and clast-dispersed bands.

Fig. 9. Irregular inclined stratification as seen in cross-sectional view, deep-sea Cap Enragé braided channel fills.. Rectangles on notebook are 5 cm long. Bic, Québec.

thinner, on the average, in the fluvial gravels. Trough cross-bed sets in the deep-sea conglomerates range from 0.3-2 m thick, with an average set thickness of 0.45 m. The rare trough cross-beds in the North Saskatchewan terrace were much thinner, with an average set thickness of 0.1 m. In addition, within the Cap Enragé conglomerates, graded beds with trough cross-stratification occurred, with the beds averaging 1.0 m thick, and trough cross-bed sets averaging 0.25 m thick. Graded/trough cross-bedded facies did not occur in the fluvial gravel terrace deposits of the North Saskatchewan River. This facies has been recognized by Baker (1973) in fluvial flood deposits. However, the proportion of this facies occurrence is probably much lower in fluvial gravels than in deep-sea gravels, owing to the greater dominance of large-scale turbidity currents in the emplacement of deep-sea clast-supported gravels. In the fluvial setting, even under flood conditions, bed load tractive processes are more dominant in the transport and deposition of gravel.

CLAST FABRIC PATTERNS

Gravel fabric patterns were measured on a grid scale from square-metre plots of upstream and downstream portions of nearly emergent, unmodified diagonal, transverse and longitudinal bar-and-channel complexes in the Kicking Horse River and Bath Creek systems. A similar study was conducted on part of the exposed terrace in the North Saskatchewan outwash. Dimensions of the three major axes were measured on each pebble; only those pebbles with long axes greater than 2 cm and those with non-spherical shapes were measured.

Rose diagrams of pebble orientations for the nearly emergent bar and channel complexes are shown in Figure 10. Diagonal, point and transverse bars show a preferred current-transverse orientation of the a-axes (long-axes) and b-axis (intermediate axis) imbrication (Figs. 10A-C, E). Channel plots have a more current-parallel orientation, with the a-axes oriented flow-parallel and the a-axes imbricate upstream (Figs. 10A, C and E). In the diagonal bar complexes (Fig. 10A) the channel plot shows a somewhat bimodal pattern which may reflect down-channel as well as cross-channel flow across the diagonal bar surface. In the swale plot, associated with the point bar (Fig. 10C), there is a multimodal pattern, reflecting complex flow patterns within the swale. The longitudinal bar plot shows a polymodal pattern, reflecting currents flowing across the bar front. All of the plots, except those with bimodality or polymodality, are significant at the 5% confidence interval (Rayleigh test, Curray, 1956). Visually most of the polymodal or bimodal plots tend to have a dominant mode; however, there is no rigid statistical test for these distributions.

Pebbles from the bar and channel complexes have similar shape distributions, with most of the clasts being elongate disks and blades, with a small percentage of rods. Similarly, grain sizes (a-axis dimensions) were the same for bar and channel plots, with an average a-axis dimension of 5 cm (range: 2-10 cm). The longitudinal orientations in the channels may reflect pivoting of the clasts around previously deposited bed material. Alternatively, the current-parallel mode in the channels may reflect intermittent saltation of clasts during bed load transport in the channels. Bar-top pebbles record the orientation of elongate pebbles that was the transport orientation, with little reworking or pivoting of the clasts after initial deposition on the bed.

The base map of the emergent bar-and-channel complex of the North Saskatchewan terrace is given in Figure 11, with rose diagrams corresponding to the pebble plots given in Figure 12 and the statistics listed in Table 2. Data from this emergent complex show a much greater amount of scatter than the nearly-emergent bar-and-channel complexes examined in the Kicking Horse River and Bath Creek (Fig. 10). This probably reflects long term reworking of exposed material on the terrace during subsequent high spring floods.

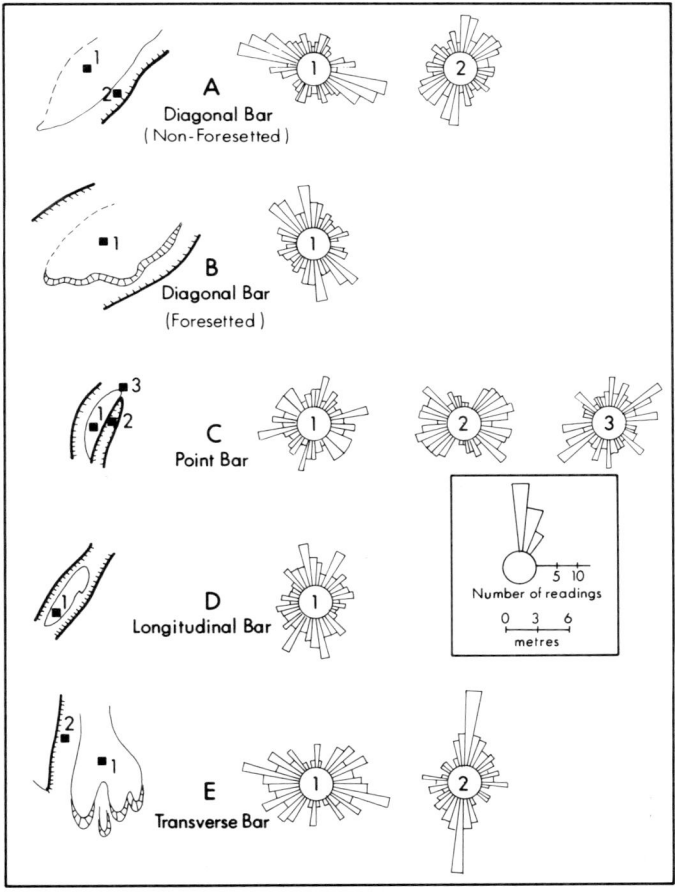

Fig. 10. Rose diagrams of the orientations of the strike of the a-b planes measured for elongate blades, disks and rods on unmodified, nearly emergent, bars. Bars were mapped by pace-and-compass techniques. Plot 1 in (**A**) and (**E**) refers to the bar-top fabric; plot 2 refers to the channel-bottom fabric. The longitudal bar (**D**) was measured in Bath Creek; other bars were from the Kicking Horse River.

Pebble plots were classified as being on bars, in channels or in topographically-lower pools. Bar plots are the most widely scattered, with most of the pebble plots showing no preferred orientation. Pools had preferred pebble orientations in directions parallel, oblique and transverse to the axis of symmetry of the pool; plots are less scattered than the bar-top plots. Channel plots showed a preferred orientation parallel to the axis of symmetry of the channel, with most plots showing significant trends. The general lack of statistically significant preferred directions on bar-tops and in the pools probably reflects complex flow patterns and rotation during successive periods of reworking.

Pebble shapes were classified and rose diagrams were replotted with respect to bar or channel axes of symmetry (Fig. 13). Flat-lying discs with no imbrication showed random a-axis orientations on bar tops and in channels. Blades in channels showed an a-axis flow-transverse, b-axis imbrication pattern. On bar tops blades had random orientations. Rods on bar tops had a large amount of scatter in orientation, but with a dominant mode parallel to the bar axis of symmetry. Disks with imbrication showed random patterns in the channels whereas on bar tops they showed a-axis orientations oblique to the bar trend and b-axes imbricate upstream.

A detailed discussion of gravel fabrics and their origins for the deep-sea conglomerates in the Cap Enragé channel fills is given by Hein (1982). In the present discussion, only the general patterns within the dominant facies of the braided complexes will be discussed. Typical fabrics of the Facies 1 coarse conglomerates are a-axis flow-parallel, and a-axis imbricate upstream. The angles of imbrication vary, and less commonly upstream and downstream

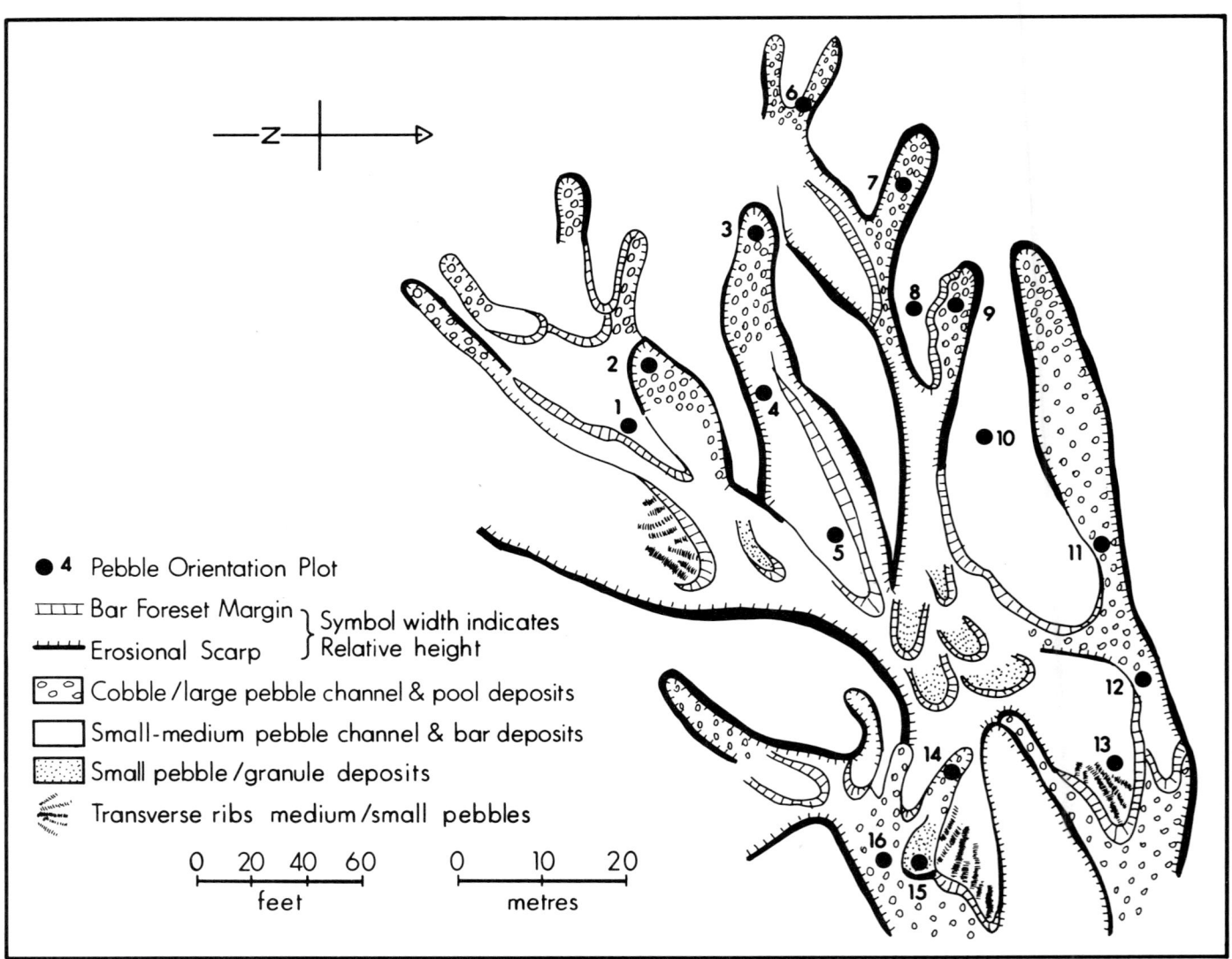

Fig. 11. Surficial plane table map of an exposed bar-and-channel complex, upper North Saskatchewan River terrace, where pebble orientations were measured. Numbers refer to the square metre plots, in which 100 pebbles were measured per plot. This site is just upstream from the terrace site that was trenched for stratification (Fig. 5).

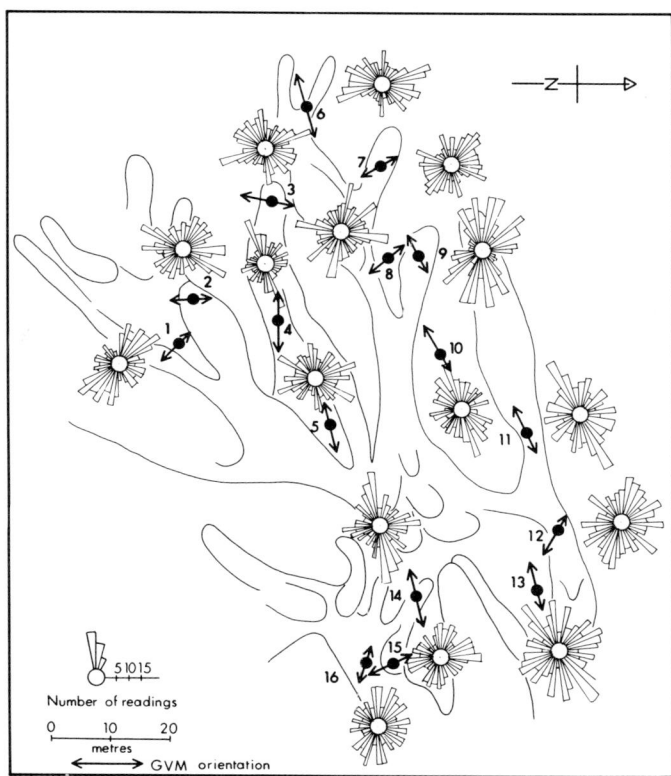

Fig. 12. Rose diagrams of the orientations of the strike of the *a-b* planes measured for elongate blades, disks and rods on the exposed upper North Saskatchewan River terrace. Numbers refer to the pebble orientation plots (located in Fig. 11). Arrows indicate the orientations of the computed grand vector means of the distributions.

(bimodal) imbrication may occur (Hein, 1979). Rose diagrams of typical bedding fabrics and imbrication patterns for the finer conglomerate and pebbly sandstone facies are shown in Figure 14. As discussed by Hein (1979) Facies 2 fine conglomerates and pebbly sandstones show a variety of fabric patterns depending upon the location in the bed. Fabric in the coarser, lower parts of beds is *a*-axis flow-parallel, *a*-axis imbricate upstream. Fabric in the upper, finer-grained bed portions is *a*-axis unimodal (61%, by number of beds) or *a*-axis bimodal pattern (30%), with less common random *a*-axis orientations (9%). Imbrications are *a*-axis unimodal upstream imbrications (47%) or *a*-axis bimodal (upstream and downstream) (53%) patterns. Facies 6 pebbly sandstones showed unimodal (31%), random (37%) or bimodal (32%) *a*-axis orientations in bedding. *A*-axis imbrications are usually unimodal upstream (78%), less commonly a bimodal (upstream and downstream) imbrication pattern (22%).

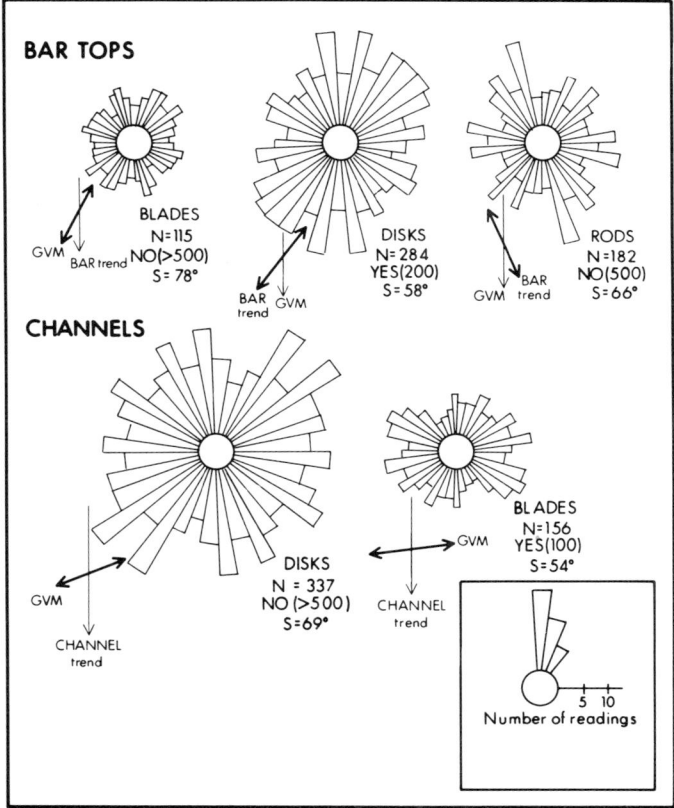

Landform	Plot	N	Rayleigh Test 5% Confidence Level	Standard Deviation	Grand Vector Mean Trend*
Bar	1	100	YES (50)	45°	transverse
	5	100	YES (50)	48°	parallel
	8	99	NO (500)	68°	transverse
	13	100	NO (200)	60°	transverse
	15	100	NO (200)	60°	transverse
Pool	2	99	NO (200)	65°	transverse
	3	99	NO (500+)	69°	transverse
	6	97	YES (50)	48°	oblique
	7	99	NO (200)	51°	oblique
	9	99	YES (50)	44°	oblique
	14	101	YES (100)	49°	oblique
	16	99	NO (200)	55°	oblique
Channel	4	100	YES (100)	50°	parallel
	10	100	YES (100)	53°	parallel
	11	100	YES (100)	54°	parallel
	12	114	NO (500)	68°	oblique

*Grand Vector Mean Trend are indicated with respect to the axis of symmetry of the respective landforms; N indicates number of data points; YES (100): Yes, it is significant, number of measurements needed; NO (500): No, it is not significant, number of measurements needed. For the non-significant trends the dominant modes are plotted in Figures 12 and 13.

Table 2. Pebble fabric statistics for gravel fabric studies, upper North Saskatchewan River terrace. See Figures 11-13 for plane table maps and rose diagrams.

Fig. 13. Rose diagrams of the orientations of the strike of the *a-b* planes of pebble distributions shown in Figure 12, corrected with respect to the axis of symmetry of the bars or channels, subdivided according to shape classes. Double headed arrows refer to the grand vector mean directions computed for the distributions; single headed arrows indicate the direction of the axis of symmetry for the bars (upper part of diagram) or the channels (lower part of diagram). N refers to the number of pebbles measured per site. Yes (50) states that the distribution is significant, with the number of measurements needed for significance given in parentheses. No (200) states that the distribution is not signigicant, with the number of measurements needed to attain significance given in parentheses (Curray, 1956). 100 pebbles were measured per site.

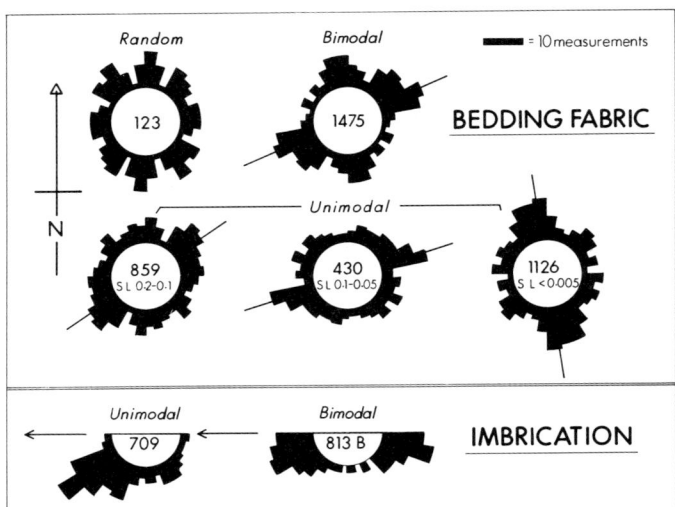

Fig. 14. Typical rose diagrams of bedding fabric and imbrications for pebbly sandstones and fine conglomerates, deep-sea Cap Enragé braided channel fills. Numbers in the centre of the rose diagrams refer to the bed numbers of Hein (1979). 100 clasts were measured in each plot. SL numbers of the unimodal bedding fabrics refer to the computed significance levels of the distributions, using the chi-squared test. Lines on bedding fabric rose diagrams refer to the computed grand vector mean directions for line-of-motion data. Arrows on imbrication rose diagrams indicate the preferred a-axis orientations in bedding. Rose diagrams were all plotted in 10° class intervals (from Hein, 1982).

In summary, gravel fabric pattern in deep-sea channel deposits is usually a-axis flow-parallel, a-axis imbricate upstream (Facies 1 is dominant). On deep-sea braid bar tops, where Facies 1, 2 and 6 commonly occur, a-axis orientations in bedding are variable, but the imbrication is generally a-axis upstream, although bimodal upstream and downstream imbrications can occur. This peculiar pattern did not occur in the fluvial gravels that were studied, the hydrodynamic implications of this are elaborated by Hein (1982). In fluvial conglomerates there is a much smaller proportion of clasts with an a-axis flow-parallel, a-axis upstream imbricate pattern. This pattern only occurs in smaller clasts within main channel deposits. Gravel fabrics, on the whole, tend to be a-axis flow-transverse, b-axis imbricate for coarser clasts within main channels and in bar cores. On the tops of exposed gravel flats and terraces this fabric may deteriorate to a polymodal or random a-axis pattern in bedding, but with a preferred b-axis upstream imbrication.

LARGER SCALE RELATIONSHIPS: MODERN GRAVEL RIVERS

In the Kicking Horse River system the different bar types were mapped in upstream, midstream and downstream sections. Sorting of bedload material was examined on transverse bars during active rising flow by bed load sampling and sieving of material in transport at upstream and downstream stations (Hein, 1974). In order to obtain a better estimate of sorting on transverse bars, sediment was sampled from square-metre plots of exposed, unmodified bars. Eight inactive and six active transverse bars were thus sampled. Sorting on diagonal bars was only examined on exposed, unmodified bars; active diagonal bars were not sampled due to the complex flows and bed load transport across these bars. On the diagonal bars the dimensions of the ten largest pebbles were measured at the upstream bar-channel contact and downstream at the riffle margin. Slopes of the bar surfaces between the sample sites was also measured. Forty-six inactive, unmodified diagonal bars were sampled on exposed flats of the Kicking Horse River, North Saskatchewan River and Bath Creek.

Bedload Sorting on Transverse and Diagonal Bars

All of the active transverse bars showed an across bar fining of the material in transport (Hein, 1974, Tables 5-10; Hein and Walker, 1977), a feature in agreement with most other studies. Cumulative frequency curves of the bed samples from the unmodified, inactive transverse bars are shown in Figure 15. Analyses of the size distributions suggest that the downstream fining in the coarser tails of the grain size distributions may reflect a partial development of a lag pavement in the upstream areas as well as differential transport of finer sediments across and onto the bar surfaces (Hein, 1974).

The most common bar form in the outwash flats of this study is the diagonal bar form, which migrates obliquely to the main current direction. Results of the grain size measurements on the exposed bar surfaces (Fig. 16) show that on diagonal bar surfaces with slopes less than 0.075, sediment becomes finer downstream; whereas, with bar surfaces slopes greater than 0.075, the sediment on the bar surface becomes coarser downflow. For the bars studied, this slope may mark the threshold at which bed load starts to avalanche down the diagonal bar surface.

Distribution of Bar Types

General comparisons of the bar-and-channel patterns developed in upstream and downstream reaches of the Kicking Horse River are given by Hein and Walker (1977), with an idealized summary diagram given in their Figure 2. Some of the original base maps from Hein (1974) are given in Figures 17 and 18. Areal percentages for the upstream and midstream sections show that the ratio of active bars and gravel sheets to active channels increases from 1:10 in upstream reaches to 1:3 in midstream reaches, over a 4 km down-valley distance. Active channel areas (without bars or gravel sheets) decrease from over 90% in the upstream reaches to 75% in midstream areas. Of bar areas in the upstream section, they were equally divided between diffuse gravel sheets and bars, each occupying 4.5% of the surficial area. In the midstream reach, 18% of the area consists of bars with angle-of-repose foresets, with 7% consisting of bars with riffle margins (without angle-of-repose foresets) and diffuse gravel sheets.

Summary: Facies and Fabric Development in Conglomeratic, Shallow Braided Rivers

Hein and Walker (1977) presented a possible model to describe the facies development in relation to bar morphologies. One can integrate that model with the observed areal percentages of bars in the Kicking Horse River outwash to predict the type of stratification or crossbedding that would occur in proximal versus distal reaches. (see Figs. 4, 17, 18).

In coarse-grained more proximal reaches, gravel is transported as individual clasts or as diffuse gravel sheets, which display very little sediment sorting and have very little topographic relief (Hein and Walker, 1977). These gravel sheets lack a downstream angle-of-repose foreset, and their progradation produces gravels which are poorly sorted and, at best, show vague horizontal stratification. Channel deposits passing laterally and downstream into deposits of diffuse gravel sheets show a slight fining trend.

Preferred fabric orientations are fairly well-developed in upstream proximal sections, with smaller clasts assuming a flow-parallel orientation in main channel areas and a more flow-transverse orientation on diffuse gravel sheets. Recognition of the gravel sheets in the geologic record would be very difficult due to their very low relief.

In finer-grained medial to distal reaches diagonal and transverse bars were common. Transverse bar deposits consist of fining-up gravels in the bar core, which grade downstream into finer-grained, planar tabular cross-bedded gravels. Diagonal bar deposits consist of fining-up gravels in the bar core, which grade downstream into either coarser-grained or finer-grained, low-angle cross-stratified gravels. Preferred fabric orientations in the bar deposits are *a*-axis flow-transverse, *b*-axis imbricate upstream. This fabric is well-developed in transverse bars, and more scattered (reflecting more complex flows) in diagonal bars. Preferred fabric patterns would decay up-section to bar-top deposits, which may have been subject to reworking processes.

Recognition of bar deposits in the ancient record, if reworking is not extensive, would be as follows. Lateral channel margin-to-bar transitions fine cross-stream, with a more pronounced trend in downstream bar margins. Pebble orientations should become more scattered from channels to bars, and there may be a higher tendency for more flow-parallel orientations for smaller clasts in the

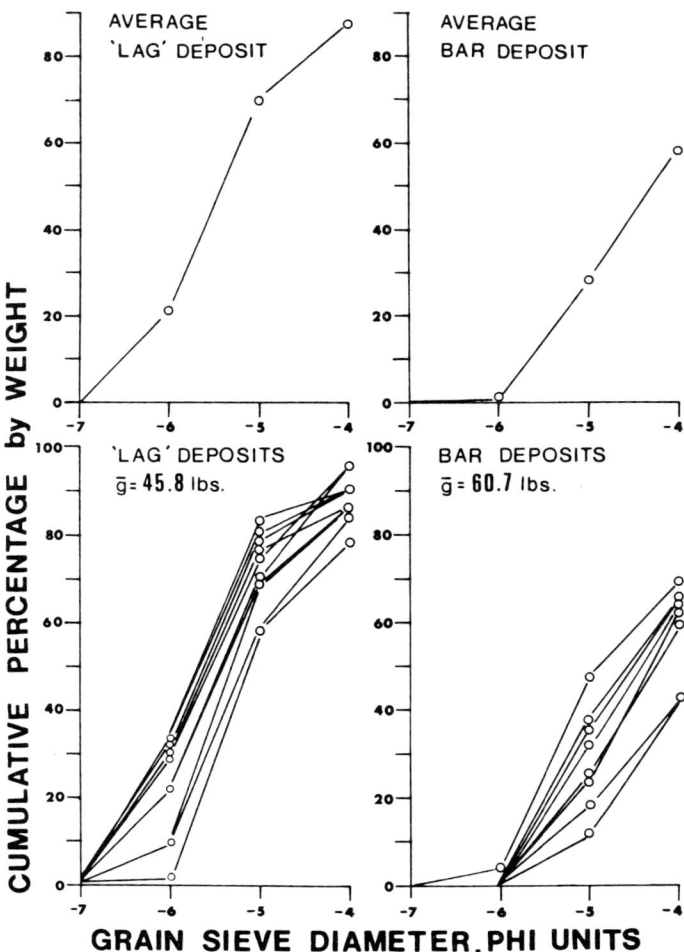

Fig. 15. Cumulative frequency curves of average bed samples from upstream ('lag') and downstream ('bar') deposits, inactive transverse bars, Kicking Horse River, Bath Creek and North Saskatchewan River. Samples were sieved at full phi intervals; $\bar{g}$ refers to the average weight of the bulk samples taken from the bar surfaces.

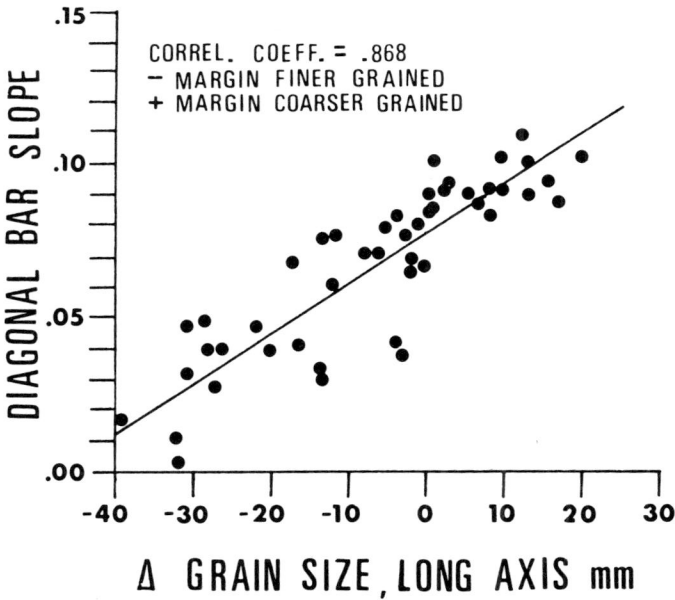

Fig. 16. Plot of diagonal bar slope versus downstream difference in grain size from upstream to downstream stations, inactive, just emergent diagonal bars, Kicking Horse River, Bath Creek and North Saskatchewan Rivers. 'Δ' Grain Size is the difference in the average *a*-axis measurements of the ten largest pebbles measured in upstream and downstream stations. Negative 'Δ' Grain Size values indicate that the downstream sediments were finer-grained; positive 'Δ' Grain Size values indicate that the downstream sediments were coarser-grained.

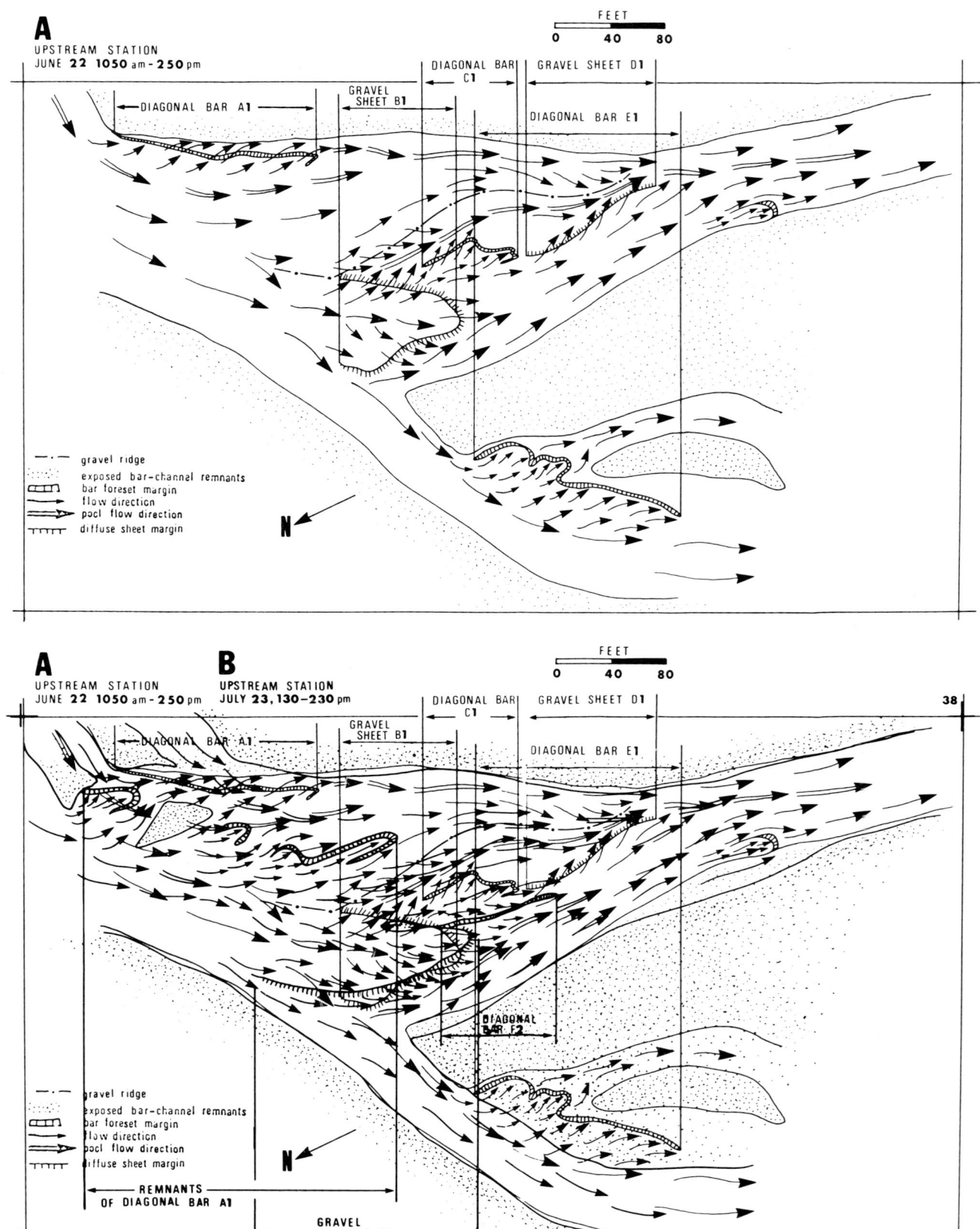

Fig. 17. Plane table maps of the upstream study area, Kicking Horse River (**A**) during rising Spring floods; (**B**) after the flood flows subsided. Note that diffuse gravel sheets were the most common geomorphological features, with bars less commonly occurring in protected lee areas. Spring melt discharge period, 1973 (from Hein, 1974).

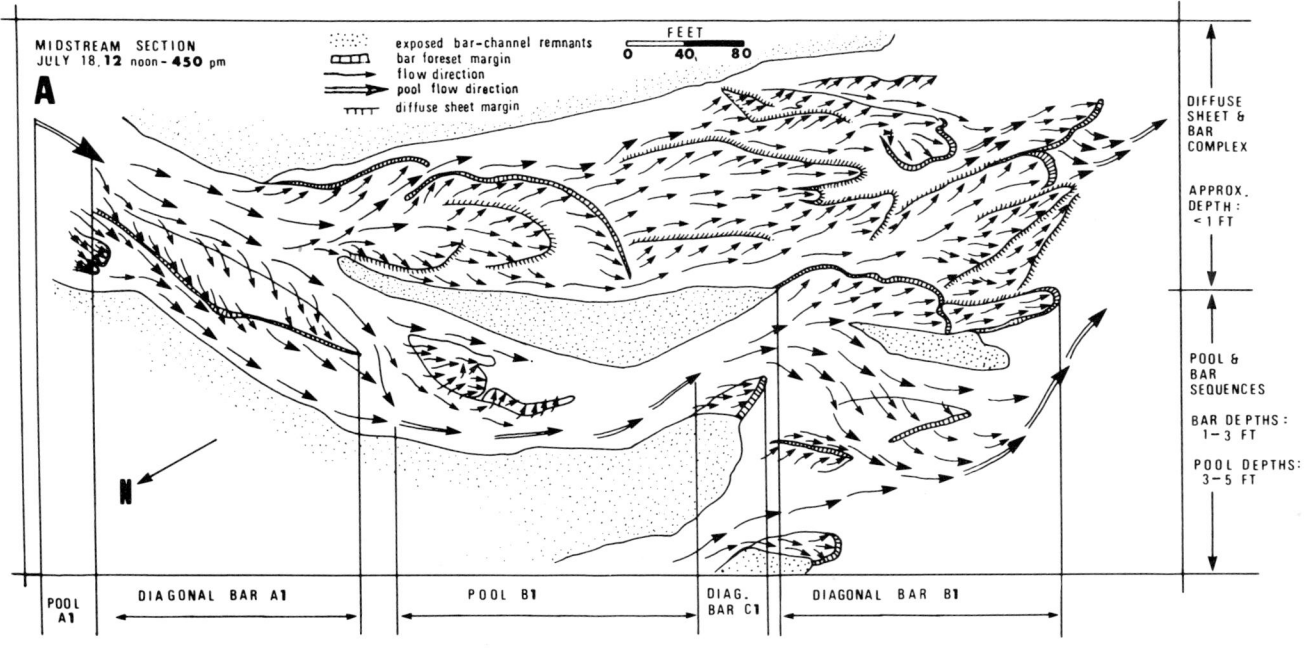
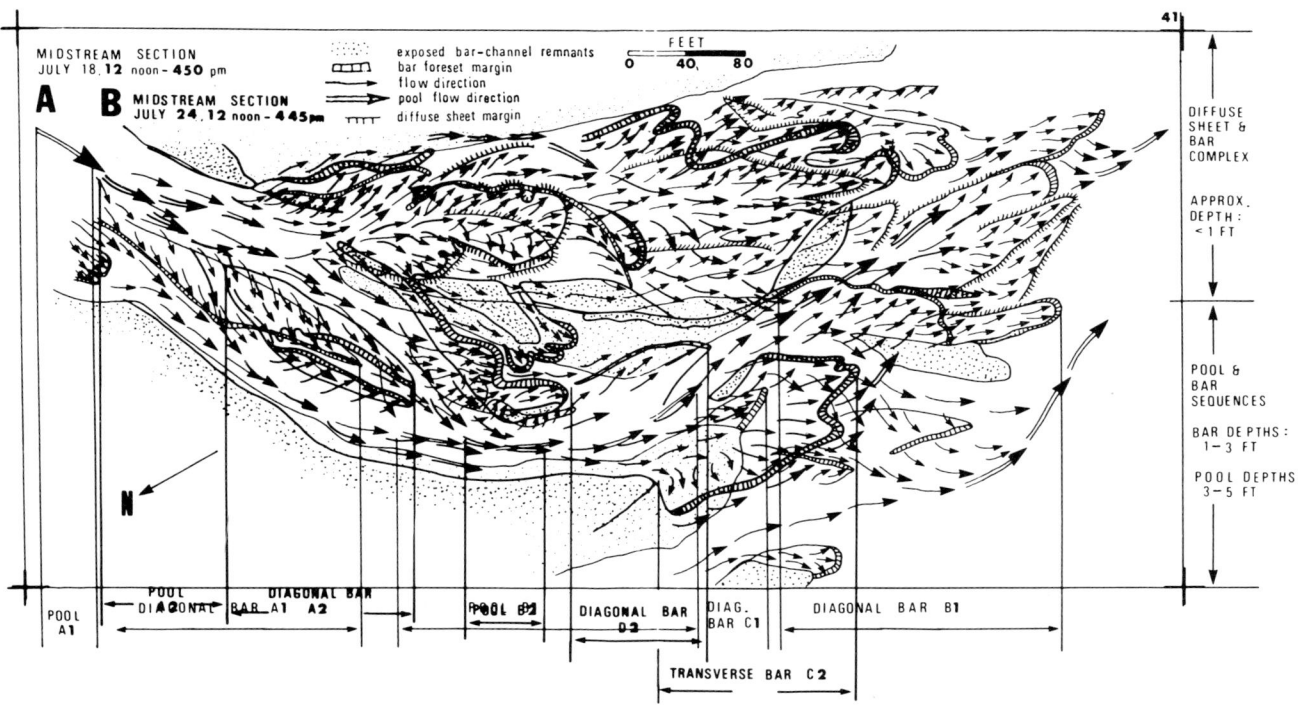

Fig. 18. Plane table maps of the midstream study area, Kicking Horse River, showing typical bar-and-channel patterns formed at different times (**A**) and (**B**). Summer diurnal discharge period (from Hein, 1974).

channels. Assuming aggradation in the system, progradation of the transverse bars would produce coarsening-up sequences of planar tabular cross-bedded gravels. Similarly, progradation of a diagonal bar would produce either fining-up or coarsening-up sequences of horizontally stratified to low-angle cross-stratified gravels. In shallow braided gravel streams these cycles would be generally on the order of 0.5-2 m thick. Any larger scale cyclicity probably records the gradual migration of individual braid channels, with coarsening-up cycles reflecting channel reoccupation and fining-up cycles recording channel abandonment. Such cyclicity was recognized by Costello and Walker (1972) and Steel and Thompson (1983) for gravelly fluvial deposits. Rapid braid channel migration or avulsion may not develop any recognizable cyclicity.

LARGER SCALE RELATIONSHIPS: CAP ENRAGÉ CHANNEL

A generalized paleotopographic reconstruction of the Cap Enragé system is given in Figure 19. Deposition within the channel system was on three major topographic levels: (1) main channels (mc, Fig. 19); (2) on bars which divided the main channels (BB, Fig. 19); and, (3) on terraces topographically above the main channel and bar complex (MT and HT, Fig. 19). A variety of different lateral and vertical facies changes occur as a result of interactions between the main channels and braid bars (mc and BB, Fig. 19); between the main channels and point bars (mc and PB, Fig. 19); and, between the main channels and terraces (mc and MT and/or HT, Fig. 19).

Summary: Facies and Fabric Development in the Deep-Sea Cap Enragé Channel System

A detailed discussion of the complete facies model for these sediments is given by Hein and Walker (1982). The following general conclusions can be drawn.

(1) In the main channels (Fig. 19, line 9), the dominant facies are very coarse (facies 1) conglomerates in the channel centres, which toward margins become finer-grained and thinner bedded (thin facies 1 and facies 2) (Fig. 19, lines 2, 3, 6 and 7) (see Table 1 for facies descriptions).

(2) In the main channel system, braid bars are mainly thin coarse conglomerates (facies 1) and graded/stratified or graded/cross-bedded fine conglomerates (facies 2), with less commonly facies 3,4,5 and 6 fine conglomerates and pebbly sandstones (Fig. 19, line 3) (see Table 1).

(3) Secondary channel fills are 1-2 m deep and show cross-cutting relationships among a wide variety of facies. Widths of the small-scale multiple channel complexes are a few metres to tens of metres (Fig. 19, lines 1, 4).

(4) Main channel conglomerates occur in concave-up, scour-based units 1-10 m deep and up to 250 m wide. Braid bar and point bar deposits occur in convex-up, flat-based (non-scoured) units 0.5-5 m high and 50-600 m wide. Point bar deposits consist of graded beds, inclined at low angles, in units up to 5 m thick. In these lateral accretion deposits of point bars, beds are usually scoured at the base and the

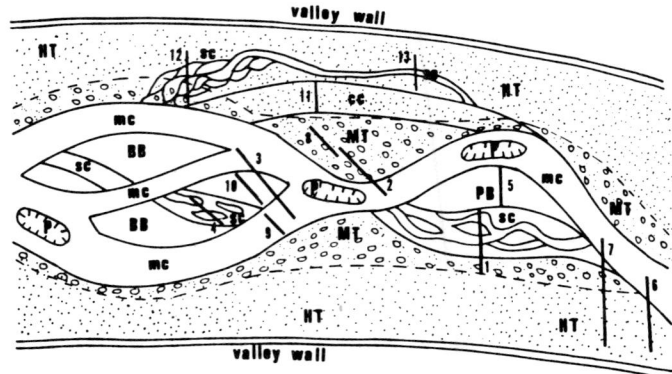

Fig. 19. Generalized paleotopographic reconstruction of the Cap Enragé deep-sea braided channel system. Numbers and heavy lines are reference sections on the diagram for discussions in the text mc: main channels; platerally thinnning and fining units within main channel conglomerates ('pods' of Johnson and Walker, 1979); cc: cut-off channel; HT: high terrace; sc: secondary braid channels. Dotted portions are mainly sandstone; dotted with open circles are pebbly sandstones; white areas are conglomeratic (from Hein, 1979).

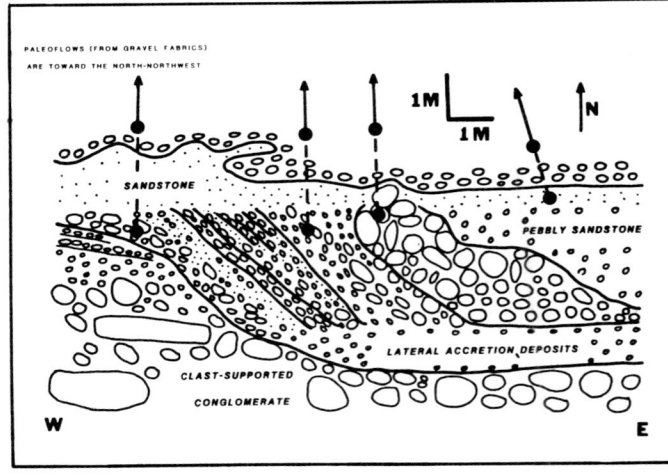

Fig. 20. Field sketch of lateral accretion surfaces of point bar, with measured preferred transport directions indicated. Beds are all graded and commmonly have scoured bases. Deep-sea Cap Enragé braided channel fills, St. Simon sur Mer section (modified from Hein, 1979).

internal fabric within individual beds indicates a transport direction perpendicular to the angle of inclination of the beds (Fig. 20). These deposits are fairly common at the sides of the main channel fills (Hein, 1979; Johnson, 1974; Johnson and Walker, 1979).

(5) Preferred fabric orientation in the main channel conglomerates is consistently a-axis flow-parallel, a-axis imbricate upstream (Hein, 1979; Johnson and Walker, 1979). Toward channel margins, with an increase in the percentage of finer (facies 2) conglomerates, the fabric pattern shows a more dominant a-axis flow-parallel mode, with a secondary flow-transverse mode. Imbrications also become

bimodal, with a secondary mode of the *a*-axes dipping downstream. Thus fabric patterns in bar-margin and bar-top deposits are quite complex and are also more scattered than those in the main channel (facies 1) conglomerates.

In the deep-sea braid bar deposits examined in this study, there is no true planar tabular cross-bedding. Bar deposits merge gradually with main channel deposits, becoming gradually coarser-grained and thicker bedded, with a change from mixed facies 2, 3 and 4 fine conglomerates and pebbly sandstones to thick facies 1 coarse conglomerates (Hein and Walker, 1982; Johnson and Walker, 1979; Hein, 1979). In contrast to the fluvial bars, which migrate downstream by bed load deposition, most of the deep-sea bars aggrade vertically by sediment gravity flow deposition, probably with very little downstream progradation.

Small-scale thinning- and fining-upward sequences (up to 15 m thick) generally involve stacked sequences of facies 1 conglomerate capped by thinner facies 1 conglomerates, with some or all of facies 2, 3, 5 and 6 sediments (Hein and Walker, 1982). These are intrepreted as representing simple, but progressive channel abandonment. Commonly there is a paleoflow divergence from one sequence to the next; and, these sequences may correlate with other types of sequences along strike. Small-scale coarsening- and thickening-upward sequences are interpreted as representing the progressive reoccupation of an area by a main channel.

Larger scale fining- and thinning-upward or coarsening and thickening-upward sequences are interpreted as due to main channel network migration and terrace progradation. Fining and thinning sequences are usually braid bar and channel deposits, overlain by terrace deposits (Fig. 21). Coarsening and thickening sequences are commonly terrace deposits overlain by braid bar and channel deposits (Fig. 21). Sequences involving marginal terrace deposits are quite complex (Fig. 21). The overall sequences from main channel to terrace sediments range from 15 m to several tens of metres thick and are much thicker than most fluvial conglomerate cycles reported in the literature.

CONCLUSIONS

A summary of the basic descriptive features of the conglomerates examined in this study is given in Table 3. The first distinction concerns separation of main channel depos-

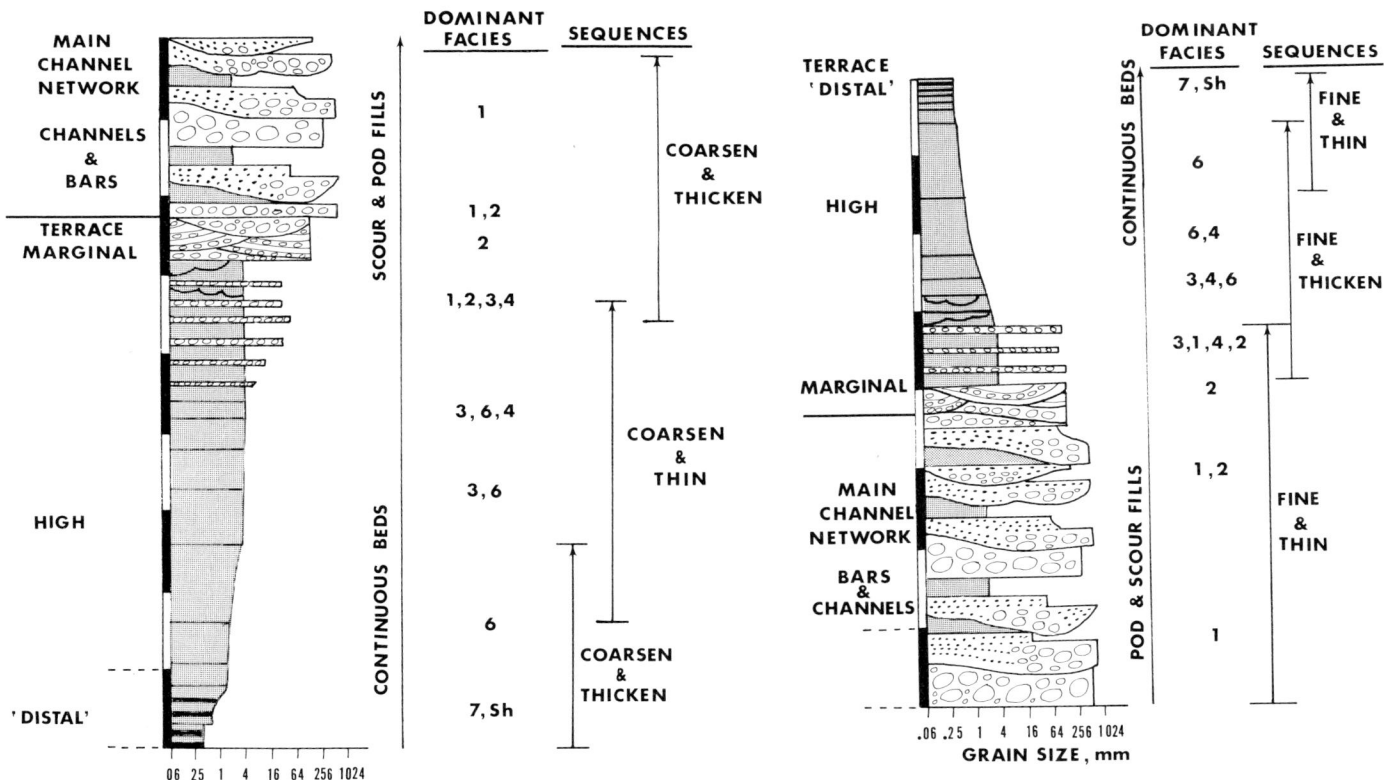

Fig. 21. Types of stratigraphic sequences generated in the deep-sea Cap Enragé braided channel system with terraces. Sequences generated by gradual channel abandonment are indicated to the right; those generated by gradual channel reoccupation are indicated to the left. Black and white scale bars were each 2 m high. 'Distal' high terrace refers to those terrace settings farthest removed from the main channel braided network of bars and channels (from Hein, 1979).

its from bar accumulations. Main channel deposits from both deep-sea and fluvial settings seem to occur in scour-based lenses, perhaps 5-10 m thick for deep-sea deposits, and somewhat thinner (?2-3 m) for fluvial conglomerates. Main channel deposits in both settings tend to be quite massive. Rare horizontal stratification or cross-bedding may occur. In directions perpendicular to the local paleoflows, the sediments of the main channel lenses become finer-grained and thinner-bedded in both the deep-water and fluvial settings.

Braid bar deposits are generally flat-based units, with only minor scouring (less than 0.5 m deep), occurring as reactivation structures. Bar deposits tend to be thinner than the main channel deposits (?1-2 m thick) and are better stratified than the relatively massive channel deposits. Sediments within the bars are flat-based, convex-up lenses, which become coarser and thicker-bedded as they grade into the main channel deposits of both fluvial and deep-sea settings.

Differences between the deep-sea and fluvial conglomerates examined in this study are mainly reflected in the type of grading, the gravel fabric and to a lesser extent on the type of stratification and cross-bedding. Grading is overwhelmingly abundant in the deep-sea bar and channel sediments, and is usually of the normal type. Grading of individual beds in fluvial settings is not so common, unless the fluvial deposits are flood-dominated. Distinction of the different bar types within either fluvial or deep-sea settings would rely on detailed paleotopographic reconstructions and paleoflow determinations.

As discussed by Harms et al., (1975) the distinctions between conglomerates deposited in deep-water and fluvial conglomerates on the basis of descriptive features within the conglomerates may be quite subtle. Obviously, the facies character of the associated deposits and their stratigraphic context must also be considered. No one physical criterion is diagnostic of either setting, with the exception of dessication cracks in shales associated with fluvial conglomerates. Marine shales may contain fossils or trace fossils suggestive of possible water depths. Stratigraphically associated sandstones with the conglomerates in deep-sea settings are usually normally graded, may show Bouma (1962) turbidite sequences, an abundance of liquefaction structures (Lowe, 1975), or be struc-

Paleotopography	Gravel Bar		Gravel Channel	
Environment	**Fluvial**	**Deep-Sea**	**Deep-Sea**	**Fluvial**
Lateral cross-stream grain size trends	Coarsening/thickening units in convex-up lenses bounded by lower flat surfaces		Fining/thinning units in concave-up lenses bounded by lower scour surfaces	
Scour surfaces	Minor <0.5 m deep as bedding contacts or reactivation structures		Major: 1-10 metres deep	Major: 1-2+ metres deep
Vertical sequences	1-2 m thick Fining-upward more common Coarsening-upward uncommon		5-10 m thick Fining-upward more common Coarsening-upward less common	2-3 m thick Fining-upward more common Coarsening-upward less common
Sedimentary structures	Dominantly stratified and/or crossbedded		Dominantly structureless, massive	
Cross-bedding:				
Low-angle	Common	—	—	—
Planar tabular	Common	Rare	—	—
Lateral accretion	Less Common	Less Common (graded)	—	—
Irregular inclined	—	Very common (graded)	Rare	—
Trough	Rare	Common (graded)	Rare	Uncommon
Grading: normal	Less common	Abundant	Abundant	Rare
inverse	Less common	Less common	Less common	Rare
complex	—	Less common	Rare	—
Gravel Fabric	$a(t)$, $b(i)$, towards base; a (bimodal, polymodal or random); $b(i)$ towards top of vertical sequences	$a(p)$, $a(i)$, towards base; a (bimodal or random);, $a(i)$ towards top of vertical sequences	$a(p)$, $a(i)$	$a(p)$, $a(i)$ smaller clasts $a(t)$, $b(i)$ larger clasts
Horizontal stratification	Layers of different clast sizes and/or layers with clast-supported, matrix-filled texture alternating with open-work texture	Layers of different clast sizes and/or layers with clast-supported texture alternating with clast-dispersed texture	Layers of different clast sizes and/or layers with clast-supported texture alternating with clast-dispersed texture	Layers of different clast sizes and/or layers with clast-supported, matrix-filled texture alternating with open-work texture

Table 3. Common associations of features seen in clast-supported conglomerates from fluvial and deep-water braided channel systems. Based upon detailed comparisons of modern fluvial conglomerates, Canadian Rockies, and an ancient deep-water braided channel system, Cap Enragé Formation (Cambro-Ordovician), Québec. This table is intended to show those features which are more common in the conglomerates examined in this study. For gravel fabrics abbreviations are as follows: $a(t)$, $b(i)$ — a-axis flow-transverse, b-axis imbricate-upstream; $a(p)$, $a(i)$ — a-axis flow-parallel, a-axis imbricate upstream.

tureless; bases of beds would be more commonly loaded (in comparison to fluvial sandstones). This is not to say that fluvial sandstones do not show grading or Bouma sequences (as shown by Baker, 1973), or liquefaction structures (as shown by Selley, 1969); but rather that these features are rare in fluvial sandstones in comparison with deep-water sandstones. Fluvial sandstones tend to be generally ungraded, and stratified or cross-bedded throughout the beds. The total assemblage of features, including lithologic or biogenic criteria, in conjunction with the stratigraphic context, must be used to develop appropriate conglomerate facies models.

REFERENCES

Baker, V.R. 1973. Paleohydrology and sedimentology of Lake Missoula flooding in eastern Washington. Geological Society of America Special Paper 144, 79p.

Bouma, A.H. 1962. Sedimentology of some flysch deposits: a graphic approach to facies interpretation. Elsevier Publishing Company, 168p.

Chough, S.K. and Hesse, R. 1976. Submarine meandering thalweg and turbidity currents flowing for 4000 km in the Northeast Atlantic Mid-Ocean Channel, Labrador Sea. Geology, v. 4, p. 529-533.

Costello, W.R. and Walker, R.G. 1972. Pleistocene sedimentology, Credit River, southern Ontario: a new component of the braided river model. Journal of Sedimentary Petrology, v. 42, p. 389-400.

Curray, J.R. 1956. The analysis of two-dimensional orientation data. Journal of Geology, v. 64, p. 117-131.

Davies, I.C. 1972. Transport of conglomerate into deep water: a study of the Cambro-Ordovician Cap Enragé conglomerate at St. Simon de Rimouski, Québec. M.Sc. Thesis, McMaster University, 91p.

Davies, I.C. and Walker, R.G. 1974. Transport and deposition of resedimented conglomerates: the Cap Enragé Formation, Gaspé, Québec. Journal of Sedimentary Petrology, v. 44, p. 1200-1216.

Fahnestock, R.K. and Bradley, W.C. 1973. Knik and Matanuska Rivers, Alaska: a contrast in braiding. In: Morisawa, Marie (Ed) Fluvial Geomorphology. Proceedings of the 4th Annual Geomorphology Symposium, State University of New York, Binghamton, N.Y., p. 220-250.

Harms, J.C., Southard, J.B., Spearing, D.R. and Walker, R.G. 1975. Depositional environments as interpreted from primary sedimentary structures and stratification sequences. Society of Economic Paleontologists and Mineralogists, Short Course Notes No. 2, 161p.

Harms, J.C., Tackenberg, P., Pickles, E. and Pollock, R.E. 1981. The Brae Oilfield area. In: Illing, L.V. and Hobson, G.D. (Eds). Petroleum Geology of the Continental Shelf of Northwest Europe, Heyden, p.352-357.

Harms, J.C. and McMichael, W.J. 1983. Sedimentology of the Brae Oilfield area, North Sea Journal of Petroleum Geology, v. 5,4, p.437-439.

Hein, F.J. 1974. Gravel transport and stratification origins, Kicking Horse River, British Columbia. M.Sc. Thesis, McMaster University, 135p.

Hein, F.J. 1979, Deep-sea valley-fill sediments: Cap Enragé Formation, Québec. Ph.D. Thesis, McMaster University, 514p.

Hein, F.J. 1982. Depositional mechanisms of deep-sea coarse clastic sediments, Cap Enragé Formation, Quebec. Canadian Journal of Earth Sciences, v. 19, p. 267-287.

Hein, F.J. and Arnott, R.W. 1983. Precambrian Miette conglomerates, Lower Cambrian Gog quartzites and modern braided outwash deposits, Kicking Horse Pass area. Canadian Society of Petroleum Geologists, Field Trip Guidebook, 46p.

Hein, F.J. and Walker, R.G. 1977. Bar evolution and development of stratification in the gravelly, braided, Kicking Horse River, British Columbia. Canadian Journal of Earth Sciences, v. 14, p.562-570.

Hein, F.J. 1982. The Cambro-Ordovician Cap Enragé Formation, Québec, Canada: conglomeratic deposits of a braided submarine channel with terraces. Sedimentology, v.29, p.309-329.

Hess, G.R. and Normark, W.R. 1976. Holocene sedimentation history of the major fan-valleys of Monterey fan. Marine Geology, v. 22, p. 233-252.

Hubert, C., Lajoie, J. and Leonard, M.A. 1970. Deep sea sediments in the Lower Paleozoic Quebec Supergroup. In: Lajoie, J. (Ed), Flysch sedimentology in North America. Geological Association of Canada, Special Paper 7, p. 103-125.

Johnson, B.A. 1974. Deep-sea fan-valley conglomerate: Cap Enragé Formation, Gaspé, Québec. M.Sc. Thesis, McMaster University, 108p.

Johnson, B.A. and Walker, R.G. 1979. Paleocurrents and depositional environments of the deep-water conglomerates in the Cambro-Ordovician Cap Enragé Formation, Gaspé, Québec. Canadian Journal of Earth Sciences, v. 16, p. 1375-1387.

Kessler, L.G., II 1979. Sedimentary structure, sequence and channel geometry in ancient submarine fan systems — examples of similarities to braided fluvial deposits. Geological Association of Canada Annual meeting, Quebec, Program with Abstracts, v.4, p. 61.

Lowe, D.R. 1975. Water escape structures in coarse-grained sediments. Sedimentology, v. 22, p. 157-204.

Nemec, W., Porębski, S.J. and Steel, R.J. 1980. Texture and structure of resedimented conglomerates: examples from Książ Formation (Famennian-Tournaisian), southwestern Poland. Sedimentology, v. 27, p. 519-538.

Rocheleau, M. and Lajoie, J. 1974. Sedimentary structures in resedimented conglomerate of the Cambrian flysch, L'Islet, Quebec Appalachians. Journal of Sedimentary Petrology, v. 44, p. 826-836.

Smith, D.G. 1973a. Aggradation and channel braiding in the North Saskatchewan River, Alberta, Canada. Ph.D. Thesis, Johns Hopkins University, 84p.

Smith, D.G. 1973b. Aggradation of the Alexandra-North Saskatchewan River, Banff Park, Alberta. In: Morisawa, Marie (Ed). Fluvial Geomorphology. State University of New York, 4th Annual Symposium on Geomorphology, Binghamton, N.Y., p. 201-220.

Smith, N.D. 1974. Sedimentology and bar formation in the upper Kicking Horse River, a braided outwash stream. Journal of Geology, v. 82, p. 205-223.

Selley, R.C. 1969. Torridonian alluvium and quicksands. Scottish Journal of Geology, v. 5, p. 328-346.

Steel, R.J. and Thompson, D.B. 1983. Structures and textures in Triassic braided stream conglomerates ('Bunter Pebble Beds') in the Sherwood Sandstone Group, North Staffordshire, England. Sedimentology, v. 30, p. 341-367.

Stow, D.A.V., Bishop, C.D. and Mills, S.J. 1982. Sedimentology of the Brae Oilfield, North Sea: fan models and controls. Journal of Petroleum Geology, v. 5, 2, p. 129-148.

Syvitski, J.P.M. and Blakeney, C.P. (compilers) 1983. Sedimentology of Arctic Fjords Experiment. HU 82-031 Data Report, v. 1, Canadian Report of Hydrography and Ocean Sciences, No. 12, 935p.

Syvitski, J.P.M. and Farrow, G.E. (in press). Structures and processes in bayhead deltas: Knight and Bute Inlets, British Columbia. Sedimentary Geology.

Walker, R.G. 1979. L'Islet Wharf: an Early Cambrian submarine channel complex. In: Middleton, G.V., Strong, P., Hein, F.J., Hendry, H.E. and Hiscott, R.N. Cambro-Ordovician submarine channels and fans, L'Islet to Sainte-Anne-des-Monts, Quebec. Geological Association of Canada Field Trip A-6, p. 4-7.

Wescott, W.A. and Ethridge, F.G. 1980. Fan-delta sedimentology and tectonic setting — Yallahs fan-delta, SE Jamaica. American Association of Petroleum Geologists Bulletin, v. 64, p. 374-399.

Winn, R.D., Jr., and Dott, R.H., Jr. 1979. Deep water fan channel conglomerates of Late Cretaceous age, southern Chile. Sedimentology, v. 26, p. 203-228.

FLUVIAL PROCESSES

CHUTES AND LOBES: NEWLY IDENTIFIED ELEMENTS OF BRAIDING IN SHALLOW GRAVELLY STREAMS

JOHN B. SOUTHARD[1], NORMAN D. SMITH[2] AND ROGER A. KUHNLE[1]

ABSTRACT

In-channel features here called *chutes and lobes* are dominant elements of braiding in the small gravel-bed outwash plain of Hilda Glacier, western Alberta, Canada. Observations of chutes and lobes were made in the field and in a laboratory flume arranged to represent a Froude scale model of the Hilda outwash plain but with steady rather than varying water discharge and sediment input rate. Chute-and-lobe behavior was closely similar between field and laboratory.

Chutes are relatively deep, narrow channels lined with relatively fine sediment, through which coarse clasts move efficiently. A lobe forms at the downstream end of a chute by stalling and jamming of coarse clasts; once initiated, the lobe grows quickly by capturing more coarse clasts until it is several clasts thick and has a downstream slope so steep (up to about 12°) that incision begins, leading to deep dissection and development of one or more new chutes. Incision leaves lobe remnants as emergent gravel-bar topography.

Coarse, poorly sorted sediment together with very shallow flow depths seem to be the essential elements for development of chute-and-lobe behavior.

RÉSUMÉ

Les principaux éléments de développement du réseau anastomosé dans la petite plaine d'épandage du glacier Hilda, ouest de l'Alberta, Canada, laquelle est composée de lits de graviers, sont les phénomènes que nous appelons ici *chutes-et-lobes* caractéristiques de l'intérieur des chenaux. Des observations de ces chutes-et-lobes ont été faites sur le terrain et également en laboratoire sur des ravins disposés de manière à représenter le modèle à échelle de Froud pour la plaine d'épandage Hilda, cependant avec un débit d'eau et un taux de production de sédiment réguliers plutôt que variables. Le fonctionnement des chutes-et-lobes sur le terrain et en laboratoire est presque identique.

Les chutes sont des chenaux relativement profonds, étroits, garnis de sédiment relativement fin, où circulent efficacement des fragments grossiers. Un lobe se forme à l'extrémité avale d'une chute par perte de vitesse et immobilisation des gros fragments; dès le début de la formation du lobe, il s'accroît rapidement en capturant d'autres gros fragments jusqu'à accumulation d'une épaisseur de plusieurs fragments et il développe une pente en aval si abrupte (jusqu'à environ 12°) que s'amorce une entaille, amenant une dissection profonde et le développement d'une ou plusieurs nouvelles chutes. L'entaille laisse apparaître des reliquats de lobes qui présentent un relief de bancs de gravier émergents.

Le sédiment grossier faiblement trié associé à des coulées peu profondes semble constituer les éléments nécessaires pour permettre le fonctionnement des chutes-et-lobes.

INTRODUCTION

The underlying causes of braided channel patterns have been discussed by many workers (for brief reviews, see Church, 1972, and Cheetham, 1979). Factors often considered to favor braiding include unstable channel banks, high valley slopes, irregular or flashy flow discharges, and abundant bed load, all of which can be considered contributive but external variables in that they do not define or regulate the actual in-channel processes that create braided patterns.

In well-sorted sand-bed channels, braiding is commonly perceived to arise mainly from low-stage distortion and incision of large tabular bed forms, solitary or repetitive, formed during high flows (Brice, 1964; Smith, 1971; Cant and Walker, 1978; Blodgett and Stanley, 1980), and resulting channel-network patterns appear to be independent of system scale (Krumbein and Orme, 1972; Hong and Davies, 1979). Because some sand is transported by even very low fluvial flows, braiding patterns in sandy streams are continuously changed by dissection and accretion. In streams with poorly sorted gravel beds, however, previous workers have tended to place more emphasis on generation and evolution of bars in the braiding process, with flow division a more or less passive result of bar emergence at lowered flow stages (Leopold and Wolman, 1957; Krigstrom, 1962; Smith, 1974; Bluck, 1974; Hein and Walker, 1977; Church, 1982). Unlike in sandy streams, bar dissection and other modifications are mainly confined to periods of high flow when the stream is competent to move coarse bed load. Bar growth and erosional sculpturing thus occur more or less simultaneously, with the result that simple

---

[1]Department of Earth, Atmospheric and Planetary Sciences, Massachusetts Institute of Technology, Cambridge, Massachusetts 02139, U.S.A.

[2]Department of Geological Sciences, University of Illinois at Chicago, Chicago, Illinois 60680, U.S.A.

This paper represents some of the results obtained during a joint field and laboratory investigation of braiding and sediment transport in gravel-bed streams. The work was sponsored by the National Science Foundation by grants to the first two authors.

We wish to thank Shaw-dyi Kang and Thomas Drake for their assistance and helpful discussions. We thank the officials of Parks Canada for permission to study in Banff National Park.

Copyright © 1984, Canadian Society of Petroleum Geologists

bar forms ('unit bars' of Smith, 1974) seldom become exposed without undergoing major modifications (see also Church, 1982; Ashmore, 1982). The usual case is that exposed gravel bars that characterize braided patterns are complex products of multiple depositional and erosional events.

In virtually all natural gravel-bed braided streams studied so far (Fahnestock, 1963, a prominent exception), the rivers have been of intermediate and large sizes, usually glacial outwash, with bars measurable in tens to hundreds or even thousands (e.g., Fahnestock and Bradley, 1973) of meters in length. In such large streams, depths of gravel-transporting channels are usually decimeters to meters, and ratios of depth to clast size (d/D) normally exceed 10 during active sediment transport. These large spatial dimensions virtually assure that most braid bars in such streams are complex multistory deposits, not the simple forms developed from a relatively straightforward series of events as envisaged by Leopold and Wolman (1957). In small gravelly streams such as that studied by Fahnestock (1963), braiding patterns may be no less complex, but the question arises as to whether the mechanisms are comparable to those of the larger streams that to date have received the most attention.

In a study of sediment transport and channel processes in a small glacial outwash stream, we have noted that in-channel features we call *chutes and lobes* are an extremely important element in the braiding process. Our purposes here are to call attention to these features from observations made both in the field and in a laboratory flume and to discuss their role in braiding.

Field Observations

SETTING

The subject of our field study was the outwash plain of Hilda Glacier, a small cirque glacier near the Columbia Icefields in the Rocky Mountains of western Alberta, Canada. The glacier is approximately 8 km southeast of the Columbia Icefields Chalet, 2 km southwest of the Icefields Highway between Lake Louise and Jasper. There is an active and rapidly aggrading outwash plain approximately 300 m long and 60 m wide about 500 m downstream from the glacier terminus. During high flows, much of the outwash plain is occupied by shallow braid channels transporting gravel. Slope decreases from 0.050 near the apex to 0.036 at the distal end. A fuller description of the area, including a discussion of the hydrology, sediment sources and yield, is contained in Hammer and Smith (1983).

Size analysis of composite bed-material samples from across the outwash showed a downstream decrease of mean grain diameter, D, from -5.0ϕ to -3.7ϕ (32-13 mm) over a distance of 200 m. Over the same distance, D decreases from -6.3ϕ to -5.1ϕ (79-34 mm). Standard deviations of composite samples ranged from 0.92ϕ to 1.23ϕ, indicating the general poor sorting of the bed material.

Water discharges are strongly diurnal on warm days and only rarely exceed $0.6 \text{ m}^3.\text{s}^{-1}$. Gravel transport is usually apparent whenever water discharge exceeds $0.25 \text{ m}^3.\text{s}^{-1}$. Well-defined braid channels are typically 1 to 3 m wide and less than 0.2 m deep (commonly 0.10-0.15 m) during gravel-transporting flows. Channel overspilling and avulsions are frequent during rising discharges, however, resulting in shallow sheet flows with ill-defined width. The characteristically shallow depths and coarse clast sizes of Hilda outwash yield low d/D ratios, rarely if ever exceeding 10; a consequence of this is that conventional bed forms with slip faces are never observed.

CHUTES AND LOBES

Chutes and lobes are active in outwash channels during intermediate and high discharges and develop most often in the proximal and medial reaches. Each feature consists of a straight and narrow gravel-transporting channel (chute) which terminates in a lobe-shaped gravel deposit (lobe) that enlarges as coarse bed load moves through the chute and comes to rest in the lobe. Chutes are typically several meters long and a meter or less wide. Fully developed lobes range from 1 to 3 m in width and are either equal or slightly greater in streamwise length. Chutes and lobes form quickly and are active for only short periods, usually no more than several minutes. Flow measurements with a direct-reading Price current meter in 17 active chutes showed depth ranges of 9 to 18 cm (average 12.9 cm) and mid-depth velocities ranging between 1.1 and 2.5 m.s^{-1} (average 1.7 m.s^{-1}).

The lobes accumulate the coarsest material in transport and form low mounds of gravel which, after exposure, are noticeably coarser than their surroundings. Lobe surfaces are slightly inclined, with downstream-dipping slope angles increasing with sediment size (Fig. 1). In no case was an

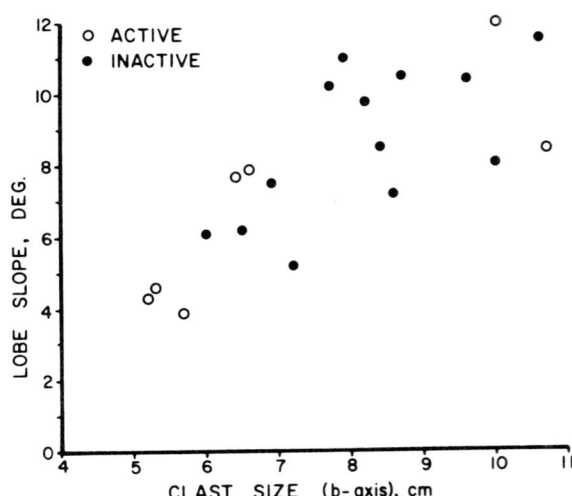

Fig. 1. Relationship between lobe slope angle and sediment clast size. Clast size represents the average of the 10 largest clasts found on lobe surface.

angle of repose ever observed or even approached; the largest measured slope angle was 13°.

To even the casual observer, the role of chutes and lobes in producing braided patterns in Hilda outwash is at once apparent. New chutes may be triggered by either local scouring in an active channel or incision of an existing lobe. Once initiated, scouring progresses quickly, forming a straight narrow chute through which gravel is rapidly transported. Chutes several meters long were observed to form in less than one minute. Eventually, the moving gravel stalls at the downstream end of the channel to form a lobe that continues to enlarge as additional gravel is supplied through the chute.

Subsequently, often as gravel supply from upstream lessens, the lobe begins to be incised and becomes partly exposed by the now sediment-deficient flow. Further incision may erode all of the lobe, often to form another one a short distance downstream, but more commonly one or more remnants are left behind as small braid bars. A typical sequence of events is shown in Figure 2. These braid bars are genetically simple features, unlike those typically found in larger braided streams. During low discharges when the outwash plain is well exposed, remnants of lobe deposits can be observed over much of the outwash surface (Fig. 3). Because they form locally positive areas, these remnants ('bars') remain exposed while adjacent low areas are gradually reoccupied during rising stages. This process of channel reoccupation, termed 'secondary anastomosis' by Church (1972), is a common cause of braided flows in the Hilda field site.

Although new chutes and lobes were often observed to form a few meters downstream from freshly dissected lobes (Fig. 3), nowhere on the outwash were we able to perceive any regular or repetitive spacing of these features. Rather, the development of a new chute followed quickly by lobe deposition appeared to be spatially random, and at no time did we feel able to predict when or where a new chute would occur. Furthermore, we were unable to recognize any particular set of 'upstream conditions' of flow, bed condition, or channel configuration that would trigger a new chute except for the fairly common (but by no means dominant) cases of newly scoured lobes at short

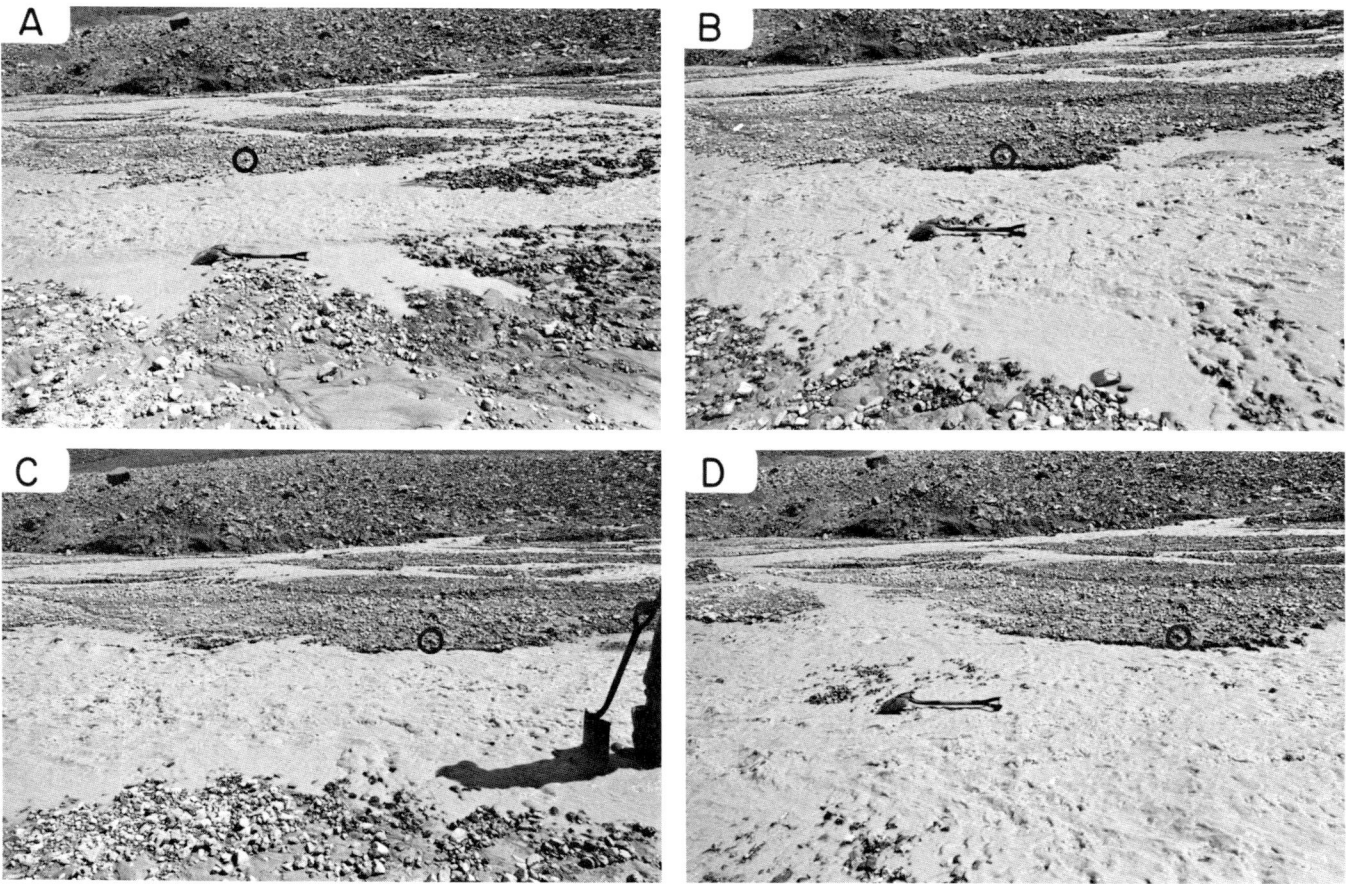

Fig. 2. Sequence of chute-and-lobe changes in a Hilda outwash channel. Note circled clast for reference point. Flow is right to left. (**A**) Active chute and lobe. Lobe position is marked by area of roiled water directly behind and to the left of shovel (1 m long). (**B**) Ten minutes later, lobe has stalled: flow has incised downward on two sides, partially exposing center portion of lobe (under shovel). Note that dissection of upper channel has eroded bank closer to reference point. (**C**) Approximately 3 minutes later, new chute forms over exposed remnant of (B), moving mass of gravel about 3 m downstream. (**D**) Five minutes after (C), new bar is formed by lateral incision and partial exposure of lobe remnant.

Fig. 3. Exposed lobe remnant (outlined) only slightly modified by lateral incision. Shovel (1 m long) lies across abandoned chute, now plugged with fine sediment. Flow was toward viewer.

distances upchannel. It seems important to note, however, that chutes and lobes are totally within-channel features and should not be confused with lobate deposits at anabranch confluences.

No more than gross observations of chute-and-lobe processes could be made in the field because of difficult working conditions: nearly opaque water, unfavorable vantage positions, unpredictability of occurrence, and the rapidity with which features are formed and destroyed. Accordingly, flume experiments were undertaken to reproduce and more closely observe chute-and-lobe behavior. We describe these experiments next.

SCALE MODELING OF BRAIDING IN GRAVELS

The irregular nature of water and sediment input at the head of the Hilda outwash stream makes it difficult to isolate the effects of these variables on braiding. Observed braiding patterns are likely to be out of equilibrium with water discharge and sediment supply, and can be viewed as a compromise over a wide range of changing values. To gain some insight into the effect of varying flow on braiding (as well as to better observe the sediment movement), we arranged a laboratory model stream. The problem in exact modeling, however, is that the scale must be far smaller than even the relatively small Hilda system. Fortunately, as shown below the coarseness of the Hilda sediment makes scale modeling a practical possibility.

In modeling a freely braiding channel system carrying a steady input of water and sediment down a broad sediment surface, seven variables can be imposed: Q_s, mass rate of sediment input at the head of the channel; Q, water discharge; D, mean or median sediment size; σ, standard deviation of the size distribution; ρ and ρ_s, liquid and sediment density; and g, acceleration of gravity. After equilibrium is reached, these variables characterize all aspects of sediment transport, channel pattern, and channel slope. Other size-distribution parameters are ignored, and clast shape is assumed not to be grossly different between the model and field.

Viscous effects are assumed to be negligible. If grain size is coarser than a few millimeters in the model as well as in the original, the flow is always fully rough even at the transport threshold, so both boundary resistance and the drag coefficient of suspended bed-material particles (if any) settling through the flow are independent of Reynolds number. Also, for sufficiently high Reynolds numbers, the overall pattern of nonuniform flow in the irregular channel geometry should also be largely independent of Reynolds number.

The seven variables above can be arranged into four dimensionless variables that completely specify the state of the system in terms of dimensionless measures of channel pattern, sediment transport, and slope (Buckingham, 1915). The following is a convenient set: $Q/(g^{1/2}D^{5/2})$ (a Froude number based on water discharge and sediment size), $Q_s/\rho Q$, σ/D, and ρ_s/ρ. For scale modeling, variables must be adjusted in model and original systems so that all four dimensionless variables have the same value in both systems. All forces and motions are then in proportion between the two systems, and the value of any dependent variable in one can be predicted from the other. This modeling approach is equivalent to that proposed by Parker (1979).

Assuming that variables have been adjusted so that the last three dimensionless variables are equal in model and original, equality of Froude number, $Q_r/(g_r^{1/2}D_r^{5/2}) = 1$ (where r denotes the ratio of any variable between original and model), constrains the relationship between scale ratio and water-discharge ratio,

$$D_r = Q_r^{2/5} \qquad (1)$$

since g is constant. We first determined the scale ratio by choosing the finest model sediment size for which effects in the model would still be comfortably independent of Reynolds number, and then obtained from (1) the necessary water discharge in the model. We chose a sediment size mix with 4 mm mean size and 3 cm maximum clast size, which when compared to the Hilda outwash sediment, with about 20 mm mean size on the proximal to medial outwash plain, gives a scale ratio of about 5. Maximum clast size then scales to 15 cm, appropriate to much of the proximal to medial Hilda outwash.

By (1), Q_r is then about 50, so Q in the model was adjusted to be about 2% of the average Q on the Hilda outwash at times of active braiding. By the constraint $(Q_s)_r/\rho_r Q_r = 1$, $(Q_s)_r$ should also be about 50. The constant $\sigma_r/D_r = 1$ was satisfied by adjusting the size distribution of the experimental sediment in accordance with that of the Hilda sediment. Finally, the constraint $(\rho_s)_r/\rho_r = 1$ was automatically satisfied by use of water and quartz-density sediment in the model.

An added advantage of modeling is that time scales are shorter. Dimensionally, $Q_r = L_r^3/T_r$, where L_r is the ratio of any length variable and T is the ratio of any characteristic time. Substituting into (1) and writing L_r for D_r, $T_r = L_r^{1/2}$; all processes are thus speeded up in the model relative to the original.

LABORATORY OBSERVATIONS

EXPERIMENTAL SYSTEM

The experimental apparatus consisted of a recirculating stream table 2 m wide by 12 m long. The channel was not ideally long, but braiding became fully developed about halfway down the channel. Early in the work, when an anabranch reached a sidewall, it would never shift away again because the smooth wall and adjacent floor provided an easy route for sediment movement. Thereafter the sidewalls and bottom were lined with glued-down gravel with a similar size to the largest clasts in the sediment.

The sediment was prepared from commercially available glacial outwash debris by screening to give the desired size distribution (Fig. 4). Coarse clasts were generally subangular to subrounded and either equant to moderately elongated or highly irregular in shape. Few clasts were discoidal. The coarser fraction was generally less well rounded than that of the Hilda outwash.

The most valuable data on braiding were obtained by qualitative visual observations of sediment transport and by visual observations and time-lapse cinematography of the channel pattern.

RESULTS

About ten runs were made specifically to study the role of chute-and-lobe activity in braiding. Experiments were made with three sediment feed rates: 2.3 kg.s^{-1}, 4.6 kg.s^{-1}, and 9.2 kg.s^{-1} (scaled). The lower two values fall into the range of day-to-day bed-load sediment discharges measured in the field, and the highest value is in the range of unusually high field measurements. Each run was started with a straight trapezoidal channel 50 cm wide and about 5 cm deep carved into a planar bed of thoroughly homogenized sediment with a given slope. After a transient stage during which a fairly regular series of channel bends developed at first and then degenerated into a highly irregular pattern of multiple braided channels (and during which the overall slope evolved toward a state of fluctuation around a long-term average value), braiding behavior was closely similar to that on the Hilda outwash plain when scaled, suggesting that the scale modeling was valid at least to a good approximation. Runs with slow net aggradation of the system showed about the same braiding behavior as equilibrium runs.

As on the Hilda outwash plain, the dominant braiding processes involved the chute-and-lobe phenomenon. The chutes, relatively narrow and deep channels with finer-sediment beds, were 20-30 cm wide, 3-4 cm deep, and 1-3 m long; the lobes were approximately equidimensional areas about half a meter across. When scaled, these dimensions are approximately the same as those on the Hilda outwash plain. In the flume, the chutes and lobes are easy to identify and watch but are not very photogenic. The rest of this section is devoted to a description of the development of chutes and lobes, based mainly on visual observations and viewing of time-lapse movies but supplemented by profiling data.

In any local area of the flume bed, there was a continuing and irregularly repetitive development of braiding features dominated by chutes and lobes. To describe this development, we focus here on a representative new chute channel. (Development of chutes themselves is addressed later in this section.) In its early stages, the chute is moderately steep and relatively narrow and deep (Fig. 5B). It has a bed of relatively fine sediment, itself in active bed-load transport, over which coarser clasts are rolled almost continuously. The chute is thus a very efficient pathway for transport of coarse clasts. As time goes on, the chute widens appreciably by bank erosion; concurrently it shoals, as its discharge is spread laterally. Cross-stream uniformity usually remains fairly high, although in some cases there tends to be development of two separate chutes separated by an indistinct longitudinal bar.

As the chute widens and shoals, the transport rate of the finer fraction decreases markedly while that of the coarser fraction, controlled more by upstream supply, decreases much less, large clasts still rolling easily down the channel. Eventually, in some region downstream in the chute, large clasts begin to come to permanent rest by lodging or jamming with adjacent large clasts to initiate deposition of a lobe. As more and more large clasts are extracted from the load in this way, a large and stable area of exclusively coarse sediment is formed. In consequence, the flow usually tends to fan out over one or both banks of the chute

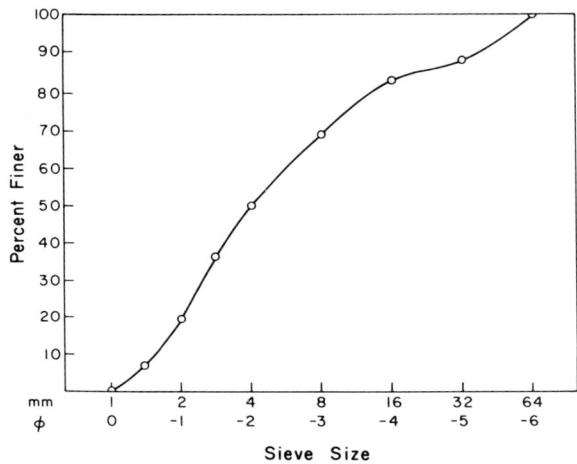

Fig. 4. Size distribution of model sediment.

Fig. 5. Two stages in the development of chutes and lobes in the model, at the intermediate sediment feed rate (4.6 kg.s^{-1}, scaled). **A.** Downstream view of an actively growing lobe. **B.** Chute incised into the lobe shown in A, leaving the lobe remnants emergent. Photograph B was taken about 10 minutes later than photograph A. Photographed width of sediment bed at downstream slipoff ramp is about 1.5 m. A water-filled footprint appears in both photographs as a point of reference.

just upstream of the growing lobe (Fig. 5A). Coarse clasts are extracted from the flow in these overbank areas in the same way, leading to development of a lobate convex-downstream body of coarse sediment. The new lobe grows downstream not by cannibalizing itself but by continuing to extract coarse clasts derived from upstream sources, and it also builds upward. At the same time, the finer fraction of the bed load aggrades the chute upstream of the lobe, and to a minor extent it is also swept onto the lobe to fill interstices. As time goes on, the upstream chute thus becomes less steeply sloping, while the lobe develops a relatively steeply sloping downstream face. The extremely shallow flow over the lobe riffles among the protruding large clasts, and newly supplied clasts skip prominently across the surface of the lobe before finding a niche.

At a certain stage the downstream face of the lobe becomes so steep (in the range of 10-15°, which is consistent with the field measurements reported above) that the flow over it begins to dislodge previously deposited clasts in one or more areas, and a phase of rapid upstream-propagating unravelling and incision (sometimes spectacularly so) begins, resulting in the formation of one or sometimes two new chutes cut through the lobe and down into the underlying, usually finer material. The newly developing chute extends itself upstream by downcutting into the finer sediment deposited upstream of the lobe. It also extends itself downstream by occupation of pre-existing largely or wholly abandoned channels as the discharge over the lobe is focused during incision. The new course of the dominant channel in the area is thus not altered greatly by the chute-and-lobe cycle; vertical changes are more spectacular than lateral changes. Some newly formed chutes evolve to carry relatively large water and sediment discharge, and it is in these that lobes are likely to form; other chutes simply become abandoned.

Because of the limited surface area of the model, we could not study directly the spatial evolution of chute-and-lobe features; however, observations on the development of new chutes are of relevance. In the model, it is clear that new chutes are typically formed by downcutting that begins at one or more points on the flanks of a lobe and then continues into an upstream reach of variable but often considerable length (up to 20 m when scaled) in which the flow is often only poorly channelized initially. Subsequently, depending apparently on supply of coarse material from upstream, a new lobe develops within the chute, as described above. This sequence of events implies an upstream progression of lobe development, in that a new lobe forms in a chute that itself was developed upstream of an earlier lobe, although the position and timing of this new lobe is governed at least in part by the behavior of lobes even farther upstream. We did not perceive this progression in the field, but it seems likely that the variability inherent in both chute development and upstream coarse-sediment supply, together with the random and rather dense arrangement of lobes at any given time, would lead to the seeming irregularity of chute-and-lobe development observed in the field.

Discussion

In both field and laboratory, primary braiding, *i.e.*, flow division, caused by local aggradation and/or dissection within active-bed-load-transporting channels, is dominated by chute-and-lobe activity. Broadly, each cycle of chute-and-lobe development goes through an aggradational stage (lobe) followed by a stage of either complete or partial degradation after the downstream-dipping surface of the lobe attains some critical slope. In both the Hilda outwash and the model, dissection is typically only partial. Depending on where the lobe is incised, the erosional remnants may remain as one or two elongate mounds resembling longitudinal bars (Fig. 6). These bars are easily recognized in the field because they form topographic highs composed of the coarsest material transported through that local area.

Lobe deposits, either whole or as remnants, are generally several clasts thick and could be expected to show low-angle cross-stratification in flow-parallel cross sections. In contrast, chutes aggrade somewhat finer grained and less well-sorted material, as it is their finer, less rough beds that permit rolling and sliding of larger clasts through the chute to the downstream lobe. During waning flows in the Hilda outwash, sand and silt frequently form thin drapes over chute-fill deposits; these are, however, only rarely preserved in section. We are as yet unable to predict the character of vertical sedimentary sequences formed by chutes and lobes in aggrading alluvium. Lobe deposits,

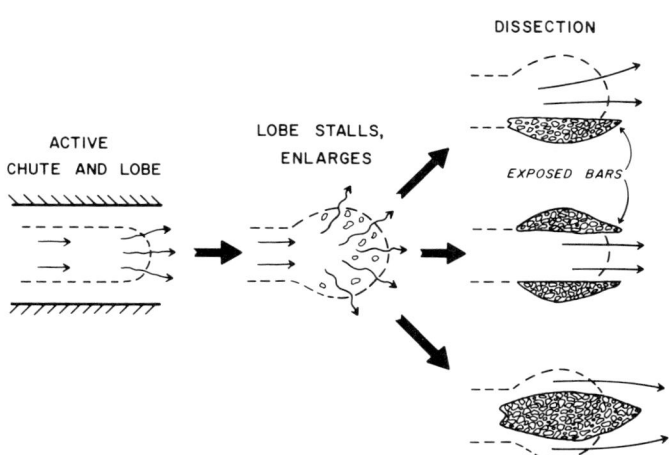

Fig. 6. Diagrammatic sequence showing how exposed bars form from chute-and-lobe activity. New chute and lobe intitially forms, either within wider channel (shown) or by incision of existing lobe. Lobe eventually stalls but continues to enlarge as gravel is transported through the chute to the lobe surface. When the downstream inclination of the lobe surface reaches some critical value, the lobe deposits begin to unravel quickly. Downward incision then exposes one or two remnants of the lobe to form elongate gravel bars.

though, should be recognizable as relatively coarse but well-sorted gravel, probably matrix-poor (open-work), forming solitary, low-angle sets no more than a few decimeters in thickness. We would expect chute deposits to be thinner, less abundant, finer grained, and less well sorted, forming horizontal beds that may be weakly graded.

The model was designed to simulate the upper outwash plain, where sediment is coarsest and discharge is greatest. Because of overall aggradation, sediment size and transport rate decrease substantially downstream. Despite these changes, chute-and-lobe behavior is qualitatively similar over the entire length of outwash, the major difference being that the tempo of braiding is slower downstream. The model did not fully address these downstream changes because sediment size and water discharge were held constant; however, the three values of imposed sediment feed rate covered most of the range of bed-load transport rates measured in the field (Kang, 1981). Chutes and lobes were most representative of the field situation at the two highest feed rates. Although sizes of chutes and lobes were about the same at these two rates (and when scaled were about the same as in the field), the tempo of chute-and-lobe activity was greatest at the high feed rate. Although we made no extensive measurements, lobes seem to be developed and incised in the field over periods of several minutes to a few tens of minutes depending on sediment transport rate, and in the model over periods of 10-15 min for the highest feed rate to some tens of minutes for the intermediate feed rate. Thus, the tempo of chute-and-lobe development is comparable between field and model, although more precise comparisons would necessitate much further work.

On the Hilda outwash, chute-and-lobe development is clearly unrelated to scour at confluences of active anabranches, and this impression was reinforced by the model study. The narrowness of the model outwash plain (10 m when scaled) limited the number of channels in any transverse section to no more than two, and seldom was there appreciable bed-load movement in more than one. That chutes and lobes were nonetheless well developed demonstrates clearly that they are independent of effects at confluences of active channels.

The similarity of braiding processes between the model and field site raises two questions: what are the underlying factors that promote chute-and-lobe behavior, and how widespread or general are these features? The salient characteristics of the Hilda outwash plain are coarse and poorly sorted bed material, shallow depths only slightly greater than the diameters of the coarser clasts in transport, high rate of sediment output, active aggradation of the outwash plain, and strongly fluctuating and laterally unconfined flows. The last two factors can be considered unessential to chute-and-lobe development because the flume experiments utilized steady water discharges with relatively confined widths. Aggradation, though essential for ultimate preservation of deposits, was shown by the model experiments not to be a factor in chute-and-lobe development, presumably because it is a far slower process. High rates of bed-load transport are probably necessary but insufficient. The rapidity at which chutes and lobes are formed, modified, and destroyed depends upon rapid movement of gravel; however, comparable bed-load discharges in other field and flume situations apparently are not associated with chutes and lobes. We have searched for chutes and lobes in two considerably larger gravel-bed outwash streams, the Kicking Horse and North Saskatchewan Rivers, both near Hilda, without success. This, plus the apparent absence of these features in published descriptions of other large gravelly streams, leads us to believe that two characteristics of the laboratory and Hilda stream beds are essential for development of chutes and lobes: poorly sorted gravel and very shallow depths.

Poor sorting seems to be essential for the development of the chute-and-lobe style of braiding in coarse sediments. As sorting becomes better, the style of braiding must become more like that of the development and modification of bars observed, for example, by Ashmore (1982) in model experiments using relatively well-sorted sediments. Both on the Hilda outwash plain and in the laboratory, sediment size varies greatly over short distances. This seems to result not so much from varying flow strengths as from the tendency for size-selective deposition of coarse and fine fractions during transport: the coarse fraction is transported easily over a bed of finer sediments in chute channels, but is moved much less readily over similar coarse material once a lobe has been initiated by the apparently random lodging of coarse clasts in an area of slightly decelerating flow.

The very shallow depths, combined with coarse clast sizes, lead to large values of relative roughness and extreme nonuniformity of flow, factors that enhance highly localized transport and deposition of varying-sized bed materials. Greater depths would spread gravel-transporting flows over relatively wider areas and tend to produce larger and more complex deposits instead of the comparatively small, simple, and short-lived chutes and lobes described here. Larger ratios of depth to grain size, say greater than 10, would likely also result in development of slip faces on some bars, as is commonly recognized in large gravel-bed braided streams. The ratio d/D_{90} in the Hilda outwash and flume rarely exceeded 3 during active transport and often was less than unity; as stated earlier, conventional bed forms or bars with slip-face margins were never observed.

We do not yet know how chutes and lobes fit into the spectrum of gravel-bed macroforms, or whether they are analogous to larger-scale lobate bar forms seen in larger rivers (e.g., Smith, 1974; Hein and Walker, 1977; Church, 1982). Model experiments as described here and by Ashmore (1982) are feasible means to investigate these further, especially if corroborated by further investigations of streams similar to or somewhat larger than the Hilda outwash plain.

REFERENCES

Ashmore, P.E. 1982. Laboratory modeling of gravel braided stream morphology. Earth Surface Processes, v. 7, p. 201-225.

Blodgett, R.H. and Stanley, K.O. 1980. Stratification, bedforms, and discharge relations of the Platte braided river system, Nebraska. Journal of Sedimentary Petrology, v. 50, p. 139-148.

Bluck, B.J. 1974. Structure and directional properties of some valley sandur deposits in southern Iceland. Sedimentology, v. 21, p. 533-554.

Brice, J.C. 1964. Channel patterns and terraces of the Loup Rivers in Nebraska. United States Geological Survey Professional Paper 422-D, p. 1-41.

Buckingham, E. 1915. Model experiments and the forms of empirical equations. American Society of Mechanical Engineers Transactions, v. 37, p. 263-292.

Cant, D.J. and Walker, R.G. 1978. Fluvial processes and facies sequences in the sandy braided South Saskatchewan River, Canada. Sedimentology, v. 25, p. 624-648.

Cheetham, G.H. 1979. Flow competence in relation to stream channel form and braiding. Geological Society of America Bulletin, v. 90, p. 877-886.

Church, M.A. 1972. Baffin Island sandurs: a study of Arctic fluvial processes. Geological Survey of Canada, Bulletin 216, 208p.

Church, M.A. 1982. Channel bars in gravel-bed streams. *In*: Hey, R. D., Bathurst, J. C. and Thorne, C. R. (Eds.), Gravel-bed Rivers. New York: J. Wiley and Sons, Ltd, p. 291-324. (See also Discussion and Reply, p. 325-338).

Fahnestock, R.K. 1963. Morphology and hydrology of a glacial stream-White River, Mount Rainier, Washington. United States Geological Survey Professional Paper 422-A, 70p.

Fahnestock, R.K. and Bradley, W.C. 1973. Knik and Matanuska Rivers, Alaska: a contrast in braiding. *In*: Morisawa, M. (Ed.), Fluvial Geomorphology. Proceedings 4th Annual Geomorphology Symposium, Publications in Geomorphology, State University of New York, Binghamton, p. 220-250.

Hammer, K.M. and Smith, N.D. 1983. Sediment production and tranpsort in a proglacial stream: Hilda Glacier, Alberta, Canada. Boreas, v. 12, p. 91-106.

Hein, F.J. and Walker, R.G. 1977. Bar evolution and development of stratification in the gravelly braided Kicking Horse River, British Columbia. Canadian Journal of Earth Sciences, v. 14, p. 562-570.

Hong, L.B. and Davies, T.R.H. 1979. A study of stream braiding. Geological Society of America Bulletin, v. 90, Part II, p. 1839-1859.

Kang, S.D. 1981. Sediment transport in a small glacial stream: Hilda Creek, Alberta. M.Sc. Thesis, University of Illinois at Chicago Circle, 267p.

Krigstrom, A. 1962. Geomorphological studies of sandur plains and their braided rivers in Iceland. Geografiska Annaler, v. 44, p. 328-346.

Krumbein, W.C. and Orme, A.R. 1972. Field mapping and computer simulation of braided stream networks. Geological Society of America Bulletin, v. 83, p. 3369-3380.

Leopold, L.B. and Wolman, M.G. 1957. River channel patterns-braided, meandering, and straight. United States Geological Survey Professional Paper 282-B, p. 39-85.

Parker, G. 1979. Hydraulic geometry of active gravel rivers. American Society of Civil Engineers, Proceedings, Journal of Hydraulics Division, v. 105, p. 1185-1201.

Smith, N.D. 1971. Transverse bars and braiding in the lower Platte River, Nebraska. Geological Society of America Bulletin, v. 82, p. 3407-3420.

Smith, N.D. 1974. Sedimentology and bar formation in the upper Kicking Horse River, a braided outwash stream. Journal of Geology, v. 82, p. 205-224.

PARTICLE INTERACTION AND ITS EFFECT ON THE THRESHOLDS OF INITIAL AND FINAL BEDLOAD MOTION IN COARSE ALLUVIAL CHANNELS

Ian Reid and Lynne E. Frostick[1]

Abstract

The data-base for initial motion of bedload is substantial, though it comes almost exclusively from laboratory studies in which practical difficulties generally preclude the use of coarse sediment. Estimates of the conditions at initial motion on coarse-grained stream beds rely on field observations, and these are rare. The lack of field data is attributed to the difficulties of registering the initial and final movement of bedload. This has been remedied by the development of two new instruments: a continuously recording bedload pit sampler, and an electro-magnetic bedload sensor which automatically registers the entrainment of individual tagged particles. A 4-year monitoring programme illustrates the short-comings of extrapolating laboratory-based predictions of thresholds of motion to larger particle sizes and applying them in the field. Extensions of the familiar curves such as those of Shields and Lane seriously underestimate the force required to initiate motion. Field values reflect the importance of microscale bedforms such as pebble clusters in delaying bedload transport. Particle interlock and mutual protection confer a stability in excess of that which would be possessed by the constituent clasts isolated on a plane bed. Perhaps more surprising is that the tractive force associated with the finish of bedload transport is shown to be one-third of that at the point of initial motion; shear stress at the bed, averaged for several floods, is 0.79 kg.m^{-2} and compares with 2.20 kg.m^{-2} for the start of bedload. This is important because bedload transport continues longer than might be predicted using standard bedload equations and because palaeohydraulic inference usually assumes the levels of tractive force involved with entrainment and not the much lower levels associated with deposition.

Résumé

La base de données sur les conditions de la mise en mouvement d'une charge de fond est considérable, quoiqu'elle repose presque exclusivement sur des études en laboratoire où l'on évite en général l'utilisation de sédiment grossier à cause des difficultés techniques inhérentes. Les données sur les conditions de mise en mouvement de lits à grain grossier de cours d'eau doivent être reliées à des observations de terrain, et elles sont rares. La pénurie de données de terrain est attribuée à la difficulté d'enregistrement du mouvement initial et final de la charge de fond. Cette difficulté a été surmontée grâce au développement de deux nouveaux instruments: un échantillonneur de trous dans la charge de fond avec enregistrement continu, et un capteur électro-magnétique de charge de fond qui enregistre automatiquement durant le transport l'embarquement de particules individuelles marquées. Un programme de surveillance électronique d'une durée de quatre années illustre les faiblesses quant à l'extrapolation pour prédiction des seuils du mouvement des particules grossières fondée sur des expériences en laboratoire et de leur application surle terrain. Les extrapolations des courbes habituelles, comme celles de Shields et Lane, sous-évaluent sérieusement la force nécessaire à la mise en mouvement. Les valeurs obtenues sur le terrain reflètent l'importance des formes de litage à micro-échelle, telles les agglomérations de galets qui entravent le transport de la charge de fond. L'enchevêtrement des particules et une protection mutuelle confèrent une stabilité supérieure à celle que possèdent les fragments isolés à la surface du lit. Ce qui surprend probablement le plus c'est que la force de traction associée à la fin du transport de la charge de fond ne correspond qu'à un tiers de celle au point du mouvement initial: la valeur moyenne de la force de cisaillement du lit de plusieurs crues est $0,79$ km.m^{-2} à la fin comparativement à $2,20$ kg.m^{-2} au départ de la charge de fond. Ceci est important car le transport de la charge de fond se poursuit sur une distance plus longue que prévue par les équations normales de charge de fond et parce que la déduction paléohydraulique présume habituellement que l'embarquement se fait dans les niveaux où s'exerce la force de traction et non à des niveaux beaucoup plus bas associés avec le dépôt.

Introduction

With increasing flood discharge the shear stress acting on the bed of a river increases to a point where it exceeds the restraining forces of gravity and particle interlock. Bedload transport commences. Experiments have traditionally acknowledged the role of particle size in determining the various restraining forces, but have generally sought to eliminate or ignore the important effects of micro-relief (for example, pebble clusters) — a factor that attains considerable importance in coarse-grained alluvial channels where bed sediments are usually extremely poorly sorted and particle size ranges widely. As a result, thresholds of motion are commonly expressed in simple form as grain diameter against some function of critical shear stress. The classical work of Shields (1936), later modified by Yalin (1972) and Miller et al., (1977) amongst others, has long been used by engineers and sedimentologists as a means of predicting and interpreting sediment transport in alluvial channels. However, the experiments on which the familiar curve of Shields is based must inevitably limit its application within the natural environment since they were carried out under conditions of steady uniform flow and over a plane bed of sediment. The data were also derived

---

[1] Departments of Geography and Geology, Birkbeck College, University of London, 7-15 Gresse Street, London W1P 1PA, England.

The Birkbeck bedload monitoring programme is partly funded by two grants from the Central Research Fund, University of London. John Layman and Andrew Brayshaw were intimately involved in the development of instrumentation and in the field programme and were each supported by a Natural Environment Research Council Training Award. Jim Best, Eddie Bates, Chris Hawkins, Derek Lee and Robert Parkinson assisted with the installation of field equipment. Access to the experimental site is by licence from the Estates and Valuation Department of the Greater London Council. We are grateful to Drs. M. G. Foley and P. D. Komar for reviewer's comments.

Copyright © 1984, Canadian Society of Petroleum Geologists

with sediment of less than 3 mm diameter, a size readily exceeded in most stream beds of coarse alluvium.

In fact, bedload equations such as that of Meyer-Peter and Müller (1948) are often found to overpredict by a considerable margin (Lauffer and Sommer, 1982). Among the factors responsible is the inevitable need to adopt a simple parameter of the size-distribution of the bed material and to assume that the critical tractive force appropriate to this size parameter will describe initial motion; this ignores the effect of protrusion of particles from the bed into the fluid stream (Fenton and Abbott, 1977), and the shielding of particles by others (Einstein, 1950; Brayshaw et al., 1983; Brayshaw, this volume). In addition, bedload equations implicitly assume that the tractive force at initial motion will be the same as at the point when bedload motion ceases. It will be shown here that this is inappropriate.

The data-base for initial motion is large, though it comes almost exclusively from flume studies conducted under controlled conditions. The range of bed-material size cannot usually be extended because of the practical difficulties of simulating coarse-grained alluvial channels. Yet there are few field observations to complement the laboratory studies since increased turbidity during flood flow precludes visual contact with the bed.

Faced with these difficulties, an approach that has been commonly adopted involves introducing marked pebbles to the bed and tracing them after flood flows (e.g., Leopold and Emmett, 1981). The largest displaced particle is *assumed* to have been moved at peak discharge, and particle size is correlated with bed-shear at the flood peak. Alternatively, if actual samples of bedload are available, the largest trapped particle is *presumed* to represent the competence of the stream at the moment of sampling (e.g., Baker and Ritter, 1975). Using a different approach, Helley (1969) devised an ingenious way of detecting the displacement of very large bed particles by observing the moments at which floats trapped under the particles are released. But, while all these methods provide useful information, they can only be considered as surrogates for the direct observation of thresholds of general motion which may, of course, involve any or all of the size fractions that constitute the stream bed.

The lack of data for gravelly sediment is attributed to the difficulty in actually registering initiation and cessation of bedload transport in the field. This has now been remedied by the development of two new devices designed to record on one hand the incidence and magnitude of bedload discharge and on the other the movement of individual particles throughout a flood wave.

Experimental Design

River Channel Character

Field experiments were carried out at Turkey Brook, Enfield Chase, 18 km north of London, England. The stream is small and therefore manageable: average channel width is 3 m. The bed is composed of rounded flint pebbles that have a consistent sphericity of 0.68, a D_{50} (median diameter) of 16 mm and a D_{90} of 35 mm. Sand generally constitutes less than 10% of the bed and, together with clay and silt particles, infiltrates the interstices of the gravel framework. The flood hydrograph is characteristically flashy and reflects the rapid rainfall-runoff response of the Eocene London Clay that underlies the drainage basin. Floods are discrete events separated by low baseflow. This is useful for experimental purposes since it allows access for servicing equipment between events. Floods are also frequent with up to 10 bedload generating events expected per year.

Birkbeck Bedload Samplers and Sensors

The channel consists of a series of straight reaches punctuated by meanders. Six Birkbeck Bedload Samplers were installed, 3 in each of 2 straight reaches. The samplers are described fully elsewhere (Reid et al., 1980), but each consists of a slot that is conformable with the bed of the stream and which leads to a box that sits in a concrete-lined pit (Fig. 1). The box is free to ride up and down and rests on a pressure pillow that registers any changes in the mass of the box and its contents. As sediment accumulates in the sampler, the change in pressure is sensed by a transducer and the output signal passed to a suitable recorder. The system is fully automatic and bedload is continuously recorded (Fig. 2). Sensitivity is good with a standard error of the estimate of only 0.3 kg. But the nature of the record means that the thresholds of both initial motion and cessation of bedload pertain to general motion on the bed and not to individual particle movement. In fact, this is useful since it removes the subjectivity that usually accompanies studies of entrainment.

Besides the Birkbeck Bedload Samplers, two Birkbeck Bedload Sensors (Reid et al., 1984) were installed in another straight reach of Turkey Brook. Each of these is a set of elongate unscreened electromagnetic coils that sit permanently in the bed of the stream and act in the same fashion as a conventional metal detector (Fig. 1). Metal tagged particles are seeded upstream, replacing like-size bed particles, and their entrainment registers on a continuous strip-chart as they move over the sensor upsetting the balance in the coils.

Results and Discussion

Threshold of Initial Sediment Motion

The results of the Birkbeck bedload monitoring programme call into question the application of laboratory-based predictions of threshold motion in the field where bed material is of coarse size. The data obtained from the Turkey Brook bedload samplers are plotted in Figure 3 along with the experimentally derived threshold curves of

Fig. 1. **A** — Birkbeck Sensor installed in the bed of Turkey Brook. **B** — Component parts of the electro-magnetic sensor. **C** — Installation of the continuously recording Birkbeck Bedload Samplers. Photographs A and B courtesy of A. C. Brayshaw.

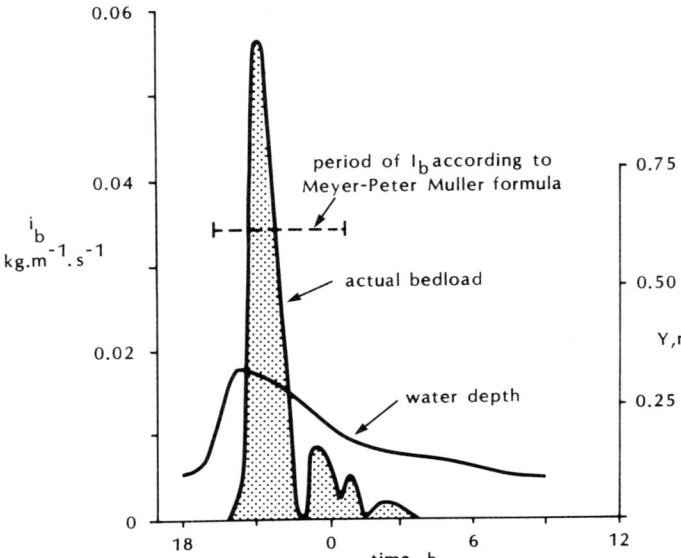

Fig. 2. Example of Turkey Brook flood wave, registered unit bedload transport and the period of bedload predicted by the Meyer-Peter and Müller (1948) formula (In this and subsequent figures, all notation is explained in the Appendix).

Shields (1936) and Lane (1955). It can be seen that the shear stress required to initiate particle motion (taking D_{50} of the sampled bedload as representative of particle size) is considerably in excess of that predicted by Shields' or Lane's curves. This might at first sight be attributed in part to the effect of channel side-wall drag — a factor of some importance in narrow channels. But the mean values of the well-tested Williams' (1970) side-wall adjustment factor at initial and final bedload transport are little different from unity (0.995 and 0.997).

An explanation for the discrepancy between actual and predicted thresholds of motion lies in the mutual interference of neighbouring particles on the bed of the stream. The use of small-sized material in flume studies and the general adoption of a plane bed as a starting condition for experimentation has meant that little attention has been paid to the influence of such interference on initial entrainment. Notable exceptions are Einstein's (1950) incorporation of a 'hiding factor' in his bedload equation, and the work of Fenton and Abbott (1977) and of Gessler (1971) on the effects of relative protrusion on the motion of individual particles. More recent work (Brayshaw et al., 1983) has provided a physical explanation for the behaviour of particles in close proximity in terms of lift and drag forces, and has added much to our understanding of grain interaction.

Particles congregate in organised clusters which are considerably more stable than the constituent pebbles would be if isolated on a plane bed (Brayshaw, this volume). Because there is both an enhanced degree of interlock and shielding of smaller pebbles by larger ones, the threshold of motion of clustered particles requires greater shear force than is predicted by the extrapolated Shields and Lane curves. Pebble clusters occupy ca. 10% of the bed of Turkey Brook and this has been shown to be an appropriate figure for streams of widely ranging clast lithology (Brayshaw, this volume). But their sphere of influence extends beyond this by substantially increasing the rugosity of the bed. Dislodgement depends upon the force required to move the most vulnerable of any one cluster's constituent particles. But once one particle has been moved the overall structure is weakened and further dislodgement is likely. The effect is that initial bedload transport is delayed.

The importance of cluster bedforms is confirmed by results obtained with the electromagnetic bedload sensor. Tagged particles were placed in both plane-bed positions and in pebble clusters and their motion during flood events recorded. For each flood, the plane-bed seeded particles, unhindered by protruding neighbours, would move earlier than their clustered counterparts, while more of the plane-bed particles would move than those seeded in cluster positions (typical ratios range from 1.3:1 to 5.4:1). Early entrainment also meant greater transport velocity per flood for plane-bed particles which would travel >10 m and compare with ca. 3 m for particles whose original position was the wake of a cluster form.

The hiding effect that is a feature of natural coarse-grained channels, and which dictates higher threshold values than are predicted for initial motion by Shields' entrainment function, is placed in perspective by the data of Coleman (1967) and of Fenton and Abbott (1977) which relate to single *exposed* particles. As would be expected their exposed particles move more readily and their points (Fig. 3) lie on the opposite side of the Shields curve relative to the Turkey Brook data.

Besides the control apparently exercised by pebble clusters over initial sediment motion, the Birkbeck bedload monitoring programme indicates the role of fine particles in binding the coarse bed material once they settle into interstices in the gravel framework. The effect is amplified when there is a prolonged period in which no floods disturb the bed. 10 December 1978 and 9 December 1979 (Table 1 and Fig. 4) were the first floods of the winter seasons of two consecutive years. Both followed autumn droughts and both floods demonstrate that the shear stress required to initiate bedload motion rises by a factor of 2 or 3 if fine particles bed themselves into interstices

FINAL BEDLOAD MOTION

Considerable attention has been paid to initial bedload transport but there is scant consideration of the forces at the point at which bedload ceases. This is somewhat surprising. Erosion is important to the engineer and to the geomorphologist and might explain the overriding interest in initial transport thresholds. In contrast the sedimentologist is primarily interested in processes of deposition responsible for aggradation of ancient alluvial successions.

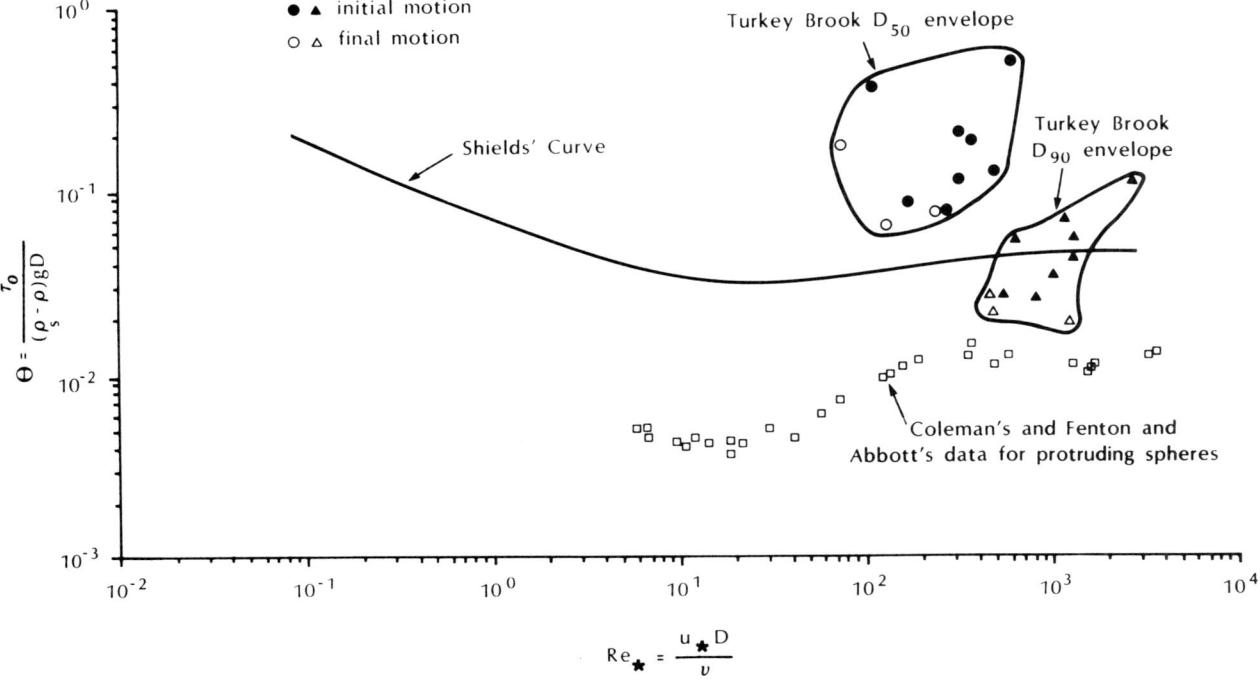

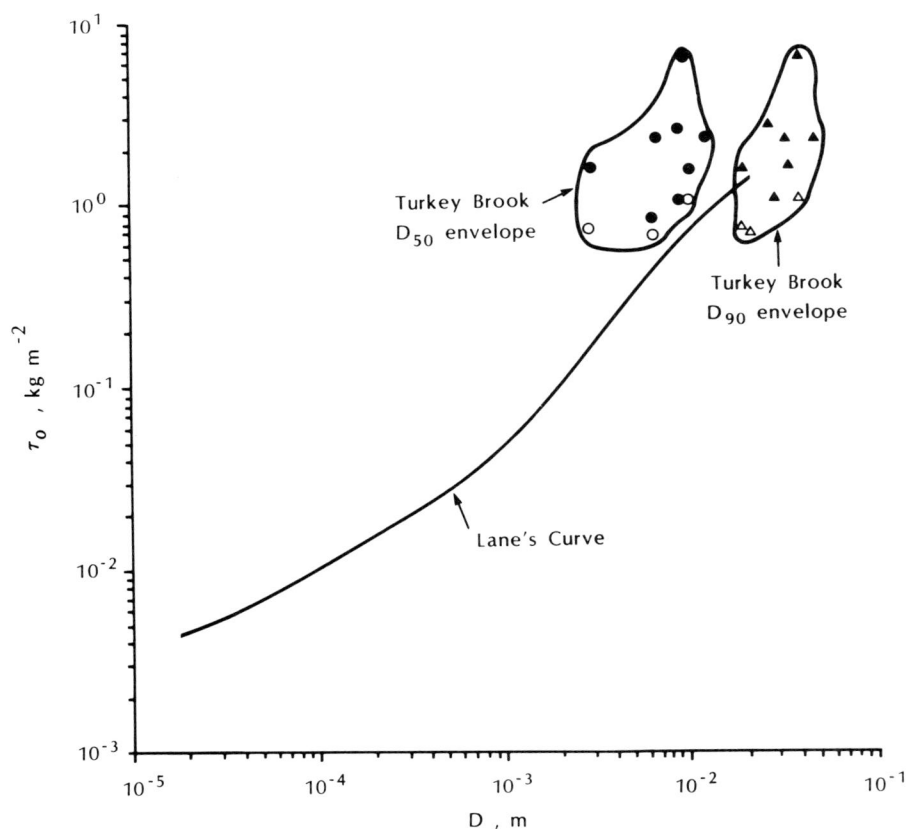

Fig. 3. **A** — Shields' threshold criterion, θ_0, plotted against grain Reynolds Number, Re_* for Turkey Brook and other data. **B** — Lane's curve of grain diameter, D, plotted against shear stress at initial motion τ_0.

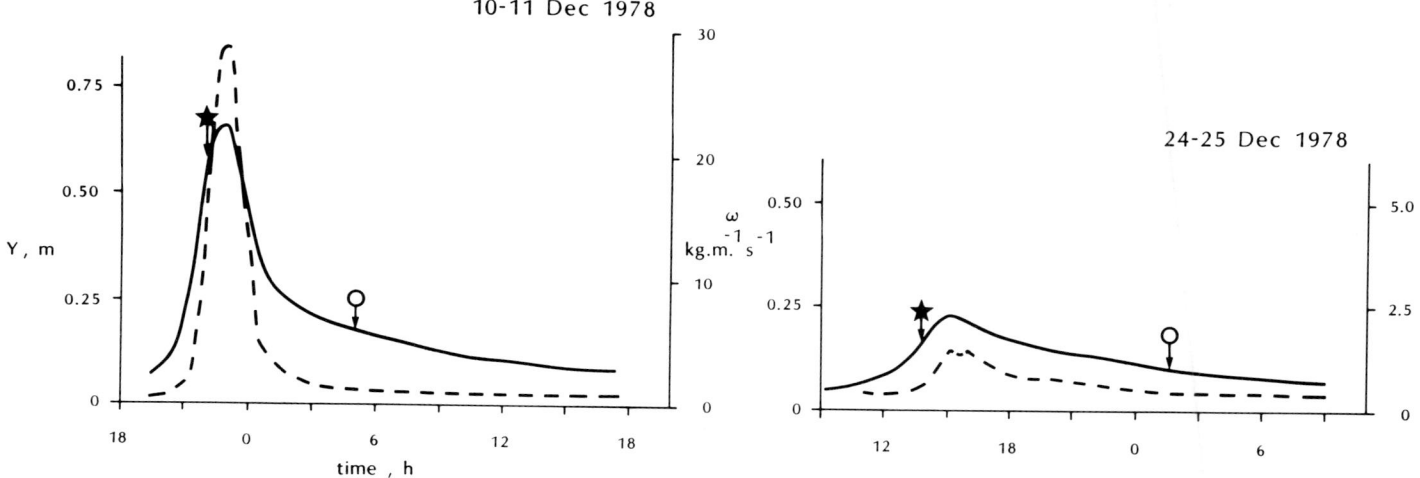
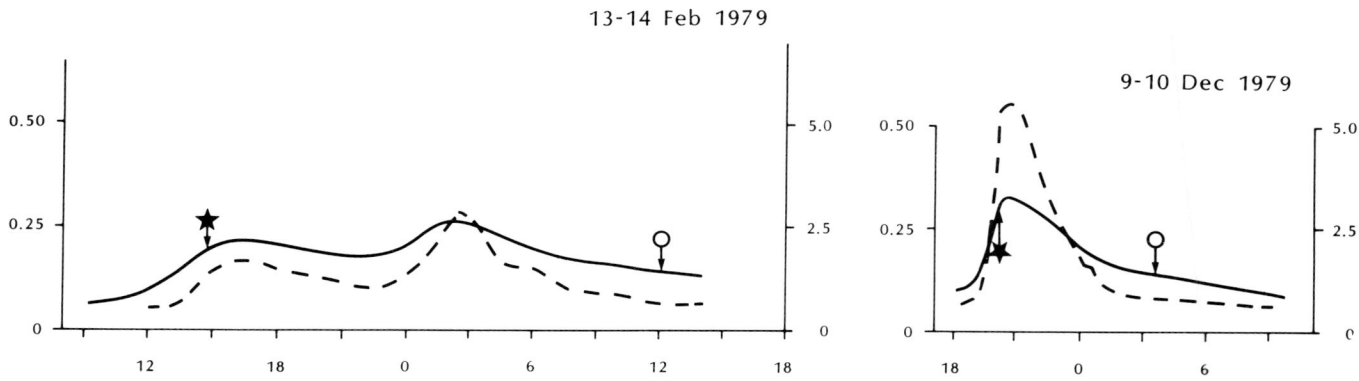

Fig. 4. Unit stream power, ω, and water depth, Y, for selected Turkey Brook floods showing initial and final motion of bedload.

Flood date	τ_o	τ'_o	ω_o	ω'_o
	kg.m^{-2}		kg.m^{-1}.s^{-2}	
10 Dec 1978	6.65	1.07	18.31	1.12
24 Dec	0.82	0.65	0.68	0.53
28 Dec	1.04	—*	0.71	—
25 Jan 1979	2.25	—	2.05	—
31 Jan	2.31	—	2.13	—
13 Feb	1.56	0.74	1.40	0.67
13 Mar	2.61	—	2.86	—
26 May	1.56	—	1.29	—
9 Dec	4.65	0.81	5.57	0.89
27 Dec 1980	1.98	—	2.26	—
12 Mar	3.78	0.72	4.13	0.78
Meyer-Peter Müller	1.56	1.22	1.60	1.27

* pit samplers full and unable to register final bedload transport.

Table 1. Bed shear and unit stream power at initial (τ_o, ω_o) and final (τ'_o, ω'_o) bedload transport for individual Turkey Brook floods and according to the Meyer-Peter Müller formula (Notation is explained in the Appendix).

The Birkbeck bedload monitoring programme clearly illustrates a distinct difference in force at the points of initial and final bedload motion, flood by flood (Fig. 4, Table 1). Values of shear stress at the start of bedload are consistently higher, ranging between 1.2 and 6.2 times the values at the moment bedload ceases.

Part of the explanation lies in the influence that pebble clusters and interstitial fines have on delaying initial motion. There is also a possibility that the size of the mobile material changes as the flood progresses. No information is available to check this since instantaneous samples of bedload were not taken. However, the nature of the material trapped by each pit sampler was deliberately noted when servicing the installation, and no vertical size grading was apparent on any occasion. In fact, the composite samples of trapped sediment and the bed material both show remarkably similar size-distributions (with mean D_{50} values of 11 and 16 mm, and mean D_{90} values of 34 and 35 mm), and it may be possible to infer from this that there is no significant change in size of the mobile sediment. There is, however, a need to consider the difference in inertia of resting and mobile particles that can be expressed as a difference between static and dynamic friction. The force required to entrain a static particle will always be greater, *ceteris paribus*, than the force prevailing at the point a mobile particle comes to rest. The process might be considered analogous to Postma's (1967) 'scour lag' for estuarine fine sediments, except that in this case grains are not loose but are subject to the extra binding force of cohesion.

Acknowledging this difference between entraining and depositing forces is not only important to the sedimentologist analysing a sedimentary sequence and inferring palaeohydraulics, but also to the engineer and geomorphologist interested in gauging bedload sediment discharge. An inspection of Table 1 will reveal that by comparison with the popular equation of Meyer-Peter and Müller (1948), bedload at Turkey Brook generally commences later (higher values of τ_0) and longer (lower values of τ_0') than would be predicted for any one flood wave. This highlights the difficulties inherent in bedload equations of characterising fluid-particle interaction using a single traction threshold. It suggests the presence of a castastrophe loop in the relationship between flow parameter and sediment transport that takes account of the difference in tractive force required to initiate and maintain particle motion. Building such a catastrophe loop into a bedload formula would undoubtedly allow better prediction, but would also complicate the equation considerably.

Conclusions

The mutual interaction of bed particles plays an important role in delaying the entrainment of coarse sediments. Pebbles involved in cluster forms resist movement even when bed shear is considerably in excess of the threshold values predicted on the basis of flume studies. The effect in Turkey Brook is that the bed reacts as though it were composed of its coarsest components. This is evident from recalculation of traction thresholds using D_{90} of the sampled bedload as the effective particle size. The values are plotted in Figure 3, and can be seen to lie in a better position relative to the curves of both Shields and Lane. The D_{90} sediment size is already incorporated in some bedload equations as an expression of bed roughness, but the Turkey Brook data clearly indicate its importance in predicting initial motion in coarse-grained channels.

Once in motion, bedload transport continues beyond the point at which the recession limb of the flood wave exerts a force equivalent to that required for initial entrainment. It also continues beyond the point predicted by equations that rely upon single thresholds which are derived from flume observations of initial motion (Fig. 2).

Estimating bedload in coarse-grained channels is therefore problematic. This has important implications for the hydraulic engineer, since rates of erosion and deposition may dictate the life or function of artificial channels, reservoirs, and so on. For the sedimentologist making interpretations about the nature of aggradational processes and palaeohydraulics and using curves such as those of Shields or Lane, there may be considerable underestimation of available stream power unless the appropriate particle size is selected. It should also be remembered while analysing a sediment that the particle size distribution reflects the hydraulic conditions of deposition and not of entrainment.

Appendix: Explanation of Notation

D	Grain diameter, m
g	Acceleration of gravity, m.s^{-2}
i_b	Unit bedload transport, submerged mass, kg.m^{-1}.s^{-1}
I_b	Bedload transport, kg.s^{-1}
Re_*	Grain Reynolds Number, dimensionless
u_*	Friction velocity, m.s^{-1}
Y	Water depth, m
θ_o	Shields threshold criterion, dimensionless
ν	Kinematic fluid viscosity, m^2.s^{-1}
ρ	Fluid density, kg.m^{-3}
ρ_s	Particle density, kg.m^{-3}
τ_o	Mean bed shear at initial motion, kg.m^{-2}
τ_o'	Mean bed shear at final motion, kg.m^{-2}
ω	Unit steam power, kg.m^{-1}.s^{-1}
ω_o	Unit stream power at initial motion, kg.m^{-1}.s^{-1}
ω_o'	Unit stream power at final motion, kg.m^{-1}.s^{-1}

References

Baker, V.R. and Ritter, D.F. 1975. Competence of rivers to transport bedload material. Geological Society of America Bulletin, v. 86, p. 975-978.

Brayshaw, A.C., Frostick, L.E. and Reid, I. 1983. The hydrodynamics of particle clusters and sediment entrainment in coarse alluvial channels. Sedimentology, v. 30, p. 137-143.

Coleman, N.L. 1967. A theoretical and experimental study of lift and drag forces. International Association for Hydraulic Research, 12th Congress Proceedings, Fort Collins, Colorado, p. 185-192.

Einstein, H.A. 1950. The bed-load function for sediment transportation in open channel flows. United States Department of Agriculture, Soil Conservation Series, Technical Bulletin 1026, 71p.

Fenton, J.D. and Abbott, J.E. 1977. Initial movement of grains on a stream bed : the effects of relative protrusion. Proceedings of the Royal Society of London Series A, v. 352, p. 523-537.

Gessler, J. 1971. Beginning and ceasing of sediment motion. In: Shen, H.W. (Ed.). River Mechanics. Fort Collins: Shen Publications, v. 1, p. 7.1 - 7.22.

Helley, E.J. 1969. Field measurement of the initiation of large bed particle motion in Blue Creek near Klamath, California. United States Geological Survey Professional Paper 562-G, 19p.

Lane, E. 1955. Design of stable channels. Transactions of the American Society of Civil Engineers, v. 120, p. 1234-1260.

Lauffer, H. and Sommer, N. 1982. Studies on sediment transport in mountain streams of the Eastern Alps. Commission Internationale des Grands Barrages, Quatorzième Congrès des Grands Barrages, Rio de Janeiro, p. 431-453.

Leopold, L.B. and Emmett, W.W. 1981. Some observations on the movement of cobbles on a streambed. In: Erosion and Sediment Transport Measurement. Proceedings of the Florence Symposium, International Association of Hydrological Sciences, p. 49-59.

Meyer-Peter, E. and Müller, R. 1948. Formulas for bed-load transport. International Association for Hydraulic Research, 2nd Congress Proceedings, Stockholm, p. 39-64.

Miller, M.C., McCave, I.N. and Komar, P.D. 1977. Threshold of sediment motion under unidirectional currents. Sedimentology, v. 24, p. 507-528.

Postma, H. 1967. Sediment transport and sedimentation in the estuarine environment. In: Lauff, G.H. (Ed.), Estuaries. American Association for the Advancement of Science, Publication 83, p. 158-179.

Reid, I., Layman, J.T. and Frostick, L.E. 1980. The continuous measurement of bedload discharge. Journal of Hydraulic Research, v. 18, p. 243-249.

Reid, I., Brayshaw, A.C. and Frostick, L.E. 1984. An electromagnetic device for automatic detection of bedload motion and its field applications. Sedimentology, v. 31, p. 269-276.

Shields, A. 1936. Anwendung der Aechnlichkeits mechanik und der Turbulenz-forschung auf die Geschiebebewegung. Mitteilung der Preussischen Versuchsanstalt für Wasserbau und Schiffbau, v. 26, p. 98-109.

Williams, G.P. 1970. Flume width and water depth effects in sediment-transport experiments. United States Geological Survey Professional Paper 562-H, 37p.

Yalin, M.S. 1972. Mechanics of Sediment Transport New York: Pergamon Press, 290p.

RELATIONSHIP BETWEEN FLOWS AND SEDIMENT SIZE IN SOME GRAVEL STREAMS OF THE ARID NEGEV, ISRAEL

Ze'ev B. Begin[1] and Moshe Inbar[2]

Abstract

The hydraulic geometry and sediment size were measured at 17 sites near hydrologic stations along five gravel streams in the arid Negev of Israel. These measurements permitted the establishment of a relationship between grain size and shear stress values of flows of different recurrence intervals. Median grain size was found to be positively correlated with shear stress arising from peak discharges having a probability of 0.9 - 0.7 to be equalled or exceeded. The linear regression of shear stresses related to these frequent flows on D_{50} and D_{60} resulted in equations which are similiar to the Shields relationship. This indicates that frequent floods, not extremely rare events, are responsible for the grain-size distribution of the sampled sites.

Résumé

La géométrie hydraulique et la granulométrie des sédiments furent mesurées à 17 emplacements près de stations hydrologiques le long de cinq rivières de gravier dans le désert du Negev, en Israël. Ces mesures ont permis d'établir une relation entre la granulométrie et les valeurs des contraintes de cisaillement des coulées, lesquelles sont répétées à différents intervalles. On a observé que la médiane granulométrique est correlée positivement avec la contrainte de cisaillement que résulte de crues maximales dont la probabilité 0,9 - 0,7 peut être atteinte ou dépassée. La régression linéaire des contraintes de cisaillement par rapport à ces coulées fréquents sur D_{50} et D_{60} se traduit par des équations analogues à la relation de Shields. Ceci révèle que les crues fréquentes, événements loin d'être extrêmement rares, sont responsables de la distribution de la grosseur des grains aux emplacements échantillonnés.

Introduction

It is well accepted in fluvial geomorphology that stream sediments are adjusted to both channel discharge and hydraulic geometry. From this assumption rises the possibility that stream discharge can be predicted from information given by the morphology of the channel and the bed material through which it flows. The reference discharge is in many cases the "dominant" discharge (Kellerhals, 1967) or the bankfull discharge (Henderson, 1961; Parker, 1979; Griffiths, 1981). However, the definition of bankfull discharge is not easy to apply in the field, and its recurrence interval may vary considerably (Williams, 1978). Moreover, it is widely considered that a channel is more likely to be the product of a wide range of flows rather than a single channel-forming discharge (Pickup and Rieger, 1979, p.41). In desert streams, which are the subject of this study, the problem of a reference discharge becomes even more acute. For the arid Negev region in Israel (Ben Zvi, 1979) it has been shown that the frequency of flows obeys the Poisson distribution (Ben Zvi and Ben Zvi, 1971; Finkel and Finkel, 1979). Most flows are of a relatively short duration — hours or days — and hydrographs are characterized by fast rising and falling limbs (Shick, 1970; Finkel and Finkel, 1979). Therefore the possible adjustment of stream bed-material to these flows is less likely than in streams where levels of flood discharge prevail over longer periods.

On the other hand, both field observations and theoretical considerations indicate that a certain regularity of sediment distribution should be found even in such desert streams. The effectiveness of a given discharge in shaping a channel is determined by both its magnitude and frequency (Andrews, 1980), and Wolman and Miller (1960) noted that the maximum amount of erosional work is done by geomorphic processes of moderate intensity but frequent occurrence. Thus it may be expected that between rare, large events a channel will be able to adjust to the lower, frequent flows, which may have a longer total duration and a cumulatively greater ability to achieve channel bed changes. It might be further expected that channel bed sediment will be in quasi-equilibrium with the more frequent flows.

The aim of this study is to relate bed-material characteristics and local hydraulic geometry of desert gravel streams to flows of certain recurrence intervals. This requires detailed information on the size-sorting properties of the channel sediment. However, obtaining a representative grain-size distribution for a given section in gravel-bed rivers is problematic. Size distributions differ according to position in the channel cross-section, but this variability depends also on the stage attained by recent flow events.

---

[1]The Geological Survey of Israel, 30 Malkhe Yisrael Street, 95 501 Jerusalem, Israel.
[2]The University of Haifa, Haifa, Israel.
 This study was carried out with funding of the Earth Science Administration of the Ministry of Energy and Infrastructure. The manuscript was typed by Mrs. V. Arieh and the line drawings were prepared by Mrs. C. Hadar. We are greatly indebted to Drs. A. C. Brayshaw and I. Reid for their most detailed review and valuable suggestions.
Copyright © 1984, Canadian Society of Petroleum Geologists

This is an important consideration given that the product of depth and local slope determines the shear stress applied to the bed and therefore the probability of particles of a certain size remaining on the bed (Gessler, 1970). Also, large local scatter in size distributions may be caused by secondary flows, channel migration, vegetation and bank slumping (Church and Kellerhals, 1978). In addition, sediment through a given channel cross-section will have been deposited by differing depths of flow. As such, careful attention needs to be given to the location of bed-material samples in the channel perimeter.

Method

Measurements were taken at 17 sites near seven hydrologic stations on five gravel-bed streams on the eastern divide of the arid Negev of southern Israel (Fig. 1, Table 1). The channel cross-section and slope were measured at each site with the aid of a theodolite (Fig. 2), and bed-material samples were collected. For coarse sediments, the size of 100 pebbles selected at random within a one meter grid square were measured (Fig. 3). For finer sediment the surface armour layer was removed and sieved in the laboratory. The position of each sample (the center point of the sampling net) within the cross-section was recorded.

For each cross-section the mean flow velocity (u) for different water elevations was calculated using Lacey's equation (Bray, 1979):

$$u = 10.8 h^{2/3} S^{1/3} \qquad (1)$$

where u is mean flow velocity (m.s^{-1}), h is mean flow depth (m) and S is bed slope. Using equation (1) with knowledge of cross-section geometry, it was possible to calculate mean flow depth and discharge for a given water elevation.

With the position of each sampling point of bed material known it was also possible to assign a water depth (y) to each point for each flood stage. Hence, with bed slope used as a surrogate for water-surface slope, the local shear stress applied to each sample point can be calculated at specific water elevations using

$$\tau = \gamma y S \qquad (2)$$

where τ is the shear stress (kg.m^{-2}) applied to the point, γ is specific weight of water, y is flow depth and S is bed slope.

With discharge calculated for each water elevation, and knowing also the shear stress applied to a point by these flows, a rating curve can be constructed for each sample

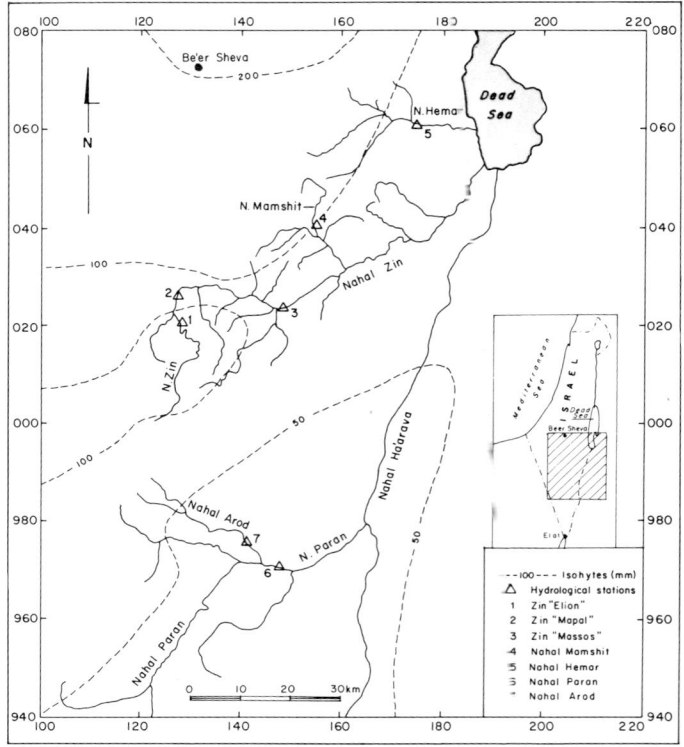

Fig. 1. Location map.

Hydrologic station	Station symbol	Coordinates Israel Grid	drainage area above site sq. km.	number of years of hydrologic measurements	peak discharge m³.s⁻¹ (*)	No. of cross-sections sampled in station
Zin "Elion"	Zel	1284/0210	120	28	6.0	3
Zin "Mapal"	Zmp	1273/0260	230	28	24.0	2
Zin "Massos"	Zms	1483/0235	660	28	44.0	1
Arod	Ard	1411/9763	160	16	6.7	1
Hemar	Hem	1745/0607	330	8	11.5	4
Paran	Par	1473/9708	3420	26	19.3	3
Mamshit	Mam	1543/0403	60	23	2.5	3

*probability of being exceeded equals 0.5

Table 1. Basic data on hydrologic stations in the Negev (Cohen and Ben Zvi, 1978) on which this study is based. Symbols of stations are those used in tables and figures.

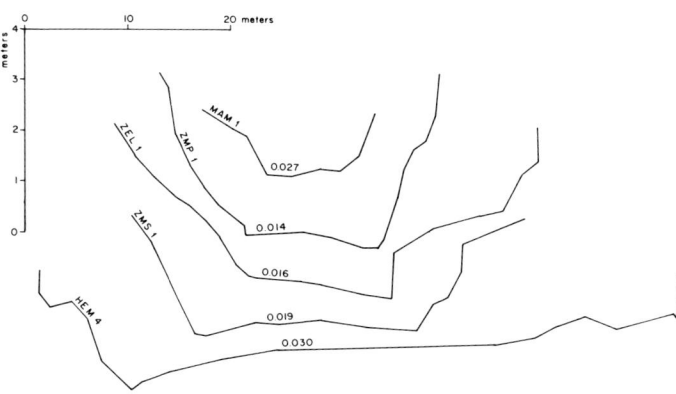

Fig. 2. Selected cross-sections of the studied streams. Numbers above cross-sections denote local down-channel slope. For station symbols see Table 1.

point relating shear stress to flow discharge (Figs. 4 and 5). Since the measurements were taken near gaging stations, the recurrence interval of each discharge value is known from discharge - probability curves (Cohen and Ben Zvi, 1978; Polak, 1981). Shear stress - discharge rating curves therefore permit recurrence intervals to be assigned to each value of shear stress. Thus, the shear stress applied to a sampled point, τ_n, on a stream perimeter is related to peak discharge Q_n, where n is the probability that Q_n and τ_n will be equalled or exceeded (Table 2).

For each sample point the grain-size distribution curve was constructed on log-probability paper and grain size percentiles D_k were determined (k is the percent of particles finer than D_k, expressed in mm; see Table 3).

Fig. 3. A. Nahal Zin "Elion" (Table 1) — characteristic view of terrain in study area. **B.** Nahal Zin "Mapal" (Table 1) — close-up view of bed-material.

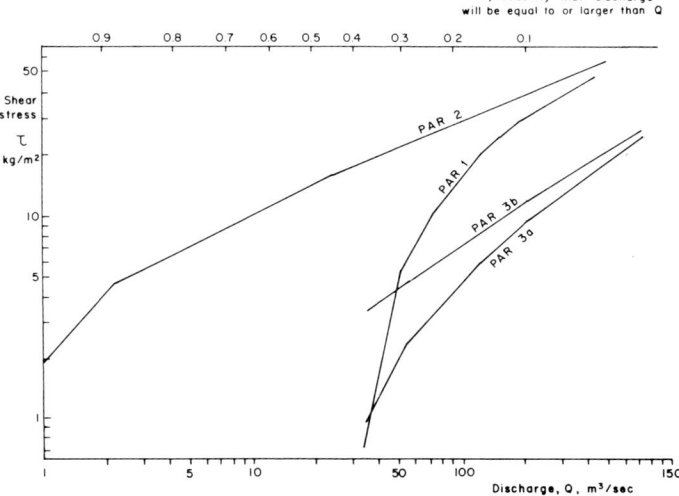

Fig. 4. Rating curves of shear stress versus discharge for some points in the Nahal Paran. The probabilities of discharge to be equalled or exceeded are taken from Cohen and Ben Zvi (1978).

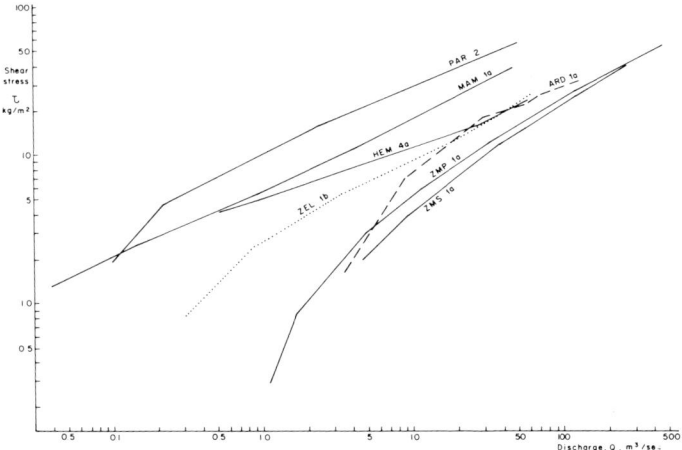

Fig. 5. Rating curves of shear stress versus discharge for selected points in the studied streams. For station symbols see Table 1.

Site	Probability n that shear stress τ_n will be equalled or exceeded.								
	0.1	0.2	0.3	0.4	0.5	0.6	0.7	0.8	0.9
Zel 1b	20.0	7.6	5.6	5.8	4.8	3.9	2.8	1.1	—
Zel 2a	9.2	4.0	3.3	2.9	2.5	2.1	1.5	0.52	—
Zel 3a	14.5	5.6	4.5	4.2	3.6	3.1	2.4	1.5	—
Hem 1a	9.8	6.6	5.2	4.3	3.5	3.2	2.9	2.8	2.6
Hem 1b	9.2	6.6	5.2	4.0	3.4	2.6	1.9	1.5	1.1
Hem 2a	16.0	11.0	8.6	6.8	5.2	4.1	3.0	2.1	1.2
Hem 2b	15.5	10.7	8.2	6.4	4.8	3.7	2.7	1.8	1.1
Hem 3	12.0	8.4	6.4	5.0	3.9	3.1	2.4	1.7	1.1
Hem 4a	23.0	17.2	14.1	12.5	11.0	9.8	8.5	7.2	5.6
Par 2	38.0	27.0	21.0	18.0	14.0	11.0	8.6	6.2	3.9
Mam 1a	33.0	24.0	17.4	12.2	8.8	5.6	2.8	1.0	—
Mam 1b	31.5	22.0	15.1	10.1	6.6	3.7	0.8	—	—
Mam 2	17.9	13.0	10.2	7.8	5.8	3.5	1.7	0.8	—
Zms 1a	26.5	21.0	18.0	15.1	13.0	10.0	7.0	3.3	—
Zms 1b	26.0	21.8	17.3	14.3	12.0	9.4	6.4	2.7	—

Table 2. Shear stress, τ_n, (kg.m^{-2}) associated with peak discharges, expected at various probability levels at the selected sites. For station symbols, see Table 1. Only data from sites with information on τ_n, with $n \geq 0.7$ are presented.

Sample site	Percent finer					
	40	50	60	70	80	90
Zel 1b	10	11	13	15	18	22
Zel 2a	14	23	33	46	69	92
Zel 3a	5.0	5.5	6	6.6	7.3	8.5
Hem 1a	14	18	20	29	39	65
Hem 1b	7	9	11	16	26	27
Hem 2a	20	31	44	54	92	120
Hem 2b	6	6.5	7	9	11	13
Hem 3	7	9	11	14	21	26
Hem 4a	26	46	55	70	86	98
Par 2	24	45	52	73	94	152
Mam 1a	0.5	3	9	26	60	92
Mam 1b	0.05	0.1	0.5	6	20	35
Mam 2	0.2	0.2	0.22	0.25	0.32	0.45
Zms 1a	9	11	17	26	31	52
Zms 1b	8	9	10	17	26	36

Table 3. Grain-size (mm) of samples taken at selected sites in the studies streams. For station symbols, see Table 1.

Results

The τ_n values (Table 2) were correlated with the D_k values (Table 3) in an attempt to identify those grain-size fractions which are most closely associated with shear stress values that have a specific recurrence interval. The result is a correlation matrix (Table 4), relating D_k to τ_n. The correlation coefficients relating grain size to shear stress values τ_n with $n<0.7$ were not found to be significant at the 0.05 level. Therefore Table 4 includes only the results of correlating D_k with τ_n for which $n \geq 0.7$. Table 4 shows that the highest correlation coefficients result from regressing shear stress on D_{50}. Bivariate diagrams of each relationship are shown in Figures 6-9.

Discussion and Conclusions

Of the correlations indicated in Table 4, the following regression equations of $\tau_{0.9}$ on D_{50} and of $\tau_{0.8}$ on D_{60} (Figs. 6 and 9) are of particular interest:

$$\tau_{0.9} = 0.09 D_{50} + 0.34 \tag{3}$$

$$\tau_{0.8} = 0.08 D_{60} + 0.76 \tag{4}$$

They have significance because of the similarity to the τ vs. D relationship anticipated from a Shields-type equation. The Shields relationship defines the threshold of motion by relating particle size to the minimum shear stress, τ_{cr}, which is required for entrainment of these particles. If it is assumed that the Shields' relation holds at near-threshold conditions (Griffiths, 1981), then

$$yS/(\gamma_s - 1)D_{50} = F \tag{5}$$

where γ_s is specific weight of the sediment. If F is assumed to be 0.06 as indicated by the Shields diagram (Graf, 1971; see also Komar, 1970; Baker, 1974; Koster, 1978) and if $\gamma_s = 2.6$ g.cm^{-3}, equation (5) becomes:

$$yS = (0.06 \times 1.6) D_{50} \tag{6}$$

Combining equations (6) and (2), the threshold shear stress τ_{cr} is:

$$\tau_{cr} = 0.096 D_{50} \tag{7}$$

By comparing this equation to the regression equations (3) and (4), it can be seen that these may be considered as defining threshold conditions. Not only are there significant statistical relationships between D_{50} or D_{60} and the bed shear of flows with specific probabilities of occurrence, but it can also be concluded that flows which have a probability of 0.8 and 0.9 to be exceeded as peak flows are directly responsible for the grain size distribution of the sampled sites. It should be noted again that these sites occur in different streams and that each site has its specific hydraulic geometry. Equations (3) and (4) represent their average tendency.

τ_n	D_{40}	D_{50}	D_{60}	D_{70}	D_{80}	D_{90}	number of observations
0.9	0.822*	0.839*	0.772*	0.792*	n.s.	n.s.	7
0.8	0.779**	0.807**	0.742**	0.738**	0.572*	0.558*	14
0.7	0.661**	0.677**	0.636*	0.655**	n.s.	n.s.	15

n.s. — not significant at the 0.05 level
* — significant at the 0.05 level
** — significant at the 0.01 level

Table 4. Correlation coefficients for the regression of τ_n on D_k. τ_n is the shear stress with a probability n to be equalled or exceeded; D_k is grain size of which k percent is finer.

As expected, the stream gravels appear to be more closely associated with the more frequent, lower flows than to the rare events. It should be remembered, though, that a frequent flow in one stream (say, $Q_{0.9} = 2.6$ m^3·s^{-1} in Nahal Paran) is quite different from another ($Q_{0.9} = 0.1$ m^3·s^{-1} in Nahal Zin-Mapal). It is assumed that the sampled sediments are related to flows through the shear stress exerted on each sample point.

It is difficult to assess the significance of the fact that no correlation was found between grain size percentiles and higher flows of lower probability of occurrence. A possible explanation may be a deficiency in large particles in the bed material: high flows move all the available particles. This is indicated by Figure 10, which shows that shear stress values associated with rare flows lie above the threshold values necessary to move most particles in the bed material of nearly all the sampled sites. This conclusion is further corroborated by a field experiment in the upper reach of Nahal Zin (Zin "Elion") (Hanokh, 1981). A flood occurring there on December 26, 1980, with a probability of occurrence estimated at about 0.1, caused the movement of the largest measured particles of up to 280 mm in *b*-axis diameter.

Based on this conclusion it becomes possible to speculate on the travel time of gravels representing the coarsest

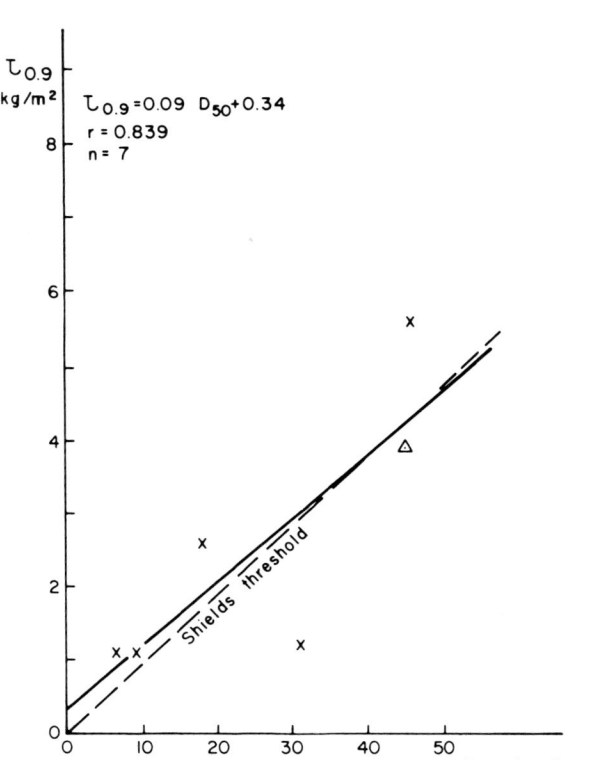

Fig. 6. Regression of $\tau_{0.9}$ on D_{50}. Note similarity between regression line and the Shields threshold line. For symbols see Figure 10.

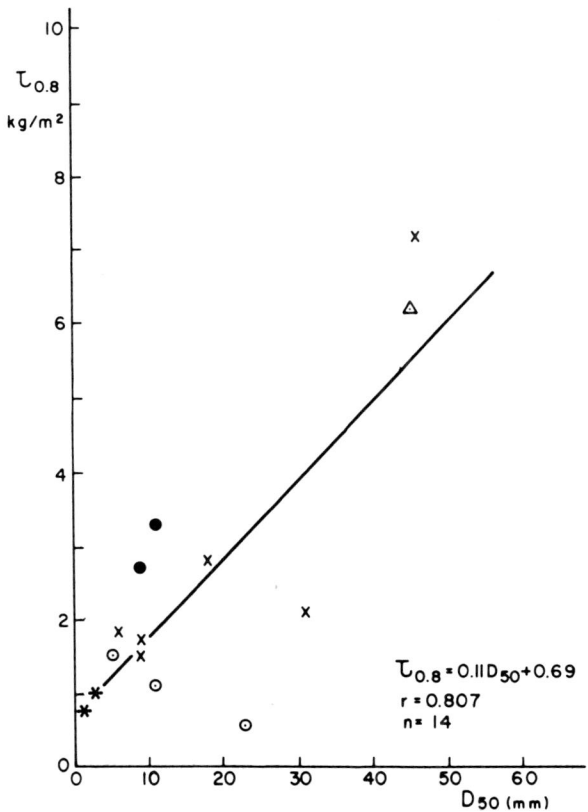

Fig. 7. Regression of $\tau_{0.8}$ on D_{50}. For symbols see Figure 10.

percentile along the stream bed. If it is assumed that particles of D_{90} size move in events with a probability of occurrence of about 0.1, and if on the average there occur about two flood events per year in the Negev streams (Cohen and Ben Zvi, 1978), then a 5-year flood should be considered as moving the fraction of D_{90}. Assuming an average length of movement in one such flood to be meters or several tens of meters, we have an order of 20 to 200 steps per kilometer. Continuing this assumption, the rate of travel of gravels in these Negev streams is therefore in the order of a hundred to a thousand years per kilometer.

Although this study is based on a relatively small sample, and although sampling took place at only one point in time (hence, no information on scour and fill in these cross-sections), these preliminary results seem to justify the conclusion that even in this flash-flood environment, bed material is distributed on the channel perimeter in accordance with the hydraulic geometry.

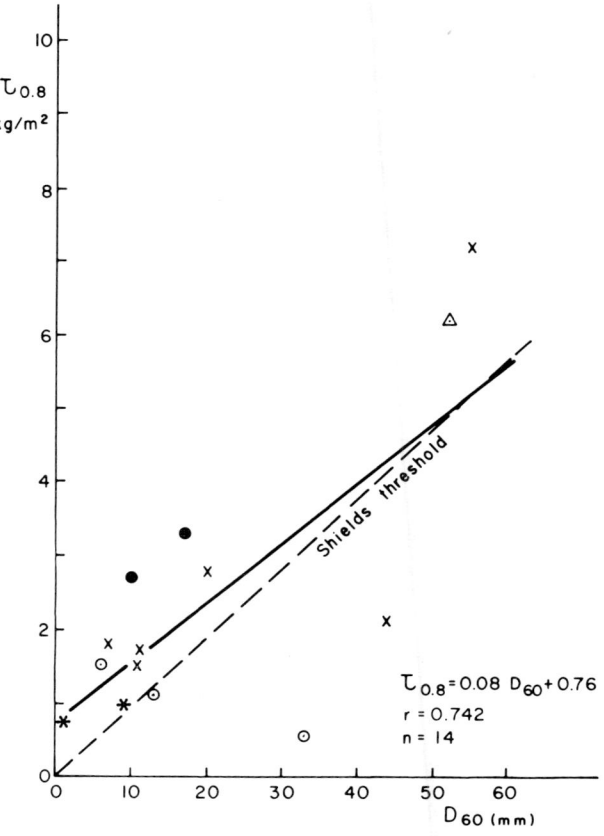

Fig. 9. Regression of $\tau_{0.8}$ on D_{60}. For symbols see Figure 10.

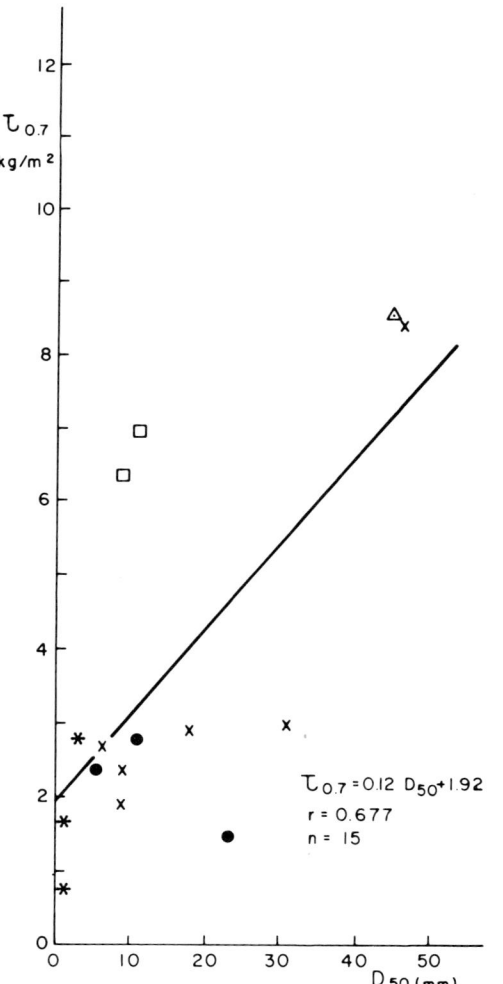

Fig. 8. Regression of $\tau_{0.7}$ on D_{50}. For symbols see Figure 10.

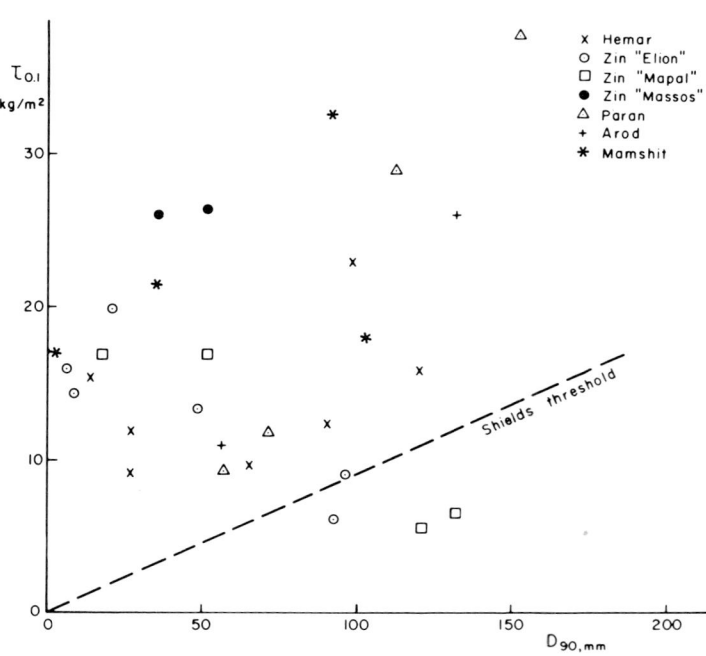

Fig. 10. Relationship between $\tau_{0.1}$ and D_{90}. Note that most data points lie above the Shields threshold line, signifying that in these rare flows most bed-material moves.

REFERENCES

Andrews, E.D. 1980. Effective and bankfull discharges of streams in the Yampa River basin, Colorado and Wyoming. Journal of Hydrology, v. 46, p. 311-330.

Baker, V.R. 1974. Paleohydraulic interpretation of Quaternary alluvium near Golden, Colorado. Quaternary Research, v. 4, p. 94-112.

Ben Zvi, M. and Ben Zvi, A. 1971. The probability distribution of flow events in the Negev. Journal of Hydrology, v. 14, p. 348-353.

Ben Zvi, A. 1979. Flows in the Negev streams. *In:* Shmueli, A. and Grados, Y. (Eds.), The Land of the Negev. Israel Ministry of Defence, p. 74-85 (in Hebrew).

Bray, D.I. 1979. Estimating average velocity in gravel-bed rivers. American Society of Civil Engineers, Journal of the Hydraulics Division, v. 105, p. 1103-1122.

Church, M. and Kellerhals, R. 1978. On the statistics of grain size variation along a gravel river. Canadian Journal of Earth Sciences, v. 15, p. 1151-1160.

Cohen, O. and Ben Zvi, A. 1978. Peak discharges to the Southern Dead Sea from regional analysis in the Negev. Israel Hydrological Service, Report Hydro/4/1978, 11p.

Finkel, H. and Finkel, M. 1979. Runoff in the Arava. *In:* Shmueli, A. and Grados, Y. (Eds.), The Land of the Negev. Israel Ministry of Defence, p. 125-139 (in Hebrew).

Gessler, J. 1970. Self-stabilizing tendencies of alluvial channels. American Society of Civil Engineers, Journal of the Hydraulics Division, v. 96, p. 235-249.

Griffiths, G.A. 1981. Stable-channel design in gravel-bed rivers. Journal of Hydrology, v. 52, p. 291-305.

Graf, W.H. 1971. Hydraulics of Sediment Transport. New York: McGraw-Hill Book Co., 513p.

Hanokh, S. 1981. The morphology of Nahal Zin above Ein-Avedat. Report, Sede Boker High School for Environmental Studies, 41p. (in Hebrew).

Henderson, F.M. 1961. Stability of alluvial channels. American Society of Civil Engineers, Journal of the Hydraulics Division, v. 87, p. 109-138.

Kellerhals, R. 1967. Stable channels with gravel-paved beds. American Society of Civil Engineers, Journal of the Waterways and Harbors Division, v. 93, p. 63-84.

Komar, P.D. 1970. The competence of turbidity current flow. Geological Society of America Bulletin, v. 81, p. 1555-1562.

Koster, E.H. 1978. Transverse ribs: their characteristics, origin and paleohydrologic significance. *In:* Miall, A.D. (Ed.), Fluvial Sedimentology. Canadian Society of Petroleum Geologists, Memoir 5, p. 161-186.

Parker, G. 1979. Hydraulic geometry of active gravel rivers. American Society of Civil Engineers, Journal of the Hydraulics Division, v. 105, p. 1185-1201.

Pickup, G. and Rieger, W.A. 1979. A conceptual model of the relationship between channel characteristics and discharge. Earth Surface Processes, v. 4, p. 37-42.

Polak, S. 1981. Nahal Zin catchment area; hydrological analysis for the period 1951-1978. Israel Hydrological Service, Report Hydro/5/1980, 39p. (in Hebrew).

Schick, A. 1970. Desert floods, interim results of observations in the Nahal Yael Research watershed, Southern Israel, 1965-1970. International Association of Scientific Hydrology — UNESCO Symposium, Wellington, New Zealand, Dec. 1970, p. 479-493.

Williams, G.P. 1978. Bank-full discharge of rivers. Water Resources Research, v. 14, p. 1141-1154.

Wolman, M.G. and Miller, J.P. 1960. Magnitude and frequency of forces in geomorphic processes. Journal of Geology, v. 68, p. 54-74.

In: Koster, E.H. and Steel, R.J. (Eds.), Sedimentology of Gravels and Comglomerates.
Canadian Society of Petroleum Geologists, Memoir 10 (1984), p. 77 - 85.

CHARACTERISTICS AND ORIGIN OF CLUSTER BEDFORMS IN COARSE-GRAINED ALLUVIAL CHANNELS

ANDREW C. BRAYSHAW[1]

ABSTRACT

Cluster bedforms, closely nested groups of clasts aligned parallel to flow, are the prevalent type of bed microtopography in poorly sorted gravel-bed streams. They have been studied in natural channels with a wide range of clast shape and size; relationships between cluster bedform geometry and properties of the channel bed material are identified. Clusters are found to range in length from 0.1-1.2 m in a streamwise direction. Using flume experiments, it has been possible to simulate the formation of clusters resembling those found in natural channels. Observations of cluster formation indicate that these bedforms are developed during falling stage from flood discharges as particles are deposited around exceptionally large clasts which act as obstacles in the flow. Cluster bedforms play a significant role in sediment transport: they delay incipient motion and limit the availability of bed material for transportation.

RÉSUMÉ

Des formes de litage en agglomérats constituées de nids très rapprochés d'amas de fragments alignés parallèlement à la direction de l'écoulement représentent le principal type de microrelief dans lits de gravier pauvrement triés des rivières. Elles ont été étudiées dans des chenaux naturels manifestant une grande variété de forme et de grosseur de fragments; les relations entre la géométrie des formes de litage en agglomérats et les propriétés des matériaux qui composent le lit du chenal sont identifiées. On observe que les agglomérations d'amas de fragments varient en longueur de 0, 1-1, 2 m dans la direction d'écoulement du cours d'eau. Grâce à des expériences sur des ravins, il fut possible de simuler la formation d'agglomérats semblables à ceux observés dans les chenaux naturels. Les observations sur la formation de ces agglomérats indiquent que les formes de litage se développent durant le stage de réduction des débits de crue lorsque les particules se déposent autour de fragments exceptionnellement gros qui servent d'obstacles à l'écoulement. Les formes de litage en agglomérats influencent considérablement le transport du sédiment; elles retardent le mouvement à son début et elles limitent la disponibilité des matériaux du lit à être transportés.

INTRODUCTION

Our understanding of sediment transport in coarse-grained alluvial channels relies heavily on appreciating the structural packing and arrangement of particles that make up the bed (Church, 1972; Laronne and Carson, 1976; Carling and Reader, 1982; Brayshaw, 1983). In particular, there is an urgent need to link hydraulic formulae, which predict thresholds of motion, to the complex bed conditions characteristic of poorly sorted gravel-bed streams where clast shape and size range widely. Thresholds of motion are derived typically from flume experiments using a plane bed of like-sized grains and uniform flow conditions (*cf.* Shields, 1936). In consequence, threshold values usually are expressed in the simple form of grain diameter against critical shear in which the forces resisting motion are primarily the submerged weights of the particles. The limited applicability of such formulae to a majority of natural channels is due to their failure to acknowledge the effects of sedimentological properties such as particle hiding, fabric and packing, and the mutual interference of neighbouring clasts. Yet all these factors are important in poorly sorted coarse-grained channels where a low-relief microtopography, superimposed on the gross morphology of the bed, influences the local arrangement of clasts and their availability for subsequent bedload transport.

Problems associated with the incipient motion of differing sizes and exposures of clasts have received attention both in theoretical treatments of sediment transport (Einstein, 1950; White and Day, 1982), and in laboratory experiments using grains of simplified geometry (Fenton and Abbott, 1977; Brayshaw *et al.*, 1983). In the field, Church (1972), Laronne and Carson, (1976) and Brayshaw *et al.* (1983) provide evidence which indicates that the movement of coarse streambed particles depends to a large extent on their stability on the bed, and the proximity of surrounding clasts. However, there is still much to learn about the highly localized structural arrangements of clasts in natural channels before spatial and temporal variations in sediment entrainment thresholds can be explained.

The small-scale bed features that comprise the major components of bed microtopography have received little attention. Where these have been dealt with, it has been mainly from the point of view of their use as palaeocurrent indicators. Bedforms which have been documented include 'pebble clusters' (Dal Cin, 1968; Teisseyre, 1977), 'imbricate clusters' (Rust, 1972; Martini, 1977), 'boulder shadows' (Laronne and Carson, 1976) and 'current shadows' (Gustavson, 1974). However, whilst such detailed descriptions have added much to our limited knowledge of small-scale gravel bedforms, there exists virtually no information on

---

[1]Sedimentology Branch, BP Exploration Company Limited, Brittanic House, Moor Lane, London EC2Y 9BU, England.
 This research was undertaken during the tenure of a Natural Environment Research Council award, and prepared for publication whilst in receipt of a Royal Society European Fellowship at the University of Florence, Italy. I should like to thank I. Reid and L.E. Frostick, who read an earlier draft of the manuscript, and M. Church who reviewed it, for constructive criticisms and helpful suggestions.
Copyright © 1984, Canadian Society of Petroleum Geologists

how they are generated and the role that they play in the relationship between the bed and the transport of bedload.

As part of a wider investigation into the effects of bed microtopography on bedload transport, the detailed characteristics of micro-relief features, encompassed under the general term *cluster bedforms* (Brayshaw, 1983), have been analysed. This paper examines some of the dimensional properties of cluster bedforms, and considers their origin in the light of field measurements and flume observations. The mechanics of cluster formation throws considerable light on the relationship between bed microtopography and sediment transport in gravel-bed streams, and casts doubt upon the simplified manner in which sediment transport theories are sometimes applied to such channels.

FIELD CHARACTERISTICS

GENERAL DESCRIPTION

Cluster bedforms consist of accumulations of bed particles, typically formed around an exceptionally large clast, above the level of an otherwise planar gravel bed (Fig. 1). Clusters incorporate a wide range of particle sizes, but have a scale of only several clast diameters. They tend to be somewhat longer in a streamwise direction than they are broad. The long axes of clusters rarely diverge by more than a few degrees from the direction of the depositing current. An assessment of a wide variety of poorly sorted, coarse-grained alluvial systems indicates that clusters usually occupy between 5 and 10% of the channel floor. However, unlike transverse ribs (Koster, 1978), where pebbles form rows normal to flow, cluster bedforms are usually non-periodic.

Field measurements of cluster bedforms were made in a number of channels of varying clast lithology ($n = 100$), deliberately chosen to provide a wide range of particle shape (sphericity, $\psi = (c^2/ab)^{1/3}$) and size sorting (S_0). Turkey Brook, Enfield Chase, England, is a small, flint gravel bed channel ($D_{50} = 14.9$ mm; $\psi = 0.68$; $S_0 = 1.37$); width at the bed is 2.45 m and bankfull depth 0.8 m. Widdale Beck, Yorkshire, is a river comprising well rounded sandstone-limestone bed material ($D_{50} = 52.0$ mm; $\psi = 0.72$; $S_0 = 1.70$); mean width is 10 m and bankfull depth 2 m. Streams from the Wye catchment, Wales, are characterized by low sphericity, slate bed particles ($D_{50} = 22.6$ mm; $\psi = 0.39$; $S_0 = 1.51$). Channel widths in the reaches studied here average 9.0 m and bankfull depths are 1.5-2.0 m.

For the purpose of describing clusters, each was subdivided into three main components — the *obstacle clast*, the accumulation of particles on the obstacle's *stoss side*, and the accretion of grains in the obstacle's downstream *wake*. For each bedform analysed, six size properties were measured (Fig. 2). The mean values for each of these are summarised in Table 1 for each channel.

Fig. 1. Examples of typical cluster bedforms from (**a**) Widdale Beck, Yorkshire, England; (**b**) Turkey Brook, Enfield Chase, England and (**c**) Afon Elan, Wye catchment, Wales.

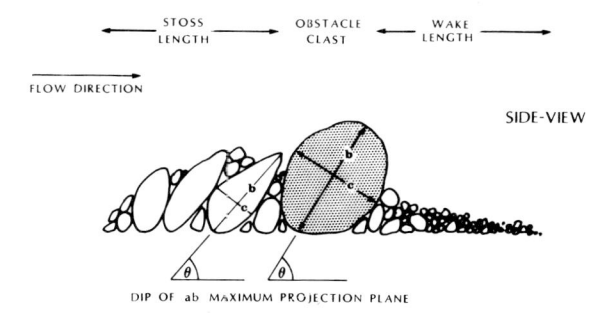

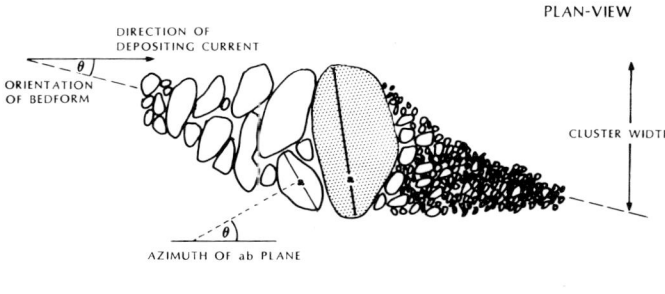

Fig. 2. Idealized diagram of a cluster bedform showing parameters used to define its characteristics.

	Wye catchment channels, Wales		Turkey Brook, Enfield Chase, England		Widdale Beck, Yorkshire, England	
	$\bar{x}$ (cm)	Range (cm)	$\bar{x}$ (cm)	Range (cm)	$\bar{x}$ (cm)	Range (cm)
Cluster Length	24.5	10-120	13.2	10-21	48.1	23-111
Cluster Width	17.1	9-40	9.2	5-13	23.1	12-48
Stoss Length	14.6	2-30	3.6	2-11	15.3	5-48
Wake Length	3.0	0-16	3.4	0-9	14.6	0-50
B-axis of stoss-side clasts	7.2	3-20	2.1	1-3	7.0	4-15
B-axis of obstacle clasts	10.3	4-28	5.6	2-8	11.4	5-21
	$n = 40$		$n = 30$		$n = 30$	

Table 1. Mean size characteristics of cluster bedforms measured in three selected streams in England and Wales.

Fig. 3. Example of an imbricate-type cluster bedform from Afon Cammarch, Wales.

GEOMETRY

Clusters vary in length from 0.1 m to 1.2 m and are roughly twice as long as they are wide. Maximum width is defined by the a-axis of the obstacle clast which is typically orientated normal to the flow. Cluster bedforms increase in both length and width with the calibre of the stream's constituent bed material and tend to be larger where channel bed sediment is both poorly sorted and negatively skewed. However, the relative lengths of stoss and wake deposits vary considerably between channels, and are shown to depend upon the shape characteristics of the sediment. Where bed material comprises well-rounded pebbles, particle interlock is limited to point contact, and stoss and wake accumulations each occupy approximately one third of the bedform's total length. Where bed particles are tabular or disc-shaped, clasts are imbricated against the upstream side of the obstacle. This leads to the formation of imbricate-type clusters in which long trains of stoss-side clasts all share a common angle of dip. However, the 'obstacle' stone need not be the largest when imbrication occurs (Fig. 3).

SIZE-SORTING CHARACTERISTICS

The distinctive size-sorting of sediment in cluster bedforms is striking (Fig. 1). Size analyses of the various components of cluster bedforms (Fig. 4) reveal that obstacle clasts represent the largest size fractions available on the bed. The median size of obstacles exceeds D_{95} for all channels. By contrast, wake deposits comprise fine calibre bed material: the median size ranges from just D_8 to D_{46}. In this respect, wake accumulations resemble coarse-grain equivalents of obstacle shadows observed in sandsize sediment by Peabody (1947) and Bagnold (1954). Clasts clustered on the stoss-side of obstacles are comparatively coarse, and represent sizes of between D_{74} and D_{94}. The actual size differentiation between the components of cluster bedforms varies, however, in accordance with the overall degree of sorting in bed material.

The coarse calibre of obstacles and stoss-side clasts conforms with our current knowledge of surface armouring: both represent the coarser size fractions of the bed material distribution and provide protection to finer underlying sediment. The calibre of sediment in wake accumulations is similar to that which occurs in the sub-surface bed material layer and which fills voids between contiguous clasts.

ALIGNMENT

The alignment of cluster bedforms in trains parallel with the flow is an important characteristic for palaeocurrent interpretation, and one which has received some attention. The current-parallel orientation of clusters recorded by Dal Cin (1968) and Teisseyre (1977) is confirmed by measurements taken here: the mean angular deviation of the long axes of bedforms from the flow direction is just 8.9°, 90% of all bedforms measured diverging by not more than

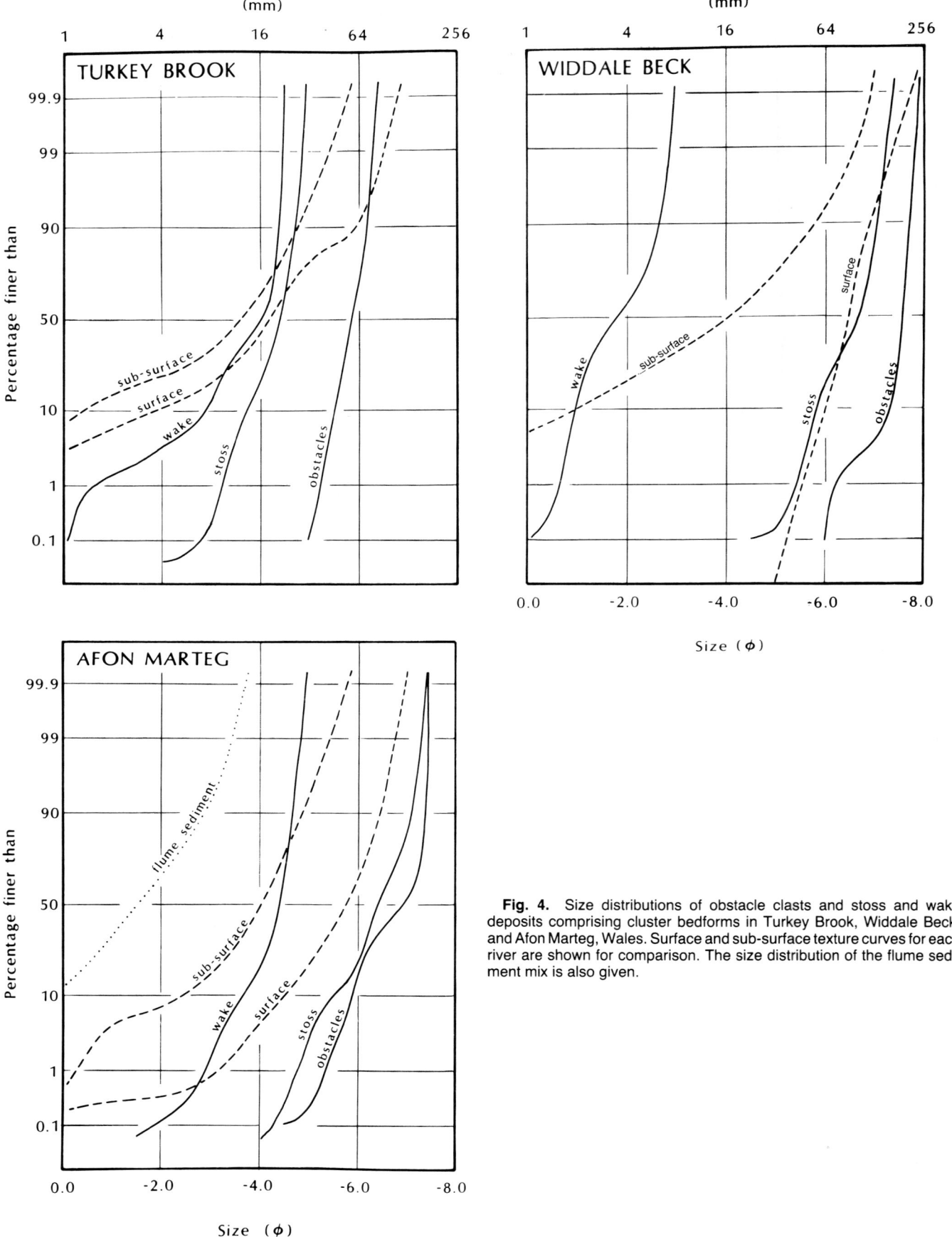

Fig. 4. Size distributions of obstacle clasts and stoss and wake deposits comprising cluster bedforms in Turkey Brook, Widdale Beck, and Afon Marteg, Wales. Surface and sub-surface texture curves for each river are shown for comparison. The size distribution of the flume sediment mix is also given.

20° from the flow direction (Table 2). This small deviation of the mean azimuth from the flow direction is illustrated in Figure 5.

Also of considerable interest is the alignment of individual clasts comprising clusters. Figure 6 compares the orientation of clasts (*ab* planes) from clusters against those sampled along transects, using the method of Wolman (1954). Transect clasts exhibit distinct patterns for each channel, attributable to their grain shape characteristics. Afon Elan and Widdale Beck show parallel-transverse and transverse patterns respectively; clasts from Turkey Brook exhibit little preferred orientation. However, clusters always display a more consistent alignment parallel with the flow direction than constituent clasts: mean divergence from flow directions are 10° and 21°, respectively. Incorporation of clasts within cluster bedforms appears to confer upon them a degree of preferred alignment which is absent where particles occupy more exposed, open plane-bed positions. A further characteristic of cluster bedforms is that constituent clasts are more steeply imbricated compared with those occupying non-clustered parts of the bed. For example, the mean dip angle of cluster clasts from the Wye catchment is 35° compared with 17° for pebbles sampled elsewhere of similar size and shape.

	Wye catchment channels, Wales	Turkey Brook, Enfield Chase, England	Widdale Beck, Yorkshire, England
Mean angular deviation of cluster bedforms	6.8°	15.3°	5.7°
Mean angular deviation of obstacles (*ab* planes)	14.2°	23.0°	19.1°
Mean angular deviation of stoss-side clasts (*ab* planes)	23.0°	27.0°	31.3°

Table 2. Mean angular deviation of cluster bedforms, obstacle clasts and clustered stoss-side clasts from the local flow direction.

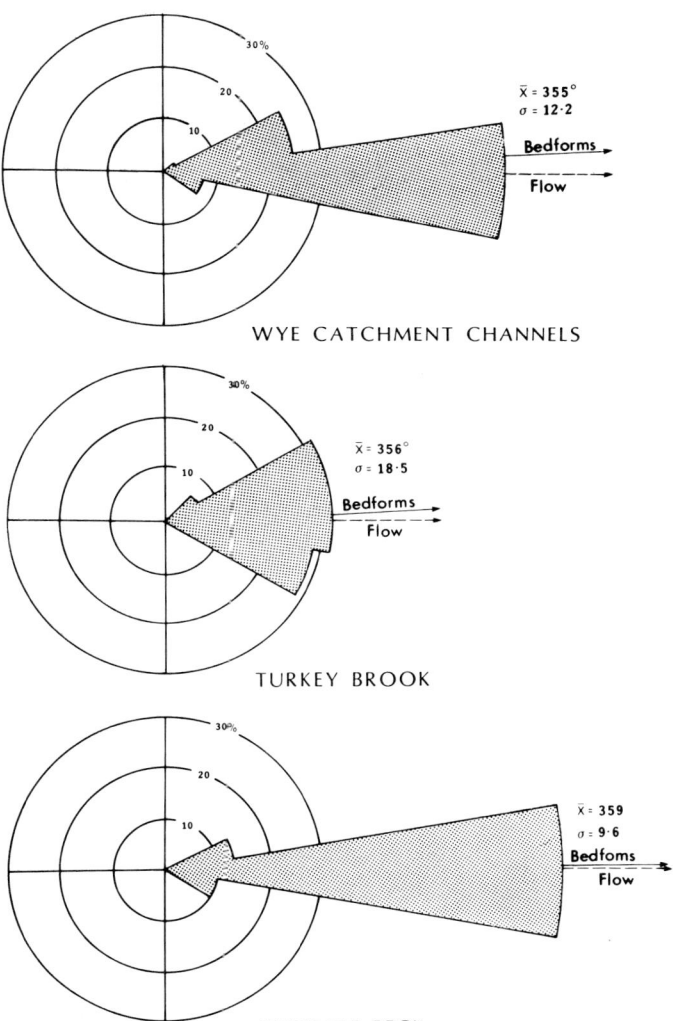

Fig. 5. Rose diagrams showing the long axis orientation of cluster bedforms from the Wye catchment channels, Turkey Brook and Widdale Beck. Reference direction for $\bar{x}$ is flow direction.

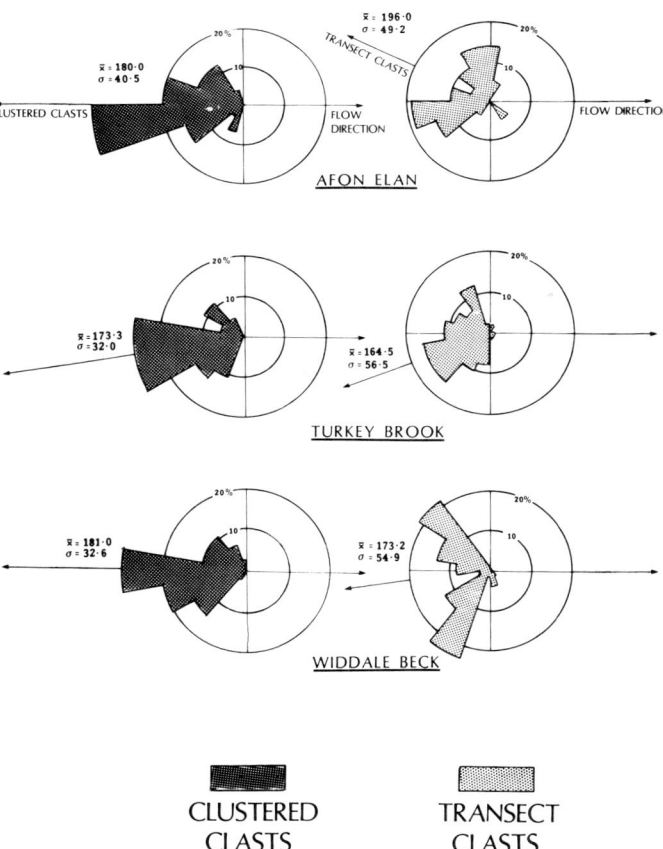

Fig. 6. Rose diagrams showing the *ab* plane orientation of clustered (stoss-side) clasts from bedforms in Afon Elan, Turkey Brook and Widdale Beck. The *ab* plane orientation of clasts taken along cross-profile transects is given for comparison.

SECTIONAL CHARACTERISTICS

In order to investigate in greater detail the internal structure of clusters, several were fixed by impregnation with resin, so that they could be removed intact from the streambed. Bedforms were then sectioned parallel to the flow direction (Figs. 7A and 7B).

Wake deposits are easily distinguishable as discrete accumulations of fine calibre sediment in the lee of obstacle clasts. These deposits appear to be marginally coarser at the surface of the streambed and fine downwards. Remnants of wake accumulations from previous depositional events are observed behind buried clasts in the sub-surface bed material.

Cut sections bring to light the surficial character of cluster bedforms. Clusters overlay a finer sub-surface bed material of similar calibre to that comprising sediment in the wake. Hydrodynamic size-sorting of sediment within the cluster is confined to only the uppermost layer of bed material — a fact which underlines the importance of clusters in exerting primary control over the availability of particles for subsequent transportation. In this respect, cluster bedforms conform with the pre-existing armour layer of channel bed material, but form distinct entities which can be clearly distinguished from the surrounding armour layer.

EXPERIMENTAL FLUME STUDY

In common with virtually all other small-scale gravel bedforms, the formation of clusters has never been witnessed in the field due to the difficulty of observing the depositional activity during waning flood flow. Dal Cin (1968) contends that pebble clusters are a feature of the selective transport of bed particles, and form by pebbles piling up against a particularly large clast which obstructs their further transport. Teisseyre (1977) considers that the formation of pebble clusters requires the same hydrodynamic conditions as imbrication, and suggests that they form only under turbulent rapid flows.

Our lack of understanding of cluster bedforms is compounded by the absence of detailed experimental studies. However, experiments reported by Brayshaw *et al.* (1983) on the effects of mutual interference between neighbouring particles provide a theoretical basis to explain the stability of closely spaced groups of pebbles. In an attempt to understand the mechanisms which produce clusters, their formation was simulated in a flume with an unsorted sediment mixture.

EXPERIMENTAL DESIGN

Experiments were conducted in a 0.3 m wide, 4 m long, recirculating tilting channel flume. At the beginning of each run, a 50 mm deep bed of sediment was laid in the channel and graded by template. A sequence of flow events with identical hydrographs was then run, each lasting for 1 hour and with a flood-peak discharge of 3.0 litres.s^{-1}

Fig. 7. (**A**) Section cut through impregnated cluster bedform fixed *in situ* in Widdale Beck, showing characteristic fine-grained wake deposit formed behind single obstacle clast. The local elevation of the wake is considerably lower than the stoss-side accumulation. (**B**) Cut section through imbricate-type cluster bedform also from Widdale Beck. Small wake deposits fill interstices between contiguous clasts, and the bedform is aligned parallel with the flow.

after 15 minutes. However, for each hydrograph, bed slope and water depth were adjusted to produce four differing values of bed shear stress at flood-peak flow. Laboratory channel slopes ranged from 0.005 to 0.010, and therefore compare favourably with the field channels (0.006-0.008). Three runs were performed for each combination of slope and depth.

During each run, sediment was fed into the upstream end of the flume by hand, so as to avoid scour of the initial bed material. At the end of the simulated flood recession limb, a gate at the flume outlet was gradually adjusted to prevent unwanted scour of bed features. The bed was then slowly drained and allowed to dry completely. Finally, all bed configurations generated within a 1.5 m long test section in the centre of the channel were impregnated with resin, removed from the bed, and sectioned parallel to the direction of flow.

The median size of sediment mix used (D_{50} = 1.55ø or 2.92 mm) gave a scale ratio of between 1/20 and 1/5 to bedload in the natural channels. To ensure similarity between flume and prototype, several dimensionless parameters were used to judge the scaling relationships. The range of Froude numbers in the laboratory experiments, 0.31 to 0.73, was essentially that covered by the studied rivers at bankfull flow (0.35 - 0.63); grain Reynolds numbers, using D_{50} as the length dimension, were around 10^3 at flood-peak flow compared with 10^4 for the natural bankfull flow; ratios of water depth to clast size, using D_{50}, at flood-peak flow in the flume were approximately 13 against 16 to 22 for the field cases.

VISUAL OBSERVATIONS

At flood-peak flow during each run all available bed material was in continuous motion, though suspension was visibly absent. Larger size fractions moved by traction, and finer clasts by saltation; suspension of clasts was not observed. Closer inspection revealed, however, that only a small proportion of the bed material was in motion at any given time, there being a continuous exchange of clasts in and out of transport.

In all runs, cluster bedforms formed within the predetermined test section. Their overall geometry displayed a striking similarity with naturally occurring counterparts, but no significant relationship was observed between the area occupied by clusters and boundary shear stress at flood-peak flow.

The first particles to come to be deposited upon falling stage are the largest clasts in the sediment mixture. These form the potential nuclei of cluster bedforms, and range in size from D_{98} to D_{99}. Soon after their deposition at relatively high flow stage, finer sediment begins to accrete on the downstream sides of obstacles. This occurs wherever the flow-paths of individual clasts are drawn into the recirculating separation zone in the obstacle's wake, the majority being deposited here. Similar observations have been reported by Parker et al. (1982), and are readily explained in terms of the modified fluid forces which grains experience in the wake of upstream particles (Mair and Maull, 1971; Brayshaw et al., 1983). Wake accumulations occur wherever stationary obstacle clasts, protruding into the flow, are present on the bed. At this stage there is no deposition of particles on the stoss-side of obstacles, where small scour holes form occasionally in response to downflow from horseshoe vortices (Breusers et al., 1977).

Stoss-side accumulations only occur at a considerably lower flow stage, near the depositional velocity for the median size of bed material: this size of grain is most easily halted by stationary obstacle pebbles. Furthermore, the deposition of one or two initial clasts on the stoss-side of an obstacle stimulates the deposition of others, since the cluster presents a larger obstruction to the flow. They are precluded from settling on the stoss-side of obstacles at higher flow velocities because they are easily carried around the obstacle's flanks by the local flow acceleration.

The size-sorting of sediment in cluster bedforms can be explained conveniently in the context of these observations. In the lee of stationary obstacles, only relatively finer grains are vulnerable to the recirculating flow which draws particles into the low pressure wake; the momentum of larger grains is such that the lateral fluid forces which attract grains towards the wake centre-line have no effect. On the upstream side of the obstacle, the same small calibre particles are easily carried around the obstacle's flanks by filaments of the horseshoe vortex, and are thus precluded from settling. It is therefore only coarser grains which are stopped on the upstream side of obstacles — the slight scour hole developed here possibly contributing to their deposition.

Size analyses of sediment deposited on the stoss and in the wake of obstacles reveal the same relationships as for naturally occurring field clusters. The median size of sediment deposited in the wake of obstacles ranges from D_{36} to D_{52}. Stoss-side deposits range in calibre from D_{58} to D_{71}.

SECTIONAL CHARACTERISTICS

Sections cut through impregnated flume clusters are shown in Figure 8. Both imbricate-type clusters (Fig. 8A) and characteristic obstacle-dominated clusters (Fig. 8B) resemble counterparts found in natural channels. However, wake deposits in flume clusters have considerably higher relief in relation to the obstacle clast than stoss-side accumulations; the converse is true for natural clusters. This reflects the finer size distribution of the flume sediment and the slightly differing grain Reynolds numbers between the laboratory and natural channels. The wake accumulation takes on the appearance of a flow-aligned ridge in the obstacle's lee (Fig. 8B) whilst the elevation of the stoss accumulation is lowered by scouring from downflow on the obstacle's upstream side.

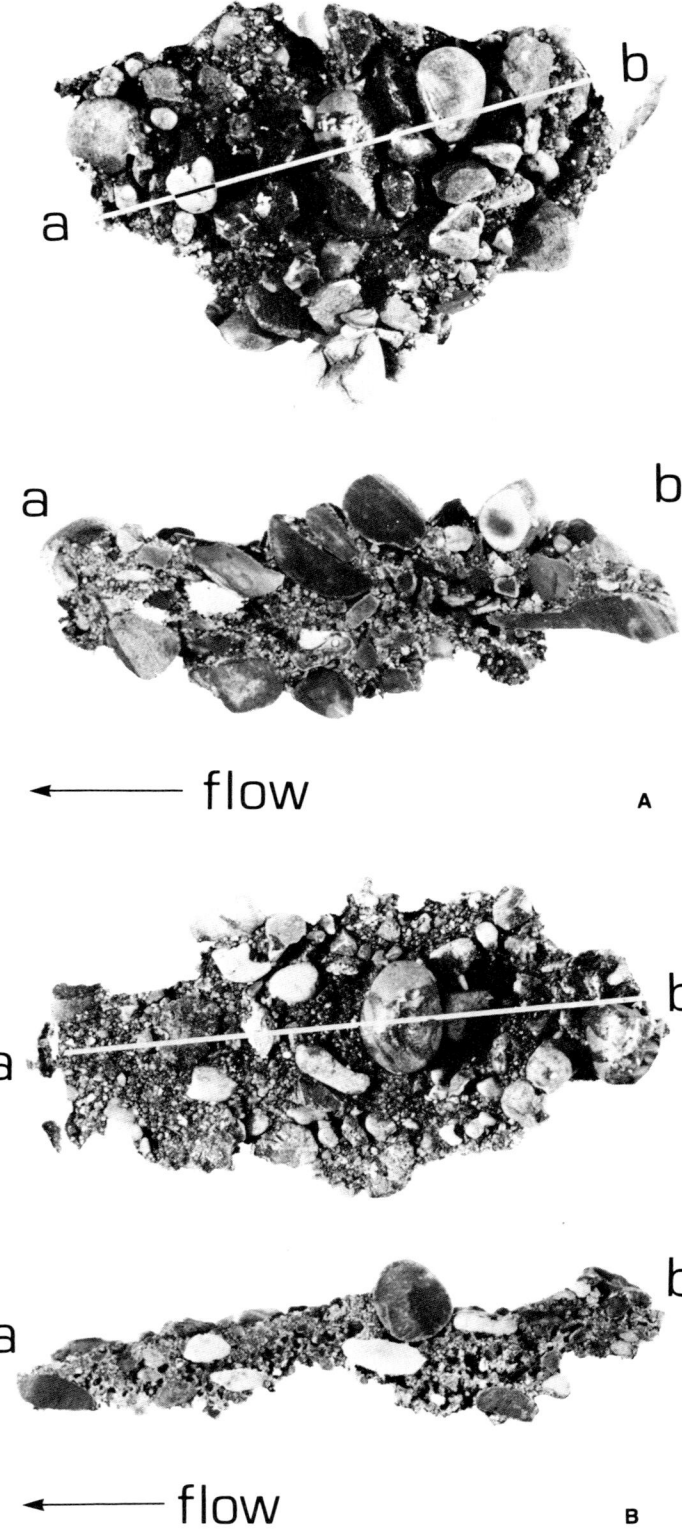

Fig. 8. (**A**) Cut section through cluster bedform showing an imbricate arrangement of particles, generated following a flood-peak boundary shear stress of 0.38 kg.m^{-2}. (**B**) Bedform generated following flood-peak shear stress of 0.33 kg.m^{-2} displaying good size-sorting characteristics. Difference in the elevation of stoss and wake deposits reflects scouring upstream from the obstacle.

IMPLICATIONS FOR SEDIMENT TRANSPORT

Cluster bedforms play a critical role in influencing the transport of coarse bed material in coarse-grained alluvial channels. Once a cluster has formed on the bed, it can be remobilized only by flows which are competent to dislodge the nucleus obstacle clast; flume observations confirm that, with increasing flow stage, there is little selective entrainment of constituent particles from either stoss or wake positions within the cluster bedform. However, some upstream stoss-side clasts are occasionally dislodged prior to the dispersal of the complete cluster bedform. This phenomenon is also witnessed in natural channels where particles seeded onto the stoss of cluster bedforms are sometimes entrained slightly earlier than well-protected wake particles (Brayshaw *et al.*, 1983). However, up to a certain threshold value on rising discharge, higher flow velocities may actually enhance the stability of pre-existing clusters: stoss-side clasts will be held more firmly against the obstacle, whilst grains in the obstacle's lee will be trapped in the low pressure separation zone. In addition, the critical entrainment threshold of obstacle clasts in cluster bedforms is above that of unobstructed counterparts by virtue of interference and interlock between neighbouring particles. In this way, cluster bedforms create a stability which both limits the availability of readily transportable particles and delays their incipient motion. This is confirmed by field measurements (Brayshaw, 1983) in which it has been shown that the threshold of motion of clasts incorporated within cluster bedforms is delayed beyond that of counterparts in open plane-bed positions.

CONCLUSIONS

A detailed examination of the structural arrangement of small-scale gravel bedforms is important to our further understanding of sediment movement in coarse-grained alluvial channels. Particle clusters consist of closely nested groups of clasts usually deposited in trains parallel with the direction of the flow. Clusters are formed by the presence of larger than average obstacle clasts, around which stoss and wake deposits accrete in response to secondary currents induced by the obstacle's presence in the flow. The structure of the flow in the vicinity of obstacles influences the trajectories of particles in such a manner that well-defined and repeatable sorting patterns prevail in cluster bedforms. As such, clusters exercise an influence over the sorting patterns of particles deposited on the streambed, and in doing so, determine the nature of contact between juxtaposed clasts at both incipient motion and in later transport.

Observations of cluster formation reveal a temporal difference in the deposition of stoss and wake deposits. This is attributable to the response of different calibre grains to the varying flow structure around an isolated clast on a streambed, and accounts for the characteristic size-sorting patterns noted in cluster bedforms. It is interesting to observe that clusters form neither as an equilibrium bedform nor quickly as a response to varied flow conditions: clusters are generated over the duration of the recession limb following bedload movement, and form subsequent to the deposition of large clasts which act as obstacles to the flow. As such, the size of the obstacle has an important influence on the geometry of the cluster bedform.

Of significance for sediment transport is that the re-entrainment of cluster constituents is a sudden event in which groups of particles are released in bursts into the flow. The threshold of incipient motion for stones in cluster bedforms depends upon the calibre of the obstacle clast and its degree of interlock with neighbouring particles. The sudden dispersal of clustered clasts on a rising flood stage may account for temporal variations in bedload transport rates (*eg.*, Einstein, 1937; Reid *et al.*, 1980; Brayshaw, 1983). It is clear that bed microtopography must be considered alongside hydraulic variables as an important factor determining sediment transport in poorly sorted gravel-bed streams.

References

Bagnold, R.A. 1954. The Physics of Blown Sand and Desert Dunes. London: Methuen and Co., 265p.

Brayshaw, A.C. 1983. Bed microtopography and bedload transport in coarse-grained alluvial channels. Ph.D. Thesis, University of London, 415p.

Brayshaw, A.C., Frostick, L.E. and Reid, I. 1983. The hydrodynamics of particle clusters and sediment entrainment in coarse alluvial channels. Sedimentology, v. 30, p. 137-143.

Breusers, H.N.C., Nicollet, G. and Shen, H.W. 1977. Local scour around cylindrical piers. Journal of Hydraulic Research, v. 15, p. 211-252.

Carling, P.A. and Reader, N.A. 1982. Structure, composition and properties of upland stream gravels. Earth Surface Processes and Landforms, v. 7, p. 349-365.

Church, M. 1972. Baffin Island sandurs: a study of arctic fluvial processes. Geological Survey of Canada Bulletin, v. 216, 208p.

Dal Cin, R. 1968. Pebble clusters: their origin and utilization in the study of palaeocurrents. Sedimentary Geology, v. 2, p. 233-241.

Einstein, H.A. 1937. Die Eichung des im Rhein verwendeten Geschiebefangers. Schweizer Bauzeitung, v. 110, p. 29-32.

Einstein, H.A. 1950. The bedload function for sediment transportation in open channel flows. United States Department of Agriculture Technical Bulletin 1026, 78p.

Fenton, J.D. and Abbott, J.E. 1977. Initial movement of grains on a stream bed: the effect of relative protrusion. Proceedings of the Royal Society of London, Series A, v. 352, p. 523-537.

Gustavson, T.C. 1974. Sedimentation on gravel outwash fans, Malaspina glacier foreland Alaska. Journal of Sedimentary Petrology, v. 44, p. 374-389.

Koster, E.H. 1978. Transverse ribs: their characteristics, origin and paleohydraulic significance. *In:* Miall, A.D. (Ed.), Fluvial Sedimentology. Canadian Society of Petroleum Geologists, Memoir 5, p. 161-186.

Laronne, J.B. and Carson, M.A. 1976. Interrelationships between bed morphology and bed material transport for a small, gravel-bed channel. Sedimentology, v. 23, p. 67-85.

Mair, W.A. and Maull, P.J. 1971. Aerodynamic behaviour of bodies in the wake of other bodies. Proceedings of the Royal Society of London, Series A, v. 269, p. 425-437.

Martini, I.P. 1977. Gravelly flood deposits of Irvine Creek, Ontario, Canada. Sedimentology, v. 24, p. 603-622.

Parker, G., Dhamotharan, S. and Stefan, H. 1982. Model experiments on a mobile, paved gravel bed stream. Water Resources Research, v. 18, p. 1395-1408.

Peabody, F.E. 1947. Current crescents in the Triassic Moenkopi Formation. Journal of Sedimentary Petrology, v. 17, p. 73-76.

Reid, I., Layman, J.T. and Frostick, L.E. 1980. The continuous measurement of bedload discharge. Journal of Hydraulic Research, v. 18, p. 243-249.

Rust, B.R. 1972. Pebble orientation in fluvial sediments. Journal of Sedimentary Petrology, v. 42, p. 384-388.

Shields, A. 1936. Anwendung der Ahnlichkeitsmechanik und der Turbulenzforschung auf Geschiebebewegung: Preussische Versuchsanstalt fur Wasserbau und Schiffbau (Berlin), Mitteil. no. 26, 26p. (Translation by W.P. Ott and J.C. Van Uchelen, Applications of similarity principles and turbulence research to bedload movement: United States Department of Agriculture, Soil Conservation Service, California Institute of Technology, 70p).

Teisseyre, A.K. 1977. Pebble clusters as a directional structure in fluvial gravels: modern and ancient examples. Geologica Sudetica, v. 12, p. 79-94.

White, W.R. and Day, T. 1982. Transport of graded gravel bed material. *In:* Hey, R.D., Bathurst, J.C. and Thorne, C.R. (Eds.), Gravel-Bed Rivers. Chichester: John Wiley and Sons, Ltd., 875p.

Wolman, M.G. 1954. A method of sampling coarse bed material. Transactions of the American Geophysical Union, v. 35, p. 951-958.

FLOOD SEDIMENTATION IN BEDROCK FLUVIAL SYSTEMS

Victor R. Baker[1]

Abstract

Bedrock fluvial systems which occur in regions of landscape degradation, are commonly ignored by sedimentologists. These systems deserve more study because (1) they are locally preserved along unconformities in the ancient record, and (2) they may contain sediments that allow paleohydraulic reconstruction of extreme flood events. Rare, high magnitude floods in narrow, deep bedrock channels produce distinctive hydraulic and sedimentary phenomena. Macroturbulent kolks, boulder transport, and hydraulic jumps occur. Very resistant rocks can induce anastomosing patterns and/or structural control of rapids, pools and riffles. Distinctive depositional macroforms include longitudinal, pendant, and expansion bars. Gravel mesoforms, such as giant current ripples, develop during deep flood flows. Preservation of these bedforms is enhanced by a rapid decrease in discharge coupled with a rapid drop in stage due to very small width/depth ratios.

Résumé

Les sédimentologues se désintéressent très fréquemment des systèmes fluviaux sur socle de régions dont le paysage est modifié par l'érosion. Ces systèmes méritent une plus grande attention car (1) ils sont préservés localement le long des discordances faisant partie du registre géologique, et (2) ils peuvent contenir des sédiments qui permettent une restitution paléohydraulique des événements qui accompagnaient les plus grandes crues. Les très fortes crues dans des chenaux étroits et profonds creusés dans le socle sont peu fréquentes, elles sont toutefois accompagnées de phénomènes hydrauliques et sédimentaires caractéristiques. Il se forme des kolks macroturbulents, un transport de gros blocs, et des ressauts hydrauliques. Les roches très résistantes peuvent occasionner l'apparition de motifs anastomosés et/ou d'obstacles structuraux contrôlant les rapides, et un relief de bancs et bassins. Les macroformes distinctives des dépôts incluent des barres longitudinales, en apophyse et d'accrétion. Des mésoformes composées de gravier, telles les rides de courant géantes, se développent lors des débits de crue puissants. La préservation des formes de litage est favorisée par une bassie brusque du débit accompagnée d'une réduction rapide des gradins résultant des très faibles rapports de largeur/profondeur.

Introduction

Fluvial studies commonly give emphasis to alluvial systems. For sedimentology this is justified because bedrock systems have a low potential for preservation in the geologic record. However, anomalies that arise in considering unusual phenomena are a critical element in the scientific process (Kuhn, 1962). Moreover, a balanced science requires the investigation of all relevant phenomena. High-energy floods through bedrock fluvial systems transport gravel and boulders when such materials are available. These sediments and, in some cases, the associated bedforms and channel patterns may be preserved along major unconformities in a stratigraphic sequence. Moreover, extremely large floods through bedrock channels provide a long-lasting record of significant, but rare geological events. The paleohydraulic reconstruction of those events provides an upper limit to the scale of sediment transport phenomena that must be understood to interpret depositional sequences. Thus, the neglect of bedrock fluvial systems by sedimentologists can only inhibit the balanced growth of sedimentology as a science.

Bedrock fluvial systems occur in regions where landscape degradation dominates over deposition. Bedrock streams flow in valleys that are incised because of base-level lowering, uplift, or climatically induced changes in water and sediment influxes. Extremely large, rare floods can be introduced by appropriate meteorological conditions, e.g., tropical cyclones, or by lake bursts. The effects of such floods are maximized in regions of extreme hydrologic variability, such as in arid and savanna landscapes (Baker, 1977). In such regions the distinction between valleys and channels is obscured by the tendency of rare floods to spill out of low-flow channels and to occupy the entire valley floor. In extreme cases floods may even spill out of valleys, forming anastomosing scour patterns over an entire landscape.

If a landscape containing bedrock fluvial systems changes from an erosional to a depositional regime, then the bedrock system may be preserved along the resulting unconformity. Buried valleys formed in this way do occur in some ancient sedimentary basin sequences. Thus, the study of erosional bedrock fluvial systems can be considered a contribution to the interpretation of paleogeomorphology (Martin, 1966).

This paper reviews the sedimentology of bedrock rivers subject to very large floods. The approach will emphasize the author's field experience with respect to the relevant

---

[1] Department of Geosciences, University of Arizona, Tucson, Arizona 85721, U.S.A.

My thinking about bedrock fluvial systems has been facilitated by research funds provided by the National Science Foundation (Grants EAR-7723025, EAR-8100391, EAR-8119981, and EAR-8300183) and by the National Aeronautics and Space Administration (Contracts NAS-913312 and NAS-115164, Grants NSG-7326 and NSG-7557). I thank Emlyn Koster for encouraging me to prepare this review. He and John Costa made numerous suggestions that greatly improved an earlier version of the paper.

Copyright © 1984, Canadian Society of Petroleum Geologists

River	Location	Flood	Recurrence Interval (years)	Contributing Drainage Area (km^2)	Peak Discharge (m^3.s^{-1})	Flood Stage (m)	Velocity (m.s^{-1})	Transported Boulder Size (m)	References
Medina	Texas	Aug. 2, 1978	500	700	6,792	15	3.72	2	This report.
Elm Creek	Texas	May 2, 1972	400	12.2	1,130	7	6.4	2	Baker, 1977
Pecos	Texas	June 31, 1954	2000	9300	27,400	30			Kochel, et al., 1982
Big Thompson	Colorado	July 31, 1976	5000	88	884	3.23	7.92	2.8	Grozier et al., 1977
Missoula Flood	Washington	Pleistocene	10000		1x10^7	100+	9-30	2-10	Baker, 1973

Table 1. Hydraulic and hydrologic properties of some rare, high magnitude floods in narrow, deep bedrock channels.

scientific literature. The selected results presented and topics discussed are meant to stimulate much-needed further research.

Flow Dynamics and Sediment Transport

Rare, high magnitude floods produce spectacular channel changes and movement of coarse sediments (Baker, 1977; Gupta, 1983). Table 1 provides some examples of these phenomena, but knowledge of such events remains scant. This is because the significant flows have recurrence intervals of hundreds of years. Measurements of flow phenomena and sediment transport are lacking because of difficulties in accessing measurement sites during extreme events and because of the effects of suspended and floating debris on instrumentation. A result of the quantitative revolution in fluvial studies, therefore, has been a neglect of these geologically significant processes. The search for quantitative data favors attention to flow events of lesser magnitude, generally for alluvial systems.

Figure 1 outlines some important flow properties and sedimentary phenomena associated with floods in narrow, deep bedrock channels. Extremely high velocities (Table 1) result from the steep gradients and deep cross-sections that characterize these systems. In alluvial systems such extreme floods would enlarge the channels conveying the flow by scour and by increasing the channel width. However, bedrock and certain other non-alluvial materials, such as till and duricrusts, can have sufficient resistance that this self-adjusting tendency is impeded. Such systems accommodate the high discharge of extreme floods by dramatically increasing flow depth (Tinkler, 1971). Where resistant banks do not readily yield sediment to an energetic flood flow, these factors lead to the development of remarkably intense turbulent phenomena, which are rarely important in alluvial systems.

This paper will emphasize sedimentation in fluvial systems that display very resistant flow boundaries and which experience intense flood flows. Although such phenomena are most common in bedrock settings, it should be noted that certain other non-alluvial materials may have sufficient bank and bed resistance to display the phenomena.

A fundamental hydraulic characteristic of very deep, high gradient flood flows is the development of secondary

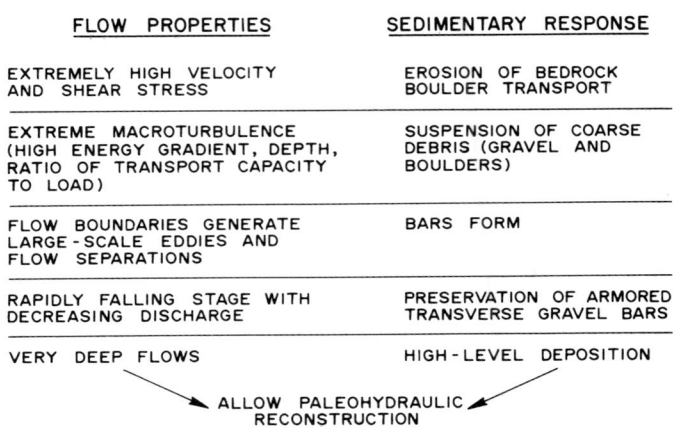

Fig. 1. Important flow properties and sedimentation phenomena associated with large floods in narrow, deep bedrock channels.

circulation, of flow separation, and of the birth and decay of vorticity around obstacles and along irregular boundaries. Such three-dimensional flow phenomena in rivers are collectively termed macroturbulence by Matthes (1947). Several forms of macroturbulence have received extensive investigation, including the flow at meander bends, separated flows at a downward step (reviewed by Allen, 1971), and helical vorticity aligned parallel to the dominant flow direction (Einstein and Li, 1958).

A much less studied form of macroturbulence is that which develops along highly irregular flow boundaries, such as behind bedrock projections into the channel, downstream of submerged large boulders or along canyon rock walls (Baker, 1978a). The macroturbulent vortices that develop at such sites were termed kolks by Matthes (1947). Kolks involve intense energy dissipation by upward vortex action. The intense pressure and velocity gradients of the vortex produce a phenomenal hydraulic lift force along the filament of the vortex. The precise magnitude of this suction effect is not known, but it greatly exceeds normal hydraulic lift forces. The conditions necessary for kolk generation according to Matthes (1947) include a steep energy gradient, a low ratio of actual sediment transported to potential sediment transport, and an irregular, rough boundary capable of generating flow separation. Jackson (1976) attributes kolks to the oscillatory growth and breakup stages of the turbulent bursting phenomenon. Bursting, as described by Offen and Kline (1975), characterizes the turbulent structure of the outer part of the turbulent boundary layer. It is not clear, however, that this mechanism as observed in smooth-walled flumes actually applies to the extremely irregular flow boundaries of bedrock channels.

Boulder transport phenomena in bedrock channel systems are especially instructive (Fig. 2). Intense floods have commonly been observed to carry 1-2 m boulders to the level of the flood high-water surface (Costa, 1983; Foley et al., 1978; Stewart and LaMarche, 1967). Boulder levees or berms may form parallel to pre-flood channelways with the largest boulders at the berm summits (Fahnestock, 1963; Scott and Gravlee. 1968). Where the possible debris flow origin of such deposits can be ruled out (Costa and Jarrett, 1981), it appears that the deposits arise either from the macroturbulent suspension of the boulders by kolks or by movement in a type of density underflow (Scott and Gravlee, 1968). Foley et al. (1978) envisage a process of kolk action in deeper channel sections, boulder entrainment, and upward transport to the berm crests. Here the boulders combine with finer load to move for short distances as subaqueous debris flows eventually lodge as lobate boulder fronts (Scott and Gravlee, 1968).

Bedrock rivers often display relatively abrupt flow expansions and constrictions, in contrast to the more uniform or gradually varying width characteristics of alluvial streams. Reasons for the latter phenomena are nicely embodied in the hydraulic geometry concept of Leopold and Maddock (1953). The anomalous behavior of bedrock rivers has important sedimentologic consequences, as recognized by Krumbein (1942) in his study of flood deposits at Arroyo Seco, near Los Angeles. He found that irregular flow boundaries and high flow velocities resulted in very coarse bed material being thrown into suspension. This suspended material was found to be deposited in the backwater zones for flood flows upstream of constrictions, while the high-velocity flows through constrictions induced scour and extremely high flow competence. The sedimentological interpretation of such situations must also consider the possibility of debris flow deposition.

Abrupt flow expansions may occur at the mouths of tributaries, especially where a deep tributary canyon enters the trunk canyon at a high angle. Very resistant rock units may also form local, narrow gorge reaches on trunk canyons that end at abrupt expansions. If flood flows become supercritical in constricted reaches, the subsequent transition to an abrupt expansion can induce a hydraulic jump. Komar (1971) reviews the sedimentologic aspects of hydraulic jumps. They consist of stationary surges or shock waves in which the speed of advance of upstream waves is balanced by the velocity of flow downstream. Kinetic energy in the constricted, high-velocity flow is transformed to potential energy in the deep, low-velocity flow down-

Fig. 2. Large boulders transported by floods in bedrock canyon environments. **A.** Boulders 2.5 x 1.5 x 0.9 m transported by the May 11, 1972, flood on Elm Creek, central Texas (Table 1). **B.** Boulders up to 5.6 x 3.7 x 3.2 m transported by late glacial flooding on the Big Lost River, east-central Idaho. **C.** Boulders up to 2.5 x 1.5 x 1.0 m transported by the March 1983 flood of the Finke River, central Australia.

stream from the jump. This transition involves an abrupt loss of mechanical energy through the generation of intense turbulence within the jump. The transition point will be the site either of erosion into bed materials or nondeposition of the coarse bedload in transport. Fan-shaped deposits accumulate immediately downstream, as illustrated in laboratory experiments by Jopling and Richardson (1966).

It is somewhat ironic that the intensive flood flow phenomena that defy direct measurement in bedrock channels also provide unique opportunities for paleohydraulic reconstruction. Two methods are applied. The first involves the fact that given a suitable supply of sediment, the extreme competence of bedrock channel flows produces a broad range of transported boulder sizes that can be related to responsible flow hydraulics (Baker, 1973; Bradley and Mears, 1980; Costa, 1983). The second arises from the rapid fall in stage that follows from the flood peak through narrow, deep bedrock channels. Suspended sand is deposited at high flood levels in slack-water areas, and preserved as the flood rapidly recedes. Such deposits allow the reconstruction of past flood water-surface profiles, in turn allowing the calculation of the flood magnitude. A stratigraphic sequence of such deposits, as preserved in certain ideal settings, allows the generation of long term flood-frequency curves for the largest ancient flow events (Baker *et al.*, 1979, 1983; Kochel and Baker, 1982; Kochel *et al.*, 1982).

Flood transported boulders and slack-water sediments may occur in the same reach or cross section of narrow deep bedrock river systems (Fig. 3). For the example shown in Figure 3 a 17 m flood stage is indicated by the slackwater deposits, and the channel has a width/depth ratio of approximately 2.2. In such a flood a considerable proportion of the total stream power goes into the generation of shear at the stream bed, rather than into channel widening. Boulders several meters in diameter can be transported by the flows when a sufficient gradient is present.

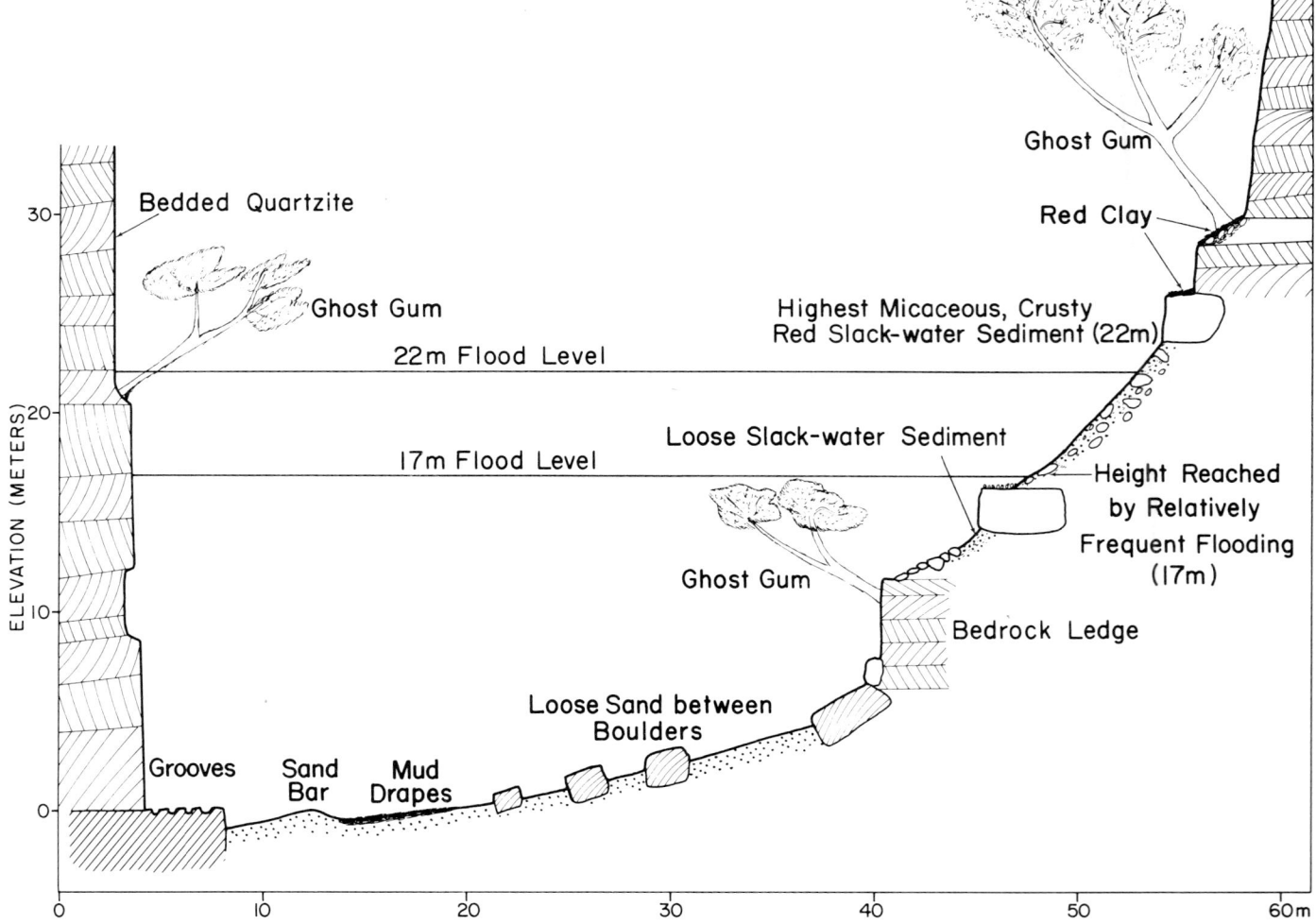

Fig. 3. Cross section of Ormiston Gorge, Northern Territory, Australia, showing flow depths of ancient floods inferred from slack-water sediment accumulations.

Flow Patterns and Channel Morphology

Alluvial rivers display channel patterns that can be systematically related to sediment types composing channel banks, to sediment loads, and to streamflow characteristics (Schumm, 1977). In bedrock rivers, the weathering and resistance characteristics of the rock banks also produce a spectrum of channel types and patterns (Shepherd, 1979). Meandering patterns occur in many rock types and in the appropriate tectonic settings (Gardner, 1975). However, streams cut into very resistant rocks will exhibit pronounced structural control. Narrow, deep fault- or joint-controlled patterns can occur (Fig. 4A). Because a high-magnitude flood cannot efficiently dissipate its energy at high angle bends, structurally controlled patterns can lead to manifestations of turbulent phenomena not usually observed in meandering systems, as already discussed.

Where flood flows exceed the capacity of bedrock canyons or gorges, the water may spill across adjacent uplands and divide producing channel anastomosis. The term 'anastomosis' is a general one, not to be confused with braiding. Braiding refers to branching and rejoining around alluvial islands or bars. Braided streams are part of a continuous series of fluvial forms that develop in quasi-equilibrium with external controls on the river systems. Anastomosis does not necessarily have a genetic connotation, although a similar term 'anastomosed river system' has been applied to interconnected alluvial networks characterized by low-gradient, deep and narrow channels with stable banks (Smith and Smith, 1980). As used in this paper the term 'anastomosis' will refer to interconnected channel systems whether in alluvial streams or in bedrock streams. Anastomosis is extensively developed in the Channeled Scabland (Baker, 1978b). It also occurs where resistant rocks so constrict a channel that extreme floods spill out of the bedrock valley (Fig. 4B). Eroded rock surfaces produced in this fashion are also termed 'scabland.'

Misfit streams are streams that are either too small or too large for the valleys in which they flow (Dury, 1964, 1977). The underfit variety is relatively common, and such streams often show smaller channel widths and meander wavelengths than the winding valleys in which they flow. The disparity between river and valley size is explained by a reduction in stream discharge, generally induced by climatic change (Dury, 1965). Although there is considerable debate over the validity of the climatic explanation of underfit streams, the concept does apply, in a general qualitative sense, to many alluvial valleys (Baker and Penteado-Orellana, 1977). An overfit stream is too large for the valley in which it flows. Dury (1964) considered overfit streams in the context of sudden increases in discharge with rapid channel enlargement. Although such a condition would not persist long in an alluvial valley, the erosion of bedrock may provide an opportunity to preserve overfit stream relationships (Fig. 4B).

Pools are topographically low areas produced by scour on river beds. They seem to have a very regular spacing in stream channels relative to topographically high areas called riffles. As pointed out by Leopold et al. (1964, p. 203), pools and riffles tend to be spaced at 5 to 7 times the channel width. Keller and Melhorn (1973) have pointed out the significance of convergent and divergent flow for the development of pools and riffles. Pools tend to occur at locations where convergent flow increases the bottom shear stress at flood stages. In alluvial rivers, the riffles also develop at high stage when divergent flow induces deposition. More frequent low flow conditions produce a velocity reversal; the water moves more rapidly over the riffles, eroding them and transporting sediment into the relatively tranquil pools (Keller, 1971; Lisle, 1979). Pool-riffle sequences comprise a self-stabilizing property of alluvial rivers, generating localized energy dissipation while transporting the sediment. The same scaling relations also occur in bedrock rivers (Keller and Melhorn, 1978) show-

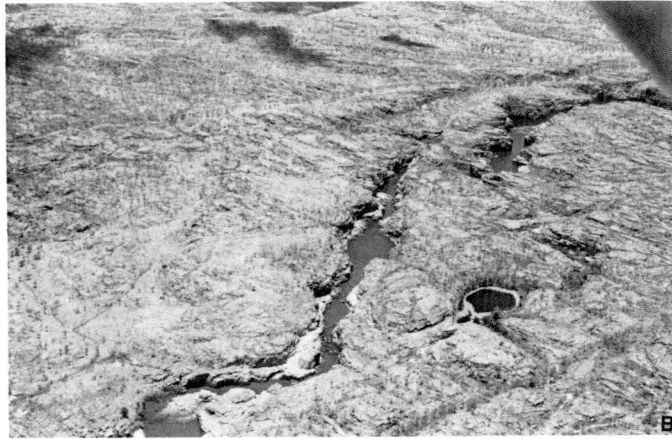

Fig. 4. Bedrock fluvial channel environments in the Northern Territory of Australia. **A.** Joint-controlled canyons at Katherine Gorge, where flood stages may exceed 15 m. **B.** Scabland developed on a high surface (pond at right center) where flood water spilled out of the Katherine Gorge.

ing that the resistance characteristics accommodated by pools and riffles are fundamental fluvial properties.

In some cases a bedrock system may offer too much resistance for self-stabilization to occur. Many of the bouldery rapids along the major canyon rivers of the Colorado Plateau are located at points of lateral influx of coarse debris, either at tributary mouths or sites of mass movement from canyon walls (Graf, 1979). However, Graf (1979) found that many other rapids are not located at sites of debris sources, nor do the rapids show a spatial scaling with discharge as observed in the pool-riffle sequences of alluvial rivers (Richards, 1976). As resistance elements in the main channel flow fields, the rapids generally show little tendency to conform to the equal spacing along canyon floors predicted by quasi-equilibrium theory (Leopold, 1969). Instead, rapids show a near random spacing (Graf, 1979) with some local control by geologic structure (Dolan et al., 1978). Individual large boulders in the Colorado Plateau rapids seem to be very stable, and many require very rare floods (recurrence interval greater than 100 years) for their entrainment (Graf, 1979). One concludes that boulder rapids neither reflect nor contribute to an overall self-adjusting regimen, but rather generate and depend on local geomorphic and hydraulic factors.

Where side-canyon fans of extremely coarse debris constrict main-channel flows, a distinctive pattern of scour and fill is induced (Howard and Dolan, 1981). Pools upstream from the boulder rapids that develop in the constriction will scour at high stage because of supercritical flow. Pools downstream of the rapids will also scour because of turbulence generated at the rapids. During low and/or falling stages, the scour pools in gravel will be partly filled by sand carried in suspension during high discharge. Thus, the pools and ripples function as dynamic elements of flow resistance, sediment transport and sediment storage. Their role is dictated by rock control and by the flood hydraulics, and either factor can dominate.

Bedforms

A fluvial bedform is defined as follows: "any deviation from a plane bed that is readily detectable by eye or higher than the largest sediment size present in the parent bed material" (American Society of Civil Engineers, 1966). Most of the interest in bedforms has focussed on relatively small primary forms developed in sand-bed alluvial rivers (Allen, 1968, 1982). Channel-scale bedforms in alluvial gravel-bed rivers have generated more recent interest (Church and Jones, 1982), but gravel bedforms in narrow, deep bedrock rivers have received relatively little study.

Jackson (1975) developed a hierarchical classification of bedforms generated by fluid shear and composed of cohesionless granular material. The classification relates to bedform size and to the time span of existence for various bed configurations. In alluvial bed rivers the hierarchy begins with macroforms, including point bars, scroll bars, alternate bars, and pool-and-riffle sequences. These bedforms do not relate to local flow conditions. They rather respond to long-term hydrologic factors. Their size is scaled to channel width. Church and Jones (1982) added a higher level 'megaforms' in which bar assemblages or sedimentation zones develop on valley wavelength or greater size scales and persist for time scales that correspond to the fluvial geomorphic regime. Lower in the hierarchy are mesoforms, which include large-scale ripples (dunes), antidunes, and large-scale lineation. The spacing of mesoforms depends on the outer zone of the turbulent boundary layer as the flow varies through a dynamic event, such as a flood. In rivers the boundary layer control is approximated by flow depth. Microforms include current lineation and small-scale ripples. Microforms respond to flow structure in the inner part of the turbulent boundary layer, and their lifetime is much shorter than the periodicity of dynamic events.

Baker (1978b) applied a modified version of Jackson's hierarchical classification to the erosional and depositional bedforms of the Channeled Scabland of eastern Washington. A more generalized scheme for gravel bedforms in narrow deep bedrock channels is shown in Figure 5. In addition to the hierarchical attributes of the bedforms, the classification includes the position in the flow field and the modification by flows subsequent to those responsible for generating the primary bar forms. Microforms, which appear as small-scale ripples and lineations in sand-bed systems, are absent in gravel deposition because of the large particle sizes.

MACROFORMS

The term *bar* has been somewhat indiscriminately applied to all large-scale depositional forms in rivers. The term originally applied to impediments to navigation within a river channel. A modern definition is as follows: "bedforms having lengths of the same order as the channel width or

	LOCATION	PRIMARY BAR FORM	SECONDARY BAR FORM
MACROFORMS (BARS)	BENDS	LATERAL POINT BAR	STREAMLINED AND DISSECTED BAR REMNANTS
	EXPANSION	EXPANSION BAR	
	TRIBUTARY MOUTH ALCOVE	EDDY BAR	
	OBSTRUCTION IN CHANNEL	PENDANT BAR	
MESOFORMS	CHANNEL (ESPECIALLY AT PROXIMAL ENDS OF RIFFLES)	LARGE-SCALE GRAVEL RIPPLES (DUNES)	ARMORED GRAVEL UNDULATIONS

Fig. 5. A preliminary generalized classification of gravel bedforms in narrow, deep bedrock stream channels.

greater, and heights comparable to the mean depth of the generating flow" (American Society of Civil Engineers, 1966). In the classification used here, bars are therefore considered equivalent to depositional macroforms. The considerable confusion in bar nomenclature is reviewed by Smith (1978).

Longitudinal bars are elongated parallel to flow directions. In braided gravelly systems they form at local expansions and display a broad, low form with an internal structure of massive or crude horizontal bedding. In very deep, narrow flood channels, however, longitudinal bars may be high mounded forms with foreset beds. Such bars are common in the Channeled Scabland (Bretz, 1928) where they are interpreted to have developed by downstream accretion (Baker, 1973, 1978b). Malde (1968) introduced the term pendant bar to refer to streamlined mounds of Bonneville Flood gravel that occur downstream from bedrock projections on scabland channel floors. Bar formation occurs by gravel deposition in flow separations that can develop downstream from a variety of flow obstructions (Fig. 6A). The pendant bars of the Channeled Scabland may be 2 km in length and 30 m high. Much smaller

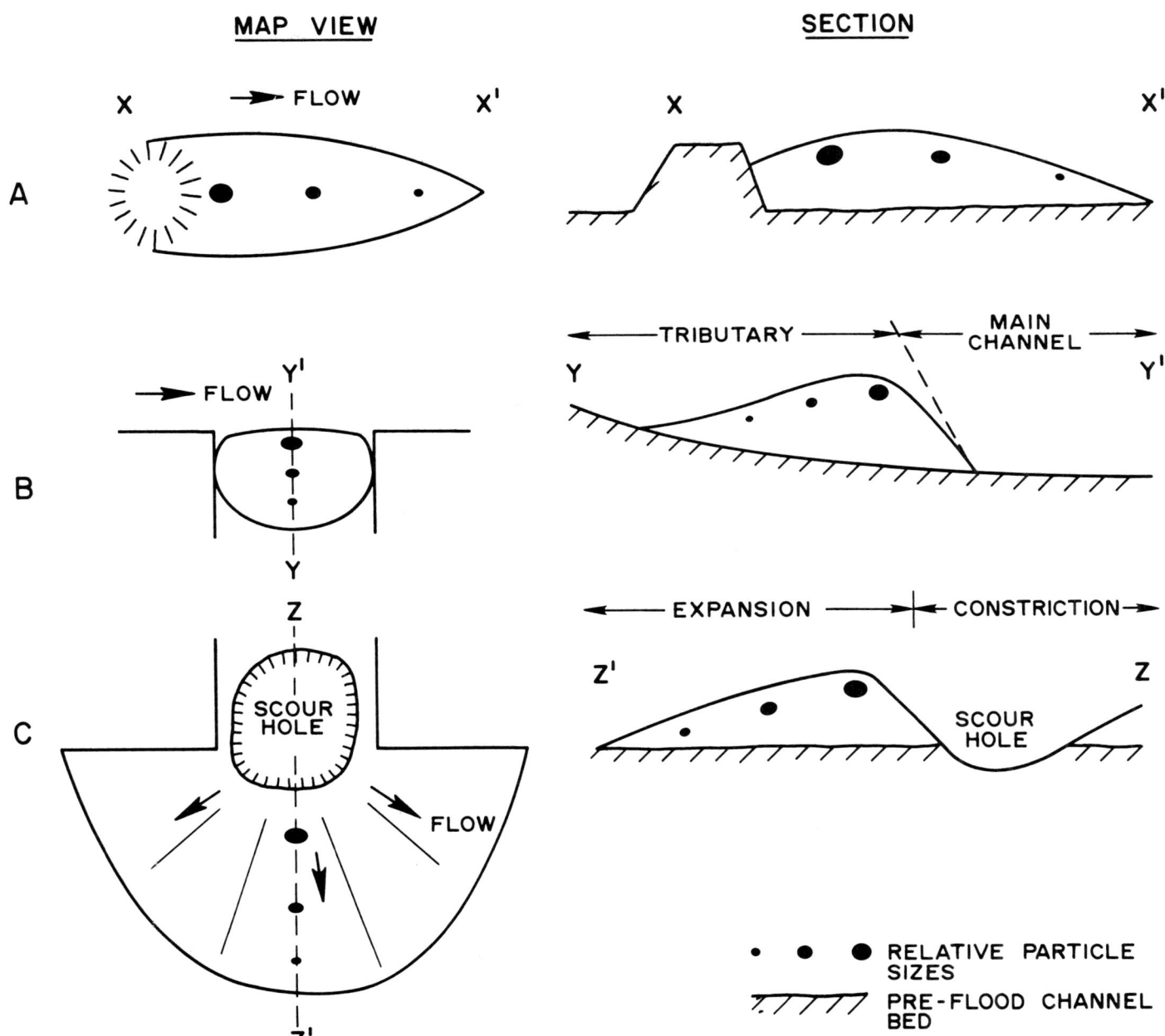

Fig. 6. Schematic diagrams of some major bar types in narrow, deep bedrock stream channels. **A.** Pendant bar. **B.** Eddy bar. **C.** Expansion bars.

analogous bars form in central Texas floods downstream from large boulders of rockfall or in the lee of barriers of stable trees and log jams (Baker, 1977).

The absence of linguoid and transverse bars from the Channeled Scabland has been discussed by Rust (1975, p. 246). Unlike most braided rivers, in which channels are exceptionally wide and shallow, the Missoula and Bonneville Flood channels were relatively constricted. The exceptionally deep flood water allowed the bars to develop prominent slip-faces. These were the stable bedforms under extreme flood conditions. The major form of modification during falling stage has been the concentration of large boulders on the bar surface by winnowing. Most braid bars, in contrast, form by deposition on sub-horizontal surfaces.

Rust (1975) observed that gravel braid bars are like the scabland bars in that they form initially as primary bedforms. They are stable under the flood flows in which all bed material is in motion. In contrast, however, many sand-bed and some gravel-bed rivers show extensive bar modification during changing river stages. These complex, modified bar forms are probably the most common forms in gravel-transporting systems (Church and Jones, 1982).

Eddy bars form at the mouths of tributaries to bedrock canyons invaded by large flood flows (Fig. 6B). They are very well developed in the Channeled Scabland (Bretz et al., 1956; Baker, 1978b) where their internal structure is distinct from that of pendant bars. The former show well developed foreset bedding. Individual foresets contain well sorted sub-rounded boulders and cobbles with an open-work structure. Other foresets are entirely made up of granules and pebbles. Eddy bars, in contrast, contain a variety of grain sizes and structures. Interfingering occurs between poorly sorted boulder gravel, laminated silts, cross-bedded granule gravel, and graded sand-silt layers. The boulders and cobbles are marked by percussion flake scars.

Foreset bedding in the pendant bars almost always dips downstream. Bedding in the eddy bars indicates varying directions of sediment transport. Crude foresets in the boulder gravels usually dip away from the main scabland channel. Frequently, however, the smaller foresets in the granule gravels dip back toward the main scabland channel. This pattern may be the result of the swirling eddies which deposited the bars. The stronger currents carried the coarsest flood debris up the tributary valley. Weaker back-flow currents then deposited the finer granule gravels.

Expansion bars (Fig. 7A) form at abrupt channel expansions (Fig. 6C). They form as unit bars at the flood peak and become considerably modified as stage drops. The dissected bars have a fan-like appearance. Sediment sizes on expansion bars show an extremely rapid proximal-to-distal fining (Baker, 1973, 1978b). For both Bonneville and Missoula flooding, the following relationship holds

$$D = ax^{-b},$$

where D is the maximum particle (m) diameter and x is an arbitrary distance (km) increasing downstream from the constriction (Baker, 1978b, Fig. 4.23). The coefficient a generally ranges from 10^2 to 10^4 and the exponent b from 1.5 to 5.0. This relationship distinguishes expansion bars from alluvial fans, which show a general adherence to Sternberg's Law for the downstream decrease in fluvial sediment size, as follows:

$$D = ae^{-bx},$$

where D and x are defined as above.

The unit bars of braided gravel-bed streams are assumed to develop from coarse sediment moving along the bed as a sheet or wave a few grain diameters thick (Hein and Walker,

Fig. 7. Bars formed by intense flooding of bedrock stream channels in central Texas (Baker, 1977). **A.** Expansion bar formed by flooding of May 11, 1972, at the mouth of a tributary to the Guadalupe River, approximately 10 km north of New Braunfels, Texas. Note that the bar has blocked and diverted the Guadalupe River. **B.** Unit bar formed by tropical storm Amelia flooding of the Medina River (August, 1978) near Medina, Texas.

1977) (Fig. 7B). The bars form where these sheets accumulate at points of flow divergence or where material is excavated from a scour pool (Church and Jones, 1982). In contrast, the large-scale gravel ripples of narrow, deep bedrock streams indicate sediment transport through the system at high stage.

MESOFORMS

The large-scale gravel ripples of bedrock streams have been extensively described for the Channeled Scabland (Bretz *et al.*, 1956; Baker, 1973, 1978b). It is clear that these features are mesoforms (Fig. 5) with their scale controlled by channel depth. In the Channeled Scabland both the ripple size and the included sediment sizes correlate well to shear stress and stream power (Baker, 1973). Observations of smaller gravel ripples in central Texas (Fig. 8) and Australia seem consistent with this trend (Table 2), but much more data will be needed to precisely establish the hydraulic significance of these interesting bedforms. The sediment comprising the giant scabland gravel ripples is some of the coarsest known to occur in large-scale depositional bed mesoforms. The largest particles may be 1.5 m or greater in diameter, and the median size is generally of pebble grade. In all observed examples less than 10% of the sediment is finer than granule gravel. Internally the ripples consist of foreset-bedded gravel with

Fig. 8. Large-scale gravel ripples developed on central Texas rivers by tropical storm Amelia flooding in August, 1978. **A.** Pedernales River near Fredericksburg, Texas. **B.** Medina River, near Medina, Texas. Ripple chord here averages 36 m. **C.** Medina River at Highland Waters, Texas. **D.** Ground photograph of the largest ripples shown in Figure 9C. Ripple chord is 87 m and flow depth here was approximately 10 m at maximum flood stage.

Flood	Height (m)	Chord (m)	Max. Size Particle (m)	Depth (m)	Velocity (m·s⁻¹)	References
Medina River, Texas (Aug. 1978)	1-2	15-25	1	10-15	3-4	This report
Pleistocene Missoula Flood	2-10	40-150	2	20-200	9-20	Baker (1973)
Finke River Australia (March 1983)	0.5	10	0.4	7	2-3	This report
Nueces River, Texas (June 1971)	2	100	0.5	5-7	3-4	Gustavson (1978)
Lake Burst Bass Lake, Minnesota	1	20	0.8	12	3	Theil (1932)

Table 2. Some properties of large-scale gravel ripples produced by flood flows.

remarkable sorting. Layers of cobbles alternate with discrete layers of granules or pebbles, which gives the gravel a distinctive open-work texture.

Boulders and cobbles generally form an armor on the stoss slopes of the ripples. This imbricate pavement probably acted to decrease flow resistance on the ripple surface during the waning stages of flood flow. A similar armoring occurred in central Texas streams after tropical storm Amelia flooding in 1978 (Fig. 9). The development of a hydraulically smoothed surface may be partially responsible for the preservation of the ripples. In most rivers mesoforms composed of sand are washed out during waning flood stages (Jackson, 1975).

Describing gravel ripples in the Nueces River, Gustavson (1978) designated the forms as 'transverse bars', not to be confused with the 'transverse bars' of braided outwash streams (Krigstrom, 1962; Smith, 1974). The latter occur at channel expansions, generally the bifurcation or confluence points of braid channels. Like the large-scale gravel ripples, Gustavson's transverse bars have a low-angle stoss side and a steep avalanche face on the lee side. However, the transverse bars of braided streams are clearly macroforms, in which the depositional system is scaled to the channel width. They constitute one of several types of unit bars that undergo progressive modification during successive flow events (Smith, 1974; Church and Jones, 1982).

Fig. 9. Armor surface formed on a gravel bar of the Pedernales River developed during tropical storm Amelia flooding near Fredericksburg, Texas. Imbricate position of surface clasts shows flow was from right to left. Internal bar texture consists of pebbles and cobbles in a sand matrix. (Photographed in October, 1978.)

CONCLUSIONS

1. Sedimentologists have generally neglected the study of rare, high magnitude flow events in bedrock systems.

2. Large floods in deep, narrow bedrock channels generate special conditions of flow dynamics and sediment transport including the following: intense macroturbulent phenomena, boulder transport, abrupt flow expansions and constrictions, slack-water sedimentation, and very high mean flow velocities and bed shear stresses.

3. Flow patterns and channel morphology, including pools and riffles, reflect structural controls in very resistant rock types, but they reflect adjustment to hydraulic factors in less resistant rock types.

4. Depositional bedforms include macroforms controlled by various flow geometries. These include expansion, eddy and pendant bars.

5. The mesoforms, given appropriate supply of gravel bedload, may be large-scale gravel ripples in narrow, deep flood channels. These ripples can be preserved because of

rapid hydrograph recession and rapid fall in stage in a narrow, deep bedrock channel.

6. The phenomena described in this paper need to be studied in more environments, utilizing paleohydraulic reconstruction of events, to understand the process-response relationships of bedrock fluvial systems.

REFERENCES

American Society of Civil Engineers, Task Force on Bedforms in Alluvial Channels 1966. Nomenclature for bedforms in alluvial channels. American Society of Civil Engineers, Journal of the Hydraulics Division, v. 92, no. HY3, p. 51-64.

Allen, J.R.L. 1968. Current Ripples: Their Relation to Patterns of Water and Sediment Motion. Amsterdam: Elsevier, 433p.

Allen, J.R.L. 1971. Transverse erosional marks of mud and rock: their physical basis and geological significance. Sedimentary Geology, v. 5, p. 167-385.

Allen, J.R.L. 1982. Sedimentary Structures, Their Character and Physical Basis. Amsterdam: Elsevier, v. 1, 594p., v. 2, 664p.

Baker, V.R. 1973. Paleohydrology and sedimentology of Lake Missoula flooding in eastern Washington. Geological Society of America Special Paper 144, 79p.

Baker, V.R. 1977. Stream-channel response to floods with examples from central Texas. Geological Society of America Bulletin, v. 88, p. 1057-1071.

Baker, V.R. 1978a. Paleohydraulics and hydrodynamics of scabland floods. In: Baker, V.R. and Nummedal, D. (Eds.), The Channeled Scabland. Washington, D.C.: National Aeronautics and Space Administration, p.59-79.

Baker, V.R. 1978b. Large-scale erosional and depositional features of the Channeled Scabland. In: Baker, V.R. and Nummedal, D. (Eds.), The Channeled Scabland. Washington D.C.: National Aeronautics and Space Administration, p. 81-115.

Baker, V.R., Kochel, R.C. and Patton, P.C. 1979. Long-term flood frequency analysis using geological data. International Association of Hydrological Sciences, Publication 128, p. 3-9.

Baker, V.R., Kochel, R.C., Patton, P.C. and Pickup, G. 1983. Paleohydrologic analysis of Holocene flood slack-water sediments. International Association of Sedimentologists, Special Publication 6, p. 229-239.

Baker, V.R. and Penteado-Orellana, M. 1977. Adjustment to Quaternary climatic change by the Colorado River in central Texas. Journal of Geology, v. 85, p. 395-422.

Bradley, W.C. and Mears, A.I. 1980. Calculations of flows needed to transport coarse fraction of Boulder Creek alluvium at Boulder, Colorado. Geological Society of America Bulletin, v. 91, Part II, p. 1057-1090.

Bretz, J H. 1928. Bars of the Channeled Scabland. Geological Society of America Bulletin, v. 39, p. 643-702.

Bretz, J H., Smith, H.T.U. and Neff, G.E. 1956. Channeled scabland of eastern Washington: new data and interpretations. Geological Society of America Bulletin, v. 67, p. 957-1049.

Church, M. and Jones, D. 1982. Channel bars in gravel-bed rivers. In: Hey, R.D., Bathurst, J.C. and Thorne, C.R. (Eds.), Gravel-bed Rivers. Fluvial Processes, Engineering, and Management. New York: John Wiley and Sons, p. 291-338.

Costa, J.E. 1983. Paleohydraulic reconstruction of flash-flood peaks from boulder deposits in the Colorado Front Range. Geological Society of America Bulletin, v. 94, p. 986-1004.

Costa, J.E. and Jarrett, R.D. 1981. Debris flows in small mountain stream channels of Colorado and their hydrologic implications. Association of Engineering Geologists Bulletin, v. 18, p. 309-322.

Dolan, R.E., Howard, A. and Trimble, D. 1978. Structural control of rapids and pools of the Colorado River in the Grand Canyon. Science, v. 202, p. 629-631.

Dury, G.H. 1964. Principles of underfit streams. United States Geological Survey Professional Paper 452-A, 67p.

Dury, G.H. 1965. Theoretical implications of underfit streams. United States Geological Survey Professional Paper 452-C, 43p.

Dury, G.H. 1977. Underfit streams. Retrospect, perspect, and prospect. In: Gregory, K.J. (Ed.), River Channel Changes. New York: John Wiley and Sons, p. 281-293.

Einstein, H.A. and Li, H. 1958. Secondary flow in straight channels. Transactions of the American Geophysical Union, v. 39, p. 1085-1088.

Fahnestock, R.K. 1963. Morphology and hydrology of a glacial stream -White River, Mount Ranier, Washington. United States Geological Survey Professional Paper 422-A, 70p.

Foley, M.G., Vessell, R.K., Davies, D.K. and Bonis, S.B. 1978. Bedload transport mechanisms during flash floods, In: Preprint volume, Conference on Flash Floods: Hydrometeorological Aspects. Los Angeles: American Meteorological Society, p.109-116.

Gardner, T.W. 1975. The history of part of the Colorado River and its tributaries: an experimental study. Four Corners Geological Society Guidebook, 8th Field Conference, Canyonlands, Utah, p. 87-95.

Graf, W.L. 1979. Rapids in canyon rivers. Journal of Geology, v. 87, p. 533-551.

Grozier, R.U., McCain, J.F., Lang, L.F. and Merriman, D.C. 1976. The Big Thompson River flood of July 31-August 1, 1976, Larimer County, Colorado. Denver: Colorado Water Conversation Board, 78p.

Gupta, A. 1983. High-magnitude floods and stream channel response. In: Collinson, J. D. and Lewin, J. (Eds.), Modern and Ancient Fluvial Systems. International Association of Sedimentologists Special Publication 6, p. 219-227.

Gustavson, T.C. 1978. Bedforms and stratification types of modern gravel meander lobes, Nueces River, Texas. Sedimentology, v. 25, p. 401-426.

Hein, F.J. and Walker, R.G. 1977. Bar evolution and development of stratification in the gravelly, braided, Kicking Horse River, British Columbia. Canadian Journal of Earth Sciences, v. 14, p. 562-570.

Howard, A. and Dolan, R.E. 1981. Geomorphology of the Colorado River in the Grand Canyon. Journal of Geology, v. 89, p. 269-298.

Jackson, R.G. II. 1975. Hierarchical attributes and a unifying model of bedforms composed of cohesionless material and produced by shearing flow. Geological Society of America Bulletin, v. 86, p. 1523-1533.

Jackson, R.G. II. 1976. Sedimentological and fluid dynamic implications of the turbulent bursting phenomenon in geophysical flows. Journal of Fluid Mechanics, v. 77, pt. 3, p. 531-560.

Jopling, A.V. and Richardson, E.V. 1966. Backset bedding developed in shooting flow in laboratory experiments. Journal of Sedimentary Petrology, v. 36, p. 821-824.

Keller, E.A. 1971. Areal sorting of bed-load material: the hypothesis of velocity reversal. Geological Society of America Bulletin, v. 82, p. 753-756.

Keller, E.A. and Melhorn, W.N. 1973. Bedforms and fluvial processes in alluvial stream channels: selected observations. In: Morisawa, M. (Ed.), Fluvial Geomorphology. Boston: George Allen and Unwin, p. 253-284.

Keller, E.A. and Melhorn, W.N. 1978. Rhythmic spacing and origin of pools and riffles. Geological Society of America Bulletin, v. 89, p. 723-730.

Kochel, R.C. and Baker, V.R. 1982. Paleoflood hydrology. Science, v. 215, p. 353-361.

Kochel, R.C., Baker, V.R. and Patton, P.C. 1982. Paleohydrology of southwestern Texas. Water Resources Research, v. 18, p. 1165-1183.

Komar, P. 1971. Hydraulic jumps in turbidity currents. Geological Society of America Bulletin, v. 82, p. 1477-1488.

Krigstrom, A. 1962. Geomorphological studies of sandur plains and their braided rivers in Iceland. Geografiska Annaler, v. 44, p. 328-346.

Krumbein, W.C. 1942. Flood deposits of Arroyo Seco, Los Angeles County, California. Geological Society of America Bulletin, v. 53, p. 1355-1402.

Kuhn, T.S. 1962. The Structures of Scientific Revolutions. Chicago: University of Chicago Press, 210p.

Leopold L.B. 1969. The rapids and pools - Grand Canyon. United States Geological Survey Professional Paper 669-D, p. 131-145.

Leopold, L.B. and Maddock, T., Jr. 1953. The hydraulic geometry of stream channels and some physiographic implications. United States Geological Survey Professional Paper 252, 57p.

Leopold, L.B., Wolman, M.G. and Miller, J.P. 1964. Fluvial Processes in Geomorphology. San Francisco: W.H. Freeman and Company, 522p.

Lisle, T. 1979. A sorting mechanism for a riffle-pool sequence-summary. Geological Society of America Bulletin, v. 90, part 1, p. 616-617.

Malde, H.E. 1968. The catastrophic late Pleistocene Bonneville Flood in the Snake River Plain, Idaho. United States Geological Survey Professional Paper 596, 52p.

Martin, R. 1966. Paleogeomorphology and its application to exploration for oil and gas (with examples from Western Canada). Bulletin of the American Association of Petroleum Geologists, v. 50, p. 2277-2311.

Matthes, G.H. 1947. Macroturbulence in natural stream flow. Transactions of the American Geophysical Union, v. 28, p. 255-262.

Offen, G.R. and Kline, S.J. 1975. A proposed model of the bursting process in turbulent boundary layers. Journal of Fluid Mechanics, v. 70, p. 209-228.

Richards, K.S. 1976. The morphology of riffle-pool sequences. Earth Surface Processes, v. 1, p. 71-88.

Rust, B.R. 1975. Fabric and structure in glaciofluvial gravels. *In*: Jopling, A.V. and McDonald, B.C. (Eds.), Glaciofluvial and Glaciolacustrine Sedimentation. Tulsa: Society of Economic Paleontologists and Mineralogists, Special Publication 23, p. 238-248.

Scott, K.M. and Gravlee, G.C., Jr. 1968. Flood surge on the Rubicon River, California - hydrology, hydraulics and boulder transport. United States Geological Survey Professional Paper 422-M, 40p.

Schumm, S.A. 1977. The Fluvial System. New York: John Wiley and Sons, 338p.

Shepherd, R.G. 1979. River channel and sediment responses to bedrock lithology and stream capture, Sandy Creek drainage, central Texas. *In*: Rhodes D.D. and Williams G.P. (Eds.), Adjustments of the Fluvial System. Iowa: Kendall/Hunt, p. 255-275.

Smith, D.G. and Smith, N.D. 1980. Sedimentation in anastomosed river systems: examples from alluvial valleys near Banff, Alberta. Journal of Sedimentary Petrology, v. 50, p. 157-164.

Smith, N.D. 1974. Sedimentology and bar formation in the upper Kicking Horse River, a braided outwash stream. Journal of Geology, v. 82, p. 205-224.

Smith, N.D. 1978. Some comments on terminology for bars in shallow rivers. *In*: Miall, A.D. (Ed.), Fluvial Sedimentology. Canadian Society of Petroleum Geologists, Memior 5, p. 85-88.

Stewart, J.H. and LaMarche, V.C. 1967. Erosion and deposition produced by the flood of December 1964 on Coffee Creek, Trinity County, California. United States Geological Survey Professional Paper 422-K, 22p.

Theil, G.A. 1932. Giant current ripples in coarse fluvial gravel. Journal of Geology, v. 40, p. 452-458.

Tinkler, K.J. 1971. Active valley meanders in south-central Texas and their wider implications. Geological Society of America Bulletin, v. 82, p. 1783-1800.

PALEOHYDROLOGIC TECHNIQUES WITH ENVIRONMENTAL APPLICATIONS FOR SITING HAZARDOUS WASTE FACILITIES

M.G. Foley, J.M. Doesburg and D.A. Zimmerman[1]

Abstract

Facilities for storing hazardous waste on or slightly below the Earth's surface must be located or designed to avoid disruption of the containment system and dispersal of the wastes by the maximum credible flood in their vicinity. Because many forms of hazardous waste must be isolated for hundreds or thousands of years, the historical record is not adequate to estimate the magnitude of the 'maximum' flood requiring consideration. Paleohydrologic techniques, based commonly on data derived from the geometry and sedimentology of alluvial gravels, allow the extension of the historical record into the Holocene and late Pleistocene. This method, unlike statistical manipulations of the historical hydrologic record, includes the manifested effects of the range of past climates on floods in the specific drainage basin. Properly interpreted, including estimation of the span of the paleohydrologic record compared to the desired containment period and the physical relation between processes leading to past extreme floods and future ones, paleohydrology is a valuable asset for aiding the assessment or design of a near-surface storage site for hazardous waste.

Paleohydrologic methods are based on the mechanics of flow and sediment transport in alluvial streams. The governing equations relate the flow width, depth, slope, and discharge, and the maximum size of sediment transported. Paleohydrologic studies involve determining or estimating as many of these variables as possible in the field, and calculating the unknowns. Different techniques are subject to varying degrees of uncertainty, but generally the most reliable estimates can be made for paleoflow surface elevation or depth, followed by width, and then slope with least reliability. Multiple, independent approaches are desirable where adequate data are available; confidence in a paleoflow estimate is increased if different approaches give similar results. Combined methods produce the least uncertainty, but data are seldom adequate to allow their use.

Résumé

Les facilités pour l'emmagasinage de déchets dangereux en surface ou enfouis dans le sol doivent être localisée ou construites de façon à éviter toute rupture du système de retenue et également la dispersion des déchets qui pourrait être occasionnée par la plus grande crue de pointe prévisible dans leurs environs. Vu que plusieurs types de déchets dangereux doivent être isolés durant des centaines ou des milliers d'années, le registre historique ne suffit pas pour évaluer le débit de pointe de la crue "maximale" dont il faut tenir compte. Les techniques paléohydrologiques, fondées généralement sur les données dérivées de la géométrie et de la sédimentologie des graviers alluvionnaires, permettent d'étendre le "registre historique" à l'Holocène et au Pléistocène tardif. Cette méthode, différente des manipulations statistiques du registre hydrologique historique, inclut les effets produits par une séquence de paléoclimats sur les crues dans un bassin de drainage donné. Interprétée convenablement, tout en considérant une évaluation de la période couverte par le registre paléohydrologique comparée à celle de la période de retenue désirée, et en plus de la relation physique entre les processus responsables des crues de pointe anciennes et à venir, la paléohydrologie apporte des enseignements très précieux pour choisir et planifier un emplacement pour l'emmagasinage de déchets dangereux près de la surface.

Les méthodes paléohydrologiques sont fondées sur la mécanique d'écoulement et le transport de sédiment de rivières alluviales. Les équations de contrôle relient la largeur, la profondeur, la pente et le débit de l'écoulement avec la grosseur maximale des particules sédimentaires transportées. Les études paléohydrologiques comprennent la détermination ou l'évaluation du plus grand nombre possible de ces variables sur le terrain, et le calcul des inconnues. Les diverses techniques sont assujetties à différents degrés d'incertitude, mais généralement les évaluations les plus fiables peuvent être faites pour l'élévation de surface ou la profondeur des paléoécoulements, suivies par la largeur et ensuite par la pente, mais avec moins de certitude. Lorsqu'il y a suffisamment de données disponibles, il est préférable d'utiliser les méthodes d'indépendants multiples; la confiance dans les résultats pour un paléoécoulement s'accroît d'autant plus que différentes méthodes fournissent des résultats senblables. Les méthodes combinées offrent le plus d'assurance, cependant les données sont rarement assez précises pour en justifier l'usage.

Introduction

Disposal of hazardous wastes (as distinct from storage) is usually regarded by regulatory agencies as 'permanent', no-maintenance isolation from the surrounding environment. The definition of permanent varies according to the type of waste; for example, it means 200 to 1,000 years for uranium mill tailings in the U.S.A. Many types of hazardous waste are disposed of in pits at, or slightly below the ground surface (Fig. 1), and usually above the permanent water table. Contemporary pits usually include relatively impermeable liners and covers designed to prevent ero-

---

[1] Battelle, Pacific Northwest Laboratories, P.O. Box 999, Richland, Washington 99352, U.S.A.

John E. Costa reviewed an earlier version of this paper. The paper has been improved considerably by his review, and those of John Shaw and an anonymous reviewer. Any errors or omissions remain the responsibility of the authors. This work was funded by a contract from the U.S. Nuclear Regulatory Commission.

Copyright © 1984, Canadian Society of Petroleum Geologists

sion and transportation of solid wastes, and migration of liquid and gaseous contaminants. Containment of the wastes can be compromised if the liner or cover is disturbed, or if the water table rises to engulf part of the pit. Therefore, to demonstrate compliance with regulatory criteria, it must be shown with reasonable assurance that future floods or changes in the water table are not likely to disrupt the waste-isolation barriers. This paper will not deal with changes in the water table, but rather will address the methods by which the extent and hydraulics of maximum credible surface-water floods can be established. For disposal periods of 10^3 to 10^4 years, these necessarily include the use of paleohydrologic techniques because climate can be expected to change over such a time period (*e.g.*, Imbrie and Imbrie, 1980) within the general bounds of late Pleistocene-Holocene climates. Statistical extrapolation of present flood magnitudes, or probable maximum flood (PMF) estimates are necessarily based on rare events in the present climate, and are not necessarily suitable for the longer containment periods. Even if containment periods are shorter than 10^3 years, paleohydrology coupled with absolute age dating is a valuable tool in estimating the maximum credible flood to be used for design purposes.

Figure 1 shows a waste disposal pit (partially above ground) protected against floods by a rock armor blanket, and protected against erosion by rainfall and runoff above the elevation of the maximum credible flood by a layer of gravel or by a vegetative cover. Cost control is critical in the construction of erosion protection for large-area waste disposal sites, and it is imperative that the rock armor blanket be designed efficiently and conservatively, but not with gross over-conservatism. Thus, estimates of the elevation and hydraulics (primarily velocity) of the maximum credible flood must be as realistic as possible so that the rock blanket extends high enough without excessive freeboard and that the rocks are large enough to prevent being dislodged by the flood flow, but not excessively so.

Techniques useful in determining the parameters of past extreme flows in river systems relevant to the design of rock armor are largely based on understanding the movement and deposition of the coarsest sediment. If it can be assumed that the coarsest clasts were the largest that the maximum flood could move, then quantitative relations between channel geometric properties, hydraulic parameters and maximum sediment size allow estimation of the magnitude of the extreme flood. This paper will discuss paleohydrologic techniques suitable for the long time periods necessary for performance analyses of hazardous waste sites.

Quantitative paleohydrology is analyzed by the same techniques used for indirect analysis of the hydrology of existing streams. A method useful for determining the flow parameters of a recent flood in an ungaged stream is also applicable to the study of flows that occurred tens of thousands of years ago. The only difference is that the features used in the determination of flow parameters tend to become obliterated over time by slope development processes and continued fluvial erosion and deposition.

Methods useful for paleohydrology are based primarily on the hydraulics of flows and sediment transport mechanics of open alluvial channels. Additional techniques, based on observational data relating channel geometric parameters instead of sedimentological ones, have been discussed in Foley *et al.* (1982). All techniques are subject to uncertainty, the degree varying with the type of field data available and the complexity of the channel being studied.

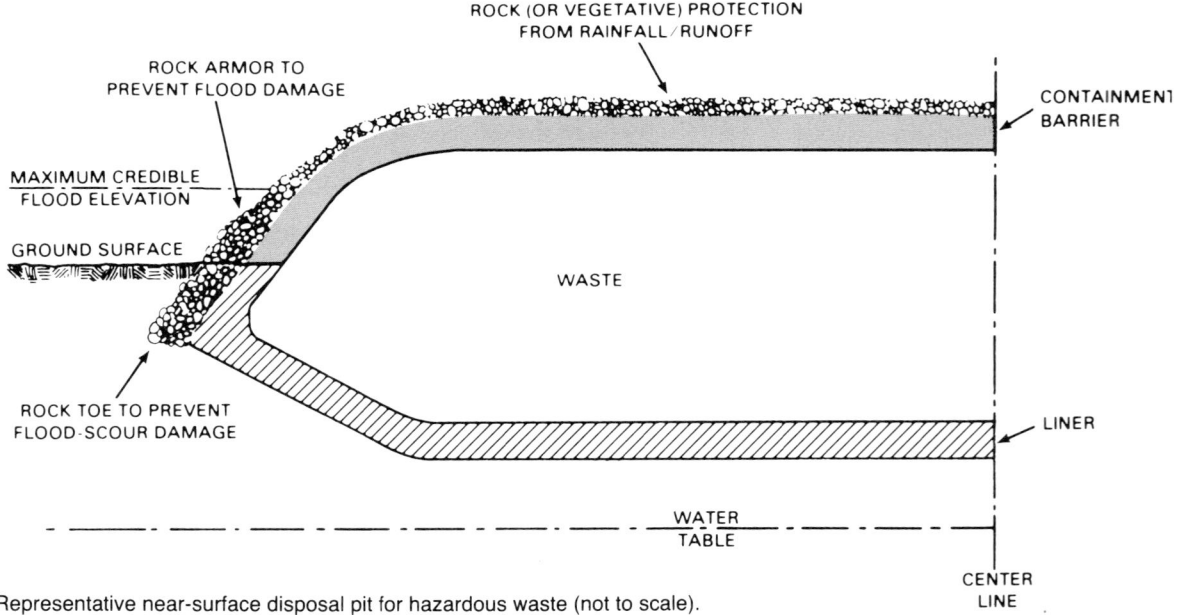

Fig. 1. Representative near-surface disposal pit for hazardous waste (not to scale).

The following sections present the physical basis of the hydraulics of flow and sediment transport in alluvial channels as a framework for discussion of specific paleohydrologic methods. Uncertainties can be mitigated by the use of several, complementary approaches and by the use of combined techniques, although sufficient data are seldom available to use more than one approach.

Modeling of Flow Systems

The hydraulic study of self-forming alluvial channels is extremely difficult in detail due to the complexity of interactions between flow, bed and banks. Further, each channel is a historical record of past erosional and depositional events, and cannot be treated in detail by a generic physical model. Foley (1980a) discussed the analysis of such channels in terms of three classes of models, which are relevant to the entire field of paleohydrology:

"a deterministic class in which known physical principles and initial and boundary conditions allow accurate description of system behavior; a stochastic class, in which systems are assumed to have inherent randomness and cannot be described by deterministic methods; and a parametric class, in which empirical relations among system variables may give some ability to predict system behavior."

Purely deterministic models of systems such as alluvial channels require overwhelming quantities of extremely detailed data, and have yet to be formulated. Parametric models, based on observations of field and laboratory channels, are the basis of 'hydraulic geometry' methods. These also require considerable data to generate a model for a specific channel; however, approximate applications are possible if the channel being analyzed is similar to the one observed. Although uncertainties increase dramatically with increasing difference between properties of the observed and modelled channels, such applications are relatively common in cases for which no other approach is available.

Foley (1980a) suggested that:

"a middle ground may exist between deterministic and parametric modeling, in which data-input requirements are minimized but output accuracy is compromised by use of empirical parameters whose details are unknown, but whose variability about some 'mean' may be estimated".

Foley further suggested that such 'quasi-deterministic' models may have some predictive capabilities outside the envelope of observed data that are lacking in parametric models.

The basis for hydraulic and sedimentological studies of alluvial channels of use in paleohydrology is a collection of quasi-deterministic flow and sediment-transport equations. These are based on empirical parameters that, in some cases, represent the collective effects of several physical processes. The equations of general interest are those relating channel geometric parameters to flow resistance and discharge, since these establish a useful relation between the variables of interest and those which can be quantified in the field. Other empirical and quasi-deterministic relations between channel sediment properties and flow or geometric parameters will be discussed in more detail below.

Shear Stress

An open-channel flow that does not vary with time (steady) or position along the channel (uniform) is such that the frictional resistance exerted by the channel perimeter is balanced by the gravitational acceleration of the flow. It can be shown that this equilibrium requires that:

$$\tau_0 = \rho g r S, \quad (1)$$

where τ_0 is the average shear stress on the bed and banks, ρ is water density, r is the flow's hydraulic radius, and S is channel slope. Henderson (1966) has shown that, for non-uniform flow, S, must be replaced by the friction slope:

$$S_f = \frac{d}{dx}\left(h - \frac{V^2}{2g}\right) \quad (2)$$

where x is the along-channel direction and V is the mean flow velocity in a cross-section, defined in relation to the flow volumetric discharge, Q, and cross-sectional area, A, as:

$$V = \frac{Q}{A} \quad (3)$$

The derivation of the equation for shear stress assumes that no non-frictional energy losses occur in the flow. Flow separations around obstacles or at sudden expansions in the channel perimeter, free overfalls, hydraulic jumps, and other energy losses will invalidate the above shear-stress relation, and must be avoided in reaches to be included in paleohydraulic analyses. Some special types of nonfrictional energy losses related to expanding channel reaches can be approximated, as discussed by Dalrymple and Benson (1967).

Discharge

The volumetric discharge, Q, or more commonly the equivalent mean cross-sectional velocity, V, can be expressed in terms of hydraulic and channel geometric parameters in the form of several semi-empirical equations. The two most widely used are the Manning equation and the Chezy equation. The Manning equation is:

$$V = \frac{1}{n} r^{2/3} S^{1/2} \quad (4)$$

where n is a dimensional, empirical frictional coefficient and all variables must be expressed in MKS units. Tables of empirical values for n may be found in texts on open channel flow (e.g., Henderson, 1966), and vary from 0.025 to 0.150 for natural channels. The USGS (Barnes, 1967) has published a well-illustrated report showing a variety of channels (different shapes, slopes, and sediment sizes) and their values of n determined by calculation from gaged velocities.

The Chezy equation
$$V = C(rS)^{1/2} \quad (5)$$
is also empirical, but is often presented with $C = (8g)^{1/2}$ because that form,
$$V = \left(\frac{8grS}{f}\right)^{1/2} \quad (6)$$
is the open-channel equivalent of the well-known Darcy-Weisbach pipe-flow equation. The resistance coefficient, f, known as the Darcy-Weisbach friction factor, is related to the roughness of the channel bed and banks and to flow parameters (Vanoni, 1975).

Values for n (Limerinos, 1970) and f (Brownlie, 1981) can be determined from measured bed roughness in natural channels. However, in applications to alluvial channels, the bed roughness measured is often the median bed-sediment size, which may not be appropriate for determining n or f (Vanoni, 1975). Further, the bed roughness even in a straight, uniform channel is a function both of sediment roughness and bedform roughness. The ratio of bed roughness, f, to flat-bed roughness, f', can be much greater than 2 for bedforms such as ripples and dunes. Specific techniques discussed below address these relations in more detail.

QUANTIFYING INDIVIDUAL PARAMETERS

Paleohydrologic studies are usually conducted for the purpose of estimating the maximum discharge (or maximum flow depth) that has occurred in a particular channel. For application to estimates of the maximum credible flood for waste disposal facilities, the identification of the paleohydrologic estimate as the maximum flood, as opposed to a morphologically more-dominant flood of lesser magnitude, must be demonstrated. However, for most applications interested in transport conditions of the maximum possible gravel size, such a demonstration is straightforward; the maximum gravel size is moved by the maximum flood and cannot be disturbed significantly by lesser flows, regardless of the relative importance of lesser and maximum flows on overall channel morphology. It is therefore recommended that paleohydrological reconstructions using channel parameters be based on maximum available clast sizes, since this avoids a possible under-estimate of the highest credible flood in situations where morphologically-dominant flows involve lesser discharges of greater frequency.

Examination of the above equations shows that paleohydrologic methods for determining the flow depth, width, cross-sectional shape, friction slope, and resistance coefficient stem from combined field and analytical techniques. Preservation of features that may be used to determine the values of these parameters is often serendipitous, and paleohydrologic investigations require a flexible, ad hoc approach for specific field conditions. An additional goal of paleohydrologic studies is to establish the ages of prehistoric maximum flows. In turn, this permits determination of the statistical distribution of expected discharges in the current climatic regime, or the expected change in peak discharges which may accompany the onset of a pluvial or glacial climate.

WATER-SURFACE ELEVATION

Depth of flow in an existing channel or paleochannel is the most important parameter, in addition to the resistance coefficient, for calculating discharge. Where the channel bed can be shown to have been stable during the flow, techniques for estimating paleodepth depend on the preservation of indicators of water-surface elevation or of sediments deposited in known relation to the water surface. Less direct estimates relate the competence of the flow to move coarse sediment to depth and these can be used independently, or jointly with high-water marks, as a test of the bed stability assumption.

High-Water Marks

Benson and Dalrymple (1967) have discussed field techniques for indirect determinations of unmeasured stream discharge. An important aspect is accurate measurement of high-water marks, and techniques for their identification carry over into paleohydrologic studies. Floating debris deposited along banks, 'trim lines' dividing water-scoured banks below and unaffected banks above, and bent grass or other vegetation are the principal indicators of water-surface elevation for studies of modern, ungaged flows. However, indicators of water-surface elevation need careful assessment to ensure that they are representative of the local reach and not the result of local perturbing influences. For example, marks on the upstream and downstream sides of obstacles will tend to be too high and low, respectively. Similarly, marks on the outsides of bends will be too high, those on the insides too low. For these reasons, Benson and Dalrymple (1967) cautioned that high-water marks be measured preferentially in relatively straight channels on stable surfaces parallel to the flow.

These techniques are not generally useful in paleohydrologic studies because high-water marks are transitory features in all climates, and especially so in humid regions. However, Stewart and Bodhaine (1961) were able to reconstruct stages for the 1815 and 1856 floods on the Skagit River, Washington, from high-water marks on trees and canyon walls and from suspended sediment lodged in tree bark and in the crevices in canyon sides. In paleohydrological work, paucity of data may require that less desirable locations for high-water marks be used, in which case Benson and Dalrymple's (1967) guidelines provide a basis for evaluating those data.

Ice-Rafted Debris

Anomalously large clasts deposited from free-floating or grounded ice can provide excellent, long-lived markers. The highest such clasts beyond glacial limits are usually accurate high-water marks if Benson and Dalrymple's (1967) evaluation criteria are met, and if the water-surface profile is such that ice jams can be discounted. Ice-rafted erratics provide a major source of high-water marks in studies of the Channeled Scablands of the Columbia Plateau (e.g., Fryxell and Cook, 1964; Baker, 1973). However, Baker (1973) pointed out that they were usually concentrated in areas where floodwaters were ponded, and are lacking in more uniform flood reaches where open-channel flow theory applies.

Eroded Channel Margins

The highest indication of erosion by a flood flow, or the highest base of a flood-cut scarp, is a minimum elevation for the water surface. For example, Bretz (1928) attempted to estimate flow depths of floods in the Channeled Scablands on the basis of scarps eroded in loess or on the highest scabland eroded in basalt. For paleohydrologic purposes, this analog of the trimlines used in estimating water-surface elevations for modern floods is of limited accuracy. In the case of Bretz's measurements, for example, it is known that the actual water surface was at least 60 m higher than the bases of the loess scarps (Baker, 1973).

Minor Divide Crossings

For the catastrophic floods of the Columbia Plateau, Baker (1973) found that the most consistent measurements of water-surface elevation were provided by the elevations of the smallest channels eroded where flood waters topped the usual drainage divides. The water surface was bracketed between elevations of the floors of the minor channels and those of the lowest nearby divides unmodified by flood flows.

Highest Flood Gravel

Baker (1973) used the tops of flood-deposited gravel bars as indicators of minimum water-surface elevation. In the case of floods on the Columbia Plateau, these gravel bars were covered by flows of unknown depth and give water-surface elevations no more reliable than those indicated by eroded channel margins.

However, investigations of less catastrophic floods have suggested that the highest flood gravels may be reliable indicators of water-surface elevation. For example, Stewart and LaMarche (1967) reported that the crests of natural, boulder-mantled levees formed during the flood of December 1964 on Coffee Creek, California, were up to 3 m above the channel bed and only 0.5 to 1 m below the flood high-water marks. Vessell (1977) found that the tops of some boulders up to 2.3 m in diameter deposited on levee or 'boulder berm' crests in floods on the Rio Guacalate, Guatemala, were exposed above flood high-water marks. Vessell also found cobbles up to 20 cm in diameter deposited on a bridge deck approximately 4 m above the bed of the Rio Guacalate and very near the water-surface elevation of the flood reconstructed from conventional type of high-water marks discussed previously.

The boulder berms on the Rio Guacalate were deposited in a diverging reach downstream of a flow constriction (Vessell et al., 1977). This suggests (Foley et al., 1978) that non-frictional energy dissipation in the form of macroturbulent kolks (Matthes, 1947) may have been the mechanism by which large boulders were deposited on the tops of the berms. G. Parker suggested (written communication, 1978), based on laboratory experiments, that the boulder berms could have been formed by gravel-bar migration. However, Vessell (1977) showed that the boulders may have been transported in the deeper channels (necessary for entrainment by kolks) but not by the flows across the berm crests. Parker's observations may apply to formation of boulder berms in more uniform reaches, and may explain the formation of some of the bars observed by Stewart and La Marche (1967), Fahnestock (1963), and Scott and Gravlee (1968). Foley et al. (1978) and Baker (1973) suggested that macroturbulent transport of boulders is likely in diverging stream reaches where non-frictional energy dissipation is important. In general, boulder berms provide a more reliable and more permanent record of maximum water-surface elevation than elevations of gravel bars in more uniform stream reaches. However, this method assumes that the flow transported a sufficient quantity of gravel to supply potential depositional sites and that its competence level was satisfied by the available sizes of gravel.

Slack-Water Deposits

Recent investigations (Patton et al., 1979; Kochel and Baker, 1982) have used slack-water deposits — relatively fine-grained sediments that accumulate in areas of reduced flow velocity during floods — to estimate water-surface elevations of flood flows. Kochel and Baker (1982) found that stream floods in west Texas often follow flood peaks on their tributaries, resulting in late-stage surges up the tributary valleys. These surges deposit the finer-grained detritus carried primarily as suspended load, and often cap such deposits with a layer rich in floating organic debris. If the slack-water environment remains one of periodic aggradation, each layer will record the passage of a trunk stream flood that overtopped the elevation of previous slack-water deposits. The organic-rich horizons allow separation of sedimentation units and provide material suitable for radiocarbon dating.

As in the case of gravel bars, the slack-water deposits are covered by an unknown depth of water during deposition. However, by comparing water-surface estimates based on slack-water deposits with gaged data for floods on the lower Pecos River in 1954 and 1974, Kochel and Baker (1982) found that flow depth estimates were only about 10% too low. That is, slack-water sediment heights were usually 2 to 3 m below the water surface for peak flood stages of approximately 30 m. This relation between slack-water sediment elevation and peak flow elevation is probably a function of flood duration, suspended sediment size, and magnitude of the flood and would not apply to channels elsewhere without independent corroboration.

Evaluation of Indicators

The preceding discussion of water-surface indicators has concentrated on their reliability and utility for paleohydrologic analyses. These range from very good for minor divide crossings and sediment lodged in canyon-wall crevices, to qualitative for the bases of flood-cut scarps.

In order that the evidence of a water-surface elevation for a paleoflood relate to the flow depth, it must be assumed that: 1) the water-surface indicator is reliable as discussed above; 2) the present channel is representative of the channel during the flood, with no significant aggradation or degradation during or after; and 3) the water-surface indicator represents flows in the present channel. Significant scour and fill along the channel bed during flood-wave passage has been recognized by, for example, Leopold, Wolman and Miller (1964). However, Foley (1978) has shown that such processes have least effect in the relatively straight, uniform reaches which are preferably used in paleohydrologic analyses. The relationship of paleoflow high-water marks to the present channel bed can be more difficult to determine. For example, a water-surface indicator above stream terraces may simply indicate a flood that overtopped the terraces, or may suggest that significant channel incision has occurred since the major flood. In such a case (*e.g.*, Foley, 1980b), careful field mapping, interpretation, and historical reconstruction must be performed before the flow geometry can be determined from remaining evidence. This process is almost mandatory for work with Quaternary paleohydrology because enough time has elapsed to render the above assumptions as seldom valid.

DIRECT DEPTH DETERMINATION

Several techniques are available to estimate depth of flow directly from features of channel deposits, although certain of the indirect techniques may have greater reliability. They can however serve as an independent check on water-surface determinations where the relation between the paleochannel and present channel is unclear. Further, the depth-estimation techniques are useful in cases in which water-surface indicators are missing or where the bounds of the former channel are uncertain.

Competence

The maximum size of sediment particle that a flow is capable of transporting can be expressed as an empirical function of the shear stress (*e.g.*, Vanoni, 1975; Blatt *et al.*, 1980). Typically, such relations provide a critical shear stress for initiation of motion on a flat, alluvial bed composed of like-sized particles. Often, relations are further simplified by assuming that the particles have the same density as quartz and are immersed in water of 20°C. Baker (1973), Baker and Ritter (1975), and Foley (1977) have suggested that large clasts exposed on a finer bed are moved at lower shear stresses than predicted by common shear-stress criteria. Baker and Ritter (1975) derived an empirical relation for particle size versus mean shear stress for coarse bed-load material transported in rivers:

$$d_{s_{max}} = 65\tau_c^{0.54}, \quad (7)$$

where $d_{s_{max}}$ is the largest intermediate diameter in mm, the critical "stress" τ_c is in kg.m^{-1} (original units), and the correlation coefficient is 0.92. Williams (1983) has recalculated the regression equation using Baker and Ritter's data to estimate depth as a function of sediment size (same units as Equation 7) where the friction slope of the flow is known or can be estimated:

$$\tau_c = 0.017 d_{s_{max}}^{1.31}, \quad (7a)$$

with a correlation coefficient of 0.49.

Use of Equation (7a) to determine paleodepth assumes that the coarsest clasts were: 1) transported as bedload, 2) the largest that the flow could transport, 3) transported across a flat bed of generally smaller particles, and 4) not moved by subsequent flow events. It further assumes that the friction slope of the paleoflow was the same as the present bed slope. Foley (1980b) used a similar relation to estimate flow depths in an abandoned glacial outwash channel. To mitigate spurious determinations caused by ice-rafted debris, the $d_{s_{max}}$ value used for each estimate was the average intermediate diameter of the ten largest boulders at locations beyond the former glacier margins. Since the sediment source was glacial debris, the flow was certainly not under-supplied with respect to quantity or calibre of its bedload. In studies of the Channeled Scabland (Baker, 1973), this assumption cannot be made even for boulders of several meters diameter.

Competence estimates in the vicinity of morainal debris from Pleistocene valley glaciers require more care in field measurements (Leopold *et al.*, 1964; Baker, 1974; Foley, 1980b). Upstream of terminal moraines, maximum boulder sizes in stream channels cannot be assumed to have been transported by fluvial processes. However, a plot of water-surface estimates made along a stream reach from all boulders larger than a given size may suggest a dividing line between fluvial bedload and residual glacial erratics. Figure 2 shows such a plot for all boulders larger than 1 m in maximum diameter in a 40 m reach of the Middle Fork of the Popo Agie River, Wyoming (M.G. Foley, 1979, unpublished data). Competence estimates were made using Baker and Ritter's (1975) equation, and the dividing line shown in Figure 2 could be interpreted as a maximum water-surface elevation.

Antidune Bedforms

Theoretically more reliable competence estimates can be made on the basis of deposits from antidune-regime flow. Sub-regularly spaced clusters of coarse, sometimes imbricate gravels in the deposits of steep streams are often relict antidune bedforms called transverse ribs (Koster, 1978; Rust and Gostin, 1981) or bedload dropout armor (Foley, 1977). Foley (1977) has shown in laboratory experiments that such deposits form from the largest bedload particles that a flow can transport by 'normal' (*i.e.*, not macroturbulent) fluvial processes over a flat bed by deposition on the upstream sides of growing antidunes. Allen (1983) has suggested a mechanism unrelated to antidunes for the formulation of features somewhat similar to bedload-dropout armor, so some care must be taken with field relations. Bedload-dropout armor related to antidunes allows an approximate determination of antidune wavelength.

The amplitude and wavelength of antidunes are closely related to flow velocity and depth (Kennedy, 1961; Vanoni, 1975). Hand (1969) has suggested that antidunes can be represented as trochoidal waves, and has related their wavelength, amplitude and maximum face slope to the mean water depth and amplitude of the accompanying water waves. This technique is promising, although antidune cross-laminae are not as common as those of ripples and dunes, and tend to be less regular or complete (Middleton, 1965).

Evaluation of Indicators

Except for antidune bedforms, all of the above techniques depend on some variant of a critical shear-stress criterion for the maximum size of transported bed material. Confidence is greatest when competence estimates are applied to the movement of large particles across a finer-grained pavement, using Baker and Ritter's (1975) criterion. Even then (Baker and Ritter, 1975, Fig. 1), shear stress (hence depth if slope is known) may be in error by as much as a factor of 7 for 10 cm cobbles, and by a factor of 2 for 1 m boulders. Baker and Ritter suggested that the effect of hydrodynamic lift was not considered explicitly in empirical determinations of critical shear stress. Thus, large particles (relative to those around them) in relatively shallow flows will have lift forces proportionately larger than in the experimental determinations, whereas those submerged in very deep flows with small velocity gradients will have proportionately smaller lift forces. Foley (1980b) found that water-surface elevations estimated from Baker and Ritter's (1975) relation at two different gravel bars on the same cross-section differed by less than 4 cm for water depths of 1 and 2 m. This precision of 2 to 4% may have been accidental, considering the relative crudity of the technique, and probably does not lend credibility to the accuracy of the method in general.

Helley (1969) derived a quasi-deterministic initiation-of-motion criterion for boulders on beds of smaller particles, based on hydrodynamic lift and drag, and tested it against field measurements. He considered that the boulders were tri-axial ellipsoids with intermediate axes inclined upstream between 0 and 25°, short axes upward, and long axes

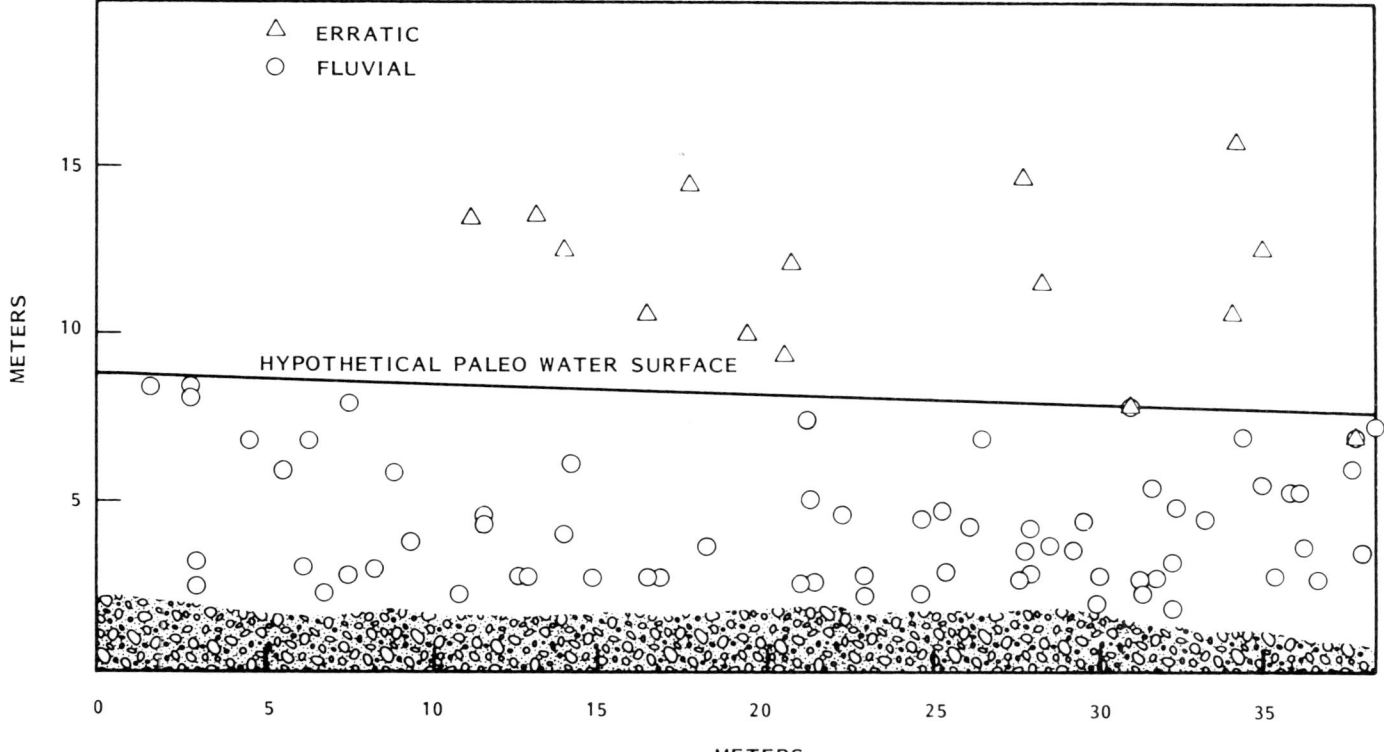

Fig. 2. Estimation of paleowater surface in the Middle Fork of the Popo Agie River, Wyoming, by calculating paleodepths competent to move all boulders larger than 1 m diameter and by discriminating scattered points (? glacial erratics) from grouped ones (? fluvial).

horizontal and normal to the flow. Critical bottom velocities (*i.e.*, at a distance of 0.63 times the short axis above the bed) calculated from the theoretical analysis and measured for 34 boulders in Blue Creek, California, were within 10% for 47.1% of the measurements, 11 to 20% for 23.5% of the measurements, 21 to 30% for 20.6% of the measurements, and greater than 30% for 8.8% of measurements. Although the details are beyond the scope of this report, Helley's ignoring the geometric effects of upstream inclination of the intermediate axis results in a 24% error in calculated drag at an inclination angle of 25°. Further, it is not clear that the same velocity is appropriate for calculating lift and drag. Finally, although Helley's analytical technique holds promise for initiation-of-motion studies, the usefulness for paleohydrology is not as direct as criteria based on a critical shear stress. As discussed below, estimation of the velocity profiles of sediment-laden flows is subject to great uncertainty. Thus, conversion of a relatively accurate critical bottom velocity based on Helley's analysis to a paleodepth or mean flow velocity may result in a potential order-of-magnitude error overall.

SLOPE

The techniques discussed so far have assumed that the channel of interest is still in existence, and that water-surface or depth indicators (or both) allow reconstruction of the cross-section of the paleoflow. Two or more reconstructed cross-sections permit both the channel and water-surface slopes to be determined. The slope-area method, outlined below, is then used to calculate friction slope and paleodischarge. However, if paleohydrologic studies are based on remnants of channel sediments, the determination of paleoslope is crucial to hydraulic calculations.

The critical shear-stress criterion for movement of large clasts yields only the product dS when S is not known. Where transverse ribs are present, the wavelength of the antidunes that produced them is also preserved (Foley, 1977; Koster, 1978; Rust and Gostin, 1981). Kennedy (1961) has shown that the wavelength, L, of antidunes is related to the mean flow velocity by the relation:

$$V = \left(\frac{gL}{2\pi}\right)^{1/2} \quad (8)$$

Foley (1977) derived the relation:

$$d = 2.5 \, k_s \exp\left[\left(\frac{V}{u_*} - a_r\right)\right] \kappa, \quad (9)$$

where k_s is the bed roughness length, κ is von Karman's constant, a_r is a roughness coefficient (Bakhmeteff, 1936), and

$$u_* = \left(\frac{\tau_0}{\rho}\right)^{1/2}. \quad (10)$$

These three equations may be solved simultaneously for d if the stream width-to-depth ratio is large. However, the 'constant' κ may vary from 0.2 to 0.4, depending on the concentration of suspended sediment (Vanoni, 1975). A 10% error in κ results in a 100% error in d as determined by this method, suggesting that it should be used for order-of-magnitude calculations only (Foley, 1977). However, Coleman (1981) has suggested, based on his experiments, that earlier findings were in error, and that the value of κ is independent of the amount of suspended sediment in open channel flow. Evaluation of Coleman's findings may prove that the technique suggested by Foley (1977) is subject to less inherent error caused by the variable κ than previously suspected.

Other techniques for determining paleoslope require the estimation of former valley cross-sections from the geometry of tributary streams, or the preserved remnants of peripheral parts of the former valley. Matthes (1930), in his classical study of the Yosemite Valley, reconstructed multiple Pleistocene positions of the Merced River by projecting the profiles of tributary streams in hanging valleys across the Yosemite Valley. Multiple reconstructions were possible because several graded tributary profiles were preserved between knickpoints in the bedrock hanging valleys. Foley (1980b) used valley remnants preserved under glacial till (undisturbed river terraces preserved nearby suggested little subglacial erosion) to reconstruct a valley profile at the time of diversion of the Dearborn River, Montana, and a later profile projected from valley remnants buried by a younger till. These techniques clearly suffer from the uncertainty associated with the projection of curvilinear features over distances of a kilometer or more. However, several determinations made along a reach of tens of kilometers should mitigate the effects of local errors.

WIDTH

The cross-section of a paleoflow in an existing channel can be determined by surveying the channel and estimating the flow depth somewhere along that cross-section, as described above. Uncertainty is introduced by post-flow changes in the banks, such as slumping and landsliding or deposition of nearbank colluvial material. Further uncertainty can result from super-elevation of flows in curved reaches. However, these uncertainties can be avoided by careful selection of cross-sections in relatively straight reaches away from areas of post-flow erosional or depositional activity.

When the paleochannel is missing, or as in the case of the Missoula floods was ill-defined when active, a quantitative estimate of paleowidth is difficult and subject to large uncertainties. Schumm (1968) estimated the widths (and bankfull depths) of buried prior stream channels of the Murrumbidgee River in Australia from detailed cross-channel stratigraphy.

Cotter (1971) estimated the width and depth of channels in which sands of the Late Cretaceous Ferron Sandstone were deposited using stratigraphic and textural data. Cotter assumed (based on Moody-Stuart, 1966) that bankfull depth in the meander bends was equal to the thickness of the intervals of epsilon cross-stratified point bar sandstones. Further (also based on work reviewed by Moody-Stuart, 1966), he estimated paleochannel width as 1.5 times the width of the point bar sandstone. Cotter then 'adjusted' paleochannel depth downward somewhat to describe the depth of a straight reach, and derived a width-depth ratio of 12 for the straight paleochannel. This value compared with a width-depth ratio of 8 determined by measuring the silt-clay content of the 'bed' and 'banks' from thin-section analysis and using Schumm's (1960) relation for alluvial channel geometry as a function of sediment type. The 50% discrepancy between Cotter's (1971) width-depth ratios calculated by independent methods is probably a good indication of the uncertainty of paleowidth determinations. Cotter's direct estimate of width may have been in error by tens of percent. His direct estimate of depth would have had a similar uncertainty, although the use, when possible, of the more accurate techniques described in a previous section could have reduced that uncertainty. In addition, the estimates of silt-clay contents of the bed and banks are prone to error due to diagenetic modifications of texture. Schumm's (1960) hydraulic geometry relation also depends on the silt-clay content, but Foley (1975) found it to be generally useful: however it would be highly unreliable in predicting individual events, especially where depth is also estimated. Cotter's (1971) work is valuable in that it quantifies the uncertainty associated with a specific type of paleohydrologic analysis, and also demonstrates the necessity for multiple, independent approaches to paleohydrology.

Combined Techniques

Individual methods for determining the 'paleovalues' of hydraulic variables are all subject to uncertainties of varying magnitude, as has been discussed. In addition, application of the flow equations introduces some uncertainty because natural alluvial systems are seldom uniform. Nonfrictional energy losses induced by changes of channel shape, flow separations, and free overfalls result in smaller discharge for the same geometric flow parameters. These uncertainties may be reduced by using a multiple, independent approach to determine values of hydraulic parameters and by making paleoflow estimates at several cross-sections.

SLOPE AREA METHOD

Dalrymple and Benson (1967) have described this technique in excellent detail, and the reader is referred to their treatment for actual application. In brief, several cross-sections are combined in this approach and treated with the Manning equation. High-water marks and *in situ* determinations of roughness coefficient are necessary input data. Each cross-section may be divided into several subsections with different roughness coefficients, and flow may be non-uniform (although converging flow is preferable to diverging flow). Given good high-water marks, the major uncertainty in this method is the estimate of roughness coefficient. Riggs (1976) presented a simplified slope-area method that requires input of only friction slope and high-water marks. Overall accuracy of these two methods is difficult to estimate for an ungaged site, although Riggs' simplified approach may be subject to slightly greater uncertainty. However, Riggs (1976, p. 285) noted:

"Opinions range from claims of high accuracy to the comment of a prominent (unnamed) hydraulic engineer who was quoted by Henry Beckman (written communication, 1924) as saying that, whenever results obtained by the slope-area method came nearer than 25 percent to the correct result, it was due either to accident or to a second choice of factors to use in the formula after the first choice had gone amiss. Fifty years later, wide differences of opinion as to the accuracy of the method still exist."

Despite differing opinions concerning its uncertainty, where data of sufficient quantity and accuracy are available the slope-area method is probably the most sophisticated technique of real use in most paleohydrologic studies. It is the basis for Baker's (1973) paleohydrologic studies of the Channeled Scablands and Vessell's (1977) study of the Rio Guacalate. In addition, it is often used in studies of maximum flood discharges in ungaged streams, or in rivers with gages that were destroyed by floods.

OTHER METHODS

Modern river studies employ more sophisticated techniques than those described above. Examples are the step-backwater analysis (Shearman, 1976) and other computerized models such as HEC-2 (United States Army, 1973). However, most paleohydrologic studies are data-limited, and the less sophisticated techniques described above are adequate.

References

Allen, J.R.L. 1983. A simplified cascade model for transverse stone-ribs in gravelly streams. Proceedings of the Royal Society of London, Series A, v. 385, p. 253-266.

Baker, V.R. 1973. Paleohydrology and sedimentology of Lake Missoula flooding in eastern Washington. Geological Society of America, Special Paper 144, 79p.

Baker, V.R. 1974. Paleohydraulic interpretation of Quaternary alluvium near Golden, Colorado. Quaternary Research, v. 4, p. 94-112.

Baker, V.R. and Ritter, D.F. 1975. Competence of rivers to transport coarse bedload material. Geological Society of America Bulletin, v. 86, p. 975-978.

Bakhmeteff, B.A. 1936. The Mechanics of Turbulent Flow. Princeton, New Jersey: Princeton University Press, 97p.

Barnes, H.H., Jr. 1967. Roughness characteristics of natural channels. United States Geological Survey Water Supply Paper 1849, 213p.

Benson, M.A. and Dalrymple, T. 1967. General field and office procedures for indirect discharge measurements. United States Geological Survey Techniques of Water-Resources Investigations, Book 3, Chapter A1, 30p.

Blatt, H., Middleton, G. and Murray, R. 1980. Origin of Sedimentary Rocks. 2nd Edition. New Jersey: Prentice Hall, 782p.

Bretz, J.H. 1928. The Channelled Scabland of Eastern Washington. Geographical Review, v. 18, p. 446-477.

Brownlie, W.R. 1981. Re-examination of Nikuradse roughness data. American Society of Civil Engineers, Journal of the Hydraulics Division, v. 107(HY1), p. 115-119.

Coleman, N.L. 1981. Velocity profiles with suspended sediment. Journal of Hydraulic Research, v. 19, p. 211-229.

Cotter, E. 1971. Paleoflow characteristics of a Late Cretaceous River in Utah from analysis of sedimentary structures in the Ferron Sandstone. Journal of Sedimentary Petrology, v. 41, p. 129-138.

Dalrymple, T. and Benson, M.A. 1967. Measurement of peak discharge by the slope-area method. United States Geological Survey Techniques of Water-Resources Investigations, Book 3, Chapter A2, 12p.

Fahnestock, R.K. 1963. Morphology and hydrology of a glacial stream — White River, Mount Rainier, Washington. United States Geological Survey Professional Paper 422-A, 70p.

Foley, M.G. 1975. Scour and fill in ephemeral streams. California Institute of Technology, Pasadena, W.M. Keck Laboratory of Hydraulic and Water Resources Report No. KH-R-33, 189p.

Foley, M.G. 1977. Gravel-lens formation in antidune regime flow — a quantitative hydrodynamic indicator. Journal of Sedimentary Petrology, v. 47, p. 738-746.

Foley, M.G. 1978. Scour and fill in steep, sand-bed ephemeral streams. Geological Society of American Bulletin, v. 89, p. 559-570.

Foley, M.G. 1980a. Hydrology. Geotimes. v. 25, p. 30-31.

Foley, M.G. 1980b. Quaternary diversion and incision, Dearborn River, Montana. Geological Survey of America Bulletin, v. 91, part II, p. 2152-2188.

Foley, M.G., Vessell, R.K., Davies, D.K. and Bonis, S.B. 1978. Bedload transport mechanisms during flash floods. Proceedings of Conference on Flash Floods: Hydrometeorological Aspects. American Meterologial Society, Boston, Massachusetts, p. 109-116.

Foley, M.G., Zimmerman, D.A., Doesburg, J.M. and Thorne, P.D. 1982. Review and evaluation of paleohydrologic methodologies. United States Nuclear Regulatory Commission, NUREG/CR-3055. National Technical Information Service, Springfield, Virginia, 87p.

Fryxell, R. and Cook, E.F. 1964. A field guide to the loess deposits and Channeled Scablands of the Palouse Area, eastern Washington. Washington State University, Pullman Anthropology Laboratory, Report of Investigations, No. 27, 32p.

Hand, B.M. 1969. Antidunes as trochoidal waves. Journal of Sedimentary Petrology, v. 39, p. 1302-1309.

Helley, E.J. 1969. Field measurement of the initiation of large bed particle motion in Blue Creek near Klamath, California. United States Geological Survey Professional Paper 562-G, 19p.

Henderson, F.M. 1966. Open Channel Flow. New York: MacMillan, 522p.

Imbrie, J. and Imbrie, J.Z. 1980. Modeling the climatic response to orbital variations. Science, v. 207, p. 943-953.

Kennedy, J.F. 1961. Stationary waves and antidunes in alluvial channels. California Institute of Technology, Pasadena, W.M. Keck Laboratory of Hydraulics and Water Resources Report, No. KH-R-2. 146p.

Kochel, R.C. and Baker, V.R. 1982. Paleoflood hydrology. Science, v. 215, p. 353-361.

Koster, E.H. 1978. Transverse ribs: their characteristics, origin and paleohydraulic significance. *In:* Miall, A.D. (Ed.), Fluvial Sedimentology. Canadian Society of Petroleum Geologists, Memoir 5, p. 161-186.

Leopold, L.B., Wolman, M.G. and Miller, J.P. 1964. Fluvial Processes in Geomorphology. San Francisco: W.H. Freeman and Company, 522p.

Limerinos, J.T. 1970. Determination of the Manning coefficient from measured bed roughness. United States Geological Survey Water Supply Paper 1898-B, 47p.

Matthes, F.E. 1930. Geologic history of the Yosemite Valley. United States Geological Survey Professional Paper 160, 137p.

Matthes, G.H. 1947. Macroturbulence in natural stream flow. American Geophysical Union Transactions, v. 28, p. 255-262.

Middleton, G.V. 1965. Antidune cross-bedding in a large flume. Journal of Sedimentary Petrology, v. 35, p. 922-927.

Moody-Stuart, M. 1966. High- and low-sinuosity stream deposits with examples from the Devonian of Spitsbergen. Journal of Sedimentary Petrology, v. 36, p. 1102-1117.

Patton, P.C., Baker, V.R. and Kochel, R.C. 1979. Slack-water deposits: a geomorphic technique for the interpretation of fluvial paleohydrology. *In:* Rhodes, D.D. (Ed). Adjustments of the Fluvial System. Dubuque, Iowa: Kendall-Hunt Publishing Company, p. 225-253.

Riggs, H.C. 1976. A simplified slope-area method for estimating flood discharges in natural channels. United States Geological Survey Journal of Research, v. 4, p. 235-291.

Rust, B.R. and Gostin, V.A. 1981. Fossil transverse ribs in Holocene alluvial fan deposits, Depot Creek, South Australia. Journal of Sedimentary Petrology, v. 51, p. 441-444.

Schumm, S.A. 1960. The shape of alluvial channels in relation to sediment type. United States Geological Survey Professional Paper 352-B, 30p.

Schumm, S.A. 1968. River adjustments to altered hydrologic regimen-Murrumbidge River and paleochannels, Australia. United States Geological Survey Professional Paper 598, 65p.

Scott, K.M. and Gravlee, G.C. Jr. 1968. Flood surge on the Rubicon River, California — hydrology, hydraulics and boulder transport. United States Geological Survey Professional Paper 422-M, 38p.

Shearman, J.O. 1976. Computer applications for step-backwater and floodway analyses. United States Geological Survey Open File Report 76-499, Washington, D.C., 119p.

Stewart, J.H. and Bodhaine, G.L. 1961. Floods in the Skagit River basin, Washington. United States Geological Survey Water Supply Paper 1527, 66p.

Stewart, J.H. and LaMarche, V.C., Jr. 1967. Erosion and deposition produced by the flood of December, 1964 on Coffee Creek, Trinity County, California. United States Geological Survey Professional Paper 422-K, 22p.

United States Army. 1973. HEC-2: Water surface profile, users manual. United States Army Corps of Engineers, Hydrologic Engineering Center, Davis, California.

Vanoni, V.A. 1975. Sedimentation Engineering. American Society of Civil Engineers Manual 54, 745p.

Vessell, R.K. 1977. Morphology, hydrology, and sedimentology of Rio Guacalate — volcanic highlands — Guatemala. M.A. Thesis, University of Missouri, Columbia, 144p.

Vessell, R.K., Davies, D.K., Foley, M.G. and Bonis, S.B. 1977. Sedimentology and hydrology of flood flows on the active Volcano Fuego, Guatemala. Geological Society of America Abstracts with Program, v. 9, p. 1210-1211.

Williams, G.P. 1983. Improper use of regression equations in Earth Sciences. Geology, v. 11, p. 195-197.

MODERN AND ANCIENT
ALLUVIAL SYSTEMS

GEOMORPHOLOGY AND SEDIMENTOLOGY OF HUMID-TEMPERATE ALLUVIAL FANS, CENTRAL VIRGINIA

R. Craig Kochel[1] and Robert A. Johnson[2]

Abstract

Alluvial fans in mountainous humid-temperate climatic regions are typically formed by mass wasting processes generated by infrequent catastrophic rainfall events. Coarse gravel alluvial fans are the dominant depositional landforms along the slopes of the Blue Ridge Mountains in central Virginia. Humid-temperate alluvial fans such as those in central Virginia are markedly different in process, geomorphology and sedimentology than alluvial fans described in arid, humid-glacial humid-tropical regions and humid-periglacial regions.

Fans along the western slope of the Blue Ridge contain clast-supported sandy gravels presumed to have been deposited by braided streams. These gravels are subrounded, well-imbricated, contain channel fills, and show occasional large-scale planar cross-bedding.

Fans east of the Blue Ridge, in Nelson County, are composed of cobbles and boulders in a fine-grained matrix and result from infrequent debris avalanche and debris flow processes. Radiocarbon dating of fan stratigraphy indicates that periods of inactivity range from 3,000 to 6,000 years. These fans appear to have been active throughout the Quaternary and most recently in 1969 due to the intense rainfall from Hurricane Camille. Recognition of individual events was based on identification of buried soil horizons, changes in matrix mineralogy, bedding character, and clast weathering characteristics. Discrete depositional events were correlated between alluvial fans in the Nelson County area by radiocarbon dating of buried soil horizons formed on fan paleosurfaces.

The Virginia alluvial fans are small, have steep segmented profiles and are less than 100 m thick. Most of the fans observed were less than 20 m thick. Soil horizons are well-developed on fans which have been inactive for thousands of years, while inceptisols occur on fans that have received recent sedimentation.

Résumé

Les cônes de déjection dans les régions montagneuses à climat humide-tempéré sont formés spécifiquement par des phénomènes de mouvement de masse engendrés par les événements catastrophiques peu fréquents d'averses de pluie violentes. Les cônes de déjection composés de gravier grossier constituent les principales formes de dépôt de terrain le long des pentes de la chaîne des Montagnes Bleues du centre de la Virginie. Les cônes de déjection formés en climat humide-tempéré comme ceux du centre de la Virginie se distinguent nettement par leur processus de formation, leur géomorphologie, et leur sédimentologie d'avec les cônes de déjection décrits dans les régions à climat aride, humide-froid, humide-tropical et humide-périglaciaire.

Les cônes le long de la pente occidentale des Montagnes Bleues renferment des graviers sableux à fragments jointifs que l'on croit déposés par des cours d'eau anastomosés. Les graviers sont subarrondis, bien imbriqués, contiennent des sédiments accumulés dans des chenaux, et exhibent sporadiquement une stratification oblique planaire à grande échelle.

Les cônes à l'est des Montagnes Bleues, dans le comté de Nelson, sont composés de galets et de blocs dans une matrice à grain fin et résultent de phénomènes de glissement de terrain et de coulées de débris. Les datations au radiocarbone de couches stratigraphiques des cônes révèlent des périodes inactives il y a entre 3 000 et 6 000 années. Ces cônes apparaissent avoir été actifs durant le Quaternaire et plus récemment en 1969 à cause des averses de pluie intenses provoquées par l'ouragan Camille. La reconnaissance d'événements distincts était fondée sur l'identification d'horizons de sol ensevelis, les changements minéralogiques dans la matrice, les marques caractéristiques du litage, et les altérations typiques des fragments. Des épisodes de formation de dépôts individuels furent corrélés entre divers cônes de déjection du comté de Nelson au moyen des datations au radiocarbone des horizons de sol ensevelis formés sur les paléosurfaces des cônes.

Les cônes de déjection de la Virginie sont de petite dimension, présentent des profils abrupts et segmentés, et ont moins de 100 m d'épais. La majorité des cônes observés avait moins de 20 m d'épais. Les horizons des sols sont bien développés sur les cônes qui sont demeurés inactifs durant des milliers d'années, tandis que les sols jeunes à horizons rapidement formés apparaissent sur les cônes qui furent recouverts récemment de sédiment.

Introduction

A review of previous studies of alluvial fans suggests that significant differences exist in morphology, processes, deposits and facies between alluvial fans formed in various climatic regions. The primary factor responsible for these differences is the nature of the dominant depositional processes contributing sediment to the fan.

Sedimentological and geomorphological models of modern and ancient alluvial fans have largely focused on fans

---

[1]Department of Geology, Southern Illinois University, Carbondale, Illinois 62901, U.S.A.
[2]United States National Park Service, South Florida Research Center, Everglades National Park, Homestead, Florida 33030, U.S.A.
We thank the following students for their assistance in the field and lab during our research: R. Broome, C. Burgess, L. Clark, T. Diggs, H. Dyson, N. Fisher, J. Flanzenbaum, A. Fromer, A. Holt, E. Koehler, L. Lisle, B. Marshall, W. Millar, J. Peatross, C. Peach, J. Piper, M. Potts, R. Savage, R. Schneider, C. Silliman, C. Wayland, and R. Wayland. We also express thanks to Mr. and Mrs. Valentino, Mr. and Mrs. Harvey, Mr. and Mrs. Matthews, and Mr. Thomas for permission to work on their property. Mr. S. Valastro Jr. of the University of Texas Radiocarbon Lab. in Austin graciously performed the radiocarbon analyses.
Copyright © 1984, Canadian Society of Petroleum Geologists

developed in arid regions (Blissenbach, 1954; Bull, 1964; Denny, 1965; Lustig, 1965; Hooke, 1967; Beaty, 1970; Bull, 1972; Nilsen, 1982), fans formed in humid-glacial outwash environments (Reimnitz, 1966; Boothroyd, 1972; Boothroyd and Ashley, 1975; Nummedal and Boothroyd, 1976), fans formed in humid-tropical environments (Gole and Chitale, 1966; Ruxton, 1970; Mukerji, 1976; Wescott and Ethridge, 1980) or fans in humid-periglacial areas (Ryder, 1971; Church and Ryder, 1972; Wasson, 1977). Little attention has been given to alluvial fans developed in humid-temperate regions, such as those in Virginia noted by Hack and Goodlett (1960) and Williams and Guy (1973). Pierson (1980) described debris flows and their deposits on humid-temperate fans in New Zealand. No sedimentological model has been developed for comparing humid-temperate fans with fans formed in other environments.

Arid alluvial fans are characterized by alternating periods of rapid deposition followed by periods of inactivity. Humid-glacial outwash fans are formed by constantly shifting braided streams on an aggrading outwash plain. Humid-tropical fans are formed largely by seasonal deposition by braided streams and/or debris flows. The humid-temperate fans in central Virginia are constructed by infrequent episodes of debris avalanching initiated by large rainstorms. Our data suggest that periods of inactivity may be on the order of several thousands of years.

In this paper we will summarize our preliminary data on the geomorphological and sedimentological characteristics of Quaternary humid-temperate alluvial fans in central Virginia, with special emphasis on alluvial fans activated during Hurricane Camille in 1969. The fieldwork for this report was undertaken during 1981 and 1982. Studies of the fans activated during Camille were limited to those not significantly modified by man after the storm. These data will be used to compare these fans with fans formed in other climatic environments. Our morphostratigraphic studies of alluvial fans and downstream floodplain stratigraphy indicate that slope and fluvial processes can be correlated. Magnitude and frequency of processes on these alluvial fans will be discussed using radiocarbon-dated buried soil horizons.

BLUE RIDGE ALLUVIAL FANS

BEDROCK GEOLOGY

Blue Ridge alluvial fans included in our study area occur along the western flank of the Blue Ridge in Rockingham and Augusta Counties, Virginia (Fig. 1). The Blue Ridge forms the western limb of the Southwestern Mountain-Blue Ridge anticlinorium which is overturned to the west. This linear arch extends from southern Pennsylvania to central Virginia. Precambrian crystalline and low-rank metamorphic rocks are exposed in the axis of the fold. The flanks are composed of a thick sequence of resistant late Precambrian and Lower Cambrian metavolcanics and clastic sedimentary rocks (Bloomer and Werner, 1955; Gathright, 1976). Folded Paleozoic rocks overlie the Blue Ridge complex to the west. Headwater regions of the alluvial fan drainage basins are underlain by Precambrian and Cambrian igneous and metamorphic rocks. Most of these areas are underlain by the Catoctin Formation which is composed of metabasalts and interbedded clastic sediments. Smaller areas of the fan headwaters are underlain by granitic and gneissic rocks of the Virginia Blue Ridge complex and metasediments of the Swift Run Formation. The Blue Ridge alluvial fans have prograded westward over Lower Cambrian clastic sediments of the Chilhowee Group (Weverton, Harpers, Antietam Formations) and Lower Cambrian carbonates of the Shady and Rome Formations (Brent, 1960).

Hillslope soils are composed of angular rock fragments in a matrix of sand, silt, and clay. Soils are highly variable depending on the underlying bedrock. Deep, highly-weathered soils occur on the Catoctin Formation and Virginia Blue Ridge complex rocks. Thinner, poorly-developed soils overlie Cambrian clastics like the Antietam Formation.

FAN MORPHOLOGY

Blue Ridge fans exhibit typical alluvial fan morphology, *i.e.*, fan shape in plan view, downslope-arcuate contours, and convex-upward cross-profiles. In most areas the fans

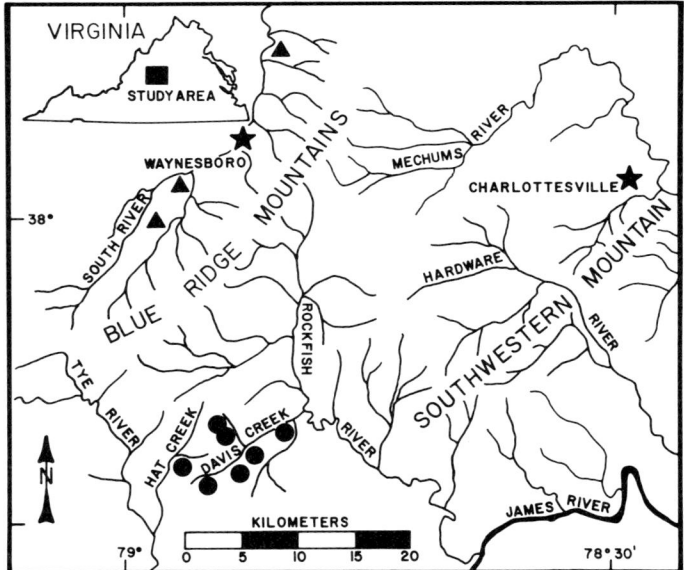

Fig. 1. Index map of the study area showing the Blue Ridge Mountains and foothills in central Virginia. The two areas of alluvial fans discussed are the Blue Ridge fans located along the western flank of the Blue Ridge Mountains and the debris fans activated by Hurricane Camille located east of the Blue Ridge in Nelson County. The triangles indicate specific sites of fans on the western slopes of the Blue Ridge studied during this research. The circles indicate specific sites of fans in Nelson County studied during this research.

have coalesced to form a gently-sloping bajada-like alluvial apron west of the Blue Ridge. The fans are recognized by slight convex-upward undulations on the bajada surface and by the occurrence of small inter-fan lakes and muds at sites of abandoned and active clay, sand, and manganese quarries. The Blue Ridge fans are forested, show no signs of recent depositional activity, and are currently drained by small boulder-floored streams that flow down the fan axes. Drill-hole data near Waynesboro (Fig. 1) indicate that these fans are between 30-110 m thick (Gathright et al., 1977). Axial fan profiles are smooth, moderately-steep and concave-upward.

Figure 2 shows the relationship between fan surface area and drainage area above the fan apex. In general, Blue Ridge fans are smaller than arid fans and show a small range in surface area with changes in drainage area (Bull, 1964; Denny, 1965; Lustig, 1965; Hooke, 1967; Bull, 1972). The Blue Ridge fans exhibit a poorly-developed relationship compared with arid fans studied in California by Bull (1964). Part of the reason for this ill-defined relationship may be due to problems in determining their areal extent because of the coalesence of these features into a broad bajada-like alluvial apron. A better relationship may exist if fan boundaries could be determined from subsurface data.

SEDIMENTOLOGY AND STRATIGRAPHY

The sedimentological characteristics of the Blue Ridge alluvial fans will be summarized briefly because they were based on only a few observations of gravel pits in several fans. Fan sediments are dominantly clast-supported with boulders, cobbles and pebbles in a sandy matrix. Proximal-distal fining from boulder-cobble grade fanhead gravels to distal pebbly, more matrix-rich deposits occurs over a distance of 2 to 3 km. Exposures of gravel in quarry walls showed well-developed cobble and pebble imbrication, large-scale foresets, fining-upward sequences, and sharp lateral and vertical variations in texture (Fig. 3).

The boulder facies generally lacks imbrication and the matrix contains greater amounts of mud than the less coarse distal sediments. Boulder-grade gravels compose less than 10% of the total section in distal fan exposures, and do not generally exceed 20% of the sediments in proximal fan areas. Most of the boulders are subrounded, resistant quartzites, metabasalts, and granitic rocks.

A cobble-sized framework with a sand matrix accounts for over 80% of the Blue Ridge fan deposits. Clasts are commonly imbricated and even the most resistant rock types are subrounded. Most imbricated cobble-sized clasts indicate a unidirectional paleoflow pattern toward the west-southwest (Fig. 3).

Bedding in the Blue Ridge fans commonly occurs as continuous sheets over tens of meters across quarry walls transverse and parallel to the fan axes. Locally, channel fills display crudely-developed upward-fining sequences with occasional large-scale planar tabular cross-bedding.

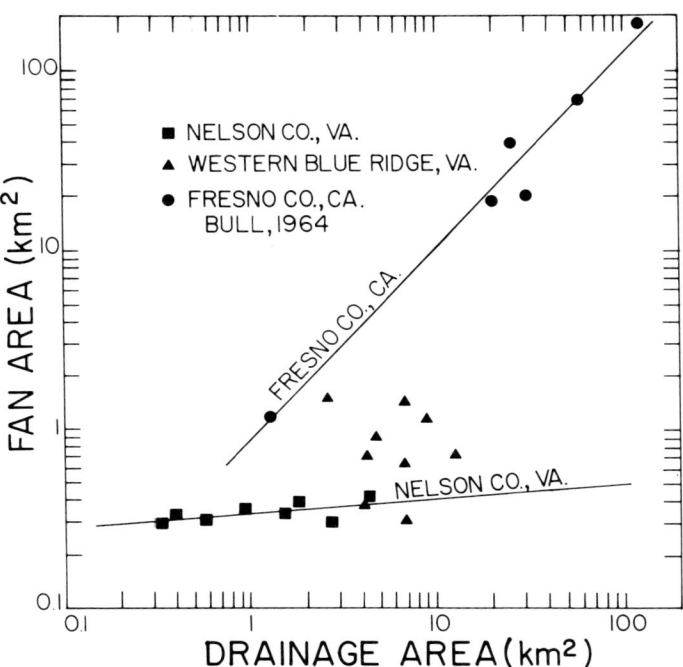

Fig. 2. Relationship between alluvial fan surface area and the area of the drainage basin above the fan apex for alluvial fans in Virginia and in California. No distinct trend line could be fitted to the Blue Ridge fans.

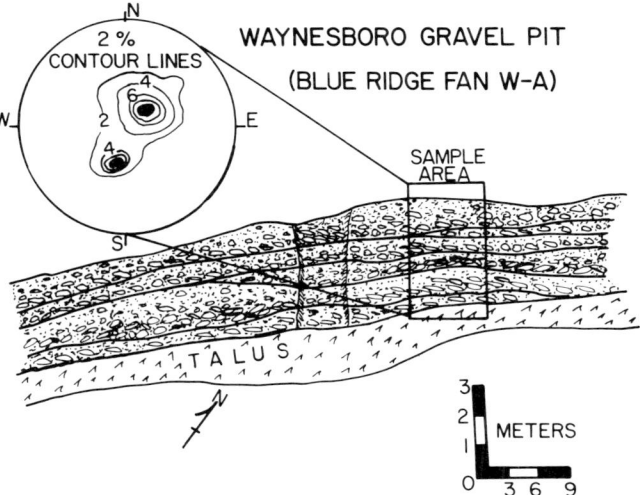

Fig. 3. Sketch of a portion of the northern wall of a small gravel pit in Waynesboro, Virginia. The pit is in the proximal region of a small Blue Ridge fan and shows the characteristic facies. Cobbles and pebbles are found in a sand matrix. Individual beds have erosional bases and fine upward. In most areas a well-developed clast imbrication occurs. The equal area plot shows the dominant orientations of the imbrication and is based on orientation of the plane containing the long (L) and intermediate (I) clast axes. The dominant orientation strikes N 44° W and dips 8° S. A secondary orientation strikes N 60° W and dips 22° SE. These observations are interpreted as indicating a southwesterly paleoflow in a braided stream depositional environment.

DEPOSITIONAL PROCESSES AND FACIES

Interpretations of depositional processes on the alluvial fans along the western flank of the Blue Ridge are based on sedimentological data because these fans have remained inactive in historic times. We interpret the dominant cobble-grade facies as the high-gradient deposits of braided streams during the Pleistocene and perhaps also the early Holocene. Clast-supported gravel facies like those in the Blue Ridge fans characterize modern and ancient braided stream deposits (Miall, 1977; Cant, 1982). Crude fining-upward sequences occur in these fan sediments similar to those described in the Donjek River braided system (Williams and Rust, 1969), but vertical cyclic patterns typical of point bar facies (Bernard et al., 1970) were not observed. Where present, cross-stratification is planar-tabular, similar to those described in modern and ancient braided stream deposits (Cant and Walker, 1976; Miall, 1977). Well-developed upslope imbrication (Fig. 3) and current-normal long axis orientation of clasts are commonly observed in braided river sediments (Boothroyd, 1972; Boothroyd and Ashley, 1975; Harms et al., 1975). Any fines that may have been present in these deposits would have been rapidly winnowed by the competent currents.

The minor mud-supported boulder facies without a preferred clast fabric occurs in thin (less than 2 m) units that appear to be most common in proximal portions of the Blue Ridge fans. We consider that these deposits were deposited by debris flow processes similar to those described in the following section.

NELSON COUNTY ALLUVIAL FANS

BEDROCK GEOLOGY

The Nelson County alluvial fans occur east of the Blue Ridge in central Virginia (Fig. 1). The fans studied in detail are located in the drainages of Davis and Hat Creeks. This area lies in the center of the Blue Ridge anticlinorium and is underlain by the Lovingston Formation. This unit was eroded to near base level before deposition of the overlying Precambrian and Cambrian metasediments (Bloomer and Werner, 1955; Nelson, 1962). The Lovingston Formation is exposed in a broad belt trending northeast-southwest bordered by the Mechum River to the west and Lewis Mountain to the east.

The Lovingston Formation is a coarse-grained crystalline rock with variable composition including granite, quartz monzonite, granodiorite and granite gneiss (Nelson, 1962). Within the Nelson County area the Lovingston is deeply weathered and contains numerous joint sets and fractures.

The hillslopes in the study basins are covered by a thin colluvial mantle of angular rock fragments in a matrix of sand and silt. The soils show very little horizon development and are shallow, ranging from a few centimeters to 2 m in thickness. Observations of the soils bordering debris avalanche scars revealed the presence of extensive blocks of detached weathered bedrock throughout the soil mantle.

FAN MORPHOLOGY

The alluvial fans in Nelson County are generally small compared to those reported in California (Bull, 1964; Denny, 1965; Hooke, 1967), or the Blue Ridge fans reported earlier (Fig. 2). The fans do not display typical fan shapes because most of them are constricted between narrow basin interfluves on steep hillslopes. This confinement is seen in the relationship of fan area to drainage area in Figure 2. There is a wide range of drainage area with respect to fan area but very little variation in fan surface areas with a mean of 0.36 km^2 and standard deviation of 0.06 km^2.

The Nelson County fans are typically elongate and irregular (Fig. 4). Only the fans formed further down-basin in expanded valley reaches are free from lateral confinement so that a true fan form can be developed (Fig. 5). The Nelson County fans are segmented and steep, with average slopes of 40 to 100 m.km^{-1}, similar to debris-flow fans formed in arid regions (Bull, 1964; Denny, 1965). Excavations, augering, and observations of cut banks in

Fig. 4. Oblique aerial photograph of alluvial fan site HA-A and associated debris avalanche in the headwaters of Davis Creek. Photograph was taken two days after the Camille flood in 1969. Note the relationships between the slope, fan and downstream river system. The coarse bouldery debris with a muddy matrix can be seen on the fan surface along with numerous trees denuded from adjacent slopes. Much of the sediment was transported across the fan and into the mainstream along the fan base. (Photograph courtesy of the Virginia Division of Mineral Resources, Charlottesville).

fan-head trenches suggest that most of these fans are thin, ranging from 5 to 20 m in thickness.

Figure 4 shows one of the study sites in the headwaters of Davis Creek. The geomorphic components of this alluvial fan system are well illustrated by this photograph taken two days after Hurricane Camille in August 1969. Two debris avalanches contributed sediment to the fan apex and there are distinct erosional channels down-fan from the point of entry of each debris avalanche (Fig. 4). The locus of deposition and erosion on the fan during any single event is controlled by the distribution and magnitude of the slope failures occurring upslope.

Figure 5 shows a typical alluvial fan in a broad valley reach downstream of the eroded hillslopes. Post-storm aerial photographs taken following the 1969 storm show that 20 to 70% of the fan surfaces were activated during this event. The shifting of the locus of deposition across the fans smoothed the surface topography by depositing sediments into fan surface depressions (Fig. 6, Fan VV-B, 1969 unit). The fan surface depressions appear to have been formed by fluvial erosion during the long periods between episodes of debris avalanching.

SEDIMENTOLOGY AND STRATIGRAPHY

All of the 1969 storm deposits observed on the alluvial fans in Nelson County were coarse gravels with extremely poor sorting of matrix. Maximum clast sizes of these gravels exceed 3 m. Field point counts of 200 clasts were made on several fans to approximate the textural characteristics of the coarse-grained fraction and to look for systematic vertical and lateral changes in mean grain size. Samples of the fine-grained matrix were collected for sieve analysis. All Nelson County fan deposits were found to be internally structureless, except for occasional poorly-developed imbrication of the coarse clasts oriented in the down-fan direction. Within a single stratum, clast sizes ranged from clay to large angular to subangular boulders. Many of the debris avalanche deposits in the study basins were observed to be inversely graded with the largest clasts occurring near the fan surface. Figure 7 illustrates typical inverse

Fig. 5. Oblique aerial photograph of a debris fan located in a distal portion of the basin. This fan was not constrained by valley walls and the typical fan-shape was developed. Note that about 50% of the fan surface was activated in 1969 by the Camille flood and that sediment from the fan progradation diverted the mid-valley stream channel. Debris avalanche scars are visible on distant valley slopes. (Photograph courtesy of the Virginia Division of Mineral Resources, Charlottesville).

grading observed in the 1969 deposits in the Nelson County area. Inverse grading is common in debris flow deposits because of the strength and buoyancy forces caused by high dispersive stresses produced in these high concentration flows (Johnson, 1970). Similar inverse grading has been described by Fisher (1971) and Costa and Jarrett (1981) in debris flow deposits.

Figure 8 is a schematic summary of the general trends of sediment texture and mineralogy based on laboratory analysis of the fine-grained matrix of five alluvial fans in the Davis Creek basin. In all cases the 1969 deposits can be clearly recognized by the dominance of coarse clasts and by the unweathered nature of the sediments. The 1969 deposits contained the largest clasts in all of the stratigraphic sections observed. Each abrupt perturbation in Figure 8a and 8b represents a discrete sedimentation unit. The sharp changes in trend lines are due to the discrete nature of each unit and to the weathering of sediments between depositional episodes. Radiocarbon-dated organic matter and paleosol horizons provided additional criteria for distinguishing the Hurricane Camille episode of activity from earlier ones. Figure 8c shows that the pre-1969 deposits had higher clay contents, greater oxidation, and higher quartz/feldspar ratios. Coarse clasts in the older units were commonly weathered to grus while the 1969 clasts were fresh. The decrease in sand concurrent with the increase in mud with depth reflects post-depositional weathering. The original granitic clasts disintegrate after deposition due to biotite hydration. Once the clasts are broken down the feldspars weather to form clay minerals while the quartz fragments remain relatively stable. The accessory minerals show a similar decrease with depth due to their unstable nature. Each of the depositional units observed was separated by an erosional contact and several of these units had truncated paleosols that could be radiocarbon dated.

DEPOSITIONAL PROCESSES AND FACIES

Field observations were made along the axis of several fans to distinguish proximal to distal variations. In addition, samples of the fine-grained matrix were collected from the 1969 deposits and at least one subsurface unit on two fans (Fig. 6) to examine textural variations in the down-fan direction. However, no significant textural variations were discernible. The 0.25 to 1 km length of these fans may be too short a distance for down-fan sorting to occur, particularly since they were deposited by debris flow processes.

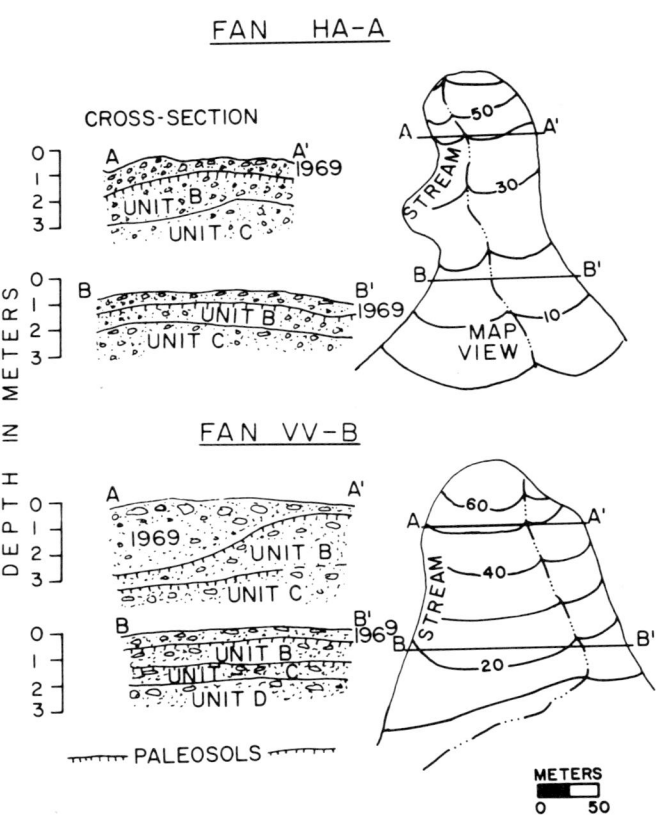

Fig. 6. Plan views and cross-profiles of two debris fans in Davis Creek watershed. Profile data was obtained by digging a series of trenches at several locations along incised stream banks and by augering along the section lines located on the maps. Only the upper few meters of fan sediment were penetrated.

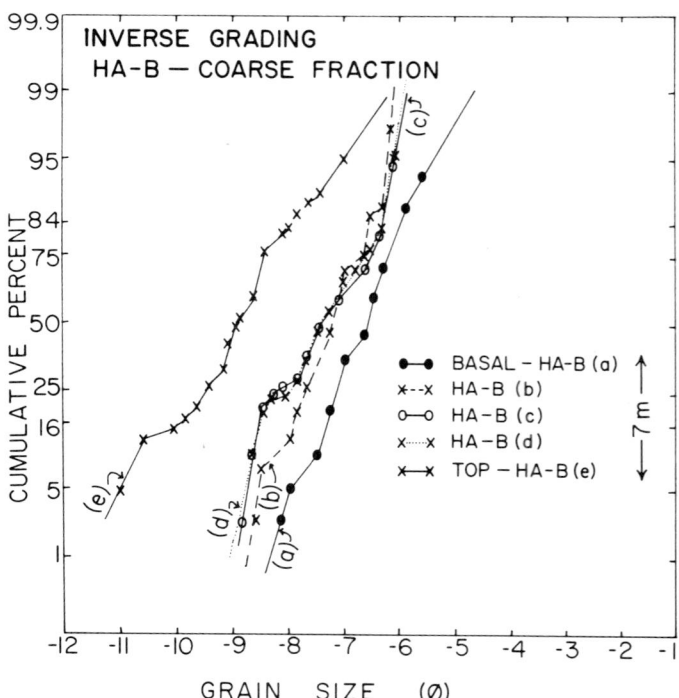

Fig. 7. Cumulative frequency curves illustrating the grain size trends and inverse grading characteristic of the 1969 deposit on alluvial fan HA-B in Nelson County, Virginia. Samples are based on field point counts of 200 clasts in each of five levels in the deposit.

Bed thickness does show a tendency to decrease in the down-fan direction from a peak at the fan apex. Lateral variations in the fan sediments are common (Fig. 6) due to the irregular lobate nature of the debris deposition on the fans. Figure 6 suggests that the surface topography of each fan had a major effect on the locus of deposition during the 1969 storm.

No debris flow levees or terminal lobes were observed on the fans in the photographs taken after the Hurricane Camille storm. Much of the morphologic evidence may have been destroyed by storm runoff following the slope failures or by recontouring by man. Levees were reported on several fans and along the margins of a number of debris avalanche scars in an area of similar terrain in the North River basin during a large rainstorm in 1949 (Hack and Goodlett, 1960).

Figure 9 is a CM diagram (Passega, 1957) commonly used to distinguish between traction fluvial and debris flow modes of deposition, based on the sorting of the coarsest half of the sediment population. This technique was used by Bull (1964, 1972) to determine the depositional processes of arid alluvial fan sediments. Most of the Nelson County alluvial fan sediments plot within the mudflow envelope as defined by Bull (1972), and all of the fan

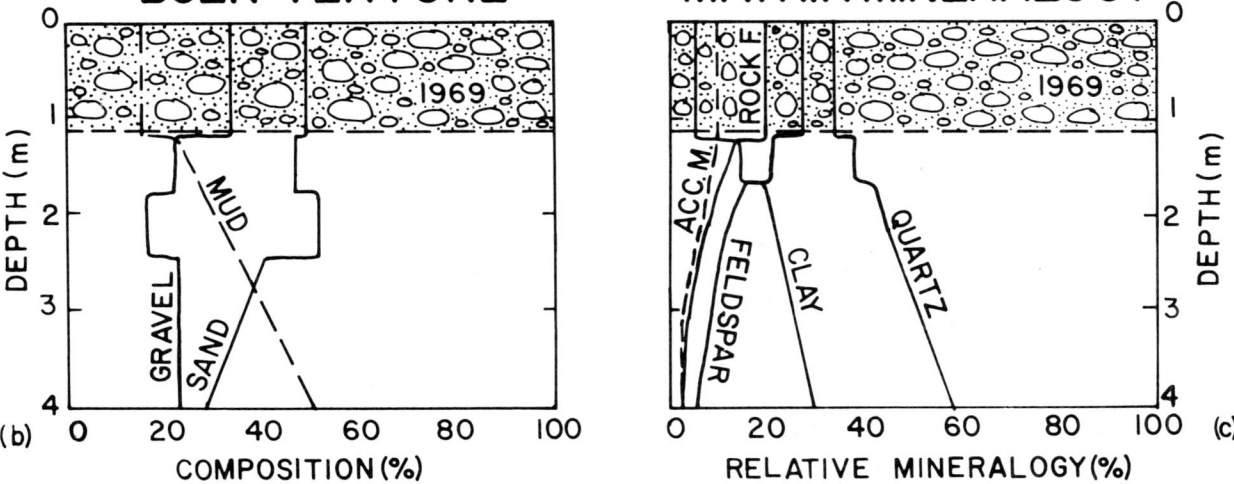

Fig. 8. Schematic illustration of the vertical changes in texture and mineralogy in Davis Creek debris fan sediments based on generalizations of observations from 5 fans. (**a**) textural variation with depth shows abrupt changes due to differences in sediment size shed onto the fan by discrete depositional events, (**b**) bulk texture shows the variations in relative amount of gravel, sand, and mud with depth, and (**c**) relative variations in matrix mineralogy with depth showing the effects of time between depositional events and weathering processes acting on the granitic sediments.

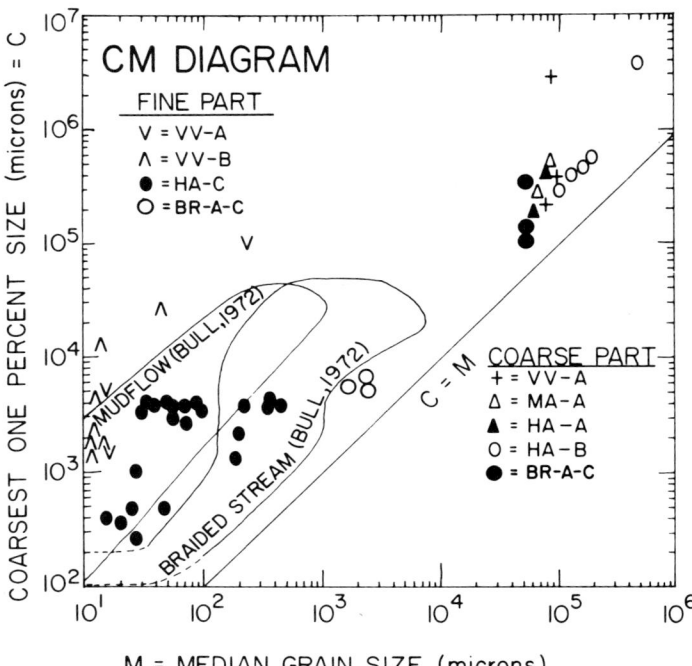

Fig. 9. CM diagram for representative samples from the debris fans and floodplain sediments in Davis Creek watershed. The envelopes of Bull (1972) for mudflow and braided stream sediments of California fans are also plotted for comparison with the Virginia data. Note that the coarse and fine fractions of the samples were plotted separately. In most cases, the Virginia fan sediments are much coarser than the arid fan sediments of Bull (1972).

sediments plot to the left of the better sorted fluvial sediments that cluster near the C = M line. Most of the alluvial fan sediments are coarser than the fan deposits described by Bull (1972).

Additional evidence supporting the importance of debris flow processes on these fans is the presence of rip-up clasts of soil and weathered granitic boulders in the alluvial fan deposits. These soil clasts and weathered boulders would not have survived the more abrasive tractive transport and indicate that they were deposited by debris flow processes. These appear to have been competent to transport the entire range of available clasts over the length of the alluvial fans. This observation is supported by post-storm photographs showing the presence of large boulders and debris along the downstream floodplains that were as coarse as the clasts found near the apex of the activated fans. To summarize, evidence supporting the occurrence of debris flows in the downstream alluvial channels includes: (1) very poor sorting and coarse texture; (2) indistinct stratification; (3) sharp basal contacts; (4) absence of current structures; (5) superelevation of the debris lines; (6) inverse graded bedding; and (7) preservation of rip-up clasts of soil and weathered boulders.

One of the alluvial fans in the Hat Creek basin (Fig. 1) showed evidence of several depositional processes during its history. At this site the fan surface activated in the 1969 storm was entrenched 3 to 5 m into older deposits. Intensely weathered clasts and well-developed soil horizons suggest that this higher fan surface is considerably older, possibly pre-Holocene in age. These coarse, clast-supported gravels are similar to the Blue Ridge fan sediments described earlier and observations suggest deposition by braided stream processes.

Correlation and Depositional Frequency

In the following section we will focus attention on the correlation and frequency of depositional episodes on the alluvial fans in Nelson County. Our work on the Blue Ridge alluvial fans is in its preliminary stages and we have not investigated potential sources for determining the depositional chronology of these alluvial fans. The geomorphic role of rare, large magnitude events is difficult to assess and quite controversial (Wolman and Miller, 1960; Wolman and Gerson, 1978). The hillslope failures and resulting deposits in the Nelson County study basins were triggered by up to 70 cm of rainfall produced during a twelve hour period beginning at 7 p.m. on August 19, 1969. Eyewitness reports indicate that the majority of the debris avalanching occurred during the later part of the storm (Simpson and Simpson, 1970).

Few detailed studies have described the erosional and depositional effects of catastrophic events of this type on hillslope and channel morphology (*cf.* Bürgisser, this volume). The major reasons for this are the relatively infrequent nature of their occurrence and the lack of detailed antecedent morphologic conditions. Recurrence intervals for rare events of this type are difficult to estimate because of the short hydrologic and meteorologic records available for most regions. Kochel and Baker (1982) and Kochel *et al.* (1982) showed that the chronology and frequency of Holocene flooding in west Texas could be determined using radiocarbon dating of organic debris and paleosols contained in slackwater sediments.

RADIOCARBON STRATIGRAPHY

No previous research on Appalachian hillslope or alluvial fan deposits has utilized radiocarbon-dated sequences to calibrate estimates of the recurrence intervals of sediment-producing events. Our preliminary estimate, based on only six radiocarbon dates, suggests that the recurrence interval for storms of a magnitude similar to the Hurricane Camille storm for any given location is on the order of 3,000 to 6,000 years. This estimate is based on dates from three fans in the Davis Creek basin. In Figure 10 one can readily see the correlation between the lower strata uncovered in fans VV-A and VV-B of an event that occurred at approximately 10,700 years B.P. Observations of fan stratigraphy suggest that there has been at least three events of a magnitude similar to the 1969 storm during the last 10,700 years. Other fan deposits in the study area also show repeated episodes of activity throughout the Holocene. Additional radiocarbon dating is needed to calibrate the recurrence interval of these events.

Estimates of the recurrence intervals of major sediment-producing events of this type are difficult if they are based solely on alluvial fan stratigraphy. Our studies of the Nelson County fans show that the number of events preserved in a given stratigraphic section can vary across the fan due to the laterally discontinuous nature of the debris flow deposits (Fig. 6). A second problem is related to the temporal ordering of major storm events. Once a slope has failed, subsequent large rainstorms may be unable to produce additional hillslope failures because of the long period of time needed to replace the colluvial mantle. For this reason, major sediment-producing events may go unrecorded in the alluvial fan sequence. Our study suggests that these problems can be reduced if floodplain deposits can be considered jointly with several alluvial fan deposits to produce a more regional estimate of the frequency of extreme events.

All of the radiocarbon dates included in Figure 10 were based on samples of buried soil horizons. The sample dated in profile HA-A was from an organic rich A horizon, while the samples dated in profiles VV-A and VV-B were from clay-rich B horizons. The low concentration of organic material in these B horizons required the collection of large (1-2 kg) samples to obtain enough carbon for accurate dating. In all cases, the samples were depth-integrated throughout the depositional unit.

Profiles HA-C and MA-A in Figure 10 are from floodplain sequences in the Davis Creek basin. HA-C is in the headwaters of the basin and shows a large number of flood generated units, although these probably represent local events limited to the headwater region of this basin. Profile MA-A is from an eroded channel bank near the mouth of Davis Creek. This profile contains three distinct flood events. The upper unit is the 1969 storm deposit and is the coarsest of the three units. The lower two units each contain well-developed buried soil horizons which indicate the infrequency of depositional events. It is possible that the lower two units correspond to the 6,300 and 10,700 year B.P. events recorded in the stratigraphy of fans further upstream (Fig. 10, profiles VV-A and VV-B). The radiocarbon date of 370 ± 70 years B.P. from the middle unit in MA-A was from a sample of wood that appears to have been a root. Additional dating is needed to develop a

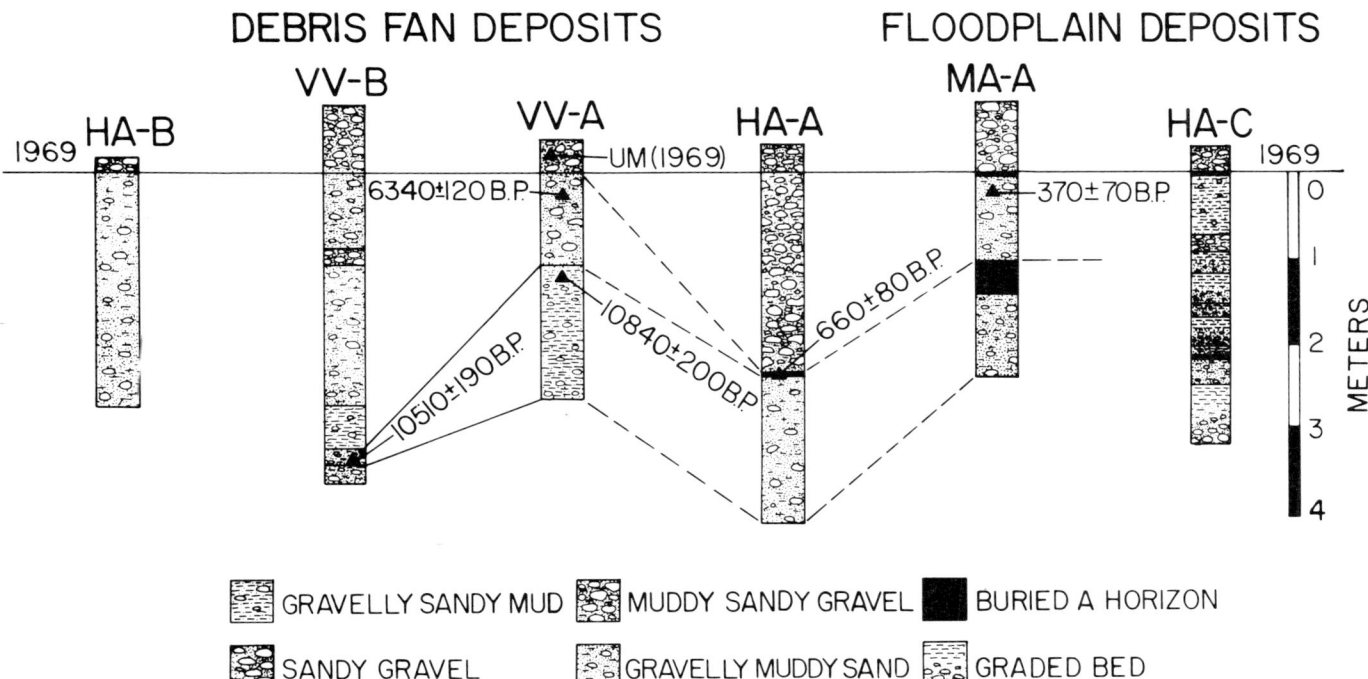

Fig. 10. Stratigraphic summary of debris fans and floodplain sites studied in the Davis Creek watershed in Nelson County, Virginia. Sites HA-B, VV-B, VV-A, and HA-A are debris fans and sites MA-A and HA-C are distal and proximal basin floodplain sites, respectively. The triangles are samples used for radiocarbon dating and the ages are years before present. The dates have been corrected for dendrochronology and for the half-life of ^{14}C. One standard deviation is listed beside each date. The sample marked UM signifies ultra-modern age (post-1945) and confirms the 1969 origin of the uppermost unit. These sections are hung along the base of the 1969 deposit because of the pronounced sedimentological contact visible in all sections.

regionalized chronology for the major sediment-producing events preserved in alluvial fan and floodplain deposits in this area.

There is some controversy over the dating of organic matter in paleosols because its accumulation during soil formation is a continuous process. For this reason, radiocarbon dates from a paleosol represent the entire period of accumulation rather than the initial period of soil accumulation (Schapenseel, 1971). Birkeland (1974) noted that the accumulation of organic matter in soils is a non-linear process. Therefore, the actual soil date is really a mean residence time (Geyh *et al.*, 1971) and probably represents a minimum age of the soil. Bloomfield and Valastro (1974) have shown that when the three soil fractions of soluble humic acid, insoluble residue, and total soil humus are dated and yield similar ages, contamination can be considered minor and accurate mean residence times are obtained. All three of these fractions were dated for each of our samples so that the dates reported represent the mean residence times for each paleosol.

HILLSLOPE-FAN-FLOODPLAIN SYSTEMS

As stated earlier, additional calibration of the frequency estimates for the alluvial fan deposits is possible by correlating these with estimates from floodplain stratigraphy downstream. For each major basin-wide event, there is a probable continuum of depositional processes and resulting landforms between the hillslope, fan, and floodplain systems. Figure 11 schematically illustrates these relationships. Above the fan, a complex set of variables (rainfall intensity and duration, hillslope gradient, vegetation patterns, and the characteristics of the soil and underlying bedrock) control the threshold of slope stability. In this way, the intensity and spatial distribution of hillslope failures (Fig. 11, areas A and B) control the morphology of the downslope fan and floodplain areas (Fig. 11, areas C and D). When a slope failure occurs, the moving debris denudes the hillslope regions below the initial failure producing large chutes through which the eroded material is transported to the depositional sites on the downslope alluvial fans and floodplains. Floodplain gravels associated with Hurricane Camille are widespread but their coarseness appears to exceed the competence level of subsequent drainage events.

The importance of hillslope processes on the downstream fluvial system was demonstrated through a quantitative analysis using stepwise multiple regression. This study showed that the amount of downstream channel bank erosion was controlled by the frequency of hillslope failures occurring upstream, the overall channel gradient, and basin shape. Using this approach, 82% of the variance in channel bank erosion can be explained at the .001 level of significance based on only these three variables (Johnson, 1983).

The results of this study suggest that in humid regions of high relief, narrow valley bottoms, and moderate to steep channel gradients the amount of downstream channel erosion is related to the frequency of major sediment-producing events. Under these conditions, the supply of coarse detritus which modifies the downstream fluvial system is closely related to the stability of the adjacent hillslopes.

THE VIRGINIA HUMID-TEMPERATE ALLUVIAL FAN MODEL

Many of the alluvial fans observed in humid-temperate regions have been considered as relict landforms. Considerable research in Appalachian geomorphology has focused on the question of whether the present landforms and deposits are active or inherited from a pre-Holocene climatic regime. The features most commonly in question are talus slopes, boulder-fields and alluvial fans (Hack and Goodlett, 1960; Hack, 1965; Mills, 1981). Many investigators believe that the Appalachian alluvial fans and boulder-fields are relict features inherited from periglacial conditions that may have existed in the Pleistocene or from climates that may have prevailed in the late Tertiary (Smith, 1949, 1953; Craig, 1969; Potter and Moss, 1968; Dunford-Jackson, 1978).

In contrast to the above studies, a number of investigators have observed signs of recent activity on these hillslope

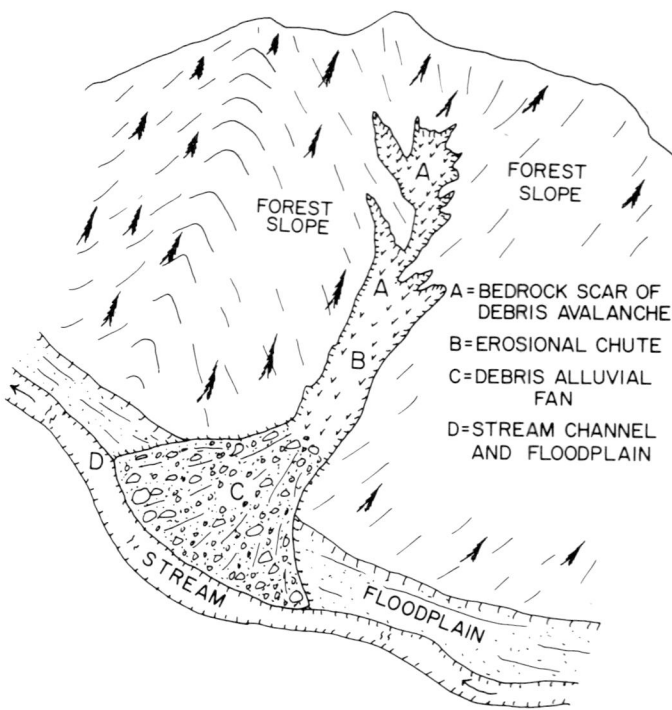

Fig. 11. Schematic illustration of an alluvial fan system in the Nelson County area. Area A is the once-forested slope now showing exposed bedrock following the slope failure. Area B shows the erosional chute cut by the debris being transported downward from the zone of failure. Area C is the debris fan and the site of the bulk of the deposition. Area D represents the downstream floodplain deposit created by the sediment-laden streams draining the failed upland areas. These areas represent a continuum of form and process and can be related through correlation of stratigraphies.

and alluvial fan landforms. Hack (1965) studied colluvial deposits throughout the Virginia Shenandoah Valley and concluded that at least some of the scree slopes showed geomorphic and botanic evidence of recent activity. Hack suggested that many of these features may become activated by catastrophic storms like the Hurricane Camille storm in 1969.

Additional support for the recent activity of scree slopes is also presented by Hupp (1983). Other studies of recent catastrophic events in the central Appalachians have shown that renewed erosion and deposition has occurred on many alluvial fan surfaces (Hack and Goodlett, 1960; Williams and Guy, 1973; Bogucki, 1976). In a recent study of hillslopes near Grandfather Mountain, North Carolina, Mills (1981) described a two-layer colluvial stratigraphy which he suggested was indicative of the transition from periglacial to more humid-temperate climatic conditions. Mills based his interpretations on weathering characteristics and the soil stratigraphy. Pollen data from the southern Appalachians are also suggestive of periglacial climatic conditions in the Pleistocene (Watts, 1979). These pollen data indicate that the early Holocene, from 11,000 to 6,000 years B.P. was somewhat warmer and drier than present. After this, the increase in bog and swamp deposits and the replacement of deciduous trees by pines suggest that the climate became wetter and cooler (Watts, 1979). These changes suggest that the Holocene from 11,000 B.P. to present was of a humid-temperate character similar to the present.

It is important to note that none of the studies of hillslope and alluvial fan landforms described above have utilized radiocarbon-dating techniques to develop a chronology of their activity. This report describes a sedimentological and geomorphological model for humid-temperate alluvial fans in the central Nelson County area of Virginia. The model is based on observations of fans that are known to have been activated repeatedly throughout the Holocene.

The Virginia alluvial fans described in this report range from broad, fan-shaped forms in open valley reaches to irregular, elongate forms in constricted headwater areas. Fan sequences are thin (generally less than 20 m), of small area (typically less than 1 km^2), and have steep, segmented axial profiles (40 to 100 m.km^{-1}). Surfaces are generally smooth due to the lateral shifting of the locus of deposition which tends to fill in surface depressions. Major storm events commonly erode the upper apex portions of these fans cutting deep fan-head trenches and depositing coarse sediment and debris across the lower fan surface.

The deposits on these alluvial fans are poorly-sorted mud-supported gravels, with inverse grading and poorly developed internal stratification. Bed thicknesses are variable, ranging from 50 cm to 5 m and show little down-fan change. Vertical variations in texture and mineralogy are sharp, and are associated with sedimentation unit boundaries. Texture within each depositional unit is a function of the intensity of hillslope failures occurring upslope. During the long periods between depositional events, unstable minerals (feldspars, micas, and accessory minerals) tend to weather, producing clay films around the weathered bedrock clasts and stable minerals.

The major depositional process on the Nelson County alluvial fans appears to be debris flow and debris avalanches triggered by intense rainfall. Observations suggest that only a portion (20 to 70%) of the fan surfaces are activated during an event. Our limited radiocarbon dating suggests that the return period for these events is in the range of 3,000 to 6,000 years for any particular fan. Additional radiocarbon dating of alluvial fan and adjacent floodplain deposits will enable us to develop a regional chronology of major geomorphic events for the local area.

Alluvial Fans Formed in Different Environments

Significant differences exist in the morphology, sedimentology, depositional processes and resulting facies of alluvial fans formed under various climatic regions. Arid alluvial fans are characterized by alternating periods of rapid deposition followed by periods of inactivity. Humid-glacial fans are formed by constantly shifting braided streams aggrading an outwash plain. Humid-tropical fans are formed by seasonal deposition by braided streams and/or debris flows. Table 1 summarizes the generalized characteristics of alluvial fans formed in various climatic environments, and is intended to provide a means of comparing the Nelson County alluvial fans with those reported by previous investigators. A table of this type necessarily separates the characteristics of the various alluvial fan types artificially. Clearly, alluvial fans have been reported that are gradational to the arbitrary groups selected for this comparison. For example, the types of glacially-associated alluvial fans reported by Ryder (1971) and Wasson (1977) have morphologic and sedimentologic characteristics that are similar to those reported for humid-glacial and humid-temperate environments.

The results of this study suggest that more attention should be given to the alluvial fans found in present humid-temperate regions. Our future plans are to expand our study of humid-temperate alluvial fans and adjacent floodplain deposits in the central Virginia area. This will include their stratigraphy, sedimentology and morphologic characteristics. We also plan to expand radiocarbon dating on a regional basis to include additional alluvial fan and floodplain deposits in order to develop a chronology of the major sediment-producing events of this area. It is hoped that studies of hillslope and fluvial deposits in other areas could adopt this approach to document the periods of activity of these landforms.

Parameter	Arid Fans	Humid-Glacial Fans	Humid-Tropical Fans	Virginia Humid-Temperate Fans
Fan Morphology				
Plan View	broad fan-like symmetrical	broad fan-like symmetrical	broad fan-like symmetrical	broad fan-like to elongate
Axial Profile	segmented (20-100m/km)	smooth (1-20m/km)	smooth	segmented (40-100m/km)
Thickness	up to 100's m	up to 100's m	up to 100's m	5 m to 20 m
Area	small	very large	large	small
Depositional Processes				
Major Processes	debris flow braided stream sheet flood sieve flood	braided stream	braided stream debris flow	debris flow (avalanche)
Return Interval	1-50 years discrete events	0-few days seasonally constant	seasonally constant to discrete	3000-6000 yrs discrete events
Fan Area Activated	10-50%	80-100%	30-70%	10-70%
Triggering Processes	heavy rain snow melt	meltwater outwash	heavy rain monsoon	heavy rain hurricane
Discharge	flashy	seasonal	seasonal	flashy
Sedimentology and Stratigraphy				
Bedding	debris flow- inverse grading braided stream (see next column)	channel fills foresets, bars normal grading	channel fills foresets, normal grading inverse grading,	inverse grading crudely bedded
Gravel %	high	high	high	high
Sand %	high	high	high	moderate
Mud %	low-moderate	very low	high-moderate	moderate-high
Matrix	sand and silt	sand	sand and mud	mud and gravel
Sorting	very poor	moderate-poor	moderate-poor	very poor
Clast Morphology	subangular to subrounded	subrounded to rounded	subrounded to rounded	subangular
Fabric	minor	imbricated	imbricated	minor
Bed Thickness	10's cm to several m	10's cm to several m	10's cm to several m	50 cm to 5 m
Proximal to Distal Fan	units remain about constant	units thin distally	units thin distally	thick proximal, then constant
Lateral Variations	variable lobate, levees	sheets, fills,	sheets, fills, lobate, levees	variable lobate
Vertical Variations	grain-size varies with each event	about constant	about constant	grain-size varies with each event
Post-Depositional Modifications				
Vegetation	sparse, scrub	sparse, forest	dense, jungle	dense, forest
Soils	thin entisols, caliche	entisols	thick, oxidized, lateritic	thick, oxidized, high organics
Bioturbation	minor	minor	variable	variable
Subsequent Events	fan-head erosion	mostly depositional	mostly depositional	fan-head erosion
Chemical	desert varnish		iron oxides	iron oxides
Paleosols	minor	minor	possible	abundant
Eolian	active	active	none	none

Generalized from the following sources:
Arid Fans — (Blissenbach, 1954; Bull, 1964, 1972; Denny, 1965; Lustig, 1965; Hooke, 1967; Nilsen, 1982)
Humid-glacial Fans — (Reimnitz, 1966; Boothroyd, 1972; Boothroyd and Ashley, 1975; Nummedal and Boothroyd, 1976; Nilsen, 1982)
Humid-tropical Fans — (Gole and Chitale, 1966; Ruxton, 1970; Mukerji, 1976)
Humid-temperate Fans — (This study)
Humid-periglacial Fans such as those described by Church and Ryder (1972), Ryder (1971) and Wasson (1977) have characteristics similar to humid-glacial and humid-temperate alluvial fans.

Table 1. Generalized characteristics of alluvial fans formed in different environments.

REFERENCES

Beaty, C.B. 1970. Age and estimated rate of accumulation of an alluvial fan, White Mountains, California, U.S.A. American Journal of Science, v. 268, p. 50-77.

Bernard, H.A., Major, C.F., Parrot, B.S. and LeBlanc, R.J. 1970. Recent sediments of south-east Texas. University of Texas, Bureau of Economic Geology, Guidebook No. 11.

Birkeland, P.W. 1974. Pedology, Weathering, and Geomorphological Research. New York: Oxford Press, 285p.

Blissenbach, E. 1954. Geology of alluvial fans in semi-arid regions. Geological Society of America Bulletin, v. 65, p. 175-190.

Bloomer, R.O. and Werner, H.J. 1955. Geology of the Blue Ridge region in central Virginia. Geological Society of America Bulletin, v. 66, p. 579-606.

Bloomfield, K. and Valastro, S. 1974. Late Pleistocene eruptive history of Nevada de Toluca volcano, central Mexico. Geological Society of America Bulletin, v. 85, p. 901-906.

Bogucki, D.J. 1976. Debris slides in the Mt. Le Conte Area, Great Smokey Mountains National Park. Geographiska Annaler, v. 58A, p. 179-192.

Boothroyd, J.C. 1972. Coarse-grained sedimentation on a braided outwash fan, northeast Gulf of Alaska. Office of Naval Research Technical Report No. 6-CRD, Charleston: University of South Carolina, 127p.

Boothroyd, J.C. and Ashley, G.M. 1975. Processes, bar morphology, and sedimentary structures on braided outwash fans, northeastern Gulf of Alaska. In: McDonald, B.C. and Jopling, A.V. (Eds.), Glaciofluvial and Glaciolacustrine Sedimentation. Society of Economic Paleontologists and Mineralogists, Special Publication 23, p. 193-222.

Brent, W.B. 1960. Geology and mineral resources of Rockingham County. Virginia Division of Mineral Resources Bulletin 76, 173p.

Bull, W.B. 1964. Geomorphology of segmented alluvial fans in western Fresno County, California. United States Geological Survey Professional Paper 352-E, p. 89-129.

Bull, W.B. 1972. Recognition of alluvial fan deposits in the stratigraphic record. In: Rigby, J.K. and Hamblin, W.K. (Eds.), Recognition of Ancient Sedimentary Environments. Society of Economic Paleontologists and Mineralogists, Special Publication 16, p. 63-83.

Cant, D.J. 1982. Fluvial facies models and their applications. In: Scholle, P.A., and Spearing, D. (Eds.), Sandstone Depositional Environments. American Association of Petroleum Geologists, Memoir 31, p. 115-137.

Cant, D.J. and Walker, R.G. 1976. Development of a braided-fluvial facies model for the Devonian Battery Point Sandstone, Quebec. Canadian Journal of Earth Sciences, v. 13, p. 102-119.

Church, M. and Ryder, J.M. 1972. Paraglacial sedimentation: a consideration of fluvial processes conditioned by glaciation. Geological Society of America Bulletin, v. 83, p. 3059-3072.

Costa, J.E. and Jarrett, R.D. 1981. Debris flows in small mountain stream channels of Colorado and their hydrologic implications. Association of Engineering Geologists Bulletin, v. 18, p. 309-322.

Craig, A.J. 1969. Vegetation history of the Shenandoah Valley, Virginia. Geological Society of America, Special Paper 123, p. 283-296.

Denny, C.S. 1965. Alluvial fans in the Death Valley region, California and Nevada. United States Geological Survey Professional Paper 466, 62 p.

Dunford-Jackson, C.S. 1978. The geomorphic evolution of the Rappohannock River Basin. M.Sc. Thesis, University of Virginia, 92p.

Fisher R.V. 1971. Features of coarse-grained, high concentration fluids and their deposits. Journal of Sedimentary Petrology, v. 41, p. 919-927.

Gathright, T.M. 1976. Geology of the Shenandoah National Park, Virginia. Virginia Division of Mineral Resources Bulletin, No. 86, 93p.

Gathright, T.M., Henika, W.S. and Sullivan, J.L. 1977. Geology of the Waynesboro East and Waynesboro West quadrangles, Virginia. Virginia Division of Mineral Resources Publication, No. 3, 53p.

Geyh, M.A., Benzler, J.H. and Roeschmann, G. 1971. Problems of dating Pleistocene and Holocene soils by radiometric methods. In: Yaalon, P.H. (Ed.), Paleopedology: Origin, Nature, and Dating of Paleosols. Jerusalem: Israel University Press and International Society of Soil Scientists, p. 63-75.

Gole, C.U. and Chitale, S.V. 1966. Inland delta building activity of the Kosi River. American Society of Civil Engineers, Journal of the Hydraulics Division, Proceedings HY2, v. 92, p. 111-126.

Hack, J.T. 1965. Geomorphology of the Shenandoah Valley, Virginia and West Virginia, and origin of the residual ore deposits. United States Geological Survey Professional Paper 484, 83p.

Hack, J.T. and Goodlett, J.C. 1960. Geomorphology and forest ecology of a mountain region in the central Appalachians. United States Geological Survey Professional Paper 347, 66p.

Harms, J.C., Southard, J.B., Spearing, D.R. and Walker, R.G. 1975. Depositional environments as interpreted from primary sedimentary structures and stratigraphic sequences. Society of Economic Paleontologists and Mineralogists, Short Course No. 2, 161p.

Hooke, R.L. 1967. Processes on arid-region alluvial fans. Journal of Geology, v. 75, p. 453-456.

Hupp, C.R. 1983. Geobotanical evidence of late Quaternary mass wasting in blockfield areas of Virginia. Earth Surface Processes, v.8, p. 438-450.

Johnson, A.M. 1970. Physical Processes in Geology. San Francisco: Freeman Co., 577p.

Johnson, R.A. 1983. Stream channel response to extreme rainfall events-Hurricane Camille in Central Nelson County, Virginia. M.Sc. Thesis, University of Virginia, 120p.

Kochel, R.C. and Baker, V.R. 1982. Paleoflood hydrology. Science, v. 215, p. 353-361.

Kochel, R.C., Baker, V.R. and Patton, P.C. 1982. Paleohydrology of south-west Texas. Water Resources Research, v. 18, p. 1165-1183.

Lustig, L.K. 1965. Clastic sedimentation in Deep Springs Valley, California. United States Geological Survey Professional Paper 352-F, p. 131-192.

Miall, A.D. 1977. A review of the braided-river depositional environment. Earth Science Reviews, v. 13, p. 1-62.

Mills, H.H. 1981. Some observations on slope deposits in the vicinity of Grandfather Mountain, North Carolina. Southeastern Geology, v. 22, p. 209-222.

Mukerji, A.B. 1976. Terminal fans of inland streams in Sutlej-Yamuna plain, India. Zeitschrift fur Geomorphologie, v. 20, p. 190-204.

Nelson, W.A. 1962. Geology and mineral resources of Albemarle County, Virginia. Virginia Division of Mineral Resources Bulletin 77, 92p.

Nilsen, T.H. 1982. Alluvial fan deposits. In: Scholle, P.A. and Spearing, D. (Eds.), Sandstone Depositional Environments. American Association of Petroleum Geologists, Memoir 31, p. 49-86.

Nummedal, D. and Boothroyd, J.C. 1976. Morphologic and hydrodynamic characteristics of terrestrial fan environments. Office of Naval Research, Technical Report No. 10-CRD, Coastal Research Division, Department of Geology, University of South Carolina, Columbia, South Carolina, 61p..

Passega, R. 1957. Texture as characteristic of clastic deposition. American Association of Petroleum Geologists Bulletin, v. 41, p. 1952-1984.

Pierson, T.C. 1980. Erosion and deposition by debris flows at Mt. Thomas, New Zealand. Earth Surface Processes, v. 5, p. 227-247.

Potter, N. and Moss, J.H. 1968. Origin of Blue Rocks Blockfield and adjacent deposits, Berks County, Pennsylvania. Geological Society of America Bulletin, v. 79, p. 255-262.

Reimnitz, E. 1966. Late Quaternary history and sedimentation of the Copper River delta and vicinity, Alaska. Ph.D. Dissertation, University of California, San Diego, 160p.

Ruxton, B.P. 1970. Labile quartz-poor sediments from young mountain ranges in northeast Papua. Journal of Sedimentary Petrology, v. 40, p. 1262-1270.

Ryder, J.M. 1971. The stratigraphy and morphology of paraglacial alluvial fans in south-central British Columbia. Canadian Journal of Earth Sciences, v. 8, p. 279-298.

Schapenseel, H.W. 1971. Radiocarbon dating of soils — problems, troubles, hopes. In: Yaalon, D.H. (Ed.), Paleopedology: Origin, Nature, and Dating of Paleosols. Jerusalem: Israel University Press and International Society of Soil Scientists, p. 77-88.

Simpson, P.S. and Simpson, J.H. 1970. Torn Land. Virginia: J. P. Bell Company, 207p.

Smith, H.T.U. 1949. Pleistocene climatic changes in non-glaciated areas: eolian phenomena, frost action, and stream terracing. Geological Society of America Bulletin, v. 60., p. 1485-1516.

Smith, H.T.U. 1953. The Hickory Run boulder field, Carbon County, Pennsylvania. American Journal of Science, v. 251, p. 625-642.

Wasson, R.J. 1977. Catchment processes and the evolution of alluvial fans in the lower Derwent Valley, Tasmania. Zeitschrift fur Geomorphologie, Supplement Band No. 21, p. 147-168.

Watts, W.A. 1979. Late Quaternary vegetation of the central Appalachians and the New Jersey coastal plain. Ecological Monographs, v. 49, p. 427-469.

Wescott, W.A. and Ethridge, F.G. 1980. Fan-delta sedimentology and tectonic setting - Yallahs Fan delta, southeast Jamaica. American Association of Petroleum Geologists Bulletin, v. 64, p. 374-399.

Williams, G.P. and Guy, H.P. 1973. Erosional and depositional aspects of Hurricane Camille in Virginia, 1969. United States Geological Survey Professional Paper 804.

Williams, P.F. and Rust, B.R. 1969. The sedimentology of a braided river. Journal of Sedimentary Petrology, v. 39, p. 649-679.

Wolman, M.G. and Miller, J.P. 1960. Magnitude and frequency of forces in geomorphic processes. Journal of Geology, v. 68, p. 54-74.

Wolman, M.G. and Gerson, R. 1978. Relative scales of time and effectiveness of climate in watershed geomorphology. Earth Surface Processes, v. 3, p. 189-203.

DEBRIS FLOWS AND FLUVIAL DEPOSITS IN SPANISH QUATERNARY ALLUVIAL FANS: IMPLICATIONS FOR FAN MORPHOLOGY

A.M. Harvey[1]

Abstract

Quaternary alluvial fans occur throughout southeast Spain. Their development was characterised by early phases of aggradation and later dissection. This paper, based on field observations on 31 fans in Alicante, Murcia and Almeria provinces, deals with the sedimentology of the aggradation phases and its implications for fan morphology.

Four types of deposit are recognised: debris flow, sheet gravel, channel gravel and silt, their relative occurrence varying both vertically and laterally within fans as well as between fans. Within the fan sequences, there is a progressive vertical decrease in debris flow and increase in channel gravel deposition, and a downfall decrease in debris flow and increase in silt and channel gravel deposition.

Variations between fans reflect source area geology and relief. On the basis of their constitutent deposits fans are grouped into debris flow, intermediate and fluvial types. Small steep watersheds especially on massive sedimentary rocks favour debris flow fans and larger, less steep watersheds especially on metamorphic rocks favour fluvial fans.

The nature of sedimentary processes strongly influences fan morphology. Multiple regression analysis demonstrates the effect of watershed size and relief on fan slopes but regression residuals reflect fan sediment types, with debris flow fans having steeper slopes in relation to drainage basin characteristics than do intermediate and fluvial fans. The implications are important for aggradation and dissection behaviour in that steeper debris flow fans show more complex patterns of dissection than do less steep, simpler fluvial fans.

Résumé

Des cônes de déjection quaternaires apparaissent dans tout le sud-est de l'Espagne. Leur développement débute toujours par des phases d'alluvionnement suivies d'une dissection par érosion fluviatile. Nous dévrivons ici nos observations de terrain portant sur 31 cônes des provinces d'Alicante, Murcia et Almeria et nous discutons de la sédimentologie des phases d'alluvionnement et des effets sur la morphologie des cônes.

Quatre types de dépôt sont identifiés: des coulées de débris, des nappes de gravier, des graviers dans des chenaux et du limon, leur occurrence varie verticalement et latéralement à l'intérieur des cônes aussi bien qu'entre les cônes. On assiste à une décroissance verticale progressive des coulées de débris et à une augmentation des dépôts de gravier dans des chenaux et de haut en bas de la pente des cônes on observe une décroissance des coulées de débris et une augmentation des dépôts de limon et de gravier dans des chenaux au sein des séquences des cônes.

Le relief et la nature des matériaux géologiques qui alimentent les cônes sont responsables des variations entre eux. Les cônes sont classifiés sur la base des constituants des dépôts comme suit: cônes de coulée de débris, intermédiaire et fluviatile. Les petits bassins versants abrupts occupant particulièrement des assises sédimentaires massives favorisent la formation de cônes de coulée de débris et les bassins versants plus étendus, moins abrupts est localisés spécialement sur des roches métamorphiques favorisent le développement de cônes fluviatiles.

La nature des processus sédimentaires influence fortement la morphologie des cônes. L'analyse par régression multiple démontre l'effet de l'extension du bassin versant et du relief sur les pentes des cônes, mais les valeurs résiduelles de régression reflètent les types de sédiment qui forment les cônes, avec les cônes de coulée de débris exhibant des pentes plus abruptes reliées aux caractéristiques de drainage du bassin que le font les cônes intermédiaires et fluviatiles. Ces implications sont importantes quant au comportement de l'alluvionnement et la dissection par érosion dans le sens que les cônes de coulée de débris à pentes abruptes montrent des motifs plus complexes de dissection que ceux exhibés par les cônes fluviatiles moins abrupts.

Introduction

Previous work on dry region alluvial fans has tended to focus on fan morphology (Bull, 1977; Denny, 1965; Wells, 1978) and surface processes (Blissenbach, 1954; Bluck, 1964; Hooke, 1967) with emphasis on the American southwest. There has been less work on the constituent deposits (Reineck and Singh, 1975; Rust, 1980) and little on the relationships between sedimentology and geomorphology of fan development.

Quaternary alluvial fans occur throughout southeast Spain, issuing from mountains of the Betic Cordillera. This paper deals with fans in Alicante, Murcia and Almeria provinces and is based on field observations of 31 fans (Fig. 1). Fan sedimentology is considered in relation to source area characteristics, examining the extent to which topography and geology influence sediment type and the implications for fan morphology and development.

The mountain ranges of the Betic Cordillera are separated from one another by downfaulted basins or by pediment surfaces cut across pre-Pleistocene rocks. The fans occur at the mountain fronts, on basin margins or mantling pediment surfaces, in some cases backfilling into mountain valleys, or in confined basins within the mountains.

---

[1]Department of Geography, University of Liverpool, P.O. Box 147, Liverpool L69 3BX, England.

I am grateful to the University of Liverpool staff research fund for a grant towards the cost of the fieldwork, and to staff of the drawing office and photographic sections of the Department of Geography, University of Liverpool, for producing the diagrams.

The stratigraphy of fans in this region has been described by Harvey (1978, 1982). Aggradation during the Quaternary culminated in formation of calcrete crusts on upper fan surfaces. Trenching has since occurred and the later dissectional history has been characterised by cut-and-fill sequences within the fan trenches. The progressive change from aggradation to dissection reflects an overall diminution of sediment availability during the Quaternary (Harvey, 1982). The dissection phases appear to date from the Würm (Harvey, 1978). This paper deals with the pre-Würm deposits of the aggradation phases.

The present climate is semi-arid with average annual rainfall ranging from ca. 350 mm in the north of the area to ca. 170 mm in the driest parts of the south, with most precipitation falling in autumn and spring (Geiger, 1970). Much of the Pleistocene was characterised by aridity (Amor and Florschutz, 1964) and seasonality of precipitation and runoff (Butzer, 1964), and irrespective of total precipitation, a similar north-south aridity gradient may be assumed.

Source area geology (Table 1) includes a wide range of rock types. High-grade metamorphic rocks of the Sierra de los Filabres and Almenara ranges are dominantly fissile micaschists; those of the Alhamilla range also include more massive rocks. Less fissile rocks occur in the lower grade metamorphic and sedimentary terrain of the Carrascoy, Almagros and Lisbona ranges. Massive Cretaceous limestones characterise the pre-Betic ranges and less massive, generally softer, Tertiary sedimentary rocks occur in the remaining areas.

Deposits

Four types of fan deposits may be recognised on the basis of field characteristics: debris flow deposits, sheet gravels, silts and channel gravels.

They partly correspond to the facies classification of Miall (1977, 1978), with the added differentiation of sheet and channel gravels. To a certain extent the debris flow, channel gravel and silts accord with Blissenbach's (1954) fan environments, namely flash flood, stream and streamflood, but the sheet gravels may have origins in more than one of these environments.

In addition to these four facies are $CaCO_3$ cemented layers as surficial and buried calcrete crusts, and near bedrock contacts as groundwater cemented zones. Their origin and stratigraphic importance has been discussed elsewhere (Harvey, 1978; 1982). These pedogenic crusts are considered to mark inactive periods during fan aggradation.

DEBRIS FLOW DEPOSITS

These are matrix-supported conglomerates (Figs. 2a, d). The clast to matrix proportions vary and the silt matrix is commonly reddish and may be partially cemented. The sub-angular clasts may be up to ca. 70 cm (b axis) but are generally less than 15 cm and are only poorly sorted. The deposits show little internal structure or preferred fabric,

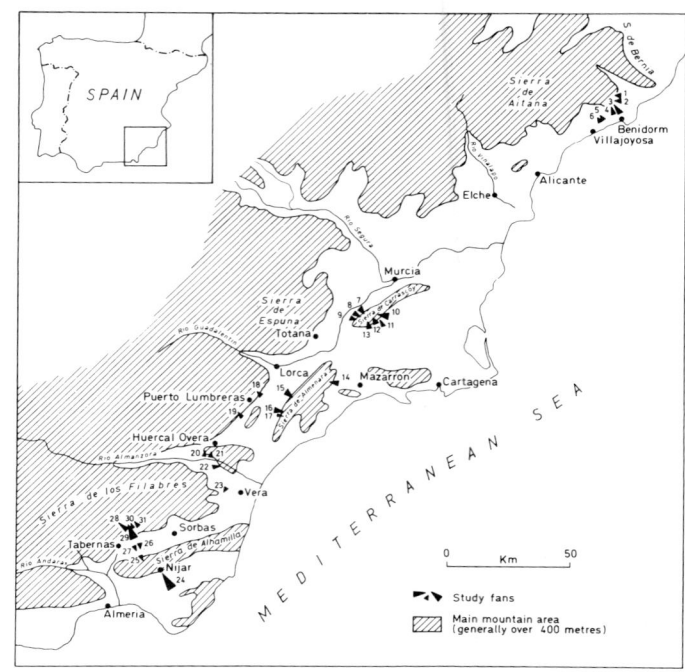

Fig. 1. Location of study fans in southeast Spain.

but occasional transverse sections reveal marginal structures identified by clast alignment parallel to that margin. The flow base may be locally irregular showing no sign of previous erosion but the basal layers are often silty. Adjacent to bedrock mountain slopes massive structureless debris flow deposits, with occasional large clasts, may reach thicknessses of 5 m. Distally from the fan apex area, debris flows occur as discrete units, normally ranging in thickness from ca. 30 cm to 1.5 m, interbedded with other deposits. They are also thinner and less common lower downfan. The debris flow deposits represent deposition by low energy viscous mudflows apparently as lobes on a fan surface or in ill-defined channels. The source of the silt matrix is likely to be former soil cover of the mountain slopes of the 'terra rossa' type (Butzer, 1964).

SHEET GRAVELS

These are almost structureless clast-supported gravels commonly with a small amount of silt-sized matrix (Figs. 2a, c). The subangular clasts are commonly up to 10 cm (b axis) and are poorly sorted, though better than those of the debris flows. These gravels show few sedimentary structures or preferred fabric apart from a common close packing (but with little or no imbrication) and occasionally a crude upward-fining tendency (Fig. 2c) within units up to 1 m thick. In upper fan locations they are widespread either as sheets or as discontinuous lobes and lenses, and interbedded with debris flows, silts and channel gravels. Lower downfan they occur in extensive sheets usually interbedded with silt units. They lack internal stratification but may overlie an erosional base. Described elsewhere as 'fan gravels' (Harvey, 1978) they may represent more than one depositional process. Some thin sheet gravels appear to grade laterally into debris flows and may locally represent matrix-poor debris flows, or debris flow deposits from which the matrix has been washed. Generally, however, they represent rapid deposition by streamflow and include both seive deposits (Bull, 1977), similar to modern seive bars in ill-defined channels and sheetflood gravels (Fig. 2b).

SILTS

Units of silt (Fig. 2b) occur mainly in midfan and distal locations. They are present as extensive 20 cm to 1 m thick beds interstratified with sheet gravels, or locally as thicker bodies up to 2 m. They are partly cemented, reddish, and structures are limited to occasional weak horizontal bedding. Their source appears to have been a pre-existing soil cover. Over distal fan surfaces they appear to represent extensive flood deposits but in upper fan locations where they are less common, their smaller extent possibly relates to suspended-sediment fallout within channels, or clast-free mudlows.

CHANNEL GRAVELS

These comprise clast-supported, bedded, generally imbricate gravels indicative of within-channel fluvial deposition (Figs. 2d, e). They differ from the sheet gravels by their greater clast roundness, better sorting and by the presence of distinct sedimentary structures. Both planar and trough cross-bedding are conspicuous in transverse or longitudinal sections, and plane bedded units commonly show strong clast imbrication and occasional local fining upwards. Channel geometry varies from small, narrow and steep-sided to wide and shallow. Some sections show repeated channel scour-and-fill during overall aggradation. Units of sand facies are locally present, especially in those fans issuing from the Sierra de los Filabres (Fig. 2e) (Fans 28-31), and these display bar-front and trough cross-bedding and plane parallel bedding. These sequences very clearly relate to within-channel deposition with an apparently wide variation in the degree of braiding.

VARIABILITY

On the 31 fans 81 outcrop sections have been examined, including 8 at fan apices adjacent to bedrock slopes, 5 in distal areas below modern intersection points and 68 within the upper fans exposed largely in fanhead trenches. The deposits were classified according to their facies, and the proportions of each in a vertical section at each site determined.

Fan Group[1]	Source area geology[2]	
Pre-Betic Group (C. Alicante) 1 Polop, 2 La Nucia, 3 Tapia, 4 Cayola, 5 Robelles N, 6 Robelles S.	Cretaceous Limestones	(Area 3, includes Eocene marls and Triassic marls locally).
Carrascoy Group (C. Murcia) 7 Roy, 8 Ginesa, 9 La Murta, 10 Herradon, 11 Carrachos, 12 Carrascoy, 13 Ros.	Permo/Trias	Sedimentaries and low grade metamorphics (sandstones, limestones, quartzites, phyllites) (Areas 10 11 also include U. Miocene sandstones Areas 10 12 13 also include basic igneous rocks)
Southern Metamorphic Group Sierra de Almenara (S. Murcia) 14 Cambron, 15 Purias, 16 Aljibejo, 17 Villaescusa,	Cambrian to Permo/Trias	High grade metamorphics (micaschists, quartzites, gneiss, marble)
Sierra de Alhamilla (S. Almeria) 24 Nijar, 25 Sierra,	Pre-Cambrian to Permo/Trias	High and low grade metamorphics (micaschists, quartzites, etc.)
Sierra de Los Filabres (S. Almeria) 28 'Cabrillo', 29 Honda, 30 'Little Honda', 31 'Little Nudos'	Pre-Cambrian to Permian	High grade metamorphics (graphite micaschists)
Southern Sedimentary Group S. Murcia 18 Las Blanquizares	Eocene, Miocene	Sedimentaries (marls, limestones and conglomerates)
19 Corascos	Pliocene	Sedimentaries (sandstones, conglomerates and silts)
N. Almeria 20 Caledron, 21 Alvantos	Pliocene	Sedimentaires (limestone, marls & sandstones)
22 Chapi (Sierra de Almagros)	Permo/Trias	Sedimentaries and low grade metamorphics (quartzites, phyllites, carbonate rocks)
23 Lisbona (Sierra de Lisbona)	Triassic	Sedimentaries and low grade metamorphics (carbonate rocks)
26 Ceporro, 27 Los Arcos (Serrata del Marchante)	Miocene	Sedimentaries (sandstones & conglomerates)

[1]Fan names mostly adopted from features on topographic maps; for locations see Figure 1.
[2]Based on IGME (Instituto Geologico y Minero de Espania) 1:50,000 geological maps and accompanying memoirs.

Table 1. Grouping of fans in study area according to source area geology.

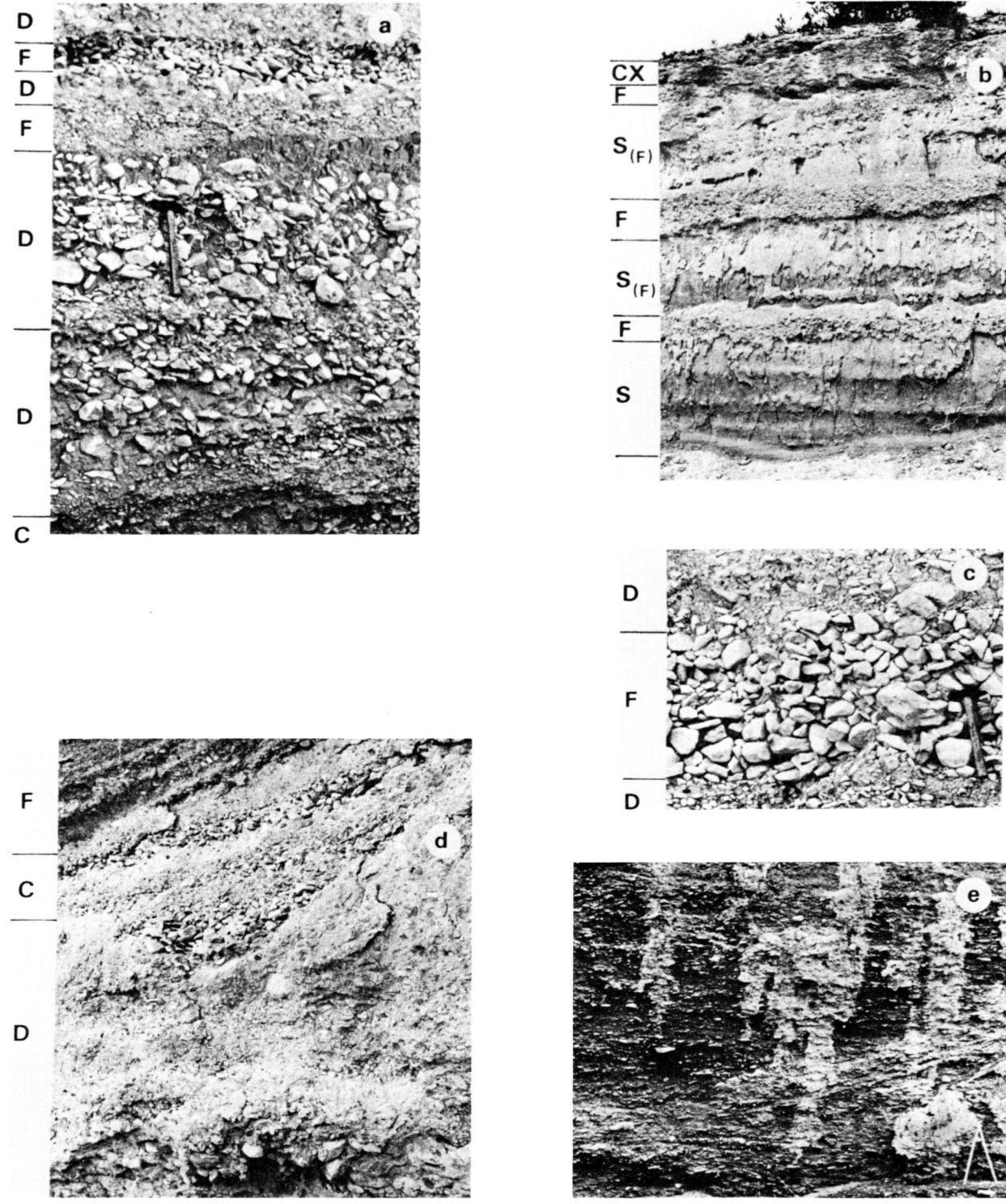

Fig. 2. Fan deposit types; **a**) Channel gravel (C), debris flow (D) and sheet gravel (F), succession at Cayola (Fan 4): length of hammer *ca.* 30 cm; **b**) Silt (S) and sheet gravel (F) sequence, capped by crusted channel gravels (CX), at Tapia (Fan 3): height of section *ca.* 8 m; **c**) Sheet gravels (F) between debris flows (D), at Cayola (Fan 3). Note fining upward tendency and packing: length of hammer *ca.* 30 m; **d**) Channel with channel gravels (C) cut into debris flows (D), sheet gravels (F) above, at Corachos (Fan 11): height of section *ca.* 3 m; **e**) Bedded channel gravels and sands, at 'Cabrillo' (Fan 28): length of marker 50 cm.

WITHIN-FAN VARIATIONS

Many fans show vertical variations in the relative importance of facies (Harvey, 1978; 1982). Debris flow deposits and silts are often more abundant basally and channel gravels more important higher up the sequences. This type of sequence resembles those described by Steel (1974) on Triassic fans in Scotland, but here appears to be the result of a progressive reduction in the availability of fines as pre-existing soils were eroded from the watersheds (Harvey, 1982).

Lateral facies variations also occur which accord with model proximal-distal facies variations (Rust, 1980). Debris flows may dominate in areas immediately adjacent to bedrock slopes but from the fan apex they become interbedded with other deposits and decrease in importance, similar to the distribution on modern fans in the American southwest (Hooke, 1967). Sheet and channel gravels occur throughout with incised channels more common upfan and wider braided channels downfan. Silt units, especially as extensive sheets, are more common downfan.

Ceporro fan (Fan 26) illustrates these within-fan variations. Figure 3 shows characteristic sections exposed at the fan apex, within the fanhead trench and in the distal area. Vertical variations are marked in this fan sequence, with debris flows dominant at the base and alluvial gravels above. Based on 9 sections, the relative abundance of debris flow deposits decreases downfan at the expense of fluvial facies. Irrespective of these trends however, the relative proportions remain fairly consistent within all but the lowest of the 7 fanhead trench sections.

BETWEEN-FAN VARIATIONS

On each fan, representative vertical fanhead trench sections have been recorded and the proportions of facies calculated (Table 2). Because of the depth of dissection, the walls of the fanhead trench commonly expose the complete sequence of fan deposits from a bedrock base. Within-fan variability, at least for proximal environments, is low when compared to between-fan variability. To facilitate comparison, each fan was classified according to deposit type. As sheet gravels and perhaps silts may be of multiple origin, the classification is based primarily on the proportions of debris flow deposits and channel gravels.

Three broad groups of fan can be recognised: debris flow, intermediate and fluvial types. The debris flow type (Group A) have more than 20% debris flows, but further subdivisions are possible using the relative abundance of debris flow and other deposits. Fans 2, 4, 7-9, 23 and 26 all have over 30% debris flows; Fans 7-9 (Northern Carrascoy fans) are high both in debris flows and channel gravels. The fluvial type (Group C) have more than 50% channel gravels; Fans 28, 29, 31 have little else. The intermediate type (Group B) includes a variety of sub-types: Fans 3, 10, 11, 22 with channel gravel maxima; Fans 19, 25 with channel gravel and silt maxima; Fans 16, 17, 24 with sheet gravels dominant and channel gravels of secondary importance; Fans 18, 21 with silt maxima; and Fan 20 with silts, sheet and channel gravels in equal proportions. Figure 4 shows a triaxial plot with combined sheet gravels and silts plotted against debris flows and channel deposits.

Three groups of factors apparently influence fan processes and therefore fan type: drainage basin size and relief, local geology and perhaps also climate, the latter being important in as much as the southward aridity gradient may have been as effective in the Quaternary as it is today. When drainage basin size is plotted against basin relief (Fig. 5), preferential plotting positions for the fan types are apparent. There is a tendency for the debris flow type (Group A) to be associated with small steep catchments and the other types with larger, lower relief basins. A regression relationship calculated between these two variables indicates that Group A has a mean positive residual for basin relief of 0.093 log units, Groups B and C together a mean negative residual of −0.072 and Group B alone −0.088: the Group A value differs significantly from the other two at the 0.1% level.

Fan	n[1]	DF	SG	Silt	Ch	Fan Type[3]
1. Polop	2	32	30	13	25	A
2. La Nucia	5	25	29	18	28	A
3. Tapia	4	7	26	29	38	B
4. Cayola	3	36	32	13	19	A
5. Robelles N	2	24	26	34	16	A
6. Robelles S	3	24	32	17	27	A
7. Roy	3	32	12	9	47	A
8. Ginesa	1	38	15	—	47	A
9. La Murta	1	36	25	—	39	A
10. Herradon	1	18	19	29	34	B
11. Corachos	3	17	9	32	42	B
12. Carrascoy	2	26	20	25	29	A
13. Ros	4	24	20	23	33	A
14. Cambron	1	9	16	20	55	C
15. Purias	1	9	11	14	66	C
16. Aljibejo	1	13	45	21	21	B
17. Villaescusa	1	15	35	20	30	B
18. Las Blanquizares	1	8	23	49	20	B
19. Corascos	1	8	15	39	38	B
20. Caledron	1	19	30	21	30	B
21. Alvantos	1	6	12	47	35	B
22. Chapi	1	11	24	28	37	B
23. Lisbona	2	40	50	—	10	A
24. Nijar	2	11	45	14	30	B
25. Sierra	1	10	23	35	32	B
26. Ceporro	7	36	26	9	29	A
27. Los Arcos	1	24	32	13	31	A
28. 'Cabrillo'	3	—	3	—	97	C
29. Honda	3	—	11	—	89	C
30. 'L. Honda'	3	17	18	—	65	C
31. 'L Nudos'	2	2	7	—	91	C

[1]Number of sections measured (upper fan).
[2]DF Debris flow, SG Sheet gravels, Ch Channel gravels.
[3]A - Debris flow type
B - Intermediate type
C - Fluvial type

Table 2. Classification of fan types.

Rather more useful than regression is a discriminant approach using the expression

$$R_b = 300\, A_d^{0.69}$$

where R_b is basin relief and A_d is drainage area, km². This may be effectively used to discriminate between Group A and Groups B and C, with Group A (debris flow fans) from smaller steeper watersheds tending to plot above the line on Figure 5 and Groups B and C below. The few anomalies may indicate geological factors in that Fans 30 and 31 from the graphitic micaschists of the Sierra de los Filabres are both of Group C but plot above the line, suggesting a dominance of fluvial processes on this lithology even from small steep drainage areas. Fans 20 and 22 both plot above the line while Fan 27 plots a little below, but as Fans 20 and 27 are near their respective class

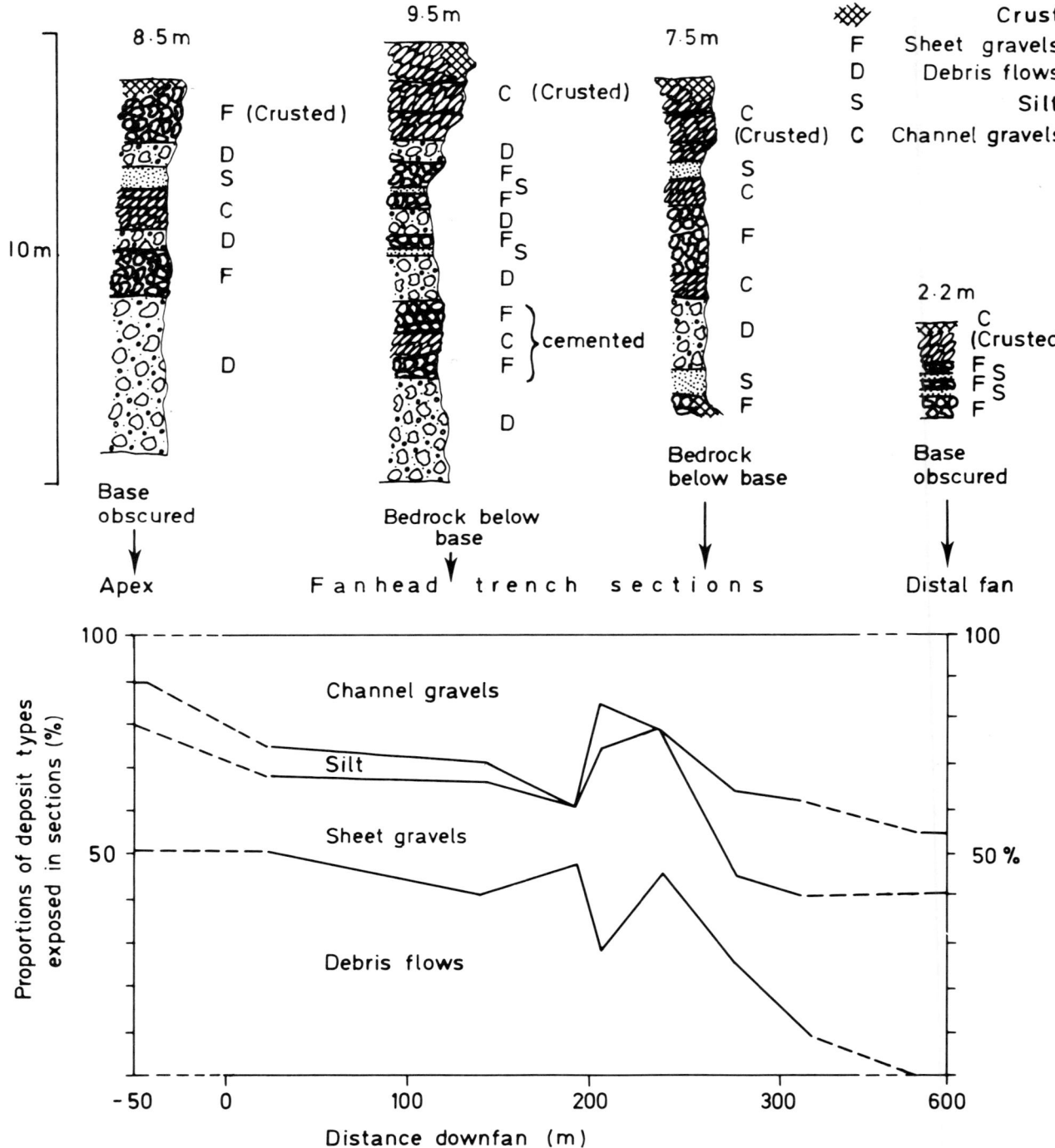

Fig. 3. Ceporro (Fan 26), within-fan variations. C — channel gravels, F — sheet gravels, D — debris flows, S — silt.

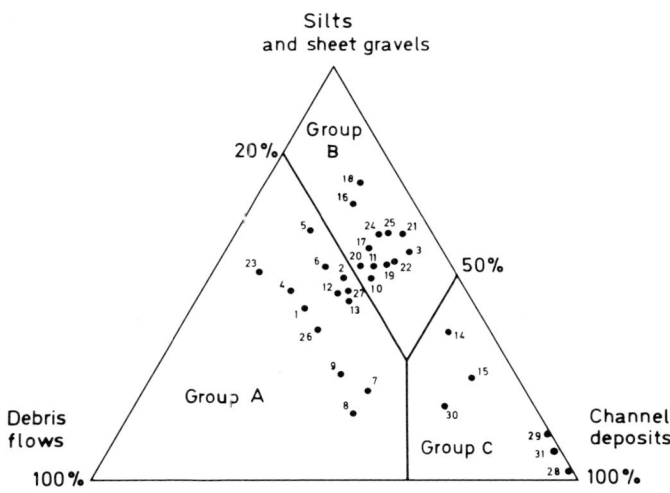

Fig. 4. Ternary classification of fans into 3 groups; A — debris flow; B — intermediate; C - fluvial.

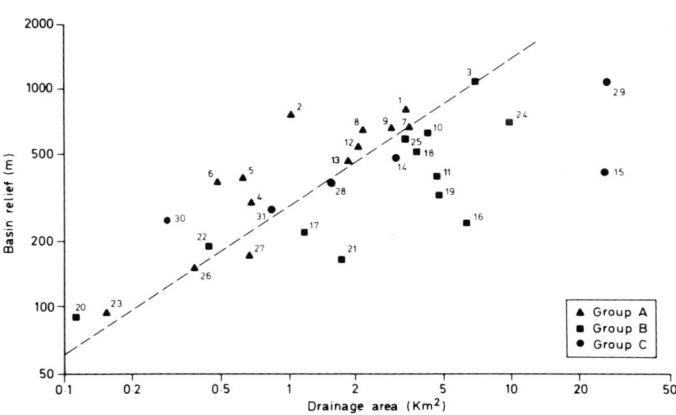

Fig. 5. Fan groups in relation to drainage area and basin relief. Dashed line separating Group A (above) from Groups B and C (below). $R_b = 300A_d^{0.69}$ (R_b in basin relief, m; A_d is drainage area, km²). For fan groups see text.

boundaries (see Table 2) these anomalies may not be important.

The importance of topographic and geological factors is highlighted when plotting positions on Figure 5 are examined for fans grouped by source area rock types (Table 1). Fans 1-6, from the limestone Pre-Betic ranges are all of Group A, except Tapia (Fan 3) of Group B which has the largest drainage area and plots well to the right. Fans 7-13, derived from the low-grade metamorphic and sedimentary rocks of the Carrascoy range are again debris flow fans except for Fans 10 and 11, of Group B, which have the largest drainage areas and plot to the right. Indeed it is possible that the drainage area to Corachos (Fan 11) may be an underestimate in that stream capture may have reduced the drainage area from a former larger extent. Most fans from the southern sedimentary groups (Fans 18-23, 26, 27) are of Group B and plot below the line. Four with smaller drainage areas plot near the line, including the three Group A fans (Fans 23, 26, 27). In the southern high-grade metamorphic areas the micaschists of the Sierra de Almenara and del los Filabres produce the only fluvial fans (Group C), some irrespective of watershed size and relief; the other fans are of Group B.

In summary, small steep basins on sedimentary and low grade metamorphic rocks produce debris flow fans but as watershed area increases or relief decreases mixed fans are the result. On high-grade metamorphic rocks fluvial fans are common, and no debris flow fans occur, irrespective of watershed size and relief. The rarity of debris flow fans in the south appears to be largely related to geology and relief.

MORPHOLOGICAL IMPLICATIONS

Two questions arise concerning the influence of fan sedimentology on fan morphology: to what extent do the constituent materials influence surface form?, and to what extent may these relationships influence fan aggradation and dissection during fan development? Much of the literature on alluvial fans deals with the influence of source area characteristics on fan morphology, particularly the relationship between drainage area and fan slope, while some studies (e.g., Bull, 1962; Hooke, 1968) have examined the influence of rock type.

For the Spanish sample of 31 fans, watershed size, relief and slope and fan slope are given in Table 3. The relationship of fan slope to drainage area may be expressed by

$$S_f = kA_d^m$$

where S_f is fan slope. For all 31 fans (Fig. 6) the relationship is

$$S_f = 0.079A_d^{-0.208}$$

with a correlation coefficient of −0.565 significant at the 1% level. There is considerable scatter on Figure 6, but when regression equations are calculated for the separate fan groups (Table 4) correlation tends to improve and standard errors decrease. However, in some cases the significance is reduced by the reduction in sample size. Northern and southern fans tend to plot separately with southern fan slopes lower than those to the north. For the separate regressions values of k and m are higher for the northern group, reflecting the steeper slopes especially of fans from small drainage areas in this group. When the groups are further subdivided according to geology, only the Pre-Betic limestones and southern high grade metamorphics yield significant relationships, again the northern group with higher values of k and m. When the grouping is on the basis of deposits, the results for the intermediate and fluvial types (Groups B and C) are generally similar to those for the southern fans as a whole, but the poor correlation for debris flow fans (Group A) is influenced by the aberrant behaviour of the southern-most three (Fans 23,

26, 27). If these are omitted from the calculations (Group A_2) the correlation improves, but because of the small sample size is still not significant. Although these results accord with previous work (*e.g.*, Bull, 1962) with fans from rock types yielding coarser sediments (and possibly, more debris flow activity having higher values of *m* than those from more fissile rocks or more fluvial deposition), they give no direct evidence as to which of source area geology, relief, or climate is the most important factor.

An alternative approach is to consider the residuals from the general regression (Table 5, Equation (a)), which reflect the scatter of points above and below the regression line on Figure 6. Northern Pre-Betic, Carrascoy groups and Groups A and A_2 (debris flow fans) have mean positive residuals, the others mean negative residuals. Tests indicate the northern group to be significantly different from the southern group, the Pre-Betic group from both southern lithological groups, and the Carrascoy group from the southern sedimentary groups. The differences between the Pre-Betic and Carrascoy, and between the southern metamorphics and both Carrascoy and southern sedimentary groups, are not statistically significant. The debris flow group, whether or not Fans 23, 26 and 27 are included, are significantly different from the intermediate and fluvial groups. Interestingly the residuals for Fans 23, 26 and 27 are more than two standard errors from the mean for the other debris flow fans, again suggesting an important difference here.

These results confirm the earlier regression results but the roles of topographic and geological factors are clarified by the following multiple regression relationship

$$S_f = 0.00092\, A_d^{-0.500} R_b^{0.748} S_b^{-0.107}$$

where S_b is basin slope. The multiple correlation coefficient is 0.836, an improvement in the coefficient of determination from 32% for the bivariate regression to almost 70%, and a reduction in the standard error of the estimate from 0.179 to 0.124 (log units). Separate multiple regressions were calculated for drainage area with basin relief and basin slope yielding multiple correlation coefficients

Fan	Fan Type[1]	Drainage Area (km²)	Basin Relief (m)	Basin Slope	Fan Slope	Channel Type[2]
1. Polop	A	3.40	810	.27	.111	X
2. La Nucia	A	1.04	790	.39	.176	X
3. Tapia	B	6.94	1118	.25	.057	X
4. Cayola	A	0.68	305	.28	.100	X
5. Robelles N	A	0.62	388	.29	.136	X
6. Robelles S	A	0.47	380	.36	.170	X
7. Roy	A	3.51	663	.23	.120	Y
8. Ginesa	A	2.16	720	.25	.110	Z
9. La Murta	A	2.91	680	.19	.110	Z
10. Herradon	B	4.13	666	.23	.061	X
11. Corachos	B	4.58	410	.16	.033	X
12. Carrascoy	A	2.01	543	.26	.096	Z
13. Ros	A	1.84	480	.20	.070	X
14. Cambron	C	3.00	498	.18	.055	Z
15. Purias	C	26.16	420	.06	.025	Z
16. Aljibejo	B	6.31	244	.07	.041	Z
17. Villaescusa	B	1.17	220	.12	.056	Z
18. Las Blanquizares	B	3.80	520	.15	.051	Z
19. Corascos	B	4.60	330	.07	.045	Z
20. Caledron	B	0.11	92	.18	.093	Z
21. Alvantos	B	1.71	166	.07	.063	Z
22. Chapi	B	0.43	195	.19	.035	Y
23. Lisbona	A	0.15	92	.20	.081	Y
24. Nijar	B	9.75	731	.18	.047	X
25. Sierra	B	3.32	632	.19	.050	Y[(3)]
26. Ceporro	A	0.38	155	.24	.081	Y
27. Los Arcos	A	0.67	170	.21	.066	Z
28. 'Cabrillo'	C	1.50	389	.26	.076	Z
29. Honda	C	27.10	1180	.11	.037	Z
30. 'L. Honda'	C	0.28	265	.48	.126	Z
31. 'L Nudos'	C	0.83	285	.21	.082	Z

Notes: Drainage area, basin relief and basin slope (basin relief/basin length) derived from topographic maps; fan slope (fan centreline slope, not necessarily identical to channel line slope; for the upper portion of the fan) measured in the field.

[1]See Table 2
[2]X - Fan channels with marked headcut development including intersection point headcuts
Y - Fan channels with limited scour, minor headcuts only
Z - Simple fan channels with no headcuts
[3]Headcut on main through channel well downstream of the fan.

Table 3. Source area and fan characteristics.

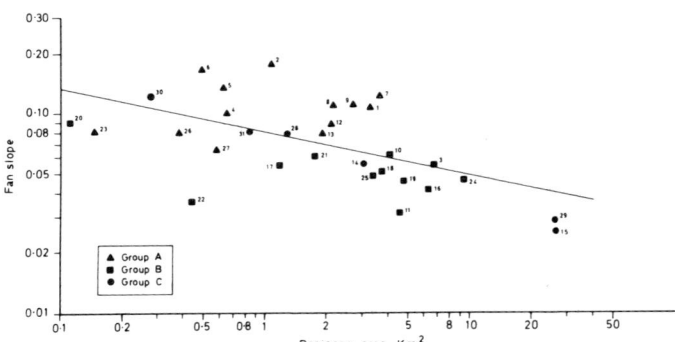

Fig. 6. Fan slope (S_f) in relation to drainage area (A_d, km²). Regression line, $S_f = 0.079 A_d^{-0.208}$; correlation coefficient -0.565, significant at the 1% level. For fan groups see text.

Fan Group[1]	n	R	k	m	Signif. level	Standard Error (log units)
All fans	31	−.565	.079	−.208	1%	.179
All N. Fans (1-13)	13	−.683	.123	−.379	5%	.155
All S. Fans (14-31)	18	−.780	.063	−.192	0.1%	.122
Pre Betic Group	6	−.772	.127	−.300	5%	.129
Carrascoy Group	7	−.519	.158	−.657	NS	.188
S Met. Group	10	−.937	−.076	−.285	0.1%	.073
S Sed. Group	8	−.605	.058	−.145	NS	.124
Group A	13	.139	.105	.043	NS	.137
Group A_2	10	−.461	.124	−.169	NS	.111
Group B	12	−.456	.055	−.098	10%	.113
Groups B + C	18	−.715	.064	−.194	0.1%	.124
Group C	6	−.973	.083	−.306	0.1%	.066

[1]Fan Groups:— Pre Betic Group, Fans 1-6; Carrascoy Group, Fans 7-13; Southern Metamorphic Group, Fans 14-17, 24, 25, 28-31; Southern Sedimentary Group, Fans 18-23, 26, 27; Group A, Debris flow group, Fans 1, 2, 4-9, 12, 13, 23, 26, 27; Group A2 as Group A less the three southern fans (23, 26, 27); Group B, Intermediate group, Fans 3, 10, 11, 16-22, 24, 25; Group C, Fluvial group, Fans 14, 15, 28-31.

Table 4. Regression relationships $S_f = k A_d^m$ (S_f is fan slope, A_d is drainage area, km²).

Fan Group[1]	n	Residuals from Equation (a)		Residuals from Equation (b)	
		mean	SD	mean	SD
All N Fans (N)	13	.144	.157	.035	.106
All S Fans (S)	18	−.099	.109	−.025	.121
Pre-Betic Group (PB)	6	.109	.123	.049	.094
Carrascoy Group (CY)	7	.105	.182	.022	.122
S Met. Group (SM)	10	−.060	.085	−.018	.085
S Sed. Group (SS)	8	−.147	.122	−.034	.162
Group A	13	.124	.169	.064	.068
Group A_2	10	.198	.106	.079	.071
Group B	12	−.109	.136	−.062	.148
Groups B C	18	−.084	.127	−.045	.125
Group C	6	−.036	.099	−.015	.054

[1] as for table 4

Note: Students t test gives the following pairs significantly different at the 5% level between their mean residuals from the regression equations above:—

Equation (a) N-S, PB-SM, PB-SS, CY-SS
A-B, A_2-B, A-BC, A_2-BC, A-C, A_2-C.

Equation (b) A-B, A_2-B, A-BC, A_2-BC, A-C, A_2-C

Table 5. Residuals from regressions
a) $S_f = 0.079 A_d^{-0.208}$
b) $S_f = 0.00092 A_d^{-0.500} R_b^{0.748} S_b^{-0.107}$

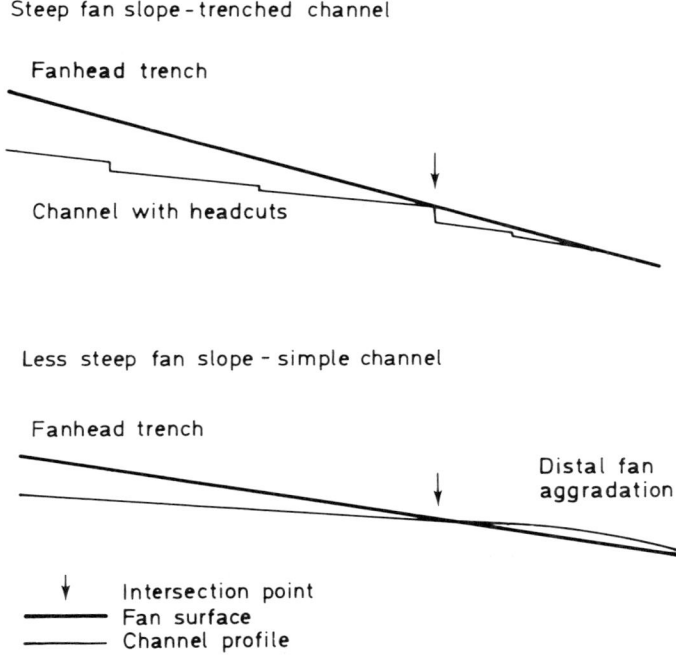

Fig. 7. Types of fan surface and fan channel profiles.

of 0.834 and 0.782, but as there is little correlation between basin slope and relief the full multiple regression with the three independent variables is used in the analysis of residuals (Table 5, Equation (b)). Mean residuals from this regression equation are markedly lower than those from the bivariate regression, but significant differences still exist between the debris flow groups and the intermediate and fluvial groups when classified on an area basis. Furthermore the residuals for the three southern debris flow fans (Fans 23, 26, 27) are still more than two standard errors from the mean of the other debris flow fans.

The implications are that drainage basin size, relief and geology all influence fan processes. Smaller steeper basins on sedimentary or low-grade metamorphic rocks yielding coarse clasts and adequate fine sediment favour debris flow deposition. Larger, less steep basins, especially on fissile high-grade metamorphic rocks that yield little fine sediment favour fluvial deposition. A relative abundance of debris flows, and possibly also the coarseness of fan debris, favours steeper fan slopes. In addition, there is some evidence that the lower fan slopes to the south may not entirely be due to topographic or geological factors in that debris flow fans in their area have low fan slopes, despite their small drainage basin areas. This may reflect the southward aridity gradient in one of two ways. In the drier areas, there is less evidence of a pre-existing soil cover to yield abundant fine sediment for debris flow processes. Secondly, the greater effectiveness of the extreme event (Baker, 1977) in more arid areas may result in higher effective discharges and consequently lower slopes per unit drainage area.

These implications are important not only for fan surface morphology but also for fan aggradation and dissection behaviour (Harvey, 1982). At the present time all the fans are incised and dominated by channel processes up-fan of intersection points with significant deposition occurring only on the distal fan surfaces. The modern trenched channels are adjusting their slopes to modern fluvial conditions, these slopes being less than the fan surface slopes. Many channels show strong headcutting especially where the lesser channel slope of the fanhead trench gives way at the intersection point to an inherited steeper fan surface slope. Headcut formation would be most likely where the discrepancy between the slopes is greatest; where it is least, simple distal fan aggradation might be more likely (Fig. 7). This is borne out by the modern channel morphology (Table 3), where although calcrete crusts may have influenced channel behaviour, there is a clear tendency for the steeper debris flow fans to have modern channel headcuts and less steep fluvial fans to have simple modern channels without headcut development. A similar behaviour in the past could have lead to a greater complexity in the development of the northern fans with more debris flows and steeper slopes than of the southern fans. This is borne out by the field evidence (Harvey, 1982).

Clearly, fan sedimentology has important implications for understanding fan morphology — but morphological aspects on the other hand, especially aggradation/trenching behaviour, are strongly related to the nature and distribution of processes and deposits.

REFERENCES

Amor, J.M. and Florschutz, F. 1964. Results of the preliminary palynological investigation of samples from a 50 m boring in Southern Spain. Boletin de la Real Sociedad Espanola de Historia Natural (Geologia), v. 62, p. 251-255.

Baker, V.R. 1977. Stream-channel response to floods, with examples from central Texas. Geological Society of America Bulletin, v. 88, p. 1057-1071.

Blissenbach, E. 1954. Geology of alluvial fans in semi arid regions, Geological Society of America Bulletin, v. 65, p. 175-190.

Bluck, B.J. 1964. Sedimentation of an alluvial fan in Southern Nevada. Journal of Sedimentary Petrology, v. 34, p. 395-400.

Bull, W.B. 1962. Relations of alluvial fan size and slope to drainage basin size and lithology in Western Fresno County, California. United States Geological Survey Professional Paper 450B, p. 51-53.

Bull, W.B. 1977. The alluvial fan environment. Progress in Physical Geography, v. 1, p. 222-270.

Butzer, K.W. 1964. Climatic-geomorphologic interpretations of Pleistocene sediments in the Eurafrican sub tropics. *In:* Howell, F.C. and Bourliere, F. (Eds.), African Ecology and Human Evolution. London: Methuen, p. 1-25.

Denny, C.S. 1965. Alluvial fans in the Death Valley region, California and Nevada. United States Geological Survey Professional Paper 466, 59p.

Geiger, F. 1970. Die Aridität in Südostspanian, Stuttgarter Geographische Studien, Band 77, 173p.

Harvey, A.M. 1978. Dissected alluvial fans in southeast Spain. Catena, v. 5, p. 177-211.

Harvey, A.M. 1982. Aggradation and dissection sequences on Spanish alluvial fans. 11th International Sedimentological Congress, Hamilton, Canada. Abstract of Papers, p. 23.

Hooke, R. le B. 1967. Processes on arid region alluvial fans. Journal of Geology, v. 75, p. 438-460.

Hooke, R. le B. 1968. Steady state relationships on arid region alluvial fans in closed basins. American Journal of Science, v. 266, p. 609-629.

Miall, A.D. 1977. A review of the braided river depositional environment. Earth Science Reviews, v. 13, p. 1-62.

Miall, A.D. 1978. Lithofacies types and vertical profile models in braided river deposits: a summary. *In:* Miall, A.D. (Ed.), Fluvial Sedimentology. Canadian Society of Petroleum Geologists, Memoir 5, p. 597-604.

Reineck, H.E. and Singh, I.B. 1975. Depositional Sedimentary Environments with Reference to Terrigenous Clastics. New York: Springer-Verlag, 439p.

Rust, B.R. 1980. Coarse alluvial deposits. *In:* Walker, R.G. (Ed.), Facies Models. Geoscience Canada Reprint Series 1, p. 9-21.

Steel, R.J. 1974. New Red Sandstone floodplain and piedmont sedimentation in the Hebridean province, Scotland. Journal of Sedimentary Petrology, v. 44, p. 336-357.

Wells, S.G. 1978. Geomorphic controls of alluvial fan deposition in the Sonoran Desert, Southwestern Arizona. *In:* Deohring, D.O. (Ed.), Geomorphology in Arid Regions. London: Allen and Unwin, p. 27-50.

SHEET DEBRIS FLOW AND SHEETFLOOD CONGLOMERATES IN CRETACEOUS COOL-MARITIME ALLUVIAL FANS, SOUTH ORKNEY ISLANDS, ANTARCTICA

Neil A. Wells[1]

Abstract

Small, steep, and laterally coalescing alluvial fans that formed under a cool and maritime climate along the sides of a 10 km wide trough are dominated by two types of planar, 1-2 m thick rudites. Sorted, gradually aggraded rudites are interpreted as traction-carpet deposits of sheetfloods. Their characteristics include clast support, moderate organization, sorted matrix, and planar stratification outlined by variable grain size and sorting; normal grading and crossbedding are rarely developed. In contrast, unorganized clast-supported conglomerates were sheet debris flows. Lack of internal partings suggest that they were instantaneously deposited. Clay is mostly less than 5% of matrix; perhaps incorporated snow was an important lubricant.

Sheetfloods and debris flows both appear to have been large, unchannelled, choked with sediment and virtually non-erosive. They seem to have carried all available material (maximum clast sizes and bed thickness are unrelated). Rapid loss of capacity maintained steep fan slopes, as shown by steep primary dips and low-angle down-fan crossbeds. The conglomerates and thin, discontinuous interbedded sandstones are primarily mid-fan deposits.

Résumé

Les cônes de déjection latéralement coalescents, de dimension restreinte et à pente abrupte, formés sous un climat froid et maritime le long des versants d'une vallée large de 10 km, sont composés principalement de deux types de rudites planaires d'une puissance de 1-2 m. Les rudites caractérisées par un triage des grains et une accrétion graduelle sont considérées comme un dépôt d'une charge de fond transportée par traction et résultant d'inondations en nappe. Leurs particularités comprennent des fragments jointifs, une organisation modérée, un triage des grains dans la matrice et une stratification planaire exprimée par l'hétérogénéité granulométrique; et le triage, le granoclassement et la stratification oblique apparaissent rarement. Au contraire, les conglomérats inorganisés à fragments jointifs sont associés à des coulées de débris en nappe. L'absence de plans de stratification indique que le dépôt fut instantané. La matrice renferme généralement moins de 5% d'argile; probablement que la neige mêlée au dépôt fut un lubrifiant important.

Il semble que les inondations en nappe et les coulées de débris ont agi toutes deux sur de grandes superficies, sans creuser de chenaux, elles étaient fortement chargées de sédiment et incapables d'éroder. Tous les matériaux disponibles semblent avoir été transporté (aucune relation entre la grosseur maximale des fragments et l'épaisseur des lits). La baisse rapide de la capacité de débit a maintenu les pentes des cônes abruptes tel que révélé par les pendages primaires forts et l'angle faible du litage oblique dans la zone basse des cônes. Les conglomérats et les couches minces et discontinues des grès interstratifiés sont essentiellement des dépôts du centre des cônes.

Introduction

This paper examines tabular, mostly 1-2 m thick, cobble and pebble conglomerates on apparently cool-climate Cretaceous alluvial fans in Antarctica. Interpreting the mode of deposition of such conglomerates is difficult because of problems of sampling and experimentation, and because aggradational processes are rarely observed. The potential for controversy was shown by Allen (1981) when he re-interpreted the debris flow conglomerates in studies by Bluck (1967), Miall (1970a) and Heward (1978) as deposits of unchannelled flowing water, in effect being broad, low longitudinal bars. This paper attempts to distinguish debris flow and sheetflood deposits.

The second point of this paper is that the South Orkney fans differ from most described fans in ways that may reflect their cool-maritime climate. They seem to have been quite steep, and their grey deposits contain few mud-rich mudflows or braided stream deposits. Unchannelled conglomerates with few sedimentary structures predominate. Models of fans rely heavily on the reddish fan deposits of braided streams and channelled mudflows in the semi-arid southwestern U.S. (*e.g.*, the Trollheim-type vertical sequence of Miall, 1977), but enough studies of different examples may eventually allow a more comprehensive understanding of fans and their processes based on climate (*cf.* Kochel and Johnson, this volume). Studies of various other settings have included those of Drew (1873),

---

[1]Department of Geological Sciences, The University of Michigan, Ann Arbor, Michigan 48109, U.S.A.
Now at Department of Geology, Kent State University, Kent, Ohio 44242, U.S.A.

This research was done while I was at the Institute of Polar Studies of The Ohio State University. It was funded by a Division of Polar Programs (National Science Foundation) Grant No. DPP 74-21509 to Dr. David Elliot, whom I had the privilege and pleasure to accompany in the field. I would like to thank Dr. Elliot for overseeing this work, and the staff of the Institute (particularly Peter Andersen, Rae Mercier, Karen Taylor, and Ron Coffman) for help during the course of the project. I would especially like to thank Captain Peter Lenie and the crew of the R/V *Hero* during cruise 77-1 for their hard work on our behalf. Reviews by Drs. A.D. Miall and R.C. Kochel and editing by Dr. E.H. Koster have considerably improved this paper.

Copyright © 1984, Canadian Society of Petroleum Geologists

Rapp (1960), Anderson and Hussey (1962), Winder (1965), Gole and Chitale (1966), Leggett *et al.* (1966), McGowen and Groat (1971), Ryder (1971), Spalletti (1972), Wasson (1977a,b) and Pierson (1980).

Field notes and an as yet unpublished report on the South Orkneys (Wells, in prep.) are on file at the Institute of Polar Studies, Ohio State University.

GEOLOGICAL AND STRATIGRAPHIC SETTING OF THE CONGLOMERATES

Figure 1 shows the location of the South Orkney Islands and their general geology. The conglomerates represent mid-fan remnants of aprons along each side of Lewthwaite Strait, which separates folded but unmetamorphosed flysch (the Greywacke-Shale Formation) and low-grade metasediments in the east from higher grade schists in the west. Wells (in prep.) suggests that the Lewthwaite Strait may have been a Late Jurassic graben that formed during fragmentation of Gondwanaland. The western conglomerates, next to the schists, are composed of clasts of mica and garnet schists, whereas those in the east consist of clasts of greywacke, shales, and metasediments up to chlorite grade. Clast lithology, therefore, indicates that there are two locally derived sets of conglomerates that face each other across Lewthwaite Strait.

The conglomerates dip away from their source rocks, and using bedding attitudes each set can be subdivided into local groups of radiating dips. Figure 2, for example, shows southward and eastward dips at the foot of Powell Island. The trends of trough axes in sandstone interbeds generally agree with dip directions of nearby beds, which would not be the case if significant folding or cross-slope tilting has occurred. Consequently, changes in dip directions are interpreted as reflecting the formation and interfingering of several fan-shaped lithosomes along both sides of Lewthwaite Strait. Early researchers named the eastern conglomerate the Powell Island Conglomerate and the western, the Spence Harbour Conglomerate (Matthews and Maling, 1967; J. Thomson, 1971, 1973, 1974).

The conglomerates are seen in isolated islands and ice-bound cliffs, so stratigraphic relationships are mostly indeterminable. However, the Spence Harbour Conglomerate is seen on schist at four sites (J. Thomson, 1974;

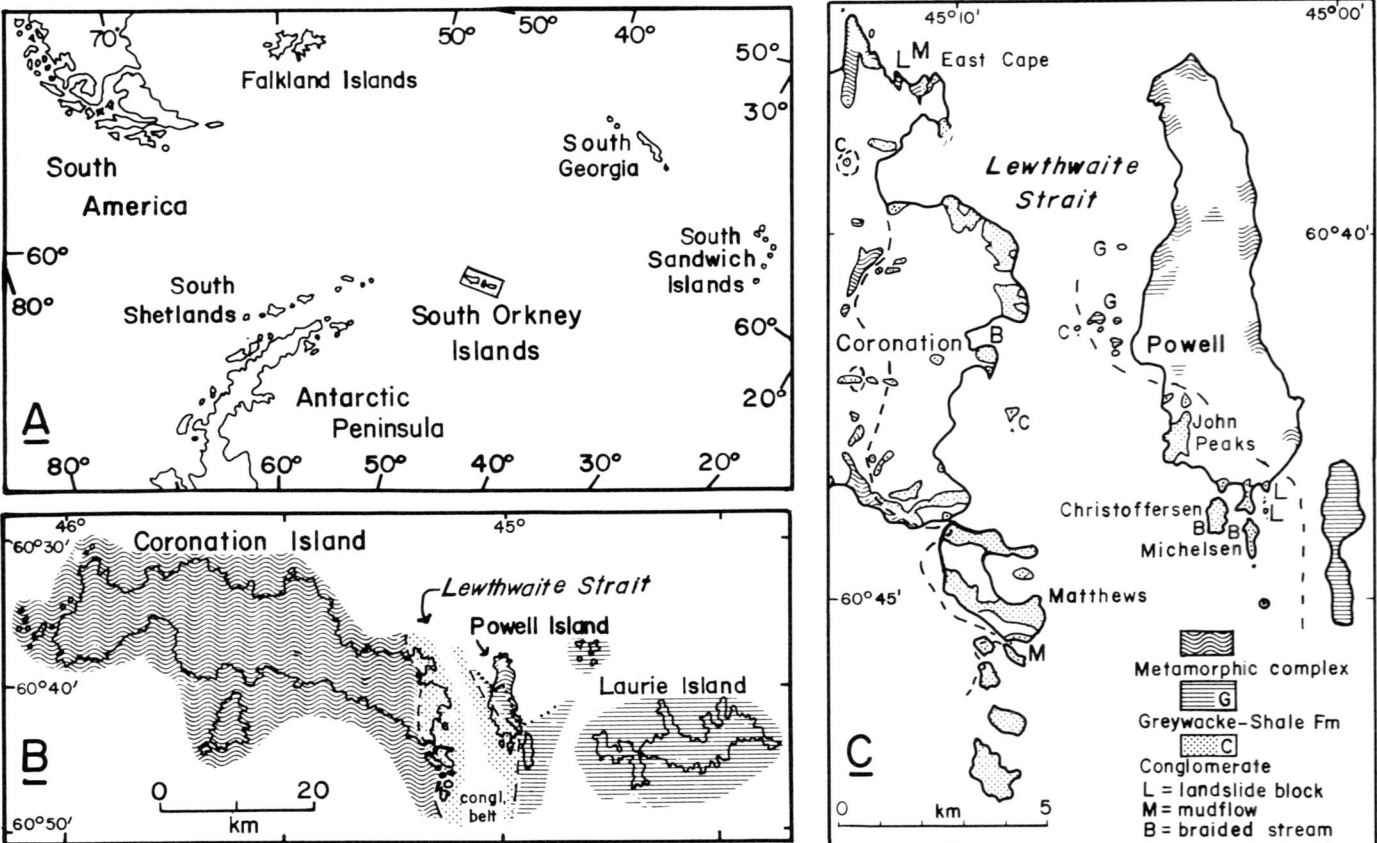

Fig. 1. Location maps for the South Orkney Islands. **A:** The Scotia arc; **B:** The South Orkney Islands (wavy lines = schist, straight lines = pre-Jurassic flysch, dots = conglomerate) (after J.W. Thomson, 1971). **C:** Geology of Lewthwaite Strait (from Dalziel et al., 1977). Note proximal positions of mudflows and landslide blocks (M and L).

Elliot and Wells, 1982), with unconformably interposed Late Jurassic — earliest Cretaceous thin marine shales, or with boulders of related marine sandstone, in three more places: fossils have been dated by M. Thomson (1975, pers. comm. in 1979) and Thomson and Willey (1975). At East Cape (Fig. 3), basal 2 m thick, mud-supported conglomerates overlie an irregular surface of weathered schist and pass up into less muddy unorganized conglomerates. The unconformity, which is believed to represent a pediment surface (cf. Williams, 1969), has some small resistant bosses and is inclined slightly to the southeast. Varied elevations of schist and conglomerate exposures around the Cape and a 30 by 15 m block of schist low in the conglomerates indicate considerable paleo-relief, perhaps even cliffs, nearby (cf. J. Thomson, 1974).

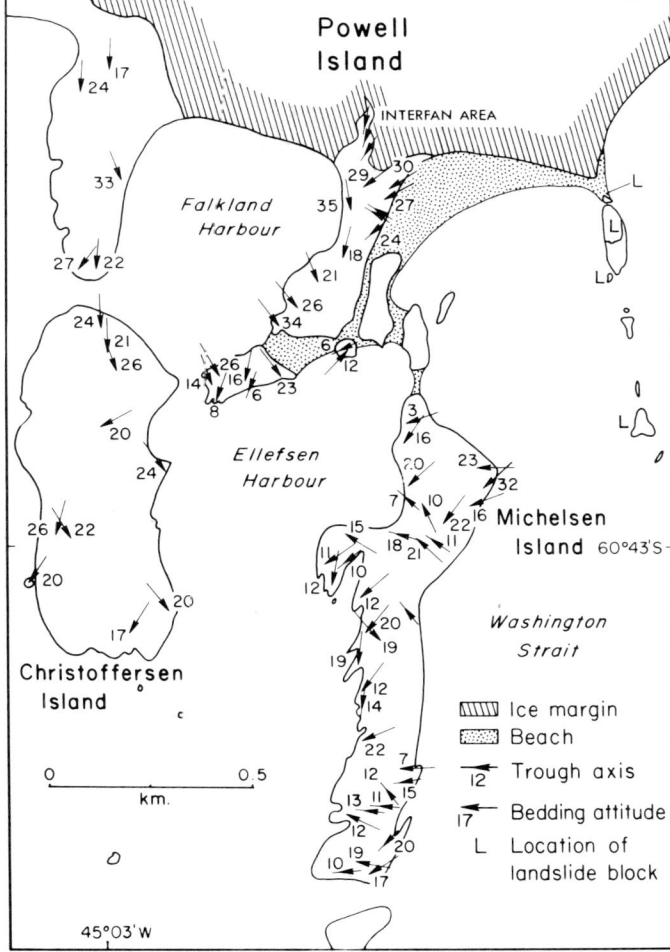

Fig. 2. Bedding attitudes, measured on sandstone interbeds, are believed to approximate radial paleoslopes. One south-facing fan can be identified over Christoffersen Island and Falkland Harbour, and a west-dipping group of perhaps three sub-fans can be identified over Michelsen Island. The fans interfinger in the area east of Falkland Harbour. Not shown are another west-dipping fan that was identified west of central Powell Island and fans along western Coronation Island that seem to have coalesced into an east-facing apron (see Elliott and Wells, 1982, Fig. 26-12).

In 414 m of strata at John Peaks, the organized conglomerate facies pass up into thinly bedded conglomerates and ultimately into channel-form sandstones and well-bedded fines with abundant (possibly Cretaceous) fossil leaves. On Matthews Island, 80 m of thin, flaggy, locally rippled shallow-marine sandstones with sparse pebbles with broken Early Cretaceous ammonites eventually coarsen upward to thinly-bedded conglomerate which then pass up into the main conglomerate facies (J. Thomson, 1974; Elliot and Wells, 1982). Michelsen Island and the Powell Islets show a kilometer-long transition from huge blocks near source rock outcrops through the predominant conglomerate facies to distal multi-channel form gravels. These several transitions are schematically depicted in Figure 3.

If the conglomerates are associated with both marine and continental beds, to which environment do they belong? Most subaerial and submarine fans are readily distinguishable by fossils, colouration or stratigraphic context, but both are fan-shaped, and mid-fan sections of either can be dominated by unfossiliferous grey conglomerates and sandstones deposited by traction and sediment-gravity flows. Therefore, grey coarse-grained and unfossiliferous remnants like the South Orkney conglomerates are not as readily interpreted as one might expect (cf. Heward, 1978; Surlyk, 1978; Allen, 1981). In the present case, evidence for a terrestrial origin for the conglomerates includes the abundance of twigs in sandstone interbeds, as well as a tree trunk and possibly in situ roots east of Falkland Harbour, the lack of marine fossils (except in the marine sandstone unit) and shales (pelagic and hemipelagic muds are commonly associated with submarine fans). The buried pediment and the transitions to non-conglomeratic facies shown in Figure 3 also support subaerial formation.

DEPOSITIONAL SLOPES

Dip magnitudes in Figure 2 presently average 19° (16° on Michelsen, 21° on Powell, and 22° on Christoffersen). They are normally distributed and range from 0° to 35°, with 78% between 10° and 27°. How much of each slope could be primary? No dips are impossibly steep, but downslope tilting (into Lewthwaite Strait) has occurred locally in the Spence Harbour Conglomerate, and could easily have occurred in the Powell Island Conglomerate as well. The amount of tilting cannot exceed the lowest accepted dip, otherwise restoration of slopes would reverse some of them. If the lowest and highest 10% of the values are discarded as erroneous, perhaps representing eroded surfaces and non-axial remnants of sand-lined channels, then the data could permit about 10° of tilting. This would reduce the average dip to 9°, which is within the following range of observed steep depositional dips.

For modern California fans, maximum slopes have been variously reported as 2-11° (Lustig, 1965), 6° (Blackwelder, 1928) and 9-13.5° (Hooke, 1967); fans produced in the laboratory by Hooke averaged 4-8° and were as steep as

7-13°. In southern Arizona, Melton (1965) observed a maximum of 10°. Small Alaskan fans observed by Anderson and Hussey (1962) slope at 10-21°. Mudflows in the Rockies and California come to rest at 7-20° (Broscoe and Thomson, 1969), 7.5-12° (Curry, 1966) and 7-15° (Fryxell and Horberg, 1943), although some locally continue to 5° (Singewald, 1928), 3° (Fryxell and Horberg, 1943) or 1° (Sharp and Nobles, 1953). For scree slopes in Pakistan, Wasson (1979) reported angles of 30-38°.

The assumption that primary Cretaceous dips have remained essentially undisturbed is reasonable because since the South Orkneys broke off Gondwanaland in the Mesozoic, they have escaped further collision or metamorphism.

Sedimentary Facies and Processes

Six partly intergradational coarse-clastic facies are recognized. Landslide blocks, mud-supported conglomerates and multi-channel form conglomerates are rare, whereas sheet-form clast-supported conglomerates of two types and sandstone interbeds are common.

LANDSLIDE BLOCKS

Clasts wider than 0.5 m are uncommon in the eastern conglomerates, and only 25 1-4 m boulders were noted in the west. However, both areas contain rare blocks of 100 to >10,000 m³.

Blocks of immense size seem likely to have been produced and deposited in a single event from cliffs, and can therefore be termed landslide blocks (*sensu* Coates, 1977). Crushing and soft-sediment deformation of conglomerates beneath most large blocks suggest high-speed emplacement, perhaps by rockfall (see Coates, 1977) or thrown by rock avalanche (see Plafker and Erickson, 1978). Crush foliation can be seen to curve around the base of a landslide block shown in Figure 26-3 in Elliot and Wells (1982). Large boulders could likewise have been created and trans-

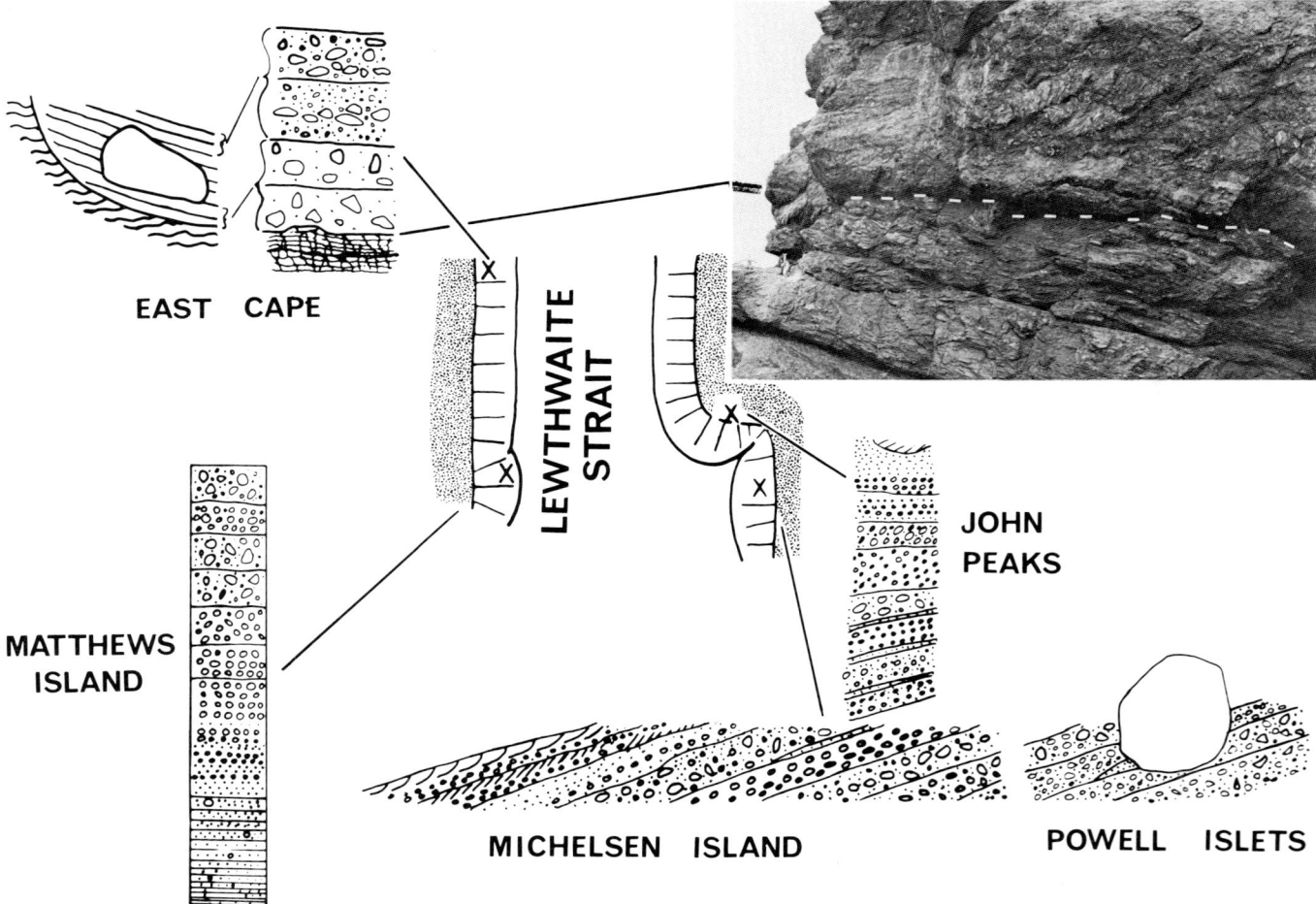

Fig. 3. Idealized cartoon of the transitions to non-conglomeratic facies, using unscaled sections. The two sets of fans bordering Lewthwaite Strait show progradation of mid-fan conglomerates over shallow-marine (fan-delta?) sandstones on Matthews Island, change from proximal conglomerates and landslide blocks around Powell Islets to distal braided stream deposits in the west and to floodplain deposits up John Peaks. At East Cape, basal and proximal mudflows near cliffs and a landslide block bury a deeply weathered schist surface (dashed lines in photo). This surface is interpreted as a terrestrial pediment. (Photo by D.H. Elliott; 35 cm penguins in lower left for scale.)

ported by rockfall (see Rapp, 1960), but most seem to be incorporated into conglomerate beds and probably represent the coarsest percentiles of 'normal' eroded and transported debris. Transport of house-sized blocks by mud-flows, debris flows and channelled floods has been documented (*e.g.*, Blackwelder, 1928; Chawner, 1935; Beaty, 1963; Broscoe and Thomson, 1969).

UNORGANIZED, MUD-SUPPORTED CONGLOMERATES

In these very poorly sorted beds, abundant clay separates the clasts, which range greatly in size (Fig. 4A). Some clasts are oriented vertically and protrude from the flow. The beds are tabular, about 2 m thick, unorganized and unstratified. Basal scouring is minimal. Unorganized fabrics and mud support identify these beds as mudflows

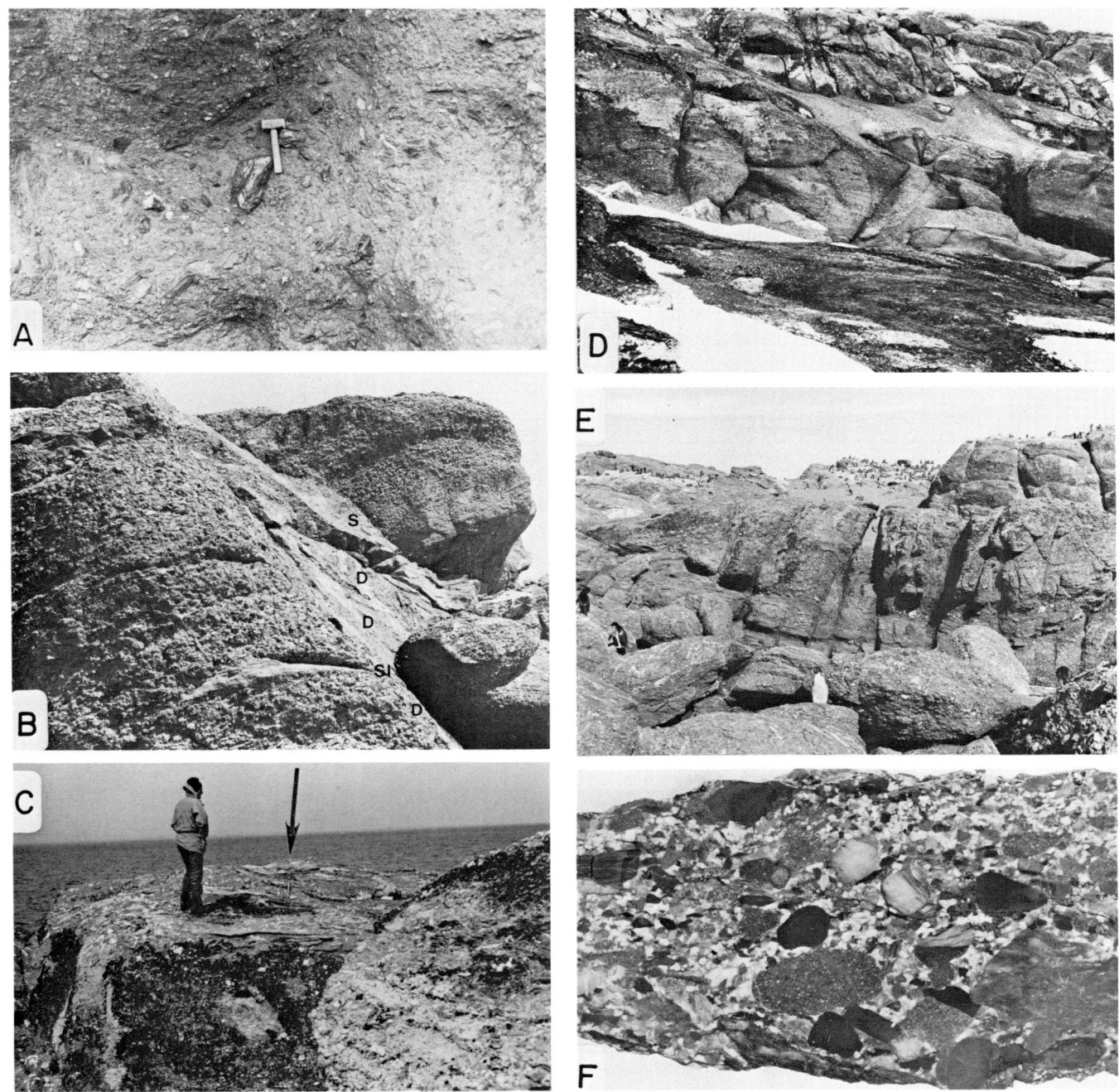

Fig. 4. **A:** Mud-supported conglomerate of mudflow origin south of Matthews Island; note the vertically oriented clast. Photo by D.H. Elliot. **B:** A laterally eroded sandstone interbed (SI) between two debris flow conglomerates (D), with S denoting a sheetflood conglomerate. Note that conglomerates are tabular and separated by thin partings or sandstone interbeds. Powell Island Conglomerate. **C:** Thin sandstone interbed infilling a shallow trough. Jacob staff (under arrow) lies along trough axis; Michelsen Island. **D:** Multi-channel form braided stream gravels, southern Christoffersen Island. **E:** Multi-channel form stream complex, western Michelsen Island. Photo by D.H. Elliot. **F:** Well-sorted, biomodal, pebbly granulestone from exposure in D.

(*sensu* Bull, 1972; Sharp and Nobles, 1953). Protruding clasts and poor sorting attest to high density and viscosity (Walker and Mutti, 1973; Walker, 1975), and vertical clasts indicate turbulence. The mudflows appear to have been broad, unchannelled and non-erosive, although modern mudflows are frequently lobate (*e.g.*, Hooke, 1967).

In a unique interfan area of muddy and root-ridden sandstones located northeast of Ellefsen Harbour at the confluence of the Michelsen and Christoffersen fans (see Fig. 2), small mudflows and sandstones are confined in 1 to 2 m deep channels. The direction and geometry of these mudflows were clearly controlled by pre-existing stream-channel relief, and there are signs that confinement caused additional scouring.

MULTI-CHANNEL FORM CONGLOMERATES

This uncommon facies in distal areas consists of well-stratified, 2 to 5 m thick fine conglomerates and conglomeratic sandstones (Figs. 4D-F). Deposition in multiple events is shown by this textural variation and the abundance of cross-cutting sets of low-angle crossbeds. Some gravelly complexes lie within broad, >30 m wide channels, although secondary cross-cutting commonly obscures the principal surface of accumulation. The apparently frequent changes in local bed topography between channels and bars suggest deposition by braided streams (Miall, 1977).

A related facies comprises well-sorted and thinly-layered fine conglomerates that lack crossbeds and cross-cutting channels. Individual beds are lenticular, ranging from 3-60 cm in thickness, and include horizons of pebbles, granules and sand (Fig. 5E). They occur between thick tabular conglomerates and marine sandstones on Matthews Island and between conglomerates and fluvial channel-form sandstones with overbank fines on John Peaks. These locations, together with their relatively thin-bedded, finer grained nature suggest a distal depositional site. Stanley (1971) and Bull (1972, Fig. 3B) attribute this kind of facies assemblage to small braided streams and small, shallow, distal sheetfloods, respectively.

CLAST-SUPPORTED SHEET CONGLOMERATES

Most South Orkney conglomerates are polymodal, clay-poor and clast-supported. They generally form well-bedded, .5 to 2 m thick, unchannelled, extensive sheets on the basis of exposures 20-200 m wide. Of these, two facies are distinguished: tabular beds that show internal organiza-

Facies	Description	Position and Abundance	Interpretation of Process
1. A) Very large blocks	100-10,000 m^3 boulders	4 in PIC, 1 in SHC* Proximal and basal	Landslide blocks from cliffs
B) Large boulders	Rounded 1-4 m boulders	25 in SHC	Some are rockfall boulders, later moved by floods
2. Unorganized mud-supported rudites	Polymodal, mud-supported, tabular, 1-2 m thick; no organized fabric	Proximal and basal; known at 7 localities, most in SHC**	Turbulent unchannelled mudflows
3. Unorganized clast-supported rudites	Polymodal, massive, clast-supported, tabular, 1-2 m thick; no internal partings with clay-poor matrix	Forms about half the conglomerates**	Sheet debris flows — instantaneous deposition by unchannelled turbulent mass-flows
4. Organized, clast-supported rudites	Clast-supported, clay-free, tabular, 1-2 m thick; heterogeneous sorting with some normal grading	Forms about half the conglomerates	Sheetfloods — aggradation by unchannelled floods
5. Sandstone interbeds	Impersistent granulestones and sandstones, 5-230 cm; very low angle crossbedding, if any	Abundant between most conglomerates, but volumetrically unimportant	Waning sheetfloods, separate small floods and dewatering of debris flows
6. A) Multi-channel form conglomerates	Thick, cross-cutting, non-tabular, crossbedded gravels in channels	Top and distal At 5 known localities most in SHC	Small braided stream complexes
B) Thinly layered gravel complexes	Flat, thinly layered, well-sorted fine gravels	Distal, 2 localities SHC and PIC	Small, shallow, and anastomosing distal floods

*PIC = Powell Island Conglomerate, SHC = Spence Harbour Conglomerate (E. and W. of Lewthwaite Strait)
**Several thousand beds were observed; specific data (sections, samples, measurements) were collected at 95 selected localities.
Remarks: Intermediates exist between facies 2 and 3; 3 and 4 are not always distinguishable; beds of 4 may grade up into 5; and facies 5 is in a sense a thin and fine version of facies 6.

Table 1. Fanglomerate Facies in the South Orkney Islands

tion are identified as sheetflood deposits, but at least as many are almost completely unorganized and they are identified as sheet debris flow deposits (Fig. 4B).

Sheetflood deposits are not necessarily well-stratified, but show some evidence of gradual accretion, such as internal partings and lateral and vertical variations in grain size and sorting (Figs. 5A-C). Crossbedding and clast imbrication are relatively uncommon. Local sorting is good: modal and maximum clast sizes are similar, and matrix is principally sand. Local shallow scouring of underlying sands is rare. In general, aggradation of thick tabular beds of sorted material implicates very large, sediment-charged and unchannelled floods. Lensing, grading, and crossbedding probably reflect anastomosing flow within the flood (McGee, 1897). These would be sheetfloods *sensu* Bull (1972) (not Davis, 1938, and others who included mudflows; see Hogg, 1982), and are similar to conglomeratic deposits described by Allen (1981).

In contrast, debris flow deposits (*sensu* Bull, 1972) are identified by their very poor sorting and unstratified nature (Figs. 4B and 5D; Fig. 26-5 of Elliot and Wells, 1982), suggesting instantaneous deposition of thick chaotic masses (Fisher, 1971). Grading is essentially absent and clasts are almost randomly oriented, as would be expected for debris flows. However, some beds have 5 cm thick sandy tops (perhaps from post-depositional dewatering or waning surface flow), and, rarely, the coarsest clasts are aggregated into a horizon low in the bed even though smaller clasts remain unsorted. Also, clast orientation in some beds tends towards subparallelism with bedding suggesting flow by laminar shear rather than turbulence (Lindsay, 1968; Enos, 1977). For ungraded types of resedimented conglomerates, Walker (1975) suggested that limited clast movement within a flow is due to low water content, in contrast to the wetter and better graded submarine debris flows. However, Hein (1982) suggested that lack of grading could also be due to reworking of deposits of the head of a flow by its more quickly moving body, syndepositional mixing and churning, and rapid deposition as flow competence declines. There is no internal stratification that might suggest that the debris flows arrived in a succession of pulses, as many have been seen to do (*e.g.*, Rickmers, 1913). The flows were almost non-erosive, like those observed by Prior and Stephens (1972, Fig. 2) and Curry (1966) and probably resembled the planar, 2 m thick, very broadly lobate, low-clay, sheet debris flows studied by Pierson (1980).

Debris flows are distinguished from mudflows by their low clay content (<5% in thin section, which is <1% in the whole bed due to sampling bias) and clast support. Clay could have been lost in dewatering, but active debris flows with very little clay have in fact been recorded. Curry (1966) observed boulder-carrying flows with a 'matrix' of only 1% clay and 9% water, and Pierson (1980) observed very fluid flows with 4% clay. Miall (1970a, p. 128; 1970b, p. 563) has identified ancient 'debris floods' with mud contents of <1% and Lustig (1965) notes many old mudflows with only 2-4% clay. Lawson (1982) notes that at water contents about 17%, lobate flows moving by basal shear begin to change into fluid, channelized, turbulent flows. Rodine and Johnson (1976) calculated that because each size fraction in very poorly sorted flows can keep clasts of the next larger size from interlocking, such flows may be 89-95% clasts. Additional support probably relates to the yield strength of the flow, upward migration of fluid, grain-to-grain dispersion (increased by turbulence), and/or buoyance (Fisher, 1971; Enos, 1977; Hampton, 1979; Pierson, 1981). Momentum would be gained from the steepness of the slopes. Buoyancy, dispersive pressure and vertical variations in yield strength and density due to changes in water content may be responsible for the above-mentioned aggregation of very coarse clasts in an otherwise homogeneous laminar flow, if the coarsest clasts alone were able to overcome the flow's yield strength and move across shearing to a level of neutral buoyancy (Sparks, 1976).

The degree of matrix sorting appeared to be a practical field criterion for distinguishing clay-free sheetflood beds from poorly sorted debris flow conglomerates (Fig. 5D). Figure 6 illustrates this tendency, although overlapping between even the selected data sets for sheetfloods and debris flows invalidates this as a discriminatory test. Analyses of packing were also attempted in order to differentiate the conglomerates and to check dispersion of like-sized grains, as required by the mass flow models of Rodine and Johnson (1976). Values of packing proximity (Fig. 7) generally suggest greater clast dispersion in debris flows (only two values are above 0.60) than sheetfloods (only one value is below 0.60). Part of the difficulty in comparing these analyses is that matrix in conglomerates is not rigorously definable, *i.e.*, a small pebble could function as interstitial matrix in a boulder conglomerate. The proportions of contacts between different sizes of clasts are highly variable and only show that, as expected, the generally better-sorted waterlain conglomerates tend to have more contacts between similarly sized clasts.

Bluck (1967), Steel (1974) and subsequently others report process-specific relationships between bed thickness and maximum clast sizes, but these variables are unrelated in the South Orkney beds (Fig. 8) using apparently non-composite beds. Debris flows and sheetfloods must, therefore, have been competent to transport all available detritus although the coarsest percentiles are relatively well-sorted. Presumably the calibre of South Orkney fan deposits mostly reflects cobble and pebble-sized detritus in the source area, with a relatively low degree of sorting accomplished by transport to the fan.

SANDSTONE INTERBEDS

The conglomerates are commonly interbedded with 5-30 cm thick granulestones and sandstones (Figs. 4B and C). Some are normally graded or massive or have very low angle crossbeds but most show horizontal stratification.

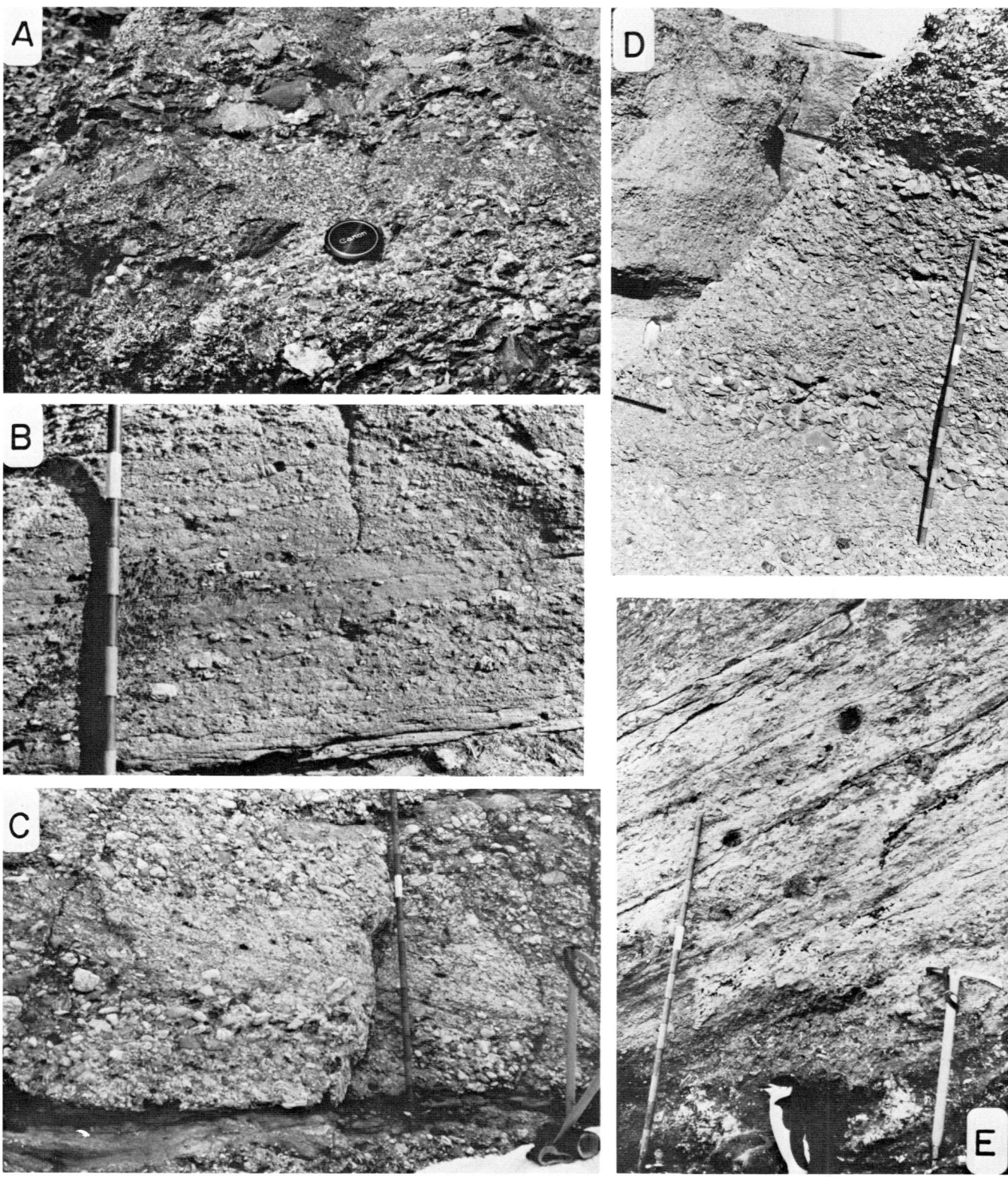

Fig. 5. Sheetflood conglomerates. **A:** Lateral and vertical variations in clast size and sorting; Michelsen Island. **B:** Tabular bed of horizontally stratified pebbly sandstone, Michelsen Island. Pebbles are oriented parallel to bedding. Divisions of Jacob staff are in decimeters. **C:** Planar crossbed in an otherwise poorly sorted and massive sheet-form conglomerate, bounded above and below by sandstone interbeds. John Peaks. **D:** Massive, clast-supported conglomerate with tightly packed clasts oriented subparallel to bedding, capped by an unorganized, poorly sorted debris flow conglomerate. Michelsen Island. Photo by D.H. Elliot. **E:** Thinly layered pebbly conglomerates above marine sandstone on Matthews Island. These beds have been tilted away from their source.

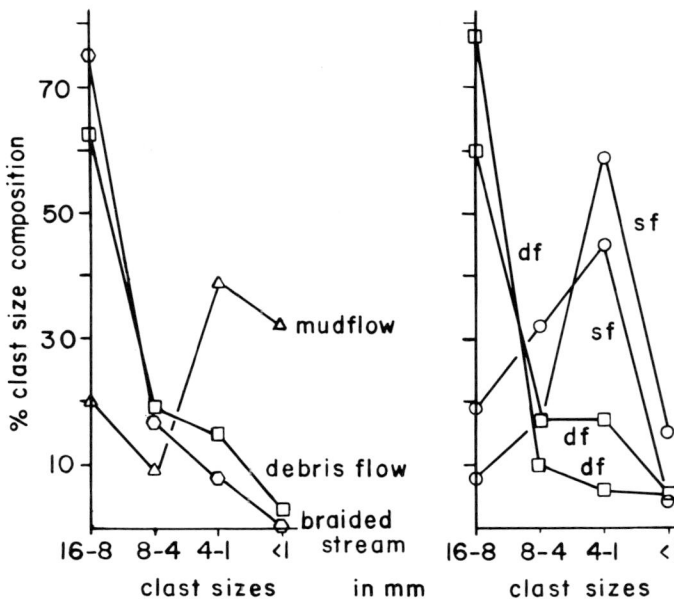

Fig. 6. Using selected data from polished slabs, the left graph shows how different processes can differently concentrate small clasts and matrix. The right graph compares debris flow (df) and sheetflood (sf) deposits. Selective concentration of mid-size fines was seen in 5 of 7 sheetfloods whereas abundance decreases with size in 7 of 11 debris flows. However, variability is high.

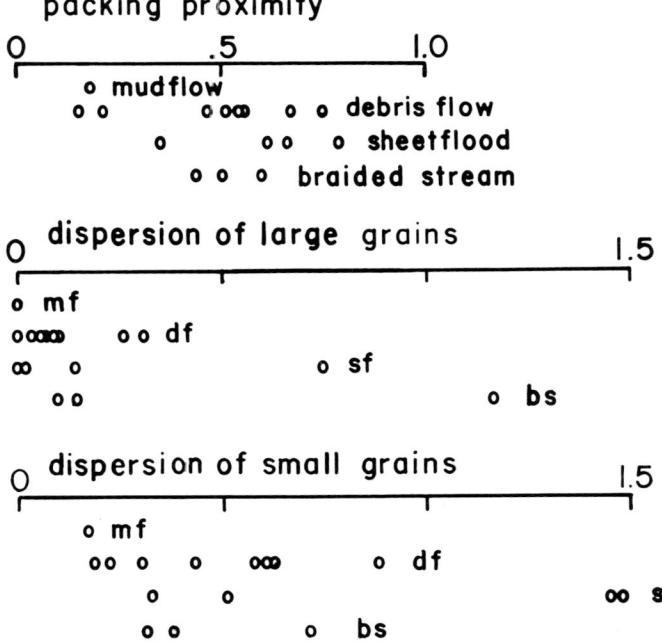

Fig. 7. Values for dispersion of grains by 'matrix' by different processes, measured on polished slabs. Packing proximity = ratio of grain-grain contacts to grain-matrix contacts (matrix = <2 mm) (adapted from Kahn, 1956). Dispersion of large grains = ratio of contacts between grains <1 cm to contacts with smaller grains and 'matrix' (<2 mm). Dispersion of small grains = similar ratio for 2-10 mm grains. Like-size grains are in contact with each other less frequently in mudflows and some debris flows than in most sheetflood and braided stream deposits.

Most are <5 m in width (a few approach 50 m) reflecting scour by the overlying conglomerate. Where sandstones are eroded, conglomerates are separated by flat partings that are level with sandstone bases (suggesting that erosion of conglomerates by conglomerates was minimal). Most sandstone bases are flat to irregular, but some occupy shallow channels and others are gradational with an underlying conglomerate.

31% of the Powell Island Conglomerate interbeds with bedding plane exposures are strewn with abundant twigs, whereas larger wood debris was found in only four conglomerates. It is envisaged that sandstones with twigs represent material washed off wooded slopes by separate small floods, whereas conglomerates must have been derived from other, virtually treeless slopes with abundant coarser debris. Some other sandstones could have formed from waning sheetfloods, by dewatering of debris flows (cf. Bluck, 1967; Miall, 1970a; Pierson, 1980), or by reworking of surficial deposits by sheetwash during storms (Moss and Walker, 1978). The variety of causes probably contributes to their abundance. Some erosion is likely because sandstones are preferentially deposited in lower areas, which coincide with the course of later debris flows and floods. Similar sandstones have been reported by Blissenbach (1954), Bluck (1967), Nilsen (1969), Wessell (1969), Steel (1974), Larsen and Steel (1978) and others. Though episodic, they are perhaps comparable 'inter-catastrophe' sediment to the pelagic muds on submarine fans.

The scarcity of bedforms other than very low angle crossbeds is doubtless partly due to the coarseness and poor sorting of the sediments (Harms et al., 1975). Low-angle avalanche faces may well reflect the steep depositional slopes of the fans, inferred to be in range of 10-15°.

Depositional Environment

The South Orkney conglomerates are interpreted as alluvial fans, because they are bodies of coarse debris with radial sediment dispersal and steep slopes that back onto their source rocks. Other evidence, cited earlier, strongly indicates a non-marine environment. It appears that two fanglomerate aprons faced each other across a possible Late Jurassic graben that contained an arm of the sea.

Initial fan deposits included debris flows, rare proximal mudflows and landslide blocks, but most deposits are mid-fan sheet debris flow and sheetflood conglomerates. Braided streams were restricted to the toes of the fans (Fig. 3). Eastern conglomerates pass locally outward from proximal landslide blocks to distal braided stream deposits, and were buried by fan-toe finer conglomerates and floodplain deposits. On Matthews Island, fanglomerates prograde over adjacent shallow-marine sands, which were possibly like the fan-deltas described by Gnaccolini (1982) and Hubert and Hyde (1982) (Fig. 3).

An interfan area identified at the confluence of the Christoffersen and Michelsen fans is characterized by rooted, muddy sandstones.

Although the bulk of the fanglomerate debris was generated nearby, the fans must have had large, complex watersheds because 1) very rare granules of marble and hornblende/actinolite schists in the Spence Harbour Conglomerate derive from at least 7 km inland (outcrop data from J. Thomson, 1968 and 1974) and 2) the presence of some anomalously well-rounded small pebbles suggests a mixture of clasts from proximal and distal sources.

PALEOCLIMATIC INFLUENCE

The type and quantity of wood in the South Orkney fans and elsewhere in the Antarctic Peninsula implies a seasonal cool-temperate climate with sufficient rainfall for forests (M. Thomson, 1982; J. M. Schopf, pers. comm. in 1977). Carbonization, rather than oxidation, of the wood, the grey colour of the conglomerates, and the absence of caliche and desert varnish all indicate a climate unlike that

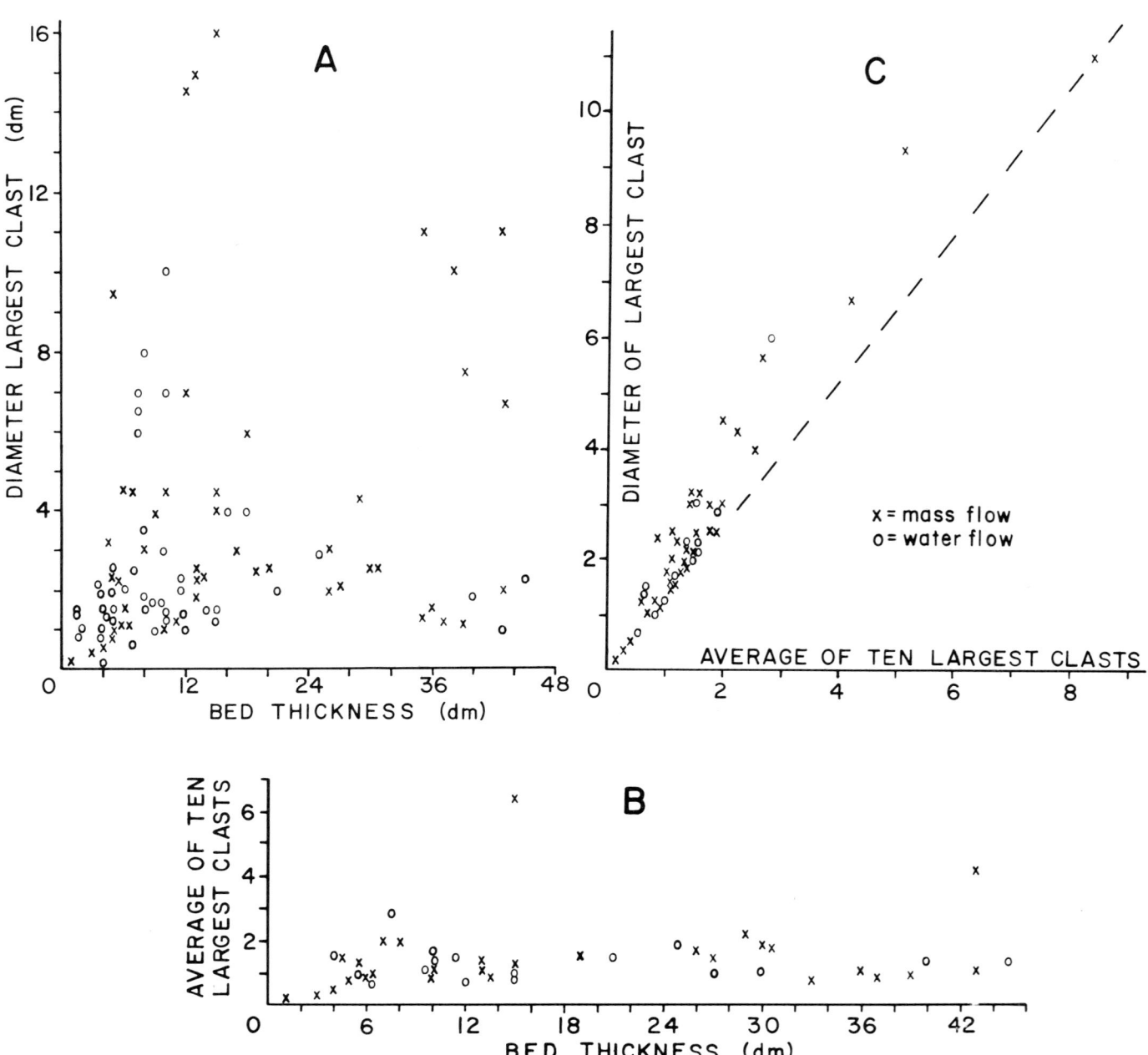

Fig. 8. A: The lack of correlation between maximum clast size and thickness of bed; as discussed in the text. O = clasts in sheetflood conglomerates; X = debris flow clasts (less securely identified). **B:** Lack of correlation between bed thickness and the average of the ten largest clasts in ca. 2.25 m². This comparison shows only that most coarse clasts were cobbles and pebbles, which was already known. **C:** Correlation between maximum clast size and the average of the ten largest clasts in the same bed shows that the occurrence of rare boulders is not completely random, which A & B might otherwise suggest.

Fig. 9. A mountain side and alluvial fan in Tierra del Fuego National Park, Argentina, that are believed to be similar to the Mesozoic fans of the South Orkneys. Features that may be analogous include the steepness of the fan, the wood on the surface (although here some of the wood could be from logging), the lower wooded slopes, and the even higher treeless slopes above. Note also the capping mantle of coarse scree, which is produced by nivational or periglacial processes and which is probably moved downslope by debris flow through the channels in the lower slope and out onto the fan.

of the modern southwestern U.S., as does paleo-magnetic positioning of the Antarctic Peninsula within 20° of the pole by late Mesozoic times (Scharnberger and Scharon, 1982; Watts, 1982).

The mountain slopes above the South Orkney fans must have been sufficiently steep to produce large rockfall blocks. The extreme rarity of tree-trunks and logs in the conglomerates suggests that fans and source areas were not extensively wooded. However, one interfan was vegetated, as were the slopes that produced all the wood in the sandstone interbeds. This suggests an alpine setting like the hillside in Tierra del Fuego shown in Figure 9, with a small, steep alluvial fan below a composite source that produces abundant dead wood and coarse detritus. Alluvial fans of coarse debris channelled through wooded slopes from cold, treeless slopes above were apparently very common in Tasmania during the Pleistocene and are still found there today (Wasson, 1977a and b).

An alpine setting, with mantles of scree produced by nivational and periglacial processes on slopes above timberline, is appealing because it allows rapid production of moderately-sized clasts and creates an unstable mantle of debris that could easily trigger debris flows or slush avalanches (Rapp, 1966), especially during storms or when snowbanks melt in spring (Rickmers, 1913, p. 193-197). The South Orkney conglomerate sheets imply generation and sudden release of large amounts of debris. Prolonged coastal rains may have triggered many local debris flows on slopes within mountain valleys (cf. Kochel and Johnson, this volume), which could be flushed out together onto the fan. Prior and Stephens (1972), Suwa and Okuda (1980) and Tufescu (1970) have found that mudflows are commonly caused by prolonged heavy rain in the cool wet climates of northern Ireland, northern Japan and in the drier Carpathians.

The variable texture and lithology of South Orkney fan deposits may have resulted from localized sources for debris flows within the watershed, and partial weathering and reworking of their deposits during temporary storage in mountain valleys. Local variability in the severity of weathering and freeze/thaw (Wasson, 1977a) may have been an additional factor. Lastly, steepness of fans is often equated with very low rainfall because fluvial processes are the only agent capable of moving sediment off fans (Bull, 1964), but steepness can also indicate rapid proximal sedimentation by short-lived and briefly competent floods and flows.

Conclusions

1) The South Orkney fans were dominated by sheetfloods and sheet debris flows.
2) Landslide blocks and mudflows occupy basal and proximal positions, whereas braided stream gravels occur distally. Both are rare compared to typical unchannelled conglomerate facies in mid-fan areas.
3) Distinguishing sheet debris flows and sheetflood conglomerates can be difficult. Both are coarse, tabular, 1-2 m thick, clast-supported conglomerates. However, sheetflood conglomerates are more highly organized with a sorted fine fraction and show evidence of aggradation during a flood, whereas unstratified and unsorted debris flows 'froze' in place as a single unit.
4) Low clay contents in debris flows could be primary, but clay may have been partly removed by dewatering, rain-wash, or reworking after deposition. If climatic inferences are correct, snow may have lubricated debris flows.
5) Conglomerates are separated by a variety of thin but distinctive sandstone interbeds, which seem to be common in many other fanglomerates.

6) Overall, the deposits are quite similar to alluvial fans described by Bluck (1967), Nilsen (1969), Miall (1970b), and Allen (1981), and they are not very different from the submarine fans described by Lewis *et al.* (1980).
7) The South Orkney fans seem to have developed in a cool wet, maritime climate, with a low tree-line and locally perennial snowbanks.
8) Features that distinguish these fans from typical semi-arid fans include grey colouration, abundant wood, steep surfaces, little channelling and few fluvial deposits.

REFERENCES

Allen, P.A. 1981. Sediments and processes on a small stream-flow dominated, Devonian alluvial fan, Shetland Islands. Sedimentary Geology, v. 29, p. 31-66.

Anderson, G.S. and Hussey, K.M. 1962. Alluvial fan development at Franklin Bluffs, Alaska. Iowa Academy of Science Proceedings, v. 92, p. 310-322.

Beaty, C.B. 1963. Origin of alluvial fans, White Mountains, California and Nevada. Annals of the Association of American Geographers, v. 53, p. 516-535.

Blackwelder, E. 1928. Mudflow as a geologic agent in semiarid mountains. Geological Society of America Bulletin, v. 39, p. 465-483.

Blissenbach, E. 1954. Geology of alluvial fans in semiarid regions. Geological Society of America Bulletin, v. 65, p. 175-190.

Bluck, B.J. 1967. Deposition of Old Red Sandstone conglomerates in the Clyde area: a study in the significance of bedding. Scottish Journal of Geology, v. 3, p. 139-167.

Broscoe, A.J. and Thomson, S. 1969. Observations on an alpine mudflow, Steele Creek, Yukon. Canadian Journal of Earth Sciences, v. 6, p. 219-229.

Bull, W.B. 1964. Alluvial fans and near-surface subsidence in western Fresno County, California. United States Geological Survey Professional Paper 437-A, 71p.

Bull, W.B. 1972. Recognition of alluvial-fan deposits in the stratigraphic record. *In:* Rigby, J.K. and Hamblin, W.K. (Eds.), Recognition of Ancient Sedimentary Environments. Society of Economic Paleontologists and Mineralogists, Special Publication 16, p. 63-83.

Chawner, W.D. 1935. Alluvial fan flooding, the Montrose, California flood of 1934. Geographical Review, v. 25, p. 225-263.

Coates, D.R. 1977. Landslide perspectives. *In:* Coates, D.R. (Ed.), Reviews in Engineering Geology: III. Landslides. Colorado: Geological Society of America, p. 1-28.

Curry, R.R. 1966. Observation of alpine mudflows in the Tenmile Range, central Colorado. Geological Society of America Bulletin, v. 77, p. 771-776.

Dalziel, I.W.D., Elliot, D.H., Thomson, J.W., Thomson, M.R.A., Wells, N.A. and Zinsmeister, W.J. 1977. Geologic studies in the South Orkney Islands. R/V Hero cruise 77-1, January 1977. Antarctic Journal of the United States, v. 12, p. 98-101.

Davis, W.M. 1938. Sheetfloods and stream floods. Geological Society of America Bulletin, v. 49, p. 1337-1416.

Drew, F. 1873. Alluvial and lacustrine deposits and glacial records of the Upper Indus Basin. Quarterly Journal of the Geological Society of London, v. 292, p. 441-471.

Elliot, D.H. and Wells, N.A. 1982. Mesozoic alluvial fans of the South Orkney Islands. *In:* Craddock, C. (Ed.), Antarctic Geoscience. Madison, Wisconsin: University of Wisconsin Press, p. 235-244.

Enos, P. 1977. Flow regimes in debris flow. Sedimentology, v. 24, p. 133-142.

Fisher, R.V. 1971. Features of coarse-grained, high-concentration fluids and their deposits. Journal of Sedimentary Petrology, v. 41, p. 916-927.

Fryxell, F.M. and Horberg, L. 1943. Alpine mudflows in Grand Teton National Park, Wyoming. Geological Society of America Bulletin, v. 54, p. 457-472.

Gnaccolini, M. 1982. Oligocene fan-delta deposits in northern Italy: a summary. Revista Italiana Paleontologia e Stratigrafia, v. 87, p. 627-636.

Gole, C.V. and Chitale, S.V. 1966. Inland delta building of the Kosi River. Proceedings of the American Society of Civil Engineers, Journal of the Hydraulics Division, HY2, v. 92, p,. 111-122.

Hampton, M.A. 1979. Buoyancy in debris flows. Journal of Sedimentary Petrology, v. 49, p. 753-758.

Harms, J.C., Southard, J.B., Spearing, D.R. and Walker, R.G. 1975. Depositional environments as interpreted from primary sedimentary structures and stratification sequences. Annual Meeting, Society of Economic Paleontologists and Mineralogists, Dallas. Short Course No. 2, 161p.

Hein, F.J. 1982. Depositional mechanisms of deep sea coarse clastic sediments, Cap Enragé Formation, Québec. Canadian Journal of Earth Sciences, v. 19, p. 267-287.

Heward, A.P. 1978. Alluvial fan and lacustrine sediments from the Stephanian A and B (La Magdalena, Cinera-Matellana and Sabero) coalfields, northern Spain. Sedimentology, v. 25, p. 451-488.

Hogg, S.E. 1982. Sheetfloods, sheetwash, sheetflow, or . . .? Earth Science Reviews, v. 18, p. 59-76.

Hooke, R.L. 1967. Processes on arid-regional alluvial fans. Journal of Geology, v. 75, p. 438-460.

Hubert, J.F. and Hyde, M.G. 1982. Sheet-flow deposits of graded beds and mudstones on an alluvial sandflat - playa system: Upper Triassic Blomidon redbeds, St. Mary's Bay, Nova Scotia. Sedimentology, v. 29, p. 457-474.

Kahn, J. 1956. The analysis and distribution of the properties of packing in sand-size sediments. 1. On the measurement of packing in sandstones. Journal of Geology, v. 64, p. 168-186.

Larsen, V. and Steel, R.J. 1978. The sedimentary history of a debris-flow dominated Devonian alluvial fan — a study of textural inversion. Sedimentology, v. 25, p. 37-59.

Lawson, D.E. 1982. Mobilization, movement and deposition of active subaerial sediment flows, Matanuska Glacier, Alaska. Journal of Geology, v. 90, p. 279-300.

Leggett, R.F., Brown, R.J.E. and Johnson, G.H. 1966. Alluvial-fan formation near Aklavik, Northwest Territories, Canada. Geological Society of America Bulletin, v. 77, p. 15-30.

Lewis, D.W., Laird, M.G. and Powell, R.D. 1980. Debris flow deposits of Early Miocene age, Deadman Stream, Marlborough, New Zealand. Sedimentary Geology, v. 27, p. 83-118.

Lindsay, J.F. 1968. The development of clast fabric in debris flows. Journal of Sedimentary Petrology, v. 38, p. 1242-1253.

Lustig, L.K. 1965. Clastic sedimentation in Deep Springs Valley, California. United States Geological Survey Professional Paper 352-F, p. 131-192.

Matthews, D.H. and Maling, D.H. 1967. The geology of the South Orkney Islands, I, Signy Island. Falkland Islands Dependencies Survey Scientific Report No. 25, 32p.

McGee, W.J. 1897. Sheetflood erosion. Geological Society of America Bulletin, v. 8, p. 87-112.

McGowen, J.H. and Groat, C.G. 1971. Van Horn Sandstone, West Texas: an alluvial fan model for mineral exploration. Bureau of Economic Geology, University of Texas at Austin, Report of Investigation 72, 57p.

Melton, M.A. 1965. The geomorphic and paleoclimatic significance of alluvial deposits in southern Arizona. Journal of Geology, v. 73, p. 1-37.

Miall, A.D. 1970a. Continental-marine transition in the Devonian of Prince of Wales Island, Northwest Territories. Canadian Journal of Earth Sciences, v. 7, p. 124-144.

Miall, A.D. 1970b. Devonian alluvial fans, Prince of Wales Island, Arctic Canada. Journal of Sedimentary Petrology, v. 40, p. 556-571.

Miall, A.D. 1977. A review of the braided-stream environment. Earth Science Reviews, v. 13, p. 1-62.

Moss, A.J. and Walker, P.H. 1978. Particle transport by continental water flows in relation to erosion, deposition, soils, and human activities. Sedimentary Geology, v. 20, p. 81-139.

Nilsen, T.H. 1969. Old Red sedimentation in the Buelandet-Vaerlandet Devonian district, west Norway. Sedimentary Geology, v. 3, p. 35-57.

Pierson, T.C. 1980. Erosion and deposition by debris flows at Mt Thomas, North Canterbury, New Zealand. Earth Surface Processes, v. 65, p. 227-247.

Pierson, T.C. 1981. Dominant particle support mechanism in debris flows at Mt Thomas, New Zealand, and implications for flow mobility. Sedimentology, v. 28, p. 49-60.

Plafker, G. and Erickson, G.E. 1978. Nevados Huascaran avalanches, Peru. In: Voight, B. (Ed.), Rockslides and Avalanches, v. 1, Developments in Technical Engineering 14A. New York: Elsevier, p. 423-437.

Prior, D.B. and Stephens, N. 1972. Some movement patterns of temperate mudflows: examples from north-eastern Ireland. Geological Society of America Bulletin, v. 83, p. 2533-2544.

Rapp, A. 1960. Recent development of mountain slopes in Karkevagge and surroundings, northern Scandinavia. Geografiska Annaler, v. 42, p. 1-197.

Rapp, A. 1966. Solifluction and avalanches: notes on mass-wasting in the Scandinavian mountains. Proceedings of the 1963 International Conference on Permafrost, National Academy of Science, p. 150-154.

Rickmers, W.R. 1913. The Duab of Turkestan. Cambridge: Cambridge University Press, 563p.

Rodine, J.D. and Johnson, A.M. 1976. The ability of debris, heavily freighted with coarse materials, to flow on gentle slopes. Sedimentology, v. 23, p. 213-234.

Ryder, J.M. 1971. The stratigraphy and morphology of paraglacial alluvial fans in south-central British Columbia. Canadian Journal of Earth Sciences, v. 8, p. 279-298.

Scharnberger, C.K. and Scharon, L. 1982. Paleomagnetism of rocks from Graham Land and western Ellsworth Land, Antartica. In: Craddock, C. (Ed.), Antarctic Geoscience. Madison, Wisconsin: University of Wisconsin Press, p. 331-338.

Sharp, R.P. and Nobles, L.H. 1953. Mudflow of 1941 at Wrightwood, southern California. Geological Society of America Bulletin, v. 64, p. 547-560.

Singewald, J.T. 1928. Discussion [of Blackwelder, 1928]. Geological Society of America Bulletin, v. 39, p. 480-483.

Spalletti, L.A. 1972. Sedimentologia de los cenoglomerados de Volcan, provincia de Jujuy. Revista del Museo de La Plata (new series), Geology section, v. 8, p. 137-225.

Sparks, R.S.J. 1976. Grain size variations in ignimbrites and implications for the transport of pyroclastic flows. Sedimentology, v. 23, p. 147-188.

Stanley, K.O. 1971. Tectonic and sedimentologic history of Lower Jurassic Sunrise and Dunlap Formations, west-central Nevada. American Association of Petroleum Geologists Bulletin, v. 55, p. 454-477.

Steel, R.J. 1974. New Red Sandstone floodplain and piedmont sedimentation in the Hebridean Province, Scotland. Journal of Sedimentary Petrology, v. 44, p. 336-357.

Surlyk, F. 1978. Submarine fan sedimentation along fault scarps on tilted fault blocks (Jurassic-Cretaceous boundary, East Greenland). Grønlands Geologiske Undersøgelse Bulletin 128, 108p.

Suwa, H. and Okuda, S. 1980. Dissection of valleys by debris flows. Zeitschrift fur Geomorphologie Supplementband, v. 35, p. 164-182.

Thomson, J.W. 1968. The geology of the South Orkney Islands, II, The petrology of Signy Island. British Antarctic Survey Reports, no. 62, 30p.

Thomson, J.W. 1971. The geology of Matthews Island, South Orkney Islands. British Antarctic Survey Bulletin, no. 26, p. 51-57.

Thomson, J.W. 1973. The geology of Powell, Christoffersen, and Michelsen Islands, South Orkney Islands. British Antarctic Survey Bulletin, no. 33 and 34, p. 137-167.

Thomson J.W. 1974. The geology of the South Orkney Islands, III, Coronation Island. British Antarctic Survey Reports, no. 86, p. 1-39.

Thomson, M.R.A. 1975. Fossils from the South Orkney Islands, II, Matthews Island. British Antarctic Survey Bulletin, no. 40, p. 75-79.

Thomson, M.R.A. 1982. Mesozoic paleogeography of western Antarctica: In: Craddock, C. (Ed.), Antarctic Geoscience. Madison, Wisconsin: University of Wisconsin Press, p. 331-338.

Thomson, M.R.A. and Willey, L.E. 1975. Fossils from the South Orkney Islands. I. Coronation Island. British Antarctic Survey Bulletin, no. 40, p. 15-21.

Tufescu, V. 1970. Mudflows in the flysch Carpathians and Bend Sub-Carpathians of Romania. Zeitschrift fur Geomorphologie, Supplementband 9, p. 146-156.

Walker, R.G. 1975. Generalized facies models for resedimented conglomerates of turbidite association. Geological Society of America Bulletin, v. 86, p. 737-748.

Walker, R.G. and Mutti, E. 1973. Turbidite facies and facies associations: In: Middleton, G.V. and Bouma, A.H. (Co-chairmen), Turbidites and Deep-Water Sedimentation. Pacific Section, Society of Economic Paleontologists and Mineralogists, Anaheim, Short Course Notes, p. 119-158.

Wasson, R.J. 1977a. Last-glacial alluvial fan sedimentation in the Lower Derwent Valley, Tasmania. Sedimentology, v. 24, p. 781-799.

Wasson, R.J. 1977b. Catchment processes and evolution of alluvial fans in the lower Derwent valley. Zeitschrift fur Geomorphologie, v. 21, p. 147-168.

Wasson, R.J. 1979. Stratified debris slope deposits in the Hindu Kush, Pakistan. Zeitschrift fur Geomorphologie, v. 23, p. 301-320.

Watts, D.R. 1982. Potassium-argon and paleomagnetic results from King George Island, South Shetland Islands. In: Craddock, C. (Ed.), Antarctic Geoscience. Madison, Wisconsin: University of Wisconsin Press, p. 255-261.

Wells, N.A. in prep. Mesozoic alluvial fanconglomerates and other sedimentary rocks of the South Orkney Islands. Institute of Polar Studies (Ohio State University), Report no. 712.

Wessell, J.M. 1969., Sedimentary history of Upper Triassic alluvial fan complexes in north-central Massachussets. University of Massachusetts, Department of Geology, Contribution no. 2, 157p.

Williams, G.E. 1969. Characteristics and origin of a Precambrian pediment. Journal of Geology, v. 77, p. 183-207.

Winder, C.G. 1965. Alluvial cone construction by alpine mudflow in a humid temperate region. Canadian Journal of Earth Sciences, v. 2, p. 270-277.

A UNIQUE MASS FLOW MARKER BED IN A MIOCENE STREAMFLOW MOLASSE SEQUENCE, SWITZERLAND

Heinz M. Bürgisser[1]

Abstract

A unique sedimentary marker bed outcrops within a Middle Miocene sequence of conglomerates and finer clastics in the North Alpine molasse basin. The sequence is interpreted as the deposits of large, gently-sloping ($\leq 1°$) alluvial fans building out from the foot of the Alps. Up to 6 m thick, the marker bed has a volume of the order of 10^9 m^3 and occurs across the entire width (65 km) of one of these fans.

Using information from 216 profiles, the marker bed is subdivided into four facies, all characterized by a more restricted, carbonate-rich composition than the enclosing fluvial deposits. One of these facies, confined to a 18 km wide mid-fan region, consists of several superimposed depositional units. These are either 0.5 - 3 m thick and consist of clast-supported rudites which are ungraded or inversely-graded at the base, or dm thick, matrix-supported rudites. Common features include relatively small, angular carbonate clasts and a carbonate mud matrix. Even in the thicker beds, maximum clast size (<10 cm) is finer than the surrounding streamflow conglomerates. These beds have a measurable width of several 100 m, and are interpreted as having been deposited from a series of small subaerial debris flows which originated from a uniquely large event. This was possibly an exceptional rockslide from the front range of the Alps. The other facies of the marker bed comprise fluvial conglomerates and detrital lacustrine limestones, both of which represent reworked material from the large event.

Résumé

Une couche repère sédimentaire unique affleure à l'intérieur d'une séquence de conglomérats et de roches détritiques plus fines d'âge Miocène moyen dans le bassin molassique des Alpes du Nord. On considère que la séquence est formée de dépôts de cônes alluviaux, à faible pendage ($\leq 1°$), édifiés aux pieds des Alpes. La couche repère d'une puissance de 6 m occupe un volume de l'ordre de 10^9 m^3 et se manifeste sur toute la largeur (65 km) d'un de ces cônes.

La couche repère a été subdivisée en quatre faciès après examen de 216 profils, tous caractérisés par une composition plus haute en carbonate que celle des dépôts fluviatiles qui les encaissent. L'un de ces faciès, d'occurrence limitée à une région de 18 km de large au milieu du cône, renferme plusieurs unités de dépôts surimposés. Ils sont composés soit d'une épaisseur de 0,3 - 3 m de rudites à fragments jointifs sans granoclassement ou à granoclassement inverse, ou d'une couche de rudites à fragment flottants dans une matrice et dont l'épaisseur se mesure en dm. Les traits communs sont les fragments angulaires et relativement plus petits de carbonate et la matrice boueuse carbonatée. Même dans les couches les plus épaisses, la grosseur maximale des fragments (<10 cm) est inférieure à celle dans les conglomérats fluviatiles. Ces couches repères sont présentes sur une largeur mesurable de plusieurs 100 m, et on croit qu'elles furent déposées par une série de petites coulées de débris subaériennes originant d'un même grand événement. Il s'agit possiblement d'un éboulement exceptionel du front de la chaîne des Alpes. Les autres faciès de la couche repère comprennent des conglomérats fluviatiles et des calcaires lacustres détritiques, tous les deux sont fabriqués de matériaux remaniés par cet important événement.

Introduction

Compared to other types of clastic sediments, alluvial and deep-sea resedimented conglomerates are deposited characteristically during brief periods of high transport energy. While subaerial mass flows and floods are often catastrophic in human terms, their deposits are not rare in the geological record because very little coarse sediment transport occurs during the lengthy 'fair-weather' periods between two such high-energy events. In contrast, an exceptional process or event produces a single deposit of *unique* character (Fig. 1). The resulting bed is so different from those produced by the aforementioned 'regular catastrophes' that it stands out, and is frequently a good stratigraphic marker (Reading, 1978, p. 10).

Exceptional events in the geologic record are preserved in a variety of ways. The probable impact of a huge celestial body at the Cretaceous-Tertiary boundary is reflected by rather gradual faunal changes and such subtleties as changes in the amounts of rare earth elements in clays (see Hsü, 1983). In contrast, the sudden, catastrophic collapse of a carbonate platform in Spain during the Eocene led to a huge sediment gravity flow; its deposit, more than 100 m thick, has been traced in the direction of transport for 75 km, and contains clasts as much as 100 m across and

---

[1]Geologisches Institut, ETH Zürich, Switzerland.
Present address: Koninklijke/Shell, Exploratie en Produktie Laboratorium, Postbus 60, NL-2280 AB Rijswijk, The Netherlands.

This paper has developed from part of a Ph.D. thesis completed at the Geological Institute, ETH Zürich, Switzerland. I am grateful to K.J. Hsü for inspiring discussions during the various stages of the research. G.V. Middleton read a first, G. Postma a first and second version of this manuscript and made useful suggestions. The paper has further benefited from constructive comments by M. Epting, B.K. Levell and M.A. Naylor. Thanks are also extended to Shell Research B.V. for permitting publication of this paper.

Copyright © 1984, Canadian Society of Petroleum Geologists

several tens of metres thick (Johns *et al.*, 1981). Another exceptional deposit, including 3 m blocks in sets of giant cross-beds, resulted from a catastrophic flood caused by the rapid emptying of a Pleistocene lake in Idaho (Malde, 1968). The sedimentary features resulting from major bedrock floods are described by Baker (this volume).

This paper examines the record of another exceptional event which occurred at the northern front of the Alps at some moment in the Middle Miocene. Its record is in the form of a thin (≤ 6 m) sedimentary marker bed with features that are unique in a 7000 m thick, predominantly alluvial, basin fill. However, in contrast to the above-mentioned olistostrome or flood deposits, large blocks, typically associated with an exceptional event, are absent. The grain size of the marker bed is even smaller than that of the surrounding streamflow rudites. A hypothesis on the origin of the marker, explaining this contrast between grain size and uniqueness of the bed, will be put forward in the final part of this paper.

Geologic Setting

The marker bed occurs in the North Alpine Foredeep in central Europe, a typical, elongate (750 x 80-150 km), late-orogenic molasse basin bordering the Alps (Fig. 2). It is the direct successor of the northernmost trough of the Tethys Sea, in which deep-sea clastic sedimentation prevailed throughout the Early Oligocene. During the early Late Oligocene, the part of this trough north of the Western and Central Alps was infilled (see Van Houten (1981) for a review of this part of the basin). The shallowing-upwards sequence, capped by coastal sandstones, is known as *Lower Marine Molasse*, the following continental sequence

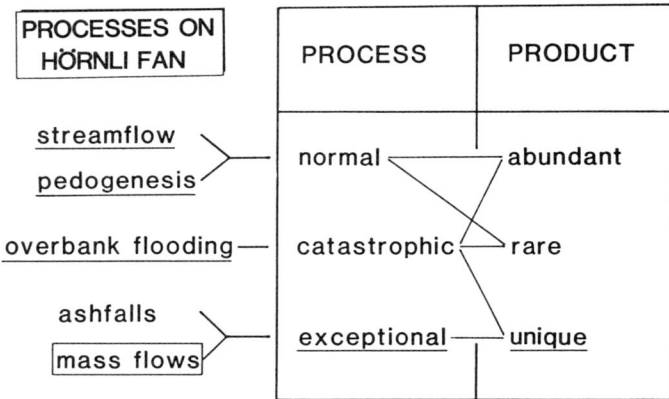

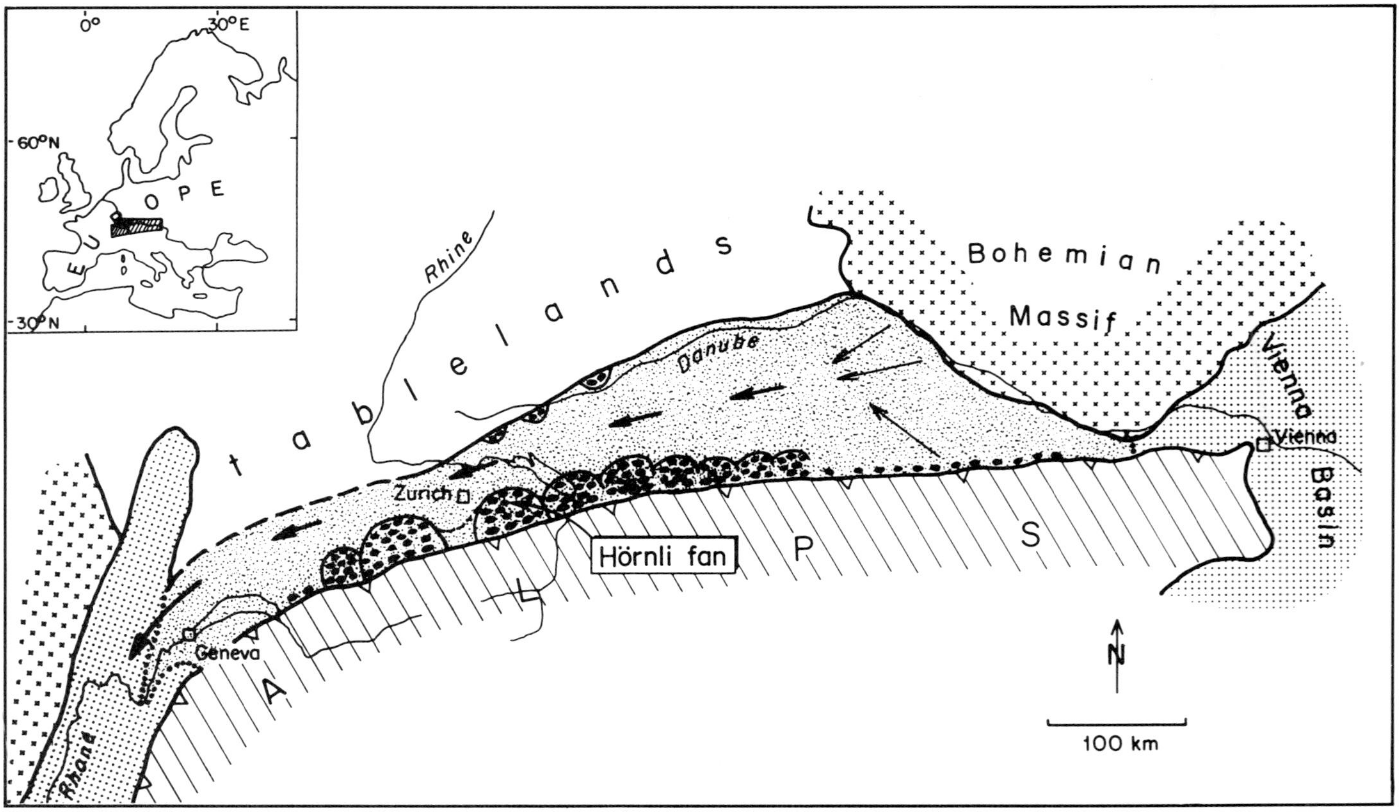

Fig. 1. Normal, catastrophic and exceptional processes on the Miocene Hörnli alluvial fan.

Fig. 2. North Alpine Foredeep during deposition of the Upper Freshwater Molasse (Middle Miocene). Arrows indicate basin drainage towards the Rhone/Saône/Bresse depression in the west. The marker bed is confined to the area of the Hörnli fan east of Zurich, one of the large alluvial fans built up by radial rivers at the foot of the Alps.

as *Lower Freshwater Molasse* (Fig. 3). Deep-sea sedimentation continued in an embayment north of the then relatively subdued Eastern Alps (Moiola and Malzer, 1982). In the Early Miocene, the sea also extended across the western part of the foredeep (*Upper Marine Molasse*), but subsequently withdrew completely from it. The following continental sequence, known as *Upper Freshwater Molasse*, is therefore developed across the entire basin (Fig. 2). The youngest preserved deposits of this series have been dated as Middle Miocene in the western part of the basin (Fig. 3) and Upper Miocene in the eastern part.

The marker bed is found in the Upper Freshwater Molasse (Fig. 3). Sediment composition and textural properties suggest that this youngest basin-filling sequence represents alluvial deposits laid down either by radial Alpine rivers, mostly on large alluvial fans building up at the front of the Alps, or by the trunk river that drained the foredeep axially towards the west (Lemcke *et al.*, 1953; Hofmann, 1955; Füchtbauer, 1967; Fig. 2). According to floral remains, the prevailing climate on the alluvial plain was warm-temperate and seasonally humid during deposition of the marker bed (Hantke, 1984).

The occurrence of the marker bed is restricted to the deposits of one of these radial rivers, the Hörnli river (see Fig. 2). These deposits crop out between the towns of Zurich and St. Gallen in northeastern Switzerland, a distance of about 65 km (Fig. 4). Radially, they extend for 40 to 55 km; however, the most proximal 10-15 km have been eroded following their tilting and uplift during the Late Miocene to Quaternary.

The bulk of the Hörnli sediments was deposited by streamflow processes in channels and during flooding in the overbank area, where they were later modified by pedogenesis (Fig. 1). Besides the unique marker bed, a few very thin bentonite layers in the Hörnli sediments can also be regarded as the products of exceptional events, *i.e.* volcanic eruptions, resulting in extensive ashfalls (Fig. 1). The following brief description of the Hörnli streamflow deposits serves to contrast them with the marker bed. Their interpretation (Bürgisser, 1981) allows a reconstruction of the foredeep morphology just prior to the deposition of the marker bed.

THE HÖRNLI STREAMFLOW DEPOSITS

The conglomerates are polymict, with well-rounded clasts of dominantly limestone and dolomite, and subordinate amounts of granite, metamorphics, quartzite, sandstone

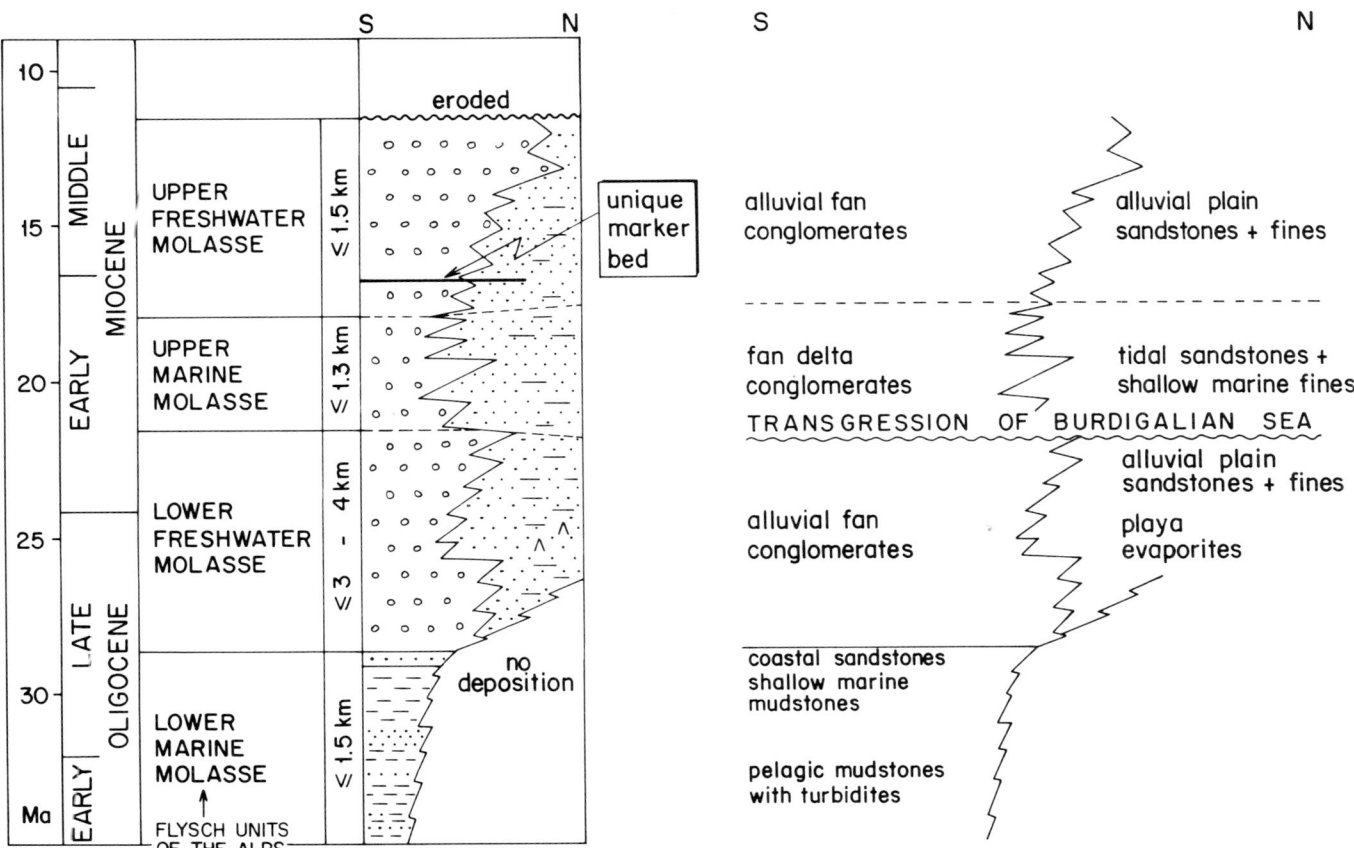

Fig. 3. Summary of the lithostratigraphic subdivision of the western part of the North Alpine Foredeep. The marker bed occurs within the alluvial Upper Freshwater Molasse.

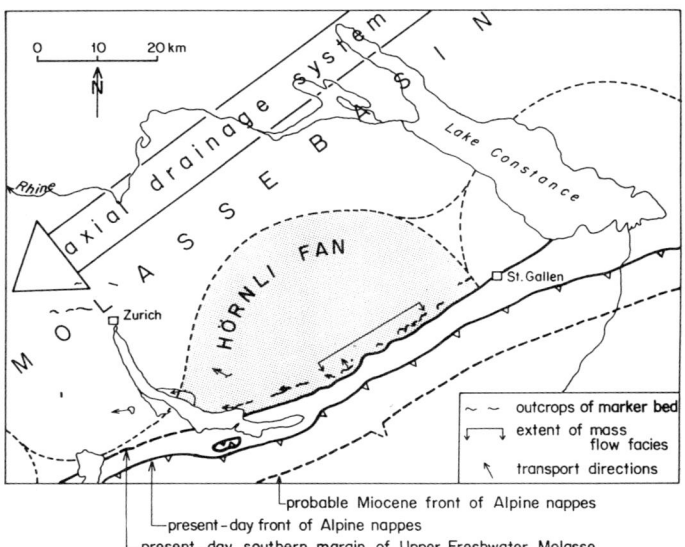

Fig. 4. Molasse Basin in northeastern Switzerland with approximate extent of the Hörnli alluvial fan at the time of deposition of the marker bed. Outcrops of the latter are found only in the proximal part of the fan deposits and in the alluvial plain sediments west of the fan. The mass flow facies is restricted to the central part of the fan.

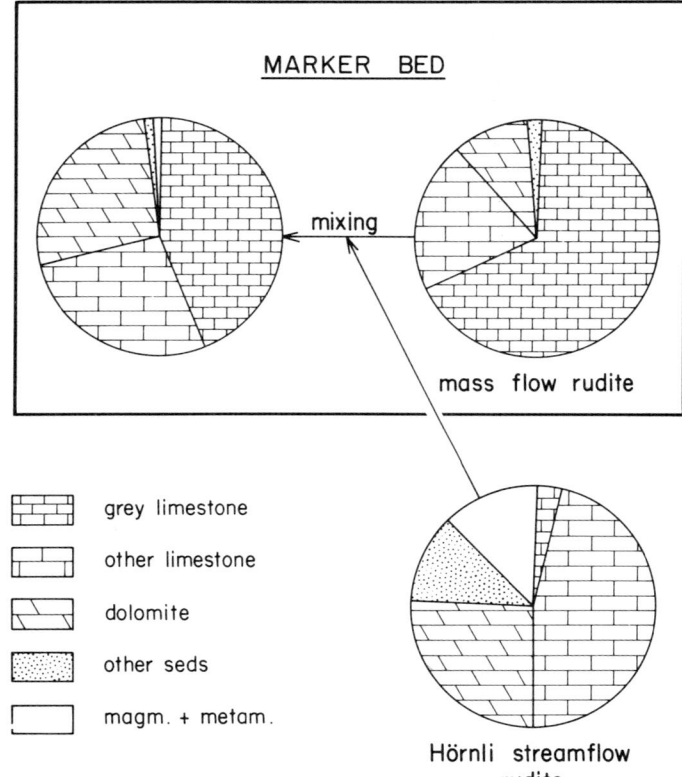

Fig. 5. Clast lithologies of the marker bed rudites. The composition of the mass flow facies deviates markedly from that of the normal Hörnli alluvial fan rudites. The streamflow facies of the marker bed show less difference because of mixing of rockslide material with normal Hörnli river gravel.

and chert (Fig. 5). Also the sandstones contain abundant carbonate grains (60-70%; calcite:dolomite 2:1), with quartz amounting to only 20-30%. This immature, lithic composition suggests that the Hörnli river must have had a drainage area with varied, but predominantly carbonate, source lithologies.

Three facies associations have been recognized (Bürgisser, 1981). The *conglomerate facies association* (H1), occurring closest to the Alps, comprises 80% cobble conglomerates, mostly in 0.6-1.4 m thick, massive to crudely horizontally bedded, amalgamated units. Structure and thickness of the conglomerate units are comparable to those of the Scott model sequence for braided stream deposits (Miall, 1977), but H1 sequences also contain some fine-grained clastics. This association is interpreted to record deposition on the proximal part of a humid alluvial fan, comprising an overbank area beside the zone of rather shallow, braided gravelly channels (Fig. 6). Comparison of pebble size with that of similar Recent environments suggests a fan slope of 7-16 m.km^{-1} (0.4-0.9°) in the proximal and 3-5 m.km^{-1} (0.15-0.3°) in the distal part of this facies association (Bürgisser, 1981).

The *conglomerate/siltstone facies association* (H2) comprises (a) various amounts of pebble to cobble conglomerate, in 1-2.2 m thick, massively bedded or up to 4 m thick cross-bedded, deeply channelled units, (b) little channel sandstone, and (c) overbank sand-siltstone. This facies association is interpreted to have been deposited on the distal part of a humid alluvial fan, with an extensive overbank area flanking a main sinuous channel several metres deep, similar to the modern Nueces River in Texas (Gustavson, 1978). Channel slope is estimated to be in the order of 0.7-4 m.km^{-1} (Bürgisser, 1981).

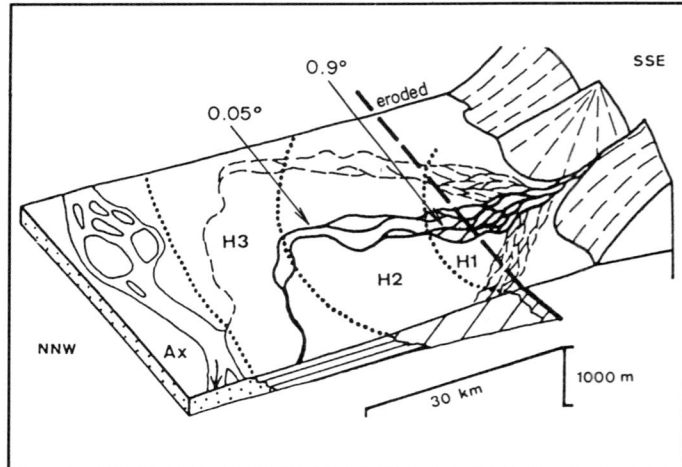

Fig. 6. Schematic reconstruction of the gently sloping Hörnli alluvial fan in the early Middle Miocene (after Bürgisser, 1981, modified).
H1 Conglomerate facies association
H2 Conglomerate/siltstone facies association
H3 Sandstone/siltstone facies association
Ax Sandstones of axial river

The *sandstone/siltstone facies association* (H3) comprises thick, variegated argillaceous siltstones with rare beds of lignite and freshwater limestone, in which ribbons of sandstones (200-500 m wide, 3-8 m thick, multiphase fill) are intercalated. This facies association and channel directions suggest deposition on the alluvial plain containing the axial trunk stream of the basin (see Fig. 6).

THE MARKER BED

SETTING AND SUBDIVISIONS

The marker bed is found in the lower part of the Hörnli deposits, in all three facies associations described above (Fig. 7). The outcrops suggest deposition not only over the entire width of the alluvial fan, with palaeocurrents diverging in accordance with the assumed fan slope (Fig. 4), but also beyond it, on the alluvial plain to the northwest. The following features characterise the marker bed:
— a pebble assemblage less polymict than the streamflow deposits (Fig. 5);
— the occurrence of matrix-supported conglomerates;
— normal and inverse grading;
— angular limestone clasts;
— bedded limestones up to 6 m thick; and
— a degree of induration which allows the rock to be used as a building stone.

The interpretations are based on descriptive parameters recorded in 216 vertical profiles through the marker bed. The most useful parameters include amount and grain size of the conglomerate matrix; maximum particle size (MPS = mean of the apparent long axis of the 10 largest clasts of a bed or part of it, measured within a lateral distance of 2 m); type of bedding; presence and type of grading; and bed geometry.

These parameters occur combined in such a way as to subdivide the marker bed into four facies which originated from different processes (Table 1, Fig. 8): (1) The Degersheim Conglomerate, interpreted as having been deposited by *mass flows*; (2) the Abtwil and (3) Hüllistein Conglomerates, deposited from traction processes in broad and narrow *fluvial channels*, respectively (see Fig. 9); and (4) the sheet-like Meilen Limestone, dominantly originating from *suspension sedimentation in a lake*. This paper concentrates on the mass flow facies (Degersheim Conglomerate), which forms the key to the understanding of the entire marker bed.

SEDIMENTOLOGY OF THE MASS FLOW FACIES

The mass flow facies (Degersheim Conglomerate) consists of one to six superimposed beds of well-indurated, oligomict pebbly rudites and arenites composed of angular to subangular carbonate clasts and fine carbonate matrix. The facies is typically 0.2 to 5 m thick (Figs. 9, 10, 11).

Occurrence and Extent of Beds

Outcrops of the mass flow facies are restricted to the southern, denuding margin of the Upper Freshwater Molasse (Fig. 4). Since beds dip 10-20° NNW (Fig. 12), the outcrops are within a virtually linear zone running in an ENE-WSW direction, perpendicular and oblique to the slope of the former Hörnli alluvial fan (Fig. 4). The distance between the westernmost and easternmost outcrops is about 18 km (Figs. 4, 9). Along the line of outcrop, exposure of the marker bed is rather discontinuous (Fig. 4). Over distances of up to 5 km, the Degersheim Conglomerate is covered by Quaternary deposits or has been demonstrably eroded shortly after deposition (Figs. 9, 13). The most continuous outcrop is approximately 2 km long.

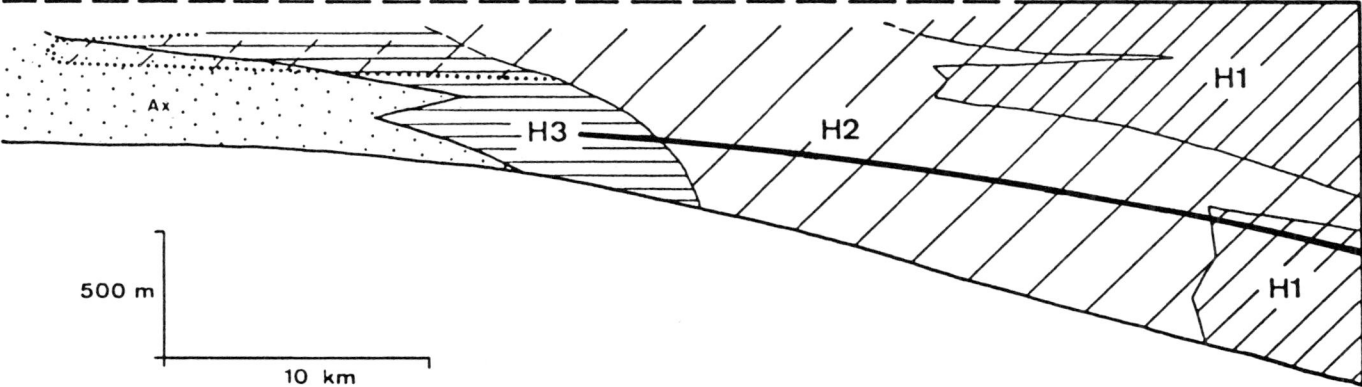

Fig. 7. Downfan profile through the Middle Miocene Hörnli deposits as preserved today, showing schematic distribution of facies associations H1-H3 and Ax (see Fig. 6) and the marker bed (bold line) (after Bürgisser, 1981, modified). The overall sedimentation pattern was not altered by the deposition of the marker bed. Fan progradation was generally ca. 10 km, but in one area 25 km (dotted line). The distance between the right-hand margin of the figure to the front of the Alps was about 10-15 km in the Middle Miocene; these fan deposits (mainly H1) have been eroded.

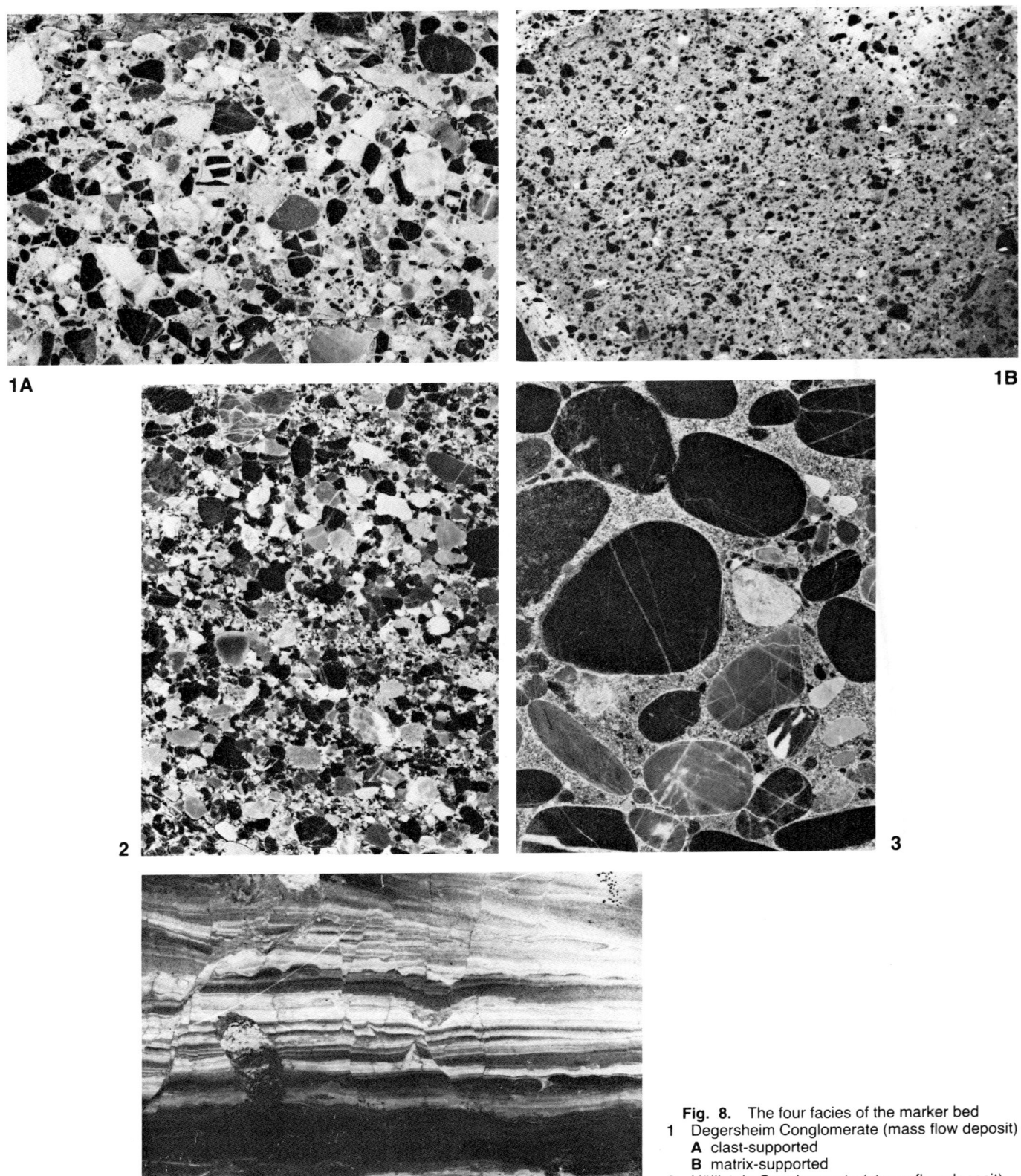

Fig. 8. The four facies of the marker bed
1 Degersheim Conglomerate (mass flow deposit)
   A clast-supported
   B matrix-supported
2 Hüllistein Conglomerate (streamflow deposit)
3 Abtwil Conglomerate (streamflow deposit)
4 Meilen Limestone (lacustrine suspension deposit)
Polished slabs, approximately 1.2x natural size.

Shape and size of outcrops limit reconstructions of bed geometry. Thicker beds possibly extend up to 1 km perpendicular to the transport direction, thin beds for at least 200 m. The orientation of the outcrop belt precludes observations as regards bed extent parallel to the direction of transport.

The Degersheim Conglomerate is largely intercalated in the conglomerate facies association (H1) of the Hörnli deposits. In the easternmost 5 km, it is within the conglomerate/siltstone facies association (H2). The outcrops of the Degersheim Conglomerate are therefore restricted to the assumed central part of the Hörnli fan (Fig. 6).

Clasts and Matrix

Calcite encrustation and moss cover often conceal internal features of the mass flow beds. On a freshly broken surface, however, the clasts which are predominantly dark, are readily distinguished from the yellowish grey matrix (Figs. 8-1, 14A, 15).

Clasts are defined for this study as all material of sand size and larger. Clast-support is assumed if less than 30% matrix is present on a surface (see Beard and Weyl, 1973). The rocks of the mass flow facies are usually clast-supported (Figs. 14A, 15). Matrix-support is confined to thin beds (thickness <0.5 m) where the matrix may comprise up to 99% of the rock (Figs. 11, 15). Clasts floating in matrix are also present at inversely graded bases and normally graded tops of beds (see below).

Clasts of the Degersheim Conglomerate are angular or subangular and fall typically in the granule to medium pebble range (2-16 mm). Maximum particle size (*MPS*) is below 9.5 cm and at the lateral margins of the outcrop belt below 1 cm (Fig. 16). A correlation between *MPS* and bed thickness could not be established because of erosion of the upper parts of beds. In the matrix-supported beds, the *MPS* value is inversely proportioned to the percent matrix.

Clasts with the mass flow beds are virtually all carbonates. Dark to medium grey, microsparitic limestones dominate over dolomites and other limestone types (Fig. 5). The matrix has a composition similar to that of the clasts in containing more calcite than dolomite. Clay minerals never exceed a few percent. The matrix is so well-indurated that the rock normally breaks through the clasts.

'Oversize' Clasts

More than 30 'oversize' clasts were observed in the area of largest bed thickness and grain size (see Fig. 16). They occur isolated at the base (Fig. 11) or within the lowermost metre of clast-supported beds. Oversize clasts are two to four times longer than the second-largest adjacent clast (the largest clast, with a long axis of 24 cm, is illustrated in Fig. 17) and are of lithologies, *e.g.,* quartz arenite, not present amongst the smaller clasts (Fig. 17). The high rounding of some of these clasts is also exceptional. Because of this different character they have not been used to calculate the *MPS* values in Figure 9. Size, roundness and

	Rudite Facies of Marker Bed			Hörnli Alluvial Fan Conglomerates
	Degersheim Conglomerate	Hüllistein Conglomerate	Abtwil Conglomerate	
Texture	clast- or matrix-supported	clast-supported	clast-supported	clast-supported
Grain size of matrix	≤silt	fine sand	fine to medium sand	fine to medium sand
Maximum particle size	<60 mm (one bed: <95 mm)	<70 mm	<110 mm	proximal: 140-205 mm distal: 65-180 mm
Pebble roundness	angular to subangular	subangular to rounded	subangular to rounded	rounded to well-rounded
Bedding types	massive	crudely horizontal and cross-bedded	crudely horizontal and cross-bedded	crudely horizontal and cross-bedded
Grading	ungraded or inversely graded (at base) or/and normally graded (at top)	ungraded	ungraded	ungraded
Geometry	sheet-like (?)	20-50 m wide channel fills	several 100 m wide channel fills	several 100 m wide channel fills
Clast composition	oligomict (60-70% dark limestone; no basement or flysch clasts)	oligomict (predominantly dark limestone; rare basement clasts)	oligomict to polymict	polymict (3-5% dark limestone)
Interpreted Depositional Process	SEDIMENT GRAVITY FLOWS	STREAMFLOW	STREAMFLOW	STREAMFLOW

Table 1. Characteristics of the three conglomerate facies of the marker bed and of normal Hörnli alluvial fan conglomerates. The fourth facies of the marker bed, the Meilen Limestone, is a sheet of detrital, fine-grained carbonate rock (see Fig. 8-4).

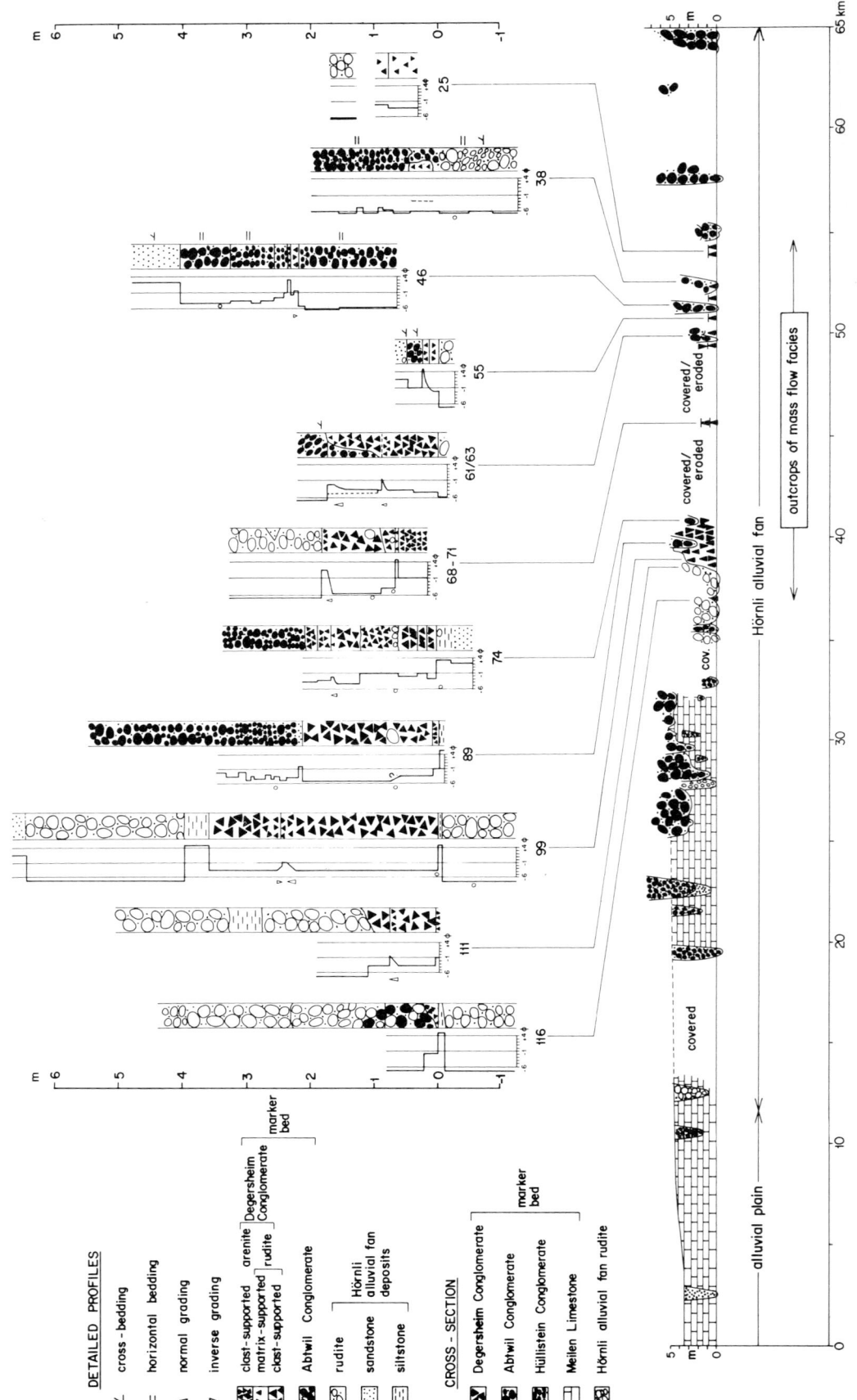

Fig. 9. Cross-section through the marker bed showing the lateral distribution and mutual position of the four marker bed facies (below) and 11 selected profiles in the area of the mass flow facies (above). Grain size curves represent maximum particle size (MPS); small circles indicate single oversize clasts.

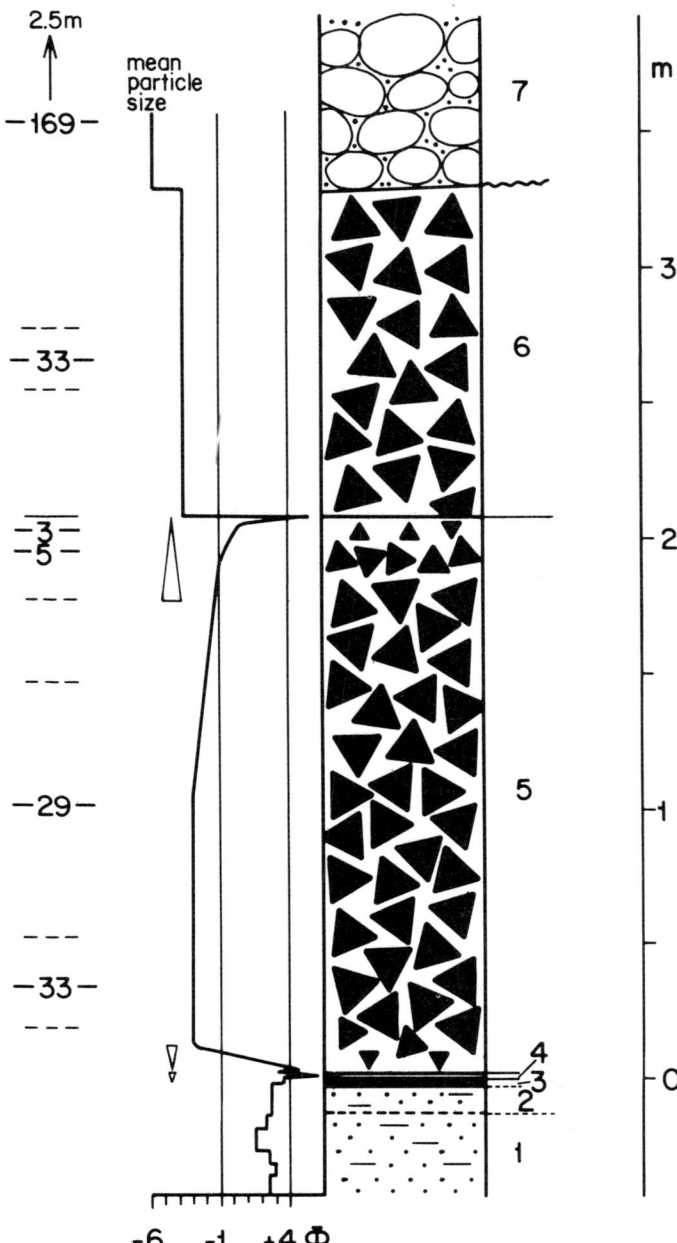

Fig. 11. Outcrop of Degersheim Conglomerate (profile 69; see Fig. 9) showing superposition of four beds: (1) clast-supported arenite; (2) bed consisting of carbonate mud with very few small clasts; (3) clast-supported rudite with large, rounded clasts (4.7 cm long) at base; (4) clast-supported rudite. Note plant cover and moss (dark patches). Scale subdivisions 1 cm.

Fig. 10. Type profile of the mass flow facies of the marker bed (Degersheim Conglomerate):

1, 2   silty calcareous sandstone with coal fragments
3   greenish grey sandy calcareous siltstone
4   inversely graded carbonate bed (lutitic ⟶ arenitic)
5   inversely and normally graded, clast-supported carbonate rudite with muddy matrix
6   ungraded, clast-supported carbonate rudite with muddy matrix
7   polymict rudite with sandy matrix
-33- maximum particle size in mm (10 largest clasts)
△   normally graded part of bed
▽   inversely graded (part of) bed

The units of the marker bed (4-6) are within the conglomerate facies association (H1) of the Hörnli deposits.

Fig. 12. Up to 2.5 m protruding base of NNW-dipping mass flow rudite of the marker bed

clast lithology of the largest oversize clasts are similar to the coarsest clasts in the normal Hörnli streamflow conglomerates.

Base of Mass Flow Beds

Hörnli deposits below the marker bed are visible at 31 of the 97 profile stations within the outcrop of the Degersheim Conglomerate. In the western half of the outcrop zone, overbank siltstones and sandstones are present at nine localities (*e.g.,* Fig. 10, beds 1-3) and channel conglomerates at four localities. In the eastern half, conglomerates dominate over overbank fines (14 versus 4 localities).

The mass flow beds have sharp, slightly undulating bases without any signs of significant erosion. Although the basal surfaces are easily discerned due to the better cohesion of this facies compared with the normal streamflow conglomerates (see Figs. 12, 13), clearly defined sole marks

Fig. 13. Erosional contact between the marker bed (mass flow facies) and a streamflow channel

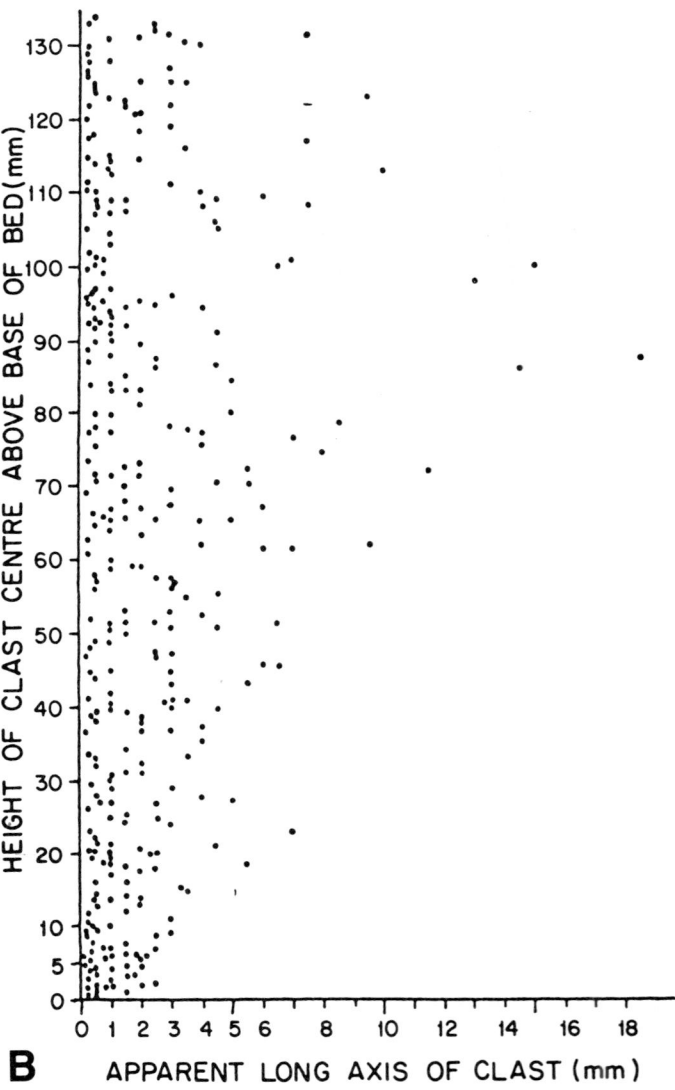

Fig. 14. A) Inversely graded, matrix-supported base of 2 m thick rudite bed (bed 5 of Fig. 10) B) Plot of apparent long clast axis vs distance of clast centre from base of bed. The amount of fine matrix increases and the size of the largest clasts decreases towards the base of the bed.

are rare. Flute-like casts of a few decimetres length, about 5 cm width and 2-3 cm depth were observed at three locations indicating transport towards the NNW.

Thin arenaceous beds of a composition similar to the conglomerates locally underlie the lowest conglomerate bed. One bed, 4-12 cm thick, traceable along 1 km of the outcrop belt, is massively bedded or shows an oblique alternation of arenaceous and carbonate mud laminae (Fig. 18) and locally fills a pre-existing, shallow relief.

Bedded and Grading

Internal bedding appears to be massive, although vague zones of finer and coarser material can be recognized. Crude horizontal bedding occurs in the upper part of thicker conglomerate beds. Of the 99 beds observed at the profile stations, two-thirds (66) appear to be ungraded (Fig. 19). Inverse grading occurs 13 times at the base of thick rudite beds (Figs. 10, 14A, 15), but lateral continuity of these thin (<10 cm) graded zones is poor. At three profile stations, one of the above-mentioned thin arenaceous beds underlies the inversely graded zone. Centimetre-thick, entirely inversely graded beds occur below the inversely graded base of a 2 m thick conglomerate bed (Fig. 10, bed 4) and below a 20 cm thick, matrix-supported layer (Fig. 20). At 15 stations, normal grading occurs at the top of thicker beds, one of which is also inversely graded at the base (Fig. 10, bed 5). Six other stations possibly expose the same 30 cm thick bed which is normally graded except at the base (Fig. 9, profile 55). All observed grading is defined by both changing *MPS* and amount of matrix; Figure 19 classifies the conglomerates according to the second parameter and grading type.

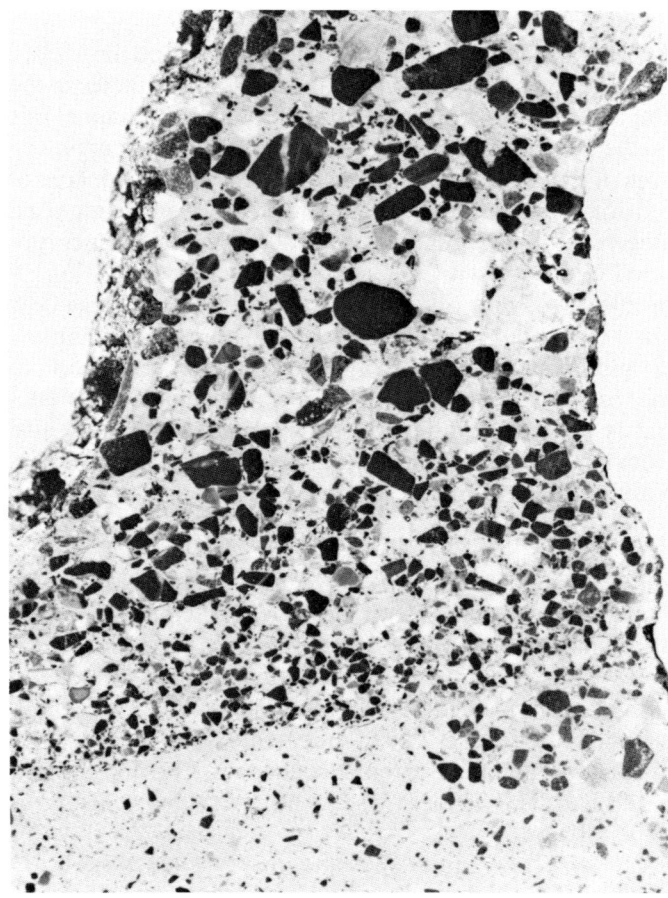

Fig. 15. Inversely graded, clast-supported rudite of Degersheim Conglomerate overlying a matrix-supported bed with sharp, wavy contact. Note pocket of coarser clasts at top of lower bed. Polished slab, width 5.5 cm.

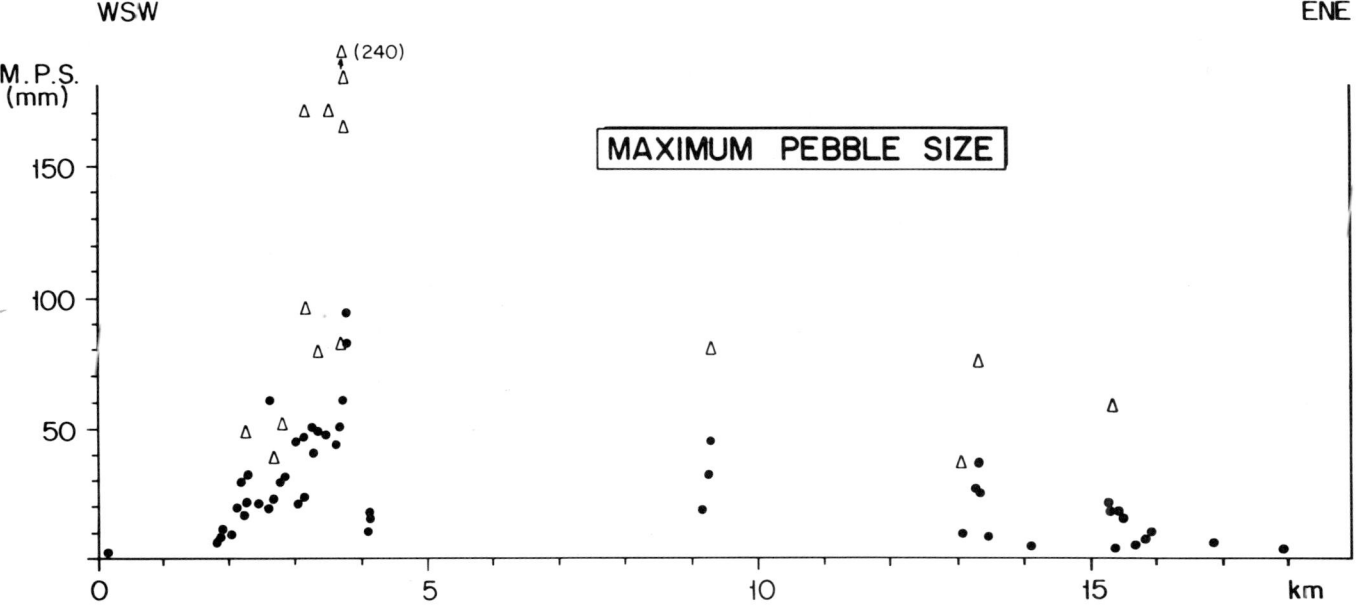

Fig. 16. Maximum particle size (*MPS*) of the Degersheim Conglomerate along its 18 km long outcrop belt (see Fig. 9). dots: angular carbonate clasts (mean of the apparent long axis of the 10 largest clasts)

Relationship to Other Facies of the Marker Bed

Outcrops of the Hüllistein Conglomerate and the Meilen Limestone are west of those of the mass flow facies in the Degersheim Conglomerate (Fig. 9). Only the channel fills of the Abtwil Conglomerate (Table 1) also occur in the area of the mass flow beds, involving a variable degree of erosion of the latter (Fig. 21). Blocks of matrix-supported Degersheim Conglomerate up to 80 cm are found incorporated in the Abtwil Conglomerate at one locality (Fig. 9, profile 38). Thin, fine-grained graded beds of mass-flow character occur rarely as interbeds within the streamflow conglomerates (Fig. 9, profile 46). Progressive dilution of the restricted mass flow lithologies by normal Hörnli material is evident in different Abtwil channel fills and also from the base to the upper parts of individual fills (see Fig. 5 and profile 116 in Fig. 9).

INTERPRETATION OF DEPOSITIONAL PROCESSES

The angularity of limestone clasts and presence of a very fine matrix suggest that the Degersheim Conglomerate resulted from gravity-driven processes known as mass flows, sediment gravity flows or sediment flows (Middleton and Hampton, 1973). The flows must have been non-channelled because their deposits invariably cover both fan channel and fan overbank sediments. The non-erosive base suggests that the flows had little turbulence just prior to deposition.

The superposition of beds and the presence of six types of beds (Fig. 19) indicate several sediment flows of different characteristics. In the *clast-supported* beds (types I-V), the absence of arenaceous beds with the well-known Bouma sequence, deposited from low-density turbidity currents, indicates deposition from flows with high particle concentrations. The low amount of matrix in bed types I-V suggests deposition from flows in which the clasts remained mostly in contact with one another. Suspension was due to buoyancy and static grain-to-grain contact rather than by cohesive strength of the mud with interstitial water. These clast-support mechanisms dominate in high-concentration gravelly sediment flows in humid mountainous terranes, loosely called debris flows (Curry, 1966; Kojan and Hutchinson, 1978; Pierson, 1981), and are considered to have been the mode of transport for the clasts of the clast-supported beds in the marker bed.

Several features point to some rheological details of these flows.

Fig. 17. Isolated, oversize clast (24 cm long, yellow quartz arenite) in conglomerate of mass flow facies. *MPS* of the other clasts is 5.1 cm only.

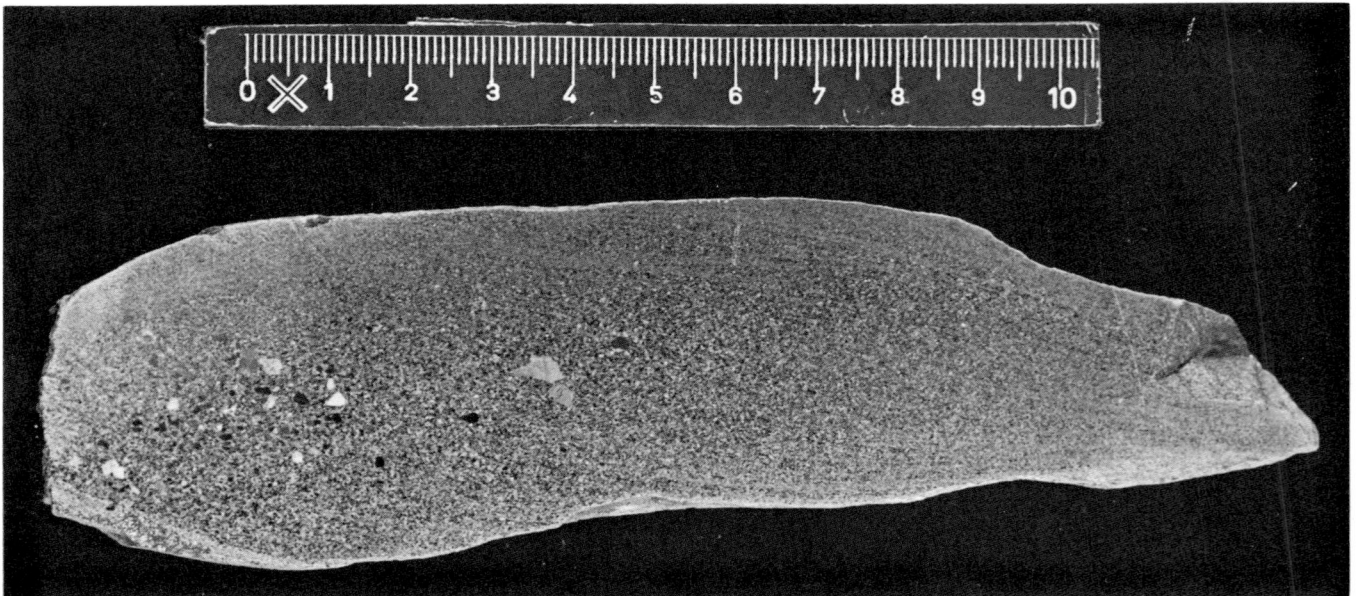

Fig. 18. Obliquely laminated arenite below thick mass flow conglomerate bed. The light laminae are identical to the carbonate mud matrix of the overlying rudite. Polished slab, scale units cm/mm.

(a) The oversized clasts were probably also supported by buoyancy and grain-to-grain contact. Their position in the lower part of beds, however, suggests that these mechanisms were not quite sufficient to fully support their weight.

(b) The absence of large grains at the top of beds, as observed in deposits of subaerial true debris flows (Rodine and Johnson, 1976), supports the assumption that strength of the mud-water mixture was low.

(c) The normal grading of only the top part of a few beds (types III and IV) may indicate a downward migration of a more cohesive rigid plug within the flow with time (Johnson, 1970).

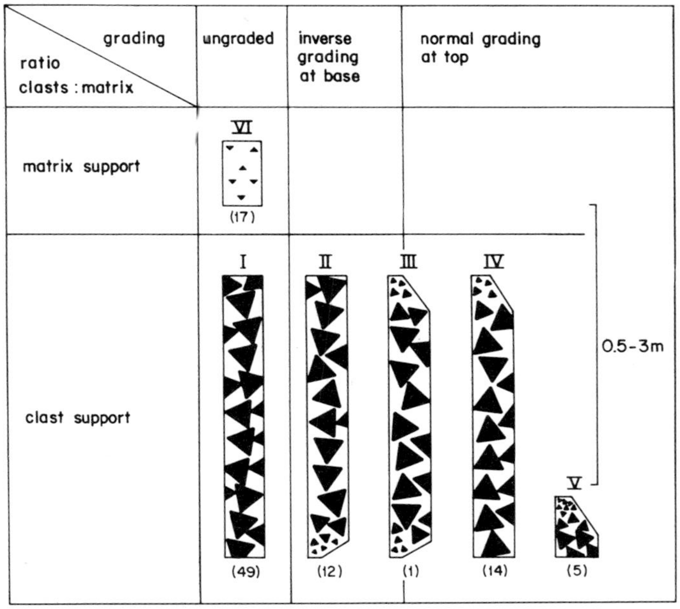

Fig. 19. Types of rudite beds in the Degersheim Conglomerate of the marker bed, according to clast-or matrix-support and grading characteristics. Figures in brackets indicate number of occurrences at the profile stations.

(d) The well-developed normal grading (type V) suggests deposition from a dense, cohesionless suspension, *i.e.* a flow with some turbulence and a lower sediment concentration than the others.

(e) Where inverse grading is present (types II and III), dispersive pressure became the support mechanism for grain-to-grain contacts. Inverse grading is known from (1) deposits of density-modified grain flows, where it is developed over the entire bed thickness (Lowe, 1976); (2) thin traction carpet layers below high-density turbidity currents (Lowe, 1982); and (3) below the rigid plug of gravelly debris flows (Naylor, 1980). The laterally discontinuous, base-only inverse grading of thick conglomerate beds (types II and III) indicates that dispersive pressure was important only locally in the basal zone of shearing of these flows, which again places them into the debris flow group. Only in the rare flows depositing the very thin, entirely inversely graded beds (Figs. 10, bed 4, and 20), was dispersive pressure probably the main clast-supporting mechanism.

(f) The thin basal sandy layers with ripple-like structures (Fig. 9, profile 111; Fig. 18) are suggestive of traction deposition, which may occur at the base of cohesionless mass flows, evidence of which is rarely preserved (Lowe, 1982). The combination of a basal sandy layer with an overlying inversely graded base of a conglomerate bed resembles both Lowe's (1982) traction and traction-carpet stage of high-density turbidity currents, and the structures at the base of low-temperature ignimbrites (Sparks, 1976).

To summarize, features indicating dispersive pressure, traction and suspension sedimentation are locally present in the boundary zones of the clast-supported conglomerates of the marker bed, but the consistent occurrence of a small amount of very fine matrix and the predominantly massive, ungraded appearance suggest that the main characteristics of these flows were those of gravelly, virtually cohesionless debris flows.

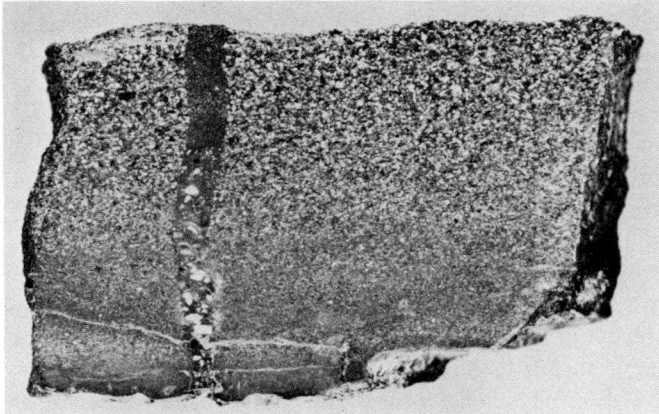

Fig. 20. Inversely graded arenitic layer with ?burrow below main mass flow bed. This layer strongly resembles Fig. 6A of Lowe (1982), showing presumable traction carpet layers. Polished slab, natural size.

Fig. 21. Steep-sided, 0.8 m deep scour demonstrating erosion of Degersheim Conglomerate (lower right) by channels filled by Abtwil Conglomerate (Fig. 9, profile 63). Note coarser grain size and poorer induration of the latter, and cross-bedding above hammer.

The *matrix-supported beds* (type VI) suggest deposition from flows in which the clasts were suspended within the mud-water mixture and supported by its cohesion. Such flows are known as mudflows, cohesive debris flows (Lowe, 1979) or 'true debris flows' (Middleton and Hampton, 1973). The cohesive strength of the mud-water mixture was only high enough to keep particles of a few mm size in suspension. The uniformly dispersed clasts (Figs. 8-1B, 15) imply a weak level of intergranular dispersive pressure (Lowe, 1976; 1979). The occurrence of this bed type at the margin of the outcrop belt of the Degersheim Conglomerate suggests that these flows moved further than those with high clast concentrations.

To summarize, gravelly debris flows with very low matrix strength and thin, cohesive true debris flows were probably the main types of sediment flows from which the beds of the Degersheim Conglomerate were deposited. Since many mass flows change their flow behaviour during movement (see Lowe, 1982, Fig. 12), the two flow types recognized may have occurred during the same flow event.

PROVENANCE

The small number of lithologies present in the clast assemblage of the mass flow deposits suggests a local, restricted source area compared with the drainage basin of the Hörnli river. Clasts from the mass flow conglomerates were compared, in thin sections, with samples from outcrops of Alpine nappes. The comparison was restricted to structurally high nappes because pebbles from the lowest nappe complex (Helvetic nappes) appear only in younger Hörnli deposits. Unfortunately, neither the dark limestones nor the dolomites are distinctive lithologies in the sequences of higher Alpine nappes, and age-diagnostic fossils have not been found. Nevertheless, the association, in the nappes, of dark limestones and dolomites with other lithologies, which are not found as clasts in the mass flow conglomerates, excludes many nappe sequences as a potential source.

The best match was achieved with the Norian to Rhaetian carbonate sequence of the Austroalpine nappes (Table 2). These nappes are not present at the Alpine front behind the outcrops of the marker bed, but the degree of metamorphism of the nappes outcropping today suggests that the Austroalpine nappes may have formed the Alpine front in the Hörnli fan area in the Middle Miocene but were eroded following Miocene to Quaternary uplift.

THE EXCEPTIONAL EVENT - DISCUSSION

The central facies of the marker bed, the Degersheim Conglomerate, represents deposition from several, successive sediment gravity flows. Flow characteristics varied but in general flows were thin and had a high concentration of solids. Could flows with these characteristics have travelled directly from the postulated source area at the Alpine front? The distribution of the preserved Hörnli sediments

Clast Lithologies of Marker Bed (mass flow facies)	Probable Source Formation
dark, microsparitic limestone, occasionally with ostracods	Plattenkalk (Norian) or Kössen Formation (Rhaetian)
light grey, homogeneous, sparry dolomite	Hauptdolomit-Formation (Norian)
oolitic limestone	Kössen Formation (Rhaetian)

Table 2. Provenance of the clasts of the marker bed. Hauptdolomit-Formation, Plattenkalk and Kössen Formation form the Upper Triassic sequence of the Austroalpine nappes of the Alps.

and tectonic considerations suggest that the Alpine front was 10-15 km south of the present-day mass flow outcrops (see Figs. 4, 6). Recent examples demonstrate that debris flows resulting from bursting of landslide- or moraine-dammed lakes may travel such distances when contained in a narrow valley (*e.g.*, Cluff, 1971), but subaerial gravelly debris flows normally flow only little beyond the confinement (see *e.g.*, Pierson, 1981). It appears therefore unlikely that thin flows with high sediment concentrations could have moved subaerially 10-15 km on a fan surface sloping 1° and less. An origin of the marker bed as lake-burst deposit has been previously suggested (Büchi and Welti, 1950), although, beside the problem of travel distance, such deposits commonly consist of a single bed with huge blocks (*cf.* Eisbacher, 1982) which testify clearly to a catastrophic event. Nevertheless, the direct superposition of all mass flow beds and the unique occurrence suggest deposition in connection with an exceptional event.

Large subaerial rockslides and debris avalanches may move further than would be expected by assuming rigid sliding of the rock mass. Several mechanisms, not to be discussed here, have been suggested to be responsible for this "excessive distance of transport" (Hsü, 1975). In the case of the Blackhawk rockslide in southern California, the rock mass spread out as a lobe by one of the above-mentioned mechanisms (see Voight and Pariseau, 1978, p. 27-32), for 7 km over the gently inclined alluvial slope at the foot of the mountain. The slide deposit is composed predominantly of angular carbonate fragments, ranging from powder size to about 25 cm in diameter, with the dominant clast size being about 2.5 cm. This small grain size has been explained by the pervasive fracturing of the source formation, which forms a thrust sheet (Shreve, 1968). In terms of clast size, petrography, nappe origin of the source formation and travel distance, the marker bed is therefore comparable to the Blackhawk slide deposit. However, the latter forms a 10-30 m thick sheet with steep, abrupt terminations and a poor internal stratification (Shreve, 1968), which contrasts to the presence of several thin beds in the Degersheim Conglomerate.

The apparent paradox between unique occurrence and non-catastrophic sedimentary features may be conceptually explained, in this case, by assuming the following

(Fig. 22): an exceptionally large mass of Upper Triassic carbonates broke loose at the Alpine front near the apex of the Hörnli fan and moved onto the proximal part of the fan. Most of the material, in particular the larger sizes, came to rest south of the present-day southern margin of the Upper Freshwater Molasse, to form a lobate, hummocky deposit, but some of the finer material moved further downfan, as relatively thin, local mass flows.

In the Blackhawk rockslide, such small flows did not develop at the front of the deposit, possibly because the amount of water involved in the slide was small under the arid climate. Transformation of a large gravity flow into a smaller flow occurred during the recent catastrophe of Nevado Huascarán in Peru on May 31, 1970. A huge mass of rock and ice, of an estimated volume of 5-10 x 10^7 m^3 and including boulders up to 2500 m^3 in size, travelled with an estimated average velocity of 270 km.h^{-1} down a valley for a distance of 16 km, where it changed into a slow (25-35 km.h^{-1}), diluted, muddy turbulent flow which was able to transport only a few clasts up to 50 cm (Plafker et al., 1971; Browning, 1973; Plafker and Erickson, 1978). At Mount St. Helens, sediment flows carrying small clasts developed in two valleys (11 and 4 cm mean size, respectively) from the huge slide triggered by the volcanic eruption of May 18, 1980 (Gilkey, 1982). Because of the topographic restriction, only one thick (up to 20 m) secondary flow originated during the Huascarán event. On an unrestricted alluvial fan surface, the creation of several, less powerful flows at various places at the lower end of the large event seems possible. The high mobility of the Huascarán debris avalanche has been ascribed to its high fluidity due to the incorporation of snow and water and to a steep, long slope for the first stretch (22°, vertical drop 3000 m). Similar properties may be assumed for the exceptional Miocene even in view of the postulated high altitude of the Alps at that time (Schaer, 1979), which, together with the above-mentioned high precipitation, makes the presence of montane glaciers possible.

Most mass flow beds of the marker horizon are therefore interpreted as the products of secondary sediment flows which repeatedly developed at the lower end of a much larger gravity flow deposit, immediately or shortly after the latter had formed. The direct superposition of beds implies little time between individual flows; locally there is evidence that underlying flow deposits were still soft when deposition from the next flow occurred (Fig. 15). Some thin beds, particularly those within the streamflow facies of the marker bed (Fig. 9, profile 46), may consist of material of the large event which was remobilized at a later time.

The facies distribution of the Hörnli fan deposits that underlie the marker bed suggests that the Hörnli river flowed in a north-north-easterly direction across the fan just prior to the event. After the catastrophe, channels combed through the deposits of the event in various places. The fine-grained Hüllistein Conglomerate (Table 1; Fig. 8-2) formed in westward flowing channels which probably eroded only the fine-grained deposits of the secondary mass flows,

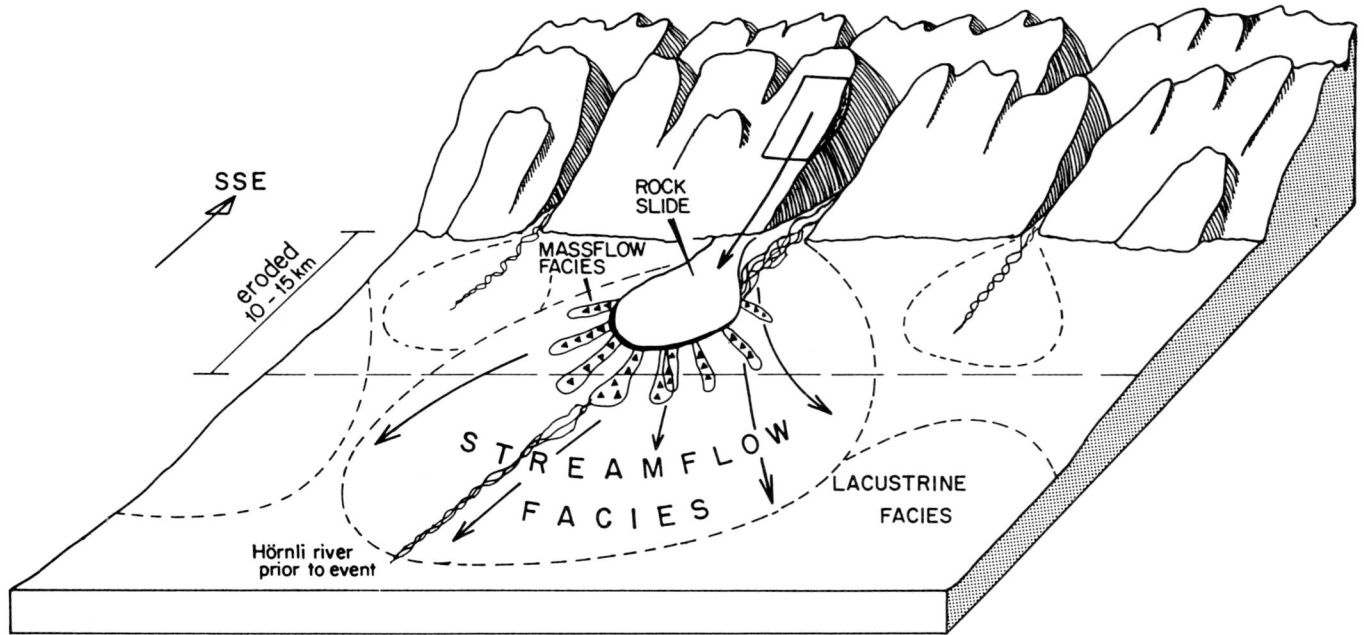

Fig. 22. Probable origin of the marker bed in connection with an exceptional event at the Alpine front. Small mass flows: Degersheim Conglomerate. Fluvial reworking of Degersheim Conglomerate: Hüllistein Conglomerate. Fluvial reworking of rockslide material: Abtwil Conglomerate. Lacustrine suspension deposit at the western toe of the fan: Meilen Limestone. Alternatively, the rockslide could have occurred on the flanks of the Hörnli river valley inside the first mountain chain.

whereas the coarser nature of the Abtwil Conglomerate (Table 1; Fig. 8-3) testifies to channels which eroded also the main rockslide deposits (see Fig. 22). Judging from the presence of channels that cut down into the deposits of the secondary mass flows (Degersheim Conglomerate), but in whose fill the event cannot be detected (Fig. 9, profile 111; Fig. 13), the fluvial reworking of the event took place during a relatively short time. The overall sedimentation pattern was not changed by the catastrophic event (see Fig. 7).

The size of the exceptional event can be estimated from the volume of the marker bed. The collective volume of the conglomerate facies can only be estimated since knowledge on at least one dimension is lacking, but the facies with the largest volume, the Meilen Limestone, can be determined rather exactly (Table 3). Because the estimated volume of the marker bed represents only a fraction of the entire rock volume involved in the exceptional event, this total volume may have been 10^{10}-10^{11} m^3. The largest landslides known on earth have volumes of this magnitude (Table 4). According to the Holocene record in the central Southern Alps of New Zealand, landslides of a volume of 10^8 m^3 occur with a frequency of 10^{-3} per year (Whitehouse and Griffiths, 1983). This compares with Hsü's (1983) worldwide frequency-size relationship for landslides, which may be written in the form,

$$f = 10^5 V^{-1}$$

where f denotes frequency (per year) and V the slide volume (m^3). According to this relationship, a 10^{10}-10^{11} m^3 slide occurs with a frequency of 10^{-5} to 10^{-6} per year. The frequency of such a slide occurring specifically at the front range of a particular mountain chain is believed to be one to three orders of magnitude smaller, which does not then contradict the uniqueness of the marker bed in the Oligocene to Miocene fill of the Molasse Basin.

Facies	Dimensions measured	Dimensions estimated	Volume
Degersheim Conglomerate (mass flow beds)	width (18 km), thickness (average 1 m)	length (1-5 km)	1.8-9.0·10^7 m^3
Hüllistein Conglomerate (streamflow channels)	channel cross-section (6-400 m^2, average 100 m^2)	length (30-100 km)	0.3-1.0·10^7 m^3
Abtwil Conglomerate (streamflow channels)	channel depth (3-8 m, average 5 m)	width (0.2-1 km), length (100-400 km)	10^8-2.0·10^9 m^3
Meilen Limestone (lacustrine)	area (>1000 km^2), thickness (4 m)	—	>4.0·10^9 m^3
			>4-6·10^9 m^3

Table 3. Volume estimate of the four facies of the marker bed. The sheet-like Meilen Limestone contributes most to the total volume, estimated to be >4-5 × 10^9 m^3.

Time of Event	Event, Location	Volume (m^3)	Largest Event Known Within
ca. 1630	Craigieburn Range slide, Central Alps, New Zealand	5·10^8	10^4 y, Central Alps, N.Z.
1974	Mayunmarca slide, Andes, Peru	10^9	ca. 10^3 y, Western Hemisphere
1915	d'Ousoi slide, Pamirs	2·10^9	ca. 10^3 y, Earth
12,000 B.P.	Flims slide, Alps, Switzerland	1.2-1.5·10^{10}	ca. 1.5·10^4 y, Alps
?	Saidmarreh slide, Zagros, Iran	2·10^{10}	?
330,000 B.P.	Mount Shasta slide, California, U.S.A.	2.6·10^{10}	?
Eocene	Heart Mountain slide, Wyoming, U.S.A.	0.9-1.4·10^{12}	10^7 - 10^8 y ?
Miocene	Hörnli Fan slide, Alps/Molasse Basin, Switzerland	10^{10}-10^{11}?	1.6·10^7 y, front range of the Alps

Table 4. Volumes and occurrence interval of some large landslides (data from Voight and Pariseau, 1978; Hsü, 1975; Crandell et al., 1983; Whitehouse and Griffiths, 1983) compared with the exceptional event of this paper.

References

Beard, D.C. and Weyl, P.K. 1973. Influence of texture on porosity and permeability of unconsolidated sand. American Association of Petroleum Geologists Bulletin, v. 57, p. 349-369.

Browning, J.M. 1973. Catastrophic rock slide, Mount Huarascan, North-Central Peru, May 31, 1970. American Association of Petroleum Geologists Bulletin, v. 57, p. 1335-1341.

Büchi, U.P. and Welti, G. 1950. Zur Entstehung der Degersheimer-Kalknagelfluh im Tortonien der Ostschweiz. Eclogae geologicae Helvetiae, v. 43, p. 17-30.

Bürgisser, H.M. 1981. Fazies and Paläohydrologie der Oberen Süsswassermolasse im Hörnli-Fächer (Nordostschweiz). Eclogae geologicae Helvetiae, v. 74, p. 19-28.

Cluff, L.S. 1971. Peru earthquake of May 31, 1970; engineering geology observations. American Association of Petroleum Geologists Bulletin, v. 61, p. 511-533.

Crandell, D.R., Miller, C.D., Glicken, H.X., Christiansen, R.L. and Newhall, C.G. 1983. Huge debris avalanche of Pleistocene age from ancestral Mount Shasta Volcano, California. Geological Society of America, Abstracts with Programs, v. 15, p. 330.

Curry, R.R. 1966. Observation of alpine mudflows in the Tenmile Range, central Colorado, Geological Society of America Bulletin, v. 77, p. 771-776.

Eisbacher, G.H. 1982. Mountain torrents and debris flows. Episodes, v. 1982/4, p. 12-17.

Füchtbauer, H. 1967. Die Sandsteine der Molasse nördlich der Alpen. Geologische Rundschau, v. 56, p. 266-300.

Gilkey, K.E. 1982. Sedimentology of debris flows generated during the 1980 eruption of Mount St. Helens. 11th International Sedimentological Congress, Hamilton, Canada, Book of Abstracts, p. 138.

Gustavson, Th.C. 1978. Bedforms and stratification types of modern gravel meander lobes, Nueces River, Texas. Sedimentology, v. 25, p. 401-426.

Hantke, R. 1984. Floreninhalt, biostratigraphische Gliederung und Paläoklima der mittelmiozänen Oberen Süsswassermolasse (OSM) der Schweiz und ihrer nördlichen Nachbargebiete. Günzburger Hefte, p. 47-55.

Hofmann, F. 1955. Neue geologische Untersuchungen in der Molasse der Nordostschweiz. Eclogae Geologicae Helvetiae, v. 48, p. 99-124.

Hsü, K.J. 1975. Catastrophic debris streams (Sturzstroms) generated by rockfalls. Geological Society of America Bulletin, v. 86, p. 129-140.

Hsü, K.J. 1983. Actualistic catastrophism. Sedimentology, v. 30, p. 3-9.

Johns, D.R., Mutti, E., Rosell, J. and Séguret, M. 1981. Origin of a thick, redeposited carbonate bed in Eocene turbidites of the Hecho Group, south-central Pyrenees, Spain. Geology, v. 9, p. 161-164.

Johnson, A.M. 1970. Physical Processes in Geology. San Francisco: Freeman, Cooper and Co., 577p.

Kojan, E. and Hutchinson, J.N. 1978. Mayunmarca rockslide and debris flow, Peru. In: Voight, B. (Ed.), Rockslides and Avalanches, v. 1. Amsterdam: Elsevier Scientific Publishing Co., Developments in Geotechnical Engineering, v. 14A, p. 315-361.

Lemcke, K., von Engelhardt, W. and Füchtbauer, H. 1953. Geologische und sedimentpetrographische Untersuchungen im Westteil der ungefalteten Molasse des süddeutschen Alpenvorlandes. Beihefte zum Geologischen Jahrbuch, v. 11, 109p.

Lowe, D.R. 1976. Grain flow and grain flow deposits. Journal of Sedimentary Petrology, v. 46, p. 188-199.

Lowe, D.R. 1979. Sediment gravity flows: their classification and some problems of application to natural flows and deposits. In: Doyle, L.J. and Pilkey, O.H. Jr. (Eds.), Geology of Continental Slopes. Society of Economic Paleontologists and Mineralogists, Special Publication 27, p. 75-82.

Lowe, D.R. 1982. Sediment gravity flows: II. Depositional models with special reference to the deposits of high-density turbidity currents. Journal of Sedimentary Petrology, v. 52, p. 279-297.

Malde, H.E. 1968. The catastrophic Late Pleistocene Bonneville Flood in the Snake River Plain, Idaho. United States Geological Survey Professional Paper 596, 52p.

Miall, A.D. 1977. A review of the braided-river depositional environment. Earth Science Reviews, v. 13, p. 1-62.

Middleton, G.B. and Hampton, M.A. 1973. Sediment gravity flows: mechanics of flow and deposition. In: Middleton, G.V. and Bouma, A.H. (Co-chairmen), Turbidites and Deep Water Sedimentation. Pacific Section, Society of Economic Paleontologists and Mineralogists, Anaheim, p. 1-38.

Moiola, R.J. and Malzer, O. 1982. Sedimentology of Oligocene sandstone and conglomerate reservoirs, Molasse Basin, Austria. 11th International Sedimentological Congress, Hamilton, Canada, Book of Abstracts, p. 152.

Naylor, M.A. 1980. The origin of inverse grading in muddy debris flow deposits — a review. Journal of Sedimentary Petrology, v. 50, p. 1111-1116.

Pierson, T.C. 1981. Dominant particle support mechanisms in debris flows at Mt. Thomas, New Zealand, and implications for flow mobility. Sedimentology, v. 28, p. 49-60.

Plafker, G. and Erickson, G.E. 1978. Nevados Huascarán avalanches, Peru. In: Voight, B. (Ed.), Rockslides and Avalanches, v. 1. Amsterdam: Elsevier Scientific Publishing Co., Developments in Geotechnical Engineering, v. 14A, p. 277-314.

Plafker, G., Ericksen, G.E. and Fernández Concha, J. 1971. Geological aspects of the May 31, 1970, Perú earthquake. Seismological Society of America Bulletin, v. 61, p. 543-578.

Reading, H.G. (Ed.) 1978. Sedimentary Environments and Facies. Oxford: Blackwell Scientific Publications, 557p.

Rodine, J.D. and Johnson, A.M. 1976. The ability of debris, heavily freighted with coarse clastic materials, to flow on gentle slopes. Sedimentology, v. 23, p. 213-234.

Schaer, J.-P. 1979. Mouvements verticaux, érosion dans les Alpes, aujourd'hui et au cours du Miocène. Eclogae geologicae Helvetiae, v. 72, p. 263-270.

Shreve, R.L. 1968. The Blackhawk Landslide. Geological Society of America Special Paper 108, 47p.

Sparks, R.S.J. 1976. Grain size variations in ignimbrites and implications for the transport of pyroclastic flow. Sedimentology, v. 23, p. 147-188.

Van Houton, F.B. 1981. The odyssey of molasse. In: Miall, A. D. (Ed.), Sedimentation and Tectonics in Alluvial Basins. Geological Association of Canada Special Paper 23, p. 35-48.

Voight, B. and Pariseau, W. 1978. Rockslides and avalanches: an introduction. In: Voight, B. (Ed.), Rockslides and Avalanches, v. 1. Amsterdam: Elsevier Scientific Publishing Co., Developments in Geotechnical Engineering, v. 14A, p. 1-67.

Whitehouse, I.E. and Griffiths, G.A. 1983. Frequency and hazard of large rock avalanches in the central Southern Alps, New Zealand, Geology, v. 11, p. 331-334.

SEDIMENTOLOGY OF A PRECAMBRIAN QUARTZ-PEBBLE CONGLOMERATE, SOUTHWEST COLORADO

FRANK G. ETHRIDGE,[1] NOEL TYLER,[2,3] AND LARY K. BURNS[1]

ABSTRACT

Sedimentological characteristics suggest that the Precambrian quartz-pebble Vallecito Conglomerate of the Needle Mountains originated as an alluvial fan complex built by high-gradient, short duration peak discharge braided streams and occasional debris flows. Scott, Donjek and South Saskatchewan-type stream deposits dominate proximal, medial and distal fan sequences, respectively. Four fining-upward megacycles, averaging 127 m thick, suggest reactivation of basin-margin faults followed by scarp retreat and lowering of relief during deposition. Thinner cycles within these megacycles, ranging from a few meters to a few tens of meters in thickness, are related to second order extrinsic controls such as flood events. Although the sedimentological setting and gross lithological composition of the Vallecito Conglomerate are similar to Precambrian uranium-bearing fossil placers, the unit is too young to contain stable primary detrital uranium minerals and pyrite.

Field and petrographic evidence indicate that the unit rests unconformably on metavolcanic and metasedimentary rocks of the Irving Formation. Compositional data and limited paleocurrent data suggest that the Vallecito Conglomerate was derived from an uplifted subduction zone complex north of the present outcrop belt and deposited in a tectonically active continental-margin basin. This margin was the southern edge of the North American Plate during mid to late Proterozoic time. This setting, coupled with its age, are in general agreement with the plate-tectonic model proposed by Condie (1982) for Precambrian rocks of the southwestern United States. The depositional-tectonic model presented here represents a first attempt to develop such a model for Precambrian metasedimentary rocks of the southern Rocky Mountains.

RÉSUMÉ

Les particularités sédimentologiques révèlent que le conglomérat à galets quartzeux de Vallecito de Needle Mountain, Précambrien, doit son origine à un complexe de cônes de déjection édifié par les débits de pointe de courte durée de rivières anastomosées et de coulées de débris occasionnelles. Les modes d'accumulation sédimentaire proximale, médiane et distale, typiques des rivières Scott, Donyek et Saskatchewan Sud, prédominent respectivement. Quatre mégacycles positifs i.e. à granoclassement normal, d'une épaisseur moyenne de 127 mètres, indiquent une réactivation des failles qui bordent le bassin suivie d'un recul des escarpements et d'un adoucissement du relief durant l'épisode de sédimentation. Des cycles d'une épaisseur de quelques mètres à quelques dizaines de mètres, à l'intérieur des mégacycles, sont rattachés à des événements de deuxième ordre extrinsèques comme les crues fluviales. Quoique le contexte sédimentologique et la lithologie générale du conglomérat de Vallecito soient semblables aux placers lithifiés porteurs d'uranium du Précambrian, l'unité est trop jeune pour contenir des minéraux d'uranium primaires détritiques et de la pyrite.

Les phénomènes de terrain et les études pétrographiques montrent que l'unité repose en discordance sur les roches métavolcaniques et métasédimentaires de la formation d'Irving. Les analyses de la composition et quelques données sur les paléocourants suggèrent comme provenance du conglomérat de Vallecito une zone complexe de subduction soulevée au nord de la ceinture actuelle d'affleurements et une sédimentation dans un bassin d'une marge continentale tectoniquement active. Ce contexte, et son âge, s'accordent dans les grandes lignes avec le modèle de tectonique des plaques proposé par Condie (1982) pour les roches Précambriennes du sud-ouest des Etats-Unis. Le modèle tectonique de sédimentation décrit ici représente une première tentative de l'application d'un tel modèle pour les roches métasédimentaires de sud des Montagnes Rocheuses.

INTRODUCTION

BACKGROUND AND OBJECTIVES

The original impetus for this study of the Vallecito Conglomerate came from the need to evaluate potential uranium resources in the United States. One of the more important classes of uranium resources is the Early Proterozoic quartz-pebble conglomerate type, bordering older Precambrian shield areas, that contains detrital uranium, gold and pyrite minerals. The most well-known are the Witwatersrand deposits in South Africa (Pretorius, 1976; Minter, 1976; Smith and Minter, 1980) and the Elliot Lake-Blind River deposits of Canada (Robertson, 1976; Roscoe, 1969; Theis, 1979). Recent discoveries of Proterozoic uranium-bearing conglomerates at the southern edge of the Archean Wyoming Precambrian Province (Houston and

---

[1] Department of Earth Resources, Colorado State University, Fort Collins, Colorado 80523, U.S.A.
[2] Bureau of Economic Geology, The University of Texas at Austin, Austin, Texas 78712, U.S.A.
[3] Publication authorized by the Director, Bureau of Economic Geology, The University of Texas at Austin.

The authors would like to acknowledge S. Abou-Zied, R.C. Horton, and F.B. Loomis of the Bendix Field Engineering Corporation (BFEC) and William L. Chenoweth of DOE for reviewing an earlier report from which this manuscript was compiled. Rex Cole, N. J. Theis, Karl Karlstrom and J.A. Robertson unselfishly gave of their time and expertise discussing many of the broad scientific aspects of Precambrian quartz-pebble conglomerates.

We would like to express our special thanks to Alice S. Gross, Conde Thorn and A.M. Campo for their unfailing support in the field investigations. Melanie Keenan, Alice Gross and Doris Rust drafted the figures. Terri Bostedt and Eileen Saracino typed the manuscript. We also wish to thank the reviewers of this symposium volume whose comments aided in bringing some clarity to this paper.

Copyright © 1984, Canadian Society of Petroleum Geologists

Karlstrom, 1979; Hills and Houston, 1979; Graff, 1979 and Karlstrom *et al.*, 1981) have resulted in an extended uranium exploration effort in Precambrian metasedimentary terranes of the United States.

The Vallecito Conglomerate is one of two formations that were considered the most likely in the Needle Mountains to contain fossil placer deposits of uranium. Detailed field and laboratory investigations of the Vallecito Conglomerate were undertaken under the National Uranium Resources Evaluation Program of the United States Department of Energy (Burns *et al.*, 1980). The research described here represents the first attempt to develop a comprehensive depositional-tectonic model for Precambrian metasedimentary rocks in the southern Rocky Mountains south of the Wyoming border. However, metamorphism and deformation obviously limit the ability to make detailed facies analyses. The Vallecito Conglomerate is unique in that it provides a record of the depositional setting at or near the southern edge of an accreting North American Plate (Condie, 1982; Burns and Wobus, 1983).

DESCRIPTION OF STUDY AREA

The Vallecito Conglomerate crops out in the southeastern part of the Needle Mountains of southwestern Colorado (Fig. 1). The Needle Mountains consist of supracrustal and intrusive Precambrian rocks that are exposed in an uplifted domal arch.

The oldest unit in the Needle Mountains is the Irving Formation. The Vallecito Conglomerate (Fig. 2) unconformably overlies the Irving (Fig. 3; Burns *et al.*, 1980;

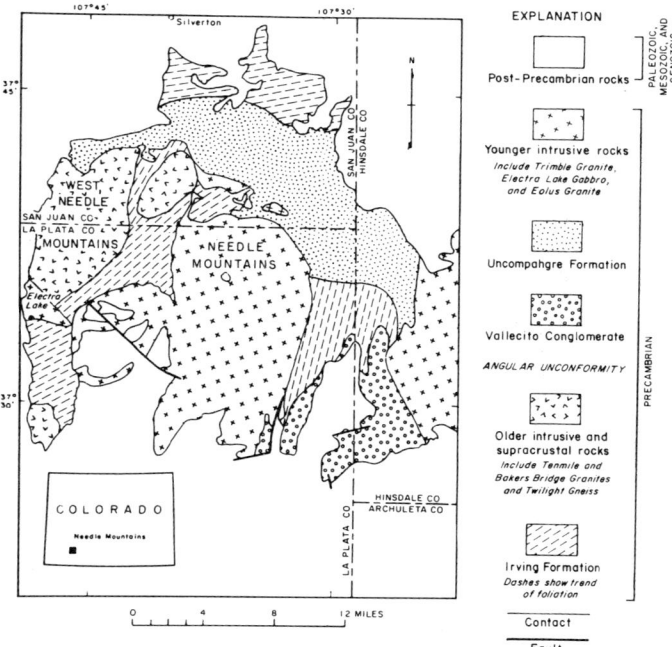

Fig. 1. Generalized sketch map of the geology of the Needle Mountains SW Colorado (modified from Barker, 1969b).

Fig. 2. Cliffs of Vallecito Conglomerate west of Runlett Peak looking northwest from the Granite Ranch in Pine River Valley near the west border of the Granite Peaks quadrangle, SW Colorado. Top of cliff is 645 m above valley floor.

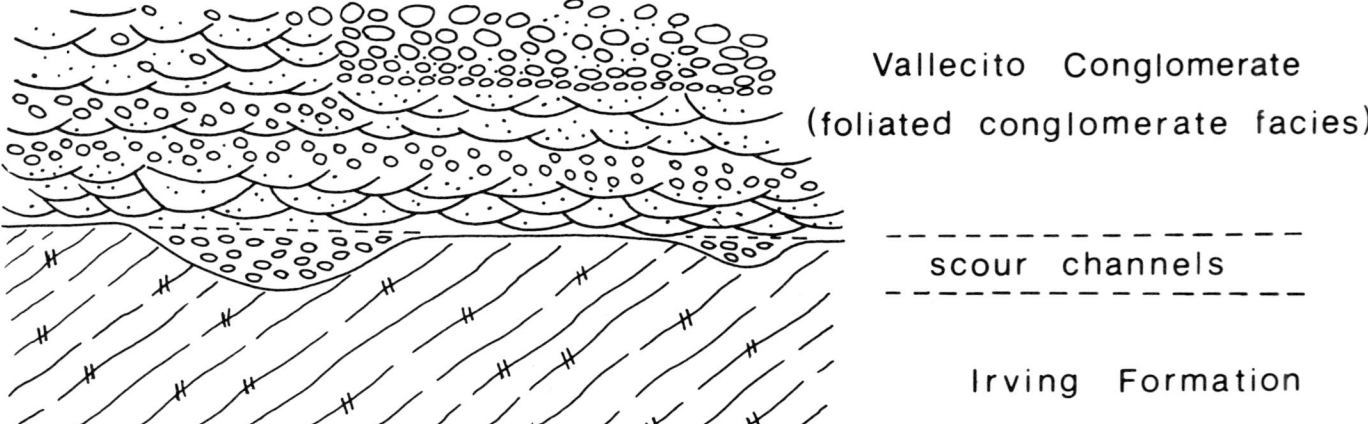

Fig. 3. Schematic sketch of contact between the Vallecito Conglomerate (foliated conglomerate facies) and the Irving Formation. Outcrop is located 1.75 km west of Dollar Lake in the headwaters of Dead Horse Creek, Hinsdale County, Colorado.

Burns and Wobus, 1983). Neither the Vallecito Conglomerate or the Irving Formation has been dated; therefore their exact ages are unknown. The oldest dates reported in the Needle Mountains are for a meta-rhyolite unit called the Twilight Gneiss. Barker (1969a) believed that the 1.7 b.y. old Twilight Gneiss is younger than the Irving Formation. Since the Vallecito Conglomerate is nowhere in contact with the Twilight Gneiss, it is not possible to determine the relative ages of the two units.

The Vallecito Conglomerate lies unconformably on the Irving Formation and contains boulder- to cobble-size clasts of amphibolite, chlorite schist, greenstone, biotite-quartz-plagioclase schist, epidote-quartz gneiss, and iron oxide-quartz rock that are similar to common rock types found in the Irving Formation (Burns et al., 1980). The Uncompahgre Formation is shown to be younger than the Vallecito Conglomerate in Figure 1; however, these two units are nowhere in direct contact with each other and may be, at least in part, contemporaneous.

The Vallecito Conglomerate is intruded by the 1.45 b.y. Eolus Granite (Silver and Barker, 1968; Bickford and et al., 1969). A comprehensive description of all the Precambrian units of the Needle Mountains is given by Barker (1969a).

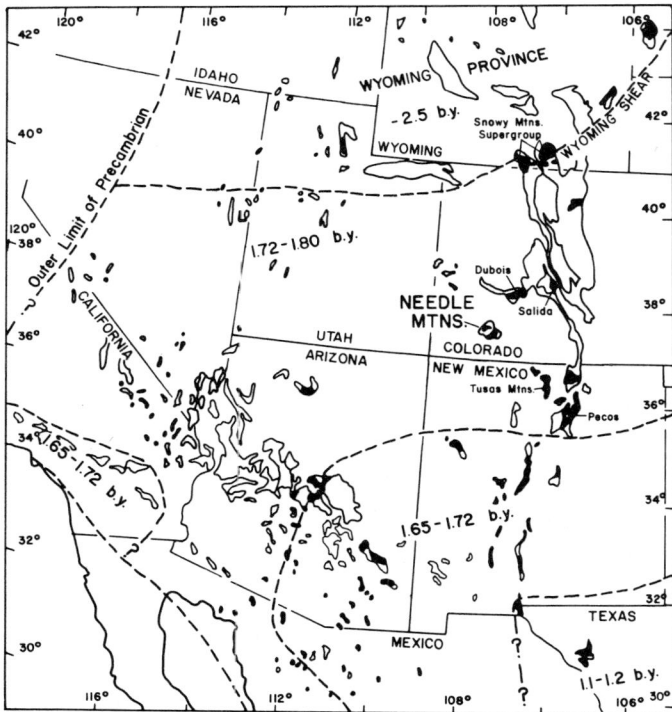

Fig. 4. Precambrian crustal provinces of the southwestern United States, defined chiefly on the basis of Rb-Sr isochron and U-Pb zircon dates from sub-crustal rocks. Major pre1.0 b.y. Proterozoic supra-crustal successions shown in black. Note: Needle Mountain study area located in southwest Colorado. (Age dates from Condie, 1981; map modified from Condie, 1982).

TECTONIC SETTING AND PROVENANCE

Condie (1982) has proposed a plate-tectonic model for Proterozoic continental accretion in the southwestern United States. Supracrustal rocks in this area define three major age provinces which are from north to south: 1.72 to 1.80, 1.65 to 1.72 and 1.1 to 1.2 b.y. (Fig. 4). Precambrian granites intrude all three provinces. Within each province the successions are similar and constitute two assemblages — namely, a bimodal volcanic assemblage overlain by a quartzite-shale assemblage (Burns and Wobus, 1983). In the Needle Mountains the first assemblage consists of the Irving Formation (basaltic) and the Twilight Gneiss (rhyolitic). This is overlain by the quartzite-dominated Vallecito Conglomerate which is in turn overlain by the quartzite-metashale dominated Uncompahgre Formation. The Vallecito Conglomerate of the Needle Mountains falls into the 1.72 to 1.80 b.y. province (Fig. 4). The observations of Condie and others can best be explained in terms of a plate-tectonic model that involved successive marginal basin closures and an Andean-type orogeny associated with southward-migrating arcs (Condie, 1982).

The exact plate-tectonic setting of the Vallecito Conglomerate, in terms of sedimentary basin type, cannot be determined with any accuracy. Compositional data and limited paleocurrent data, however, suggest that the Vallecito Conglomerate was derived from an uplifted subduction complex located north of the present outcrop belt, and that it was deposited in an actively subsiding continental margin basin. The key signal of sediments having such a derivation is an abundance of chert grains, which exceed combined quartz and feldspar grains by a factor of two or three (Dickinson and Suczek, 1979). This ratio is based on data from Phanerozoic rock sequences and may not be totally applicable to Proterozoic sequences. In the examples reviewed by Dickinson and Suczek the principal lithic types include chert, argillite and greenstone, suggesting subduction complexes starved for clastic sediments. The chemical sediment signal is prevalent in the Vallecito Conglomerate although somewhat dampened by at least two subsequent metamorphic events and by the presence of metamorphosed lithic rock fragments derived from the uplifted trench-fill deposits.

The proposed plate-tectonic setting and source terrane coupled with the age of the Vallecito Conglomerate are in agreement with the generalized plate-tectonic model of Condie (1982) (Fig. 4).

LITHOFACIES

All the rocks of the Vallecito Conglomerate show some degree of metamorphism, however, their sedimentary features have been well preserved and they have generally been given sedimentary rather than metamorphic rock names (e.g., Larsen and Cross, 1956; Barker, 1969a). The formation as a whole shows a general upward-fining trend. Conglomerate beds are thicker and make up a larger proportion of the lower part of the formation. Quartzites and quartz-pebble conglomerates are the dominant rock types in the middle of the formation. In the upper part of the sections, quartzites with intercalated thin beds of pelites are the most common lithologies.

LITHOFACIES SUBDIVISION

The Vallecito Conglomerate has been subdivided into four lithofacies based on data collected from four detailed stratigraphic sections and examination of numerous other exposures within the outcrop belt. The lithofacies include: (1) a foliated conglomerate facies, (2) a boulder conglomerate facies, (3) a pebble conglomerate facies, and (4) a quartzite facies. The subdivision is based on diagnostic rock features such as clast size in the coarser-grained units, relative amounts of different rock types, stratification types, and foliation. The foliated conglomerate facies is found only at the base of the formation. The boulder conglomerate facies is not found in direct contact with the foliated conglomerate, however, reconnaissance mapping indicates that it overlies the foliated conglomerate facies.

The pebble conglomerate facies is dominant in section intervals above the boulder conglomerate facies and the quartzite facies is intercalated with the pebble conglomerate facies. Quartzite with minor pelitic interbeds becomes dominant in the upper part of the formation.

The Foliated Conglomerate Facies

The foliated conglomerate facies crops out in an area of 1.6 km² located 1.75 km to the west of the Dollar Lake section (Fig. 5) where it lies unconformably on the Irving Formation. This facies is characterized by the alignment of deformed clasts, by a distinct foliation of matrix material between the clasts and by a schistosity in the quartzite units that are interbedded with the conglomerates (Fig. 6A). The foliation is parallel to bedding in most of the facies, especially in the alignment of the deformed pebble, cobble and boulder clasts. The clasts of the foliated conglomerate facies (schist, phyllite, gneiss, amphibolite, and greenstone) appear to have been much softer and more

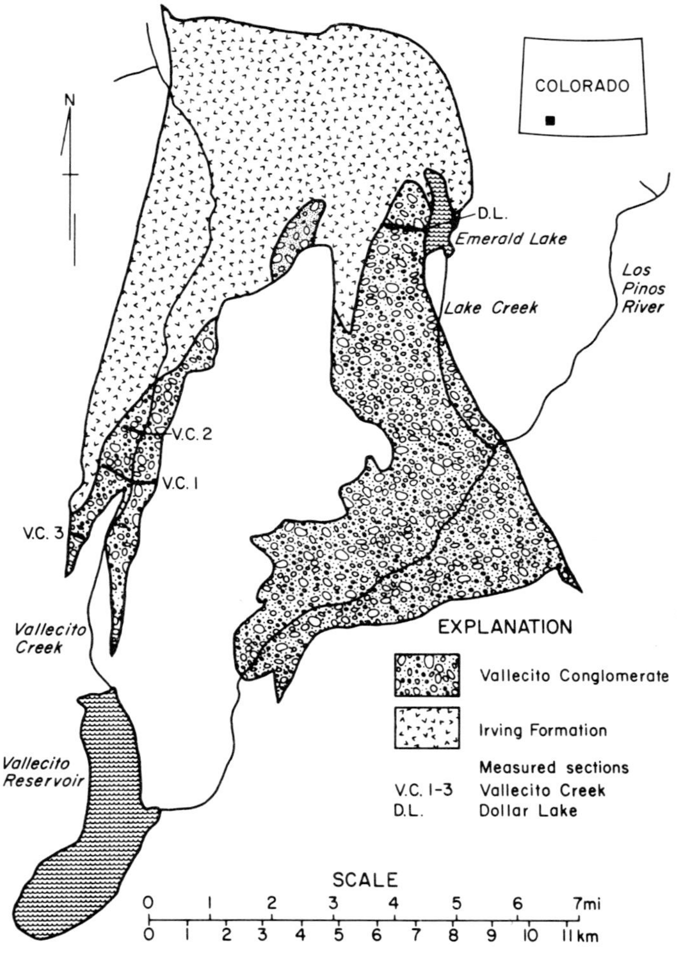

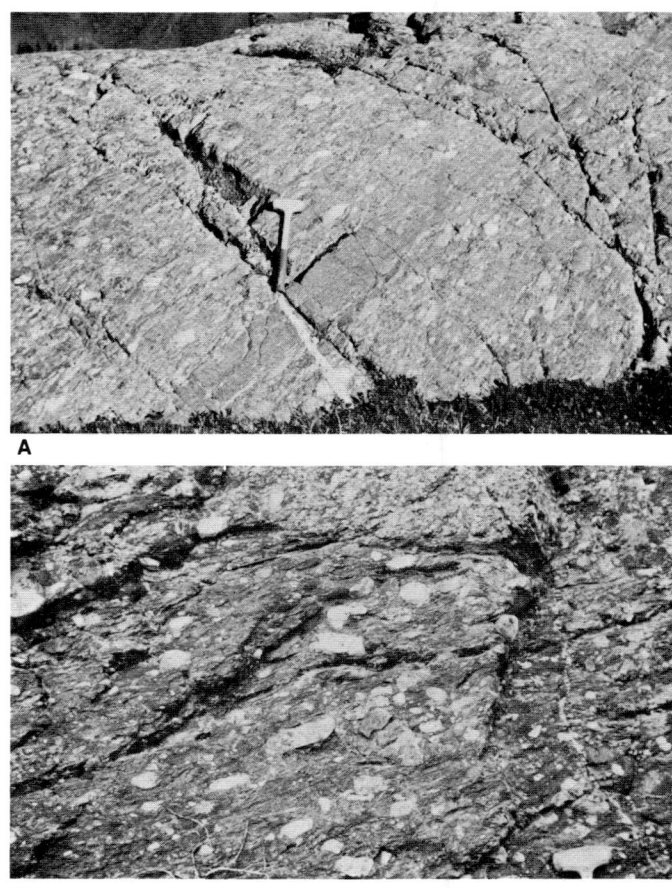

Fig. 5. Location map of detailed measured sections (geology after Steven et al., 1974).

Fig. 6. A. Foliated conglomerate facies of Vallecito Conglomerate near Table Mountain in the headwaters of Dead Horse Creek in the Emerald Lake quadrangle, SW Colorado. Hammer is 42 cm long. **B.** Massive matrix-supported boulder conglomerate, near Table Mountain in the headwaters of Dead Horse Creek in the Emerald Lake quadrangle, SW Colorado. Hammer head is 13 cm wide.

readily deformed than the more resistant silica-rich clasts (vein quartz, quartzite, jasper, chert and banded iron formation clasts) that are dominant in the non-foliated facies. Many of the clasts of schist, amphibolite and greenstone are identical to common lithologic types of the underlying Irving Formation. Minor quantities of vein quartz, quartzite and jasper-bearing rock clasts found in this facies are identical to those found in the other facies of the formation.

Stratification types include massive matrix-supported (Fig. 6B) and clast-supported cobble- to boulder-size conglomerates. The matrix-supported conglomerate beds range up to 6.7 m in thickness. Clast size within these units, which in some instances show inverse grading, reaches a maximum of 70 cm. Parallel laminated and contorted siltstones occur rarely as thin units interbedded with conglomerates in this facies (Fig. 7). Some of these units appear to be rippled and possible burrow structures were observed in one unit.

The Boulder Conglomerate Facies

The conglomerates of this facies which are exposed principally near Dollar Lake (Fig. 5), are only locally foliated. Most are undeformed. The boulder conglomerate facies is characterized by poorly-sorted conglomerates containing subrounded to rounded, occasionally well-imbricated clasts larger than 18 cm (Fig. 8A), but not exceeding boulders of *ca.* 1 m. Clasts are composed of a higher percentage of resistant rock types than those of the foliated conglomerate facies. On the average, about 70% consist of vein quartz, dark fine-grained quartzite, hematite-stained reddish quartzite, jasper and banded iron formation. Only 30% of the clasts are schist, amphibolite and gneiss. Despite the coarseness of this facies, quartzite lenses and pebble conglomerates are intercalated with the boulder beds.

The dominant stratification type is massive clast-supported cobble to boulder conglomerate (Fig. 8B) in indistinct to crudely horizontal beds. Minor stratification types include massive matrix-supported cobble to boulder conglomerates, crude horizontally-bedded to massive pebbly quartzites and trough cross-bedded pebbly quartzites. Beds of massive clast-supported cobble to boulder conglomerates range from 1-4.5 m, and have scoured bases. The interbedded thin lenticular pebbly quartzites are trough cross-bedded or horizontally laminated. Aggregate thickness of this facies approaches 50 m.

The Pebble Conglomerate Facies

The pebble conglomerate facies is the most extensive in the study area, underlying the entire Vallecito Creek area,

Fig. 7. Parallel laminated and contorted siltstones interbedded with crudely bedded pebbly conglomerates, near Table Mountain in headwaters of Dead Horse Creek in the Emerald Lake quadrangle. Hammer is 42 cm long.

Fig. 8. **A.** Boulder conglomerate facies of Vallecito Conglomerate, Dollar Lake Section (Fig. 5). Knife is 9 cm long. **B.** Massive clast-supported cobble to boulder conglomerate, Dollar Lake section (Fig. 5) Hammer is 42 cm long.

as well as most of the area on the Los Pinos River and Lake Creek drainages (Fig. 5).

The pebble conglomerate facies is differentiated from the previous facies on the basis of a maximum clast size of 18 cm, although a diameter of 6 cm is rarely exceeded. Unlike the boulder conglomerate facies, this facies is not predominantly composed of massive gravelly units. Instead, alternating units of crudely horizontally-bedded pebble conglomerates and trough cross-bedded, pebbly, very coarse-grained quartzites (Fig. 9A) are the most common stratification type. Pebble conglomerate beds are generally less than 1 m thick, have scoured bases, and are better sorted than the boulder conglomerate units. Clasts are generally rounded and display crude imbrication. Pebbles in the trough cross-bedded pebbly quartzites are generally well-rounded and elongate and are either scattered randomly or are aligned along bedding planes. Large-scale, planar cross-bedded sets up to 1 m thick of medium- to coarse-grained quartzites without pebbles and rare, thin beds (generally less than 30 cm) of horizontally laminated, medium-grained quartzites (Fig. 10) are found in the upper portion of the pebble conglomerate facies (Fig. 9B).

Fining-upward sequences that grade from pebble conglomerates to trough cross-bedded pebbly quartzites are therefore defined.

The Quartzite Facies

The eastern part of the Los Pinos River/Lake Creek area is underlain by the quartzite facies (Fig. 5). The quartzite facies is characterized by a lack of conglomerate units, and the presence of discontinuous pelites. Planar and trough cross-beds (Figs. 11A and B) are the dominant stratification types.

INTERPRETATION OF DEPOSITIONAL SYSTEM AND CYCLES

DEPOSITIONAL SYSTEM

The coarse-grained nature of the sediments, limited paleocurrent data, and the preferred vertical distribution and rapid lateral changes in stratification types suggest that the Vailecito Conglomerate was deposited as an alluvial fan system (Fig. 12) that prograded southward from a highland source area located immediately north of the present outcrop belt. Deposition of fan sediments occurred primarily by vertical aggradation of longitudinal bars and progradation of dune bed forms in high-gradient, short-duration peak discharge braided streams and by rare debris flows.

At least three types of preferred vertical sequences of stratification types are present in the Vallecito Conglomerate. These preferred vertical sequence types, which can be equated with the vertical profile models erected by Miall (1977, 1978), represent deposits of Scott, Donjek and South Saskatchewan-type braided streams. The Scott type (Scott fan of Boothroyd and Ashley, 1975) is characteristic of proximal, gravelly rivers. It consists of gravelly cycles of

Fig. 9. A. Alternating units of crude horizontally-bedded pebble conglomerate and trough cross-bedded pebbly, very coarse-grained quartzite, Dollar Lake section. Hammer is 42 cm long. B. Large-scale planar cross-bedded quartzites lacking pebbles below trough cross-bedded pebbly, very coarse-grained quartzite, Dollar Lake section. Canteen to right is 21 cm high.

Fig. 10. Horizontally laminated medium-grained quartzites interbedded with trough cross-bedded and crude horizontally bedded pebble conglomerates, Dollar Lake section. Field case is 30 cm high.

Fig. 11. **A.** Planar cross-bedded quartzites, along Vallecito Creek near V.C. 1 (Fig. 5). **B.** Trough and planar cross-bedded quartzites with hematite staining along foreset beds, south of Dollar Lake section (Fig. 5). Pencil is 14 cm long.

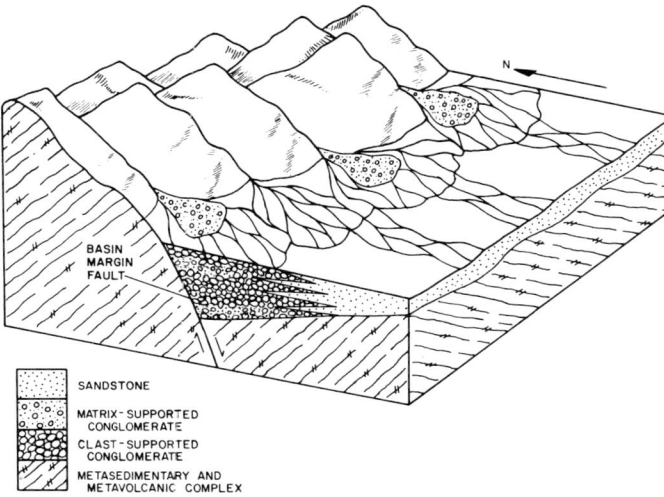

Fig. 12. Block diagram illustrating inferred paleogeography of study area.

waning flood origin, and intervals of superimposed longitudinal bar deposits (Miall, 1977). Figure 13 illustrates thin fining-upward cycles of this type that are characteristic of the middle and upper portions of the boulder conglomerate facies in the Vallecito Conglomerate (113 to 327 m, in megacycles 2 and 3; and 360 to 408 m, in megacycle 4). Cycles in the interval between 360 and 408 m range from 0.86 to 4.3 m and average 1.9 m in thickness.

The Donjek-type braided system (Williams and Rust, 1969; Rust, 1972) is characteristic of more distal gravelly rivers. It is the most varied of the models described by Miall (1977, 1978) and consists of fining-upward cycles of several scales. The thicker of these cycles may reflect sedimentation at different topographic levels within a channel system or successive events of vertical aggradation followed by channel switching (Miall, 1977). In the Vallecito Conglomerate these sequences are characteristic of the lower and upper portions of the Dollar Lake section in the pebbly conglomerate facies (0 to 113 m, megacycle 1; and 413 to 512 m, megacycle 4; Fig. 13). Cycles within these intervals have a much wider range in thickness than those of the Scott type, ranging from 0.3 to 7.3 m and averaging 1.5 m.

The South Saskatchewan type (Cant and Walker, 1978) is characteristic of sandy braided rivers, and analogous deposits occur in the upper portions of the pebble conglomerate facies and in the quartzite facies. Insufficient quantitative data are available to establish the exact nature of cycles within this sequence. The principal stratification type is trough cross-bedding, and minor bed types include thin lenticular pebble conglomerates and horizontally-bedded and planar cross-bedded quartzites.

In general, the three types of braided river deposits discussed above reflect variations from proximal to distal portions of the fan system. Proximal fan deposits consist mostly of cobble to boulder conglomerates (Scott type); mid-fan deposits of interbedded pebble conglomerates and pebbly quartzites (Donjek type); and distal fan deposits of pebbly quartzites and quartzites (South Saskatchewan type). Proximal to distal trends are in general reflected by the overall fining-upward succession of deposits within the Vallecito Conglomerate. This simplified overall trend is disrupted by external controls as suggested by the presence of megacycles.

MEGACYCLES

Four megacycles, based on vertical changes in maximum clast size, are recognized within the 512 m of exposed Vallecito Conglomerate at Dollar Lake (Fig. 13). These megacycles range from 111-153 m in thickness and average 127 m.

The first megacycle begins below the base of the Dollar Lake section and terminates at 113 m above the base (Fig. 13). This megacycle corresponds to the alternating sequence of horizontally-bedded pebble conglomerates and trough cross-bedded pebbly quartzites. Maximum clast size is commonly less than 25 cm.

The second megacycle begins with the boulder conglomerate at 113 m (Fig. 13). Maximum clast size in this unit ranges from 53 cm at the base to 13 cm near the top of the cycle at 238 m. Internally the megacycle consists of thick boulder to pebble conglomerate units and thin lenticular pebbly quartzites which were deposited in proximal Scott type braided streams.

The third megacycle begins just above the covered interval at 238 m and extends to 360 m above the base of the section. Maximum clast size ranges from 1 m near the base to less than 40 cm in thinly bedded conglomeratic units near the top. Internally, this megacycle is the most complex as it consists of matrix-supported conglomerates, clast-supported conglomerates, large-scale planar cross-bedded pebble conglomerates, and thin lenticular pebbly quartzites. Maximum clast size varies greatly over short vertical intervals, although some show coarsening-upward sequences.

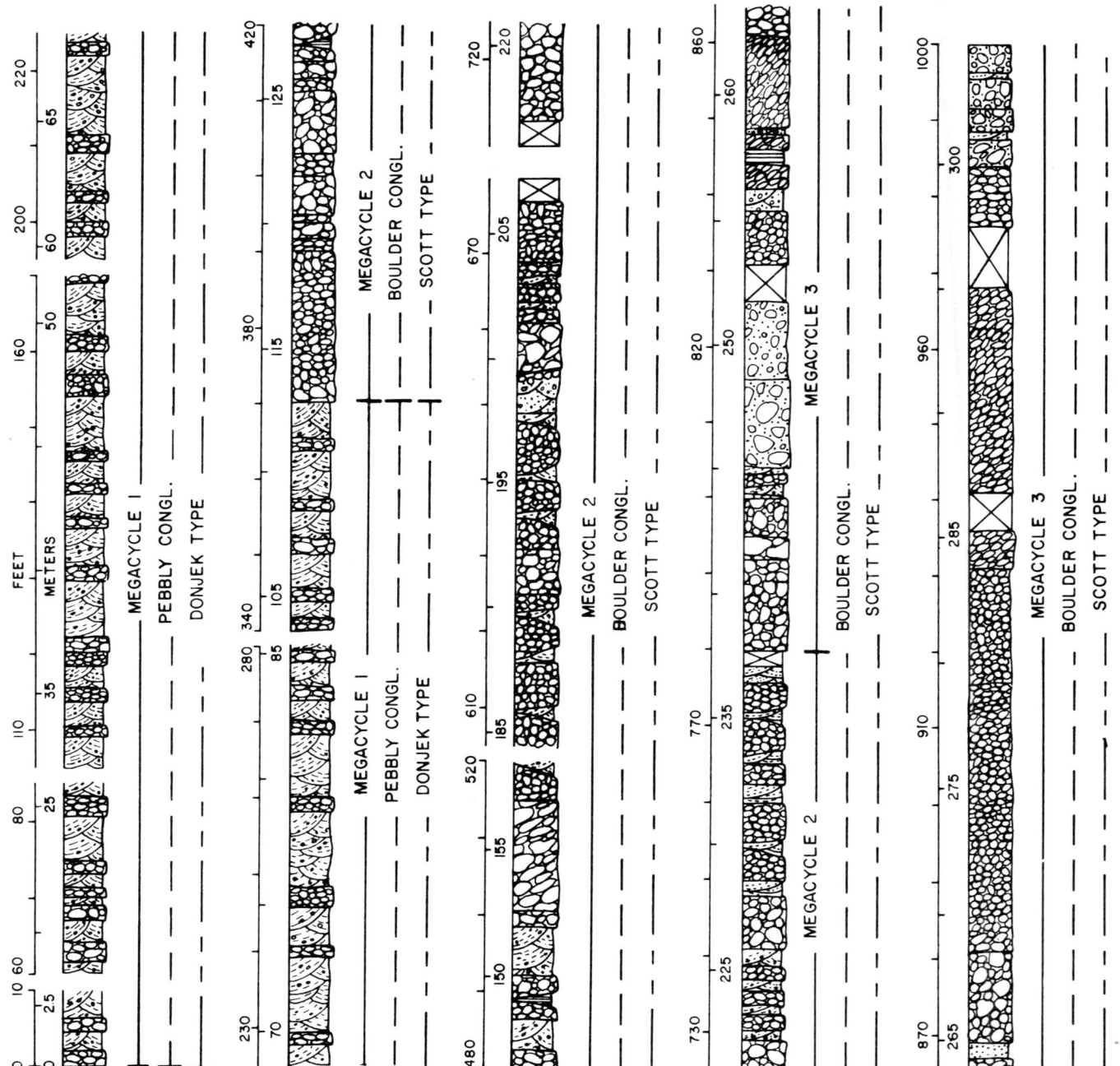

The final megacycle begins with the basal boulder conglomerate at 360 m and extends to the top of the measured section (Fig. 13). Maximum clast size decreases somewhat systematically from 68 cm at the base to 7.5 cm in thin pebbly conglomerates near the top of the megacycle. This sequence resembles deposits of both the Scott and Donjek-type braided streams.

The megacycles within the Vallecito Conglomerate are of the same order of magnitude and are similar to the fining- and thinning-upward megacycles recorded in the Upper Carboniferous of northern Spain (Heward, 1978). From base to top, these megacycles probably record deposition at progressively more distal locations relative to the source area, caused by scarp retreat and/or lowering of relief following reactivation of basin margin faults (Heward, 1978).

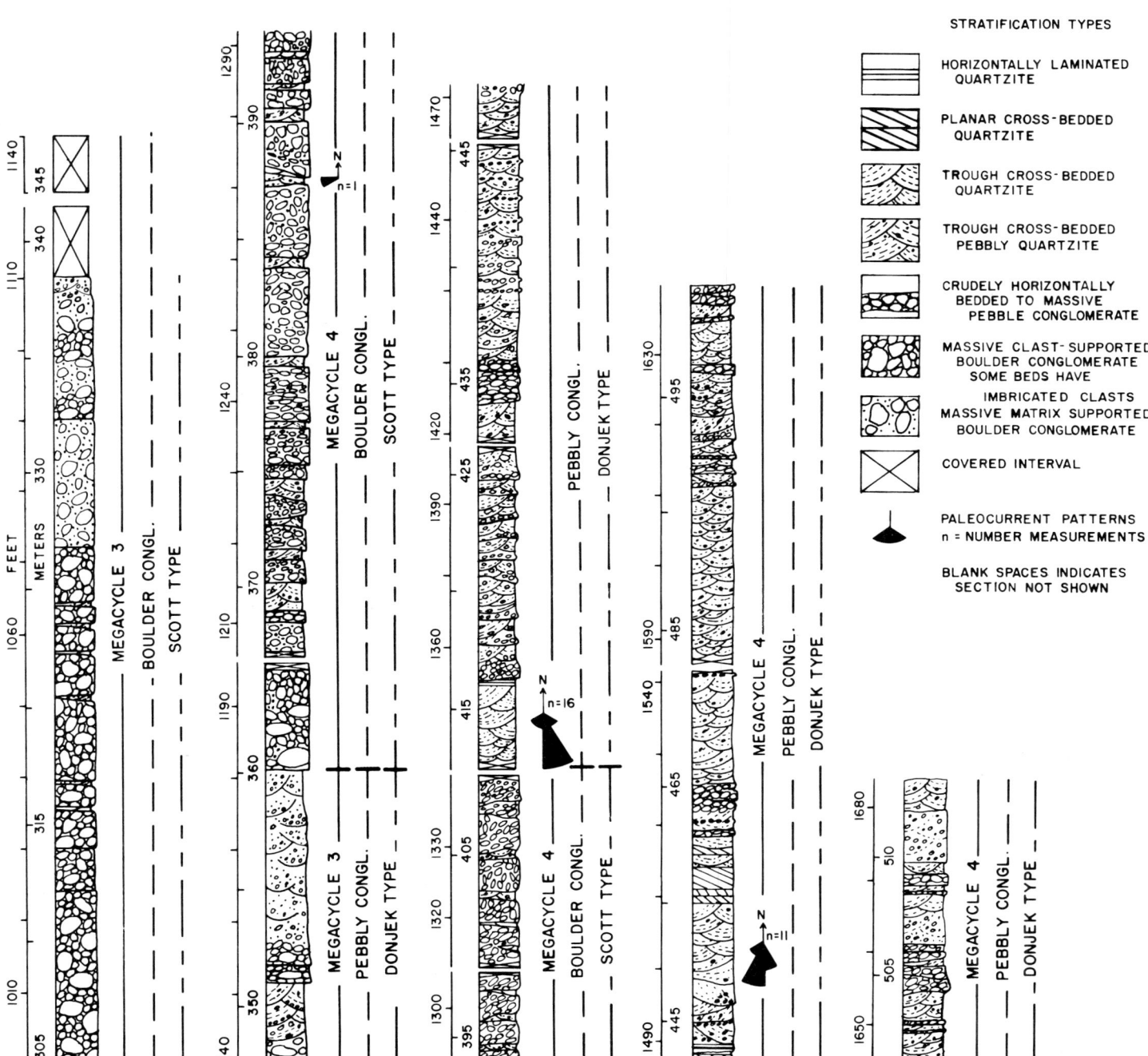

Fig. 13. Dollar Lake stratigraphic section illustrating sedimentary structures and relative grain size through 512 m of Vallecito Conglomerate, grouped into four megacycles. Facies types and inferred braided stream types are also shown.

Summary and Conclusions

The Precambrian Vallecito Conglomerate, located in the Needle Mountains of southwestern Colorado, is similar in terms of composition and depositional setting but younger than Precambrian uranium-bearing fossil placers. The unit was deposited on metavolcanic and metasedimentary rocks of the Irving Formation, and sedimentological data suggest that it was derived from an uplifted subduction zone complex located north of the present outcrop belt.

The lateral and vertical changes in grain size and stratification types suggest that the Vallecito Conglomerate originated as an alluvial fan complex built by braided stream and occasional debris flow deposits. The detailed vertical sequence illustrated in Figure 13 and others discussed in the text show cyclical deposits that can be related to the Scott, Donjek and South Saskatchewan-type braided stream deposits. These sequences represent proximal, mid, and distal fan deposits respectively.

Extrabasinal controls on deposition, such as reactivation of basin-margin faults followed by scarp retreat and lowering of relief during deposition, are reflected by large-scale, fining-upward megacycles that average 127 m in thickness. Autocyclic controls such as flood events are reflected by small-scale cycles ranging from a few meters to a few tens of meters in thickness and by debris flow deposits.

The inferred source area, depositional system and penecontemporaneous faulting coupled with the age and regional plate tectonic setting suggest that the Vallecito Conglomerate was deposited in a tectonically active continental-margin basin along the southern edge of the North American Plate during mid to late Proterozoic time.

References

Barker, F. 1969a. Precambrian geology of the Needle Mountains, southwestern Colorado. United States Geological Survey Professional Paper 644-A, 35p.

Barker, F. 1969b. Gold investigations in Precambrian clastic and pelitic rocks, southwestern Colorado and northern New Mexico. United States Geological Survey, Bulletin 1272-F, 22p.

Bickford, M.E., Wetherill, G.W., Barker, F. and Lee-Hus, C.N. 1969. Precambrian Rb-Sr chronology in the Needle Mountains, southwestern Colorado. Journal of Geophysical Research, v. 74, p. 1660-1676.

Boothroyd, J.C. and Ashley, G.M. 1975. Processes, bar morphology, and sedimentary structures on braided outwash fans, northeastern Gulf of Alaska. In: Jopling, A.V. and McDonald, B.C. (Eds.), Glaciofluvial and Glaciolacustrine Sedimentation. Tulsa, Society of Economic Paleontologists and Mineralogists, Special Publication 23, p. 193-222.

Burns, L.K. and Wobus, R.A. 1983. Correlation and revision of Precambrian stratigraphy, Needle Mountains, southwest Colorado and Tusas Mountains, north-central New Mexico (Abstract). Geological Society of America, Rocky Mountain and Cordilleran Section, Abstracts with Programs, v. 15, p. 424.

Burns, L.K., Ethridge, F.G., Tyler, N., Gross, A.S. and Campo, A.M. 1980. Geology and uranium evaluation of the Precambrian quartz-pebble conglomerates of the Needle Mountains, southwest Colorado. United States Department of Energy, Open-file Report GJBX-118(80), 161p. + 4 plates.

Cant, D.J. and Walker, R.G. 1978. Fluvial processes and facies sequences in the sandy, braided South Saskatchewan River, Canada. Sedimentology, v. 25, p. 625-648.

Condie, K.C. 1981. Precambrian rocks of southwestern United States and adjacent areas of Mexico. New Mexico Bureau of Mines and Mineral Resources, Map 13.

Condie, K.C. 1982. Plate-tectonics model for Proterozoic continental accretion in the southwestern United States. Geology, v. 10, p. 37-42.

Dickinson, W.R. and Suczek, C.A. 1979. Plate tectonics and sandstone compositions. American Association of Petroleum Geologists Bulletin, v. 63, p. 2164-2182.

Graff, P. 1979. A review of the stratigraphy and uranium potential of Early Proterozoic (Precambrian X) metasediments in the Sierra Madre, Wyoming. University of Wyoming, Contributions to Geology, v. 17, p. 149-157.

Heward, A.P. 1978. Alluvial fan sequence and megasequence models: with examples from Westphalian D-Stephanian B coalfields, Northern Spain. In: Miall, A.D. (Ed.), Fluvial Sedimentology. Canadian Society of Petroleum Geologists, Memoir 5, p. 669-702.

Hills, F.A. and Houston, R.J. 1979. Early Proterozoic tectonics of the central Rocky Mountains, North America. University of Wyoming, Contributions to Geology, v. 17, p. 89-109.

Houston, R.S. and Karlstrom, K.E. 1979. Uranium-bearing quartz-pebble conglomerates: exploration model and United States resource potential. United States Department of Energy, Open-File Report GJBX 1(80), 510p.

Karlstrom, K.E., Houston, R.S., Flurkey, A.J., Coolidge, C.M., Kratochvil, A.L. and Sever, C.K. 1981. A summary of the geology and uranium potential of Precambrian conglomerates in southeastern Wyoming. United States Department of Energy, Open File Report DJBX-139 (81), v. 1, 541p. + 7 plates.

Larsen, E.S. and Cross, W. 1956. Geology and petrology of the San Juan region and southwestern Colorado. United States Geological Survey Professional Paper 258, 303p.

Miall, A.D. 1977. A review of the braided-river depositional environment. Earth Science Reviews, v. 13, p. 1-62.

Miall, A.D. 1978. Lithofacies types and vertical profile models in braided river deposits: a summary. In: Miall, A.D. (Ed.), Fluvial Sedimentology. Canadian Society of Petroleum Geologists, Memoir 5, p. 597-604.

Minter, W.E.L. 1976. Detrital gold, uranium and pyrite concentrations related to sedimentology in the Precambrian Vall Reef placer, Witwatersrand, South Africa. Economic Geology, v. 71, p.157-176.

Pretorius, D.A. 1976. Nature of the Witwatersrand gold-uranium deposits. In: Wolf, K.H., (Ed.), Handbook of Strata-bound and Stratiform Ore Deposits. New York: Elsevier Publishing Company, v. 7, p. 29-88.

Robertson, J.A. 1976. The Blind River uranium deposits: the ores and their settings. Ontario Division of Mines, Miscellaneous Paper 65, 45p.

Roscoe, S.M. 1969. Huronian rocks and uraniferous conglomerates in the Canadian Shield. Geological Survey of Canada, Paper 68-40, 205p.

Rust, B.R. 1972. Structure and process in a braided river. Sedimentology, v. 18, p. 221-245.

Silver, L.T. and Barker, F. 1968. Geochronology of Precambrian rocks in the Needle Mountains of southwestern Colorado: Pt. 1 U-Pb Zircon. Geological Society of America, Special Paper 115, p. 204-205.

Smith, N.D. and Minter, W.E.L. 1980. Sedimentological controls of gold and uranium in two Witwatersrand paleoplacers. Economic Geology, v. 75, p. 1-14.

Steven, T.A., Lipman, P.W., Hail, W.J., Jr., Barker, F. and Luedke, R.G. 1974. Geologic map of the Durango quadrangle, southwestern Colorado. United States Geological Survey, Miscellaneous Geological Investigations Map I-764.

Theis, N.J. 1979. Uranium-bearing and associated minerals in their geochemical and sedimentological context, Elliot Lake, Ontario. Geological Survey of Canada, Bulletin 304, 50p.

Williams, P.F. and Rust, B.R. 1969. The sedimentology of a braided river. Journal of Sedimentary Petrology, v. 39, p. 649-679.

SEDIMENTOLOGY AND HYDROCARBON DISTRIBUTION OF THE LOWER CRETACEOUS CADOMIN FORMATION, NORTHWEST ALBERTA

CHRISTOPHER J. VARLEY[1]

Abstract

The Cadomin Formation is the basal unit of the Lower Cretaceous Blairmore Group in northwestern Alberta, and is a wedge-shaped conglomeratic molasse deposit associated with the Columbian Orogeny. Laramide thrusting and subsequent erosion exposed the thickest sections of Cadomin in the Foothills of the Rocky Mountains. Beyond the eastern limit of overthrusting the deposit has been deeply buried in the Alberta Deep Basin. There it forms a thin, laterally extensive sheet-like unit which overlies a regional unconformity of low topographic relief.

The Cadomin was deposited in an alluvial fan - braidplain environment under humid climatic conditions. Braided river systems reworked the fan detritus laterally and distally to produce a thin, yet widespread distribution. A Scott to Donjek type braided-river deposit is inferred on the basis of systematic textural variations in facies assemblages. Proximal to distal trends include decreasing clast size and increasing sand content, with a resulting drop in abundance of bimodal textures.

In the subsurface, the unit forms a gently southwest dipping monocline in which a Deep Basin type hydrocarbon reservoir has developed. Porous and permeable water-bearing strata pass downdip into poor reservoir quality gas-saturated strata. Significant gas reserves have been established in the Elmworth Area but production is limited to higher permeability zones. Permeability and ultimate reservoir potential is a dual function of primary depositional texture and diagenetic modifications. Moderately to poorly-sorted conglomerate textures generally have higher porosity and permeability than bimodal textures. A variation in diagenetic modification exists in the Cadomin because of a geothermal gradient and an increasing depth of burial towards the southwest of the Deep Basin area. Permeabilities are reduced in zones of quartz overgrowth cementation and added authigenic clay matrix. Secondary porosity is developed and permeability enhanced in zones of dissolution and leaching by pore fluids. Successful exploitation of these conglomerates depends on knowledge of reservoir facies textures and diagenetic controls.

Résumé

La formation de Cadomin est l'unité basale du groupe Blairmore, Crétacé inférieur, du nord-ouest de l'Alberta. C'est un dépôt de molasse conglomératique en forme de biseau associé à l'orogénèse du Columbien. Les failles de chevauchement de l'orogénèse du Laramide et l'érosion subséquente ont exposé les sections les plus épaisses de la formation de Cadomin dans les Foothills des Montagnes Rocheuses. Au-delà de la limite orientale des nappes de chevauchement, la formation se trouve enfouie sous d'épais sédiments dans le Deep Basin de l'Alberta. En ce lieu, elle constitue une nappe de grande étendue latérale laquelle recouvre une discordance régionale de faible relief.

Les sédiments de la formation de Cadomin se sont accumulés dans un complexe cône alluvial-plaine anastomosée sous des conditions climatiques humides. Les systèmes des rivières anastomosées ont remanié les particules détritiques des cônes et les ont redistribuées latéralement et à distance éloignée pour construire une nappe mince mais très répandue. Les variations des textures dans les assemblages des faciès permettent de déduire que le dépôt correspond au même type que ceux développés par les rivières anastomosées de Scott à Donjek. La transition du dépôt proximal à distal est caractérisée par une diminution de la grosseur des fragments et une augmentation du contenu en sable, ce qui entraîne une réduction des valeurs de la proportion des textures bimodales.

L'unité forme en subsurface un monoclinal à faible pendage sud-ouest où s'est développé un réservoir d'hydrocarbures du type Deep Basin. Les strates poreuses et perméables gorgées d'eau se modifient le long du pendage en profondeur en strates magasins de mauvaise qualité renfermant du gaz. Des réserves importantes de gaz ont été établies dans la région d'Elmworth mais la production est limitée aux zones de perméabilité maximum. La perméabilité et le potentiel ultime du réservoir dépendent tous les deux de la texture primaire du dépôt et des modifications diagénétiques. Les textures des conglomérats modérément ou pauvrement triés possèdent généralement des porosités et des perméabilités supérieures à celles de sédiments à textures bimodals. La formation de Cadomin fut affectée par une variation due à des changements diagénétiques causés par le gradient géothermique et une augmentation de l'épaisseur des sédiments susjacents en direction sud-ouest de la région Deep Basin. Les perméabilités sont réduites dans les zones de cimentation par l'excroissance des quartz et par une addition de la matrice argileuse authigénique. Dans les zones de dissolution et de lessivage par les fluides interstitiels il s'est développé une porosité secondaire et des perméabilités accrues. Le succès de l'exploitation de ces conglomérats dépend de la connaissance des textures des faciès et des contrôles diagénitiques de ce réservoir.

Introduction

The conglomeratic Cadomin Formation is the basal unit of the Lower Cretaceous Blairmore Group in the Foothills of the Canadian Rocky Mountains, and lies unconformably on the Upper Jurassic to Lower Cretaceous Nikanassin Formation and equivalent formations (Fig. 1). It is overlain conformably by sediments of the Gladstone Formation, Gething Formation and equivalents. The Cadomin is widely distributed and extends from the border with the United

---

[1]Shell Canada Resources Limited, P.O. Box 100, Calgary, Alberta T2P 2H5, Canada.

I would like to sincerely thank Dr. J. C. Hopkins for supervising the Masters Thesis from which most of this work is taken. Canadian Hunter Exploration Ltd. provided financial support and technical assistance for which I am very grateful. Drs. J.R. McLean, M.R. Leeder and F.G. Ethridge helped by reviewing earlier versions of this manuscript. Dr. E.H. Koster suggested many editorial changes which improved the final manuscript. Shell Canada Resources Ltd. gave permission to publish.

Copyright © 1984, Canadian Society of Petroleum Geologists

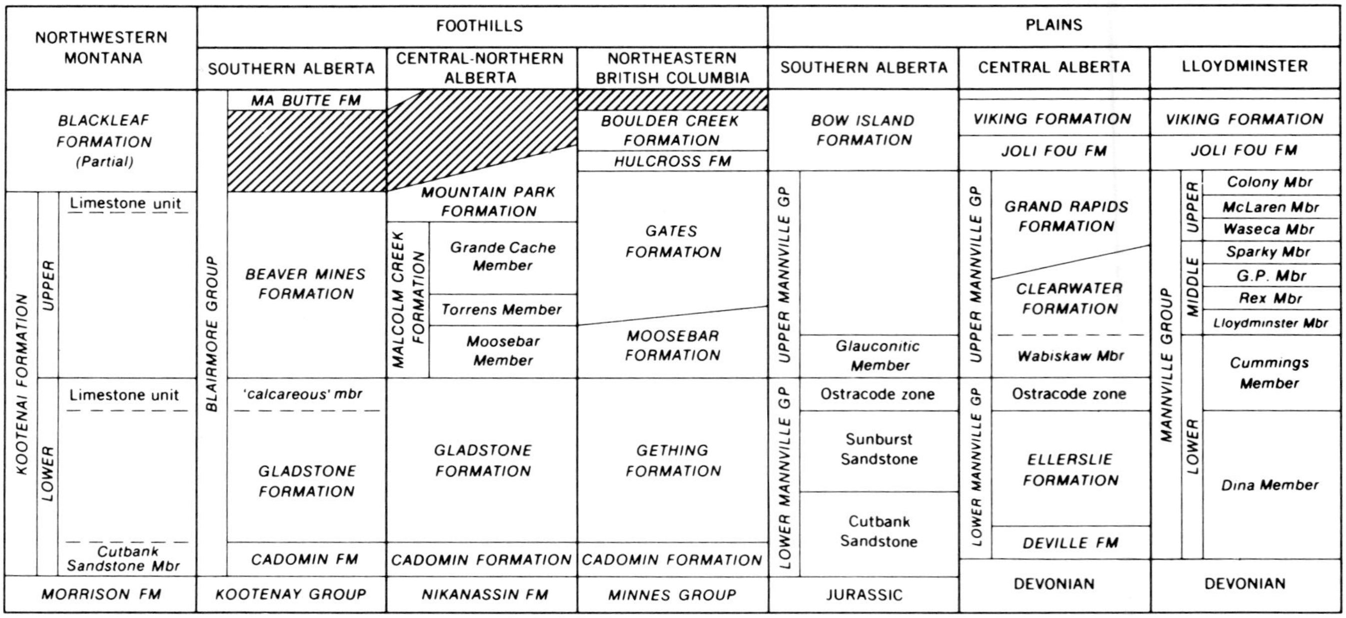

NORTHWESTERN MONTANA
FOOTHILLS
PLAINS
SOUTHERN ALBERTA
CENTRAL-NORTHERN ALBERTA
NORTHEASTERN BRITISH COLUMBIA
SOUTHERN ALBERTA
CENTRAL ALBERTA
LLOYDMINSTER
BLACKLEAF FORMATION (Partial)
KOOTENAI FORMATION
UPPER
LOWER
Limestone unit
Limestone unit
Cutbank Sandstone Mbr
MORRISON FM
MA BUTTE FM
BLAIRMORE GROUP
BEAVER MINES FORMATION
'calcareous' mbr
GLADSTONE FORMATION
CADOMIN FM
KOOTENAY GROUP
MOUNTAIN PARK FORMATION
MALCOLM CREEK FORMATION
Grande Cache Member
Torrens Member
Moosebar Member
GLADSTONE FORMATION
CADOMIN FORMATION
NIKANASSIN FM
BOULDER CREEK FORMATION
HULCROSS FM
GATES FORMATION
MOOSEBAR FORMATION
GETHING FORMATION
CADOMIN FORMATION
MINNES GROUP
BOW ISLAND FORMATION
UPPER MANNVILLE GP
Glauconitic Member
LOWER MANNVILLE GP
Ostracode zone
Sunburst Sandstone
Cutbank Sandstone
JURASSIC
VIKING FORMATION
JOLI FOU FM
UPPER MANNVILLE GP
GRAND RAPIDS FORMATION
CLEARWATER FORMATION
Wabiskaw Mbr
LOWER MANNVILLE GP
Ostracode zone
ELLERSLIE FORMATION
DEVILLE FM
DEVONIAN
VIKING FORMATION
JOLI FOU FM
MANNVILLE GROUP
UPPER
MIDDLE
LOWER
Colony Mbr
McLaren Mbr
Waseca Mbr
Sparky Mbr
G.P. Mbr
Rex Mbr
Lloydminster Mbr
Cummings Member
Dina Member
DEVONIAN

Fig. 1. Correlation of stratigraphic units in the Foothills with areas in the Alberta Plains, Canada.

States to beyond the Peace River area of northeastern British Columbia, a distance of over 1000 km. Correlative units include the conglomerate of the Ephraim Formation in northcentral Utah (McGookey *et al.*, 1972) and the Basal Kootenai conglomerate of northwestern Montana (Mudge and Sheppard, 1968).

The Cadomin is a wedge-shaped foreland molasse deposit associated with the Columbia Orogeny. The most probable source areas were thrust-sheets within or near the eastern part of the Rocky Mountain Main Ranges (Schultheis and Mountjoy, 1978). Laramide thrusting and subsequent erosion exposed the thickest sections of Cadomin in the Foothills of the Rocky Mountains where it forms a prominent and distinctive, conglomeratic unit. Beyond the eastern limit of overthrusting, the deposit has been deeply buried in the Alberta Deep Basin. Cadomin outcrops have been examined in the Mount Belcourt-Mount Torrens field area (Fig. 2), and the subsurface form of the deposit in the Deep Basin has been deduced. A maximum thickness of 155 m for the Cadomin Formation was determined from outcrops at Mount Belcourt in northeastern British Columbia. At Mount Torrens in northwestern Alberta (32 km from Mount Belcourt), only 10 m of Cadomin is observed. In the subsurface, the thickness ranges from 1.2 to 18.8 m and has a mean thickness of 7.7 m. At the time of Cadomin deposition the unconformity surface had a very gentle regional slope towards the north-northeast with minimum internal relief. The Cadomin formed as a wedge-shaped deposit that was thickest to the southwest; to the northeast, the Formation occurs as a relatively thin, sheet-form deposit above the peneplained unconformity surface.

120°
116°
57°
55°
53°
51°
49°
ELMWORTH
ALBERTA
STUDY AREA
EASTERN DEEP BASIN LIMIT
FOX CREEK ESCARPMENT
THRUSTING
MT.BELCOURT
MT.TORRENS
GRANDE CACHE
EDMONTON
CALGARY
BRITISH COLUMBIA
100 km

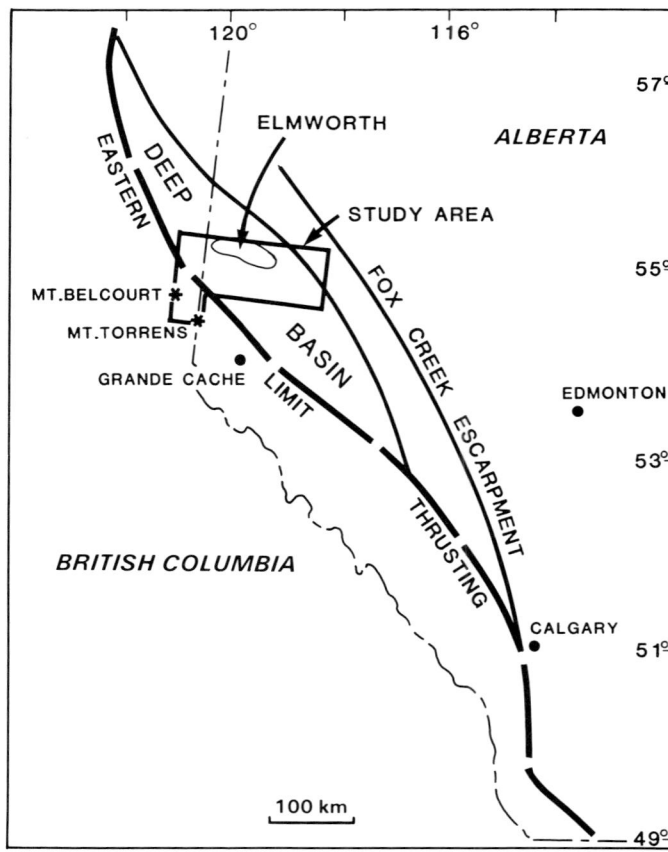

Fig. 2. The study area located in the Deep Basin in northwestern Alberta.

Study Area

The study area is located in the Deep Basin in northwestern Alberta (Fig. 2). The term Deep Basin is applied to the deepest part of the Alberta syncline and its extension into British Columbia (Masters, 1979). The study area includes foothills' outcrops at Mount Belcourt and Mount Torrens, and a subsurface part of the Deep Basin including the Elmworth field, from Townships 64-70, and from Range 1 West of the 6th Meridian to the Alberta/British Columbia border. In this area there are 309 wells and 40 cores penetrating the Cadomin Formation. The Fox Creek Escarpment (McLean, 1976) in the Jurassic subcrop forms the northeastern boundary to Cadomin deposition (Fig. 2).

Lithofacies Assemblages

INTRODUCTION

The Cadomin Formation, although characterized by conglomerate, also contains beds of conglomeratic sandstone, coal, shale and claystone. Observations from outcrops combined with a study of cores from the subsurface have revealed a well-developed textural variation in a northeast direction across the Deep Basin. On these grounds five lithofacies assemblages, A through E, can be defined (Fig. 3; for further details see section on Depositional Setting). Assemblages C, D and E occur only in the subsurface and are described from cores. The term lithofacies as definition of Bates *et al.* (1980) to define a laterally mappable subdivision of a designated stratigraphic unit, distinguished from adjacent subdivisions on the basis of lithological characteristics, including all mineralogical and petrographic characters that influence the appearance, composition or texture of the rock.

Since textural characteristics within the Cadomin change in a progressive manner, the lithofacies assemblage boundaries cannot be rigidly described. Hence a generalized association of textural characteristics and facies must be used to define an assemblage; transitional deposits share characteristics of adjacent assemblages.

Facies have been described using the nomenclature proposed by Miall (1977). The main facies are:

Facies *Gm* — massive or crudely stratified, clast-supported gravel (generally bimodal or poorly-sorted texture),

Facies *Gms* — massive, poorly-sorted, matrix-supported gravel with a high argillaceous content,

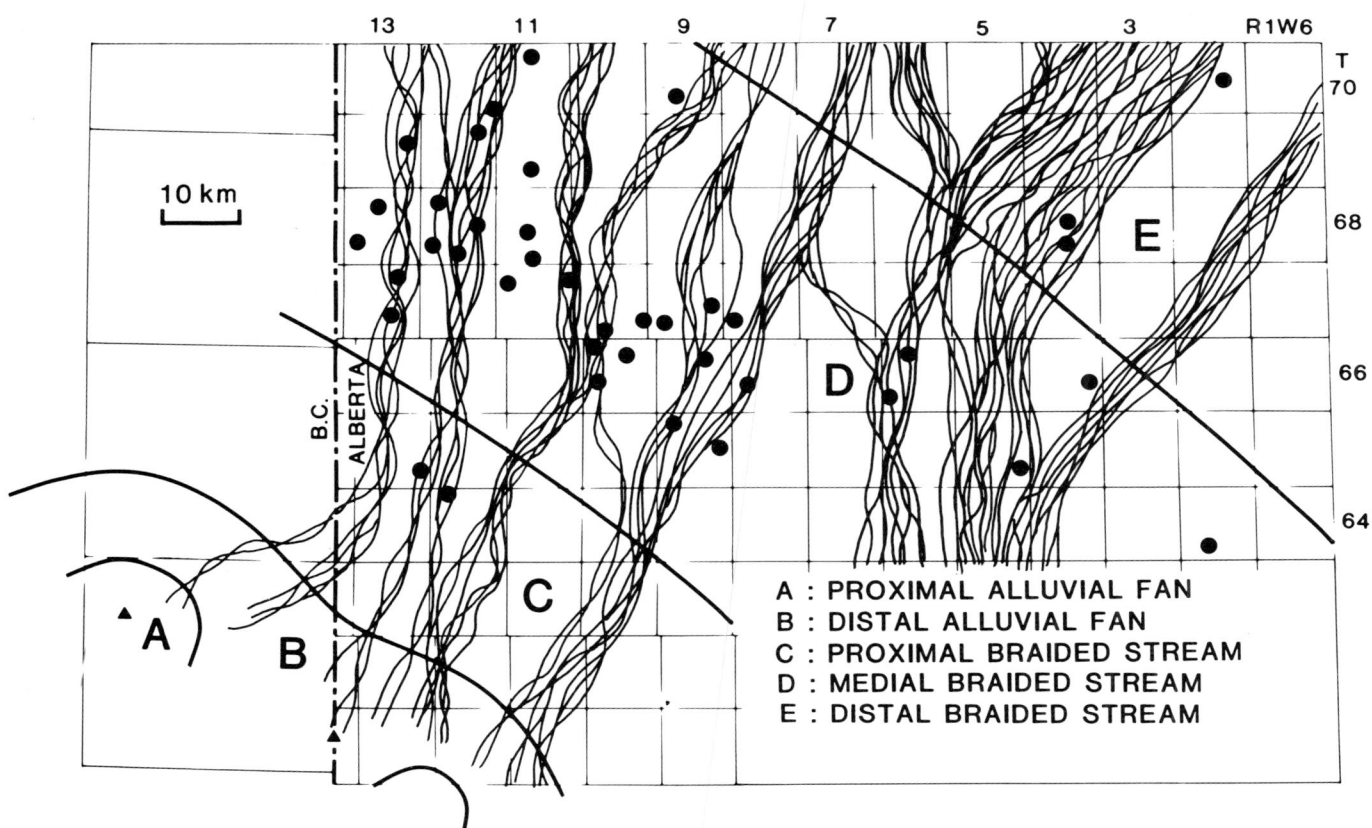

Fig. 3. The distribution of lithofacies assemblages and the interpretation of corresponding paleoenvironments: the environments are in their present location (*i.e.*, not palinspastically reconstructed) but the overall distribution of assemblages with A and B to the southwest and C, D and E progressively to the northeast is correct both for the time of deposition and the present. Locations of cores penetrating the Cadomin Formation are shown by dots. The two triangles show the position of Mount Belcourt and Mount Torrens (see Fig. 2).

Facies *Gp* and *Sp* — planar cross-stratified gravel or sand,
  Facies *St* — trough cross-stratified sand,
  Facies *Sh* — horizontally-stratified sand.
Based on visual determination, conglomerate textures are either unimodal and moderately sorted, unimodal and poorly sorted, or bimodal. The first two types are divided using a standard deviation of 0.5 ɸ. Bimodal conglomerates have two moderately-sorted modes (but are poorly-sorted overall) of gravel and coarse-grained or finer sand.

ASSEMBLAGES

Assemblage A is represented by the Cadomin strata at Mount Belcourt in British Columbia where the maximum thickness is 155 m (Fig. 4A). Conglomerates occur in a series of stacked units which are lense-shaped, 50 to 100 m across and 1 to 5 m thick. The sequences of stacked lenses are generally 10-20 m thick and do not exceed 30 m (Fig. 4B). They are separated by recessive intervals of thinly bedded sandstones and shales with coals which range from 0.1-5.0 m thick. Conglomerate lenses have two distinctly different forms:

1. Broadly upward-fining sequences of bimodally-sorted conglomerates commonly involving a scoured base below clast-supported, generally imbricate conglomerate, followed by matrix-supported conglomerate, in turn capped by pebbly sandstone (Fig. 5A).

   Facies *Gm* is prevalent, and all levels in the cycle display crude horizontal or massive stratification.

2. Poorly defined lenses of facies *Gms*, ranging from boulders to clay-sized particles (Fig. 5B). The boulders can be angular but rounded cobbles are most common; preferred clast fabric or other sedimentary structures are not present. This second type of unit was only observed at Mount Belcourt where it formed a minor component of the Cadomin deposit.

Associated with these lense-shaped units are other minor conglomerate or sandstone deposits. Wedge-shaped planar cross-stratified conglomerates and sandstones (facies *Gp/Sp*) in units up to 1.5 m thick have been recognized located laterally adjacent to, and inter-bedded with lenses of facies *Gm* (Fig. 6A). Trough cross-stratified units of sandstone (facies *St*) ranging in thickness up to 0.5 m (Fig. 6B), and minor intervals of mudstone and shales with thin discontinuous coal beds are also found interbedded with conglomerates.

Assemblage B is characterized by the Cadomin strata at Mount Torrens in Alberta. The Formation is here comprised of a single conglomeratic unit of facies *Gm*, 10 m thick (Fig. 7). Imbrication is well-developed, and the unit displays a crude upward fining from boulder-sized clasts at the base. Lithofacies B is characterized by coarse facies *Gm* with minor interbeds of conglomeratic sandstone.

Assemblage C comprises clast-supported conglomerates which attain cobble grade with a fine- to medium-grained subangular sand matrix up to 15% by volume (Fig. 8A). The main facies is *Gm* with minor facies *Gp*; sandstone beds are absent. Bimodal conglomerate textures predominate (80%) but the pebble mode is usually poorly sorted. Crude horizontal bedding in facies *Gm* is outlined by variations in sorting or clast shape (Figs. 8B, 8C). Planar or trough cross-stratification is difficult to discern in core, except where horizons of elongate pebbles are orientated oblique to the core (Fig. 8D).

Assemblage D deposits generally have very large pebbles as the maximum clast size set in a medium- to coarse-grained sandy matrix of moderately-rounded grains (Fig. 9A). Clast or matrix-supported conglomerates form about 90% of this assemblage with pebbly sandstones the remaining 10%. Textural categories include bimodal (55%) (Fig. 9A), poorly-sorted (35%) (Fig. 9B), and moderately-sorted (10%) (Fig. 9C). Facies *Gm* is most common, but normal

Fig. 4. Proximal alluvial fan deposits, Mount Belcourt, British Columbia: lithofacies assemblage A. **A)** Northeast face of Mount Belcourt, showing proximal alluvial fan deposits of the Cadomin Formation. C illustrates stacked conglomeratic lenses; G shows recessive intervals of thinly bedded sandstone and shales with coal. The two arrows show the upper (A) and lower (B) contacts of the Cadomin Formation which has a maximum thickness of 155 m. **B)** Facies *Gm*; crudely horizontally-stratified conglomerate. Clast-supported bimodal conglomerates (C) are interbedded with matrix-supported pebbly sandstones (S). The section is 2.5 m thick.

and crude inversely-graded units up to 1 m thick occur. In the former case, openwork conglomerates overlie an erosive base and are overlain by bimodal conglomerates with poorly-sorted pebbly sandstones at the top of the unit (Fig. 9D). In other examples, the basal horizon has been infilled by matrix to some degree, and is therefore classed as a bimodal conglomerate. The cores from wells 10-8-70-9W6 and 10-4-67-10W6 contain particularly good examples of these upward-fining units. Paleosol zones sometimes directly overlie these sequences.

Assemblage E occurs farthest to the northeast in the subsurface study area. It consists of mainly matrix-supported pebbly conglomerates (60%) (Fig. 10A), and pebbly sand-

Fig. 6. Proximal alluvial fan deposits, Mount Belcourt, British Columbia: lithofacies assemblage A. **A)** Facies *Gp/Sp*; planar cross-stratified conglomerate and sandstones. **B)** Facies *St*; trough cross-stratified sandstones.

Fig. 5. Proximal alluvial fan deposits, Mount Belcourt, British Columbia: lithofacies assemblage A. **A)** A broadly upward-fining sequence of bimodally sorted conglomerates commonly involving a scoured base (A) below clast-supported generally imbricate conglomerate (B), followed by matrix-supported conglomerate (C), in turn capped by pebbly sandstone (D). Overlying is another sequence of similar deposits, and the entire deposit may represent a stacked sequence of waning-flow cycles. **B)** Facies *Gms* (debris-flow deposit); extremely poorly-sorted agglomeration of boulders through to clay-sized particles illustrated by (D). A shaley mudstone unit (S) is underlain by Facies *Gm* conglomerates (C). Note hammer below (D).

Fig. 7. Distal alluvial fan deposits, Mount Torrens, Alberta: lithofacies assemblage B. A single 10 m unit of Facies *Gm* conglomerate.

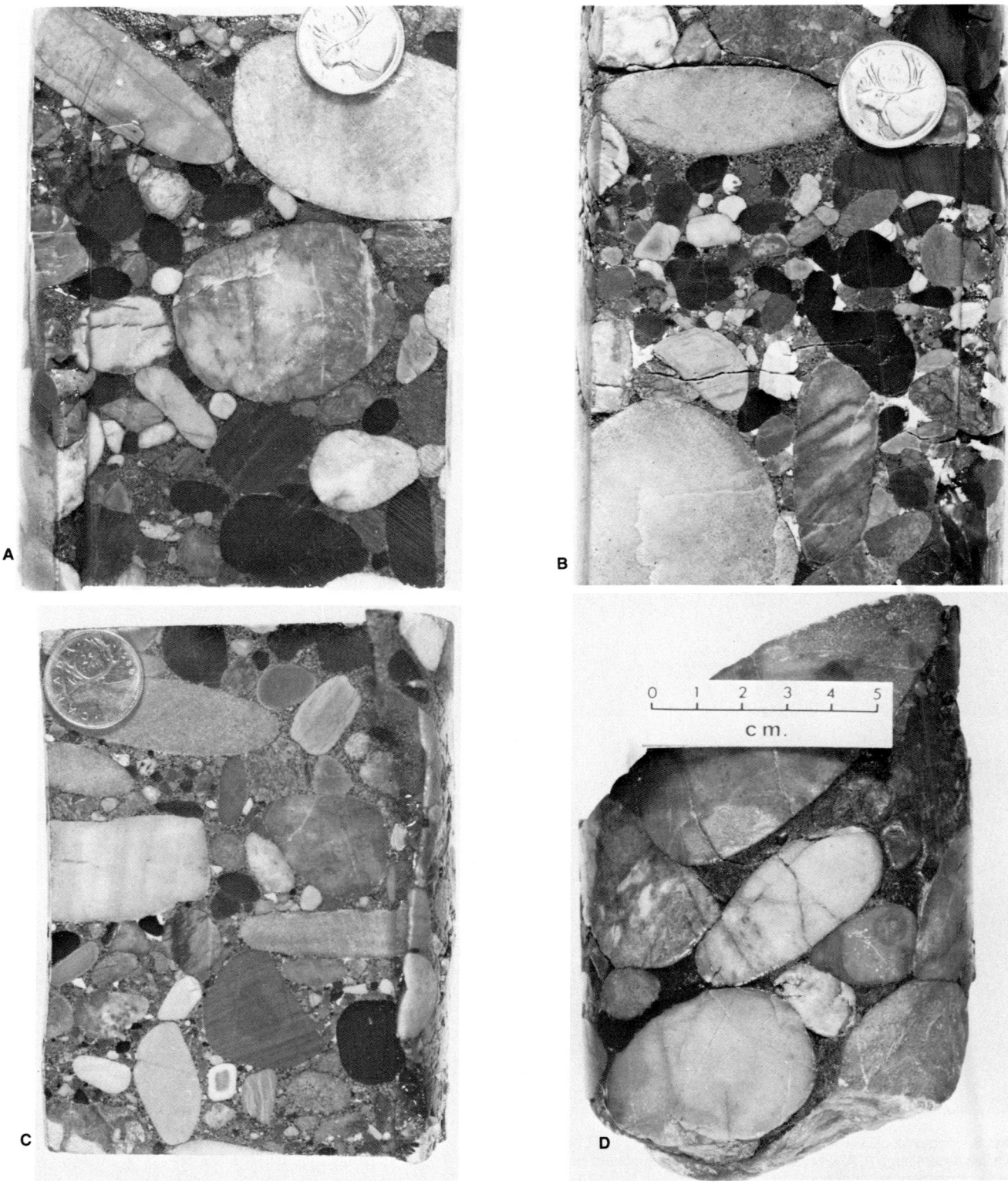

Fig. 8. Core-slab photographs from the subsurface study area (Coin = 2.4 cm dia.) showing representative conglomerate textures from lithofacies assemblage C. **A)** 16-31-64-12W6; depth 3320 m. Bimodal clast-supported conglomerate; medium pebbles to small cobbles with a fine to medium sand matrix. **B)** 8-11-65-13W6; depth 3250 m. Poorly-sorted clast-supported conglomerate; medium sand to small cobbles. Abundant well-rounded and ellipsoidal pebbles and cobbles with a sandy matrix. The central zone of smaller pebbles crudely defines horizontal stratification. Authigenic kaolinite infills where no matrix sand is present. Horizontal fractures cross-cut the core. **C)** 8-11-65-13W6; depth 3248.7 m. Facies *Gm*; crudely horizontally-stratified, clast-supported bimodal conglomerate; small pebbles to small cobbles in a fine to medium sand matrix. **D)** 8-11-65-13W6; depth 3251.1 m. Cross-bedded clast-supported conglomerate consisting of very large pebbles and small cobbles with a fine to medium sand matrix. Each mode is moderately sorted.

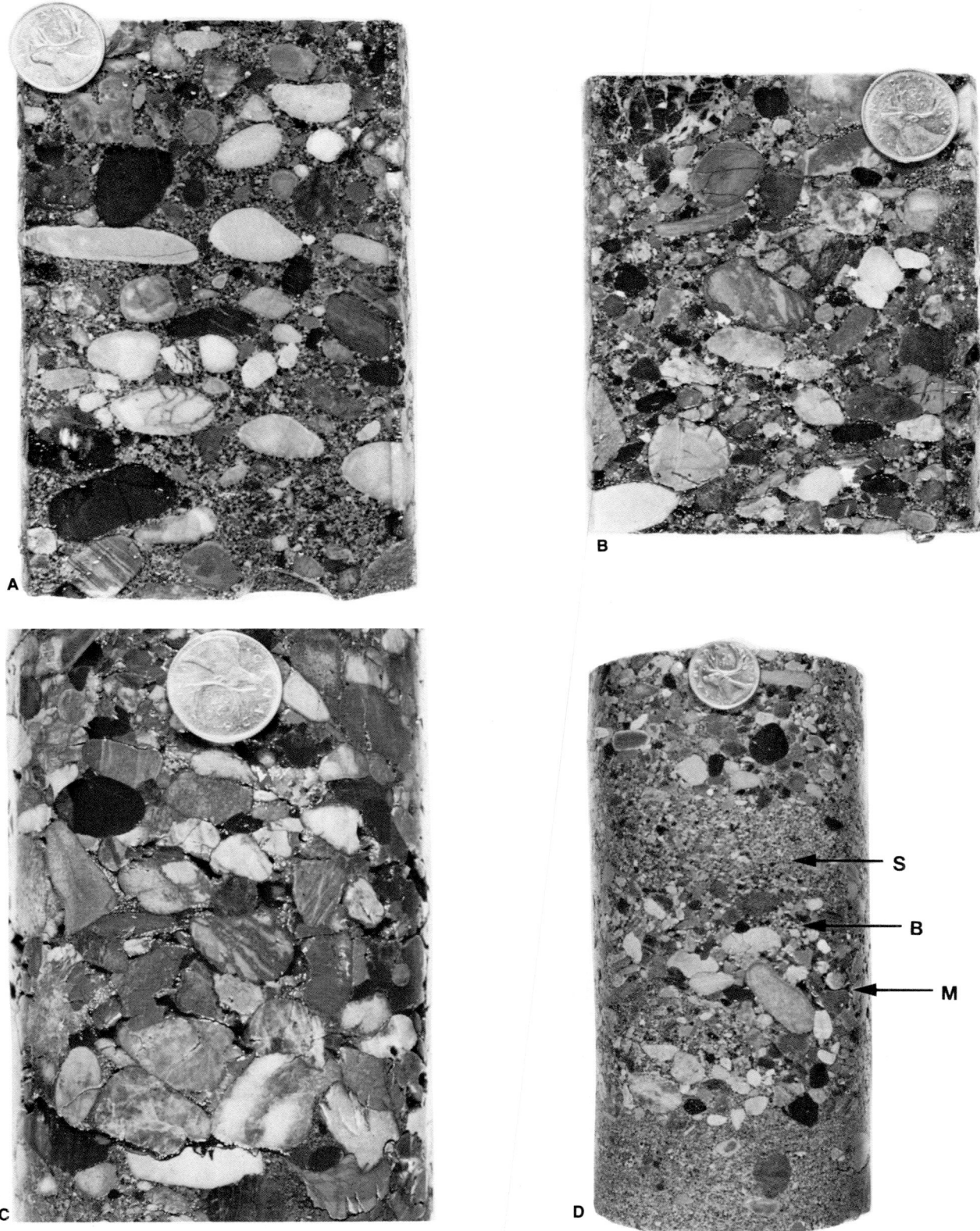

Fig. 9. Core-slab photographs from the subsurface study area (Coin = 2.4 cm dia.) showing representative conglomerate textures from lithofacies asssemblage D. **A)** 10-14-67-9W6; depth 2645.2 m. Bimodal conglomerate; small to very large pebbles with a medium to coarse sand matrix; sand content 20-25%. **B)** 10-14-67-9W6; depth 2645.85 m. Very poorly-sorted conglomerate with medium sand to very large pebbles; sand content 20%; porosity 6%; permeability 21 mD. **C)** 10-4-67-10W6; depth 2824.7 m. Moderately-sorted very large pebble conglomerate, in parts lacking a sandy matrix; porosity 5.3%; permeability 80 mD. The conglomerate is from the basal part of a fining-upwards cycle that is perhaps a waning-flow deposit. **D)** 10-8-70-9W6; depth 2195.2 m. Upward-fining from matrix-free conglomerate (M), to bimodally-sorted conglomerate (B), to sandstone (S); perhaps indicative of waning-flow; porosity 6%; permeability 8 mD.

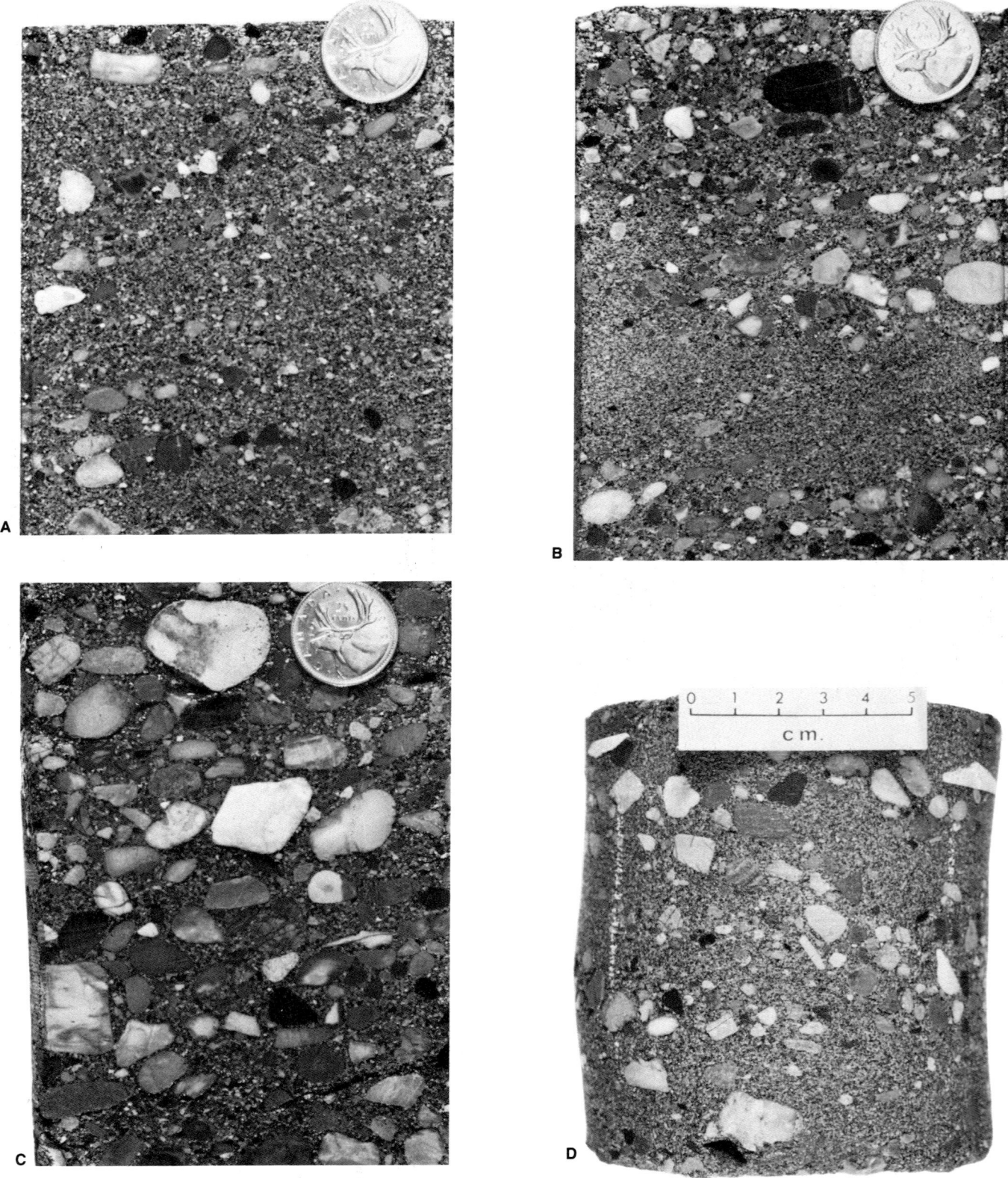

Fig. 10. Core-slab photographs from the subsurface study area (Coin = 2.4 cm dia.) showing representative conglomerate and sandstone textures from lithofacies assemblage E. **A)** 10-21-68-4W6; depth 2011.5 m. Poorly-sorted pebbly sandstone; small to medium pebbles in a fine to very coarse sand matrix; sand content is 85%. **B)** 10-21-68-4W6; depth 2012.3 m. Poorly-sorted matrix-supported sandy conglomerate; fine sand to medium pebbles; sand content 60%. **C)** 10-21-68-4W6; depth 2015.8 m. Matrix-supported sandy conglomerate; medium to large pebbles in a well-sorted coarse sand matrix; sand content 40%. **D)** 10-21-68-4W6; depth 2014.7 m. Matrix-supported sandy conglomerate; small to medium pebbles in a fine to medium sand matrix.

stones (40%) (Fig. 10B). The clasts are generally medium to large pebble grade, fairly well-rounded, and set in a medium to very coarse sandy matrix. Poorly-sorted textures form 80% of this assemblage and bimodal textures 20% (Fig. 10C). Units are generally massive or crudely horizontally stratified, but can be cross-bedded (Fig. 10D).

TRENDS

In summary, distinct textural variations have been noted from lithofacies assemblage A to E, in a northeastward direction across the Deep Basin study area. They consist of:

1. Decreasing maximum clast size from boulders to large pebbles;
2. Decreasing average clast size from cobbles to medium pebbles;
3. Increasing average size of matrix sand;
4. Increasing average matrix sand content from 5 to 60%;
5. Increasing abundance of sandstone facies in assemblages C to E from 0 to 40%;
6. Decreasing ratio of clast-supported to mainly matrix-supported textures;
7. A gradation from predominantly bimodal to increasingly poorly-sorted textures;
8. Increasing roundness of pebble-grade clasts; and
9. Increasing roundness of sand-sized matrix grains.

The following generalized areal trends in facies have also been noted:

1. Facies *Gm* is widespread, and forms the main facies in all five assemblages.
2. Facies *Gms* was only observed in assemblage A.
3. Sandy facies *Sp*, *St* and *Sh* occur as minor interbeds in assemblage A, B and C, and increased in abundance through assemblages D to E.
4. Facies *Gm* units mainly contain interbedded bimodal and poorly-sorted conglomerates; moderately-sorted conglomerates do not exceed 10%.

DEPOSITIONAL SETTING

The geometry and lithofacies assemblages of the Cadomin in the study-area all favour an alluvial fan and extensive braidplain setting. McLean (1977) suggested that humid conditions can be inferred from the presence of coal in the overlying Gething Formation. A number of other factors support an interpretation of fairly humid conditions. Bull (1972) suggested that semiarid alluvial fans characteristically have abundant debris-flow deposits and few systematic sorting patterns. Few debris-flow deposits were observed in assemblage A at Mount Belcourt. In addition many cherts have a weathered outer rim observable by colour variations which probably resulted from subaerial exposure on the braidplain under humid conditions.

McLean (1976) described the Cadomin as having a fluvial origin with low-sinuosity streams depositing sheet-like units of coarse sediment during flood stages. Contrary to this earlier paper, McLean (1977) suggested that the Cadomin is a fluvially-modified pediment gravel. It is difficult to interpret whether the Cadomin accumulated over a long period as a fluvially-modified pediment gravel, or was dispersed by fluvial processes subsequent to a period of peneplanation which formed the unconformity surface. However, both situations involve fluvial processes as the agent responsible for preserved facies. The interpreted depositional environments with the lithofacies assemblage distribution are illustrated in Figure 3. The environments are in their present location with the foothills outcrops not palinspastically restored to their true location at the time of deposition. However the overall distribution of assemblages with A and B to the southwest and C, D, and E progressively to the northeast is correct both for the time of deposition and the present. The distribution of channel systems on Figure 3 is based upon 1) valley trends on the pre-Cadomin unconformity surface, as interpreted from isopach mapping and 2) thick trends on an isopach map of the Cadomin Formation. However, location of specific channels is schematic and illustrates the probable distribution at one point in time during deposition.

A diagrammatic representation of the general Cadomin depositional setting is illustrated in Figure 11. To the southwest was a rising mountain front formed by the Columbian Orogeny, which resulted in deformation and emplacement of thrust-sheets of Precambrian and Paleozoic strata. Erosion of these thrust-sheets provided the abundant chert and quartzite-rich coarse detritus which comprises the Cadomin Formation (Schultheis and Mountjoy, 1978). Alluvial fans formed where rivers emerged from restricted mountain valleys onto a piedmont area. Braided rivers on

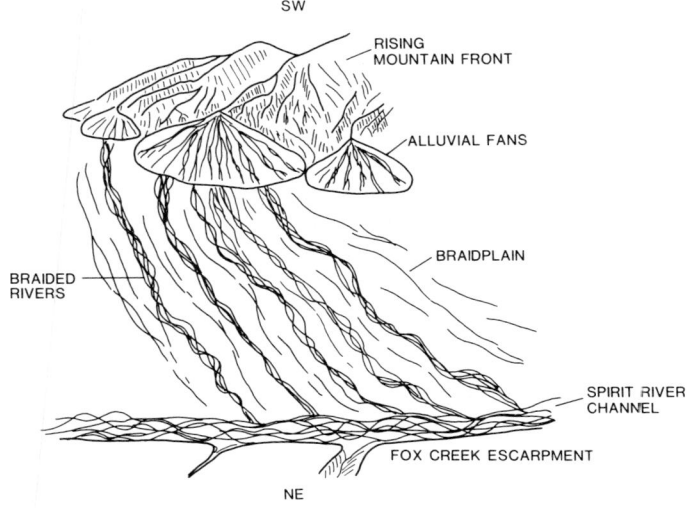

Fig. 11. Depositional Setting of the Cadomin Formation.

the wedge-shaped fan bodies then flowed north-eastwards onto a low-relief peneplain. The thin, widespread nature of the braidplain sequence strongly suggests multiple reworking by shifting channel systems. Eventually this braidplain merged with the main Spirit River Channel which flowed north-westwards and was bounded to the east by the Fox Creek Escarpment (McLean, 1976 and 1977).

McLean (1977) suggested the Kosi River in India as a modern analogy. After flowing through a gorge in the Himalayas, the Kosi River empties onto the Gangetic plain and has built an alluvial fan and plain covering 7800 km^2. The Kosi is a braided river with channels spread over a width of 6 to 16 km, which have shifted position in an east to west direction over 110 km in the last 230 years, thus forming an enormous alluvial deposit. Gole and Chitale (1966) reported that the process of channel avulsion in the Kosi system results from locally aggraded bed levels. This type of process could have resulted in the formation of the sheet-like geometry of the Cadomin Formation. Uplift and deformation in the rising mountain front must have increased initial stream gradients and provided an abundant supply of detritus, resulting in a change from an erosional to an aggradational system in the piedmont area. Braided rivers then reworked the detritus laterally and distally and a thin, yet widespread distribution resulted. The Cadomin probably represents the type of thin deposit that is preserved when conditions are marginally aggradational on the braidplain.

Although glacial melt waters are probably not related to Cadomin deposition, braided-river processes still apply and broad braidplains may result. On the Skeidara outwash plain in southeastern Iceland (Nummedal et al., 1974), there are two inter-related downslope trends — increased dominance of lower order braids, and increased braiding parameter. Rust (1977) considered that the downstream trend towards lower order braids and increased braiding parameter is common in braided systems that change from gravel- to sand-dominant distally. An additional factor related to an increased braiding parameter on outwash systems like the Skeidara is the space available for the active tract. In the middle and lower reaches, the Skeidara outwash plain is 6 km wide and almost entirely covered with braid channels. The Cadomin Formation has a similar gravel to increasingly sandy distal trend and very low topography over which the channels could spread. It is therefore suggested that the Cadomin braidplain system was increasingly dominated downstream by lower order braids with a high braiding parameter. The style of Cadomin deposition was probably similar to Recent glacial outwash plains.

The facies assemblages and sedimentary structures for humid-region coarse-grained alluvium have been described by Boothroyd and Nummedal (1977). Their lithofacies assemblages P1, P2 and M form a proximal to medial trend which corresponds to the proximal to distal trend observed in the Cadomin from lithofacies A to E.

Facies Interpretation

Interpretation of facies in the Cadomin Formation follows Miall's (1977) synthesis for modern and ancient braided-river deposits. Massive or crudely-stratified conglomerates (Facies Gm) are abundant and were deposited as longitudinal bars and channel-infills in a braided network. Facies Gm deposits comprising a unit of bimodal clast-supported conglomerates overlain by finer-grained poorly-sorted, clast or matrix-supported conglomerates are interpreted to be cyclical deposits formed under waning flow conditions. Units of facies Gms are debris-flow deposits. Facies Gp/Sp, planar cross-stratified conglomerates and pebbly sandstones formed as wedge-shaped deposits adjacent to longitudinal bars. Trough cross-stratified sandstones (Facies St) are generally poorly-sorted and represent within-channel deposits. Facies Sh, horizontally-stratified sands, were formed under planar bed flow conditions insufficiently competent for gravel transport.

The lithofacies assemblages form a trend from proximal alluvial fan deposits to distal gravel-dominated braidplain deposits. A comparison is made between the Cadomin lithofacies assemblages and the types described by Miall (1977) and Rust (1977). Lithofacies assemblage A (proximal alluvial fan) is very similar to Miall's Scott type assemblage (Rust's G_{II} assemblage) in which Facies Gm predominates with associated minor facies Gp, Gt/St and Sh. Lithofacies assemblages B and C (distal alluvial fan and proximal braided-river) are also Scott type, but with sand forming a very minor component in this area. Lithofacies assemblage D (medial braided-river) contains 90% conglomerate and 10% sandstone units and is transitional to Miall's Donjek type assemblage (Rust's G_{III} assemblage). Lithofacies assemblage E (distal braided-river) has 30-40% sandstone units and can be classified as a Donjek type deposit. The abundance and size of the clasts and sand grains in the lithofacies assemblages significantly affect the overall textural distribution in the Cadomin. Distal trends include a decreasing clast size and an increasing abundance and coarseness of sand. In proximal braidplain areas bimodal conglomerates tend to predominate. This is due to the deposition of a clast-supported framework of cobbles and pebbles into which only the finer sand-sized grains were able to filter and form a matrix. Most of the coarser sand-sized grains bypassed the proximal braidplain areas and were deposited in the medial to distal areas. In this area the deposition of smaller clast sizes and coarser sand grains, accompanied by a greater overall proportion of sand, resulted in poorly-sorted conglomerates, matrix-supported conglomerates and pebbly sandstones. There is therefore a significant textural variation from mainly bimodal conglomerates in proximal braidplain areas to mainly poorly-sorted conglomerates and pebbly sandstones in distal areas. This textural distribution is important when considering the porosity and permeability in the Cadomin as poorly-sorted conglomerates tend to be slightly more porous and permeable than bimodal conglomerates.

Diagenesis, Porosity and Permeability

Uplift and eastward overthrusting associated with the Columbian Orogeny resulted in the initiation and infilling of the Western Canadian molasse basin. The Mesozoic rock section, only 300 m thick on the shelf in eastern Alberta, thickens westwards to over 4570 m in the more greatly subsided Deep Basin in front of the Foothills' overthrusts (Masters, 1979). A structure contour map for the top of the Cadomin Formation defines a south-westward dip into the Deep Basin. Wells in the northeast and southwest of the study area penetrate the Cadomin at approximately 1000 m and over 2100 m below sea level, respectively. The top of the Cadomin forms a monocline that dips basinward at 0.75° (Varley, 1983).

A diagenetic gradient potentially exists in the Cadomin Formation because of a geothermal gradient associated with increasing depth of burial towards the southwest.

The main diagenetic characteristics are:

1. Compaction, causing long contacts and suturing which reduce the porosity and permeability, increases downdip.
2. Quartz and calcite cementation and authigenic clay mineral crystallization, (mainly kaolinite and illite) which reduce the porosity and permeability, increase downdip.
3. Secondary porosity development, due to the leaching action of hydrodynamic pore-fluids is well developed at shallower depths in the areas of lithofacies assemblage E and the updip parts of D.

The textural features of the conglomerates and diagenetic modifications jointly affect the resultant porosity and permeability distribution in the Cadomin Formation. As much as possible, these effects should be considered separately. The relationship of primary texture to porosity and permeability was investigated for conglomerate samples from a similar depth to minimize differential diagenetic effects. Figure 12 shows a porosity and permeability cross-plot for 121 unstressed core sample analyses.

Poorly-sorted conglomerates generally have slightly higher porosity and permeability than bimodal textured conglomerates derived from a similar depth of burial. Moderately-sorted conglomerate textures, although rare, generally have the highest porosity (averaging 8%) and permeability (averaging 80 mD). Poorly-sorted conglomerates averaged 6% porosity and 20 mD permeability while bimodal conglomerates averaged 4% porosity and 5 mD permeability. This somewhat surprising result is mainly due to irregular packing in the sand matrix of poorly-sorted conglomerates, as compared to the matrix sands of bimodal conglomerates. The former are also more prone to secondary porosity development while bimodal conglomerates are more prone

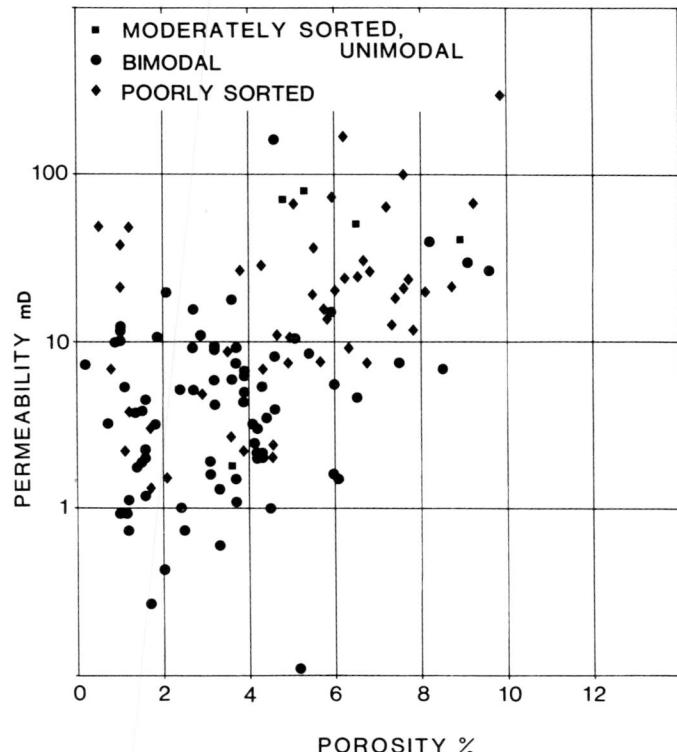

Fig. 12. Porosity/permeability cross-plot for congomerate textural types.

to porosity destruction by quartz overgrowth cementation. If therefore poorly-sorted conglomerates preserve their porosity and permeability to a greater extent than bimodal conglomerates, then the relative abundance of these various textures will directly affect the porosity/permeability distribution. Comparing lithofacies assemblages C, D, and E, it is evident that the abundance of conglomerates with poorly-sorted texture generally increases in a distal direction. Therefore the average porosity and permeability should increase in the downstream (or updip) direction.

The intensity of diagenetic modification also significantly affects the petrophysical properties of the Cadomin Formation in the subsurface. Compaction, quartz overgrowth cementation and authigenic clay mineral crystallization which reduce porosity and permeability all increase in abundance in a downdip direction. Secondary porosity development has been recognized in the shallow updip portions of the deposit. These diagenetic modifications therefore complement the influence of the texture in causing a downdip decrease in porosity and permeability. Overall, the reservoir characteristics of the Cadomin Formation apparently involve an average porosity decrease downdip, from 10% in shallow zones to less than 2% in the deepest parts of the basin. The permeability which varies from 1 to 400 mD in updip zones, decreases downdip to less than 0.1 mD at maximum depths. These porosity and permeability measurements have been derived from analyses of unstressed core samples. Jones and Owen (1980) have stated that unstressed permeabilities may be greater than those present under reservoir conditions by more than a hundred-fold because of the great relief of stress, absence of connate water and increased gas slippage.

Hydrocarbon Distribution

DEEP BASIN GAS ACCUMULATION

The Cadomin Formation is a major gas reservoir in the Elmworth area and is a good example of the Deep Basin trap type. The trap mechanism of the Deep Basin gas accumulation has puzzled explorationists since its discovery in 1976. Many have speculated how gas could be trapped in the downdip basinal portion of a clastic wedge, while the same units were water-bearing on the northeast flank of the basin. Masters (1979) recognized that the porous generally water-saturated sands in the eastern portion of the Deep Basin became less porous and permeable westwards and downdip, passing from a water-bearing area with local gas traps, through a transition zone to a gas-bearing area. Many of the gas reservoirs are in unconventional or 'tight' rocks which McMaster (1981) defines as having permeabilities less than 0.5 mD. Welte et al. (1982) stated that an enormous gas generating ability of the coals which are interbedded with the reservoir rocks has resulted in the deep portion of the basin becoming widely saturated with gas. The Cadomin Formation has been used as a model for the Deep Basin type gas trapping mechanism (Varley, 1984). A low rate of migration from the zone of gas accumulation resulting from stratigraphic, diagenetic and dynamic factors was found to be the critical trapping mechanism for the Cadomin Deep Basin type gas reservoir.

CADOMIN GAS ACCUMULATION

The Cadomin Deep Basin gas accumulation covers an area of 9,838 km^2 and reservoir pressure analysis reveals this to be single pool with approximately 420×10^9 m^3 (15×10^{12} ft^3) of gas in place (Gies, 1982). The author estimates the Cadomin gas reserves within the study area to be approximately 45.6×10^9 m^3 (1.63×10^{12} ft^3) of gas in place. In December 1981, the Alberta Energy Resources Conservation Board (1981) recognized the development of a multi-field Cadomin Formation reserve in the Elmworth, Sinclair and Wapiti fields.

Figure 13 illustrates the Cadomin pore-fluids distribution and the gas well control. An updip water-bearing

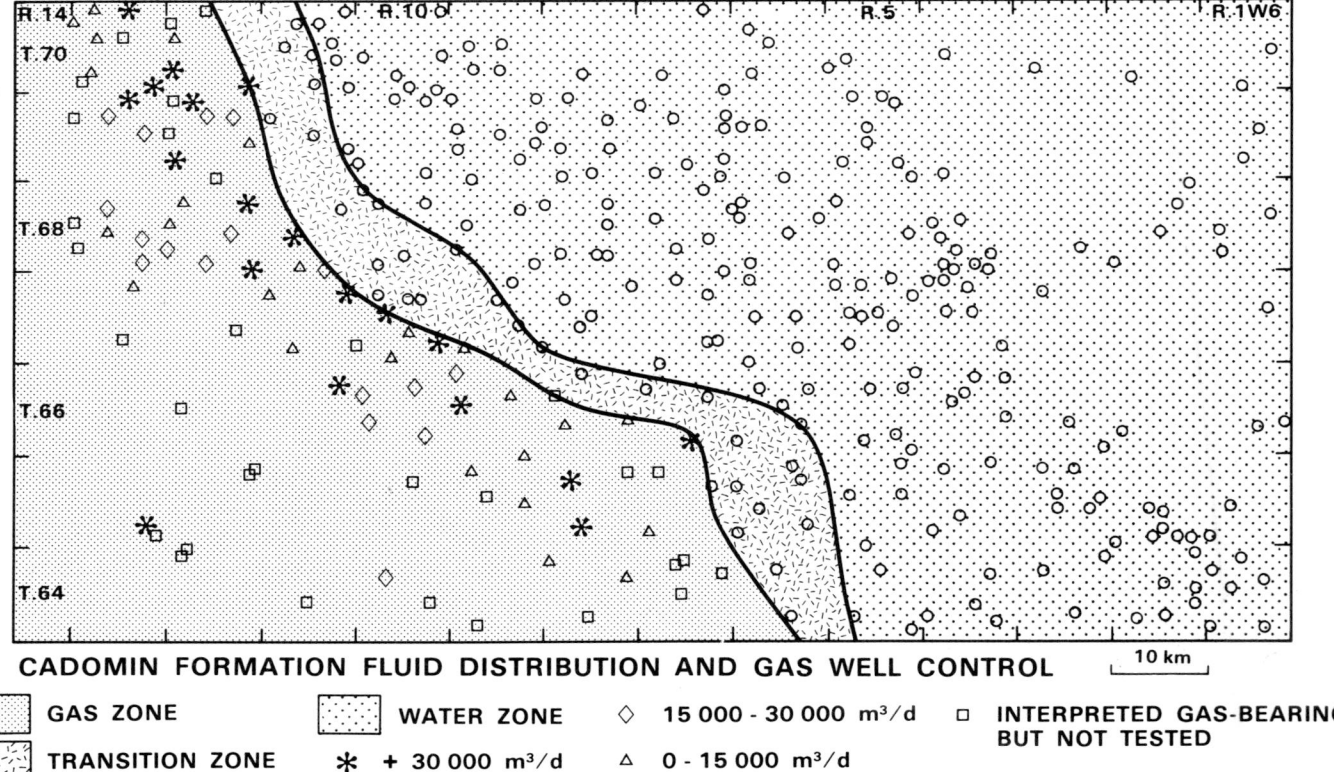

Fig. 13. The configuration of pore-fluids consists of an updip water-bearing section, a transition zone and a downdip gas-bearing zone. The most productive gas wells with flow rates >30 000 m³/day are generally found in the most updip parts of the gas-bearing zone where a higher porosity and permeability trend is located. In the updip portion of the gas zone where more distal deposits are located, poorly-sorted conglomerates are more abundant, and adverse diagenetic modifications are not as intense as deeper in the basin.

section, transition zone, and downdip gas-saturated zone with productive gas wells are shown. The gas-bearing zone corresponds with lithofacies assemblage C and the deeper parts of lithofacies assemblage D. The rocks can here be considered to be characteristically 'tight' due to the greater proportion of poor reservoir (bimodal) textures, extensive porosity destruction by diagenetic modification, and minimal or no secondary porosity. The water-bearing zone lies updip across a transition zone, and consists of the updip portions of lithofacies assemblage D and all of E. Resembling conventional reservoir rocks, they have a greater proportion of the better reservoir-forming conglomerates with poorly-sorted textures, and only minor adverse diagenetic changes. The configuration of reservoir rocks therefore consists of gas-saturated generally 'tight' conglomerates downdip and conventional conglomerates and sandstones updip that are water-bearing.

Figure 13 illustrates that the most productive Cadomin gas wells with the best flow rates are generally found in the updip portions of the gas-saturated zone. This is due to the greater preservation of porosity and permeability in this updip area due to the stratigraphic and diagenetic factors outlined. Therefore a knowledge of facies distribution, in particular their textural characteristics, and the nature and extent of diagenetic modification can assist gas exploration in the Cadomin.

To conclude, a successful exploitation of the Cadomin gas reservoir will result from:

1. An understanding of the reservoir textural characteristics and the effects of diagenesis;
2. Accurate mapping of the pore-fluid distribution to ensure exploration wells penetrate the updip portions of the gas-saturated area where a higher porosity and permeability trend is located; and
3. Isopach mapping of the Cadomin thickness to ensure exploration wells are located on thick trends which will maximize the net pay thickness and productivity from the strata.

References

Alberta Energy Resources Conservation Board, 1981. Alberta's reserves of crude oil, gas, natural gas liquids, and sulfur, at 31 December 1981. Alberta, Canada: Energy Resources Conservation Board.

Bates, R.L. and Jackson, J.A. (Eds.) 1980. Glossary of Geology. Falls Church, Virginia: American Geological Institute, 749p.

Boothroyd, J.C. and Nummedal, D. 1977. Proglacial braided outwash: a model for humid alluvial-fan deposits. In: Miall, A.D. (Ed.), Fluvial Sedimentology. Canadian Society of Petroleum Geologists, Memoir 5, p. 641-668.

Bull, W.B. 1972. Recognition of alluvial-fan deposits in the stratigraphic record. In: Rigby, J.K. and Hamblin, W.K. (Eds.), Recognition of Ancient Sedimentary Environments. Society of Economic Paleontologists and Mineralogists, Special Publication 16, p. 63-83.

Gies, R.M. 1982. Origin, migration, and entrapment of natural gas in the Alberta Deep Basin: Part 2. American Association of Petroleum Geologists, Annual Convention, Calgary, Handout to accompany poster-session, p. 16.

Gole, C.V. and Chitale, S.V. 1966. Inland delta building activity of Kosi River. Proceedings of the American Society of Civil Engineers, Journal of the Hydraulics Division, v. 92, p. 111-126.

Jones, F.O. and Owen, W.W. 1980. A laboratory study of low permeability gas sands. Journal of Canadian Petroleum Technology, v. 32, p. 1631-1640.

Masters, J.A. 1979. Deep Basin gas trap, Western Canada. American Association of Petroleum Geologists Bulletin, v. 63, p. 152-181.

McGookey, D.P., Haun, J.D., Hale, L.A., Goodell, H.G., McCubbin, D.G., Weimer, R.J. and Wulf, G.R. 1972. Cretaceous System. In: Mallory, W.W. (Ed.), Geological Atlas of the Rocky Mountain Region, U.S.A. Rocky Mountain Association of Geologists, p. 190-228.

McLean, J.R. 1976. Cadomin Formation: eastern limit and depositional environment. Geological Survey of Canada, Paper 76-18, p. 323-327.

McLean, J.R. 1977. The Cadomin Formation: stratigraphy, sedimentology and tectonic implications. Bulletin of Canadian Petroleum Geology, v. 25, p. 742-827.

McMaster, G. 1981. Gas reservoirs, Deep Basin, Western Canada. Journal of Canadian Petroleum Technology, v. 20, p. 62-66.

Miall, A.D. 1977. Lithofacies types and vertical profile models in braided river deposits: a summary. In: Miall, A.D. (Ed.), Fluvial Sedimentology. Canadian Society of Petroleum Geologists, Memoir 5, p. 597-604.

Mudge, M.R. and Sheppard, R.A. 1968. Provenance of igneous rocks in Cretaceous conglomerates in northwestern Montana. United States Geological Survey Professional Paper 600-D, p. 137-146.

Nummedal, D., Hine, A.C., Ward, H.G., Hayes, M.O., Boothroyd, J.C., Stephens, M.F. and Hubbard, D.K. 1974. Recent migrations of the Skeidararsandur shoreline, southeast Iceland. Final Report of Contract No. N60921-73-C-0258 to United States, Naval Ordnance Laboratory, Department of Geology, University of South Carolina, 183p.

Rust, B.R. 1977. Depositional models for braided alluvium. In: Miall, A.D. (Ed.), Fluvial Sedimentology. Canadian Society of Petroleum Geologists, Memoir 5, p. 605-625.

Schultheis, N.H., and Mountjoy, E.W. 1978. Cadomin conglomerate of western Alberta — a result of early Cretaceous uplift of the Main Ranges. Bulletin of Canadian Petroleum Geology, v. 26, p. 297-342.

Varley, C.J. 1983. The sedimentology and diagenesis of the Cadomin Formation, Elmworth area, northwestern Alberta. M.Sc. Thesis, University of Calgary, Alberta, 173p.

Varley, C.J. 1984. The Cadomin Formation: a model for the Deep Basin type gas trapping mechanism. In: Stott, D.F. and Glass, D.J. (Eds.), The Mesozoic of Middle North America. Canadian Society of Petroleum Geologists, Memoir 9, p. 471-484.

Welte, D.H., Schaefer, R.G., Radke, M. and Weiss, H.M. 1982. Origin, migration and entrapment of natural gas in the Alberta Deep Basin: Part 1 (Abstract). American Association of Petroleum Geologists Bulletin, v. 66, p. 642.

SEDIMENTOLOGY AND DEPOSITIONAL SETTING OF THE UPPER PROTEROZOIC SCANLAN CONGLOMERATE, CENTRAL ARIZONA

LARRY T. MIDDLETON[1] AND ALAN P. TRUJILLO[2]

ABSTRACT

The Upper Proterozoic Apache Group crops out throughout central Arizona. Radiometric dating of underlying granites and intrusive diabase sills bracket deposition between 1.45 and 1.15 b.y. The basal unit of the Apache Group, the Scanlan Conglomerate, unconformably overlies older Precambrian igneous and metamorphic rocks. Relief on the basement is as great as 30 m and defines paleovalleys into which Scanlan sediments were shed.

Facies analysis of the Scanlan indicate deposition in an alluvial fan/proximal braided stream complex. Two generalized facies sequences are common. Facies sequence 1 comprises interbedded matrix-supported, disorganized conglomerate and crudely to horizontally bedded clast-supported conglomerate which represent, respectively, debris flows on alluvial fans and in-channel sheet deposits and longitudinal bars. Facies sequence 2 consists of clast-supported, imbricate conglomerate and large sets of planar-tabular cross-bedded conglomerate. Associated facies include trough cross-bedded conglomerate and cross- and horizontally stratified sandstone. The second facies sequence formed in a proximal braided fluvial system and reflects in-channel deposition and both longitudinal and transverse bar development. Paleocurrent and mineralogic data indicate a south-sloping paleoslope away from a large Precambrian highland. Local variations in flow are related to structurally controlled paleovalleys.

RÉSUMÉ

Le groupe Apache du Protérozoïque supérieur affleure dans tout le centre de l'Arizona. Les datations radiométriques des granites sous-jacents et des filons-couches de diabase intrusive encadrent la période de dépôt entre 1,45 et 1,15 G.a. Le conglomérat de Scanlan constitue l'unité basale du groupe Apache et il repose en discordance sur des roches ignées et métamorphiques précambriennes plus anciennes. Le relief au niveau de socle peut atteindre 30 m et définit les paléovallées au fond desquelles furent abrités les sédiments.

Une analyse des faciès du conglomérat de Scanlan révèle que le dépôt doit son origine à un complexe cône alluvial/rivière anastomosée proximale. Les séquences généralisées définissent deux faciès d'occurrence fréquente. La séquence du faciès 1 inclut des couches interstratifiées à fragments flottants dans la matrice, un conglomérat inorganisé, et des couches conglomératiques plus ou moins horizontales à fragments jointifs qui représentent, respectivement, des coulées de débris sur des cônes alluviaux et des dépôts en nappes dans des chenaux et des barres longitudinales. La séquence du faciès 2 est constituée d'un conglomérat à fragments jointifs, de couches imbriquées et d'une importante série de corps planaires - tabulaires conglomératiques à stratification oblique. D'autres faciès associés incluent un conglomérat à stratification oblique et un garès à stratification oblique et horizontale dans un fossé tectonique. La séquence du faciès numéro 2 s'est développée par un système de ruissellement anastomosé proximal et représente des dépôts dans des chenaux et une édification de barres longitudinales et transversales. L'étude des paléocourants et les déterminations minéralogiques indiquent une paléopente inclinée vers le sud qui s'éloigne d'une vaste région de hautes terres précambriennes. Les variations locales de l'écoulement dépendent des accidents structuraux dans les paléovallées.

INTRODUCTION

The depositional and tectonic settings of Middle and Upper Proterozoic (terminology of Harrison and Peterman, 1980) basins in Arizona are poorly known. These sequences comprise two thick successions of sedimentary and volcanic rocks: the Grand Canyon Supergroup in northern Arizona and the Apache Group in central Arizona (Fig. 1). Although lithologic correlations between these units are difficult to establish, radiometric and paleomagnetic data suggest that the basal part of the Grand Canyon Supergroup, the Unkar Group, is correlative with part of the Apache Group to the south (Young, 1981; Elston and McKee, 1982).

The intent of this study is to examine the conglomerate deposits that comprise the basal unit of the Apache Group, the Scanlan Conglomerate, in the northernmost part of the Apache basin of deposition. These deposits formed during the initial stages of younger Precambrian basinal development in southwestern North America and as such are of paramount importance in understanding the sedimentologic and structural evolution of the western margin of the North American plate.

---

[1] Department of Geology, Northern Arizona University, Flagstaff, Arizona 86011, U.S.A.

[2] Union Oil Company, 461 South Boylston Street, Los Angeles, California 90017, U.S.A.

Field discussions with Drs. Mary J. Kraus and Ronald C. Blakey were helpful in formulating many aspects of this study. Constructive reviews by Drs. A.B. Hayward, N.D. Smith, and Emlyn Koster greatly improved the manuscript. Typing and drafting were done by Sandy Nuneviller and Greg Rossel, respectively. Debbie Meier of the Bilby Research Center (Northern Arizona University) assisted with the photographic plates. Partial support for this study was provided by a Sigma Xi Grant-In-Aid of Research to Trujillo and an Organized Research Grant (Northern Arizona University) to Middleton.

Copyright © 1984, Canadian Society of Petroleum Geologists

Lateral and vertical facies analyses were used to evaluate transport and depositional processes associated with accumulation of coarse silici- and volcaniclastic sequences within the Scanlan Conglomerate. Facies analysis was combined with petrologic data based on clast and matrix compositions, as well as paleocurrent information, to establish the provenance of the Scanlan Conglomerate and to delineate sediment dispersal patterns.

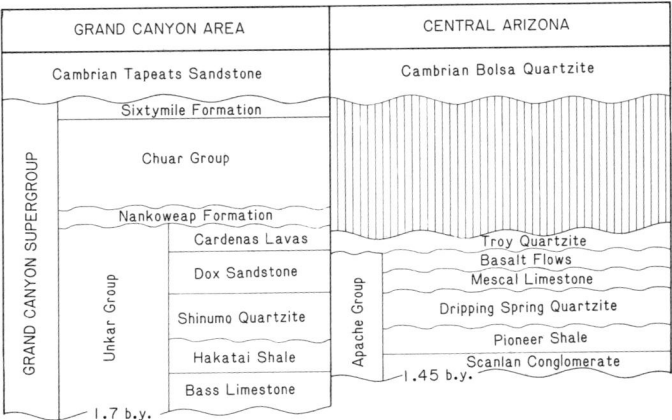

Fig. 1. Stratigraphic nomenclature and tentative correlation of Upper Proterozoic sequences in northern and central Arizona.

Geologic Setting

The Apache Group crops out throughout central and southern Arizona in a fairly continuous belt between 35 and 110 km wide and 365 km long from the Mogollon Rim on the north to the Santa Catalina Mountains on the south; the best exposures of the entire group are in the Sierra Anchas Range between Young, Arizona and Roosevelt Lake (Fig. 2). The group consists of, in ascending order, Scanlan Conglomerate, Pioneer Shale, Dripping Spring Quartzite, Mescal Limestone, and several unnamed basalt flows (Fig. 1). Aggregate thickness for the group is approximately 485 m although thickness trends within each formation are variable (Shride, 1967). Radiometric dating of underlying granites has yielded a maximum age of 1.42 b.y. for the group (Silver, 1964 in Livingston and Damon, 1968). The entire group is intruded by a number of diabase sills which have been dated by Livingston and Damon (1968) at 1.15 b.y. thus establishing a minimum age for the Apache Group.

The depositional settings of the Apache Group are poorly known. Previous studies have suggested the presence of tidal and shallow marine environments, intertidal to supratidal carbonate flats, and fluvial complexes (Gastil, 1953; Granger and Raup, 1964; Shride, 1967; McConnell, 1972, 1975a, 1975b; Trujillo et al., 1983).

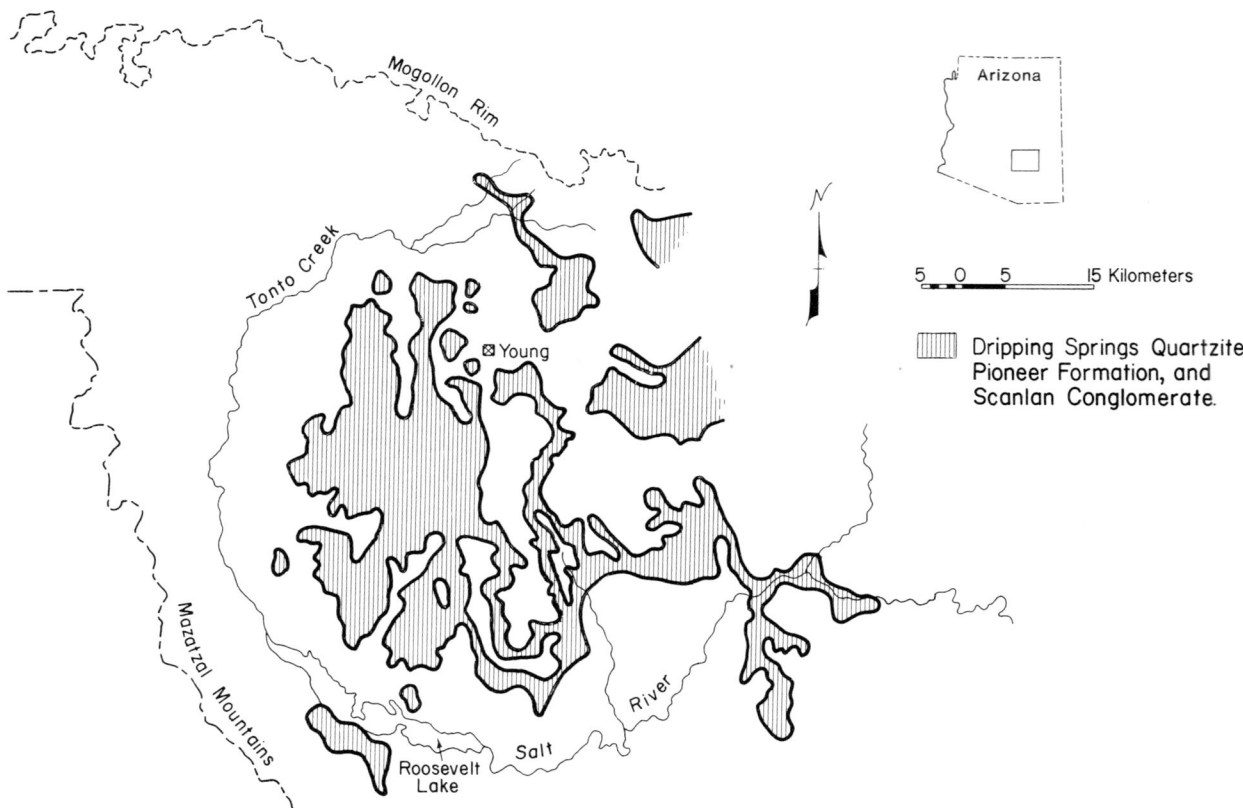

Fig. 2. Location map of study area and outcrop belt of Apache Group exclusive of Mescal Limestone (modified from Shride, 1967).

Throughout the study area these strata unconformably overlie older Precambrian granites, metavolcanics, and metasedimentary rocks (Gastil, 1953). All older Precambrian rocks have been regionally metamorphosed to greenschist facies. Near large intrusions of granite, metamorphic grade can be high, and in areas near major faults, older Precambrian rocks have been cataclastically deformed. Structurally, the older Precambrian rocks within the study area constitute a northeast-southwest anticlinorium, and major faults in the area tend to parallel the major structural grain (Gastil, 1958).

Just north of the study area the Apache Group onlaps an older Precambrian positive area (Shride, 1967). This Precambrian high area is considered to represent an extensive north to northeast-trending mountain range produced during the Mazatzal Revolution which occurred at the close of older Precambrian deposition (Wilson, 1939; Silver, 1978; Granger and Raup, 1964). Stoyanow (1936) termed this mountain system Mazatzal Land, and Anderson and Wirth (1980) referred to it as the Central Arizona Arch. The importance of older Precambrian structures and the Central Arizona Arch on younger Precambrian sedimentation will be discussed in a later section.

The Apache Group is generally flat lying and unmetamorphosed throughout its extent. Although block faulting has displaced the Apache Group to different topographic levels, dips of beds rarely exceed 20° except in the northern portion of the study area where steeply dipping strata occur. Following Apache deposition and before deposition of the overlying Precambrian Troy Quartzite, broad regional folding deformed Apache Group strata.

SCANLAN CONGLOMERATE

The Scanlan Conglomerate lies with pronounced angular unconformity upon older Precambrian granites, schists, metavolcanics and metasediments (Fig. 3). Relief on the older Precambrian surface is as great as 30 m and in several areas the Scanlan pinches out against 'hills' of older Precambrian basement with the overlying Pioneer Shale resting directly on basement. These areas represent paleovalleys into which coarse clastics of the Scanlan were deposited. Thickness varies from 0 to 53 m, although it is generally less than 35 m. As discussed later, the position of older Precambrian positive areas affected Scanlan depositional styles as well as clast composition. In general, the Scanlan thickens away from these highs.

The lithology of the Scanlan varies markedly over the study area as a function of older Precambrian lithology and structure. Dominant clasts include white and red quartzite, white vein quartz, and various rhyolites. Subordinate clasts include a variety of older Precambrian metasediments and metavolcanics. Matrix composition and texture are variable and also reflect basement rock compositions. These include coarse-grained feldspathic sandy matrix, quartz-rich sandy matrix, fine-grained rhyolitic sand, and mud-rich matrix. Lithologic variations of conglomerate clast types and matrix are used to delineate petrofacies within the Scanlan.

In general, units within the Scanlan fine upward although inversely graded and ungraded beds are common near the base of the unit. Conglomerate clasts are poor to moderately sorted within beds and matrix ranges from poor to well sorted. Very angular to rounded clasts occur. Clast size is generally in the coarse pebble to cobble range although boulders up to 3.2 m occur in the northern part of the study area.

The unit is characterized by both clast- and matrix-supported conglomerates that exhibit both trough and planar-tabular cross-stratification, horizontal stratification, or appear structureless. Intercalated lithologies include trough, planar-tabular, and horizontally stratified sandstone, horizontally laminated siltstone and thin shale drapes. The finer grained lithologies are more abundant upward in the unit.

The coarse-grained texture of the Scanlan contrasts markedly with the fine-grained nature of the overlying Pioneer (very fine grained sandstone and siltstone) with which it is gradational, attesting to a rapid change in depositional setting. Although both units have long been considered to represent marine sedimentation (Shride, 1967), facies analyses of the Scanlan Conglomerate (Trujillo et al., 1983; this study) and the Pioneer Shale (McConnell, 1975b) indicate deposition in a variety of continental depositional systems.

FACIES DESCRIPTION

Facies within the Scanlan Conglomerate are based on varieties of type of framework support, types of sedimentary structures, and general lithologies. Terminology used in facies classification is similar to that of Miall (1977,

Fig. 3. Angular unconformity separating older Precambrian schistose basement (Board Cabin Formation) and younger Precambrian Scanlan Conglomerate along Colcord Mesa approximately 12 km northwest of Young, Arizona.

1978) and Rust (1978, 1979). Supplementary characteristics used in facies designation include conglomerate fabric, sorting, and compositional variation and also unit geometries. Facies descriptions are summarized in Table 1 and described below.

FACIES Gms — STRUCTURELESS, MATRIX-SUPPORTED CONGLOMERATE

This facies consists primarily of cobble-size clasts although boulders up to 3.2 m occur locally in the northern part of the study area near the Mogollon Rim (Fig. 2). Clasts are angular to subrounded, with angularity increasing with decreasing clast size. The cobble and boulder-size clasts are moderately to poorly sorted. Matrix is either silt and clay or moderate to poorly sorted coarse-grained sand and granules (Fig. 4). Conglomerates containing the two different types of matrix are intimately interbedded and grade into one another.

Conglomerate fabric is disorganized (Fig. 5). Although imbrication is rare, some elongate clasts are oriented with their long axes horizontal to sub-horizontal and intermediate axes inclined. Sedimentary structures are absent and distinct bedding planes difficult to recognize (Fig. 5). Grading is relatively uncommon although both normal and inverse coarse-tail graded sequences do occur.

Where fabric and/or textural changes allow discrimination of individual sedimentation units, average bed thickness is 1 to 2 m and these occur as stacked units or are interbedded with clast-supported conglomerates of Facies Gm. Units within this facies tend to be tabular and can be traced 100 to 200 m where they typically grade into other conglomeratic facies. This facies is most common near the base of the Scanlan and increases in thickness to the north.

Fig. 4. Matrix-supported conglomerate (coarse-grained sand matrix) directly overlying older Precambrian basement. Note inverse coarse-tail grading and absence of imbrication. Clast in the center (A) is 45 cm long dimension.

Facies Code	Lithofacies description	Sedimentary Structures or Features	Interpretation
Gms	Massive, matrix-supported conglomerate.	Unstratified.	Debris and mass flows.
Gm	Massive or crudely bedded, bimodal to polymodal clast-supported conglomerate.	Horizontal stratification, imbrication	Longitudinal bars and diffuse gravel sheets.
Gt	Trough cross-stratified clast-supported conglomerate.	Trough cross-stratification.	Channel fills.
Gp	Planar-tabular cross-stratified clast-supported conglomerate.	Planar-tabular cross-stratification	Linguoid or transverse bars.
St	Fine to very coarse trough cross-stratified sandstone.	Trough cross-stratification.	In-channel dunes.
Sp	Fine to very coarse planar-tabular cross-stratified sandstone.	Planar-tabular cross-stratification.	Linguoid bars and dunes.
Sh	Very fine to very coarse horizontally stratified sandstone	Horizontal stratification, parting lineations.	Bar top sand sheets.
F	Laminated very fine sandstone, siltstone, and shale.	Fine laminations.	Waning flood deposits.

Table 1. Lithofacies codes, descriptions, sedimentary structures, and interpretations for the Scanlan Conglomerate (after Miall, 1977, 1978; Rust, 1978, 1979).

Fig. 5. Matrix-supported conglomerate with disorganized fabric. Although it appears structureless, the unit consists of a number of sedimentation units.

FACIES *Gm* — HORIZONTAL TO CRUDELY BEDDED CLAST-SUPPORTED CONGLOMERATE

Pebble to cobble-size clasts in this facies are subangular to rounded and often have small, less than 1 cm in diameter, surface indentations. Both bimodally sorted (a moderately sorted matrix mode and a moderately sorted clast mode) and polymodally sorted (a continuum of sand through pebble sizes) occur with the latter being more common (Fig. 6). Sand-sized matrix is usually present but rarely comprises more than 20% of any unit.

Crudely developed bedding and horizontal stratification are ubiquitous (Fig. 7). Clast imbrication is locally abundant with intermediate *b* axes dipping as steeply as 20°. Normal grading is very common (Fig. 8) although inversely graded sequences occur. A distinguishing feature of this facies is its organized fabric (Steel and Thompson, 1983), that is, a fabric with some degree of order, preferred orientation, or stratification.

Bedding is up to 1 m thick for pebble-sized conglomerates and up to 2 m for cobble-dominated beds. This facies typically forms multistorey sequences several tens of meters thick. Contacts between beds are sharp and planar to highly irregular and scoured. It is interbedded with all other conglomerate facies as well as cross-stratified sandstone. Lateral extent of individual units is up to several hundred meters, and most have widths of less than 25 m. This facies constitutes the bulk of each measured section.

FACIES *Gt* — TROUGH CROSS-STRATIFIED CONGLOMERATE

Cross-bedded sets of this facies range from 15 to 100 cm thick with most less than 50 cm and consist of clast-supported pebble conglomerates. Clasts are moderately to poorly sorted and sub-rounded to rounded. Crossbeds consist of alternating fining-upward foresets of pebbles and very coarse-grained sand.

Fig. 6. Poorly sorted, clast-supported conglomerate exhibiting a continuum of clast sizes from very coarse-grained sand to boulders. Hammer is 39 cm long.

Fig. 7. Horizontally stratified conglomerates of Facies *Gm*. At least five sedimentation units are present and there is a weakly developed coarsening-upward trend.

Fig. 8. Upward-fining deposit at base overlain by an upward-coarsening sequence-Facies *Gm*. Scale is in inches (approximately 20 cm).

These units occur as single sets or in cosets of up to six sets. Lenticular geometries are the norm with most sets being several tens of meters long and only a few meters wide. Conglomerates of this facies typically overlie a scoured basal surface developed on facies *Gm* or *Gp* and are overlain by finer grained facies.

FACIES *Gp* — PLANAR CROSS-STRATIFIED CONGLOMERATE

This facies comprises single sets of planar-tabular cross-stratified pebble conglomerate ranging from 50 to 240 cm in thickness (Fig. 9). Clasts are moderately sorted and sub-rounded. The foresets consist of alternating fine and coarse pebbles that become finer grained up the foreset.

These sets can be continuous laterally for about 50 m and grade into facies *Gm* or thin cross-stratified sandstone. A thin, horizontally bedded conglomerate typically underlies planar-tabular sets with either a sharp or gradational contact (Fig. 9). Facies *Gp* is overlain by a sandstone facies (Fig. 10) or is sharply truncated and overlain by Facies *Gt* or *Gm* (Fig. 9).

FACIES *St* — TROUGH CROSS-STRATIFIED SANDSTONE

This facies consists of fine- to very coarse-grained, moderately sorted sandstone in sets between 15 and 45 cm thick (Figs. 11). Cosets up to 1.5 m thick are common. Rounded granules and pebbles occur along some foresets and also as thin, discontinuous stringers in trough axes.

Individual trough cross-stratified sets have lengths of over 15 m and widths less than 5 m. The basal contact of most cosets is scoured and veneered with a granule to pebble lag. This facies overlies Facies *Gp*, *Gt* or *Gm* and is commonly overlain by horizontally laminated sandstone (Facies *Sh*) or a stratified conglomerate facies. In the latter case the contact is typically scoured. This facies becomes more abundant toward the top of the Scanlan.

Fig. 9. Two large-scale sets of planar-tabular cross-stratified pebble conglomerates (Facies *Gp*) overlain by low-angle trough cross-bedded facies (*Gt*) and underlain by clast-supported conglomerate (*Gm*). Foresets consist of alternating fine and coarse pebbles.

Fig. 10. Two large-scale planar-tabular cross-bedded conglomerate of Facies *Gp* overlain and underlain by thin, lenticular sandstone beds.

FACIES Sp — PLANAR-TABULAR CROSS-STRATIFIED SANDSTONE

Planar-tabular cross-stratified fine- to very coarse-grained sandstone characterize this facies (Fig. 12). Most sandstones are moderately sorted and composed of alternating foresets of fine- to medium-grained sand and coarse-grained sand to very fine pebbles. Although single sets between 5 and 15 cm thick are common, cosets of up to seven sets occur.

Sandstones of this facies are continuous laterally for several tens of meters, overlie sharp or scoured surfaces, and are overlain by horizontally bedded sandstone (Sh) or thick gravel beds (Facies Gt and Gm). This facies is relatively rare and occurs primarily near the top of the formation.

FACIES Sh — HORIZONTALLY STRATIFIED SANDSTONE

These sandstones are very fine- to very coarse-grained and moderately sorted. A thin pebble layer is present at the base of most units. Individual horizontally stratified sets are less than 10 cm and the thickest continuous development of this facies is about 1 m. Parting lineation is locally abundant.

This facies typically overlies trough or planar-tabular cross-stratified sets. Lateral extent is of the order of 10 m, and Facies Sh is generally overlain by cross-bedded sandstone or clast-supported conglomerate.

FACIES F — LAMINATED SILTSTONE AND SHALE

This is a very minor facies and usually occurs as beds less than 5 cm thick or as thin, lenticular bodies less than a meter in extent. This facies is interbedded with sandstone facies and rarely with conglomerate beds.

FACIES ASSOCIATIONS AND INTERPRETATIONS

Lateral and vertical facies associations coupled with analysis of paleocurrent trends and isolith-isopleth maps indicate that deposition of the Scanlan Conglomerate occurred within alluvial fan and braided fluvial systems.

The coarse-grained texture of the unit, general absence of fines, and mature lithologies suggest vigorous bedload transport and extensive winnowing of fine-grained detritus. Absence of any features or association of features suggestive of marine transport and sedimentation further substantiate a continental setting. Many textural and sedimentary structural features within the Scanlan are reported from both modern (Boothroyd and Ashley, 1975; Bull, 1964; Ore, 1964; Rust, 1972; Smith, 1971, 1974; and Williams and Rust, 1969) and ancient (Bluck, 1967, 1980; Bull, 1972; Eriksson and Vos, 1979; Larsen and Steel, 1978; and McGowen and Groat, 1971) alluvial fan/braided stream settings. The following discussions center on documenting processes of sediment transport and reconstructing the depositional systems.

FACIES SEQUENCE 1

Although variations in the vertical juxtaposition of facies occur throughout the study area, two recurring styles clearly dominate the Scanlan. The first sequence is shown schematically in Figure 13.

Fig. 12. Planar-tabular cross-stratified sandstone facies (Sp) near top of Scanlan. Two stacked sets are shown (contact at hammer head). Underlying bed consists of clast-supported conglomerate (Facies Gm). Hammer head is 19 cm across.

Fig. 11. Large-scale trough cross-stratified sandstone near the top of the Scanlan Conglomerate overlying clast-supported conglomerates of Facies Gm.

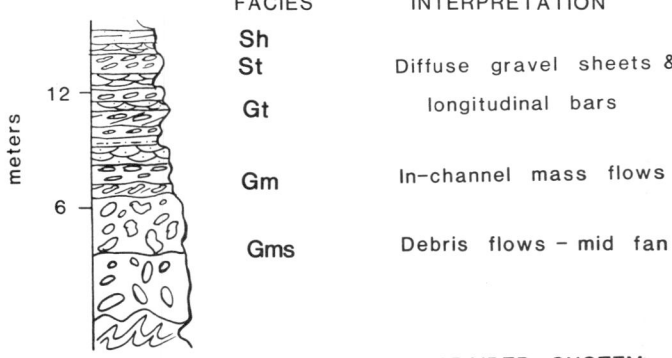

Fig. 13. Measured vertical profile through facies sequence 1. Symbols used to designate facies are explained in the text.

The fundamental motif of sequence 1 involves an interbedding of the coarsest Facies *Gm* and *Gms* (Fig. 13). Subordinate facies include trough cross-bedded conglomerate and sandstone (Facies *Gt* and *St*) and minor horizontally bedded sandstone (Facies *Sh*). Contacts between these facies are usually planar or less commonly scoured. This association is most prevalent near the base of the formation and is more abundant in the northern part of the study area.

Important characteristics of Facies *Gms* with respect to depositional processes include the coarse-grained texture, angularity of many clasts, poor sorting, sandy and/or muddy matrix support, disorganized fabric, and absence of sedimentary structures (Table 1). Although the coarseness of the deposits suggests high-energy flow conditions, the textural immaturity of the unit (poor sorting, high matrix content and angularity) indicates that a) few grain-to-grain bedload collisions occurred and b) the transporting medium was incapable of winnowing fines and sorting the clasts. In this case, therefore, both the matrix and clast were transported and deposited together.

High sediment-concentrated flows such as debris flows on alluvial fans are characterized by many of the features described above (Bull, 1964, 1972). The ability of debris flows to move on relatively gentle slopes and with a minimum of very fine matrix (clay and silt grades) has been documented experimentally (Rodine and Johnson, 1976) and in a number of modern flows (*e.g.*, Pierson, 1981). Highly viscous debris flows are characterized by a muddy matrix and often occur in the proximal reaches of alluvial fans whereas sandy matrix is typical of less viscous, distal fan debris flows.

In the study area the mud content of Facies *Gms* increases toward the north, and here this facies is interpreted to represent viscous debris flows on the upper reaches of alluvial fans. Where the sand content of the conglomerate matrix is high (compared to mud matrix), this facies is more extensively interbedded with the more organized Facies *Gm*. In these instances Facies *Gms* is interpreted to be the product of less viscous debris or mass flows. Such flows are reported from distal areas of alluvial fans (Miall, 1970; Rust, 1978) where they are commonly associated with water-laid deposits of a braided fluvial origin. As discussed below this is the interpretation of Facies *Gm*.

The characteristics of Facies *Gm* strongly support an active bedload system and one in which the transporting medium was capable of molding the sediment into broad sheet-like bodies. The clast-supported framework indicates that currents were capable of winnowing finer grained sediments and, therefore, the matrix of these conglomerates likely represents a late-stage infiltrate. A fluvial origin for this facies is supported by the presence of imbricated clasts with intermediate axis (*b*) imbricated and long axis (*a*) transverse to flow and the presence of normally graded beds indicating waning flow conditions (Fig. 14).

The horizontally stratified conglomerates of this facies (Fig. 7) were probably emplaced following deposition of diffuse gravel sheets during high-water stages (Hein and Walker, 1977). In their model high sediment and water discharges promote downstream growth of horizontally stratified gravel sheets. With continued vertical and lateral growth the sheets develop into longitudinal and/or diagonal bars. During waning-flow conditions finer grained material is deposited on top of the bars and also infiltrates into the open framework of the gravels. As pointed out by Rust (1978), however, this model might not be applicable to major flood deposits. Similar interpretation for this facies are reported from proximal reaches of modern braided outwash streams (Boothroyd and Ashley, 1975; Scott model of Miall, 1977) and from ancient proximal braid plains (Eriksson and Vos, 1979).

The lateral migration of braided channels with abundant longitudinal bars could result in the generation of sheet-like gravels (*cf.* Coleman, 1969). Alternatively, these could result from emplacement of gravel sheets during major

Fig. 14. Portion of gravel sheet (facies sequence 1) in Facies *Gm*. Two crudely fining-upward deposits are separated by thin sandstone horizon. Scale is in inches (approximately 20 cm).

floods on distal fan surfaces. The general absence of channeled bases in these units accords with the latter interpretation.

The association of finer facies (*Sh*) and occasionally trough cross-stratified conglomerates (*Gt*) is documented in distal fan/proximal braided reaches (Rust, 1978). The former results during falling stages where the water is not flowing rapidly enough to transport the gravel load but still is capable of traction transport of coarse sand and granules. Such conditions are common on bar tops as well as in channels (Bluck, 1979). The trough cross-stratified gravels in the Scanlan are typically low-angled and represent small channel-fills following low-stage dissection of the bar top and/or front and subsequent in-filling with rising stage.

FACIES SEQUENCE 2

This sequence (Fig. 15) occurs in the upper parts of the Scanlan Conglomerate and is most abundant in the southern part of the study area (Fig. 2). The majority of this package is composed of clast-supported conglomerate of Facies *Gm* with interbedded units of planar cross-bedded conglomerate (*Gp*) and minor trough cross-stratified conglomerate and sandstone of Facies *Gt* and *St*. Sets of planar cross-bedded sandstone are locally abundant (Fig. 12) as are fine-grained sandstones and siltstones of Facies *F*.

Processes associated with deposition of Facies *Gm* are discussed in the previous section. What is important to discuss here is the association of cross-stratified conglomerates and sandstones with Facies *Gm* and how these are related to the evolving character of the Scanlan depositional systems.

The large-scale planar-tabular crossbeds are typically enclosed by horizontally to crudely bedded conglomerates (Fig. 9). Thick sets of cross-bedded conglomerate have been reported from a few well-studied braided sequences (e.g., Bluck, 1979; Steel and Thompson, 1983; Kraus, this volume). Their origin requires water depths considerably greater than foreset thickness and quite probably persistent, non-ephemeral flow. Cross-stratification develops through repeated avalanching of debris over the bar crest down the front of linguoid or transverse bars. In most cases sediment is delivered to the bar front by movement of gravel across the upper bar surface (Bluck, 1974; Steel and Thompson, 1983). The alternating coarse and fine foresets of large sets may have resulted from the passage of well sorted, finer grained bar-tail gravels down the bar front followed by avalanching of coarser grained bar-head deposits. A similar process (although involving passage of bar top dunes) has been demonstrated for sandy braided systems (Smith, 1970, 1971).

The vertical juxtaposition of horizontally bedded conglomerate overlain by planar crossbeds could be due to

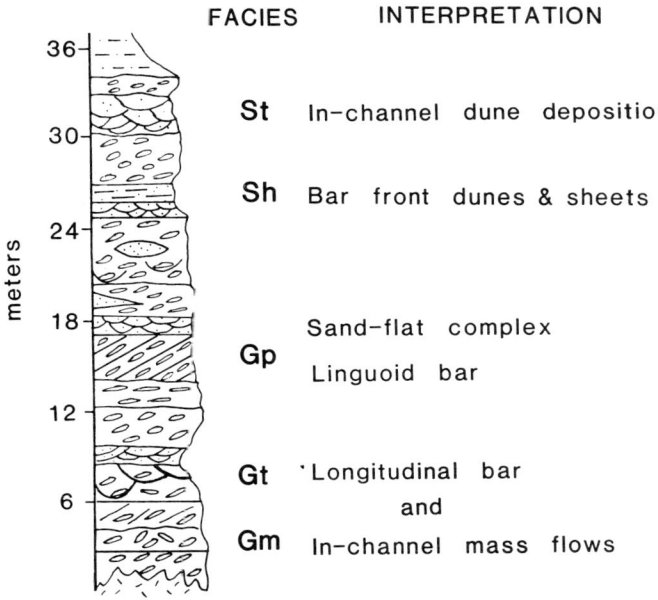

Fig. 15. Measured vertical profile through facies sequence 2. Symbols used to designate facies are given in the text.

Fig. 16. Vertical section through a portion of facies sequence 2 showing interbedded sandstone facies (*St*) and crudely to horizontally bedded clast-supported gravels (*Gm*). Sandstone at the top is 1.3 m thick.

increasing water depths which allowed formation of thick transverse bars followed by downstream migration of these features over Facies *Gm*. Hein and Walker (1977) proposed an alternative hypothesis. In this model with decreasing sediment and water discharges, vertical growth of diffuse gravel sheets proceeds faster than downstream growth resulting in development of bed forms with a downstream declivity. With continued bar-top sediment transport, well-defined bar-front foresets form. Evidence for the latter is present where the crossbeds grade laterally (upstream direction) into progressively lower angle cross-strata within a few tens of meters.

Interbedded with *Gm* and *Gp* Facies are a number of trough and horizontally bedded units (Fig. 16). These discontinuous bodies can be traced laterally (upstream direction) into Facies *Gm* and *Gp*. In both cases these represent low-stage migration of small bedforms on the bar tails and/or the foresets. Bluck (1974) among others has described this association, and such deposits are common on modern longitudinal bars in the study area.

Abundant trough and horizontally stratified sandstone occur near the top of the sequence (Fig. 16). Facies *St* is interpreted to represent both in-channel dune migration and in some cases dune migration across the tops of longitudinal bars. Facies *Sh* formed during plane bed transport of sand over the bars. In both instances the absence of sediment coarser than granules implies decreasing flow discharges.

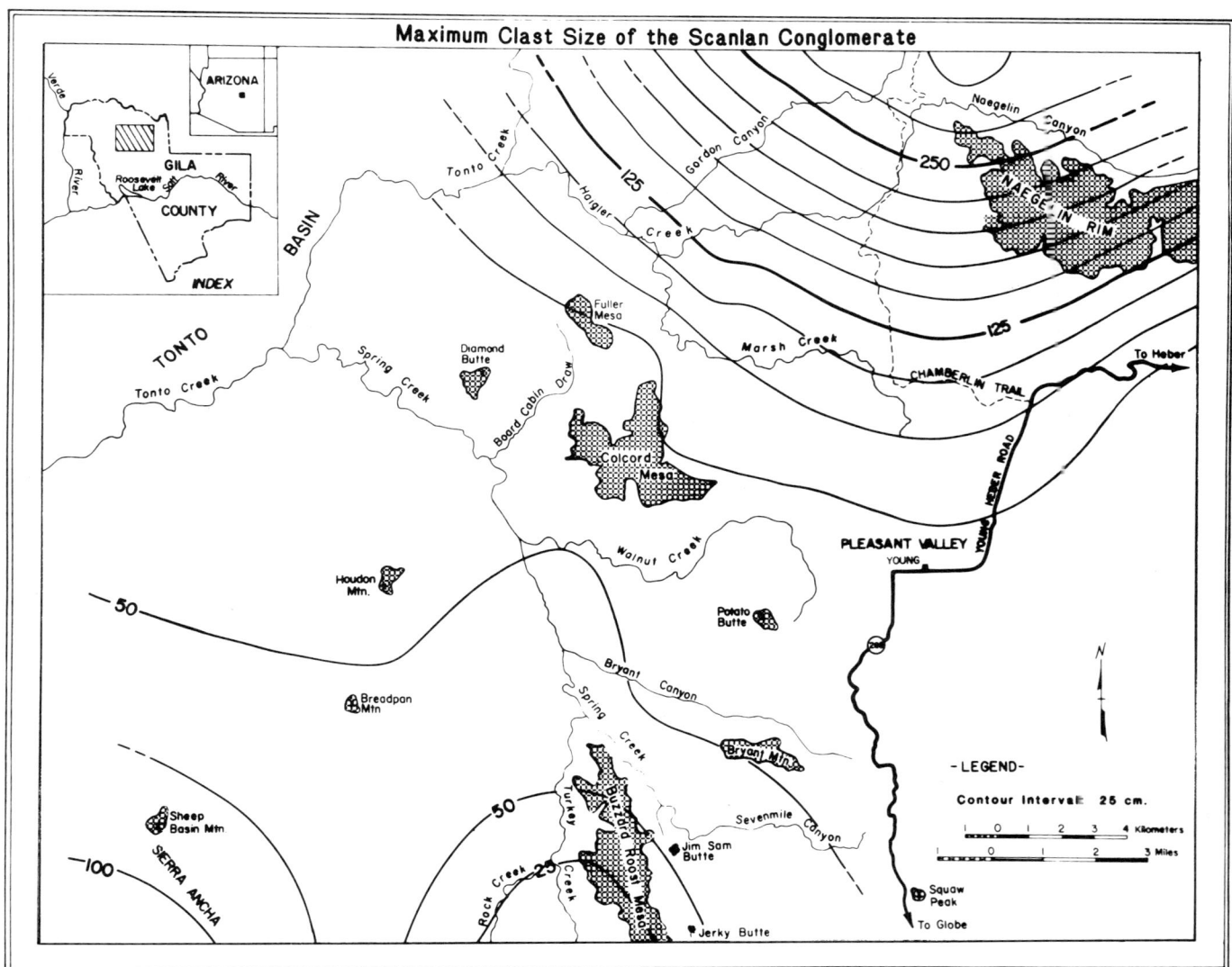

Fig. 17. Computer-generated maximum clast size map of the Scanlan Conglomerate in the study area showing systematic south to southwestern decrease in maximum clast size. Data were collected by measuring 100 clasts per square grid the size of which was dictated by clast size and sorting. Thirty-six localities were sampled.

Depositional Synthesis

Facies analysis of the Scanlan Conglomerate in the northern part of the Apache basin of deposition has documented a high-energy alluvial setting consisting primarily of distal alluvial fan to proximal braided stream systems (Figs. 13 and 15). Petrologic and paleocurrent analyses indicate a south-sloping paleoslope although, as discussed below, there are a number of local perturbations.

Figure 17 depicts variations in maximum clast size throughout the study area. In the northern part of the study area, north of Colcord Mesa, there is a systematic decrease in maximum clast size from clasts greater than 3 m on Haigler Creek to an average maximum clast size of 50 cm near Colcord Mesa. Such a distribution of decreasing clast size is consistent with an alluvial fan interpretation (Bull, 1964; Rust, 1978). As previously stated, Facies *Gms*, which is interpreted to represent subaerial debris flows, is common north of Colcord Mesa and appears to define the extent of the Scanlan fan(s). The interbedding of Facies *Gm* increases to the south reflecting the interplay of debris flow and proximal braided systems along the fan perimeter.

South of Colcord Mesa, facies interpreted as debris flows form only a minor part of the succession. In these areas the Scanlan is clearly dominated by Facies *Gm*, most of which was generated through migration of longitudinal bars in a proximal braided system (Fig. 15). The coarseness of these deposits coupled with stratification types indicate a high-energy fluvial system. While reconstruction of paleoflow velocities is tentative at best, a number

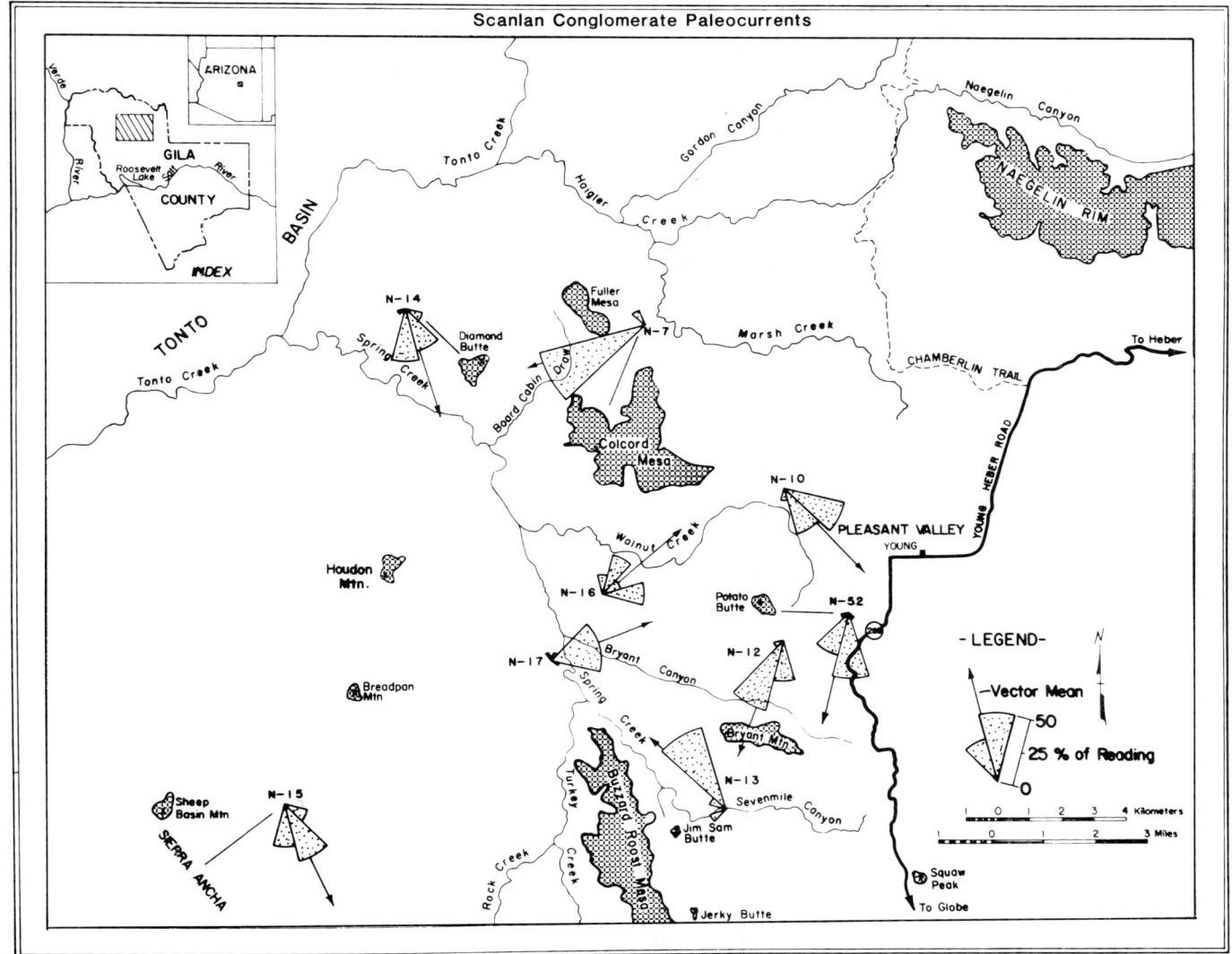

Fig. 18. Paleocurrent map showing large-scale trough axis orientations in Scanlan Conglomerate. Vector mean for each outcrop shown by bold arrow. Only those areas where data are significant at 95% confidence level are plotted. *N* equals number of readings.

of empirical relationships have been used to estimate critical shear velocity necessary for initiation of particle movement (Helley, 1969; Koster, 1978; Bradley and Mears, 1980). Using an average intermediate axis b length of 12 cm as determined by clast measurements on the outcrop, a mean flow velocity of between 1.1 to 2.2 m.sec^{-1} is indicated (Koster, 1978, Fig. 10).

A vertical change from proximal to more distal fluvial settings is suggested by the increase in deposits interpreted to represent transverse or linguoid bars and the increasing amount of coarse-grained sand toward the top of the Scanlan. This change is likely to be the result of decreasing stream gradient with decreasing source area relief.

Paleocurrent analyses of trough axis orientations and imbrication data define a complicated drainage system (Fig. 18). Between Colcord and Buzzard Roost Mesas current directions from facies sequence 2 define a crude centripetal flow pattern (Fig. 18). This is a common feature in other areas as well and is related to paleotopography developed on the underlying older Precambrian. In these areas the Scanlan pinches out against 'hills' of older Precambrian rocks and the overlying Pioneer Shale rests directly on the older basement (Trujillo et al., 1983). Scanlan deposition in these areas was confined to a series of paleovalleys, the orientation of which was controlled by structural and lithologic patterns of underlying basement (Middleton and Trujillo, in prep.).

Lithologic differences between these paleovalleys are often marked. There are numerous instances where Scanlan outcrops only a few kilometers apart have appreciable differences in clast as well as matrix compositions. In most cases the lithologic differences can be traced to local source rock control and confined paleovalley fluvial systems.

REGIONAL SYNTHESIS

Evolution of younger Proterozoic basins in southwestern United States is poorly known (Stewart and Poole, 1974). Theories ranging from failed-arm (aulacogens) rift valleys to simple passive continental margins have been proposed (Stewart and Poole, 1974; Young, 1981). Although sedimentologic studies to date have been few, data from the Scanlan Conglomerate indicate that a least part of the Apache Group was deposited in a high-energy continental setting characterized by alluvial fans and coarse-grained braidplains. The association of alluvial fans with the position of the Central Arizona Arch (Anderson and Wirth, 1980) just north of the study area, suggests that a structurally and topographically high area existed to the north. Studies by Gastil (1953, 1954, 1958) and others confirm the active structural and volcanic setting of parts of the Apache basin of deposition. Placement of this basin in a precise sedimentary tectonic framework necessitates detailed sedimentologic and structural analysis of the rest of the Apache Group. Analysis of the Scanlan Conglomerate, however, clearly points to a syntectonic origin for at least part of the Apache Group.

REFERENCES

Anderson, P. and Wirth, K.R. 1980. Uranium potential in Precambrian conglomerates of the Central Arizona Arch. Tucson: Wallaby Enterprises, 121p.

Bluck, B.J. 1967. Deposition of some Upper Old Red Sandstone conglomerates in the Clyde area: a study in the significance of bedding. Scottish Journal of Geology, v. 3, p. 139-167.

Bluck, B.J. 1974. Structure and directional properties of some valley sandur deposits in southern Iceland. Sedimentology, v. 21, p. 533-554.

Bluck, B.J. 1979. Structure of coarse grained braided stream alluvium. Transactions of the Royal Society of Edinburgh, v. 70, p. 181-221.

Bluck, B.J. 1980. Structure, generation and preservation of upward fining, braided stream cycles in the Old Red Sandstone of Scotland. Transactions of the Royal Society of Edinburgh, v. 71, p. 29-46.

Boothroyd, J.C. and Ashley, G. M. 1975. Processes, bar morphology, and sedimentary structures on braided outwash fans, northeastern Gulf of Alaska. In: Jopling, A.V. and McDonald, B.C. (Eds.), Glaciofluvial and Glaciolacustrine Sedimentation. Society of Economic Paleontologists and Mineralogists, Special Publication 23, p. 193-222.

Bradley, W.C. and Mears, A.I. 1980. Calculations of flows needed to transport coarse fraction of Boulder Creek alluvium at Boulder, Colorado. Geological Society of America Bulletin, v. 91, p. 1057-1090.

Bull, W.B. 1964. Alluvial fans and near surface subsidence in western Fresno County, California. United States Geological Survey Professional Paper 437-A, 71p.

Bull, W.B. 1972. Recognition of alluvial fan deposits in the stratigraphic record. In: Rigby, J.K. and Hamblin, W.K. (Eds.), Recognition of Ancient Sedimentary Environments. Society of Economic Paleontologists and Mineralogists, Special Publication 16, p. 63-83.

Coleman, J.M. 1969. Brahmaputra River: channel process and sedimentation. Sedimentary Geology, v. 3, p. 129-239.

Elston, D.P. and McKee, E.H. 1982. Age and correlation of the late Proterozoic Grand Canyon disturbance, northern Arizona. Geological Society of America Bulletin, v. 93, p. 681-699.

Eriksson, K.A. and Vos, R.G. 1979. A fluvial fan depositional model for Middle Proterozoic red beds from the Waterberg Group, South Africa. Precambrian Research, v. 9, p. 169-188.

Gastil, R.G. 1953. Geology of the eastern half of the Diamond Butte Quadrangle, Gila County, Arizona. Ph.D. Thesis, University of California, Berkeley. 150p.

Gastil, R.G. 1954. Late Precambrian volcanism in southeastern Arizona. American Journal of Science, v. 252, p. 436-440.

Gastil, R.G. 1958. Older Precambrian rocks of the Diamond Butte Quadrangle, Gila County, Arizona. Geological Society of America Bulletin, v. 69, p. 1495-1513.

Granger, H.C. and Raup, R.B. 1964. Stratigraphy of the Dripping Spring Quartzite, southeastern Arizona. United States Geological Survey Bulletin 1168, 119p.

Harrison, J.E. and Peterman, Z.E. 1980. A preliminary proposal for a chronometric time scale for the Precambrian of the United States and Mexico. Geological Society of America Bulletin, v. 91, p. 337-380.

Hein, F.J. and Walker, R.G. 1977. Bar evolution and development of stratification in the gravelly, braided, Kicking Horse River, British Columbia. Canadian Journal of Earth Sciences, v. 14, p. 562-570.

Helley, E.J. 1969. Field measurement of the initiation of large bed particle motion in Blue Creek near Klamath, California. United States Geological Survey Professional Paper 562-G, 19p.

Koster, E.H. 1978. Transverse ribs: their characteristics, origin, and paleohydraulic significance. In: Miall, A.D. (Ed.), Fluvial Sedimentology. Canadian Society of Petroleum Geologists, Memoir 5, p. 161-186.

Larsen, V. and Steel, R.J. 1978. The sedimentary history of a debris flow-dominated, Devonian alluvial fan — a study of textural inversion. Sedimentology, v. 25, p. 37-59.

Livingston, D.E. and Damon, P.E. 1968. The age of stratified Precambrian rock sequences in central Arizona and northern Sonora. Canadian Journal of Earth Sciences, v. 5, p. 763-772.

McConnell, R.L. 1972. The Apache Group (Proterozoic) of central Arizona, with special reference to the paleoecology of the Mescal Formation. Ph.D. Thesis, University of California, Berkeley, 170p.

McConnell, R.L. 1975a. Biostratigraphy and depositional environment of algal stromatolites from the Mescal Limestone (Proterozoic) of central Arizona. Precambrian Research, v. 2, p. 317-328.

McConnell, R.L. 1975b. Stratigraphy and depositional history of the Pioneer Formation (late Precambrian) of central Arizona. Geological Society of America, Abstracts with Programs, v. 7, p. 344.

McGowen, J.H. and Groat, C.G. 1971. Van Horn Sandstone, West Texas: an alluvial fan model for mineral exploration. Austin: Bureau of Economic Geology, Report of Investigations, v. 72, 57p.

Miall, A.D. 1970. Devonian alluvial fans, Prince of Wales Islands, Arctic Canada. Journal of Sedimentary Petrology, v. 40, p. 556-571.

Miall, A.D. 1977. A review of the braided-river depositional environment. Earth Science Reviews, v. 13, p. 1-62.

Miall, A.D. 1978. Lithofacies types and vertical profile models in braided river deposits: a summary. In: Miall, A.D. (Ed.), Fluvial Sedimentology. Canadian Society of Petroleum Geologists, Memoir 5, p. 597-604.

Ore, H.T. 1964. Some criteria for recognition of braided stream deposits. University of Wyoming, Contributions to Geology, v. 3. p. 1-14.

Pierson, T.C. 1981. Dominant particle support mechanisms in debris flows at Mt. Thomas, New Zealand, and implications for flow mobility. Sedimentology, v. 28, p. 49-60.

Rodine, J.D. and Johnson, A.M. 1976. The ability of debris, heavily freighted with coarse clastic materials, to flow on gentle slopes. Sedimentology, v. 23, p. 213-234.

Rust, B.R. 1972. Structure and process in a braided river. Sedimentology, v. 18, p. 221-245.

Rust, B.R. 1978. Depositional models for braided alluvium. In: Miall, A. D. (Ed.), Fluvial Sedimentology. Canadian Society of Petroleum Geologists, Memoir 5, p. 605-625.

Rust, B.R. 1979. Coarse alluvial deposits: In: Walker, R. G. (Ed.), Facies Models. Geoscience Canada Reprint Series 1, p. 9-21.

Shride, A.F. 1967. Younger Precambrian geology in southern Arizona. United States Geological Survey Professional Paper 566, 89p.

Silver, L.T. 1978. Precambrian formations and Precambrian history in Cochise County, southeastern Arizona. New Mexico Geological Society, 29th. Field Conference Guidebook, p. 157-163.

Smith, N.D. 1970. The braided stream depositional environment; comparison of the Platte River with some Silurian clastic rocks, north-central Appalachians. Geological Society of America Bulletin, v. 81, p. 2993-3014.

Smith, N.D. 1971. Transverse bars and braiding in the lower Platte River, Nebraska. Geological Society of America Bulletin, v. 82, p. 3407-3420.

Smith, N.D. 1974. Sedimentology and bar formation in the upper Kicking Horse River, a braided outwash stream. Journal of Geology, v. 82, p. 205-233.

Steel, R.J. and Thompson, D.B. 1983. Structures and textures in Triassic braided stream conglomerates ('Bunter' Pebble Beds) in Sherwood Sandstone Group, North Staffordshire, England. Sedimentology, v. 30, p. 341-367.

Stewart, J.H. and Poole, F. G. 1974. Lower Paleozoic and uppermost Precambrian Cordilleran miogeocline, Great Basin, western United States. In: Dickinson, W.R. (Ed.), Tectonics and Sedimentation. Society of Economic Paleontologists and Mineralogists, Special Publication 22, p. 28-57.

Stoyanow, A.A. 1936. Correlation of Arizona Paleozoic formations. Geological Society of America Bulletin, v. 47, p. 459-540.

Trujillo, A.P., Middleton, L. T. and Best, D. M. 1983. Sedimentologic and tectonic setting of the Scanlan Conglomerate (Upper Proterozoic), central Arizona. Geological Society of America, Abstracts with Programs, v. 15, p. 392.

Williams, P.F. and Rust, B.R. 1969. The sedimentology of a braided river. Journal of Sedimentary Petrology, v. 39, p. 649-679.

Wilson, E.D. 1939. Precambrian Mazatzal revolution in central Arizona. Geological Society of America Bulletin, v. 50, p. 1113-1162.

Young, G.M. 1981. Upper Proterozoic supracrustal rocks of North America: a brief review. Precambrian Research, v. 15, p. 305-330.

SEDIMENTOLOGY AND TECTONIC SETTING OF EARLY TERTIARY QUARTZITE CONGLOMERATES, NORTHWEST WYOMING

MARY J. KRAUS[1]

ABSTRACT

Quartzite cobble and boulder conglomerates are widely scattered through the Paleocene-early Eocene stratigraphic record of the western Bighorn Basin, Wyoming. Paleocurrent indicators, in conjunction with the timing of tectonic events in northwest Wyoming, demonstrate that accumulation of these conglomerates was instigated by episodic orogenic activity in the Jackson Hole area, over 100 km west of the Bighorn Basin. Deformation resulting from intrabasinal tectonic pulses also controlled the periodicity of gravel influxes to the Bighorn Basin as well as the depositional settings in which the quartzite conglomerates formed. Different modes of deposition are distinguished on the basis of conglomerate outcrop pattern and distinctive suites of depositional facies.

The oldest (earliest Paleocene) conglomerate is dominated by massive or horizontally bedded cobble conglomerate that was deposited by a braided river that flowed through an area of dissected, upturned strata along the western margin of the Bighorn Basin. Unusually thick sets of planar cross-stratified conglomerate are also common and are attributed to deep and prolonged floodstages, reflecting humid conditions and valley confinement of the stream system. In contrast, the youngest (early Eocene) conglomerate, though dominated by massive or horizontally bedded conglomerate, lacks abundant large scale cross-bedded conglomerate and was deposited on an extensive braidplain. Facies change markedly over short distances due to lateral spread of flow on the braidplain and decrease in paleoslope toward the Tertiary structural axis of the Bighorn Basin.

RÉSUMÉ

Les conglomérats à blocs et à galets de quartzite sont largement répandus dans tout le registre stratigraphique du Paléocène à l'Eocène inférieur du bassin occidental Bighorn, au Wyoming. Les révélations des paléocourants, conjuguées avec les épisodes d'événements tectoniques du nord-ouest du Wyoming, démontrent que l'accumulation de ces conglomérats fut provoquée par une activité orogénique épisodique dans la région de Jackson Hole, au-delà de 100 km à l'ouest du bassin Bighorn. La déformation produite par les contraintes tectoniques de l'intérieur du bassin a également déterminé la périodicité des apports en gravier dans le bassin Bighorn tout autant que le contexte des lieux de sédimentation où se formait le conglomérat à galets de quartzite. On distingue les différents modes de dépôt par le motif des affleurements du conglomérat et les séquences particulières du faciès sédimentologique.

Le plus ancien conglomérat (Plaéocène hâtif) est massif ou composé de couches horizontales de galets résultant de l'accumulation des sédiments laissés par une rivière anastomosée, laquelle traversait une région de strates retournées et disséquées qui longeaient la marge occidentale du bassin Bighorn. Des séquences exceptionnellement épaisses, de corps planaires de conglomérat à stratification oblique se rencontrent fréquemment et selon toute apparence elles doivent leur origine à des épisodes prolongées de crues, reflétant des conditions humides accompagnées d'un système fluviatile confiné à la vallée. Au contraire, le plus jeune conglomérat massif ou stratifié horizontalement, ne renferme pas ces abondantes couches à stratification oblique à grande échelle et fut plutôt déposé sur une vaste plaine anastomosée. Les faciès varient de façon très marquée sur de courtes distances à cause d'une propagation latérale du ruissellement sur la plaine anastomosée et une diminution de la paléopente en direction de l'axe structural tertiaire du bassin Bighorn.

INTRODUCTION

Over the last decade, studies of the nature and origin of alluvial gravels and conglomerates have intensified. Whereas much of the earlier work on coarse alluvial sediments was directed to semi-arid alluvial fan and fluvioglacial sediments, interest has now expanded to embrace coarse alluvial deposits in a broad spectrum of climatic and tectonic settings (*e.g.*, McLean, 1977; Abbott and Peterson, 1978; Rust, 1978; Parkash *et al.*, 1980). Because it has proved difficult to completely reconcile mechanisms for the formation of ancient conglomerates with models developed from coarse fluvial sediments in modern systems (*e.g.*, Eynon and Walker, 1974; Rust, 1978), additional studies of ancient sequences are essential for a better understanding of gravel environments.

The present study examines conglomerate horizons that occur sporadically throughout the Paleocene Fort Union and lower Eocene Willwood Formations of the Bighorn Basin of northwest Wyoming. Though the conglomerates are coarse, with boulders up to 40 cm long, they contain distinctive and exotic metaquartzite clasts that demon-

---

[1]Department of Geological Sciences, The University of Colorado, Boulder, Colorado 80309, U.S.A.

This paper represents part of the author's Ph.D. thesis written under the direction of T.R. Walker at the University of Colorado. J.D. Love encouraged the project and provided generous consultation and informative discussion. Critical review by T.R. Walker, G.M. Ashley, T.M. Bown, and H.E. Clifton substantially improved the manuscript. Dr. Bown also contributed informative discussion and valuable ideas and suggestions from his field experience in the Bighorn Basin. Biostratigraphic data were generously supplied by T.M. Bown, J.A. Lillegraven, K.D. Rose, and J. Zimmerman. I thank D.A. Lindsey for locality information and access to his field data and C.M. West, B.E. Kraus, and M. Barton for assistance in the field.

This report is the result of research supported by the Geological Society of America, Sigma Xi Grants-in-Aid of Research, the University of Colorado Van Riper Fund, and a University of Colorado Doctoral Fellowship.

Copyright © 1984, Canadian Society of Petroleum Geologists

strate derivation from a distant western source. The principle aims of this paper are to describe facies sequences from the two major Fort Union and Willwood conglomerate bodies and to compare and contrast their depositional and tectonic settings. Factors that controlled the episodic nature of cobble and boulder influxes to the Bighorn Basin are also discussed.

Geologic Setting

The Fort Union and Willwood Formations are early Tertiary alluvial sequences exposed throughout much of the Bighorn Basin of northwest Wyoming (Fig. 1). Deposition of these units accompanied Laramide structural development of the basin which began during latest Cretaceous time (Love, 1960; Love et al., 1963). Structural elevation of the Beartooth Mountains started during latest Cretaceous-earliest Paleocene time (Thom, 1952) and major uplifts of the Beartooth area also occurred during middle and late Paleocene times. Love et al. (1963) and Keefer (1965) suggested that minor anticlinal folding also began at the present site of the Owl Creek Range at the advent of the Paleocene.

Laramide orogenic activity continued through the early Eocene, as indicated by distinct unconformities separating Willwood sediments (some as young as latest early Eocene) from all older deposits in the western Bighorn Basin. The Owl Creek Mountains, as well as the southern Bighorn Mountains, were arched and thrust to the south along major reverse faults during early Eocene time, and the Beartooth Mountains were thrust eastward. Volcanic activity in the Yellowstone-Absaroka region, which started in the Paleocene (Love et al., 1976), dominated that region from the late early Eocene through the remainder of the epoch, resulting in the mass of stratified volcanic and volcaniclastic debris now known as the Absaroka Range (Love, 1939; Love et al., 1963; Love and Keefer, 1975; Bown, 1982).

Sediments of the principally alluvial Fort Union Formation, deposited in this area during Paleocene time, are dominated by sandstones and drab mudstones with minor thin coals and carbonaceous shales. Fort Union strata rest unconformably on the Upper Cretaceous Lance Formation or on other, older Montana Group rocks at many places in the Bighorn Basin. Erosional and/or angular unconformities are especially well developed at the western, eastern, and southeastern margins of the basin (Hewett, 1926; Bown, 1979). The base of the Fort Union Formation in the western Bighorn Basin is either a conglomerate or a massive buff sandstone which is locally pebbly (Hewett and Lupton, 1917; Hewett, 1926).

Strata of the overlying alluvial lower Eocene Willwood Formation are distinguished from those of the Fort Union by the absence of coals and the development of brightly variegated, rather than drab, mudstones (Van Houten, 1944). In the western part of the basin, Willwood sediments dip gently basinward and overlap all underlying rocks with angular or erosional unconformities (Hewett, 1926; Van Houten, 1944; Bown, 1982). On the eastern flank of the Absaroka Range, the Willwood rests on rocks as old as Lower Cretaceous Cloverly sediments (Pierce and Andrews, 1941; Pierce, 1978).

On the basis of paleofloras in Fort Union sediments of the northern Bighorn Basin, Hickey (1980) concluded that this region experienced climatic cooling from Maestrichtian through Tiffanian (early late Paleocene) time. The Tiffanian flora is dominated by deciduous forms that suggest mean annual temperatures of only 10°C and marked seasonality. Subtropical genera, susceptible to frosts, reappeared in Clarkforkian (latest Paleocene-earliest Eocene) time and heralded a return to significantly warmer conditions. Hickey (1980) computed a mean annual temperature of 13.5°C for the Clarkforkian age with a mean temperature above freezing for the coldest month.

That this warming trend persisted through the early Eocene is suggested by an increase in floral diversity, through the introduction of new subtropical and tropical forms (Wing, 1980). Early Eocene climatic conditions are substantiated by alluvial paleosol studies (Bown and Kraus, 1981; Kraus and Bown, 1982). Most Willwood paleosols resemble soils that develop today under warm-temperate to subtropical climates with alternating wet and dry seasons.

Quartzite Conglomerates

Six distinct quartzite conglomerate horizons or lithosomes occur at irregular intervals through the Fort Union-Willwood section of the west-central Bighorn Basin (Fig. 2). Reconnaissance studies indicate that no quartzite conglomerate

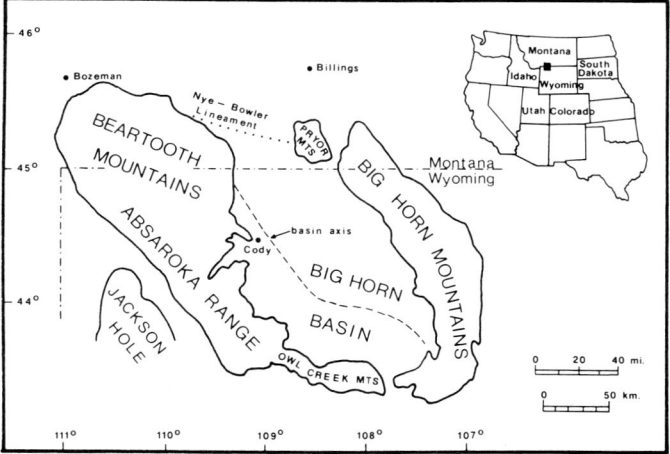

Fig. 1. Map of northwestern Wyoming showing principal structural features of the Bighorn Basin region. Tertiary structural axis of the basin follows that of Osterwald and Dean (1961). (Map modified from Eckelmann and Poldervaart, 1957, p. 1227).

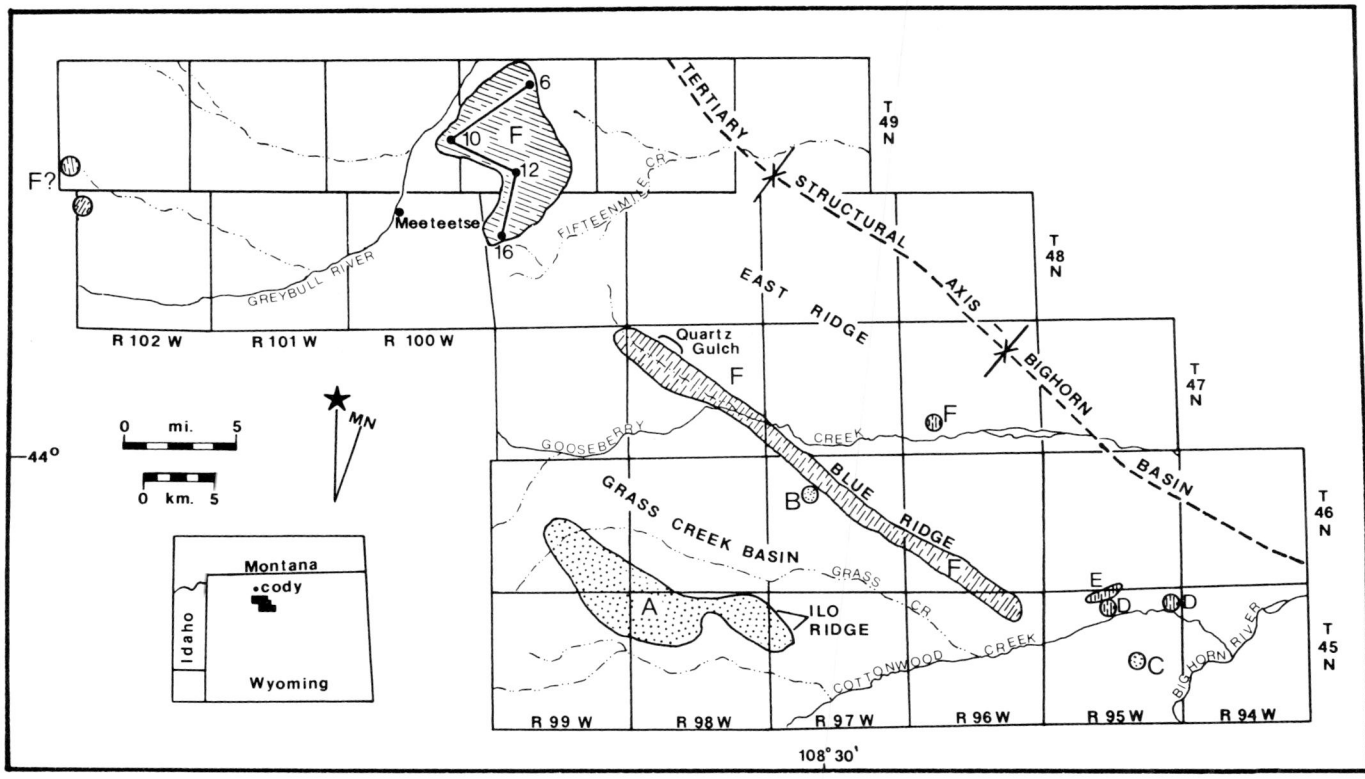

Fig. 2. Map of part of west-central Bighorn Basin showing areas of exposure of the Fort Union and Willwood quartzite conglomerate lithosomes. Lithosomes A, B, and C (stippled pattern) occur in the Fort Union Formation; lithosomes D, E, and F (lined pattern) occur in the Willwood Formation. Line of section in Lithosome F with locality numbers 6, 10, 12, and 16 refers to cross-section illustrated in Figure 4.

occurs in either formation east of the Tertiary structural axis of the basin, as delimited by Osterwald and Dean (1961).

Lithosome A delimits the base of the Fort Union Formation in the Grass Creek area (Fig. 2), where it overlies the Cretaceous Lance Formation with apparent unconformity. In most of its area of distribution, the basal Fort Union quartzite conglomerate maintains a relatively uniform thickness of approximately 15 m (Fig. 3). Its base is a sharp, irregular scour surface, whereas its upper boundary is sharp and planar. Palynomorphs from surrounding mudstones indicate that Lithosome A might be of either latest Cretaceous or earliest Paleocene age. Lithosomes B and C represent two additional Fort Union quartzite conglomerates. Both are younger (though certainly Paleocene) gravel deposits than Lithosome A.

Three similar quartzite conglomerate horizons are recognized in the Willwood Formation. Lithosomes D and E, which are poorly exposed and laterally impersistent, are very early Eocene (earliest Wasatchian) in age on the basis of fossil mammals. The youngest quartzite conglomerate (Lithosome F) occurs in the upper part of the Willwood Formation (Fig. 2). There, two to four conglomerate horizons typically appear within a 50-80 m thick zone of sandstone and conglomerate near the locally unconformable

Fig. 3. Basal Fort Union quartzite conglomerate (Lithosome A) resting unconformably above late Cretaceous Lance sediments on north side of Spring Gulch; NE¼, sec. 2, T. 45 N., R. 99 W. View to the southeast.

base of the Willwood Formation (Fig. 4). This lithosome is exposed along strike in a northwest-southeast direction over 50 km. Fossil mammals are well known from mudstones enveloping the conglomerate complex and these demonstrate that Lithosome F was deposited during the late early Eocene (late Wasatchian).

Further discussion is restricted to Lithosomes A (latest Cretaceous-earliest Paleocene) and F (early Eocene) because they are laterally the most extensive of the quartzite conglomerate bodies and they possess the best exposed vertical sequences.

CONGLOMERATE LITHOLOGY AND TEXTURE

Fort Union and Willwood conglomerates are lithologically distinctive. At least 65% of the cobbles and pebbles are quartzite and the gravel fraction in some samples is nearly 95% quartzite (Kraus, 1983). Petrographic studies by Young (1972) demonstrated that 90% of the quartzite clasts from a random sampling of Willwood conglomerate localities showed stretched quartz grains with sutured boundaries. Though grey quartzite dominates, there is a wide variety of colors including pink, brown, green, white, black, and yellow. The conglomerates are also characterized by minute quantities of detrital gold occurring as matrix particles (Antweiler and Love, 1967).

Fort Union and Willwood quartzite conglomerates are generally clast supported with a medium to very coarse sand matrix. Maximum clast size was determined for randomly selected localities by averaging the a-axis lengths from the ten largest clasts (Pettijohn, 1975). Maximum grain size in the basal Fort Union conglomerate (Lithosome A) ranges from 15 to 33 cm, however, no clear geographic

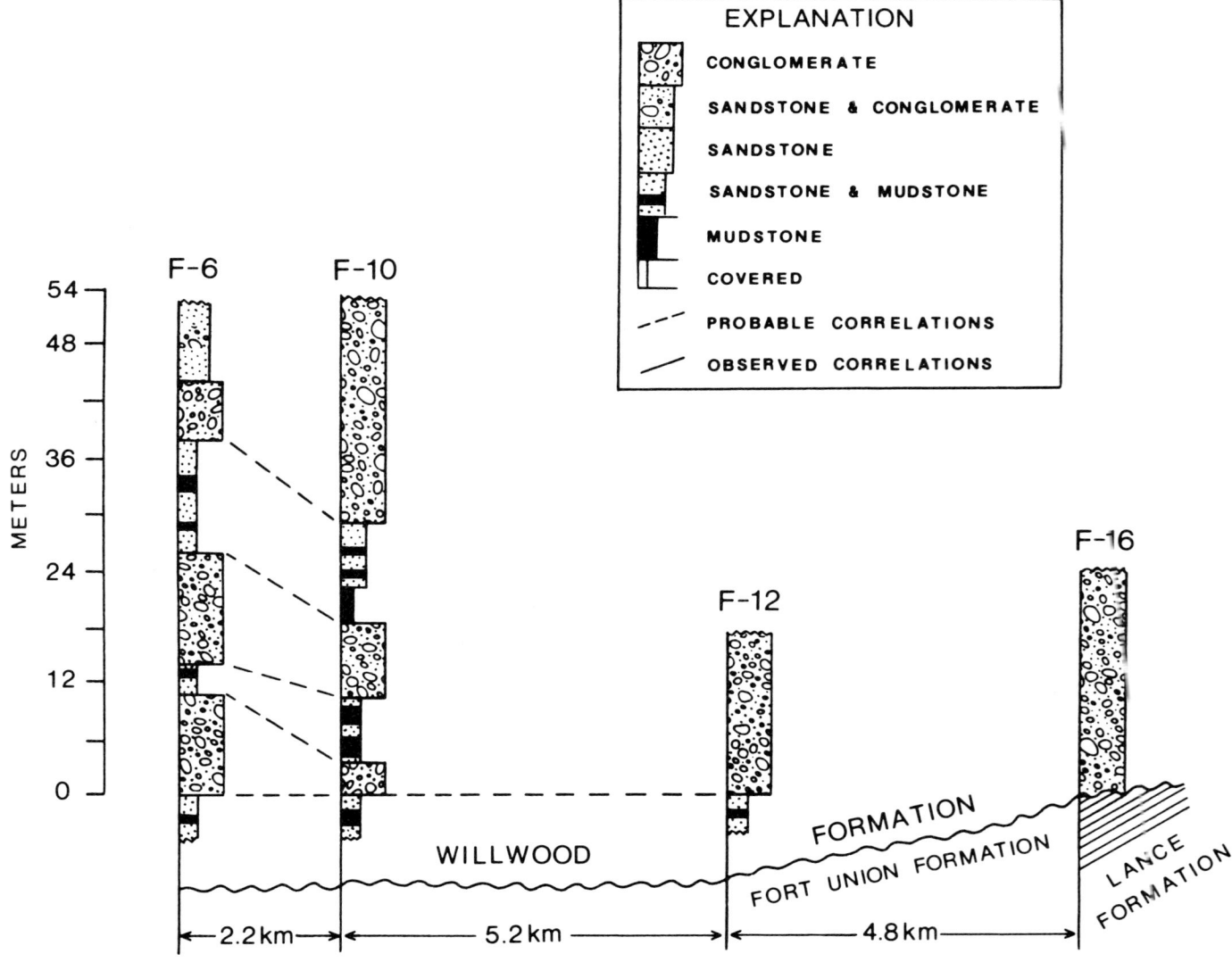

Fig. 4. Cross-section through Lithosome F in the area northeast of Meeteetse, showing vertical stacking of lithofacies. Location of measured sections and line of section shown in Figure 2. Note that at locality F-16, the conglomerate lies unconformably above the Fort Union and Lance Formations.

grain size patterns emerge nor do any vertical trends within lithosomes. In Lithosome F, maximum clast size varies from 15 to 39 cm for the localities sampled. The largest clasts occur west of the town of Meeteetse and a west to east fining trend is observable in exposures in the region of Meeteetse. No other regular patterns of grain size change (for example, northwest to southeast) occur in Lithosome F.

PROVENANCE

Paleocurrent patterns, based primarily on cobble imbrication, are illustrated in Figure 5. Large scale cross-strata in sandstones and pebbly sandstones provide a secondary means and, in some areas of the Willwood conglomerate, the sole means of paleocurrent determination available. Grand vector mean azimuths, calculated only from imbrication, are 106° for the basal Fort Union conglomerate and 057° for the Willwood quartzite conglomerate, demonstrating that early Tertiary quartzite gravel stream systems flowed eastward from a western source.

Lithologically similar conglomerates are present in other Upper Cretaceous and lower Tertiary sequences in northwest Wyoming, northeast Idaho, and south-central Montana (Fig. 6). Examples include the Upper Cretaceous Harebell and Upper Cretaceous and Paleocene Pinyon Formations in the Jackson Hole area of northwest Wyoming (Love, 1956, 1973; Lindsey, 1972), and the Divide quartzite conglomerate lithosome of the Beaverhead Formation (Ryder, 1967; Wilson, 1970; Ryder and Scholten, 1973), which apparently spans latest Early Cretaceous to mid-Paleocene time (Ryder and Ames, 1970). These conglomerates are all characterized by abundant well-rounded quartzite gravel and by small amounts of detrital gold in the matrix (Antweiler and Love, 1967). Compositional similarities, as well as regional paleocurrent trends, suggest a common stratigraphic source for all of these quartzite conglomerates, as well as for the Bighorn Basin quartzite conglomerates (e.g., Love, 1956, 1973; Antweiler and Love, 1967; Lindsey, 1972).

Kraus (1983) concluded that the Harebell and Pinyon conglomerates in the Jackson Hole area were derived directly from thick Precambrian quartzite sources in thrust sheets emplaced in east-central Idaho during late Cretaceous-Paleocene time (Ruppel, 1975, 1978). At or near the end of the Cretaceous and prior to Pinyon deposition, parts of the Harebell conglomerate rose and were deeply eroded in response to vertical movement of the Basin Creek uplift situated in southern Yellowstone National Park (Love and

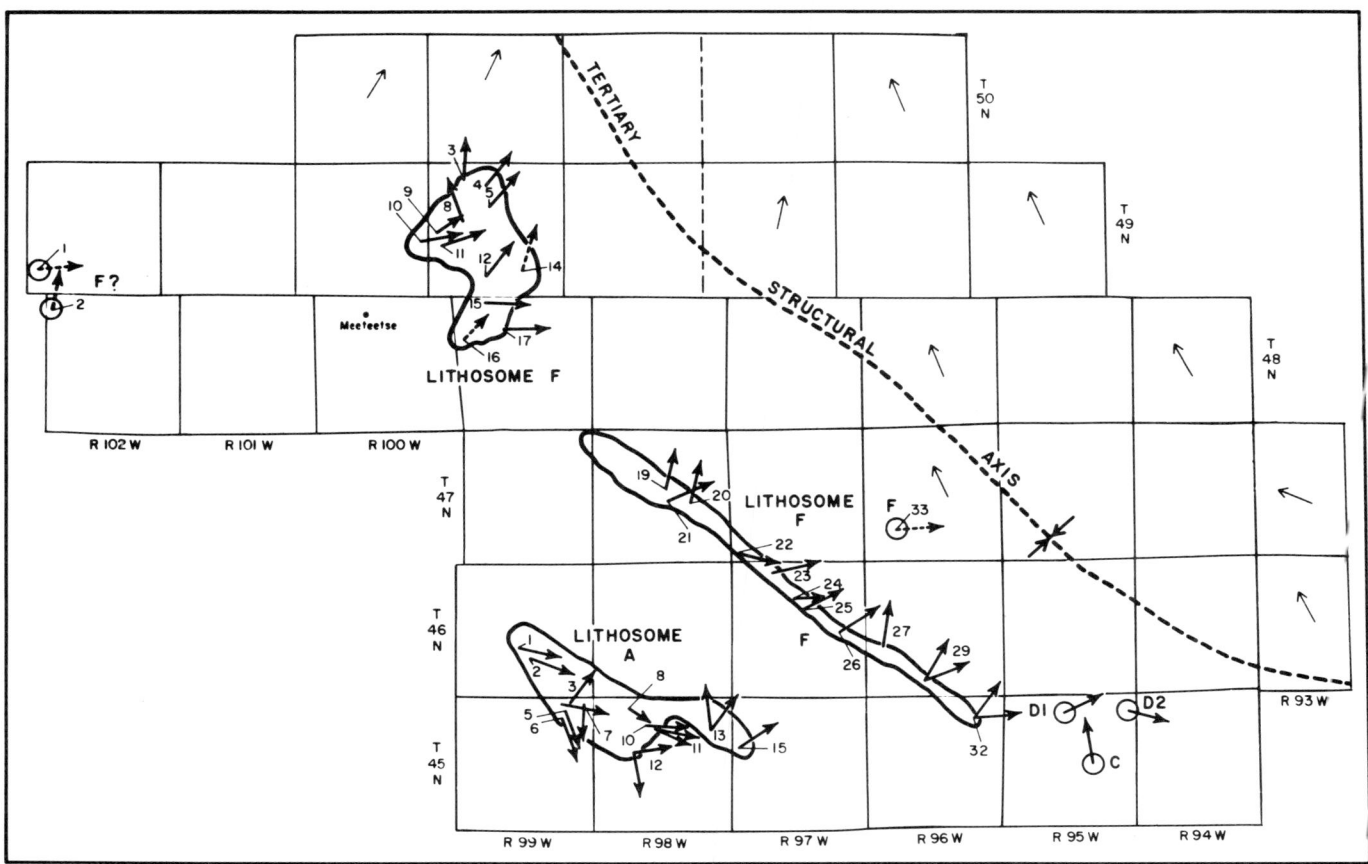

Fig. 5. Quartzite conglomerate paleocurrents based on imbrication (heavy solid arrows) and cross-bedding (heavy dashed arrows). Eocene paleocurrents from non-conglomeratic portions of the Willwood Formation (from Neasham, 1970) are shown by lighter solid arrows.

Keefer, 1969, 1975; Fig. 6). The recycled quartzite clasts were transported at least 100 km farther eastward and deposited as the basal Fort Union conglomerate (Kraus, 1983). Following this earliest Paleocene episode, a long pause ensued in the history of quartzite gravel influx to the Bighorn Basin. During this lull, which reflects a period of tectonic quiescence in the Jackson Hole area, the Pinyon conglomerate accumulated there. At least five subsequent episodes of late Paleocene-early Eocene quartzite gravel deposition occurred along the western flank of the Bighorn Basin, signaling renewed tectonic activity in Jackson Hole. The final and volumetrically most significant influx of exotic quartzite clasts occurred during the late early Eocene and was probably instigated by uplift and erosion of Pinyon conglomerate exposed on the margins of the rising Washakie Range (Kraus, 1983). Though the possibility that some Willwood quartzite cobbles were recycled from Fort Union exposures cannot be ruled out, it is unlikely that they served as a major source for Willwood conglomerates because few Fort Union conglomerate-bearing sediments were exposed to erosion in early Eocene time.

FACIES ASSEMBLAGES

Measured stratigraphic sections in Lithosomes A (Fort Union) and F (Willwood) were subdivided into depositional facies and vertical sequences of facies were analyzed by Markov chain analysis. Though the results of Markov analysis were not positive, and in fact indicate that facies transitions appear to be ordered randomly, the transition matrices aided in establishing three facies assemblages (Kraus, 1983). Facies terminology follows the code outlined by Miall (1977) and expanded by Miall (1978a) and Rust (1978).

Planar Cross-Stratified Conglomerate (Gp) Assemblage

This facies assemblage, illustrated in Figure 7, is restricted to the basal Fort Union quartzite conglomerate west and northwest of Ilo Ridge (Fig. 2). Here approximately 70% of all conglomerate is massive or crudely horizontally stratified (*Gm* facies) and the remaining 30% is planar cross-stratified (*Gp* facies). Trough cross-stratified conglomerate (*Gt* facies) was not observed in this assemblage. Though the basal Fort Union conglomerate is enveloped by sandstone and/or mudstone, the lithosome itself usually contains a mere few percent of interbedded sandstone (Figs. 7A, 7D), either between facies or as lenses within the *Gm* facies.

Though the assemblage is dominated by massive or horizontally stratified conglomerate, its most striking attribute is the presence of very large scale planar cross-stratified conglomerate (Fig. 7). Individual sets are rarely

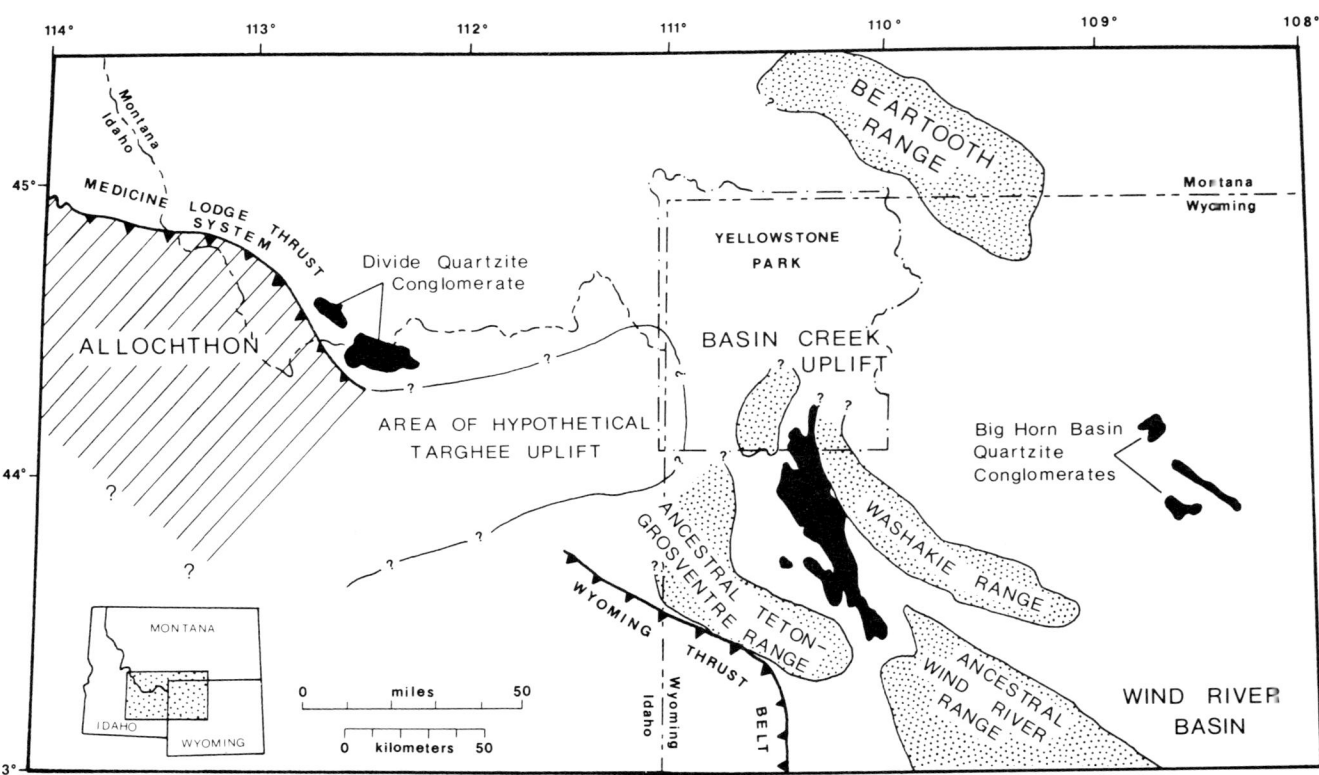

Fig. 6. Paleogeographic setting for Late Cretaceous-early Tertiary quartzite conglomerates in northwest Wyoming and vicinity. (Modified after Lindsey, 1972, Fig. 43).

less than 2 m in thickness and 50% of the 15 sets recorded in deposits of this assemblage exceed 3.5 m. Recognition of cross-stratification, which attains dips of 30°, is enhanced by the alternation of coarse and fine grained layers. Thicker foresets (25 to 50 cm thick) are dominated by cobbles, whereas foresets consisting of pebbles are usually 15-20 cm thick, but rarely reach 40 cm in thickness. Planar cross-sets are remarkably extensive in directions both perpendicular and parallel to paleoflow. A single set located in sec. 22, T. 46 N., R. 99 W. can be traced approximately 450 m in a direction perpendicular to the general paleoflow for the exposure. At a second locality, a planar cross-set whose cross-section is oriented parallel to paleoflow stretches laterally nearly 500 m.

Massive or Horizontally Stratified Conglomerate (Gm) Assemblage

This assemblage is confined to Lithosome F and is typical of exposures in Quartz Gulch and Blue Ridge (Fig.

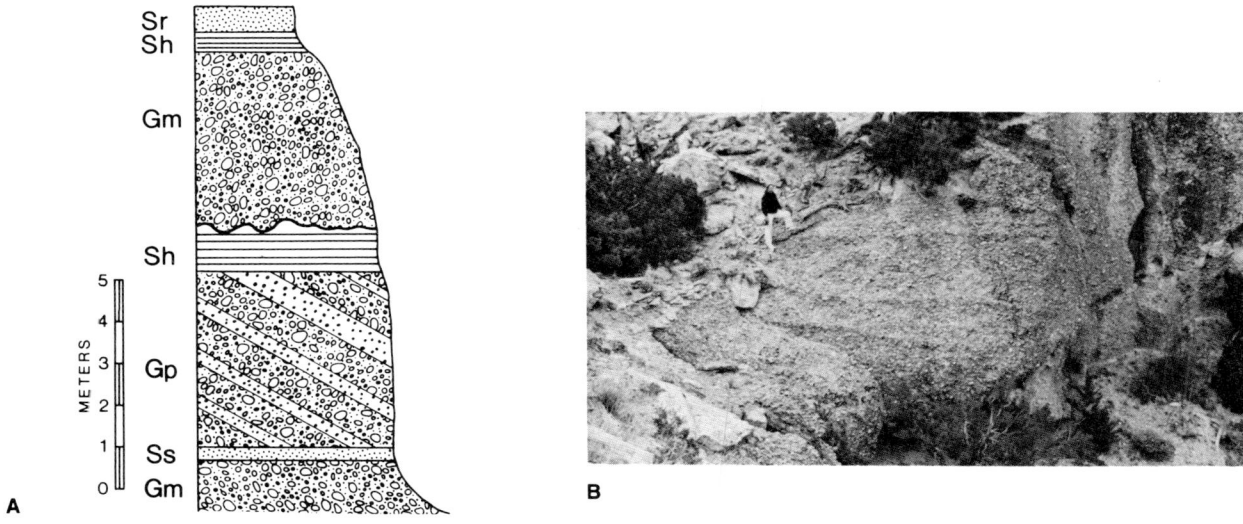

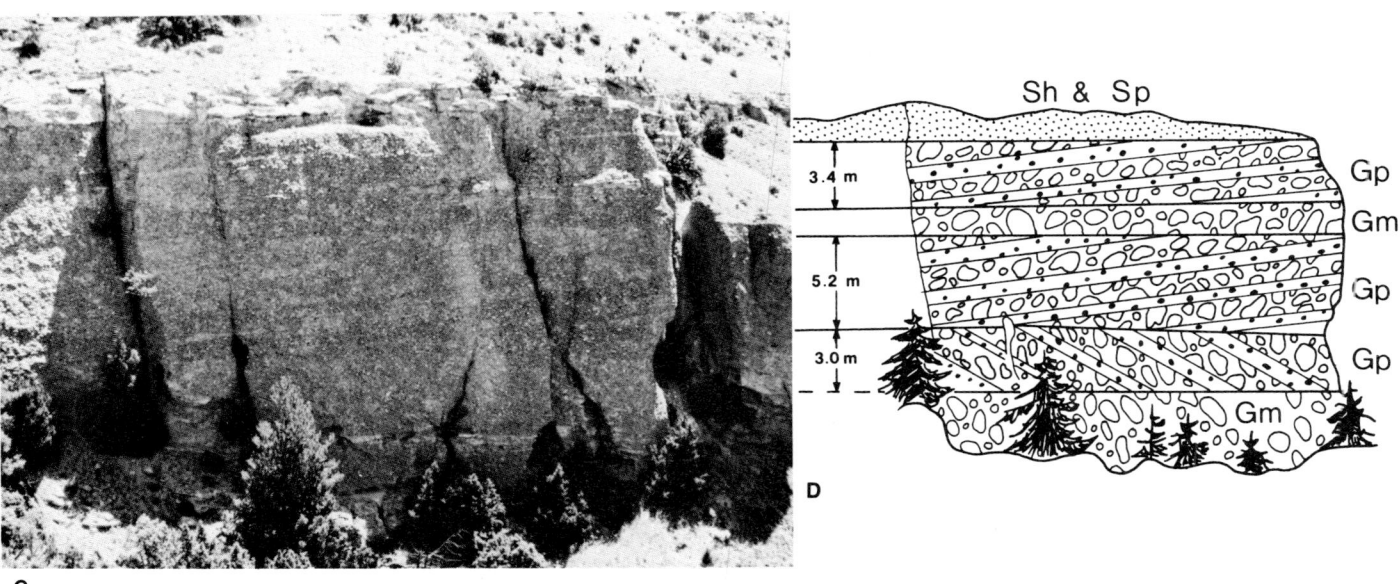

Fig. 7. Planar cross-stratified conglomerate (*Gp*) facies assemblage. **A-** Stratigraphic section of part of the basal Fort Union conglomerate, illustrating the *Gp* assemblage. The heavy line denotes an erosion surface. **B-** This solitary cross-set is 5.9 m thick and occurs at the top of the basal Fort Union conglomerate which is dipping 17° (E½, NE¼, sec. 8, T. 45 N., R. 98 W., view to the north). **C-** *Gp* assemblage showing three planar cross-sets; sec. 2, T. 45 N., R. 99 W., view to the east. **D-** Outcrop sketch of 7C showing the facies sequence.

2). Quartzite conglomerate is dominantly massive or horizontally stratified and consists of cobble conglomerate (Fig. 8). Horizontal stratification is distinguished by fine/coarse layers, sandstone lenses, or zones with well developed imbrication (Fig. 8A). Interbedded sandstone lenses that are horizontally stratified or planar cross-stratified (*Sp* facies) are locally common (Fig. 8B). Sandstone bodies range in thickness from several cm to approximately 1 m and have strongly scoured upper boundaries. In vertical section, massive conglomerate with minor sandstone is typically succeeded by sandstone rich areas of the *Gm* facies and those are in turn capped by sandstone or trough cross-stratified conglomerate (Fig. 8C).

Planar cross-stratified conglomerate is present in this assemblage, however, it is neither common nor does set thickness exceed approximately 2 m. Trough cross-stratified conglomerate is rare.

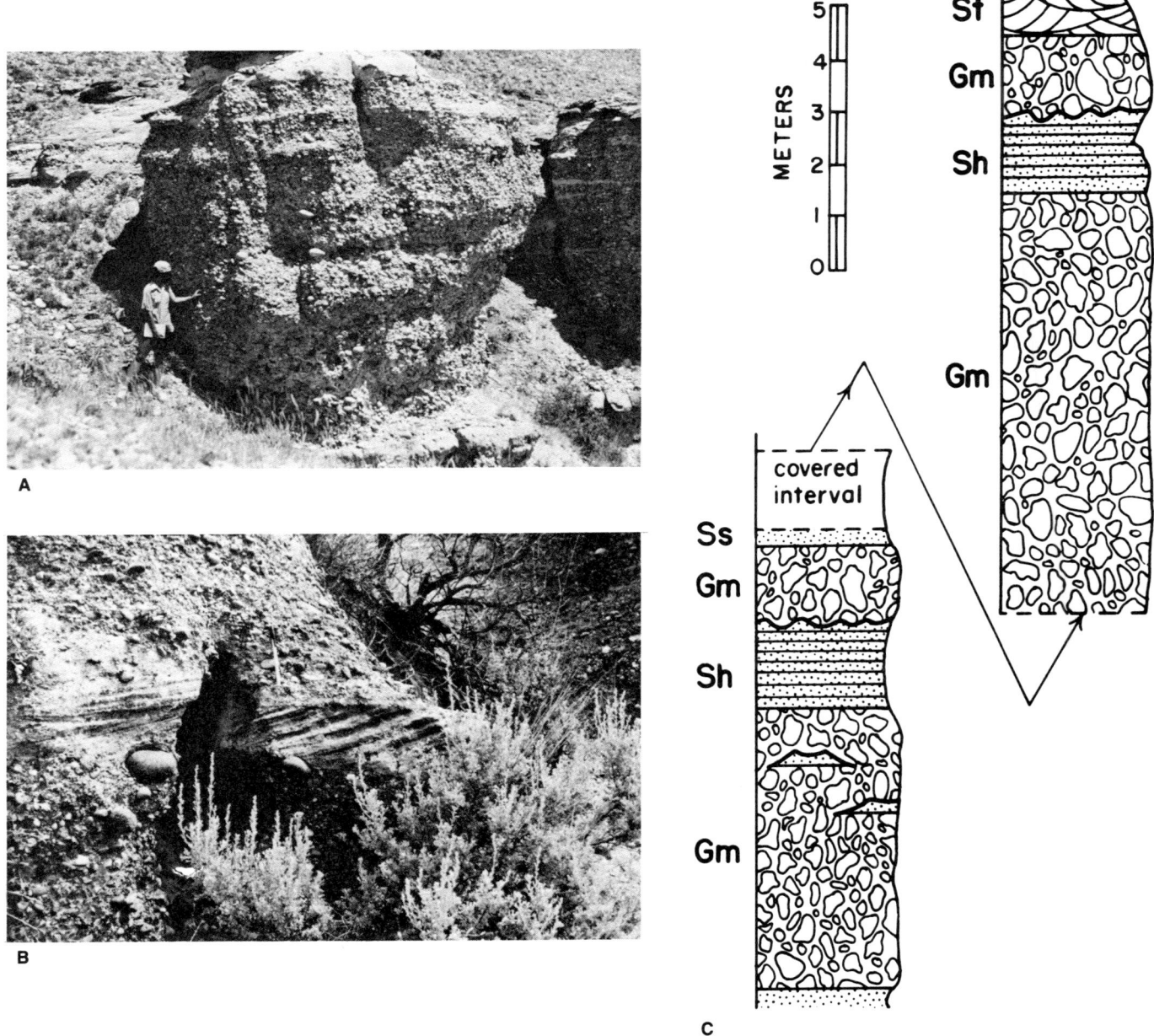

Fig. 8. Massive or horizontally stratified conglomerate (*Gm*) facies assemblage. **A-** Crude horizontal stratification in the *Gm* facies. **B-** Large scale planar cross-stratified sandstone (*Sp*) facies as lens in the *Gm* facies. 15 cm pen gives scale. **C-** Stratigraphic section of part of Lithosome F, illustrating the *Gm* assemblage. Heavy lines delimit erosion surfaces.

Trough Cross-Stratified Conglomerate (Gt) Assemblage

The third facies assemblage, the trough cross-stratified conglomerate assemblage, possesses a variety of well represented facies which complexly interchannel and crosscut one another (Fig. 9A, 9B). This assemblage is most

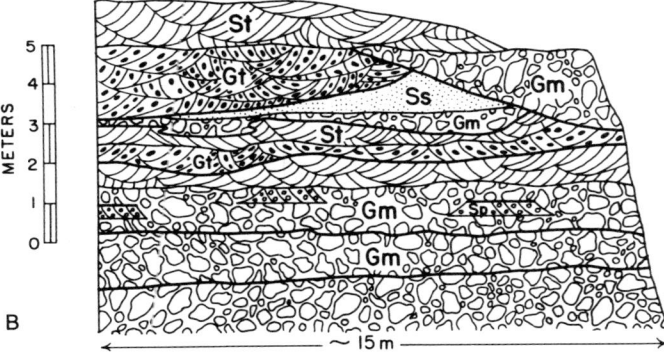

Fig. 9. Trough cross-stratified conglomerate (*Gt*) facies assemblage. **A-** Example of the *Gt* assemblage showing complex relations among facies (sec. 33, T. 49 N., R. 99 W., northeast of Meeteetse; view to the east). **B-** Outcrop sketch of 9A, illustrating the facies. Heavy lines denote erosion surfaces. **C-** Trough cross-stratified conglomerate facies, showing coarse lag above basal scour surface. Lens cap gives scale.

characteristic of Willwood conglomerate units throughout the Meeteetse area and also occurs at some localities along Blue Ridge (Fig. 2). Massive or horizontally bedded conglomerate is abundant (40% of the assemblage) and sandstones, especially pebbly sandstones, are also common (20% of the assemblage). Planar cross-stratified conglomerate is present, but is much less common than in the *Gp* facies assemblage and sets never exceed 2 m in thickness.

Trough cross-stratified conglomerate (*Gt* facies) constitutes 25% of the *Gt* assemblage and is more abundant in this assemblage than in either the *Gp* or *Gm* facies assemblages. The *Gt* facies is usually developed in pebble conglomerate, though lag deposits of somewhat coarser gravel or of sandstone intraclasts may occur above the scoop shaped, erosive base of trough sets (Fig. 8C). Set thickness ranges from 75 to 150 cm and the alternating fine and coarse foresets attain individual thicknesses of 15 cm and dips of 26°. Individual foresets may be normally graded. Probability transition matrices indicate that this facies is generally developed in the upper, finer portions of facies sequences where it is generally associated with cross-stratified sandy conglomerate and cross-stratified pebbly sandstones (Kraus, 1983).

DEPOSITIONAL SYSTEMS

The nature of depositional facies in the Bighorn Basin quartzite conglomerates, lateral and vertical transitions in those facies, and lithosome geometries are indicative of braided stream or braidplain deposition. The quartzite conglomerates strongly resemble modern glacial outwash and gravelly braided stream sediments and the three facies assemblages reflect deposition under different conditions of stream energy and flow depth. Differences in facies assemblage distribution between Lithosomes A and F, in conjunction with lithosome outcrop patterns, indicate contrasting depositional settings for the two lithosomes.

Lithosome A

The basal Fort Union quartzite conglomerate (Lithosome A) is dominated by the *Gp* facies assemblage, which bears greatest similarity to the Devonian Malbaie Formation described by Rust (1978) and the Triassic 'Bunter' Pebble Beds of Steel and Thompson (1983). The Malbaie conglomeratic succession was interpreted as an ancient analog of the proximal Donjek River deposits (Rust, 1972, 1975) or the Scott outwash sediments (Boothroyd and Ashley, 1975), in which the abundant *Gm* facies represents longitudinal bar deposits. However, the Malbaie Formation, the 'Bunter' Pebble Beds, and the basal Fort Union quartzite conglomerate differ significantly from either the Donjek or Scott sediments and from the *Gt* or *Gm* facies assemblages in that the former sequences contain relatively abundant planar cross-stratified conglomerate. The *Gp* facies, with mean set thickness of approximately 1 m, comprises 20% of the conglomeratic sequence in the Malbaie Formation (Rust, 1978) and nearly 35% of one locality in the 'Bunter'

Pebble Beds consists of cross-stratified conglomerate with set thicknesses up to 4 m (Steel and Thompson, 1983).

Rust (1978) concluded that planar cross-stratified conglomerate in the Malbaie Formation, with the possible exception of the largest sets, resulted from falling stage modification of the downstream margin of longitudinal bars. Conversely, in the 'Bunter' Pebble Beds, sequences of cross-stratified conglomerate (facies B of Steel and Thompson, 1983) overlain by horizontally bedded conglomerate (their facies A) were judged similar to the bipartite morphology characteristic of the medial bars described by Bluck (1971, 1976). These mid-channel bars contain cross-stratified bar platform deposits capped by horizontally bedded supra-bar gravels which were emergent at low flows.

Large scale conglomerate cross-sets in Lithosome A could have originated either as falling stage modifications of longitudinal bars (e.g., Eynon and Walker, 1974; Rust, 1978) or as the internal structure of migrating bedforms that were stable under flood conditions (e.g., Smith, 1974; Rust, 1975; Steel and Thompson, 1983). The great lateral extent of cross-bedding, in directions parallel to and especially perpendicular to paleoflow, in addition to the thickness of most sets, suggests that this facies resulted from the downstream migration of transverse or medial bars during floodstage. Additional evidence favoring this interpretation includes: (1) the absence of any observable lateral transition between the Gm and Gp facies; (2) sequences in which the Gp facies is overlain by the Gm facies (Fig. 7D); (3) the relative volumetric abundance of the Gp facies in the basal Fort Union conglomerate; and (4) cross-stratal angles as steep as 30°. Nonetheless, the Fort Union conglomerate is dominated by the Gm facies, most of which probably reflects deposition by longitudinal bars.

The abundance and magnitude of planar cross-sets in the Gp facies assemblage is unusual, especially in comparison with deposits described from modern gravel streams. The apparent paucity of this facies in modern systems reflects, at least partly, the difficulties of studying the internal structure of large gravel bedforms which may be submerged for long periods. It probably also reflects the fact that so many of the gravel streams described in the literature are glacially influenced. The relatively small maximum discharges and flow depths characteristic of glacial outwash environments are considered ill-suited to Gp facies genesis (Boothroyd and Nummedal, 1978); and Smith (1974) ascribed absence of the Gp facies in deposits of the Kicking Horse River to rapid changes in bar morphology arising from the diurnal as well as seasonal stage fluctuations typical of fluvioglacial settings. Cross-set thicknesses in the basal Fort Union conglomerate demonstrate that flows were relatively deep. At the very minimum, floodstage depths must have equalled cross-set thickness; thus the thickest sets reflect flow depths of at least 6 m (Fig. 7B). Rust (1978) concluded that large scale cross-stratified conglomerate in the Malbaie Formation resulted from humid conditions producing floodstages that were not only deeper but also more frequent and of longer duration than in modern proglacial areas. He further suggested that a relatively high proportion of planar cross-stratified conglomerate may be more characteristic of Devonian and older sedimentary sequences as a result of increased discharge under conditions of little or no land vegetation (Rust, 1979). This idea is contradicted by the abundance and magnitude of conglomerate cross-sets in the Fort Union basal conglomerate, which formed during earliest Paleocene time in a vegetated region. Genesis of the Gp facies in Lithosome A is attributed to both climatic and geomorphologic factors. Independent climatic indicators attest to abundant precipitation during the early Tertiary in the Bighorn Basin and probably in all of northwest Wyoming. The long distance transport of coarse detritus to the Bighorn Basin and deep floodstages are attributed to abundant run-off. Deeper and more prolonged flows also imply more stable conditions for bedform genesis. The presence of conglomerate cross-sets that are persistent in directions perpendicular and parallel to paleoslope probably resulted from deep floodstages of longer duration than those yet observed for modern systems.

The outcrop pattern of Lithosome A, in conjunction with independent evidence for deep flood-stage flows, indicates that flow was laterally confined by valley walls. The Lance-Fort Union contact, though well exposed in many areas bordering the western Bighorn Basin, is delimited by quartzite conglomerate only in the Grass Creek area. Consequently, the outcrop pattern of Lithosome A (Fig. 5) is judged to be a relatively accurate depiction of the original depositional extent of the conglomeratic body. The outcrop pattern is elongate in a northwest-southeast direction, a trend which parallels the mean paleoflow determined for Lithosome A of 106°. The essentially linear outcrop pattern is consistent with deposition in a valley confined braided river. Pre-Tertiary folds that were breached by erosion during the Paleocene are present along the eastern margin of the Absaroka Range and continue westward for an unknown distance (Bown, 1982). Streamflow was probably confined by strike valleys along these structures, and flow depth and strength were thus maintained.

Long distance transport of coarse gravel into the western Bighorn Basin probably resulted from this valley confined transport system. The decrease in maximum clast size between the quartzite gravel source in Jackson Hole and the Fort Union conglomerate is not large (36 cm to 24 cm, Lindsey, 1972; Kraus, 1983). For comparison, the Donjek River has a maximum clast diameter in excess of 1.5 m at its glacial source and 25 cm approximately 75 km farther downstream (data from Rust, 1972, Figs. 1 and 13).

Lithosome F

Lithosome F in the upper part of the Willwood Formation is composed of the Gm and Gt facies assemblages.

Though the deposits of the *Gm* assemblage are situated approximately 100 km from quartzite gravel sources in Jackson Hole, the assemblage resembles deposits of the Scott glacial outwash model (Boothroyd and Ashley, 1975), established by Miall (1977, 1978a), and the proximal Donjek River sediments described by Rust (1972, 1975). Sedimentary sequences in those modern examples are dominated by massive or horizontally stratified gravel with only minor, laterally impersistent, interbedded sand lenses.

By comparison with these and other modern examples, the *Gm* facies is interpreted as the internal structure of longitudinal bars. During periods of low water, the gravel bars were emergent and laterally impersistent, cross-stratified sandstone wedges (*Sp* facies) built out from lateral bar margins (Fig. 8B). Gravelly bar tops were also mantled by sandy deposits which are typically horizontally bedded or rippled (*e.g.*, Rust, 1972, 1978; Boothroyd and Ashley, 1975). Preservation potential of low water sands is generally poor, and this is reflected by their scarcity in the *Gm* facies in the quartzite conglomerate and in most proximal gravelly deposits.

The *Gt* facies assemblage is texturally and structurally more varied and complex than either the *Gm* or *Gp* assemblages. With its abundant fine grained facies (*Sp*, *St*, and *Sh* facies), cross-stratification, and complex channeling, the *Gt* assemblage resembles sediments deposited in areas more distal to the source than the environment represented by the *Gm* or *Gp* assemblages.

The *Gt* assemblage bears greatest resemblance to the upper midfan sediments of the Scott glacial outwash (Boothroyd and Ashley, 1975). Sequences of massive gravel subdivided by erosion surfaces with a veneer of mudstone and sandstone clasts (Fig. 9B), represent superimposed longitudinal bar deposits. Partly eroded bars, as well as those with preserved bar top deposits (Fig. 9B), are usually no thicker than approximately 1 m. This observation, in conjunction with the absence of planar cross-stratified conglomerate with sets thicker than 2 m, demonstrates that flow depths were relatively shallow.

Cross-stratified fine conglomerates and sandstones are abundant in the *Gt* assemblage. They were generated at low stages of flow in channels or on the upper surfaces of bars prior to their emergence. Vertical sequences, such as that shown in Figure 9A, record gradual channel filling; cross-stratified conglomerate and sandstones were deposited during waning flow above floodstage massive gravels. Shallow channeling represents falling stage episodes of erosion and subsequent infilling of the scours. Complex channeling among facies is prevalent and reflects rapid channel switching.

In the area northeast and east of Meeteetse (Fig. 2), Lithosome F is dominated by the *Gt* assemblage, which represents deposition under conditions of lower flow strength than does the *Gm* assemblage. Farther south in Quartz Gulch and along Blue Ridge (Fig. 2), the *Gm* assemblage is typical. Because Lithosome F was generated by flows to the northeast (Fig. 5), deposits northeast of Meeteetse were apparently more distal to the source than deposits along Quartz Gulch and Blue Ridge. Furthermore, the thick, coarse conglomerates pinch out east of the Blue Ridge-Quartz Gulch area and no quartzite conglomerate is known east of the Tertiary structural axis of the basin. These observations indicate that proximal-distal facies transitions from conglomerates to sandstones and mudstones occur over relatively short distances east of the main body of Lithosome F.

Lithosome F contrasts strongly with Lithosome A in terms of its outcrop pattern and facies assemblages. The characteristics of assemblages *Gm* and *Gt*, especially the absence of planar cross-stratified conglomerate with sets thicker than 2 m, indicate that flows were considerably more shallow than those depositing the basal Fort Union conglomerate (Lithosome A). The outcrop pattern of Lithosome F reflects extensive erosion of Willwood sediments along the western flank of the Bighorn Basin. Preserved and exposed portions of the lithosome form a belt that trends northwest-southeast, or approximately perpendicular to the mean paleocurrent direction determined for Lithosome F (57°). The orientation of the outcrop pattern of Lithosome F relative to paleoslope and the relatively shallow depths are consistent with deposition on an extensive braidplain.

Though evidence indicates that Lithosome F accumulated on a braidplain, the coarse nature of the *Gm* facies assemblage and its resemblance to proximal braided stream and outwash sediments suggest that valley confined streams transported the cobbles and boulders from source areas in Jackson Hole across the vast area now covered by the Absaroka Range volcanics (Fig. 1). At the western margin of the Bighorn Basin, the gravel systems debouched onto an extensive depositional plain that was developed in late early Eocene time. Downstream decreases in flow depth and flow strength were rapid and resulted from lateral spread of flow on the broad alluvial surface and slopes that decreased rapidly towards the Tertiary structural axis of the basin. Lateral shifting of the main channels on the broad alluvial plain left extensive floodplain areas stable for long periods during which overbank and near channel sediments accumulated and produced the mudstones and muddy sandstones interbedded with conglomerate horizons in Lithosome F.

SUMMARY AND THE NATURE OF EPISODIC GRAVEL ACCUMULATION

Quartzite conglomerate lithosomes are widely spaced through the Tertiary stratigraphic column in the western Bighorn Basin. The conglomerates represent quartzite cobble and boulder influxes from the Jackson Hole area over 100 km to the west. The final and volumetrically most significant influx of exotic clasts occurred during the late Wasatchian (late early Eocene) and was instigated by uplift and erosion of Pinyon Conglomerate exposed on the margins of the rising Washakie Range.

The preserved portions of the youngest quartzite conglomerate, Lithosome F, differ significantly from preserved exposures of the basal Fort Union conglomerate (Lithosome A). Whereas Lithosome A was deposited in a braided river system, considerable evidence indicates that Lithosome F accumulated on an extensive, relatively featureless braidplain. The different depositional settings are expressed in lithosome outcrop patterns and facies assemblages. Lithosome F extends at least 50 km perpendicular to paleoslope; in contrast, the outcrop pattern of Lithosome A is restricted to a small area and is elongate parallel to paleoflow. Lithosome A, though dominated by massive or horizontally stratified conglomerate, contains abundant, unusually thick and laterally persistent sets of planar cross-stratified conglomerate. These large scale sets were probably deposited by transverse or medial bars that migrated during deep and prolonged floodstages, resulting from humid conditions and valley confinement of the stream. Those climatic and geomorphologic factors enabled the stream to transport cobbles and boulders over 100 km to the western Bighorn Basin.

Lithosome F is characterized by the Gm or Gt facies assemblages and lacks the abundant large scale conglomerate cross-sets which typify Lithosome A. Sedimentary structures in Lithosome F and the absence of large scale conglomerate cross-sets demonstrate that flow depths on the braidplain were considerably more shallow than in the Lithosome A stream system. The rapid grain size and facies changes associated with Lithosome F are attributed to lateral spread of flow on the extensive braidplain and rapid decrease in paleoslope toward the Tertiary structural axis of the basin.

Possible controls on the episodic nature of quartzite gravel deposition in the Bighorn Basin include tectonic pulses, climatic fluctuations, or a combination of those factors. Change in the lithology of source rocks in Jackson Hole is improbable because vastly thick and areally widespread quartzite conglomerates were available there (the closest available source) from the latest Cretaceous until well into the Eocene. High precipitation was certainly important for generation of the conglomerates. As described above, humid climates have been documented for the Bighorn Basin during the early Tertiary. There is no independent evidence to support any precipitation changes in northwestern Wyoming during this period. Furthermore, there is some evidence (for example, paleosols, plants) against varying precipitation during the early Tertiary. Though climatic fluctuations cannot be entirely eliminated, they do not appear to have been a major control on the episodic nature of quartzite deposition in the western Bighorn Basin.

However, deformation resulting from both extrabasinal and intrabasinal tectonic pulses definitely influenced punctuated influx of exotic gravels into the western Bighorn Basin. Paleocurrent indicators, in conjunction with the dates established for tectonic events in the Jackson Hole region, demonstrate that accumulation of the basal Fort Union conglomerate (Lithosome A) and Lithosome F were related in time to vertical movements on the Basin Creek uplift and the Washakie Range, respectively (Kraus, 1983).

In addition to extrabasinal tectonic events, intrabasinal tectonic activity (specifically rates of local basin subsidence) partly controlled quartzite gravel ingress and accumulation in the Bighorn Basin. A major syntectonic unconformity lies subjacent to Lithosome F. Though in some localities the conglomerate rests directly on the unconformable surface, in most areas the conglomerate is separated from the unconformity, which diminishes in magnitude and eventually dies out to the east (Fig. 10). Bown (1980, 1982) attributed development of the unconformity to an increased rate of subsidence in the basin center with respect to the basin margin, and thus to differential structural elevation of the western basin margin. Those processes resulted in more or less uninterrupted fluvial deposition in the basin center but warping and erosion of the western border. Absence of Bighorn Basin quartzite conglomerates of middle Wasatchian age (middle early Eocene) indicates that the differential warping of the western border of the basin (marked there by a syntectonic unconformity) provided a structural baffle preventing ingress of any quartzite gravel from western sources in Jackson Hole. Reworked Cretaceous palynomorphs in Willwood sediments west of Blue Ridge (D. Nichols, 1982, written communication to T.M. Bown; Fig. 2) demonstrate local epiclastic (intrabasinal) sources for Willwood sediments in this area prior to accumulation of Lithosome F. Due to the occurrence of numerous reworked palynomorphs, this provenance was almost certainly the most proximal Cretaceous rocks available, that is, those in the warped western borderland of the Bighorn Basin.

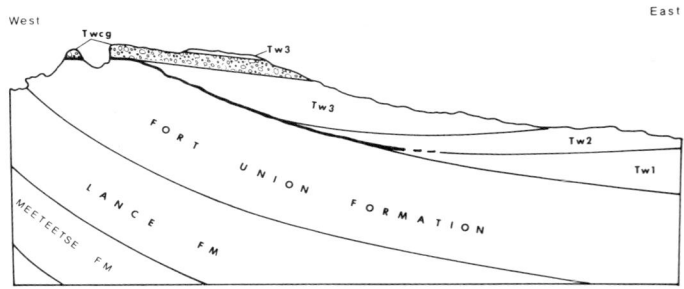

Fig. 10. Schematic cross-section of the western Bighorn Basin illustrating stratigraphic relations of Upper Cretaceous-lower Tertiary sediments below and lower Tertiary sediments above the syntectonic unconformity (heavy line) discussed in text. The beveled, upturned edges of the Lance and Fort Union Formations were overlapped by either Willwood mudstones and sandstones or quartzite conglomerates. Tw1- early Wasatchian Willwood sediments; Tw2- middle Wasatchian Willwood sediments; Tw3- late Wasatchian Willwood sediments; Twcg- late Wasatchian Willwood quartzite conglomerate (Lithosome F).

Subsidence of the central Bighorn Basin slowed relative to the western margin during late Wasatchian time. Younger Willwood deposits filled the basin center and eventually overlapped the upturned western margin (Fig. 10). Strike valleys between hogbacks of upturned Paleocene and older rocks were filled with mudstones and sandstones, and a broad alluvial plain developed (Fig. 10). Renewed depression of the entire Bighorn Basin and concomitant increased uplift of the Washakie Range developed steeper paleoslopes along the western margin of the basin, allowing, once again, ingress of extrabasinal gravel. The quartzite gravel was deposited on an extensive braidplain above valley fill mudstones and sandstones of the Willwood Formation, or in some cases, directly on the beveled surface of Paleocene and older sediments.

Syntectonic unconformities (like that in the western Bighorn Basin) have commonly been attributed to uplift of a basin margin with contemporaneous sedimentation in the undisturbed basin center (*e.g.*, Riba, 1976; Miall, 1978b). Various combinations of active uplift and subsidence can generate syntectonic unconformities, including (1) stability of the basin margin and basin center subsidence, (2) basin margin uplift with basin center stability, and (3) margin uplift accompanied by basin center downwarp. Each permutation results in differential elevation of the basin margin and continued sedimentation in the basin center; however, the unconformities that develop will have different geometry. In the case of the Willwood syntectonic unconformity, younger sediments, including the quartzite conglomerate of extrabasinal provenance, overlap this unconformity along the western border of the basin (Fig. 10). This geometry is ascribed to an increase in the rate of subsidence of the basin interior compared to the western basin margin (Kraus and Bown, in prep.), followed by quiescence, alluvial fill and overlap. Consequently, the Willwood example documents a syntectonic unconformity that developed as a result of differential rates of basin subsidence in different areas of the basin, rather than as a result of active margin uplift.

References

Abbott, P.L. and Peterson, G.L. 1978. Effects of abrasion durability conglomerate clast populations; examples from Cretaceous and Eocene conglomerates of the Sand Deigo area, California. Journal of Sedimentary Petrology, v. 48, p. 31-42.

Antweiler, J.C. and Love, J.D. 1967. Gold-bearing sedimentary rocks in northwest Wyoming — a preliminary report. United States Geological Survey Circular 541, 12p.

Bluck, B.J. 1971. Sedimentation in the meandering River Endrick. Scottish Journal of Geology, v. 7, p. 93-138.

Bluck, B.J. 1976. Sedimentation in some Scottish rivers of low sinuosity. Transactions of the Royal Society of Edinburgh, v. 69, p. 425-456.

Boothroyd, J.C. and Ashley, G.M. 1975. Processes, bar morphology, and sedimentary structures on braided outwash fans, northeastern Gulf of Alaska. *In*: Jopling, A.V. and McDonald, B.C. (Eds.), Glaciofluvial and Glaciolacustrine Sedimentation. Society of Economic Paleontologists and Mineralogists, Special Publication 23, p. 193-222.

Boothroyd, J.C. and Nummedal, D. 1978. Proglacial braided outwash: a model for humid alluvial fan deposits. *In*: Miall, A.D. (Ed.), Fluvial Sedimentology. Canadian Society of Petroleum Geologists, Memoir 5, p. 641-668.

Bown, T.M. 1979. Geology and mammalian paleontology of the Sand Creek Facies lower Willwood Formation (lower Eocene), Washakie County, Wyoming. Geological Survey of Wyoming, Memoir 2, 186p.

Bown, T.M. 1980. Summary of latest Cretaceous and Cenozoic sedimentary, tectonic and erosional events, Bighorn Basin, Wyoming. *In*: Gingerich, P.D. (Ed.), Early Cenozoic Paleontology and Stratigraphy of the Bighorn Basin Wyoming, 1880-1980. University of Michigan Papers on Paleontology No. 24, p. 25-32.

Bown, T.M. 1982. Geology, paleontology, and correlation of Eocene volcaniclastic rocks, southeast Absaroka Range, Hot Springs County, Wyoming. United States Geological Survey Professional Paper 1201-A, 75p.

Bown, T.M. and Kraus, M.J. 1981. Lower Eocene alluvial paleosols (Willwood Formation, northwest Wyoming, U.S.A.) and their significance for paleoecology, paleoclimatology, and basin analysis. Palaeogeography, Palaeoclimatology, Palaeoecology, v. 34, p. 1-30.

Eckelmann, F.D. and Poldervaart, A. 1957. Geologic evolution of the Beartooth Mountains, Part 1: Archean history of the Quad Creek area. Geological Society of America Bulletin, v. 68, p. 1225-1262.

Eynon, G. and Walker, R.G. 1974. Facies relationships in Pleistocene outwash gravels, southern Ontario: a model for bar growth in braided rivers. Sedimentology, v. 21, p. 43-70.

Hewett, D.F. 1926. Geology and oil and coal resources of the Oregon Basin, Meeteetse, and Grass Creek Basin quadrangles Wyoming. United States Geological Survey Professional Paper 145, 107p.

Hewett, D.F. and Lupton, C.T. 1917. Anticlines in the southern part of the Big Horn Basin, Wyoming. United States Geological Survey Bulletin 656, p. 1-192.

Hickey, L.J. 1980. Paleocene stratigraphy and flora of the Clark's Fork Basin. *In*: Gingerich, P.D. (Ed.), Early Cenozoic Paleontology and Stratigraphy of the Bighorn Basin, Wyoming, 1880-1980. University of Michigan Papers on Paleontology, No. 24, p. 33-50.

Keefer, W.R. 1965. Geologic history of Wind River Basin, central Wyoming. American Association of Petroleum Geologists Bulletin, v. 49, p. 1878-1892.

Kraus, M.J. 1983. Genesis of Early Tertiary exotic metaquartzite conglomerates in the western Bighorn Basin, northwest Wyoming. Ph.D. Thesis, The University of Colorado, Boulder, 158p.

Kraus, M.J. and Bown, T.M. 1982. Alluvial paleosols: recognition and significance for paleoenvironmental reconstruction and basin analysis. 11th International Sedimentological Congress, Hamilton, Canada, Abstracts of Papers, p. 13.

Lindsey, D.A. 1972. Sedimentary petrology and paleocurrents of the Harebell Formation, Pinyon Conglomerate, and associated coarse clastic deposits, northwestern Wyoming. United States Geological Survey Professional Paper 734-B, 68p.

Love, J.D. 1939. Geology along the southern margin of the Absaroka Range, Wyoming. Geological Society of America, Special Paper 20, 133p.

Love, J.D. 1956. New geologic formation names in Jackson Hole, Teton County, northwestern Wyoming. American Association of Petroleum Geologists Bulletin, v. 40, p. 1899-1914.

Love, J.D. 1960. Cenozoic sedimentation and crustal movement in Wyoming. American Journal of Science, v. 258-A, p. 204-214.

Love, J.D. 1973. Harebell Formation (Upper Cretaceous) and Pinyon Conglomerate (uppermost Cretaceous and Paleocene), northwestern Wyoming. United States Geological Survey Professional Paper 734-A, 54p.

Love, J.D. and Keefer, W.R. 1969. Basin Creek uplift and Heart Lake Conglomerate, southern Yellowstone National Park, Wyoming. United States Geological Survey Professional Paper 650-D, p. 122-130.

Love, J.D. and Keefer, W.R. 1975. Geology of sedimentary rocks in southern Yellowstone National Park, Wyoming. United States Geological Survey Professional Paper 729-D, 60p.

Love, J.D., McGrew, P.O. and Thomas, H.D. 1963. Relationship of latest Cretaceous and Tertiary deposition and deformation to oil and gas in Wyoming. *In*: Childs, O.E. (Ed.), Backbone of the Americas. American Association of Petroleum Geologists, Memoir 2. p. 196-208.

Love, L.L., Kudo, A.M. and Love, D.W. 1976. Dacites of Bunsen Peak, the Birch Hills, and the Washakie Needles, northwest Wyoming, and their relationship to the Absaroka volcanic field, Wyoming - Montana. Geological Society of America Bulletin, v. 87, p. 1455-1462.

McLean, J.R. 1977. The Cadomin Formation: stratigraphy, sedimentology, and tectonic implications. Bulletin of Canadian Petroleum Geology, v. 25, p. 792-827.

Miall, A.D. 1977. A review of the braided-river depositional environment. Earth Science Reviews, v. 13, p. 1-62.

Miall, A.D. 1978a. Lithofacies types and vertical profile models in braided river deposits: a summary. *In*: Miall, A.D. (Ed.), Fluvial Sedimentology. Canadian Society of Petroleum Geologists, Memoir 5, p. 597-604.

Miall, A.D. 1978b. Tectonic setting and syndepositional deformation of molasse and other nonmarine-paralic sedimentary basins. Canadian Journal of Earth Sciences. v. 15, p. 1613-1632.

Neasham, J.W. 1970. Sedimentology of the Willwood Formation (lower Eocene): an alluvial molasse facies in northwestern Wyoming, U.S.A. Ph.D. Thesis. Iowa State University, Ames, 98p.

Osterwald, F.W. and Dean, B.G. 1961. Relation of uranium deposits to tectonic pattern of the central Cordilleran foreland. United States Geological Survey Bulletin 1087-I, p. 337-390.

Parkash, B., Sharma, R.P. and Roy, A.K. 1980. The Siwalik Group (molasse) — sediments shed by collision of continental plates. Sedimentary Geology, v. 25, p. 127-159.

Pettijohn, F.J. 1975. Sedimentary Rocks (3rd. edition). New York: Harper and Row, 628p.

Pierce, W.G. (compiler) 1978. Geologic map of the Cody 1° x 2° quadrangle, Wyoming. United States Geological Survey Miscellaneous Field Studies Map MF-963.

Pierce, W.G. and Andrews, D.A. 1941. Geology and oil and coal resources of the region south of Cody, Park County, Wyoming. United States Geological Survey Bulletin 921-B, 180p.

Riba, O. 1976. Syntectonic unconformities of the Alto Carderner, Spanish Pyrenees: a genetic interpretation. Sedimentary Geology, v. 15, p. 213-233.

Ruppel, E.T. 1975. Precambrian Y sedimentary rocks in east-central Idaho. United States Geological Survey Professional Paper 889-A, 23p.

Ruppel, E.T. 1978. Medicine Lodge thrust system, east-central Idaho and southwest Montana. United States Geological Survey Professional Paper 1031, 23p.

Rust, B.R. 1972. Structure and process in a braided river. Sedimentology, v. 18, p. 221-245.

Rust, B.R. 1975, Fabric and structure in glaciofluvial gravels. *In*: Jopling, A.V. and McDonald, B.C. (Eds.), Glaciofluvial and Glaciolacustrine Sedimentation. Society of Economic Paleontologists and Mineralogists, Special Publication 23, p. 238-248.

Rust, B.R. 1978. Depositional models for braided alluvium. *In*: Miall, A.D. (Ed.), Fluvial Sedimentology. Canadian Society of Petroleum Geologists, Memoir 5, p. 605-625.

Rust, B.R. 1979. Coarse alluvial deposits. *In*: Walker, R.G. (Ed.), Facies Models. Geoscience Canada Reprint Series 1, p. 9-22.

Ryder, R.T. 1967. Lithosomes in the Beaverhead Formation, Montana-Idaho: a preliminary report. Montana Geological Society, 18th Annual Field Conference, Guidebook, p. 63-70.

Ryder, R.T. and Ames, H.T. 1970. Palynology and age of the Beaverhead Formation and their tectonic implications in the Lima region, Montana-Idaho. American Association of Petroleum Geologists Bulletin, v. 54, p. 1155-1171.

Ryder, R.T. and Scholten, R. 1973. Syntectonic conglomerates in southwestern Montana — their nature, origin, and tectonic significance. Geological Society of America Bulletin, v 84, p. 773-796.

Smith, N.D. 1974. Sedimentology and bar formation in the upper Kicking Horse River, a braided outwash stream. Journal of Geology, v. 82, p. 3407-3420.

Steel, R.J. and Thompson, D.B. 1983. Structures and textures in Triassic braided stream conglomerates ('Bunter' Pebble Beds) in the Sherwood Sandstone Group, North Stafforshire, England. Sedimentology, v. 30, p. 341-367.

Thom, W.T. 1952. Structural features of the Big Horn Basin rim. Wyoming Geological Association, 7th Annual Field Conference Guidebook, p. 15-17.

Van Houten, F.B. 1944. Stratigraphy of the Willwood and Tatman Formations in north-western Wyoming. Geological Society of America Bulletin, v. 55, p. 165-210.

Wilson, M.D. 1970. Upper Cretaceous-Paleocene synorogenic conglomerates of southwestern Montana. American Association of Petroleum Geologists Bulletin, v. 54, p. 1843-1867.

Wing, S.L. 1980. Fossil floras and plant-bearing beds of the central Bighorn Basin. *In*: Gingerich, P.D. (Ed.), Early Cenozoic Paleontology and Stratigraphy of the Bighorn Basin. Wyoming, 1880-1980. University of Michigan Papers on Paleontology, No. 24, p. 119-126.

Young, M. 1972. Willwood metaquartzite conglomerate in a southwestern portion of the Bighorn Basin, Wyoming. M.S. Thesis, Iowa State University, Ames, 71p.

ANCIENT FAN-DELTA SYSTEMS

TECTONIC SETTING, RECOGNITION AND HYDROCARBON RESERVOIR POTENTIAL OF FAN-DELTA DEPOSITS

Frank G. Ethridge[1] and William A. Wescott[1]

Abstract

Holocene fan-delta systems form along narrow coastal plains in close proximity to high-relief, relatively young mountains. Fan morphology is best developed along microtidal coastal zones characterized by abundant rainfall. Fan-delta deposits are fault-bounded on their proximal margins and consist of gravel and gravelly sands that are usually immature to submature texturally and mineralogically.

Fan-deltas are common Holocene depositional systems along divergent plate margins, convergent plate margins, and strike-slip margins. Ancient fan-delta deposits are generally found along the margins of tectonically active basins analogous to Holocene settings as well as along the margins of some intracratonic basins.

The differentiation of fluvial and beach conglomerates, conglomeratic sandstones, and sandstones is crucial to the recognition and paleogeographic reconstruction of ancient fan-delta systems. A series of criteria useful in core and outcrop have been developed for this purpose.

In general, distinguishing features of channel deposits are poor sorting, less well segregated/more lenticular bedding, erosional contacts, small-scale fining-upward sequences, clayey and coaly laminae, gravelly sandstones, coarser clast size with higher sphericity, lower roundness and upcurrent dipping imbrication. Horizontal beds and swash laminae are common in beach deposits, and large-scale trough and/or planar crossbeds and horizontal beds are common in braided fluvial deposits. Also important is the differentiation of fan-delta and submarine fan deposits. In fan-delta deposits conglomerates are usually more abundant, nongraded, and clast supported. Nonmarine fan-delta deposits are more poorly sorted, have less well developed bedding continuity and regularity, and lack graded beds, Bouma sequences, and slump fold strata.

Models for recognition of fan-delta deposits are designed to differentiate slope, shelf, and Gilbert-type sequences. Slope-type fan-deltas are truncated by the shelf-slope break and have poorly developed coarsening-upward trends. Shelf-type fan deltas are more fully developed and display well developed coarsening-upward trends. Gilbert-type fan-deltas have gravelly foreset beds and are known only from some ancient intracratonic basins.

The importance of fan-delta deposits as hydrocarbon reservoirs is just being realized. Productive reservoirs in fan-delta deposits have variable porosity and permeability and may be difficult to evaluate with conventional wireline logging tools. They are found in divergent plate tectonic and foreland basin settings where combination structural-stratigraphic hydrocarbon traps are common.

Résumé

Des systèmes de cônes de déjection holocènes forment le long de plaines littorales étroites bordant un haut-relief, des montagnes relativement jeunes. Les formes de cônes de déjection les mieux développées se manifestent le long de zones littorales microtidales affectées par une pluviosité abondante. Les dépôts des cônes de déjection ont une marge proximale limitée par des failles et sont composés de gravier et de sable graveleux dont les textures et la minéralogie sont en général immatures ou submatures.

Les cônes de déjection constituent des systèmes de dépôts fréquents de l'Holocène le long des marges de plaques divergentes, des marges de plaques convergentes, et des bordures de décrochements horizontaux. Les dépôts anciens de cônes de déjection sont généralement distribués le long de marges de bassins tectoniquement actifs similaires aux ensembles structuraux de l'Holocène et aussi le long des marges de certains bassins intracratoniques.

La différenciation des conglomérats de plage et fluviatiles, des grès conglomératiques, et des grès est fondamentale pour retracer et reconstituer la paléogéographie d'anciens systèmes de cônes alluviaux. Une gamme de critères utlies appliquée aux carottes de sondages et aux affleurements a été établie à cette fin.

En général, les critères distinctifs des dépôts dans les chenaux présentent un faible triage, un litage lenticulaire plus abondant avec ségrégation atténuée près des contacts d'érosion, des séquences à granoclassement normal à petite échelle, des laminations argileuses et charbonneuses, des grès graveleux, des fragments plus grossiers à plus grande sphéricité, avec un indice d'émoussé plus faible et une imbrication inclinée vers l'amont. Des lits horizontaux et des rides de courant sont fréquents dans les dépôts de plage, et des sillons profonds et/ou des corps tabulaires à stratification oblique ou des lits horizontaux sont fréquents dans les dépôts des rivières anastomosées. Il est important de distinguer entre les dépôts de cônes de déjection et ceux de deltas de canyons sous-marins. Dans les cônes de déjection les conglomérats sont généralement plus abondants, non-classés et à fragments jointifs. Les dépôts non-marins sont habituellement caractérisées par un faible triage, un litage moins bien développé, plus discontinu et irrégulier, une absence de granoclassement, des séquences de Bouma, et un plissement intraformationnel des strates.

[continued]

---

[1]Department of Earth Resources, Colorado State University, Fort Collins, Colorado 80523, U.S.A.
[2]Amoco Production Company, P.O. Box 3092, Houston, Texas 77253, U.S.A.

Many of the ideas presented in this review paper were formulated during a study of the Holocene Yallahs delta in Jamaica. Research on the Yallahs delta was funded by the Division of Earth Sciences, National Science Foundation (EAR-76-22749). Special thanks are extended to Malcolm Hendry and Grenville Draper who acquainted us with the Yallahs fan delta and the Wagwater and Richmond outcrops in the Blue Mountains of Jamaica. Mr. R.H. Griffen (Chevron Oil Company) and Gerald N. Craig (Amoco Production Company) kindly permitted us to use original photographs shown as Figures 1 and 9B, respectively. William C. Krueger and Richard K. Vessell reviewed an earlier version of the manuscript. Terri Bostedt, Heidi Derr and Sylvia Murphy typed the manuscript and Melanie Keenan drafted the illustrations.

Copyright © 1984 Canadian Society of Petroleum Geologists

Les modèles qui permettent de reconnaître les dépôts de cônes de déjection visent à distinguer les types suivants: de pente, de plate-forme, et à séquences de Gilbert. Les cônes de déjection de type de pente sont tronqués par la pente du rebord continental et exhibent des tendances vers un granoclassement inverse peu apparent. Les cônes de déjection du type plate-forme sont bien développés et montrent des granoclassements inverses bien nets. Les cônes de déjection du type Gilbert possèdent des lits frontaux graveleux et ces cônes ont été observés seulement dans quelques anciens bassins intracratoniques.

L'importance de ces dépôts de cônes de déjection comme réservoirs d'hydrocarbures vient à peine d'être réalisée. Les réservoirs exploités de dépôts de cônes de déjection possèdent des porosités et des perméabilités variables souvent difficiles à mesurer avec les outils conventionnels de carottage par les tiges. Il existe dans des contexte structuraux de plaque tectonique divergente et de bassin frontal où se retrouvent conjuguées une structure et une stratigraphie capables de piéger des hydrocarbures.

INTRODUCTION

Fan-deltas have been described as alluvial fans that prograde into standing bodies of water from adjacent highlands (Holmes, 1965; McGowen, 1970). The essential elements necessary for the development of fan-deltas are high relief adjacent to the coastal zone and high-gradient, bed-load streams that are usually braided to the coast, resulting in a fan-shaped sedimentary deposit comprising subaerial, transitional and subaqueous components (Wescott and Ethridge, 1980; Fig. 1). The classic fan-shaped morphology is best developed in areas where coastal processes are classified as microtidal (Wescott and Ethridge, 1980). However, fan-deltas also occur in macrotidal settings such as along the south coast of the Bristol Channel, North Devon, Great Britain (Holmes, 1965) and along both sides of lower Cook Inlet, Alaska (Hayes and Michel, 1982). Fan-delta deposits are composed of gravel and coarse sand-size sediments that are usually immature mineralogically and submature to immature texturally.

Other terms that are more or less synonymous with the term fan-delta include: sea-marginal fans (Friedman and Sanders, 1978) and coastal alluvial fans (Rust, 1979). These terms have not, however, received the wide acceptance of the term fan-delta and should be discarded.

We propose that fan-deltas are common depositional systems in certain tectonic settings and that ancient fan-delta deposits contain significant hydrocarbon reserves (Ethridge and Wescott, 1981). Although a great deal of additional research is needed, some Holocene fan-deltas and fan-delta deposits have been studied sufficiently to provide models for recognition of ancient fan-delta deposits and for inferring depositional processes (Gilbert, 1890; Holmes, 1965; McGowen, 1970; McGowen and Scott, 1974; Boothroyd, 1976; Galloway, 1976; Gvirtzman and Buchbinder, 1978; Wescott, 1979; Wescott and Ethridge, 1980 and 1982; Hempton et al., 1983; Hayes and Michel, 1982; Hayward, 1982; and Hendry, 1982). Documented ancient fan-delta deposits range in age from Pleistocene to Precambrian, and re-examination of ancient coarse-grained, clastic wedges might reveal fan-delta deposits previously interpreted as submarine fans or alluvial fans.

These reassessments are essential for accurate reconstructions of paleogeography and for efficient hydrocarbon exploration and exploitation. The remainder of this paper will, therefore, be devoted to a review of the tectonic setting, recognition, depositional models and hydrocarbon potential of fan-delta deposits.

TECTONIC SETTING

The tectonic and geologic setting of Holocene fan-deltas are reviewed by Wescott and Ethridge (1980) in terms of major types of tectonic coastlines (Inman and Nordstrom, 1971; Davies, 1973). In this paper we will attempt to establish the tectonic settings of ancient and Holocene fan-deltas in terms of a tectonic framework of depositional basins (Dickinson, 1974; Mitchell and Reading, 1978; Miall, 1981).

Fan-deltas are located principally along or near the margins of plate boundaries and along the margins of fault-bounded intracratonic seas and lakes. In general, these settings are illustrated by two basin-fill models (Fig. 2; Miall, 1981). In both models fan-deltas are characterized as proximal, transverse elements. Rivers that supply sediment to the fan systems occupy small drainage basins where river length is measured in tens of kilometers or less, drainage patterns are transverse to the structural grain and local relief adjacent to standing bodies of water is high. The first model (Fig. 2A) is characterized by a series of transverse fan-deltas while the second model (Fig. 2B) illustrates a longitudinal trunk stream and a transverse fan-delta. In terms of plate tectonic terminology these depositional systems are common along divergent plate margins, convergent plate boundaries, strike-slip plate margins and one type of intracratonic basin margin.

Fig. 1. Holocene fan-delta, southern coast of Espiritu Santo, New Hebrides Islands (photo courtesy of R.H. Griffen, Chevron Oil Company, Denver).

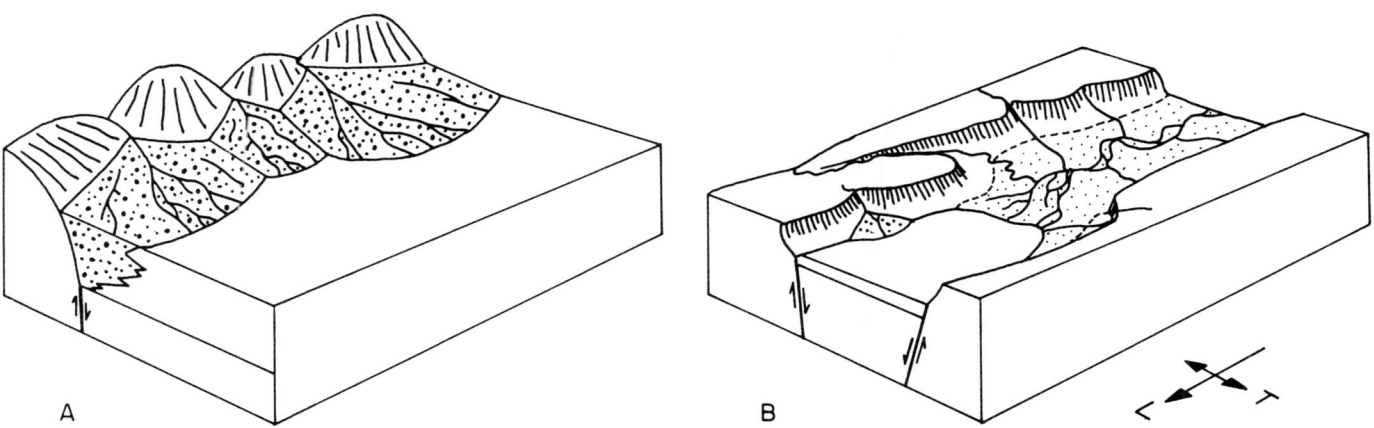

Fig. 2. Generalized alluvial basin-fill models (modified from Fig. 2 of Miall, 1981)
(A) Miall's model number 2. Proximal deposits are transverse fan-deltas and distal deposits are lake margin or sea coast.
(B) Miall's model number 6. Proximal deposits are transverse fan-deltas or rivers, medial deposits are longitudinal rivers and distal deposits are lake margin or lacustrine.

Divergent plate margins include pre-drift rift grabens and aulacogens. Holocene fan-deltas in predrift rift grabens include gravel fans on the northeast coast of Baja, California (Thompson, 1968; Meckel, 1975) and fans along the margins of the Red Sea (Hayward, 1982). Ancient pre-drift rift basin fan-delta deposits are described from Miocene deposits of southwestern Turkey (Hayward, 1982 and 1983) and from late Pleistocene deposits along the western margin of the Dead Sea (Sneh, 1979). Two well known examples of aulacogens that contain fan-delta deposits are the southern Oklahoma and North Sea (Becker, 1977; Dutton, 1982b; Harms et al., 1981; and Harms and McMichael, 1983; Webster, 1980). In the southern Oklahoma aulacogen fan-delta deposits are reported from the northeast and southwest margins of the Anarillo-Wichita uplifts (Fig. 3).

Convergent plate tectonic settings include forearc, interarc, backarc, and retroarc or foreland basins. Fan-deltas along Holocene convergent plates margins are found along the southern coast of Alaska (Reimnitz, 1966; Boothroyd and Ashley, 1975; Boothroyd, 1976; Galloway, 1976; Boothroyd and Nummedal, 1978; Hayes and Michel, 1982) and along the north coast of Alaska (McDonald and Lewis, 1973). In lower Cook Inlet, Alaska (Fig. 4), a forearc basin, fan-deltas have varying morphologies which are systematically distributed along this rock enclosed embayment (Hayes and Michel, 1982). Fan-delta deposits along ancient convergent plate boundaries are described by Brown et al. (1973), Erxleben (1975), Flores (1975, 1978), Hanford and Dutton (1980); Harbaugh and Dickinson (1981), Ricci Lucchi et al. (1981a, b), Ogliani (1981), Ori and Ricci Lucchi (1981), Rainone et al. (1981), Cotter (1983), Blakey and Gubitosa (1983), and Wescott and Ethridge (1983).

Strike-slip fault systems have elements of both compression and extension in close proximity, resulting in adja-

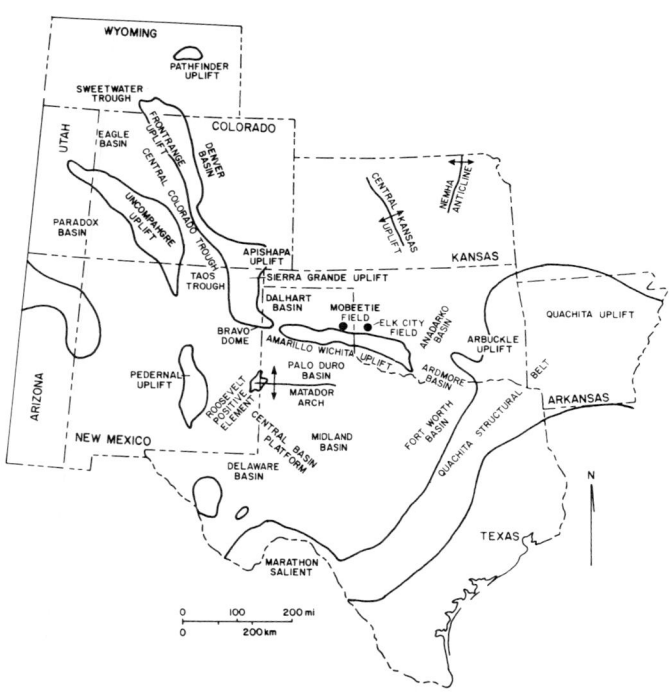

Fig. 3. Mid-Pennsylvanian (Desmoinesian) structural elements and basins of west-central United States (modified from Fig. 2 of Dutton, 1982b, and McKee and Crosby, 1975).

cent uplifted source terrains and subsiding pull-apart basins (Mitchell and Reading, 1978). Holocene fan-delta deposits within the continental lacustrine, Lake Hazar pull-apart basin in Turkey are described by Hempton et al. (1983). An oceanic pull-apart basin is characterized by the Yallahs basin, southeastern Jamaica (Burke, 1967; Wescott, 1979; and Wescott and Ethridge, 1980 and 1982; Fig. 5). Examples of fan-delta deposits associated with ancient continental or continental-margin strike-slip settings are reported by Crowell (1975); Link and Osborne (1978); Howell and Link (1979); Squires (1981); Long (1981); Gloppen and Steel (1981); Pollard et al. (1982); McLaughlin and Nilsen (1982).

Fan-delta deposits, although not generally found in most types of intracratonic basin-fill sequences, are common in one type of basin associated with block uplifts (e.g., the Pennsylvanian ancestral Frontrange and Uncompahgre uplifts of Colorado and adjacent states; Fig. 3). These basins and adjacent uplifts, bounded by narrow zones of faults, are enigmatic in terms of plate tectonic theory because they are located in an intracratonic and interplate setting some 1500 km from any contemporaneous plate margin (Kluth and Coney, 1981). Fan-deltas along the margins of these intracratonic basins have been described by Fredrickson (1978), Casey (1979, 1980), Casey and Scott (1979), Walker and Harms (1980), Langford (1982), Phillips (1982), McGowen et al. (1983), Millberry (1983), Carr and Scott (1983).

A review of the literature listed above provides a few generalizations concerning the nature of fan-delta deposits and their relation to adjacent deposits in various tectonic settings.

Fan-delta deposits:

(1) are common depositional systems along active plate margins and fault-bounded intracratonic seas and lakes and are not common along inactive continental margins or along the margins of most other types of intracratonic, intraplate basins;

(2) along the margins of pre-drift rift basins are commonly interbedded with shallow marine reef carbonate deposits;

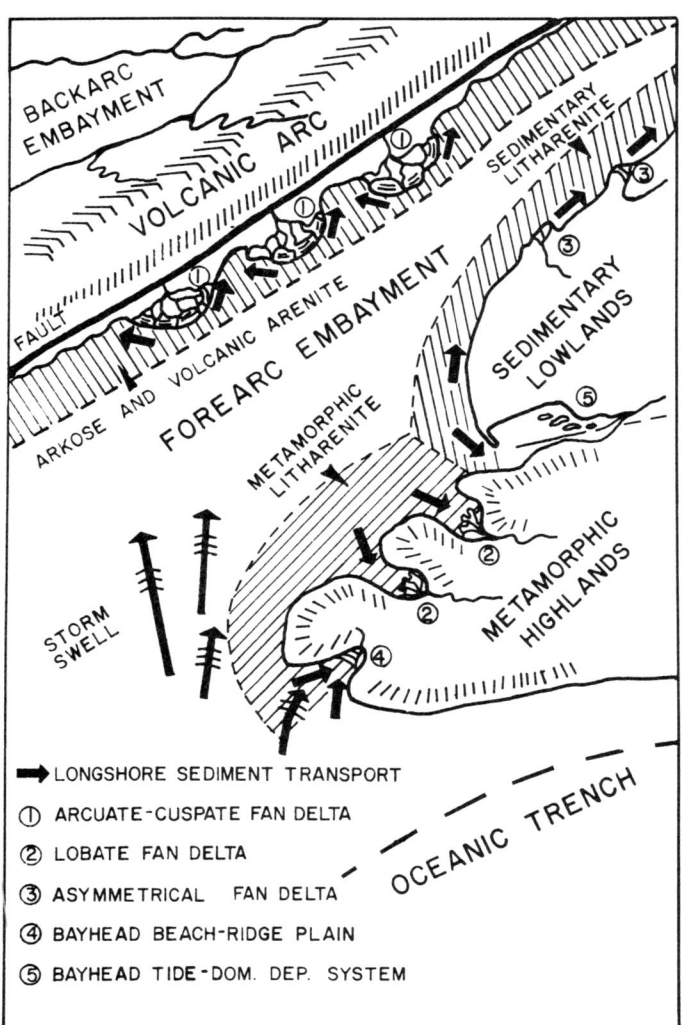

Fig. 4. Generalized depositional setting of the shoreline of lower Cook Inlet, a forearc embayment affiliated with the northeasterly extension of the Aleutian volcanic arc (after Hayes and Michel, 1982).

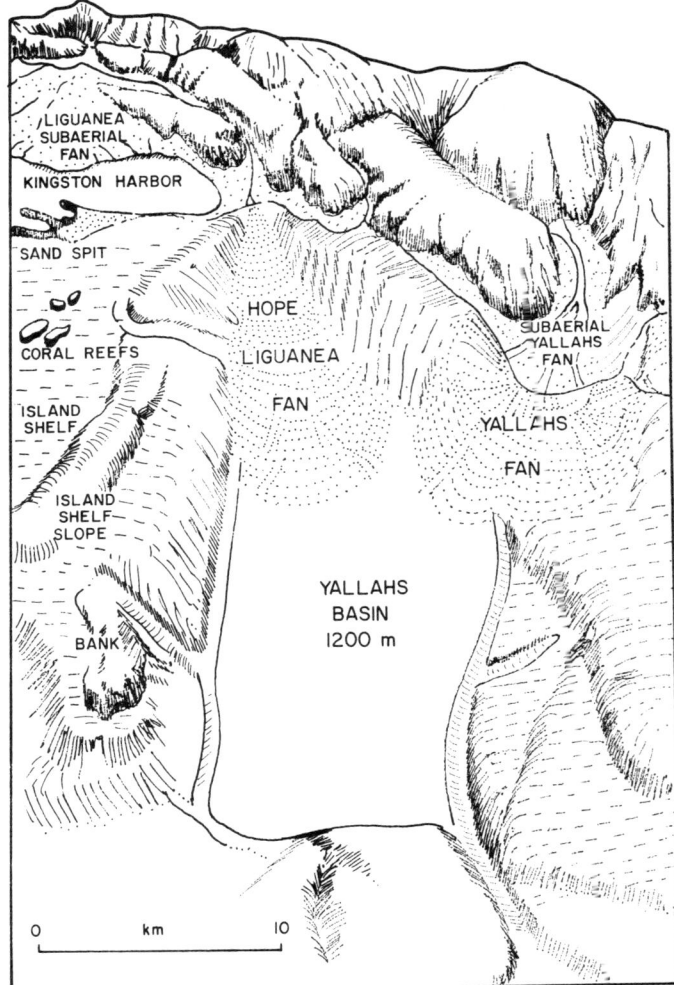

Fig. 5. Sketch map of Yallahs basin and adjacent southeastern Jamaica showing the subaerial Yallahs fan-delta and adjacent submarine Yallahs fan and subaerial Liguanea fan-delta and adjacent island shelf (modified from Burke, 1967).

(3) along the margins of aulacogens, convergent plate boundaries and oceanic or continental margin strike-slip systems may grade seaward over relatively short distances (<5km) into submarine fan and turbidite deposits or may grade laterally into shallow water shelf deposits;

(4) on the margins of continental, strike-slip basins commonly grade into and are interbedded with lacustrine deposits;

(5) along fault-bounded, intracratonic, intraplate basins (such as those associated with the ancestral uplifts in Colorado and adjacent states) are often characterized by large-scale, gravel foreset beds.

RECOGNITION

INTRODUCTION

McGowen and Scott (1974) reviewed some important aspects in the recognition of fan-deltas in the rock record. They emphasized that alluvial fans and the subaerial parts of the fan-deltas formed by similar processes and that both could be classified as arid- or humid-region fans. Arid-region fans are constructed by combinations of debris flow (Gnaccolini, 1982; Hayward and Ballance, 1982), sieve deposition and fluvial deposition while humid region fans are dominated by fluvial processes. Fan-delta deposits can be distinguished from alluvial fan deposits by a distinctive distal facies resulting from the interplay of fluvial and marine processes (Houseknecht and Ethridge, 1978). Finally, they recognized that this diagnostic distal facies can vary significantly with basin configuration, intensity of wave and tidal currents and nature of associated depositional facies. For accurate reconstruction of the paleogeography of conglomeratic sequences in the rock record, criteria must be established for differentiating subaerial alluvial fan deposits, subaerial and subaqueous fan-delta deposits, and submarine fan deposits.

BEACH AND FLUVIAL GRAVELS

One important aspect to the differentiation of alluvial fan and fan-delta deposits is the recognition of beach and braided fluvial channel gravels in the latter and only braided fluvial channel gravels in the former. Various criteria for distinguishing fluvial and beach gravels are reviewed by Dobkins and Folk (1970), Clifton (1973), Wescott and Ethridge (1980), Ethridge and Wescott (1981), Leckie and Walker (1982), Maejima (1982), and Bluck (1982) and are summarized in Table 1. Some of these criteria are useful only in outcrop exposures. In general, beach deposits are better sorted (Fig. 6), have better developed bed segregation (Fig. 7) and individual beds are more continuous laterally (Fig. 8A). Fluvial channel gravels are commonly interbedded with gravelly sandstones (Fig. 8B), and contain numerous erosional basal contacts (Fig. 9A) and small-scale, fining upward sequences (Fig 9B). Maximum clast size is generally larger in fluvial gravels, and clayey or coaly laminae may be common. Beach gravels tend to exhibit well-developed seaward dipping imbrication (Fig. 10A), and low sphericity and high roundness (Fig. 10B). Horizontal beds and swash laminations are common in beach deposits and large-scale and planar crossbeds, and horizontal beds may be common in braided fluvial deposits. Gravel texture is probably the single most obvious feature that can be used to differentiate beach and fluvial gravels. Processes responsible for producing the distinct differences in texture observed in recent deposits are reviewed by Dobkins and Folk (1970) and by Bluck (1982).

FAN-DELTA VERSUS SUBMARINE FAN DEPOSITS

In contrast to the well established criteria that have been proposed for the differentiation of channel and beach gravels, little attention has been given to the important problem of differentiating submarine fan and fan-delta deposits. Four notable exceptions are recent papers by Howell and Link (1979), Harbaugh and Dickinson (1981), Wescott and Ethridge (1983), and Harms and McMichael (1983).

In their study of Mississippian foreland deposits of Nevada, Harbaugh and Dickinson review existing depositional models for submarine fans and fan-deltas. An updated version of their review is presented in tabular form (Table

		Beach Gravels	Fluvial Gravels
*	1.	Well sorted	Poorly sorted
*	2.	Well segregated beds	Poorly segregated beds
*	3.	Continuous beds	Lenticular beds
*	4.	Gravels interbedded with gravelly sandstone - *rare*	Gravels interbedded with gravelly sandstone — *common*
*	5.	Erosional basal contacts - *rare*	Erosional basal contacts - *common*
*	6.	Repeated small-scale fining-up sequences - *rare*	Repeated small-scale fining-up sequences - *common* to *rare*
*	7.	Maximum clast size smaller than in adjacent channels	Maximum clast size larger than in adjacent beach
*	8.	Clayey or coaly laminae - *rare*	Clayey or coaly laminae - *common* to *rare*
	9.	Imbrication - *seaward dipping*	Imbrication - *landward dipping*
	10.	Sphericity - low	Sphericity - High
		Roundness - high	Roundness - low
*	11.	Horizontal beds of different size gravels or swash laminations	Horizontal beds or high angle trough cross-beds

* denotes criteria that can be identified in core

Table 1. Criteria for distinguishing beach and fluvial channel gravels (based on part on data from Dobkins and Folk, 1970, and Clifton, 1973; after Ethridge and Wescott, 1981).

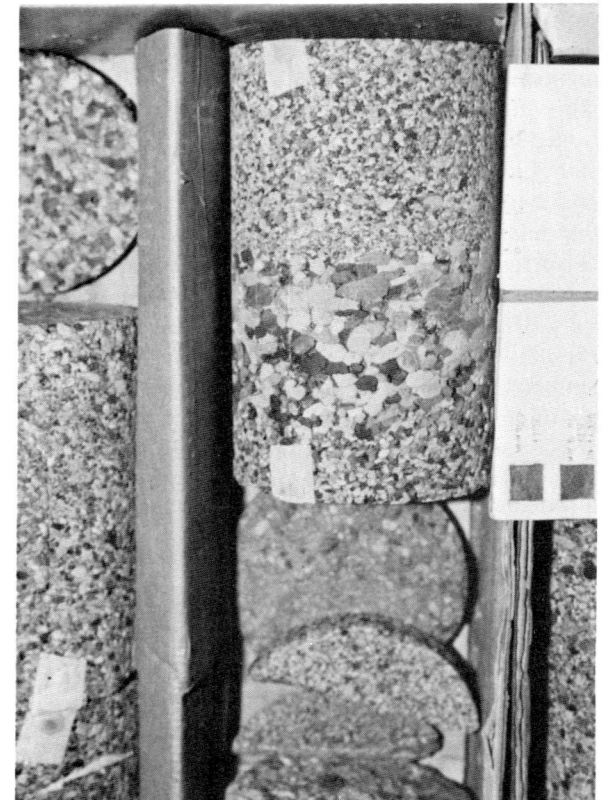 

Fig. 6. **(A)** Well sorted and segregated beach conglomerates and sandstones in Cretaceous core, western Canada. **(B)** Poorly sorted and segregated fluvial sandy conglomerate in Cretaceous core, western Canada.

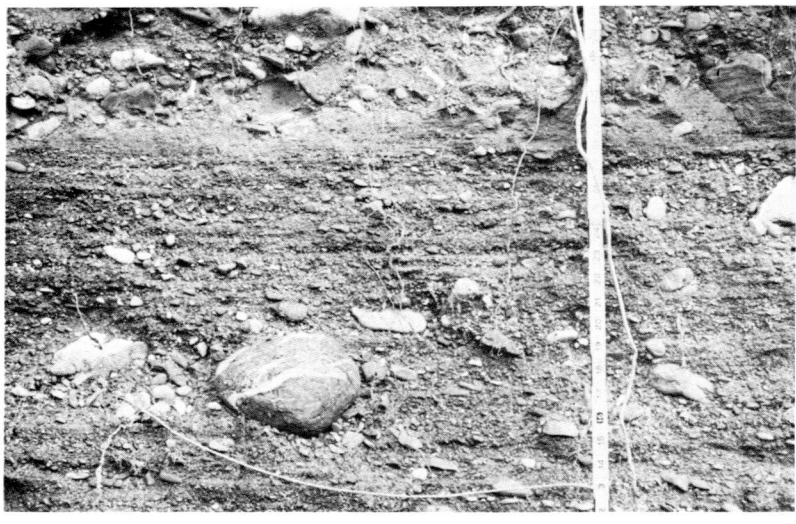 

Fig. 7. **(A)** Poorly sorted and segregated horizontally bedded, longitudinal bar deposits of Yallahs River, Yallahs fan-delta, Jamaica. **(B)** Well segregated and sorted, horizontally bedded beach sands and gravels of the Yallahs fan-delta front, Jamaica (after Wescott and Ethridge, 1980).

2). At first glance the differences between these two depositional models appears to be so great that confusion seems unlikely. On the other hand, the fact that both systems contain poorly-sorted, lenticular conglomerates and sandstones in proximal areas, better sorted and more sheet-like conglomerates and sandstones in distal areas and usually display overall coarsening upward sequences can make interpretation difficult. This difficulty is increased in the subsurface where only limited data may be available. Criteria for differentiating these two depositional systems are summarized in Table 3 using information provided by Howell and Link (1979) from the Southern California Borderland, Harms and McMichael (1983) from the Brae field area, North Sea, and Wescott and Ethridge (1983) from the Wagwater Trough, east-central Jamaica.

MODELS

GENERAL

Wescott and Ethridge (1980) published two stratigraphic models for coarse-grained, humid-region, dominantly progradational, collision-coast fan-deltas. A number of recent papers on fan-delta deposits suggest that two models are inadequate to explain all of the variations found in fandelta deposits. As previously discussed, fan-delta settings include divergent plate margins, strike-slip margins

Fig. 8. (A) Continuous and well-segregated beds of gravel and sand, Yallahs fan-delta front, Jamaica. (B) Discontinuous and poorly segregated braided fluvial deposits, Yallahs fan-delta, Jamaica.

Fig. 9. (A) Erosional contacts in overturned, braided fluvial deposits of the Wagwater (Eocene) fan-delta, Jamaica (after Wescott and Ethridge, 1983). (B) Typical small-scale fining-up cycles in braided stream deposits, Tertiary, Wyoming (photograph courtesy of Gerald N. Craig, II).

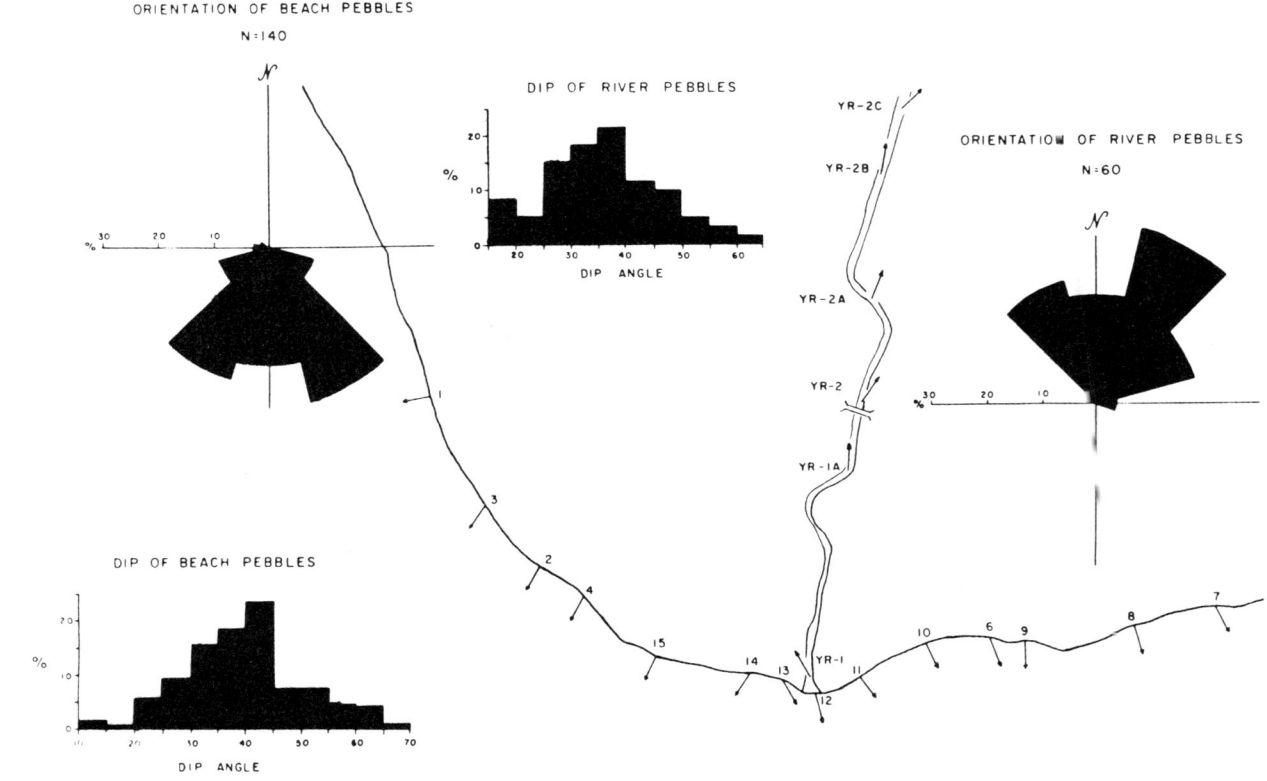

Fig. 10. (A) Sketch map of lower Yallahs fan-delta showing orientations and distributions of imbricated beach and river pebbles. Orientation and distribution shown for 10 pebbles at each sample location (after Figure 29 of Wescott, 1979). (B) Beach gravels exhibiting low roundness and seaward imbrication, southern California coast (photo courtesy of S.A. Schumm).

Submarine Fan	Fan-delta
1. *Slope deposits* - mainly hemipelagic lutites; rare lenticular, massive channel conglomerates and sandstones or olistostromal mudstones and sedimentary breccias.	1. *Subaerial delta plain deposits* resemble alluvial fan deposits and usually consist of braided stream conglomerates and sandstones and/or coarse-grained meanderbelt deposits. Fine-grained floodplain deposits are rare. In arid regions debris flow and/or sieve deposits may constitute a significant fraction of the proximal deposits.
2. *Inner fan deposits* - one or a more discrete channel conglomerates and sandstones with erosional bases and fining and thinning-upward sequences grading into overbank turbidite sandstones and lutites; bulk of deposits consist of hemipelagic lutities interbedded with thin turbidite sandstones.	2. *Delta front and shallow submarine fringe deposits* - depending upon the interplay of fluvial and marine processes and climatic factors, deposits may consist of beaches, spits, tidal flats, shallow marine bars, delta foresets and/or carbonate reefs or algal mounds.
3. *Mid-fan deposits* - bulk of deposits consists of repeated thinning-upward cycles on turbidite sandstones resulting from deposition in multiple shallow braided channels.	3. *Shelf and slope deposits* - in shallow seas shelf deposits consist of muds interbedded with shallow nearshore submarine bar sandstones. Fans that build to the shelf edge overlie slope deposits similar to those described in (1) under submarine fans.
4. *Outer fan deposits* - characteristic deposits consists of hemipelagic lutites and repeated thickening-upward cycles of turbidite sandstone beds which have sheet-like geometries.	

Table 2. Comparison of submarine fan and fan-delta deposits (modified after Harbaugh and Dickinson, 1981).

Criteria		Nonmarine Fan-delta	Shallow Marine	Submarine Fan
1.	Relative proportion of conglomerate to sandstone, mudstone, and shale	100% ———————————————————————————————————— <1%		
2.	Grading in conglomerates	Commonly ungraded ————————————————————————		Commonly normal to inverse grading.
3.	Texture of conglomerates	Clast supported (maybe matrix supported only in some arid region fans).		Matrix supported in slope and very proximal submarine feeder channels, otherwise clast supported.
4.	Imbrication	Landward dipping imbrication (well developed).	Seaward dipping in beach conglomerates.	Landward dipping (some beds show no imbrication).
5.	Orientation of long axes	Random, parallel and perpendicular.		Mostly parallel with some perpendicular.
6.	Rip - up clasts a. size b. shape	Small Well rounded		Variable Angular
7.	Sedimentary Structures	Mostly massive to crude horizontal bedding in coarse conglomerates; plannar or horizontal beds in sandy conglomerates and conglomeratic sandstones.	Beach conglomerates and sandstones have horizontal beds and swash laminae.	Slump and fold strata in proximal slope and feeder channel deposits along with massive to graded conglomerates; Bouma sequences, in mid to distal fan sequences.
8.	Fan slopes		Fan-delta fringe deposits composed of finer-grained material deposited on prodelta slopes. Depending upon the dominant transport mechanism, slope angles can range from a few degrees or less to very steep angle of respose slopes.	No mechanism for building and prograding steep slopes. Modern submarine fans have verly low average gradients; slopes of only a few minutes for larger fans to only slightly over one degree for smaller fans are common.
9.	Other criteria	Evidence of subaerial exposure including; mudcracks, roots, eolian deposits, etc.; mudstones w/caliche horizons (a function of climate); common occurrence of plant fragments; absence of marine fossils in coarser grained units.	Marine fossils and abundant vertical and/or inclined burrow structures in sandstones interbedded with conglomerates.	Abundant deep water micro-fauna; abundant grazing traces on bedding plan exposures; transported fauna in coarser grained units.

Table 3. Major sedimentologic differences between nonmarine fan-delta, shallow marine and deep-water submarine fan deposits. Generalized from data and discussion by Howell and Link (1979), Harms and McMicheal (1983) and Wescott and Ethridge (1983).

and intracratonic rifts as well as convergent plate margins. A minimum of three basic models are needed to explain the various types of fan-delta deposits. These are the slope-, shelf-, and Gilbert-type models. Furthermore, the shelf model shows several variants depending upon the interaction of fluvial and marine or lacustrine processes, climate and associated depositional systems. All fan-deltas, regardless of tectonic setting, have certain characteristics in common. Individual fan-deltas are relatively small physiographic features with the subaerial fan usually occupying an area covering only tens of square kilometers. The landward margin of fan-delta deposits is almost always fault bounded, with proximal deposits unconformably overlying bedrock. The overall geometry of a stratigraphic sequence of fan-delta deposits is a wedge or prism of coarse-grained conglomeratic sediments that thins away from a mountain front. Texturally and mineralogically, fan-delta deposits are immature to submature, reflecting the close proximity of bedrock source areas. Depositional sequences usually coarsen upward, although coarsening-upward is better developed in shelf- and Gilbert-type than in slope-type fan-deltas. The thickness and areal extent of fan-delta deposits depend upon the complex interplay of mountain-front uplift, sediment supply, and basin subsidence. The thickness of individual fan-delta deposits may be on the order of tens of meters; however, thick fan-delta clastic wedges developed along the margins of plate boundaries over considerable periods of geologic time may have cumulative thicknesses on the order of several thousands of meters and extend up to several tens of kilometers from the mountain front.

SLOPE MODEL

A generalized slope model, based on studies of the Holocene Yallahs fan-delta in southeastern Jamaica (Wescott, 1979 and Wescott and Ethridge, 1980) is illustrated by the idealized vertical sequence in Figure 11. This model is characteristic of fan-deltas that prograde onto an island or continental slope. The overall coarsening-upward sequence may not always be as well developed in slope-type fans because: 1) they are truncated by the shelf/slope break; and 2) coarse-grained slump deposits associated with the heads of submarine canyons may be locally developed just seaward of the nonmarine to marine transition zone.

Proximal fan deposits are characterized by poorly sorted, massive conglomerates that result from surge flows in confined channels (McGowen and Groat, 1971). These are interbedded with crudely horizontally bedded, landward-imbricated conglomerates and conglomeratic sandstones that result from deposition on longitudinal bars in shallow braided streams. Distal subaerial fan deposits consist of crudely horizontally bedded, landward-imbricated conglomerates and conglomeratic sandstones, and rare trough cross-bedded conglomeratic sandstones of braided stream origin. Minor amounts of fine-grained, organic deposits may represent deposition in isolated coastal lagoons and/or ponds. The nonmarine-marine transition zone is characterized by well-sorted and segregated conglomerates and sandstones that display horizontal bedding, swash laminations and seaward-dipping clast imbrication. Fossils and bioturbation structures are often abundant in marine sandstones. Slope deposits consist of marine muds and matrix-supported conglomerates which are frequently distorted by slumping. Processes active in these steep slopes include debris flows, mud flows, liquifaction transport and submarine sliding (Prior et al., 1981).

Slope-type fan-delta sequences are characteristic of aulacogens, pre-collision convergent plate margins, and oceanic strike-slip margins (i.e., where continental shelves are narrow). Ancient examples of slope-type fan-delta deposits are the lower Eocene Wagwater Richmond deposits of the Wagwater Trough, eastern Jamaica (Fig. 12; Wescott and Ethridge, 1983), Eocene conglomeratic strata of the Southern California Borderland (Howell and Link, 1979), the Archean Timiskaming Group, northeast Ontario, Canada (Hyde, 1980), the Saint Antoin Conglomerate in the Maritime Alps (Stanley, 1980), and Middle Ordivician strata northeast of Quebec City (Belt and Bassieres, 1981).

SHELF MODEL

Figure 13 is a generalized shelf model based on numerous studies of fans and fan-deltas along the southern Alaska shoreline (Reimnitz, 1966; Boothroyd and Ashley, 1975; Galloway, 1976 and Hayes and Michel, 1982). A typical proximal to distal succession of sedimentary structures on the subaerial portion of a small glacial outwash fan-delta is: (1) poorly bedded, well-imbricated, poorly-sorted, coarse-grained gravels; (2) imbricated, fine-grained gravel, and planar laminated and planar cross-bedded sand; and (3) interbedded planar cross-laminated sand and ripple-drift sand (Boothroyd, 1976). Fans that have not built directly onto the coast often have a distal facies consisting of marsh and meandering stream deposits. The distal margins of fan-deltas that prograde directly into the Gulf of Alaska are reworked into a series of barrier spits. Data from the Copper River fan-delta reveal that the coastal zone is characterized by a broad sand-rich tidal lagoon characterized by active tidal channels and subaqueous to intertidal sand flats. Sediments seaward of this tidal lagoon were deposited in marginal island, beach, barred shoreface

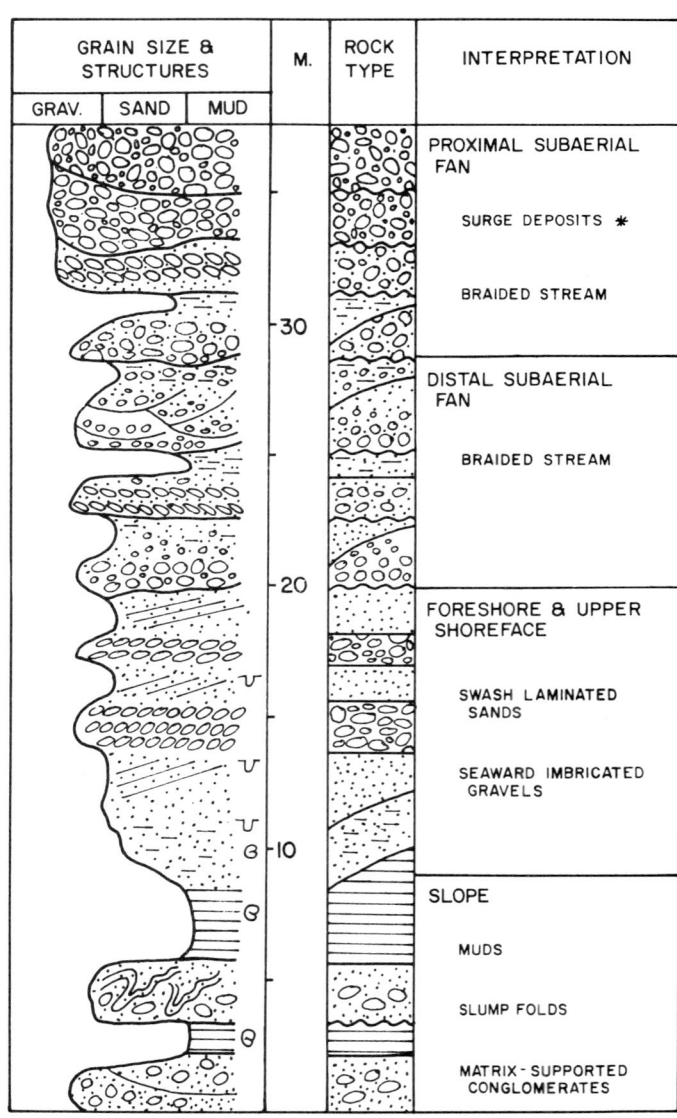

*DEBRIS FLOW OR SIEVE DEPOSITS ON ARID FANS; RIPPLE DRIFT; BURROWS; FOSSILS; ORGANICS

Fig. 11. Vertical sequence in a hypothetical slope-type fan-delta based on studies of the Yallahs fan-delta and slope, Jamaica (Wescott, 1979 and Wescott and Ethridge, 1980 and 1982).

and mid- to lower delta front environments. The entire sequence is presumably underlain by prodelta and shelf muds (Reimnitz, 1966; Galloway, 1976).

Shelf fan-deltas often have better developed coarsening-upward sequences than do slope-type fans (Fortunato and Ward, 1982). Because they are not truncated by the shelf-slope break, they can prograde for longer distances and, therefore, tend to develop sandy braided, coastal lagoon and barrier-spit deposits on their distal margins.

Numerous variants on the shelf-model presented above exist. Holocene and ancient fan-deltas along divergent plate margins and intracratonic rifts often show a close association between coarse-grained clastic delta deposits and carbonate reefs or carbonate algal mounds. These fan-deltas usually build into lagoons fringed by carbonate deposits. An excellent model for this type of shelf fan-delta is given by Brown (1979) for Pennsylvanian fan-delta deposits of the mid-continent (Fig. 14).

Examples of shelf-type fan-deltas in ancient intracratonic basins (Fig. 15A) and lacustrine fan-deltas associated with strike-slip margins (Fig. 15B) are presented by Casey and Scott (1979) and Pollard et al. (1982), respectively. The absence of beach deposits in both models is notable, as is the abundance of trough cross-bedded conglomeratic sandstones in the braided subaerial delta plain deposits. Shelf-type fan-deltas can occur along the margins of all of the plate tectonic settings listed above.

GILBERT-TYPE MODEL

Gilbert-type fan-deltas with large-scale, gravelly foresets have been reported from intracratonic basins in Colorado and New Mexico (Casey and Scott, 1979; Walker and

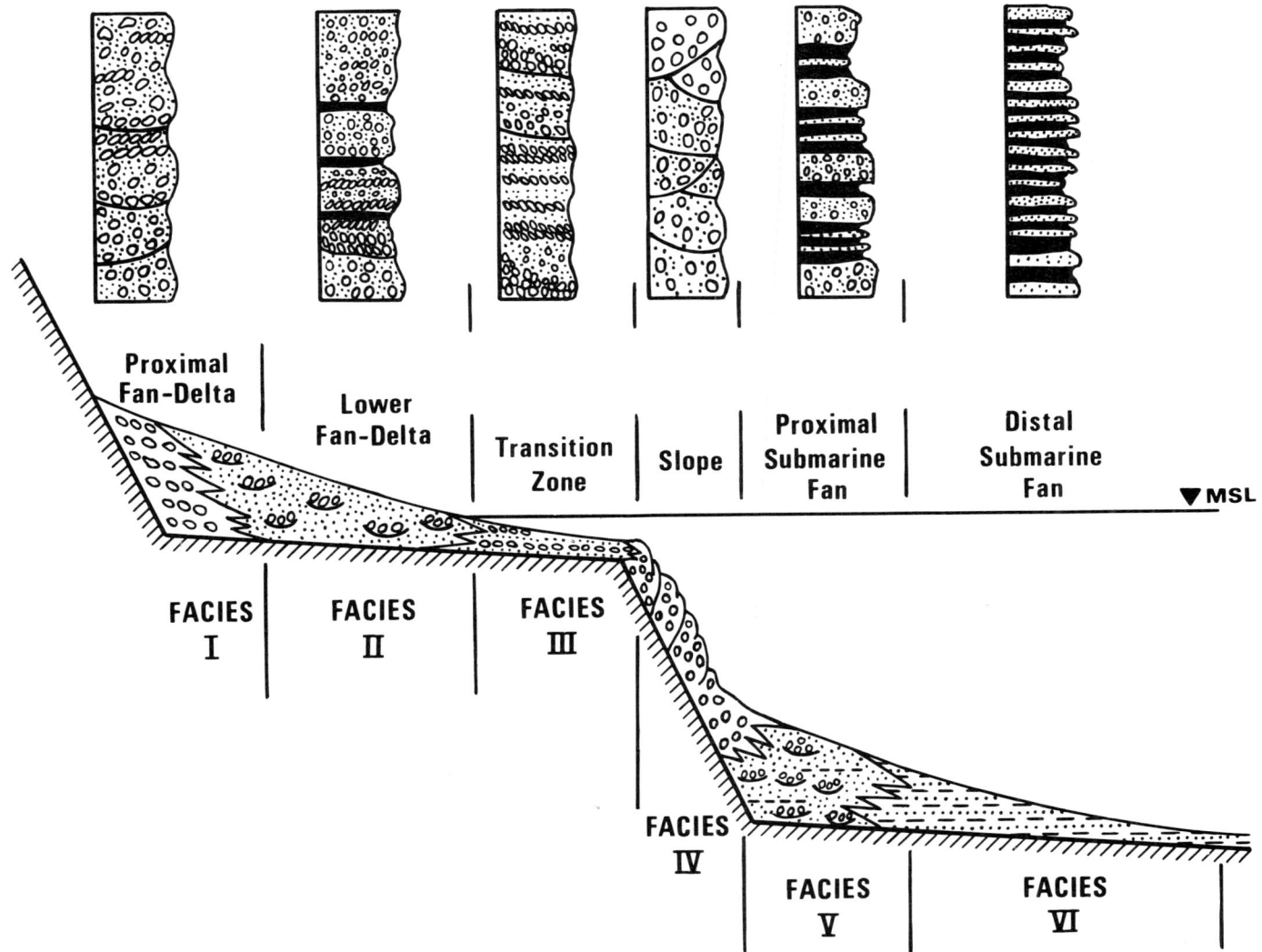

Fig. 12. Depositional model for lower Eocene Wagwater and Richmond Formations, Wagwater Trough, eastern Jamica. Generalized facies relationships and inferred depositional environments are illustrated (after Wescott and Ethridge, 1983).

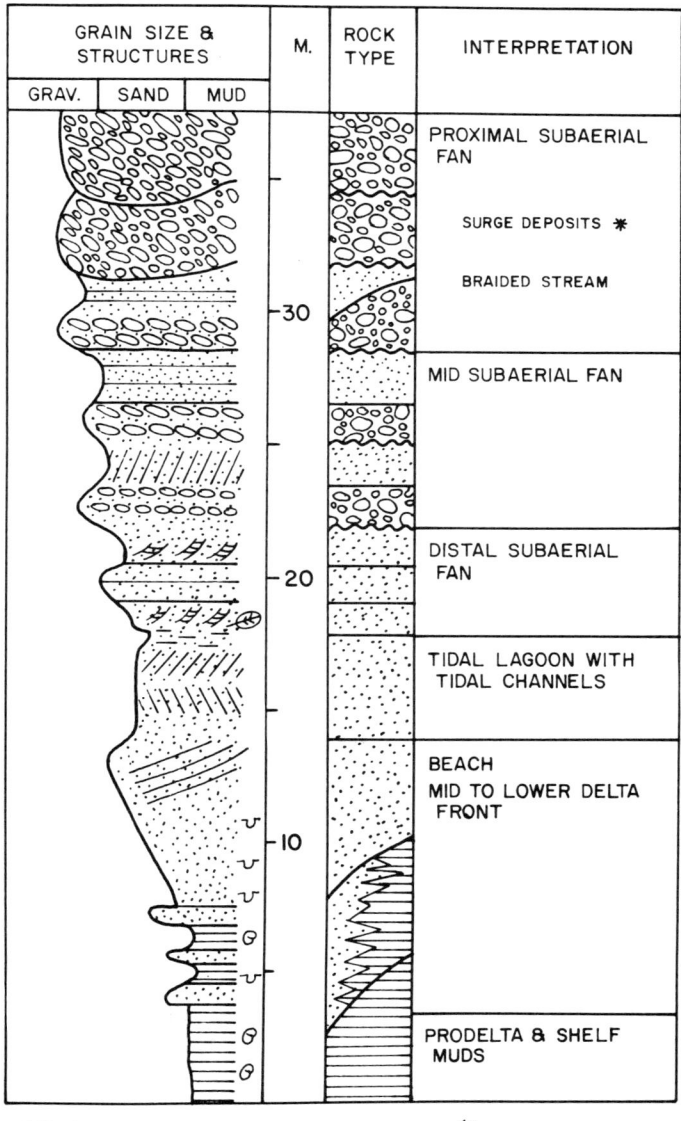

Fig. 13. Vertical sequence in a hypothetical shelf-type fan-delta based on studies of fans and fan-deltas along the southern Alaska coast between 1966 and 1982 (see text for references).

Casey and Scott (1979) believe that the development of foreset beds is related to water depth. When delta progradation occurred over shallow water platforms, delta foresets developed and sheetflooding was common. Foresets also developed in shallow water interdeltaic embayments. During floods, the saline water in these embayments might be displaced, resulting in homopycnal flow. Gilbert-type fan-delta deposits have been reported only from lacustrine and intracratonic basins.

HYDROCARBON POTENTIAL

The importance of fan-delta deposits as hydrocarbon reservoirs is just being realized. Most fields in fan-delta deposits have been found on the basis of structural plays; however, stratigraphic trapping usually plays an important role in individual reservoirs, due in large part to the rapid vertical and lateral facies changes associated with these depositional systems. In tectonically active basins that fill asymmetrically, syndepositional faulting and tilting of the basin fill combined with rapid facies changes may produce structural-stratigraphic traps. This type of trap is documented in fan-delta deposits of the Fort Worth basin (Tai Wai Ng, 1979).

Reservoir evaluation and quality are important considerations in ancient fan-delta deposits. Reservoir evaluation by conventional wireline log suites is, for example, a problem in Pennsylvanian granite wash reservoirs in Oklahoma and the Texas Panhandle, where the usual shaliness indicators are of little value. The conventional gamma ray log is affected by the presence of radioactive minerals and the SP curve is generally poorly developed. Under these conditions, spectral gamma ray logging has proven to be a reliable tool for determining reservoir shaliness and the presence of radioactive minerals (Schenewerk et al., 1981; Frost et al., 1982).

Reservoir quality in many fan-delta deposits is also affected by the bimodal conglomeratic texture of these rocks and the type and distribution of diagenetic cements. Clarke (1979) and Varley (this volume) review the reservoir properties of conglomerates and conglomeratic sandstones. Clarke (1979) demonstrated that the primary porosity of bimodal reservoir rocks may vary by a factor of 3 or 4, depending on the packing and relative proportion of sand and pebble-size grains, and that relations between permeability and porosity are usually quite different from those found in conventional reservoirs with unimodal grain size distributions.

With the exception of two excellent field studies, little is known about the quality of fan-delta reservoirs in relation to specific depositional environments and diagenesis. Elk City field, Oklahoma (Fig. 3) is a typical granite wash, fan-delta reservoir along the southern margin of the Anadarko basin. In a detailed reservoir study, Sneider et al. (1977) divided the field into a number of individual reservoirs based on studies of conventional wireline logs and

Harms, 1980; and Millberry, 1983). Figure 16A illustrates typical large-scale foresets from Pennsylvanian sequences near Talpa, New Mexico and Figure 16B shows an outcrop exposure of similar gravelly foresets in Pennsylvanian rocks from the McCoy, Colorado area. Based on faunal evidence, both of these Gilbert-type fan-delta sequences were prograding into a shallow epicontinental sea. In the Talpa, New Mexico area two types of fan-deltas are recognized. The first type, illustrated by Figure 15A, shows the development of lenticular channel deposits while the second type, illustrated in Figure 16A shows the development of high-angle foresets with sheetflood topset beds.

continous cores. They recognized barrier (delta front?), channel and delta-fringe deposits. An analysis of permeability and porosity patterns revealed that the delta front rocks were characterized by relatively high porosities and that permeabilities increased uniformly upward, parallel to depositional strike. In contrast, channel deposits contained relatively low porosity and high permeability. Rapid grain size changes and lateral variations in rock type, however, resulted in extreme vertical and lateral variations in permeability. Delta-fringe deposits, although somewhat similar to barrier bar deposits, contain thin, widespread siltstone/shale beds that cap regressive cycles and are effective barriers to vertical flow.

Dutton (1982a, b) in a detailed study of Mobeetie field, a granite wash producer in the Texas Panhandle (Fig. 3), determined that fluvial and delta-front deposits have good porosity and permeability and produced oil, while reworked fan-delta deposits are not productive because calcite cementation reduces porosity. Secondary porosity resulting form the dissolution of feldspars and rock fragments constitutes a significant proportion of existing porosity.

Hydrocarbons in ancient fan-delta deposits or deposits that can be interpreted as fan-deltas have been reported from such diverse areas as the southern Oklahoma aulacogen, the North Sea, the eastern coast of Brazil, Libya, Australia, the Northwest Territories of Canada, and the deep basin of Alberta and British Columbia in western Canada. Probably the best known fan-delta reservoirs are those of the Pennsylvanian granite wash and chert conglomerates along the southern margin of the Anadarko basin (Fig. 3; Buckthal, 1977; Cast, 1977; Cambridge, 1977; Dutton, 1981a, b, and 1982a, b; Henslee, 1977; Parker and Gibson, 1977; Sahl, 1970; Shelby, 1976, 1979, 1980; Sneider et al., 1977; and Tyler et al., 1983). Granite wash deposits are an important potential reservoir facies in the Palo Duro basin south of the Amarillo-Wichita uplift (Fig. 3; Dutton, 1979a, b, 1980a, b; Hanford, 1980; Hanford and Dutton, 1980; Hanford et al., 1980; and McCaslin, 1982). Oil is also produced from distal fan-delta deposits of Pennsylvanian age along the Midland basin (Erxleben, 1975) and in the Fort Worth basin (Tai Wai Ng, 1979), Texas. In the Viking Graben area of the North Sea, hydrocarbon accumulations are

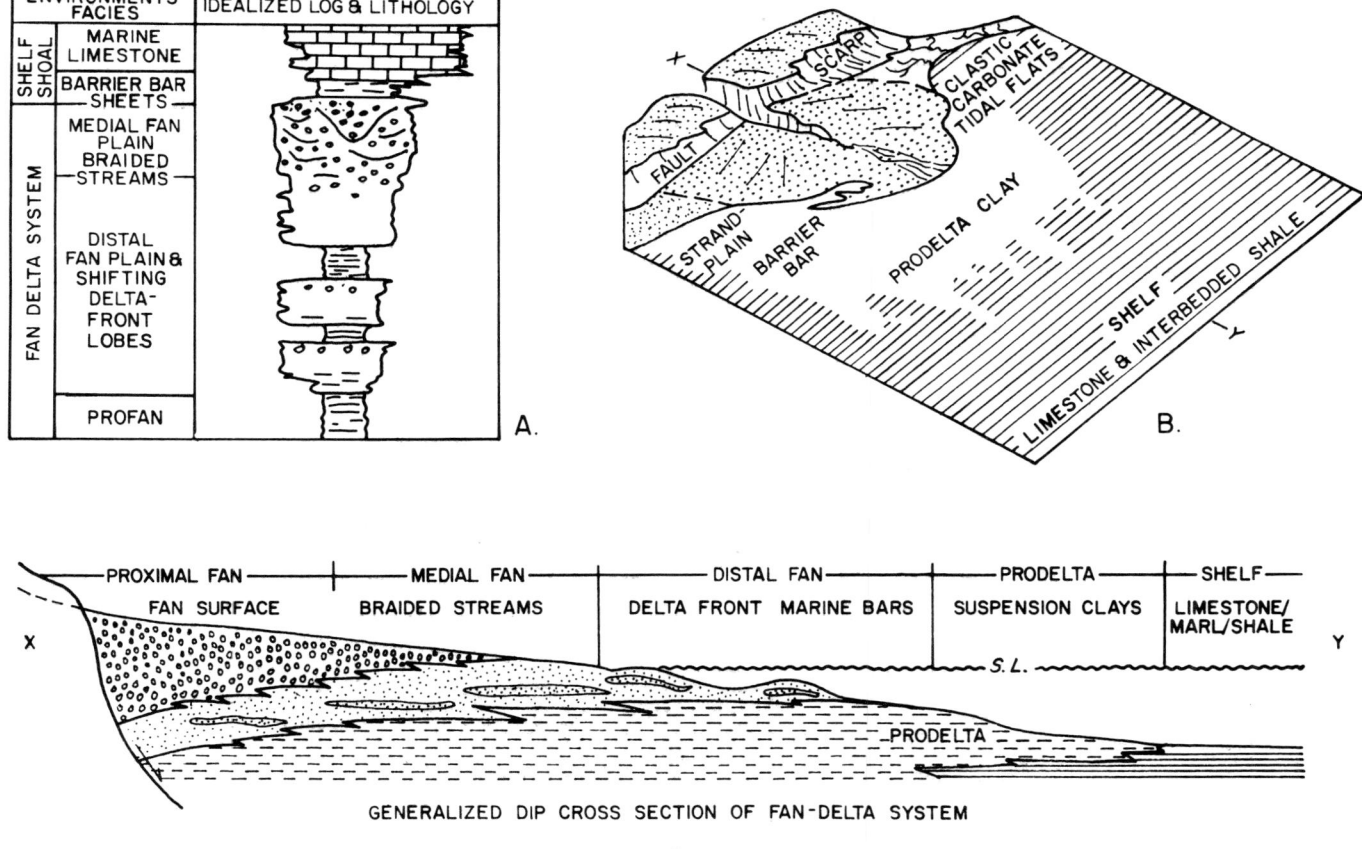

Fig. 14. Idealized vertical sequence (A), paleogeographic reconstruction (B), and cross-section (C) of shelf-type fan-deltas. This figure is based on data from mid-continent Pennsylvanian-Permian granite wash deposits and studies by McGowen (1970) (after Figure 5 of Brown, 1979).

reported from Jurassic fan-delta deposits in Brae field (Harms et al., 1981). Commerical hydrocarbon accumulations are present in fan-delta conglomeratic sequences of Early Cretaceous age in some Brazilian marginal basins (Ojeda, 1982; Meister and Aurich, 1972). Lower Cretaceous braided stream/fan-delta deposits in the offshore Gabes-Sabratha basin, northwest Libya have been described by Cable (1982) as a future petroleun exploration target.

Mackerel field, Gippsland basin, Australia produces from a massive sandstone section of Paleocene-Eocene age. The geologic model proposed for this field involves a prograding marginal marine sandstone which interfingers with continental braided stream sandstones in two nearby fields just to the north (Maughan et al., 1981), suggesting a fan-delta setting.

An oil discovery in Lower Cretaceous conglomeratic sandstones of the Tuktoyaktuk Peninsula, N.W.T. have proven the existence of a reservoir in these fan-delta deposits (Dixon, 1979); however, no commercial production is anticipated because of the inferred small reservoir size and the remoteness of the area.

Gas-bearing conglomerates of the Deep basin (Wilrich-Falher interval) in western Canada probably also represent fan-delta sequences. The conclusion is based on our observations of core from the area, descriptions by Cant (1982) and detailed descriptions and interpretations by Leckie and Walker (1982) of outcrop equivalents of the deep basin gas producing interval. Their model (Leckie and Walker, 1982, fig. 22) bears a striking similarity to other fan-delta models, and their criteria for recognition of

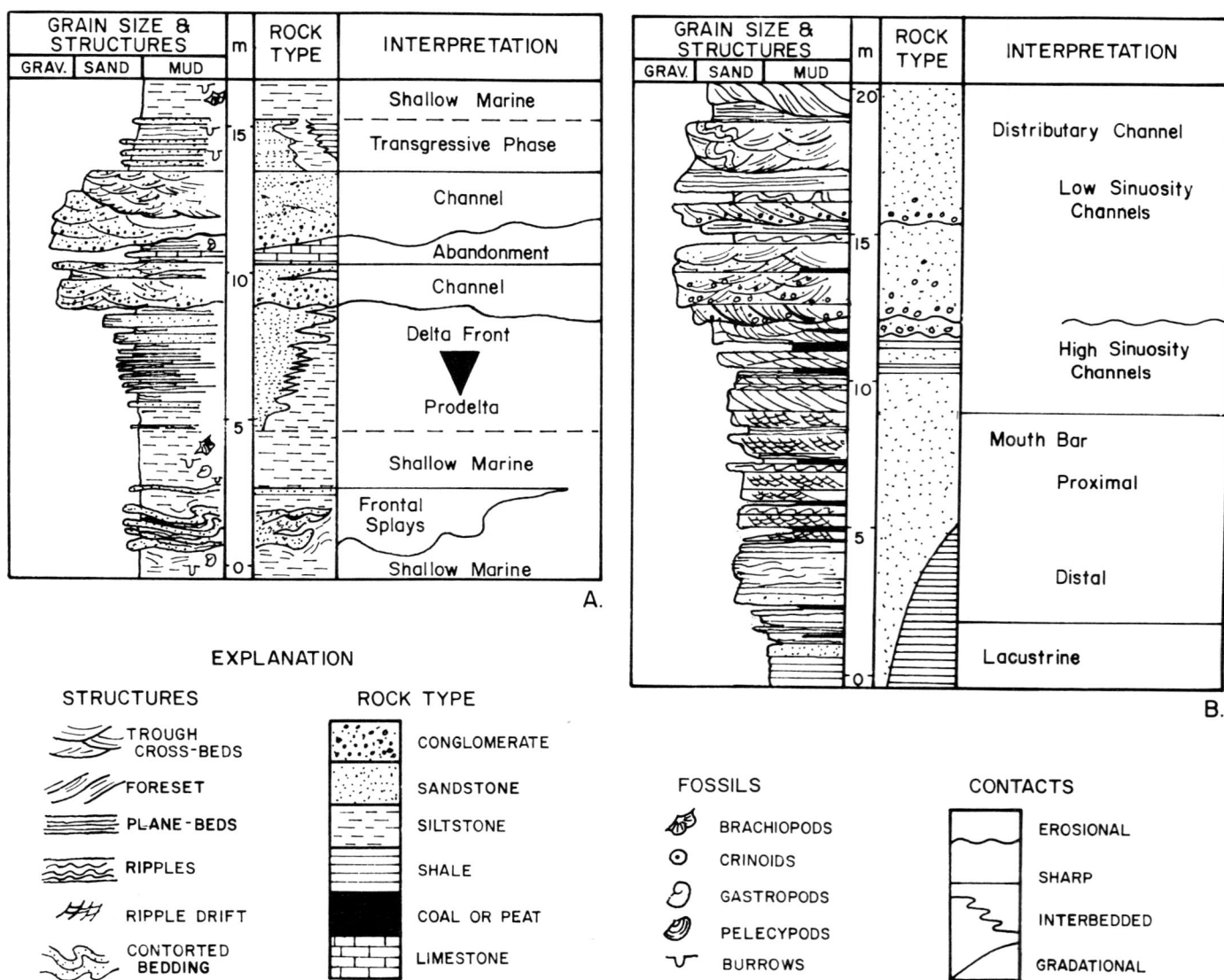

Fig. 15. Detailed vertical sequences through shelf-type fan-delta deposits. (A) Pennsylvanian deposits of the Taos Trough, New Mexico (after Figure 3 of Casey and Scott, 1979). (B) Devonian, lacustrine fan-delta deposits, Hornelen Basin, Western Norway (after Figure 7A of Pollard, Steel and Undersrud, 1982).

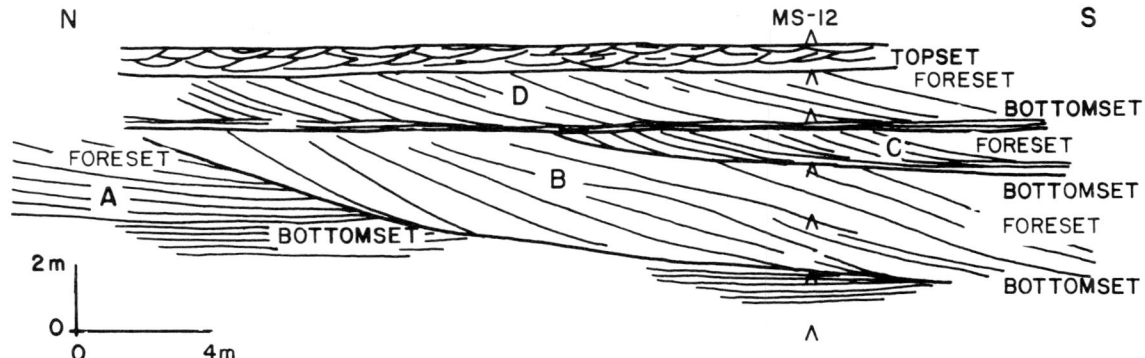

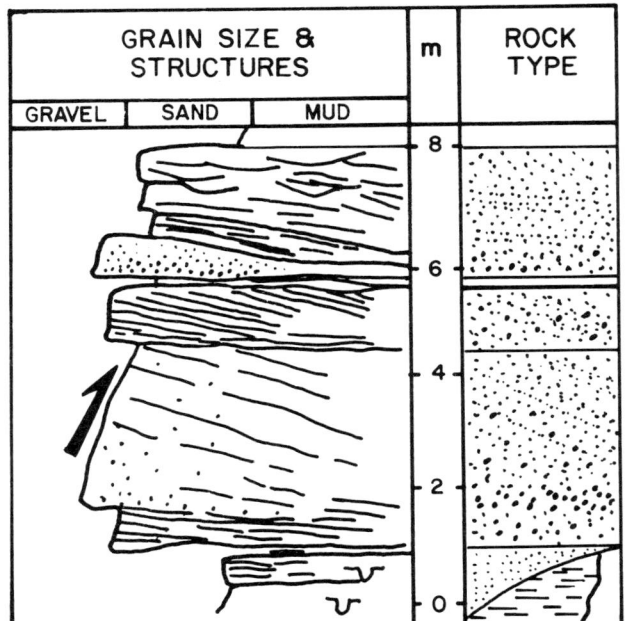

Fig. 16. (A) Outcrop sketch and measured section of topset, foreset and bottomset beds of Gilbert-type, marine fan-delta deposits, Pennsylvanian, Taos Trough, New Mexico (after Figure 5 of Casey and Scott, 1979). (B) Photo of outcrop exposure of topset (T), transgressive marine (TM) and foreset (F) beds in Gilbert-type, marine fan-delta deposits, Pennsylvanian, McCoy area, western Colorado (Reference Walker and Harms, 1980; field trip guide notes).

fluvial, beach and offshore gravel deposits (*ibid.*; Table 1) should prove useful in reservoir evaluation in the subsurface.

An evaluation of the relative importance of fan-delta deposits as hydrocarbon reservoirs is impossible at this early stage in our understanding of these deposits. However, several facts are clear: 1) fan-delta deposits are much more important as hosts of hydrocarbons than previously realized; 2) barring adverse diagenetic factors, delta front conglomerates and sandstones with better sorting and increased bed segregation and continuity make better reservoirs than braided channel conglomerates and conglomeratic sandstones; 3) reservoir potential is usually difficult to evaluate with conventional wireline logging devices; 4) known hydrocarbon-bearing fan-delta deposits occur principally along divergent plate margins (Anadarko basin, Brazilian marginal basins) and in foreland basin settings (Deep basin gas trap (?), western Canada; and Fort Worth basin, Texas); and 5) traps are mostly of the structural-stratigraphic type.

Summary and Conclusions

Fan-delta deposits are common depositional systems in the Holocene and throughout the rock record. They occur principally in active tectonic basins associated with divergent plate margins, convergent plate margins strike-slip margins and intracratonic basins. Most fan-delta deposits are composed of coarse-grained conglomerates and conglomeratic sandstones. Accurate paleogeographic reconstruction of such depositional settings requires the recognition of fluvial and beach conglomerates and the differentiation of alluvial fan, fan-delta and submarine fan sequences.

Three different types of fan-delta sequence occur repeatedly throughout the geologic column. Slope-type fan-deltas, which are characteristic of aulacogens, convergent plate margins and oceanic strike-slip margins, develop at the edge of a continental or island slope; they are truncated and have poorly developed coarsening-upward sequences. Shelf-type fan-deltas build onto continental or island shelves or into intracratonic marine and lacustrine basins, they may be interbedded with carbonate reefs and algal mounds, have well developed coarsening-upward sequences, and can be found along the margins of all active tectonic basins. Gilbert-type fan-deltas contain coarse gravelly foresets and develop on shallow water platforms in intracratonic basins.

Fan-delta deposits form hydrocarbon reservoirs in diverse areas. Documented commercial reservoirs are, however, principally associated with divergent plate margins and foreland basins. Because of better sorting, bed segregation and continuity, delta front or beach deposits usually make better reservoirs than associated channel deposits. Most traps are a combination of structural and stratigraphic controls.

References

Becker, B.D. 1977. Reciprocity of clastic and carbonate sediments, Pennsylvanian, Missourian Series, Wheeler County, Texas. M.S. Thesis, University of Texas at Austin, 115p.

Belt, E.S. and Bassieres, L. 1981. Upper Middle Ordovician submarine fans and associated facies, northeast of Quebec City. Canadian Journal of Earth Sciences, v. 18, p. 981-994.

Blakey, R.C. and Gubitosa, R. 1983. Late Triassic paleogeography and depositional history of the Chinle Formation southern Utah and northern Arizona. *In*: Reynolds, M.W. and Dolly, E.D. (Eds.), Mesozoic Paleogeography of West-Central United States. Rocky Mountain Section, Society of Economic Paleontologists and Mineralogists, p. 57-76.

Bluck, B.J. 1982. Gravel assemblages in beach and fluvial sediments. 11th International Sedimentological Congress, Hamilton, Canada, Book of Abstracts, p. 53.

Boothroyd, J.C. 1976. A model for alluvial fan-fan delta sedimentation in cold temperate environments. *In*: Miller, T.P. (Ed.), Recent and Ancient Sedimentary Environments in Alaska. Proceedings of the Alaska Geological Society Symposium, Anchorage, Alaska, p. N1-N13.

Boothroyd, J.C. and Ashley, G.M. 1975. Processes, bar morphology and sedimentary structures on braided outwash fans, northeastern Gulf of Alaska. *In*: Jopling, A.V. and McDonald, B.C. (Eds.), Glaciofluvial and Glaciolacustrine Sedimentation: Society Economic Paleontologists and Mineralogists, Special Publication No. 23, p. 193-222.

Boothroyd, J.C. and Nummedal, D. 1978. Proglacial braided outwash: a model for humid alluvial fan deposits. *In*: Miall, A.D. (Ed.), Fluvial Sedimentology. Canadian Society of Petroleum Geologists, Memoir 5, p. 641-668.

Brown, L.F., Jr. 1979. Deltaic sandstone facies of the mid-continent. *In*: Hyne, N.J. (Ed.), Pennsylvanian sandstones of the mid-continent. Tulsa Geological Society Special Publication 1, p. 35-63.

Brown, L.F., Cleaves, A.W. and Erxleben, A.W. 1973. Pennsylvanian depositional systems in north-central Texas. Bureau of Economic Geology, University of Texas, Guidebook 14, 120p.

Buckthal, W.P. 1977. Shrickey field, Texas Panhandle. Selected gas fields of the Texas Panhandle, Panhandle Geological Society, p. 67-70.

Burke, K. 1967. The Yallahs Basin: a sedimentary basin southwest of Kingston, Jamaica. Marine Geology, v. 5, p. 45-60.

Cable, D.G. 1982. Lower Cretaceous braided stream-fan delta deposition, northwest Libya: A future petroleum exploration target, offshore Gabes-Sabratha Basin (Abstract). American Association of Petroleum Geologists Bulletin, v. 66, p. 555-556.

Cambridge, T.R. 1977. Parsell field, Texas Panhandle. *In*: Selected gas fields of the Texas Panhandle. Panhandle Geological Society. p. 61-66.

Cant, D.J. 1982. Sedimentology and petroleum geology, Spirit River Formation (Lower Cretaceous), Deep Basin, Alberta (Abstract). American Association of Petroleum Geologists Bulletin, v. 66, p. 824.

Carr, D.L. and Scott, A.J. 1983. A spectrum of Late Paleozoic siliciclastic shelf bars, Sacramento Mountains, New Mexico (Abstract). American Association of Petroleum Geologists Bulletin, v. 67, p. 436-437.

Casey, J.M. 1979. Basin evolution of the Late Paleozoic Taos Trough, northern New Mexico (Abstract). American Association of Petroleum Geologists Bulletin, v. 63, p. 824.

Casey, J.M. and Scott, A.J. 1979. Pennsylvanian corse-grained fan deltas associated with the Uncompahgre uplift, Talpa, New Mexico. New Mexico Geological Society Guidebook, 30th Field Conference, Santa Fe County, p. 211-218.

Cast, R.F. 1977. Hemphill County granite wash field, Texas Panhandle. *In*: Selected Gas Fields of the Texas Panhandle. Panhandle Geological Society, p. 31-34.

Clarke, R.H. 1979. Reservoir properties of conglomerates and conglomeratic sandstones. American Association of Petroleum Geologists Bulletin, v. 63, p. 799-809.

Clifton, H.E. 1973. Pebble segregation and bed lenticularity and wave-worked versus alluvial gravel. Sedimentology. v. 20, p. 173-187.

Cotter, E. 1983. Shelf, paralic and fluvial environments and eustatic sea-level fluctuations in the origin of the Tuscarora Formation (Lower Silurian) of central Pennsylvania. Journal of Sedimentary Petrology, v. 53, p. 25-49.

Crowell, J.C. 1975. The San Gabriel fault and ridge basin, southern California. In: Crowell, J.D. (Ed.), San Andreas Fault in Southern California. California Division Mines and Geology, Special Report 118, p. 208-233.

Davies, J.L. 1973. Geographical variation in coastal development. New York: Hafner Publishing Company, 204p.

Dickinson, W.R. 1974. Plate tectonics and sedimentation. In: Dickinson, W.R. (Ed.), Tectonics and Sedimentation. Society of Economic Paleontologists and Mineralogists, Special Publication 22, p. 1-27.

Dixon, J. 1979. The Lower Cretaceous Atkinson Point Formation (new name) on the Tuktoyaktuk Peninsula, N.W.T.: a coastal fan-delta to marine sequence. Bulletin of Canadian Petroleum Geology, v. 27, p. 163-182.

Dobkins, J.E., Jr. and Folk, R.L. 1970. Shape and development on Tahiti Nui. Journal of Sedimentary Petrology, v. 40, p. 1167-1203.

Dutton, S.P. 1979a. Facies patterns and depositional models, Pennsylvanian System, Palo Duro Basin, Texas Panhandle (Abstract). American Association of Petroleum Geologists Bulletin, v. 63, p. 442-443.

Dutton, S.P. 1979b. Pennsylvanian fan-delta sandstones of the Palo Duro Basin, Texas. In: Hyne, N. J. (Ed.), Pennsylvanian sandstones of the Midcontinent. Tulsa Geological Society, Special Publication No. 1, p. 235-245.

Dutton, S.P. 1980a. Depositional systems and hydrocarbon resource potential of the Pennsylvanian System Palo Duro and Dalhart Basins, Texas Panhandle. Bureau Economic Geology, The University of Texas Geologic Circular 80-8. 49p.

Dutton, S.P. 1980b. Source-rock quality and thermal maturity, Palo Duro Basin, Texas (Abstract). American Association of Petroleum Geologists Bulletin, v. 64, p. 702.

Dutton, S.P. 1981a. Pennsylvanian fan-delta deposition, Mobeetie field, Texas Panhandle (Abstract). American Association of Petroleum Geologists Bulletin, v. 65, p. 922-923.

Dutton, S.P. 1981b. Diagenesis of Pennsylvanian fan-delta sandstones and interbedded shelf carbonates, Texas Panhandle. Geological Society of America, Abstracts with Programs, v. 13, p. 443.

Dutton, S.P. 1982a. Facies control of cementation and porosity, Pennsylvanian fan-delta sandstones, Texas Panhandle (Abstract). American Association of Petroleum Geologists Bulletin, v. 66, p. 565.

Dutton, S.P. 1982b. Pennsylvanian fan-delta and carbonate deposition, Mobeetie field, Texas Panhandle. American Association of Petroleum Geologists Bulletin, v. 66, p. 389-407.

Erxleben, A.W. 1975. Depositional systems in Canyon Group (Pennsylvanian System) North Central Texas. Bureau of Economic Geology, The University of Texas, Report of Investigations No. 82, 15p.

Ethridge, F.G. and Wescott, W. A. 1981. Tectonic setting, recognition and hydrocarbon potential of fan-delta deposits (abstract). In: Sedimentary Tectonics: Principles and Applications. Laramie, Wyoming; Department of Geology, University of Wyoming, Wyoming Geological Association and Geological Survey of Wyoming, p. 11.

Flores, R.M. 1975. Short-headed stream delta: model for Pennsylvanian Haymond Formation, West Texas. American Association of Petroleum Geologists Bulletin, v. 59, p. 2288-2301.

Flores, R.M. 1978. Braided fan-delta deposits of the Pennsylvanian Upper Haynond Formation in the northeastern Marathon Basin, Texas. In: Muzzulo, S.J. (Ed.), Tectonics and Paleozoic Facies of the Marathon Geosyncline, West Texas, Permian Basin Section. Society of Economic Paleontologists and Mineralogists, Publication 78-17, p. 149-159.

Fortunato, K.S. and Ward, W.C. 1982. Upper Jurassic-Lower Cretaceous fandelta complex- La Casita Formation of Saltillo area, Coahuila, Mexico. American Association of Petroleum Geologists Bulletin, v. 66, p. 1429.

Fredrickson, J.A. 1978. Petrology and depositional environments of the Fountain and Ingleside Formations, Owl Canyon, Colorado. M.Sc. Thesis, University of Wyoming, 100p.

Friedman, G.M. and Sanders, J.E. 1978. Principles of sedimentology. New York: John Wiley and Sons, 792p.

Frost, E., Jr., Allen T., and Fertl, W.H. 1982. Formation evaluation in granite wash reservoirs. World Oil, September, 1982, p. 121-132.

Galloway, W.E. 1976. Sediments and stratigraphic framework of the Copper River fan delta, Alaska. Journal of Sedimentary Petrology, v. 46, p. 726-737.

Gilbert, G.K. 1890. Lake Bonneville. United States Geologial Survey Monograph, v. 1, 438p.

Gloppen, T.G. and Steel, R.J. 1981. The deposits, internal structure and geometry of six alluvial fan - fan delta bodies (Devonian, Norway) — a study in the significance of bedding sequences in conglomerates. In: Ethridge, F.G. and Flores, R.M. (Eds.), Recent and Ancient Nonmarine Depositional Environments: Models for Exploration. Society of Economic Paleontologists and Mineralogists, Special Publication 31, p. 49-69.

Gnaccolini, M. 1982. Oligocene fan-delta deposits in northern Italy: a summary. Rivista Italiana Paleontologia e Stratigrafia, v. 87, p. 627-636.

Gvirtzman, G. and Buchbinder, B. 1978. Recent and Pleistocene coral reefs and coastal sediments of the Gulf of Elat. 10th International Sedimentological Congress, Jerusalem, Israel, Post-Congress Fieldtrip Guidebook, p. 161-191.

Hanford, C.R. 1980. Lower Pernian facies of the Palo Duro Basin, Texas: depositional systems, shelf-margin evolution, paleogeography and petroleum potential. Bureau of Economic Geology, The University of Texas at Austin, Report of Investigation No. 102, 31p.

Hanford, C.R. and Dutton, S.P. 1980. Pennsylvanian-Lower Permian depositional systems and shelf margin evolution, Palo Duro Basin, Texas. American Association of Petroleun Geologists Bulletin, v. 64, p. 88-106.

Hanford, C.R., Presley, M.W. and Dutton, S.P. 1980. Depositional and tectonic evolution of a basement bounded, intracratonic basin, Palo Duro Basin, Texas (Abstract). American Association of Petroleun Geologists Bulletin, v. 64, p. 717.

Harbaugh, D.W. and Dickinson, W.R. 1981. Depositional facies of Mississippian clastics, Antler Foreland Basin, central Diamond Mountains, Nevada. Journal of Sedimentary Petrology, v. 51, p. 1223-1234.

Harms, J.C. and McMichael, W.J. 1983. Sedimentology of the Brae oilfield area, North Sea. Journal of Petroleum Geology, v. 5, p. 437-439.

Harms, J.C., Tackenberg, P., Pickles, E., and Pollock, R.E. 1981. The Brae oilfield area. In: Illing, L.W. and Hobson, G.D. (Eds.), Petroleum Geology of the Continental Shelf of Northwest Europe. Heyden, p. 352-357.

Hayes, M.O. and Michel, J. 1982. Shoreline sedimentation within a forearc embayment, lower Cook Inlet, Alaska. Journal of Sedimentary Petrology, v. 52, p. 251-263.

Hayward, A.B. 1982. Coral reefs in a clastic sedimentary environment: fossil (Miocene, S.W. Turkey) and modern (Recent, Red Sea) analogues. Coral Reefs, v. 1, p. 109-114.

Hayward, A.B. 1983. Coastal alluvial fans and associated marine facies in the Miocene of S. W. Turkey. In: Collinson, J.D. and Lewin, J. (Eds.), Modern and Ancient Fluvial Systems. International Association of Sedimentologists, Special Publication 6, p. 323-326.

Hayward, B.W. and Ballance, P.F. 1982. Possible alluvial fans of basic to intermediate volcanic rudites with their feet in deep water: Piha Formation (Abstract). 11th International Sedimentological Congress, Hamilton, Canada, Book of Abstracts, p. 175.

Hempton, M.R., Dunne, L.A. and Dewey, J.F. 1983. Sedimentation in an active strike-slip basin, southeastern Turkey. Journal of Geology, v. 91, p. 401-412.

Hendry, M. 1982. Field guide to the Yallahs fan-delta complex. The Geological Society of Jamaica, 10p.

Henslee, H.T. 1977. Mobeetie field, Texas Panhandle. In: Selected gas fields of the Texas Panhandle. Panhandle Geological Society, p. 43-54.

Holmes, A. 1965. Principles of Physical Geology, 2nd Edition. New York: The Roland Press Co., 1288p.

Houseknecht, D.W. and Ethridge, F.G. 1978. Depositional history of the Lamotte Sandstone of southeastern Missouri. Journal of Sedimentary Petrology, v. 48, p. 575-586.

Howell, D.G. and Link, M.H. 1979. Eocene conglomerate sedimentology and basin analysis, San Diego and the southern California borderland. Journal of Sedimentary Petrology, v. 49, p. 517-540.

Hyde, R.S. 1980. Sedimentary facies in the Archean Timiskaming Group and their tectonic implications, Abitibi Greenstone Belt, Northeastern Ontario, Canada. Precambrian Research, v. 12, p. 161-195.

Inman, D.L. and Nordstrom, C.E. 1971. On the tectonic and morphologic classification of coasts. Journal of Geology, v. 79, p. 1-21.

Kluth, C.F. and Coney, P.J. 1981. Plate tectonics of the ancestral Rocky Mountains. Geology, v. 9, p. 10-15.

Langford, R.P. 1982. Depositional systems and geologic history of the Lower part of the Fountain Formation, Manitou Embayment, Colorado. M.A. Thesis, Indiana University, 200p.

Leckie, D.A. and Walker, R.G. 1982. Storm- and tide-dominated shorelines in Cretaceous Moosebar-Lower Gates interval- outcrop equivalents of Deep Basin gas trap in Western Canada. American Association of Petroleum Geologists Bulletin, v. 66, p. 138-157.

Link, M.H. and Osborne, R.H. 1978. Lacustrine facies in the Pliocene Ridge Basin Group: Ridge Basin, California. *In*: Matter, A. and Tucker, M. (Eds.), Modern and Ancient Lake Sediments. International Association of Sedimentologists, Special Publication 2, p. 167-185.

Long, D.G.F. 1981. Dextral strike-slip faults in the Canadian Cordillera and depositional environments of related fresh-water intermontane coal basin. *In*: Miall, A.D. (Ed.), Sedimentation and Tectonics in Alluvial Basins. Geological Association of Canada, Special Paper 23, p. 153-186.

McCaslin, J.C. 1982. Palo Duro Basin drawing new interest. Oil and Gas Journal, March 22, 1982, p. 201-202.

McDonald, B.C. and Lewis, C.P. 1973. Geomorphic and sedimentologic processes of rivers and coast, Yukon coastal plain. Environmental Social Program, Northern Pipelines, Task Force on Northern Oil Development, Government of Canada Report No. 73-39, 245p.

McGowen, J.H. 1970. Gum Hollow Fan Delta, Nueces Bay, Texas. Bureau of Economic Geology, The University of Texas at Austin, Report of Investigation 69, 91p.

McGowen, J.H. and Groat, C.G. 1971. Van Horn Sandstone, West Texas: an alluvial fan model for mineral exploration. Bureau of Economic Geology, The University of Texas at Austin, Report of Investigation 72, 57p.

McGowen, J.H. and Scott, A.J. 1974. Fan-delta deposition: processes, facies and stratigraphic analogues (Abstract). American Association of Petroleum Geologists/Society of Economic Paleontologists and Mineralogists, Annual Meeting, Abstracts, p. 60-61.

McGowen, J.H., Granata, G.E. and Seni, S.J. 1983. Depositional setting of the Triassic Dockum Group, Texas Panhandle and eastern New Mexico. *In*: Reynolds, M.W. and Dolly, E.D. (Eds.), Mesozoic Paleogeography of the West-Central United States. Rocky Mountain Section, Society of Economic Paleontologists and Mineralogists, p. 13-38.

McKee, E.D. and Crosby, E.J. (coordinators) 1975. Paleotectonic investigations of the Pennsylvanian System in the United States. United States Geological Survey Professional Paper 853, 349p.

McLaughlin, R.J. and Nilsen, T.H. 1982. Neogene non-marine sedimentation and tectonics in small pull-apart basins of the San Andreas fault system, Sonoma County, California. Sedimentology, v. 29, p. 865-867.

Maejima, Wataru, 1982. Texture and stratification of gravelly beach sediments, Enju beach, Kii Peninsula, Japan. Journal of Geosciences, Osaka City University, v. 25, Art. 3, p. 35-51.

Maughan, D.M. Mebberson, A.J. and Morton, D.J. 1981. The definition and development of the Mackerel field, Gippsland Basin. Oil and Gas Journal, June, 1981, p. 175-180.

Meckel, L.D. 1975. Holocene sand bodies in the Colorado Delta area, northern Gulf of California. *In*: Broussard, M.L. (Ed.), Deltas — models for exploration. Houston Geological Society, p. 239-265.

Meister, E.M. and Aurich, N. 1972. Geologic outline and oil fields of Sergipe Basin, Brazil. American Association of Petroleum Geologists Bulletin, v. 56, p. 514.

Miall, A.D. 1981. Alluvial sedimentary basins: tectonic setting and basin architecture. *In*: Miall, A.D. (Ed.), Sedimentation and Tectonics in Alluvial Basins. Geological Association of Canada, Special Paper 23, p. 1-33.

Millberry, K.W. 1983. Tectonic control of Pennsylvanian fan-delta deposition, southwestern Colorado (Abstract). American Association of Petroleum Geologists Bulletin, v. 67, p. 514.

Mitchell, A.H.G. and Reading, H.G. 1978. Sedimentation and tectonics. *In*: Reading, H.G. (Ed.), Sedimentary Environments and Facies. New York: Elsevier, p. 439-476.

Ogliani, F. 1981. Transgressive-regressive phases in the proximal part of a fan-delta system, Early Pliocene, Intrapenninic Basin, Bologna (Abstract). International Association of Sedimentologists, 2nd European Meeting, Bologna, p. 126-129.

Ojeda, H.A.O. 1982. Structural framework, stratigraphy and evolution of Brazilian marginal basin. American Association of Petroleum Geologists Bulletin, v. 66, p. 732-749.

Ori, G.G. and Ricci Lucchi, F. 1981. Giant epsilon bedding in coarse-grained point bars of late Pliocene fan-delta systems, Bologna (Abstract). International Association of Sedimentologists, 2nd European Meeting, Bologna, p. 137-140.

Parker, R.L. and Gibson, C.R. 1977. Saint Clair field Texas Panhandle. *In*: Selected Gas Fields of the Texas Panhandle, Panhandle Geological Society, p. 71-74.

Phillips, S.T. 1982. Fan-delta sedimentation, Waltman Member, Fort Union Formation, Wind River Basin, Wyoming (Abstract). American Association of Petroleum Geologists Bulletin, v. 66, p. 617-618.

Pollard, J.E., Steel, R.J. and Undersrud, E. 1982. Facies sequences and trace fossils in lacustrine/fan delta deposits, Hornelen Basin (M. Devonian), Western Norway. Sedimentary Geology, v. 32, p. 63-87.

Prior, D.B., Wiseman, W.J. and Gilbert, R. 1981. Submarine slope processes on a fan delta, Howe Sound, British Columbia. Geo-Marine Letters, v. 1, p. 85-90.

Rainone, M., Nani, T., Ori, G.G., Lucchi, F. 1981. A prograding gravel beach in Pleistocene fan-delta deposits of Ancona, Italy (Abstract). International Association of Sedimentologists, 2nd. European Meeting, Bologna, p. 155-156.

Reimnitz, E. 1966. Late Quaternary history and sedimentation of the Copper River Delta and vicinity, Alaska. Ph.D. Thesis, University of California, San Diego, 160p.

Ricci Lucchi, F. Colella, A., Ori, G.G. and Ogliani, F. 1981a. Pliocene fan deltas of the IntraAppenninic Basin, Bologna. *In*: Ricci Lucchi, F. (Ed.), Excursion Guidebook with Contributions on Sedimentology of Some Italian Basins. International Association of Sedimentologists, 2nd European Meeting, Bologna, Excursion No. 4, p. 81-162.

Ricci Lucchi, F., Ogliani, F., Ori, G.G. and Colella, A. 1981b. Braided versus meandering channel deposits in Pliocene fan-delta systems, Bologna Intra-Appenninic Basin (Abstract). International Association of Sedimentologists, 2nd European Meeting, Bologna, p. 161-163.

Rust, B.R. 1979. Coarse alluvial deposits. *In*: Walker, R.G. (Ed.), Facies Models. Geoscience Canada, Reprint Series 1, p. 9-21.

Sahl, H.L. 1970. Mobeetie field, Wheeler County, Texas. Shale Shaker, v. 20, p. 107-115.

Schenewerk, P.A., Sethi, D.K., Fertl, W.H. and Lochmann, M. 1981. Natural gamna ray spectral logging aids granite wash study. Oil and Gas Journal, August, 1981, p. 180-190.

Shelby, J.M. 1976. Report: Oklahoma and Texas Panhandles (Abstract). American Association of Petroleum Geologists Bulletin, v. 60, p. 322.

Shelby, J.M. 1979. Upper Morrow fan-delta deposits of Anadarko Basin (Abstract). American Association of Petroleum Geologists Bulletin, v. 63, p. 2119.

Shelby, J.M. 1980. Geologic and economic significance of the Upper Morrow chert conglomerate reservoir of the Anadarko Basin. Journal of Petroleum Technology, March, 1980, p. 489-495.

Sneh, A. 1979. Late Pleistocene fan-deltas along the Dead Sea Rift. Journal of Sedimentary Petrology. v. 49, p. 541-552.

Sneider, R.M., Richardson, F.M. Paynter, D.D., Eddy, R.E. and Wyant, I.A. 1977. Predicting reservoir rock geometry and continuity in Pennsylvanian reservoirs, Elk City field, Journal of Petroleum Technology, v. 29, p. 851-866.

Squires, R.L. 1981. A transitional alluvial to marine sequence: the Eocene Llajas Formation, southern California. Journal of Sedimentary Petrology, v. 51, p. 923-938.

Stanley, D.J. 1980. The Saint-Antonin conglomerate in the Maritime Alps: A model for coarse sedimentation on a submarine slope. Smithsonian Contributions to the Marine Sciences, No. 5, 25p.

Tai Wai Ng, D. 1979. Subsurface study of Atoka (Lower Pennsylvanian) clastic rocks in part of Jack, Palo Pinto, Park and Wise Counties, north-central Texas. American Association of Petroleum Geologists Bulletin, v. 63, p. 50-66.

Thompson, R.W. 1968. Tidal flat sedimentation on the Colorado River delta, northwestern Gulf of California. Geological Society of America, Memoir 107. 133p.

Tyler, N., Galloway, W.E., Garrett, C.M., Ewing, T.E. and Posey, J.S. 1983. Anatomy of Texas oil (Abstract). American Association of Petroleum Geologists Bulletin, v. 67, p. 560-561.

Walker, R.G. and Harms, J.C. 1980. Fan-delta deposition, Minturn Formation, McCoy area, Colorado. Unpublished field trip notes, Rocky Mountain Section, Society of Economic Paleontologists and Mineralogists Fall Field Conference.

Webster, R.E. 1980. Evolution of south Oklahoma aulacogen. Oil and Gas Journal, February, 1980, p. 150-172.

Wescott, W.A. 1979. The Yallahs fan delta, southeastern Jamaica: a depositional model for active tectonic coastlines. Ph.D. Thesis, Colorado State University, 178p.

Wescott, W.A. and Ethridge, F.G. 1980. Fan-delta sedimentology and tectonic setting-Yallahs fan delta, southeast Jamaica. American Association of Petroleum Geologists Bulletin, v. 64, p. 374-399.

Wescott, W.A. and Ethridge, F.G. 1982. Bathymetry and sediment dispersal dynamics along the Yallahs fan-delta front, Jamaica. Marine Geology, v. 46, p. 245-260.

Wescott, W.A. and Ethridge, F.G. 1983. Eocene fan delta-submarine fan deposition in the Wagwater Trough, east-central Jamaica. Sedimentology, v. 30, p. 235-245.

MASS-FLOW CONGLOMERATES IN A SUBMARINE CANYON: ABRIOJA FAN-DELTA, PLIOCENE, SOUTHEAST SPAIN

GEORGE POSTMA[1]

ABSTRACT

In the Early - Middle Pliocene, a conglomeratic fan-delta (Abrioja Formation) prograded approximately 15 km southeastwards in a confined canyon in the Almeria Basin. The canyon was 2 - 4 km wide and 100 - 200 m deep. It probably originated from rifting of the area between the metamorphic basement blocks of the Sierra de Gador and the Sierra Alhamilla.

The Abrioja fan-delta sequence, which filled the canyon is as follows (from top to bottom): 1) Alluvial fan and coastal plain deposits with paleosols; 2) Sediments of a transition zone (beach, nearshore and delta front); 3) Upper delta slope sediments, with tangential foresets of pebbly sandstones; 4) Lower delta slope sediments, with parallel bedded conglomerates, sandstones and pebbly mudstones; 5) Pro-delta sediments, with heavily bioturbated muds.

Detailed study of the subaqueous part of the fan-delta has shown that the gravel and sand were transported mainly by sediment gravity flows. The various facies described from the subaqueous fan-delta reflect different kinds of mass flow behaviour. Slumping in the upper part of the delta slope was probably the main sediment source for the conglomeratic beds of the lower delta slope, which were deposited in troughs downslope from the slump scar. The slump scars have been filled again with sediment from the transition zone.

From several conglomerate beds, some important structures have been described: a) Asymmetric conglomerate beds show a type of back-set bedding. These 'backsets' are indicated by either a diffuse, centimetre thin sandy zone or by a preferred pebble alignment. They are interpreted as shear zones due to downslope transmitted compression. The compression is probably due to resedimentation at the upslope-side of the gravel body. b) The head of a structureless, polymict gravel flow deposit shows straight troughs, of *ca.* 50 cm in cross-section filled with pebbly sandstone, issuing from this body. c) At the base and on the top of some conglomerate beds, evenly-stratified sandstones may have originated from flow surges, rather than from (basal) shearing. The sediment for the flow surges may have been progressively winnowed by liquefaction and gravity from the margins of a gravel flow, a process which was studied from experiments described here, and has been named gravity-winnowing.

Conglomerate beds in the canyon are covered by (pebbly) mudstones which were probably deposited during periods of 'normal' sedimentation conditions.

RÉSUMÉ

Un cône de déjection conglomératique (formation d'Abrioja) a avancé progressivement vers le sud-ouest d'environ 15 km dans un canyon étroit du bassin Almeria durant le Pliocène inférieur-moyen. Le canyon avait une largeur de 2-4 km et une profondeur de 100-200 m. Son origine est possiblement reliée à la création d'un fossé tectonique dans la région située entre les blocs métamorphiques du socle de la Sierra de Gador et de la Sierra Alhamilla.

La séquence formant ce cône de déjection, laquelle a rempli le canyon, est composée du toit vers la base: 1) de dépôts provenant d'un cône alluvial et d'une plaine littorale avec développement de paléosols; 2) de sédiments d'une zone de transition (plage, proximité de rivage et front de delta); 3) de sédiments de haut de pente du delta accompagnés de couches frontales tangentielles de grès caillouteux; 4) de sédiments de bas de pente de delta incluant des conglomérats en couches parallèles, des grès et des mudstones caillouteux; 5) de sédiments pro-deltaïques, avec des boues fortement bioturbées.

L'examen détaillé de la partie sous-aquatique du cône de déjection a démontré que le gravier et le sable furent transportés principalement par un écoulement par gravité de sédiments. Les divers faciès sous-aquatiques du cône de déjection reflètent différents types de comportement d'écoulement en masse. Un mouvement de masse à la partie supérieure de la pente du delta constitue probablement la principale source des matériaux des couches de conglomérat du bas de pente du delta qui ont été déposés dans les dépressions localisées le long de la pente à partir des cicatrices d'arrachement. Les cicatrices d'arrachement furent subséquemment remplies à nouveau de sédiments appartenant à la zone de transition.

L'étude de plusieurs couches de conglomérat a permis de décrire quelques structures importantes: a) des couches conglomératiques assymétriques montrant un litage de type 'backset'. Ces 'backsets' sont reconnus ou par la présence d'une zone sableuse et diffuse d'épaisseur de l'ordre d'un centimètre ou par un alignement préférentiel des galets. On les considère comme étant des zones de cisaillement développées par la transmission d'une compression dans le sens de la pente. La compression résulte probablement d'une resédimentation d'un corps graveleux du côté de la remontée de la pente. b) Le front d'un dépôt de gravier polygénique, sans litage, dérivant d'un écoulement en masse, présente des dépressions rectilignes d'environ 50 cm en coupe transversale. Elles sont remplies de grès caillouteux émanant de ces dépôts. c) A la base et au sommet de quelques lits de conglomérat, on observe des grès uniformément stratifiés formés probablement par des flots plutôt que par un cisaillement basal. Les sédiments de ces flots furent possiblement vannés progressivement par liquéfaction et gravité des rebords d'une coulée de gravier, ce processus fut étudié dans les expériences décrites ici et appelé gravité-vannage.

Les couches de conglomérat dans le canyon sont recouvertes de mudstones (cailouteux) probablement accumulés durant des périodes de conditions 'normales' de sédimentation.

---

[1]University of Amsterdam Geological Institute, N^w Prinsengracht 130, 1018 V2 Amsterdam, The Netherlands.

This paper is an updated version of the author's Ph.D. thesis at the University of Amsterdam, The Netherlands, for which fieldwork was carried out in the summer months of 1979 - 1981, and which has benefited from discussions in the field by Drs. Th. B. Roep, D.J. Beets, L.M.J.U. van Straaten. Earlier versions of the manuscript were read by the former persons and K.L. Kleinspehn, W. Nemec and R.J. Steel. F.H. Kievits did the drafting, J.J. Wiersma prepared the photographs and C. Groen constructed the accessoires for the experiments. The updated version was improved by later discussions and suggestions of Drs. K.L. Kleinspehn and W. Nemec at the University of Bergen (Norway). I thank them all most kindly.

Copyright © 1984, Canadian Society of Petroleum Geologists

INTRODUCTION

Sedimentological studies on modern and ancient shallow marine fan-deltas are scarce and little information is available on their internal structures and framework. Fan-deltas are defined as alluvial fans that prograde into a standing body of water from an adjacent highland (Holmes, 1965; McGowen, 1970). A fan-delta thus consists of a subaerial and a subaqueous part, both characterized by a higher-gradient slope than on fine grained deltas. The subaqueous part of the fan-delta has been subdivided into: 1) a delta front; the zone above wave base and the highest energy zone of the delta; 2) a delta slope; defined here as a high-gradient part of the delta with active resedimentation of sand and gravel; 3) a prodelta; the part of the delta lying beyond the delta slope which dips gently to the basin floor.

Concepts of modern and ancient fan-deltas have been summarized by Wescott and Ethridge (1980) in connection with their study of the Recent Yallahs fan system on Jamaica (Wescott and Ethridge, 1980, 1982). Most of the studies on Recent fan-deltas have concentrated on sedimentation in the alluvial fan and coastal environments, probably because of their easy accessibility. In these environments, the sediment is transported both by stream flow and mass flow processes as seen in modern and ancient examples (McGowen, 1970; Wescott and Ethridge,

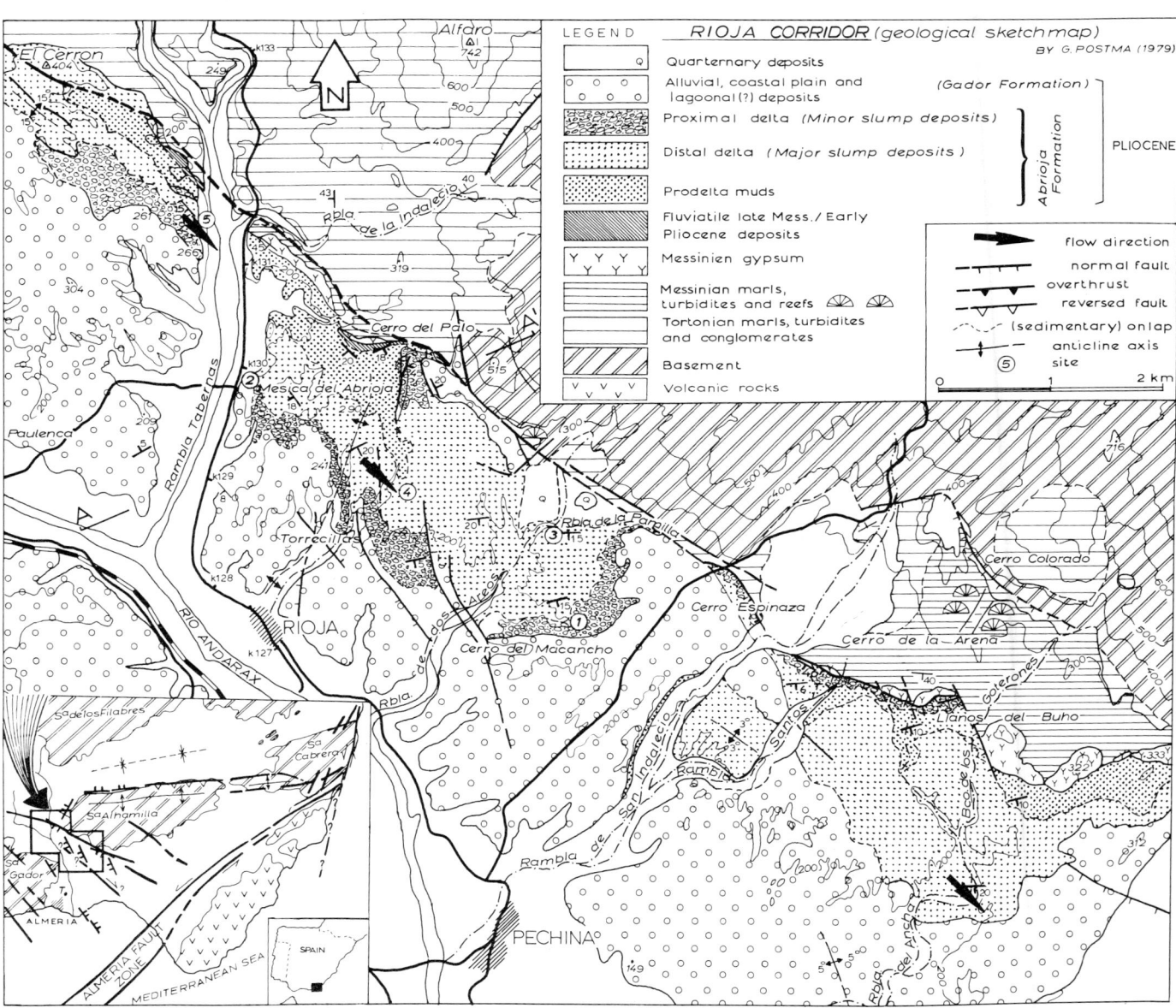

Fig. 1. Geological map of the Rioja corridor. The inset figure in the lower left corner shows some structural trends of the area on a larger scale (*cf.* Bousquet *et al.*, 1976). For the geological map, data are used from Van den Eeckhout (1980), Iaccarino *et al.* (1975), Postma (1978) and I.G.M.E. (1982). T = Torre Cardenas.

1980; Larsen and Steel, 1978; Gnaccolini, 1981; Howell and Link, 1979), and these processes may also control delta front deposition (cf. Prior et al., 1981b; Kleinspehn et al., this volume).

From the subaqueous part of modern fan-deltas, little information is available. By means of SCUBA diving, Wescott and Ethridge (1982) studied processes in nearshore canyons and canyon heads. They found that sediment, transported alongshore by waves and longshore currents, was trapped in the nearshore canyon heads. Continuous sediment accumulation, overloading and liquefaction were the main processes which caused slumping of these trapped sediments, which were then transported *en masse* farther downslope.

Studies on ancient subaqueous fan-deltas (Nemec et al., 1980; Stanley, 1980; Howell and Link, 1979; Postma, 1979, 1984; Postma and Roep (in prep); Völk, 1966; Surlyk, 1978; Porębski, 1981) and on Recent ones (Prior et al., 1981a, b) indicate major sediment emplacement by sediment gravity flows. However, little is known about the organisation of these flows within fan bodies.

This paper analyses the sedimentary setting and tectonic evolution of a Pliocene canyon in southeastern Spain (Fig. 1), which is dominantly filled by prograding sequences of sandy and gravelly sediment gravity-flows from the Abrioja fan-delta (Abrioja Formation). Sediment gravity flow deposits are described in terms of grain fabric, geometry and their downslope and lateral changes. Their depositional mechanisms are discussed in the light of experimental modelling. Further, a sedimentation model has been derived for the Abrioja fan-delta.

REGIONAL SETTING

STRATIGRAPHY

The age of the Tabernas Basin (between the Sierra de los Filabres and the Sierra Alhamilla) and Almeria Basin (Fig. 1) is Late Miocene (Late Serravalian/Early Tortonian), based on foraminifera (Jacquin, 1970; Iaccarino et al., 1975).

During the Tortonian, a submarine-fan system with feeder channels from the Sierra de los Filabres prograded southwards (Kleverlaan, 1980) filling the Tabernas and Almeria basins with gravel, turbidites and marls. The Tabernas and Almeria Basins were probably not separated by the Sierra Alhamilla at that time (Weijermars et al., in prep.).

Uplift of a metamorphic basement block in Messinian time, the present Sierra Alhamilla, is indicated by angular unconformities on its south flank (Cerro de la Arena). There, pebbly sands of mass flow origin cover tilted Tortonian sediments (Postma, 1978). Basinwards, the supply of terrigenous detritus decreased and only white marls were deposited. Rhythmites of marls, turbidites, limestone and diatomite overlying these white marls preceed the gypsum deposits in the basins (Dronkert, 1976; Postma, 1978). At some localities, barrier reefs developed along the flanks of the Sierra Alhamilla and the Sierra de Gador. Sediments overlying the gypsum deposits generally indicate further shallowing of the basins (Roep et al., 1979) and are finally covered by continental deposits.

From Pliocene time onwards, the Abrioja fan-delta prograded southeastward from the uplifted Tabernas Basin into the Almeria Basin and present Gulf of Almeria, filling these basins with mainly coarse grained detritus derived from the surrounding basement of the Sierra de los Filabres, the Sierra de Gador, and the Sierra Alhamilla (Figs. 1, 2A) and from the older basin fill. The fan-deltaic deposits of the Abrioja Formation have their type locality in the Mesica del Abrioja (Postma, 1979).

STRUCTURAL EVOLUTION OF THE CANYON DURING PLIOCENE TIME

The northern side of the canyon is well defined by steep, normal faults parallelling the Sierra Alhamilla and extending further in a NW-SE direction. Large slides of Miocene

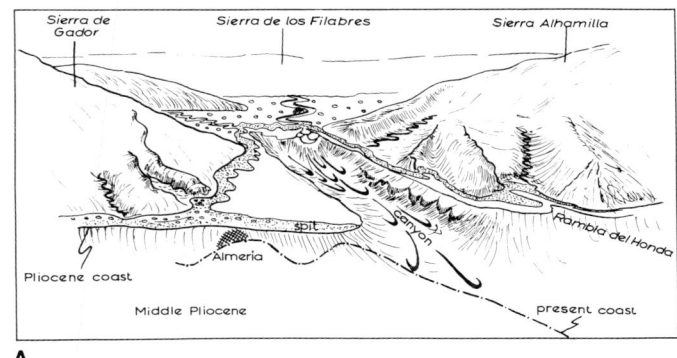

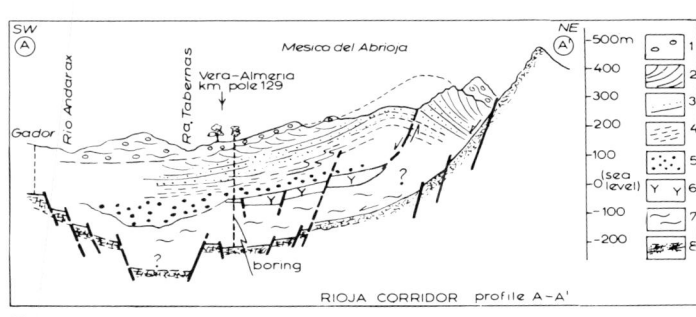

Fig. 2. A. Paleogeographic sketch from the Rioja corridor (Middle-Pliocene) with alluvial fans and shallow marine fan-deltas filling a graben between the Sierra de Gador and the Sierra Alhamilla. **B.** Section A-A' through the Mesica del Abrioja (Fig. 1). 1. alluvial fans, 2. proximal delta sediments, 3. distal delta sediments, 4. prodelta and shelf pelites, 5. Messinian/Pliocene alluvial fan serie, 6. gypsum, 7. Messinian and Tortonian sediments (mainly marls), 8. basement. Rifting probably started in the Late Messinian time, after the Miocene period of dessication in the Mediterranean region.

and Pliocene sediments and metamorphic basement material (Llano del Buho, Mesica del Abrioja, see Fig. 1) occurred locally along this fault zone as a response to uplift of the Sierra Alhamilla. These slides constitute intraformational unconformities mainly in the vicinity of the fault zone (Fig. 2B; see also Van den Eeckhout, 1980).

The southern side of the canyon is less well-defined, because it is covered by the Gador Formation. Normal faults along and across the Sierra de Gador, also trending NW-SE with a NE dip (mapped by I.G.M.E., 1982), indicate rifting of the corridor between the Sierra de Gador and the Sierra Alhamilla (Rioja Corridor). The vertical displacement in the resultant graben is 300 - 500 m, according to borings (I.G.M.E., 1974).

The rifting in the Rioja Corridor was probably a local event due to individual movements of the large basement blocks, the Sierra de Gador and the Sierra Alhamilla. Displacement probably occurred along a sinistral strike-slip fault zone, indicated on the tectonic map of Figure 1 as the Almeria fault zone (*cf.* Green *et al.*, 1977). The gap between the two basement blocks may have been caused in part by rotation of the Alhamilla block. The rotation is suggested by the presence of reverse and normal faults along the north flank of the Sierra Alhamilla (see inset on Fig. 1), recording an anti-clockwise rotation. Syn-sedimentary faults in the fan-delta sediments indicate that the tectonic movements continued during the Pliocene.

Fan-Delta Sequence

The stratigraphy of the Pliocene succession (Fig. 3) of the studied area shows a consistent, simple (deltaic) sequence, with the following units from top to bottom:

1. Alluvial fan, coastal plain with lagoon(?) — Gador Formation
2. Transition zone (coast, delta front)
3. Upper delta slope
4. Lower delta slope
5. Prodelta

— Abrioja Formation

This sequence is well exposed at site 5, in the Rambla de Tabernas (Fig. 4), and is described below:

1. Alluvial Plain, Coastal Plain with Lagoon (?) - Gador Formation

Gravels, sands and muds mainly of fluvial origin erosively overlie subaqueous sediments of fan-deltaic origin. The fluvial sediments were deposited from braided streams,

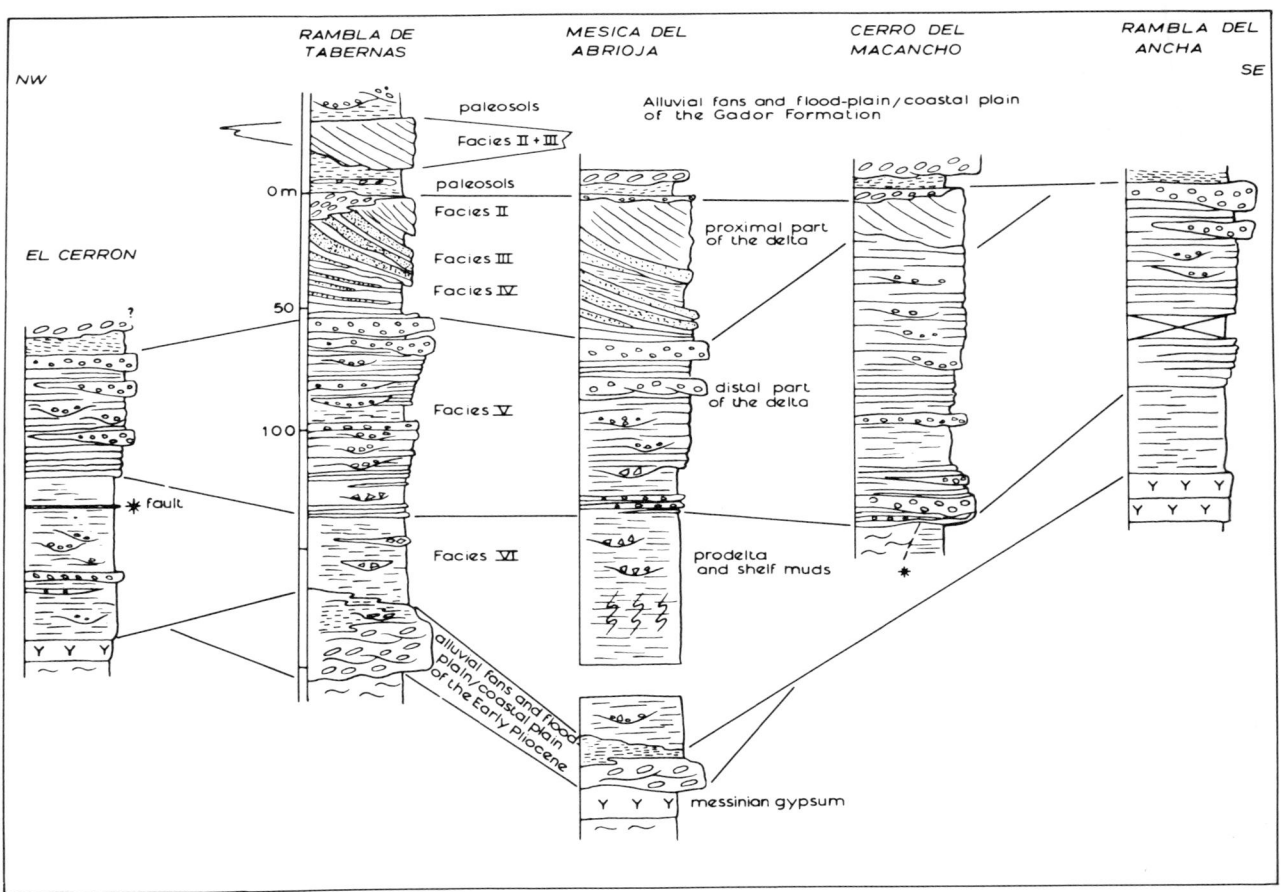

Fig. 3. Stratigraphic columns from the Rioja corridor.

and are interlayered with sediments of mass flow origin. They are further associated with finer grained floodplain deposits including paleosols with caliche nodules.

Over these paleosols, mottled sands and silts occur onlapping in a northwesterly direction. The sands are exposed near Rioja and Pechina where they contain reworked, as well as *in situ*, caliche nodules. The mottling is probably due to reduction around plant roots, remnants of which have been preserved. Scour-and-fill structures and channels as wide as 10 - 25 m are rather common in these sands and silts (Fig. 5A). Vague, hummocky-type cross-stratification is present on a small-scale. Lenses of poorly-sorted, massive conglomerates are embedded in the mottled sand and silt facies. Although more study is necessary to produce a detailed facies interpretation, the presently available data suggest a coastal plain origin with the subordinate development of lagoonal environments, with several subaqueous and subaerial stages related to sea level changes.

2. Transition Zone (Coast, Delta Front) - Abrioja Formation

Deposits representing the transition zone display great variability throughout the studied area. At many locations, the alluvial fan sediments overlie delta slope sediments of the Abrioja Formation without intervening nearshore sediments. At other sites, these two environments are separated by 'transitional' deposits consisting of a few decimeters of yellow (marine ?), pebbly sandstones and/or a few meters of greyish, fine sands and mud. The commonly poor exposure of these fine grained sediments (due to overcrusting) did not permit study of sedimentary structures. In other outcrops, thick chaotic, debris flow deposits occur directly above delta slope sediments with an erosive contact (Fig. 6). At location 'Torre Cardenas',

Fig. 5. A. Scour and fill in mottled layers with plant roots. These layers alternate with brown sands and poorly-sorted conglomerate lenses, probably representing coastal plain and alluvial fan deposits. **B.** Nearshore environment on top of deltaic sediments at location Torre Cardenas. Note cross-stratified mega-structures and flaser-type bedding in conglomerates and sandstones. The giant pillar under the arrow may have been produced by dewatering (hammer for scale).

Fig. 4. Section along the West-side of the Rambla de Tabernas (site 5), showing foresets of the proximal delta (2) and the distal delta (3). The delta section is overlain by grey coloured paleosols, from the Gador Formation (1). Details of the foresets are shown in Figure 13, the location of which is indicated by the white arrow. Foresets are approximately 30 m in height.

Fig. 6. Chaotic conglomerates (B) of Facies *I* truncate beds of the upper delta slope (A). The top of the exposure consists of stratified Facies *I* (C). Man for scale.

3 km north of Almeria (Fig. 1), the sediments are characterized by hummocky cross-stratification and thin pebble lenses. Giant pillar structures (Fig. 5B), possibly due to dewatering cut through these pebble layers. At site 2 (Fig. 1), a final type of transition is seen where a Gilbert-type delta records sediment discharge from the alluvial fan onto the delta front, and is described below as Facies *II*.

From the described lithologies alone, the depositional environment is not always clear. However, the sedimentary setting and the occurrence of wave-generated structures indicate a subaqueous environment above wave base.

3. Upper Delta Slope - Abrioja Formation

Bedding in the uppermost part of the delta slope is steeply inclined (6 - 25°) relative to paleosols of the Gador Formation, and is formed by giant, tangential foresets 30 m in height, up to 5 m thick and approximately 100 in length (Fig. 4). The sandstones and pebbly sandstones of these foresets represent mainly sediment gravity flow deposits and contain reworked nearshore fossils. These deposits, further described below in Facies *III* and *IV* (see Fig. 9), are indicated on the map of Figure 1 as 'proximal delta' deposits (together with sediments from the transition zone).

Abundant water escape structures such as pockets (bowl-shaped structures, partly filled with pebbles) and pillars (elongate structures) characterize these proximal deltaic sediments (Postma, 1983). Numerous slump scars are present in this part of the fan-delta.

4. Lower Delta Slope - Abrioja Formation

Beds of the lower delta slope are inclined up to about 6° relative to the paleosols. Conglomerate beds may form thickening- and coarsening-upward sequences (Rambla de Tabernas section, Fig. 4) totalling 30-50 m, with the uppermost strata as 8 m thick, lens-shaped conglomerate beds. Many beds can be described as bipartite beds and tripartite beds. In the case of a bipartite bed, a conglomeratic lower part and a pebbly mudstone upper part are distinguished, whereas in tripartite beds an extra sandy basal layer of variable thickness is present. Sedimentary structures of the conglomeratic (and sandy) lower part of these beds indicate deposition from gravel flows originating from the fan-delta. These sediments are described below as Facies *V* and indicated on the map of Figure 1 as 'distal delta'. They fill troughs which have steep, indented sides (Fig. 7) and which have a width of 10-100 m (as for instance the troughs in the Rambla de dos Areos and the Rambla del Ancha, Fig. 1 near arrows). Large troughs show composite infilling by gravel flow deposits and may show superimposed, smaller sized troughs separated by erosional remnants of mud and sand up to several meters in width, as shown by the dipslope view in Figure 8. Thus, within the canyon formed by rifting (Rioja corridor), there exists a second order of depressions being the large-sized troughs and a third order of depressions being the 10-20 m sized troughs.

5. Prodelta - Abrioja Formation

Heavily bioturbated muds with an abundant macro- and micro-fauna (Iaccarino *et al.*, 1975; Addicott *et al.*, 1978, see also Völk, 1967, the Cuevas Formation) dominate in the eastern part of the Almeria basin. The muds transgressively overlie older Miocene sediments, and interfinger with the conglomerates of the lower delta slope. They are interpreted as prodelta muds.

A more detailed description and interpretation of the subaqueous part of the fan-delta (Abrioja Formation) follows.

Fig. 7. Indented trough wall: Indentations are clearly following different lithologies, and are probably due to erosion in a channel. Mudstone layers are resistant and protrude from the wall. The height of the exposure shown on the photograph is *ca.* 5 m.

Fig. 8. Arrays of elongate gravel flow deposits (arrows), exposed on a dip-slope (site 4) are separated by fine sediments, . The gravels are probably deposited in troughs running downslope from the slump scars. Note the diffuse stratification upslope under the thin arrow indicating sediment aggradation at the tail of the flow. Paleo-slope is from right to left. Flows are 6-10 m in width.

ABRIOJA FORMATION

GENERAL

The general absence of structures indicative of stream flow (*e.g.*, cross-stratification or channelization), the steep canyon slope, the presence of slump scars and other sedimentary structures suggest that sediment transport occurred dominantly 'en masse' (*cf.* Carter, 1975; Middleton and Hampton, 1973, 1976; Nardin *et al.*, 1979; Lowe, 1979, 1982) due to slope failure.

In areas with relatively rapid sediment accumulation, such as fan-deltas, slope failure is common. In particular, when the slope is affected by extreme tides, waves or earth tremors, liquefaction may cause part of the slope to slide or slump downwards. Submarine slumping has been observed from onshore (*e.g.*, Andresen and Bjerrum, 1967) and in nearshore canyon heads of the Yallahs fan-delta (Wescott and Ethridge, 1982), and has been inferred from side-scan sonar images from deltas in fjords (Prior *et al.*, 1981a, b; 1982).

A comparison of the fan-delta models of the Abrioja Formation (Fig. 9) and the Pliocene Espiritu Santo Formation (Postma and Roep, in prep.) with side-scan sonar images of some Recent fan-deltas (Prior *et al.*, 1981a, b; 1982) resulted in a working model for mass flow-dominated fan-deltas as shown in Figure 10 (Postma, 1984). Both Figures 9 and 10 show the facies distribution in the Abrioja Formation.

A 'complete' deltaic sequence is easily accessible at site 5 (Fig. 1) at the west side of the Rambla de Tabernas.

SEDIMENT SOURCE

Throughout the entire delta sequence the main rock components are quartz and quartz aggregates (40%), garnetiferous micaschists (30%), detrital limestone, fragments of phyllosilicates, and (lime) mud. There is no significant change in the composition, and thus in sediment source, throughout the delta section (Postma, 1978). The clasts are derived from the surrounding metamorphic basements and from erosion of older basin sediments.

PROXIMAL DELTA SEDIMENTS
(DELTA FRONT AND UPPER DELTA SLOPE)

The proximal portion of the Abrioja Formation is divided into the following four facies (see Fig. 9):

Facies I — Chaotic conglomerates which are probably deposited from true debris flows passing, in part, over the entire delta into the canyon, and are therefore preserved throughout the delta sequence;

Facies II — Thinly-bedded, planar conglomerate foresets;

Facies IIIA — Giant tangential foresets of massive pebbly-sandstones which are probably deposited from liquefied flows (interbedded with *Facies IIIb*);

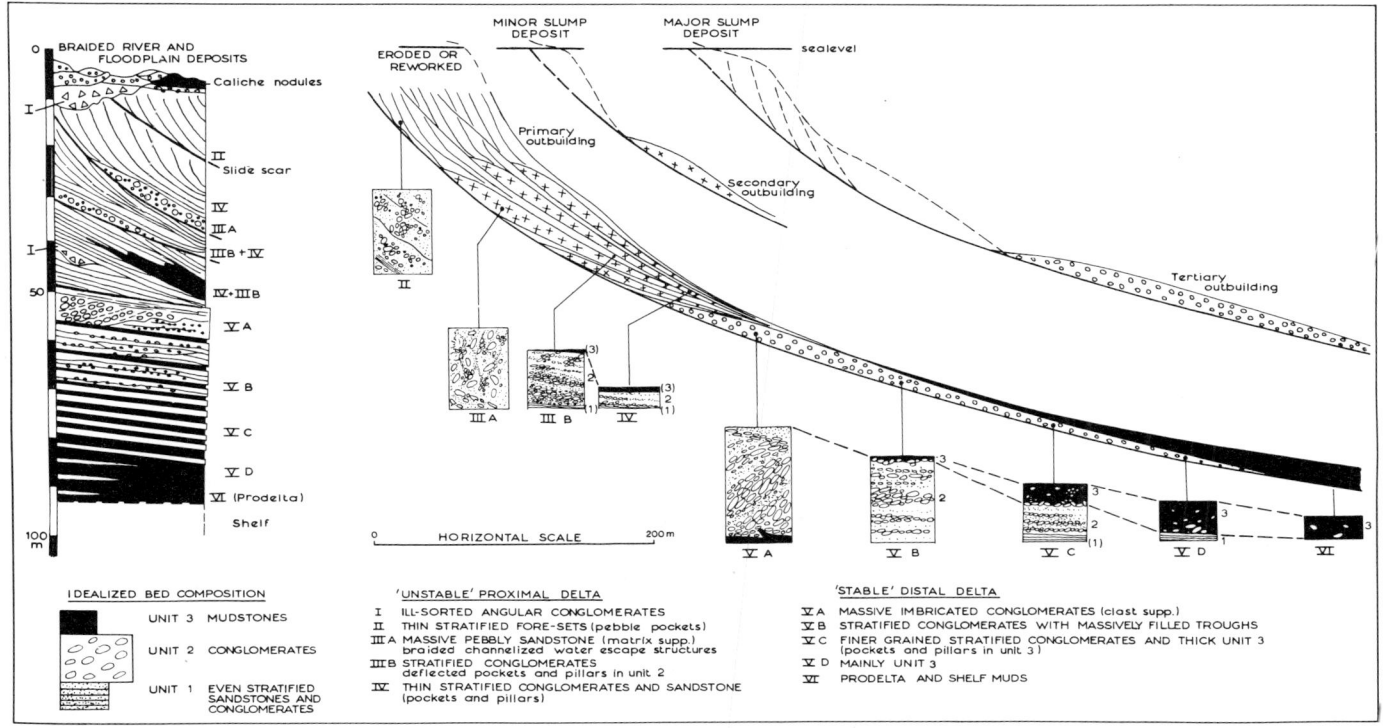

Fig. 9. Schematized model of the Abrioja fan-delta. Beach, nearshore, delta front and upper delta slope environments have been grouped as 'proximal delta sediments'; lower delta and prodelta as 'distal delta sediments' (*cf.* Fig. 1).

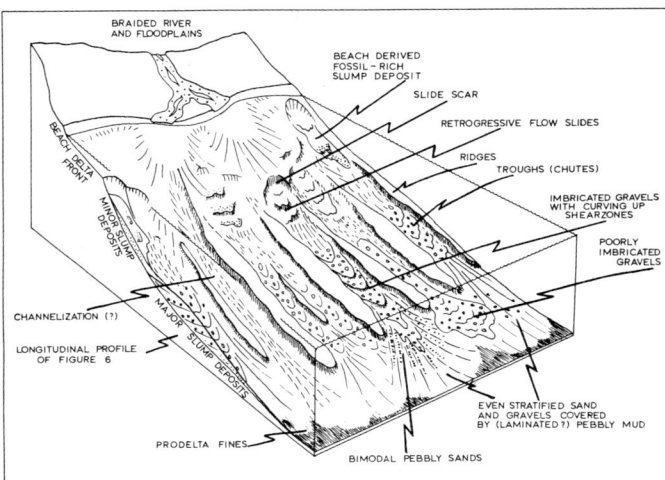

Fig. 10. Block diagram of the Abrioja Fan-Delta, a synthesis of the Abrioja Formation (Fig. 9) and side-scan sonar images of modern fan-deltas (Prior et al., 1981a, b; 1982).

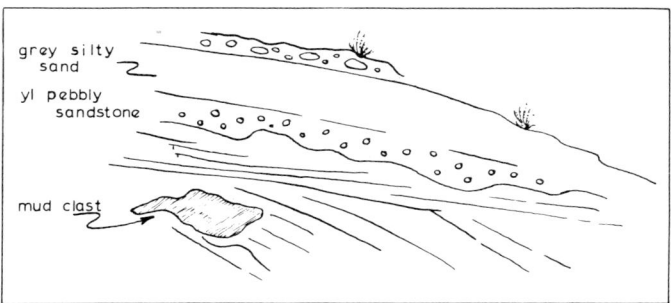

Fig. 11. Field-sketch of site 1. The structure is interpreted as the top of a Gilbert-type delta. Mud clast is approximately 75 cm. The overlying truncating sediments belong to the Gador Formation (site 1).

Fig. 12. Detail of delta foresets of Facies *II* of Figure 11. Lens cover diameter is 5,5 cm.

Facies IIIB — Giant tangential foresets of stratified, well-sorted conglomerates, probably deposited from high-density turbulent flows; and

Facies IV — Thinly-bedded bottomset beds of *Facies III*.

In addition, mudstones rarely cover the sediments deposited on the steep depositional slope, but cover the beds of *Facies IV*.

DESCRIPTION AND INTERPRETATION

Facies I

Beds of poorly sorted, disorganized conglomerates are up to 6 m thick. Conglomerates are mostly clast-supported, and consist of angular boulder-, cobble- and pebble-sized clasts. The interstices are filled with a poorly-sorted matrix composed of virtually all finer size fractions. Intraformational (sub-angular) mud clasts of various sizes may occur at any level in the beds. At a few locations, a cm-thick basal unit with inverse grading is present.

The characteristics of this facies suggest deposition from debris flows having cohesive as well as frictional strength (Fisher, 1971; Lowe, 1979). The lack of any size sorting suggests that grain interaction was unimportant, possibly damped by a dense, viscous matrix. However, contemporaneously derived mud clasts present throughout the bed suggest that the material within the flow was slowly churned.

Facies II

Thinly-bedded planar foresets, dipping 25°, show inverse and normal grading, with grain sizes ranging from fine sand to small pebbles. The pebbles may be weakly imbricated and matrix-supported (Fig. 11) in the planar foresets. At site 1, foresets are transitional upward into nearly horizontal beds (Fig. 12) similar to the topsets of a Gilbert-type delta. Some of these steeply dipping beds show evidence of fluid escape by pocket- and pillar structures and hydroplastic deformation. The location of this facies within the fan-delta sequence and its Gilbert-type delta form suggest a river outlet in the transition zone. The limited occurrence of the facies can be ascribed to its poor preservation potential due to slumping and reworking of sediment (wave action) on the delta front, or to erosion by prograding alluvial systems during a lowering of the sea level.

Facies IIIA

Giant foresets, dipping up to 25°, have tangential bases (Fig. 13). The foresets are up to 30 m high, up to 100 m long and up to 5 m thick. The beds are rarely intersected by thin (*ca.* 5 cm), finer-grained, (?) shear zones. The beds of pebbly sandstones are non-graded containing pockets and pillars indicating early post-depositional dewatering (Postma, 1984). Along the bases of these beds, pockets filled with gravel occur preferentially above fluid escape structures present in the underlying beds.

Such structures are probably related to squeezing of the latter beds during deposition of the overlying pebbly

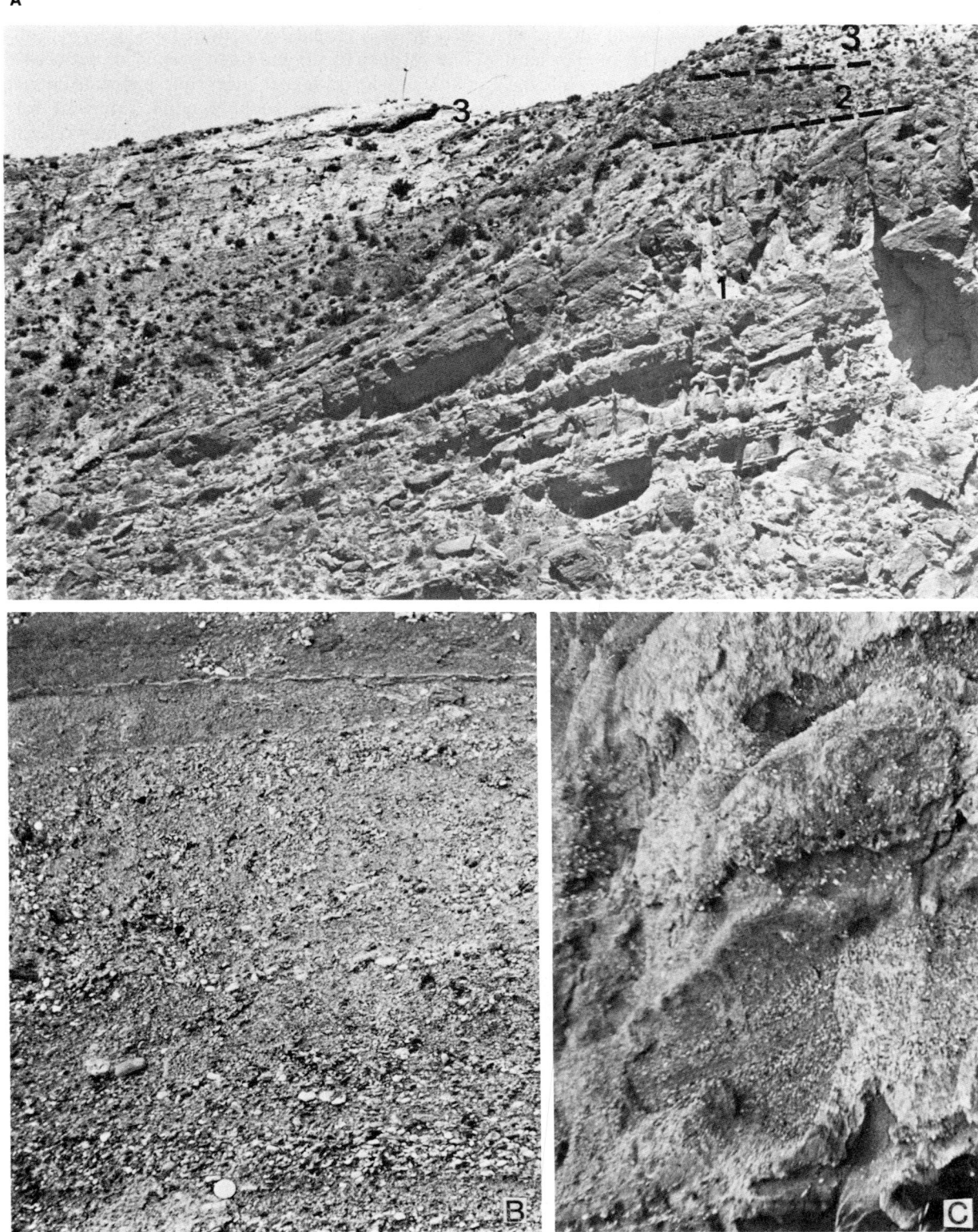

Fig. 13. **A.** Concave foresets of Facies *III* (1), are truncated by Facies *I* (2) conglomerates, which are covered by (3), the Gador Formation (site 5). Height of section is about 30 meters. The sets protruding from the outcrop are the pebbly sands of Facies *IIIA*. **B.** Detail of the upper unit of Facies *IIIB*. Note some outsized pebbles, undisturbed stratification under the coin, and disturbed stratification higher up, probably due to liquefaction and dewatering in the upper unit (coin for scale). **C.** Detail of the foresets in A. Thin, tongue-shaped pebble layers form a non-graded, stratified bed covered by pebbly sandstones of Facies *IIIA*. Stratified bed is 1,20 m. thick.

sandstones. On the other hand, the fluid escape structures isolated within the pebbly sandstones together with the structureless nature of the foresets, indicate mixing of water saturated sediment during flow. The mixing may have been due to slow churning within the flow, while the flow itself behaved plastically. Most of these sediments were deposited on a steep slope indicating rapid freezing of the flow. Based on these features, it may be concluded that the pebbly sandstones originated from liquefied flows. The upper part of such flows may remain liquefied longer than the base (Middleton, 1967, p. 495), and therefore could flow farther downslope which resulted in foresets with a tangential base.

Facies IIIB

Well-sorted layers of rounded granules and pebbles are inversely-to-normally graded or non-graded. The layers probably have a tongue-like geometry averaging 50-200 cm in width and 10 cm in height and show a well-organised lateral stacking similar to the compensation cycles of Mutti and Sonnino (1981). Beds of these facies may also show an overall inverse-to-normal trend by thickening then thinning upward, together with a corresponding coarsening and fining upward of the grain size. Such beds generally start with a sandy base. If beds thin and fine upwards, water escape structures may be present in the upper pebble layers (Fig. 13B). In some of these pocket structures outsize-clasts are present. At one locality, an overall normally graded bed was found to be composed of pebble sheets which thin upwards (Fig. 14). The fabric within the layers shows a tight packing of the pebbles and granules, and a weak imbrication of the pebbles. The size and sphericty of the pebbles did not permit measurement of *a*-axis orientation. The layers are dominantly massive, but inverse-to-normal and normal grading occurs within the layers.

The overall normal grading in many beds suggests waning flow conditions with a sufficiently low concentration to permit grain-size sorting. The fluid escape structures in the top of the bed record processes of liquefaction which may be due to rapid deposition of the lower part of the bed. The absence of silt and finer grain sizes between the pebble layers also suggest short time intervals between their deposition. Hence, such sequences may be formed by deposition from one gravel flow. They may resemble traction carpets at the base of a high-density turbulent flow, as described by Lowe (1982). Furthermore, the gravelly 'tongues' may record flow separation indicative of lateral variations in sediment concentration. However, the only evidence for turbulence in these gravel flows is the overall normal grading of the complete bed. Such grading may indicate waning flow conditions of flows with a fluidal behaviour such as high-density turbulent flows, but may also be due to, for instance, a lateral migration of the flow path.

Non-graded beds lacking the normally graded upper part (see for example Fig. 13C) are particularly difficult to explain as traction carpets. Alternatively, they may have resulted from periodic slumping of relatively small masses of sediment. Finally, the absence of muds can be ascribed to winnowing. More fieldwork and experimental work are needed in order to be explicit about the origin of these beds.

Facies IV

The above described foresets of Facies *III* are transitional into thinly-bedded, pebbly sandstones and laminated sandstones, which are interbedded with beds consisting of amalgamations of thin, chaotic gravel layers with both matrix and clast-supported fabrics (Fig. 15). Beds contain isolated clasts up to boulder size, and alternate commonly with homogeneous fine-medium sand and (pebbly) mud. Dewatering structures, mainly of post-depositional origin, are present and have been described as gravel-filled pocket and pillar structures (Postma, 1983). The fluid escape structures are in some cases aligned horizontally, and are not necessarily limited to the same bed. Muddy beds and mud drapes display bioturbation structures.

The laminated sandstones may originate either from low-density turbulent flows surging ahead of a liquefied flow of Facies *IIIA*, or from post-depositional winnowing of sediments on the upper delta slope and front. This process of winnowing may be accomplished by either water currents, or, as shown experimentally, gravity-winnowing (Postma, 1984). The latter process is reviewed in more detail at the end of this paper. The occurrence of the larger outsized clasts at the base of the steep slope is probably due to one of the two winnowing processes, and possibly due to a sorting process within the liquefied flow; large clasts tend to concentrate in the head of a flow due to their greater density.

The predominance of mud layers within this facies probably decreased the permeability, and may have promoted the build-up of sufficiently high pore-fluid pressures to liquefy and fluidize conglomeratic sediments.

Fig. 14. Stratified conglomerate bed, consisting of well-sorted graded pebble layers, which are thinning upwards (Facies *IIIB*). Because of the overall normal grading and coarse grain size, the bed may have originated from a high-density turbulent flow. Note the slight normal grading of the individual pebble layers. Lens-cap is 3,5 cm.

Fig. 15. Beds, consisting of amalgamations of thin, chaotic conglomerate layers (Facies *IV*) with outsized clasts are interbedded with pebbly-sandy toesets of Facies *IIIA* which are just visible at the top of the photograph. Flow is from right to left. The arrows indicate the location of a water escape pillar. Hammer for scale.

Fig. 16. Slump scar (arrows) truncates beds of the transition zone and upper delta slope. It is filled with foresets of Facies *III* and *IV*. Exposure is approximately 80 m in height.

SLUMP SCARS

The previously described facies of the upper delta slope are often truncated internally by smooth concave-upward surfaces. In longitudinal cross-sections, these truncation surfaces dip downslope at an angle that proximally exceeds the angle of slope, but distally converges to the latter (Fig. 16). They are draped by either thin laminae of mud and/or fine sand (sometimes bioturbated) or by thick (*ca.* 50 cm) pebbly sandstones, the base of which may be loaded into the underlying sediments. At some localities, water escape structures clearly increase below such surfaces.

The surfaces are interpreted as slump scars, mainly because of their geometry, smoothness (McCabe, 1978), setting, and the frequent similarity of sediments both above and below the scar (Laird, 1968; Clari and Ghibaudo, 1979).

Gravel flows derived from the slump scars in the upper delta slope have been transported farther downslope and are described hereafter. The slump scars have been refilled by the above described sand and gravel flows which have a volume of approximately 10,000 m^3. Homogeneous pebbly sands (Facies *IIIA*) containing *Balanus* and *Ostrea* sp. and well-rounded pebbles indicate that parts of a beach were also subject to slumping (*e.g.*, Buisonjé, 1974, p. 193; MacGillavry, 1970). These fossil-rich beds dominate along the northern side of the canyon.

DISTAL DELTA SEDIMENTS
(LOWER DELTA SLOPE)

The distal portion of the Abrioja Formation is divided into Facies *V* and *VI*. These facies consist of one or more subfacies, which are below described as units 1, 2 and 3 representing:

unit 1 — stratified, well-sorted sandstones and conglomerates

unit 2 — dominantly massive conglomerates

unit 3 — (pebbly) mudstones

The general location of the units is indicated in Figures 9 and 10. As defined, Facies *VI* consists of amalgamations of unit 3 only.

UNIT 1

Unit 1 consists of evenly bedded strata of well-sorted sediments lying parallel to the depositional slope. These strata are represented by:

a) fine to coarse sand with parallel laminae and occasional discoid pebbles lying parallel to the lamination (Fig. 17);

b) centimetre thick layers of either inversely graded, inversely-to-normally graded or normally graded sands (Fig.18);

c) thick layers up to 10 cm, of pebbly sands, both matrix- and clast-supported;

The most common types of unit 1 beds are mixtures of type (a) and (b) strata, which form beds of approximately

Fig. 17. A tripartite bed (Facies V), with evenly stratified sands (unit 1) at the base, inversely-to-normally graded gravels in the center, (unit 2) and pebbly mudstone at the top (unit 3). Lens cap is 3,5 cm.

Fig. 18. Detail of an evenly stratified unit 1, as exposed at the base of a massive conglomerate bed, with weakly developed coarsening upward cycles. The photograph shows two of such cycles above the hammer.

10-30 cm thick. Individual layers truncate each other at low angles. In rare cases, thin veneers of mud occur between the individual centimetre-layers.

In units up to a metre thick these strata types may form several coarsening upward cycles as shown in Figure 18, and may grade with diffuse boundaries into pebbly-sand layers, whereby the top of the sequence may be interrupted by small-sized troughs (Fig. 19). The troughs have a stratified conglomerate fill with upslope-dipping cross-stratification. As inferred from the general slope direction, these foresets do not reflect slipface sedimentation, but are evidently produced by upslope aggradation of sediment. A cross-section through these troughs (Fig. 19A) shows pebble layers, transitional into granule and sand layers, thinning and wedging out laterally over a distance of 2 m.

Repeated deposition of small quantities of sediment, can be inferred from mud veneers, truncations and isolated pebbles lying parallel to the laminae. The grading of the cm-layers suggests deposition from grain flows (inversely graded layers), and small turbulent flows (normally graded layers). Generation of these small turbulent flows and grain flows may be by surging flows and will be discussed later.

UNIT 2

Two types of massive, clast-supported conglomerates have been distinguished. The first type has a weak imbrication and the second type has a strong imbrication. Both are poorly-sorted, have a matrix consisting of all sand sizes, and contain a clay and lime mud component of generally less than 5% volume. The average grain size is 2-4 cm with a maximum clast size generally not exceeding 10 cm.

Type 1 Conglomerates

In this type the largest clasts are evenly dispersed throughout the structureless unit. Some of the clasts have a-axes imbricated upslope. Laterally these units may show gradational development into very diffuse, non-graded, horizontal banding.

The fabric of these conglomerates closely resembles that of debris flows (Fisher, 1971; Lewis *et al.*, 1980) and density-modified grain flows (Middleton, 1970; Lowe, 1976) with plastic flow rheology (Johnson, 1970; Hampton, 1975); the low mud content in these deposits suggests that they had a dominant frictional, but low cohesional strength.

Type 2 Conglomerates with Sigmoidally Curved Shear-zones

The second, more strongly-imbricated conglomerate type has a texture similar to the first, but its asymmetric gravel body (Fig. 20) is cut by sigmoidally curved surfaces dipping upslope (Fig. 21). These 'backsets' are demarcated by either a centimetre-thick, sandy zone or a thicker zone with pervasive, intense pebble alignment. These zones are interpreted as shear zones. At the base of some type 2 beds, ripped-up mudlayers are complexly folded into the overlying conglomerate bed, but have one end still attached

Fig. 19. Facies V, unit 1 **A.** Section normal to the slope showing even strata of well-sorted sand, which are truncated by massive and stratified gravels which fill small troughs. **B.** Longitudinal section with parallel bedded gravel layers (at the top of the photo) truncating a peculiar kind of cross-stratification, which is typified by structureless sands between the pebble layers and total absence of structures indicating slipface sedimentation. These sets are dipping upslope and are interpreted to be formed by upslope sediment aggradation (flow to the left). Sequences of this type may reflect sedimentation from a prograding front of a gravelly flow with a low plasticity. Site 3, in the Rambla de la Palmilla. Hammer is in both photographs at the same location.

to the mud substrate (Fig. 22). Associated with both type 1 and 2 conglomerates are peculiarly folded mud layers and hydroplastically-deformed underlying strata. In the downslope direction, type 2 conglomerates are transitional into type 1 conglomerates.

The folded mud clasts and rip-up structures suggest slow movement of the overriding flow. The sigmoidally curved zones occur above the rip-up structures and suggest firm shearing within a relatively 'dry' (already deposited?) sediment mixture. Both folding and shearing may have been a response to compression transmitted in a downslope direction. Such compression may have been produced by sediment accumulation at the tail of the flow.

Arcuate ridges running transverse to the slope have been observed on the surface of a modern subaqueous debris-flow deposit and have been related to compression within the flow (Prior *et al.*, 1982); although internal structures are not visible on the sonar images, it was reasoned that the only possible way to form these ridges was to thrust the sediment in a downslope direction. Internally, the result will be concave-upwards shear zones (*cf.* the shear zones in the terminal regions of glaciers, Shaw and Archer, 1979). Intensive or repeated shearing will probably result in size segregation, with either diffuse sandy zones or pervasive alignment of pebbles along the shear zones. However, similar 'shear zones' may develop by sediment aggradation at the upslope side of a stagnant gravel body. Experimentally, both kinds of shear zone have been reproduced.

UNIT 3

This unit consists dominantly of homogeneous pebbly mudstones which vary in thickness. The granules and pebbles are commonly scattered in the mud, but are also present in small concentrations forming pockets and/or pillars. The term mud is used here to denote a mixture of medium and fine grained sand, silt, clay, and large mica flakes. Mud sequences up to 30 cm thick may consist of thin, normally graded sand and silt layers and rare cross-bedding. At some other locations, ghost-signs of parallel laminae, single pebble layers and cross-stratification are observed.

Based on its primary and semi-obliterated sedimentary structures, it appears that mud units are an amalgamation of small, probably turbulent flows (normally graded sand and silt layers) and traction flows (cross-bedding). Their present homogeneity is probably due to later modification by bioturbation, pore-water expulsion (fluid escape structures) and finally, but most importantly, due to liquefaction and downslope creep of the initially stratified mud. Downslope movement is suggested by downslope verging, bent pillar structures and by hydroplastic folding and fragmentation of the laminated muds. Locally, even true debris flows (Middleton and Hampton, 1976) may have developed and transported the sediment further downslope along the canyon axis, enhancing total homogenization.

Contacts between the conglomerates and sandstones of unit 1 and 2, and the pebbly mudstones of unit 3 often show signs of reworking. Such contacts are variable and may be either sharp or gradual, and are either undulatory or straight (Fig. 23). The contacts show that the debris flows modified the top of the underlying gravelly sediments by plucking and incorporating pebbles in the overlying moving mud.

At the base of the pebbly mudstone, outsized and imbricated clasts occur (Fig. 24). This was probably the result of settling of these clasts through a flow in which its strength had been greatly reduced by water saturation. These clasts were also imbricated by the motion of the overriding pebbly mudflow. The smaller clasts remain scattered throughout the mud. Localized pebble pockets may be due to early post-depositional fluid escape.

Pebbly muds are present in Recent canyons, as for example, in the Wilmington Canyon. According to visual

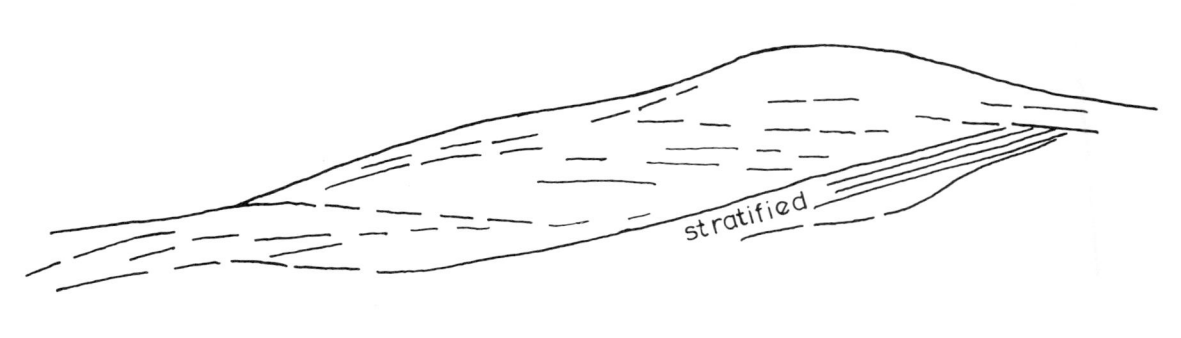

Fig. 20. Sigmoidal, upwards-curving shear zones ('backsets') in an asymmetric type-2 conglomerate lens. Pebble imbrication and general flow direction is to the left (site 5, Rambla de Tabernas). Height of flow is *ca*. 3 m.

Fig. 21. A detail of an upwards curving shear zone, as present in type-2 conglomerates (unit 2), is shown just above the hammer head (site 3, in the Rambla de la Palmilla). The shear zone is finer grained and dips slightly upslope. Flow toward the left.

Fig. 22. Rip-up structures in which a complexly folded mud layer is still attached to the mud substrate. Flow was toward the left. These structures probably record a slow, final-stage flow of the debris (Facies V). Hammer is 35 cm.

Fig. 23. Wavy contact between pebbly mud (light grey) and gravelly sediments (Facies V). At the contact the gravelly strata are truncated indicating erosion. Just below the pebbly mud, the gravelly unit shows a massive pebble lens of unit 2 sediments, which truncates evenly strata of unit 1 sediments. On top of the pebbly mud, another bed of type 1 conglomerates can be seen.

Fig. 24. Outsized clasts at the base of a pebbly mud (light grey) deposited on top of the evenly stratified sands of unit 1, which are deposited on another pebbly mud layer. Note the cm-stratification in unit 1 and discoid pebbles paralleling the stratification.

observations from a research submersible (Stanley, 1974), mud and pebbles were winnowed from the canyon walls and adjacent shelf. Mud formed the dominant surficial blanket in the canyon and became enriched in pebbles toward the canyon axis. The texture and fabric of the pebbly muds from the Wilmington Canyon closely resemble the pebbly mudstone of unit 3. Stanley (1974) discusses further the probability that the pebbly mud accumulations in the canyon axis moved farther downslope as debris flows. Slumps from the canyon wall accumulated on the axial fill and may have triggered such debris flows in the canyon axis.

The pebbly mud unit of the Abrioja Formation, therefore, may have originated during periods in which 'normal' sedimentation conditions prevailed on the fan-delta and in the canyon. Muds were deposited in the canyon due to winnowing from the canyon wall, higher shelf, and may have originated from suspension clouds accompanying the gravelly flows. Later remobilization of these muds as debris flows reworked the primary bedding structures into pebbly muds. Therefore, these layers can be used as time lines, 'sandwiching' catastrophical episodes of gravel and sand sedimentation.

COMPOSITE CONGLOMERATE BEDS

Mud layers are significant in a mass-flow dominated subaqueous environment, because they reflect periods of 'rest' or periods during which 'normal' sedimentation conditions have been maintained. Unit 3 confines the catastrophic events which brought large amounts of gravel and sand into the canyon. Such catastrophic bodies are often amalgamated and this may be recognized by mud remnants of unit 3. In the absence of such mud layers, other features are utilized. Fining- and thinning-upward stratification, water escape structures in the top of beds (Postma, 1983), or sole-marks are generally helpful to define the boundaries of one depositional event.

The downslope and lateral relationships between unit 1 and 2 are evaluated below. They form, together with unit 3, bipartite and tripartite beds.

BIPARTITE BEDS

Bipartite beds are formed by massive, weakly imbricated conglomerates (unit 2, type 1) and pebbly mudstones (unit 3). They are commonly found in beds of 0.4 -2 m thickness. Such beds have a flat lens-shape with sedimentary structures depicted in Figure 25. The base of the bed consists of a thin, inversely graded zone, and may show injection and flame structures. On the sole of some beds, both grooves and load casts are present. At some locations, the top of the conglomeratic unit is characterized by diffuse stratification, either composed of poorly sorted gravel or pebbly sand. Figure 26 shows a dipslope view of shallow, straight-sided troughs filled with pebbly sand issuing downward from a massive, polymict conglomerate body inferred as the head of a gravelly flow. This configuration corresponds to point A in Figure 25 and to point B in Figure 27.

Another type of bipartite bed is composed of unit 1 and 3 (Fig. 23), probably representing the distal parts of Facies V gravel flow deposits (Fig. 9, profile Vc). Its origin will be discussed later.

TRIPARTITE BEDS

Some of the beds of the Abrioja Formation show tripartitions and are composed of 1) a thin, stratified basal unit, 2) a massive conglomeratic central unit which is at some

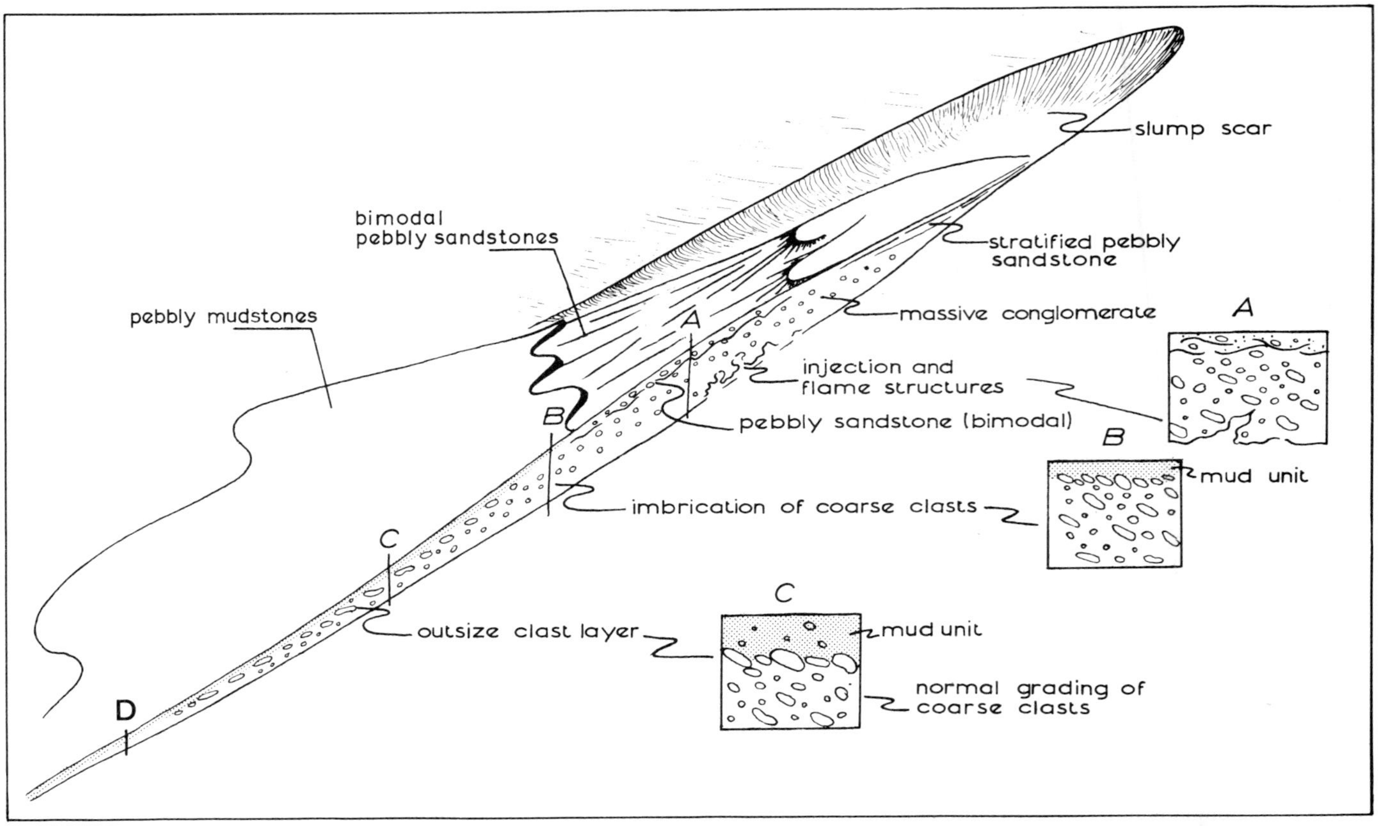

Fig. 25. Schematic drawing of a bed with bipartition. Note that stratified sediments occur here in the proximal part of the gravel flow deposit (see Fig. 8). This stratification is probably due to retrogressive flow sliding.

Fig. 26. **A.** Inferred head of a gravel flow deposit (center of photo) is surrounded and partly covered by pebbly sandstones, which have been deposited in small, straight troughs issuing from the head. **B.** Close-up view of small, straight troughs. Locality is indicated on A.

locations again covered by unit 1-type sediments, and 3) a pebbly mudstone upper unit (Fig. 17). The total thickness of units 1 and 2 ranges between 0.5 - 8 m, measured between two mud layers. The association of tripartite beds is depicted in Figure 27. Contacts between units 1 and 2 are free of mud, undulatory (with a wavelength length of *ca.* 20 m), and either sharp or gradational, without a change in matrix texture. The common superposition of unit 2 over unit 1 strongly suggests a genetic relationship.

At the top of the bed, upwards-curving shear zones are either truncated or show a downslope bend to become sigmoidal and transitional into diffuse stratification which thins and fines upwards. Downslope, at point B and C of Figure 27, laminated, well-sorted sand of unit 1 may cover the conglomerate, or are directly exposed on pebbly mud. The units wedge out downslope and laterally, covered by pebbly muds.

ORIGIN OF COMPOSITE CONGLOMERATE BEDS

A composite bed of mass flow origin as described above may, depending on characteristics of its sedimentary units 1 and 2, originate from the following processes:

1) flow segregation
2) flow surges
3) combinations of above mentioned processes.

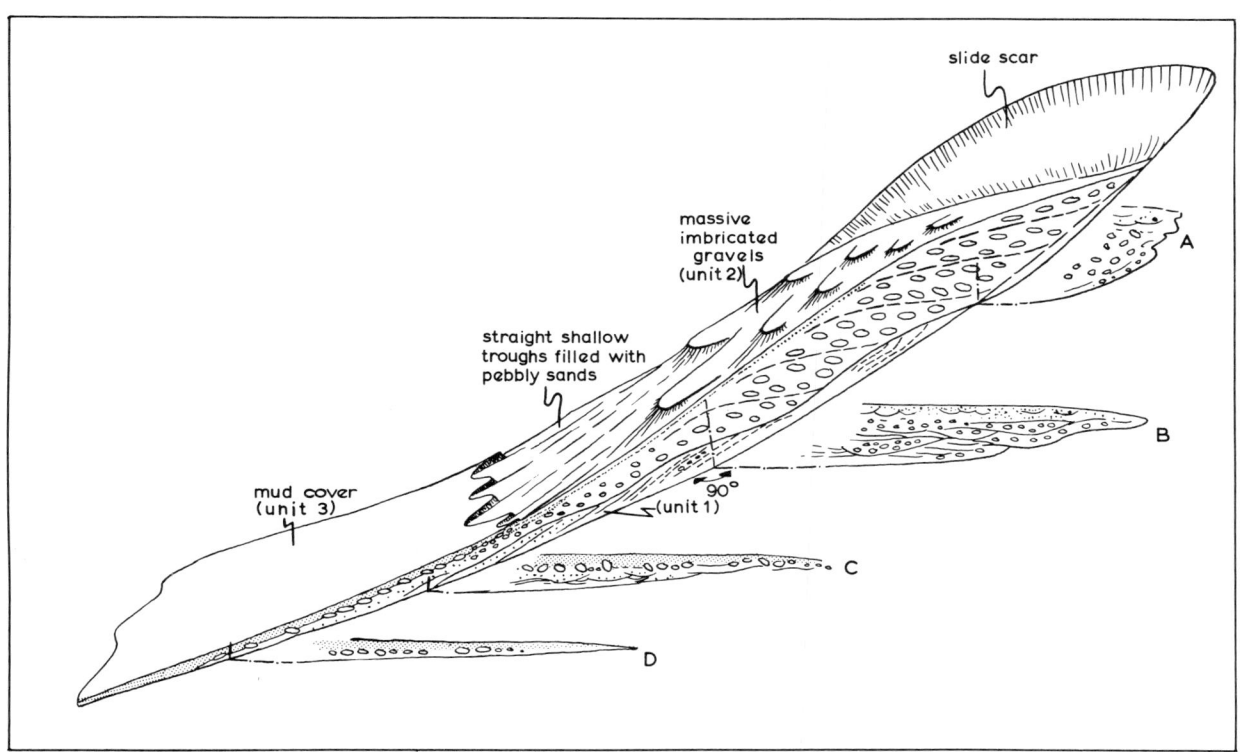

Fig. 27. Schematic drawing of a debris flow deposit, with tripartite beds (units 1-3). Evenly stratified sediments at the base of these deposits do not reflect basal shearing, but are interpreted as representing thin grain flows and turbulent flows due to surging flows issuing from the unstable margins of a stagnant (?) polymict gravelly flow.

1) It has been proposed that basal segregation of a flow due to basal shearing and laminar flow occurs in flows of a high sediment concentration exibiting a plastic flow behaviour (Johnson, 1970; Fisher, 1971; Middleton and Hampton, 1973; Aalto, 1976; Walker, 1978; Hiscott and Middleton, 1979; Lowe, 1982; Postma *et al.*, 1983). Corresponding sedimentary features are inverse grading and, if present, diffuse stratification at the base of the deposit. Alternatively, high-density turbulent flows are thought to form an inversely graded series of beds due to spasmodic sedimentation of traction carpets at the base of a highly concentrated suspension during waning flow conditions (Carter, 1975; Walker, 1978; Hiscott and Middleton, 1979; Lowe, 1982).

2) The term flow surge refers to periodic surging of sediment suspensions producing stratified deposits. Surges may originate from i) fluidization in the head of a high-velocity pyroclastic flow (Wilson and Walker, 1982); ii) retrogressive flow slides which may build up a sequence of thinning and fining upward strata due to an upslope migration of the slump scar (*cf.* Postma *et al.*, 1983); iii) gravity-winnowing at the head of a stagnant (or creeping) gravelly flow (Postma, 1984), whereby surges of thin turbulent flows and grain flows are winnowed from its unstable margins by liquefaction and the action of gravity. Experimentally, such flow surge deposits result in stratified sediments (unit 1-type) which may be covered by sediments of the main flow.

3) Combinations of processes are possible, and are discussed by Wilson and Walker (1982) and suggested by gravel flow experiments described in this paper.

In conclusion there seem to be three important, closely related units in gravel flows of the Abrioja Formation: —

1) a basal unit, due to flow segregation or flow surge;

2) a central unit, due to a dominantly viscous main flow, whose top is modified by either flow segregation of a high-density turbulent flow (Walker, 1978; Lowe, 1982), flow surges from retrogressive flow slides, or sedimentation from fine grained suspension flows (the latter may be incorporated in '3'); and

3) an upper unit, due to 'normal' conditions of sedimentation, which has been modified into homogeneous pebbly muds via redeposition by true debris flows.

ORIGIN OF TRIPARTITE BEDS

Basal Zone

Evenly stratified basal zones, described here as unit 1-type sediments at the base of tripartite beds, have been described by Fisher and Mattinson (1968), Fisher, 1971 (Fig. 9), Schmincke (1967), who described the base of a lahars, and Aalto (1976). Speculations on the origin of such a basal zone were concentrated on possible effects of flow segregation due to grain interactions at the base of highly concentrated flows. However, similar zones that develop from basal shearing are expected to be inversely graded due to the effect of dispersive pressure (Bagnold, 1954) and/or kinetic filtering (Middleton, 1970).

Although the unit 1-type sediments resemble deposits from traction carpet deposits, they are, however, never covered by an assemblage of finer grained sediments to complete a sedimentary sequence for high-density turbulent flows.

At the base of a high-speed pyroclastic flow, a basal layer, comparable to that described here, was explained by flow segregation (Wilson and Walker, 1982). Alternatively, a similar basal layer was attributed to high shear velocities by Moss et al. (1980). In flume studies, they demonstrated the existence of a rheological layer (ballistic dispersion). Such a layer was formed during supercritical flow over sand containing small amounts of clay, silt and pebbles. The result was a plane-stratified basal layer of fine and medium-grained sand covered by a 'pseudo-planar' coarser-grained layer. It is, however, rather unlikely that gravelly flows of the Abrioja Formation assumed velocities comparable to those of pyroclastic flows. And according to Wilson and Walker (1982), these basal zones will not form at velocities that are much lower. This is because a high velocity is needed to form a ballistic dispersion. Hence, the basal unit in the Abrioja tripartite beds (see unit 1 of Fig. 25) requires another explanation.

Considering the geometry of the two types of gravel flow deposits described here (see Figs. 25 and 27), a water-rich sediment mixture is inferred for a flat lens-shaped and a relatively dry sediment mixture for an asymmetric lense-shaped unit 2 forming, together with unit 1 and 3 the bipartite and tripartite beds respectively. With this in mind, experiments were carried out in an attempt to reproduce the sedimentary features of such gravelly flows.

Experiments

A tilted tank 4 m in length and 25 cm in width was filled with water of approximately 15°C. A sediment mixture with grain sizes similar to those of the gravelly beds of the Abrioja Formation was used including a lime mud matrix (volume, 5%), but having a maximum pebble size of only 2.5 cm. Subaqueous mass flows were produced by opening a valve at the top of the slope which decreased from 28° to 5° distally. Because of the limited length of the tank, observations were confined to the upper 2 m of the slope where sediment transport was gravity dominated and not influenced by unwanted water circulation. The experiments were filmed.

The plasticity of the sediment mixtures was measured according to a standard method used for plasticity measurements of concrete-cement mixtures. In this method, the plasticity scale range from 1 to 4, with the latter representing extremely wet mixtures. In the experiments, high-plasticity mixtures and low-plasticity mixtures had value 2 and 3 respectively.

After the valve was opened, the high-plasticity mixtures accelerated and rapidly became turbulent. Flow thicknesses ranged from 5 to 10 cm. After less then 2 m of downslope flow it became internally stratified into two parts: a basal, slower moving traction carpet of clast supported gravel, and a faster moving upper portion of pebbly sand, with the two zones being separated by a density/velocity boundary (Postma et al., 1984). At and somewhat above this boundary the suspended pebbles assumed a slightly imbricated configuration. With 'freezing' of the basal part, a new traction carpet of the pebbly sand above was developed which also became finally frozen.

In contrast, the low-plasticity mixtures did not move downslope after the valve was opened. The steep fronts of the stagnant sediment body were subject to a specific mode of resedimentation. In this process, lime mud matrix and fine sand fractions were elutriated from the front of the stagnant sediment mass, probably due to the settling of coarser grains and accompanying pore fluid expulsion, and were transported farther downslope as thin, surging turbulent flows. With increasing escape of the fines, successively coarser grains became undercut and unsupported, and these then started to move downslope as thin grain flows and turbulent flows. The winnowing of progressively coarser fractions was probably enhanced by the lower frictional resistance of fine grain sizes. Because this selective 'winnowing' involved progressively coarser fractions, the resultant deposits are rather well-sorted and evenly stratified and slightly coarsening upwards as the winnowing progressed, until even pebbles and some small chunks of sand containing suspended pebbles were subjected to resedimentation and produced the stratified sediments (see Fig. 28). It is worth noting that the entire process occurred on the steep static front of a stagnant gravelly body, and that no eroding currents were involved. Because of being solely governed by gravity, the entire process described above has been named 'gravity-winnowing'.

Addition of a new sediment portion at the back of the flow, to increase the instability and to simulate a retrogressive sliding process, resulted in post-depositional thrusting within this gravelly sediment. This process produced offsets within the main sediment body. The thrusting occurred along shear zones dipping upslope, which were subsequently deformed if the entire sediment mass was sliding further downslope (Figs. 25 and 30; see also experiments of Yoxall, 1983, p. 178-186; and Schwarz, 1983) over the stratified sediment deposited ahead of it by gravity-winnowing. Otherwise, undeformed 'shear zones' were preserved if the newly added sediment was insufficient to produce some major sliding of the gravelly mass. A portion of this sediment aggraded upslope of the stagnant sediment body, whereas the rest of the sediment flowed over its top.

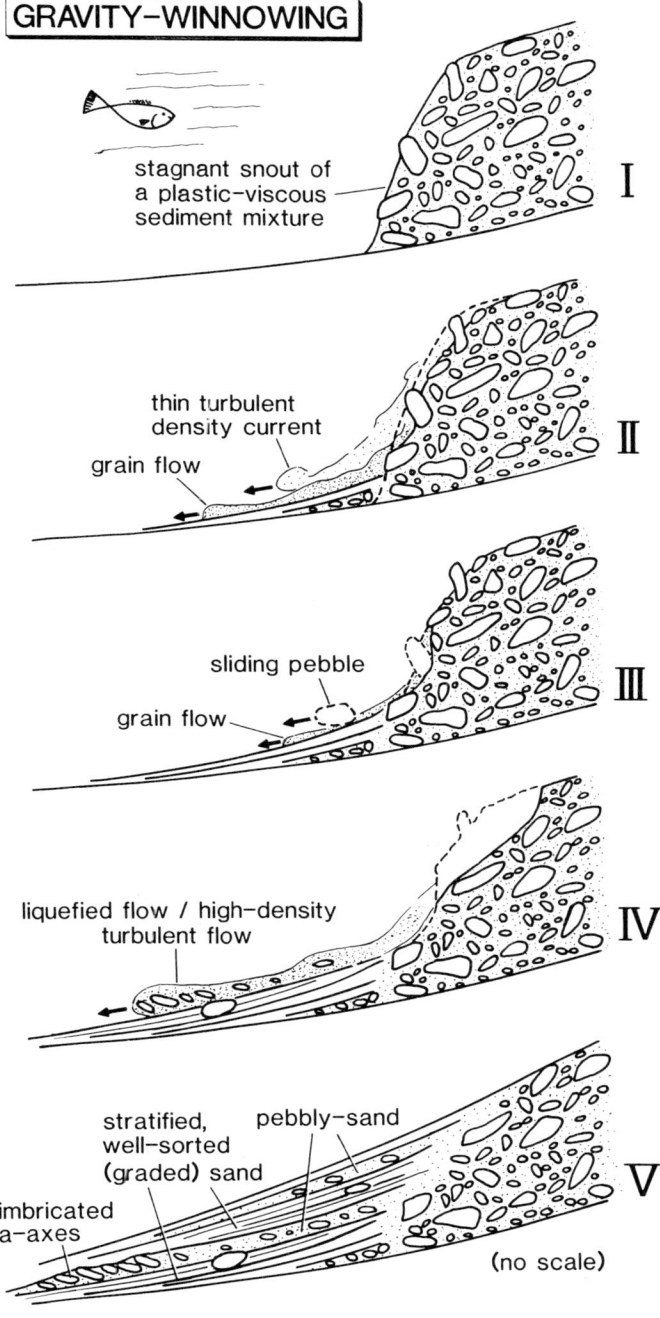

Fig. 28. Schematic illustration of the gravity-winnowing process from the snout of a stationary, but unstable, plastic-viscous sediment mass, as observed from filmed experiments: **I.** A stationary, unstable sediment mixture with a yield-strength which is proportional to the height of the snout in a static body water; **II.** Elutriated mud and fine solids, probably due to liquefaction (settling of pebbles), form thin density currents (clouds). Due to a loss of matrix support, sand-sized grains flow down in thin turbulent flows and grain flows; **III.** An undercut pebble slides down within or on top of a thin grain flow and assumes a flat, or slightly imbricated position on the bed surface; **IV.** Failure of a part of the snout causes a large sediment portion to slide downward. This portion is quickly transformed into a liquefied flow and/or a high-density turbulent flow. This type of flow produces an upslope-dipping imbrication of pebbles, a-axes and a concentration of pebbles at its front. **V.** Snout with a smoothed, decreased slope and a diffuse boundary between its massive texture and its gravity-winnowed, rather well-sorted, stratified margin.

Discussion

Although these experiments somewhat simplify natural processes (flume scale), a few observations are important:

1) Turbulent gravelly flows transported their largest pebbles in an imbricated manner within a sandy matrix.

2) Evenly stratified sediments ahead of a massive gravel mixture can be deposited by thin grain flows and turbulent flows due to gravity-winnowing (see Nemec et al., this volume).

With respect to the tripartite beds (Fig. 27), the relationship between unit 2 with upwards curving shear zones and the well-sorted, evenly stratified unit 1 at its base is inferred to have been due to gravity-winnowing, and subsequent overthrusting as described above from the experiments. The post-depositional thrusting in the upslope portion, as recorded by the shear zones, may also have transmitted compression downslope and enhanced instability in the frontal part of the gravel flow. This may have initiated fluidization and/or liquefaction elutriating the fine grain sizes and starting a gravity-winnowing process from the head of the flow. As depicted in Figure 29, remobilization and further downslope movement of the main flow body by, for example retrogressive sliding, may account for unit 1 being overridden and preserved as a basal unit. The observed coarsening upward cycles in unit 1 may indicate a prograding sediment source.

Alternatively, it is possible that unit 1-type sediments have been winnowed by traction currents, and that they represent thin sand flows in the canyon (see Lewis et al., 1980). However, the interpretation of the pebbly mudstones representing 'normal' conditions of sedimentation in the canyon would not meet this explanation satisfactory.

In the bipartite beds the conglomeratic part (unit 2) may have been deposited from a) dominantly high-density turbulent flows or b) from density-modified grain flows or debris flows with dominantly frictional strength as already discussed earlier. The lack of grading in the upslope portion of the deposits versus the coarse-tail normal grading present further downslope (Fig. 25) suggests a downslope change in sediment concentration and hence, in flow behaviour. Therefore, a downslope change from a plastic into a high-density, fluidal flow behaviour (*cf*. Lowe, 1979) is likely. The grains in such flows may have been initially supported by matrix density (buoyancy), dispersive pressure (Hampton, 1979), and static grain contacts (Pierson, 1981), but farther downslope mostly by matrix turbulence.

ABRIOJA FAN-DELTA MODEL

Numerous slump scars in the proximal part of the Abrioja fan-delta indicate that slope failure was common, and consequently the sediment was transported by sediment gravity flows of various types, resedimenting as Facies III, IV and V. Based on the increase of water escape structures near some scars, sediment failure was probably due to liquefaction. This process may be caused by contin-

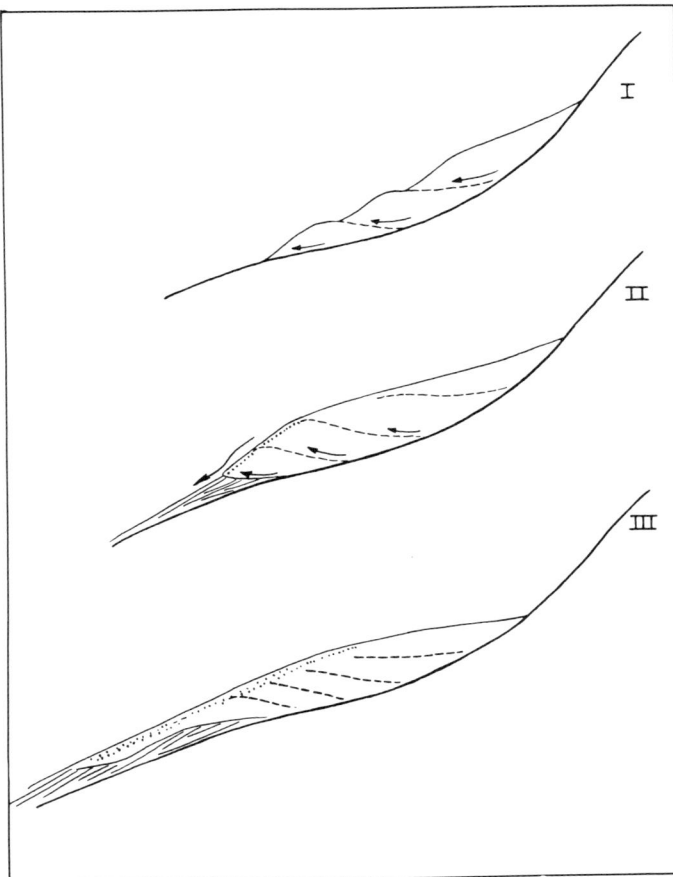

Fig. 29. The origin of an evenly laminated basal zone due to flow surge is envisaged as follows: I. Downslope transmitted compression within a gravel flow causes thrusting of the gravel body and oversteepening of its front, which will result in liquefaction and gravity-winnowing (II) from its unstable margins, producing well-sorted, evenly stratified sediments ahead. III The evenly stratified deposits may be truncated by the gravel body after renewed thrusting.

Fig. 30. Experiment during which thrusting of a stagnant gravel body occurred along upwards curving shear zones, which were partly deformed by continued sliding of the gravel in the tank (arrows indicate major shear surfaces).

uous sediment accumulation and a related increase of pore fluid pressure by loading (Chamberlain, 1964; Wescott and Ethridge, 1982). In addition, seismic activity or storm or tide-induced waves may have contributed to the initiation of slumps.

The fan-delta sequence displays cyclic sedimentation, possibly controlled by the amount of supply of sediment. The process has been demonstrated experimentally (Fig. 30) by piling up a heap of sand under subaerial conditions. Sand accumulated on an angle of repose slope would fail when piled to a height of approximately 5 m. Sand at the top of the sediment heap was the first to fail and slumped to a resting point approximately one third of the way down the slope. These small slumps were followed by massive failure in which a larger volume of sand, originating at the top of the sediment pile, was transported to the toe of the sediment mass. The process was repetitive. The volume of slumped sediment apparently controlled flow behaviour by constraining the thickness of the flow, and hence the duration of the flow. Therefore, thicker flows will flow over more gentle slopes than thinner ones. This is especially true if flow behaviour approaches that of a Bingham plastic, characterized by relatively greater strength.

The following depositional model of the Abrioja fan-delta is proposed (Fig. 9). Seasonal deposition produced a Gilbert-type delta in a fan-delta front environment (primary outbuilding). Liquefaction and sediment failure on the delta front produced small slumps, resedimenting quickly on a relatively steep slope (secondary outbuilding). Finally, overloading of these minor slump deposits resulted in a major slump which resedimented on a relatively gentle slope. The slump scar was eventually filled with minor slumps again, probably partly by retrogressive flow slides, and the process was repeated. The ideal situation was probably disturbed by unusual catastrophic events as inferred from the chaotic debris flow deposits of Facies *I*.

Slumping of the delta front resulted in arcuate scars, partly filled by slump material, which did not move farther than the lower part of the slump scar. The slumped material was confined to downslope-trending troughs, separated by erosional remnants (Fig. 7). It is suggested that the paleo-environment of the Abrioja fan-delta, as summarized in Figure 7, may be similar to some of the modern fan-deltas described by Prior *et al.* (1981a, b; 1982) and the Yallahs fan-delta system (Wescott and Ethridge, 1980).

Conclusions

1) The Abrioja Formation is interpreted as the subaqueous part of a fan-delta, which prograded southeastward in a confined submarine canyon.

2) Sedimentation on this fan-delta was dominated by processes such as overloading, slumping and resedimentation.

3) Conglomerate beds on the lower delta slope are covered by pebbly mudstones forming bipartite beds and tri-

partite beds. In the latter case, conglomerate beds show a basal unit of well-sorted, evenly stratified sandstone. This basal unit is interpreted as representing surging flows due to gravity-winnowing as shown experimentally, and is considered not to be a basal shear zone. *Gravity-winnowing* refers to a process of selective winnowing of grains from the unstable margins of sediment bodies solely by gravity. Because of the lower frictional resistance of fine grain sizes compared to coarser ones, the fine grains escape first from such bodies leaving coarser grains unsupported, so that progressively coarser fractions become unstable and are winnowed due to gravity.

4) Massive conglomerates may contain internal, upslope dipping zones which are demarcated by either diffuse sandy zones or by a preferred pebble orientation. Such zones have been interpreted, in accordance with experiments, in terms of shearing due to compression transmitted downslope or by sediment aggradation upslope of a stagnant gravel body. This may have been caused, for example, by retrogressive flow slides.

References

Aalto, K.R. 1976. Sedimentology of a mélange: Franciscan of Trinidad, California. Journal of Sedimentary Petrology, v. 46, p. 913-929.

Addicott, W.O., Parke, D.S. Jr., Bukry, D. and Poore, R.Z. 1978. Neogene stratigraphy of Southern Almeria Province, Spain: an overview. Geological Survey of America Bulletin, v. 1454, 59p.

Andresen, A. and Bjerrum, L. 1967. Slides in subaqueous slopes in loose sand and silt. *In:* Richards, A.F. (Ed.), Marine Geotechnique. Urbana: University of Illinois Press, p. 221-239.

Bagnold, R.A. 1954. Experiments on a gravity-free dispersion of large solid spheres in a Newtonian fluid under shear. Proceedings of the Royal Society of London, Series A, v. 225, p. 49-63.

Bousquet, J.C., Montenat, C. and Phillip, H. 1976. La evolución tectonica reciente de las Cordilleras Béticas Orientales. Reunion sobre la geodinamica de la Cordillera Betica y Mar de Alboran, 1976, University of Granada, Spain, p. 59-78.

Buisonjé, P.H. de 1974. Neogene and Quaternary Geology of Aruba, Curacao and Bonaire. Thesis, University of Utrecht, 293p.

Carter, R.M. 1975. A discussion and classification of subaqueous mass-transport, with particular application to grain flow, slurry-flow and fluxoturbidites. Earth Science Reviews, v. 11. p. 145-177.

Chamberlain, T.K. 1964. Mass transport of sediment in the heads of Scripps Submarine Canyon, California. *In:* Miller, R.L. (Ed.), Papers in Marine Geology. New York: MacMillan and Company, p. 42-64.

Clari, P. and Ghuibaudo, G. 1979. Multiple slump scars in the Tortonian type area (Piedmont Basin, northwestern Italy). Sedimentology, v. 26, p. 719-730.

Dronkert, H. 1976. Late Miocene evaporites in the Sorbas basin and adjoining areas. Memorie della Società Geologica Italiana, v. 16, p. 341-362.

Fisher, R.V. 1971. Features of coarse-grained, high concentration fluids and their deposits. Journal of Sedimentary Petrology, v. 41, p. 916-927.

Fisher, R.V. and Mattinson, J.M. 1968. Wheeler Gorge turbidite conglomerate series, California; inverse grading. Journal of Sedimentary Petrology, v. 38, p. 1013-1023.

Gnaccolini, M. 1981. Oligocene fan-delta deposits in northern Italy: a summary. Rivista di Italiana Paleontologia e Stratigrafia, v. 87, p. 627-636.

Green, H.G., Snavely, P.D. Jr. and Lucena, J.G. 1977. Neogene tectonics of the Gulf of Almeria, Almeria Province, SE Spain. Messinian Seminar, International Geological Correlation Program Conference Proceedings, Abstracts of Project No. 6 — Messinian correlation, Malaga, Spain, p. 26-27.

Hampton, M.A. 1975. Competence of fine grained debris flows. Journal of Sedimentary Petrology, v. 45, p. 834-844.

Hampton, M.A. 1979. Buoyancy in debris flows. Journal of Sedimentary Petrology, v. 49, p. 753-793.

Hiscott, R.N. and Middleton, G.V. 1979. Depositional mechanics of thick-bedded sandstones at the base of a submarine slope, Tourelle Formation (Lower Ordovician) Quebec, Canada. *In:* Doyle, L.J. and Pilkey, O.H. (Eds.), Geology of Continental Slopes. Society of Economic Paleontologists and Mineralogists, Special Publication 27, p. 307-326.

Holmes, A. 1965. Principles of Physical Geology. New York: The Roland Press Company, 2nd Ed., 1288p.

Howell, D.G. and Link, M.H. 1979. Eocene conglomerate sedimentology and basin analysis, San Diego and the southern California borderland. Journal of Sedimentary Petrology, v. 49, p. 517-540.

Iaccarino, S., Morlotti, E., Papini, G., Pelosio, G. and Raffi, S. 1975. Litostratigrafia e biostratigrafia di alcune serie neogeniche della provincia di Almeria (Andalusia orientale-Spanga). Estratto da "l'Ateneo Parmense". Acta Naturalia, v. 11, no. 2, p. 237-313.

Instituto Geologico y Minero de Espagna (I.G.M.E.). 1974. Estudio Hidrogeologico y de ordenacion del Campo de Nijar Tomo I y II. Servicio de Publicaciones — Claudio Coello 44, Madrid (Deposito Legal M. 23082), 99p.

Instituto Geologico y Minero de Espagna (I.G.M.E.). 1982. Mapa geologia Espana, Almeria and Tabernas 1:50:000, 23-43 and 23-42.

Jacquin, J.P. 1970. Contribution à l'étude géologigue et minière de la Sierra de Gador (Almeria — Espagne). Thésis, Nantes, 501p.

Johnson, A.M. 1970. Physical Processes in Geology. San Francisco: Freeman, Cooper and Company, 577p.

Kleverlaan, N.L.C. 1980. Stratigrafie van het bekken van Tabernas (SE Spanje). Internal report, University of Amsterdam, 329p.

Laird, M.G. 1968. Rotational slumps and slump scars in Silurian rocks, western Ireland. Sedimentology, v. 10, p. 111-120.

Larsen, V. and Steel, R.J. 1978. The sedimentary history of a debris-flow dominated, Devonian alluvial fan — a study of textural inversion. Sedimentology, v. 25, p. 37-59.

Lewis, D.W., Laird, M.G. and Powell, R.D. 1980. Debris flow deposits of Early Miocene age, Deadman Stream, Marlborough, New Zealand. Sedimentary Geology, v. 27, p. 83-118.

Lowe, D.R. 1976. Grain flow and grain flow deposits. Journal of Sedimentary Petrology, v. 46, p. 188-199.

Lowe, D.R. 1979. Sediment gravity flows: their classification and some problems of application to natural flows and deposits. *In:* Doyle, L.J. and Pilkey, O.H. (Eds.), Geology of Continental Slopes. Society of Economic Paleontologists and Mineralogists, Special Publication 27, p. 75-82.

Lowe, D.R. 1982. Sediment gravity flows II: Depositional models with special reference to the deposits of high-density turbidity currents. Journal of Sedimentary Petrology, v. 52, p. 279-298.

McCabe, P.J. 1978. The Kinderscoutian Delta (Carboniferous) of northern England: a slope influenced by density currents. *In:* Stanley, D.J. and Kelling, G. (Eds.), Sedimentation in Submarine Canyons, Fans, and Trenches. Stroudsburg: Dowden, Hutchinson and Ross, p. 116-126.

McGillavry, H.J. 1970. Turbidite detritus and geosyncline history. Tectonophysics, v. 9, p. 365-393.

McGowen, J.H. 1970. Gum Hollow fan delta, Nueces Bay, Texas. Bureau of Economic Geology, Texas University, Report of Investigations, No. 69, 91p.

Middleton, G.V. 1967. Experiments on density and turbidity currents III. Deposition of sediment. Canadian Journal of Earth Sciences, v. 4, p. 475-505.

Middleton, G.V. 1970. Experimental studies related to problems of flysch sedimentation. *In:* Lajoie, J. (Ed.), Flysch Sedimentology in North America. Geological Association of Canada, Special Publication 7, p. 253-272.

Middleton, G.V. and Hampton, M.A. 1973. Sediment gravity flows: mechanics of flow and deposition. *In:* Middleton, G.V. and Bouma, A.H. (Co-chairmen), Turbidites and Deep-Water Sedimentation. Pacific Section, Society of Economic Paleontologists and Mineralogists, Anaheim, Short Course Notes, p. 1-38.

Middleton, G.V. and Hampton, M.A. 1976. Subaqueous sediment transport and deposition by sediment gravity flows. *In:* Stanley, D.J. and Swift, D.J.P. (Eds.), Marine Sediment Transport and Environmental Management. New York: Wiley and Sons, p. 197-218.

Moss, A.J., Walker, P.H. and Hutka, J. 1980. Movement of loose sandy detritus by shallow water flows: an experimental study. Sedimentary Geology, v. 25, p. 43-66.

Mutti, E. and Sonnino, M. 1981. Compensation cycles: a diagnostic feature of turbidite sandstone lobes. International Association of Sedimentologists, 2nd European Regional Meeting, Bologna, Abstracts, p. 120-123.

Nardin, T.R., Hein, F.J., Gorsline, D.S. and Edwards, B.D. 1979. A review of mass movement processes, sediment and acoustic characteristics, and contrasts in slope and base-of-slope systems versus canyon-fan-basin floor systems. *In:* Doyle, L.J. and Pilkey, O.H. (Eds.), Geology of Continental Slopes. Society of Economic Paleontologists and Mineralogists, Special Publication 27, p. 61-73.

Nemec, W., Porębski, S.J. and Steel, R.J. 1980. Texture and structure of resedimented conglomerates: examples from Ksiaz Formation (Famennian — Tournaisian), southwestern Poland. Sedimentology, v. 27, p. 519-538.

Pierson, T.C. 1981. Dominant particle support mechanisms in debris flows at Mt. Thomas, New Zealand, and implications for flow mobility. Sedimentology, v. 28, p. 49-60.

Porębski, S.J. 1981. Swiebodzice Succession (Upper Devonian - Lowest Carboniferous; Western Sudetes): a prograding, mass flow dominated fan-delta complex. (English abstract only). Geologie Sudetica, v. 16, p. 102-192.

Postma, G. 1978. Stratigrafie van het bekken van Almeria. Internal report, University of Amsterdam. 86p.

Postma, G. 1979. Preliminary note on a significant sequence in conglomeratic flows of a mass transport dominated fan-delta (Lower Pliocene, Almeria basin, SE Spain). Koninklijke Nederlandse Akademie van Wetenschappen serie B, v. 82, p. 465-471.

Postma, G. 1983. Water escape structures in the context of a depositional model of a mass flow dominated conglomeratic fan-delta (Abrioja Formation, Pliocene, Almeria Basin, SE Spain). Sedimentology, v. 30, p. 91-103.

Postma, G. 1984. Slumps and their deposits in delta fronts and slopes. Geology, v. 12, p. 27-30.

Postma, G., Roep, Th.B. and Ruegg, G.J.H. 1983. Sandy-gravelly mass flow deposits in an ice-marginal lake (Saalian, Leuvenumsche Beek Valley, Veluwe, The Netherlands), with emphasis on plug-flow deposits. Sedimentary Geology, v. 34, p. 59-82.

Postma, G., Kleinspehn, K.L. and Nemec, W. 1984. Outsized clasts in high-density turbity currents: a mechanism for their transport. International Association of Sedimentologists, 5th European Regional Meeting, Abstracts, p. 366-367.

Prior, D.B., Wiseman, W.J. Jr. and Bryant, W.R. 1981a. Submarine chutes on the slopes of fjord deltas. Nature, v. 290. p. 326-328.

Prior, D.B., Wiseman, W.J. Jr. and Gilbert, R. 1981b. Submarine slope processes on a fan delta, Howe Sound, British Columbia. Geo-Marine Letters, v. 1, p. 85-90.

Prior, D.B., Bornhold, B.D., Coleman, J.M. and Bryant, W.R. 1982. Morphology of a submarine slide, Kitimat Arm, British Columbia. Geology, v. 10, p. 588-592.

Roep, Th.B., Beets, D.J., Dronkert, H. and Pagnier, H. 1979. A prograding coastal sequence of wave-built structures of Messinian age, Sorbas, Almeria, Spain. Sedimentary Geology, v. 22, p. 135-163.

Schmincke, H.U. 1967. Graded lahars in the type section of the Ellensburg Formation, south central Washington. Journal of Sedimentary Petrology, v. 37, p. 438-448.

Schwarz, H.V. 1983. Subaqeous slope failures — experiments and modern occurrences. Contributions to Sedimentology 11, Stuttgart, v. 11, 115p.

Shaw, J. and Archer, J. 1979. Deglaciation and glaciolacustrine conditions, Okanogan Valley, British Columbia, Canada. *In:* Schluchter, Ch. (Ed.), Moraines and Varves. Proceedings, International Quaternary Association Symposium, Genesis and Lithology of Quaternary Deposits, Zurich, 1978, Rotterdam: Balkema Press, p. 347-355.

Stanley, D.J. 1974. Pebbly mud transport in the head of the Wilmington Canyon. Marine Geology, v. 16, p. Ml-M8.

Stanley, D.J. 1980. The Saint-Antonin Conglomerate in the Maritime Alps: a model for coarse sedimentation on a submarine slope. Smithsonian Contribution to the Marine Sciences, No. 5, 25p.

Surlyk, F. 1978. Submarine fan sedimentation along fault scarps on tilted fault blocks (Jurassic - Cretaceous boundary, East Greenland). Grønlands Geologiske Undersøgelse, Bulletin 128, 108p.

Van den Eeckhout, B. 1980. Geologie van de Westelijke Sierra Alhamilla (Zuidoost Spanje). Internal report, University of Amsterdam, 212p.

Völk, H.R. 1966. Zur Geologie und Stratigrafie des Neogenbecken von Vera, Südost Spanien. Thesis, University of Amsterdam, 164p.

Völk, H.R. 1967. Aggradational directions and biofacies in the youngest postorogenic deposits of southeastern Spain. A contribution to the determination of the age of the east mediterranean coast of Spain. Paleogeography, Paleoclimatology, Paleoecology, v. 2, p. 313-331.

Walker, R.G. 1978. Deep-water sandstone facies and ancient submarine fans: models for exploration for stratigraphic traps. American Association of Petroleum Geologists Bulletin, v. 62, p. 923-966.

Wescott, W.A. and Ethridge, F.G. 1980. Fan-delta sedimentology and tectonic setting — Yallahs fan-delta, southeast Jamaica. American Association of Petroleum Geologists Bulletin, v. 64, p. 374-399.

Wescott, W.A. and Ethridge, F.G. 1982. Bathymetry and sediment dispersal dynamics along the Yallahs fan delta front. Jamaica. Marine Geology, v. 46, p. 245-260.

Wilson, C.J.N. and Walker, G.P.L. 1982. Ignimbrite depositional facies: the anatomy of a pyroclastic flow. Journal of the Geological Survey, v. 139, p. 581-592.

Yoxall, W.H. 1983. Dynamic Models in Earth-Science Instruction. Cambridge: Cambridge University Press, 210p.

RESEDIMENTED CONGLOMERATES OF A MIOCENE FAN-DELTA COMPLEX, SOUTHERN ALPS, ITALY

Francesco Massari[1]

Abstract

An Upper Serravallian - Tortonian regressive sequence, representing the submarine part of a fan-delta complex, developed in the rapidly subsiding Vittorio Veneto area concurrently with uplift of the Southern Alps. It consists of four members, the first three of which include channelized and trough-filling sandstones and conglomerates deposited in a neritic environment, dominantly by highly concentrated turbidity currents. The importance of subaqueous mass-gravity processes suggests that the fan-delta complex prograded directly onto a submarine slope. The first member consists of nested channel-fills enclosed in outer-neritic mudstones. The second member consists of a series of large composite bodies thought to represent the fills of large slide scars and/or fault-controlled troughs trapping gravity-displaced sediments at high levels on the slope. The third member is made up of closely spaced ribbon-like small channel-fills probably representing a system of gullies generated by retrogressive slumping; this system headed in a sub-littoral environment, as judged by abundance of wave ripples and hummocky cross-bedding in the associated sandstones. Finally the fourth member consists of upward-coarsening littoral sequences and is bounded at the base by an angular unconformity.

A modification of Lowe's (1982) model for deposition from highly concentrated turbidity flows is proposed: each sedimentation wave is thought to evolve from an initial inertia-flow carpet, through suspension sedimentation, to a final stage of 'traction' sedimentation. Proximal-to-distal changes include increasing sand matrix content and development of inverse and normal grading in pebble-cobble conglomerates from essentially ungraded units, the latter commonly showing steep imbrication. In addition to ungraded beds, proximal resedimented conglomerates show abundance of two characteristic units: 1) stratified layers of small-pebble conglomerate showing repetitive inverse grading and strongly anisotropic fabric thought to be deposited by basal inertia-flow carpets, and 2) antidune-like bedforms.

Résumé

Une séquence régressive du Serravallien-Tortonien, représentant la partie sous-marine d'un 'fan-delta', a été déposée au cours d'une phase de subsidence rapide dans la région de Vittorio Veneto en même temps qu'un soulèvement des Alpes du Sud. Elle comprend quatre membres, dont les trois premiers incluent des grès et des conglomérats comblant des chenaux et des dépressions en milieu néritique et déposés par des courants de turbidité de charge élevée. L'importance des processus sous-marins de déplacement en masse par gravité suggère que l'ensemble du 'fan-delta' a avancé progressivement sur une pente sous-marine. Le premier membre se compose de chenaux mutuellement emboîtés et creusés dans des mudstones de la zone néritique externe. Le deuxième membre comprend une série de grands corps à structure complexe qui sont interprétés comme des remplissages de niches d'arrachement de grands glissements et/ou de dépressions engendrées par faille; des creux pouvaient piéger les sédiments transportés par gravité dans la partie supérieure de la pente sous-marine. Le troisième membre comprend un réseau de petits chenaux très serrés entre eux, représentant probablement un système de ravins engendrés par des phénomènes de glissement rétrogressif; ce système pouvait arriver jusqu'au milieu sublittoral, comme l'abondance des rides de vagues et la stratification entrecroisée du type 'hummocky' dans les grès associés l'indiquent. Le quatrième membre enfin est constitué par des séquences littorales et il est limité à la base par une discordance angulaire.

Nous proposons une modification du modèle de Lowe (1982) concernant les dépôts de courants de turbidité fortement chargés: chaque 'onde sédimentaire' semble avoir suivi les étapes suivantes: une phase initiale de sédimentation à partir d'une nappe d'écoulement avec un support donné par la pression dispersive, ensuite une sédimentation de la suspension, et enfin un stade final de "traction". La transition à partir des dépôts proximaux jusqu'aux distaux s'accompagne de l'augmentation de la teneur en sable dans la matrice et du développement d'un granoclassement inverse et normal dans les conglomérats à partir d'unités essentiellement non-granoclassées, ces dernières présentant fréquemment des angles très élevés d'imbrication des cailloux. En plus des lits non-granoclassés, les conglomérats proximaux révèlent deux types caractéristiques d'unités: 1) des couches stratifiées d'un conglomérat à grain fin intéressé par un granoclassement inverse cyclique et par une fabrique fortement anisotropique, et considéré comme un dépôt de nappes d'écoulement avec un support donné par la pression dispersive, et 2) des couches avec des structures ressemblant à des antidunes.

Geological Setting

The studied sequence is part of the south-alpine molasse outcropping at the transition from the Eastern Prealps to the Veneto plain (Fig. 1). The entire molasse sequence, from Chattian to Messinian, can be considered a first-rank, tectonically controlled transgressive-regressive cycle, comprising in its younger part two second-rank cycles:

a) an Upper Serravallian-Tortonian cycle (the subject of this paper) with poor lateral continuity and limited to the Vittorio Veneto area; and

---

[1]Istituto di Geologia, Via Giotto 1, 35137 Padova, Italy.

The work was financed by the "Consiglio Nazionale delle Ricerche", grant n.Ct. 81.01946.05. I am grateful to the student Alberto Innocente for his assistance in the field, generous help in the elaboration of data, companionship and good humour. Particular thanks go to Emiliano Mutti, who provided helpful comments on the manuscript and contributed to great improvement of it. Ron Steel, George Postma and Frances J. Hein reviewed the manuscript and gave many helpful suggestions for its improvement. Finally I am indebted to Mr. F. Todesco and Dr. C. Brogiato for the careful execution of drawings and photographs, and to Mrs. M. Prosperi Flaviani for checking the English.

Copyright © 1984, Canadian Society of Petroleum Geologists

b) an uppermost Tortonian-Messinian cycle recording fan-delta progradation along the entire eastern Prealpine belt, followed by alluvial fan - lacustrine sedimentation. Strong downwarping of the molasse depocenter was more than compensated for by a very high sedimentation rate during the Tortonian-Messinian with a clastic wedge more than 2000 m thick forming parallel to the regional strike. Angular or progressive unconformities (Riba, 1976), implying synsedimentary deformation, have been found at different levels of this clastic wedge (Massari et al., 1976) and emphasize the tectonic control on sedimentation.

The studied complex was probably the easternmost of a belt of fan-deltas which developed during the Tortonian along the southern margin of the rapidly rising southern Alps (Fig. 1). Uplift of the source area and rapid downwarping of the basin margin probably led to narrowing and even suppression of coastal shelves. A direct connection with the turbidite fans of the Marnoso-arenacea basin, through the Padan slope (Rizzini and Dondi, 1978; Ricci Lucchi, 1975), is likely for the western complexes but is unlikely for the eastern one, due to the presence of an important intermediate structural element, the Lessini-Berici-Euganei high (Fig. 1).

The Sedimentary Sequence

The Upper Serravallian-Tortonian regressive sequence consists of the following informal units: M. Bala (oldest), M. Altare, C. Posoccon and Val Sian members (Figs. 2 and 3). Channelized conglomeratic bodies related to large-scale resedimentation processes and enclosed in outer- to inner-neritic sediments can be recognized in the first three members. The fourth (uppermost) member consists of coarsening-upward sequences recording shoreline progradation and is bounded at the base by an angular unconformity. The Tortonian sequence of the Eastern Prealps is thickest in the Vittorio Veneto area. Channelized bodies, known from there only, pass laterally into a monotonous muddy and silty sequence with sparse sandstone interbeds. Rapid subsidence and high sedimentation rates probably resulted in local overthickening and poor lateral continuity of the complex. All the observed features suggest that sediment supply to the study area during the Tortonian was controlled by a point-source, and that the Vittorio Veneto sequence can be considered the sub-

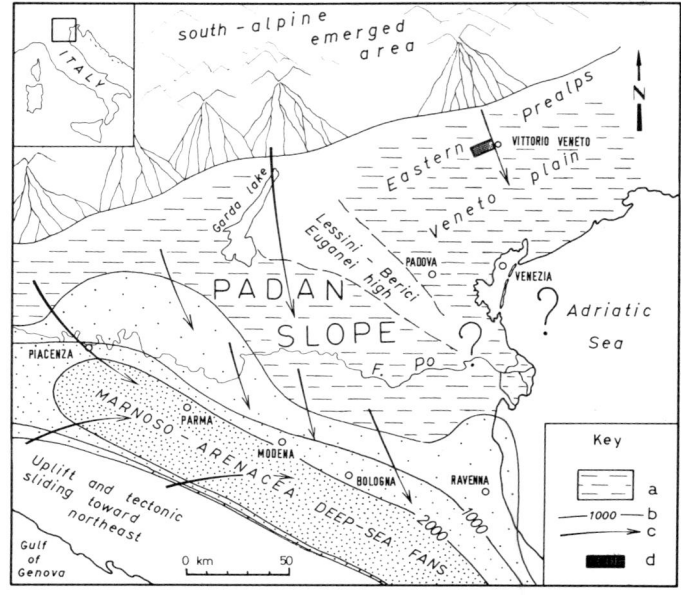

Fig. 1. Simplified paleogeographic sketch-map of the Periadriatic area during Tortonian time (after Ricci Lucchi, 1975 and Dondi et al., 1982, slightly modified). The direct connection of the Vittorio Veneto area with the Marnoso-arenacea basin is unlikely; a: Verghereto marls (Serravallian-Lower Messinian); b: isopachs of the Marnoso-arenacea Formation (Langhian-Lower Messinian); c: inferred directions of detrital influx; d: study area.

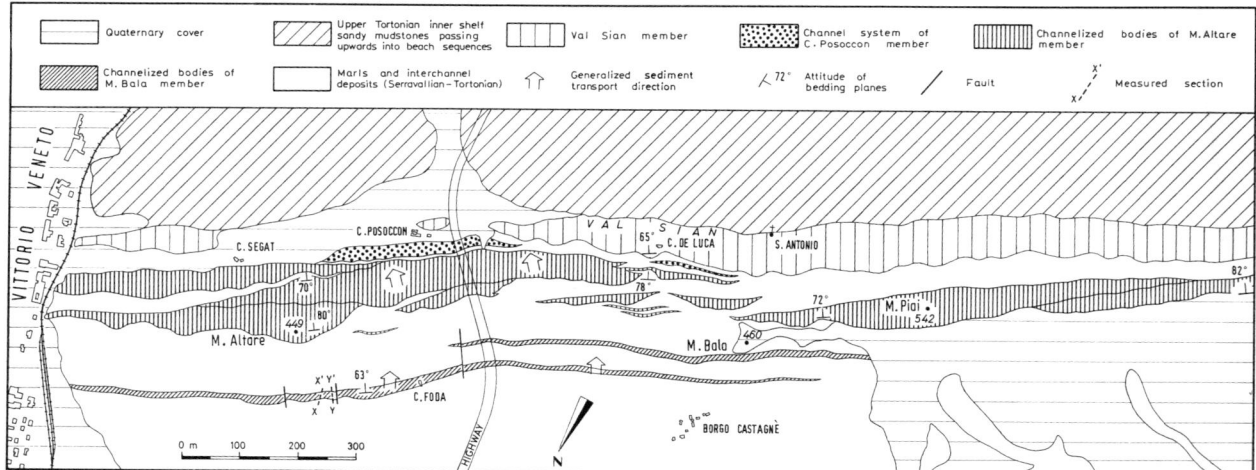

Fig. 2. Geological map of the Vittorio Veneto area. Capital letters show locations of sections represented in Figure 6. Location of the study area is indicated in Figure 1.

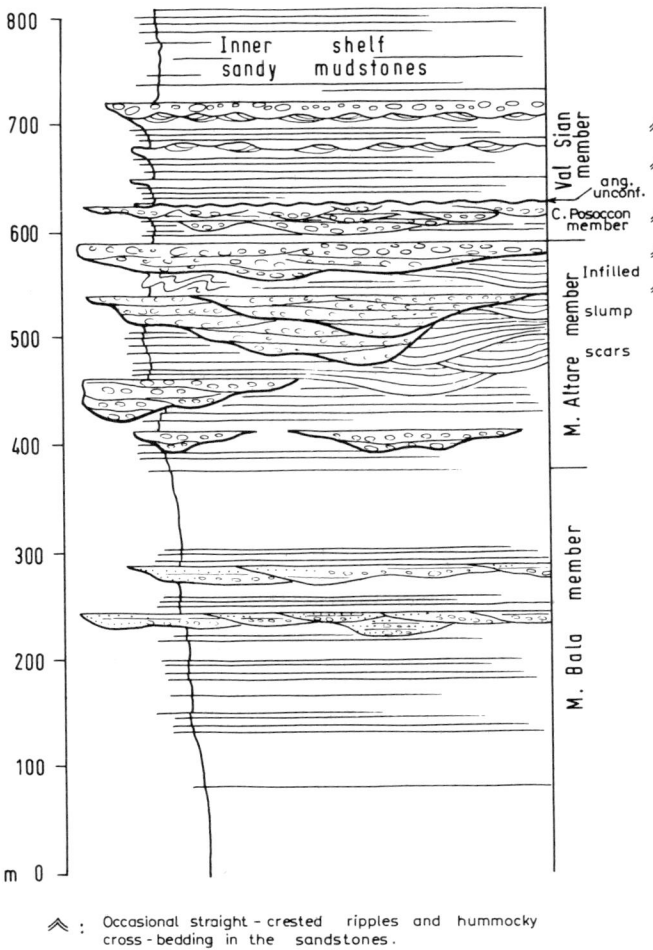

Fig. 3. The general vertical sequence of the study area.

marine part of a fan-delta complex. The sands and gravels of this complex, rapidly accumulating in coastal areas, were remobilised by a variety of subaqueous mass-movements. A north-western source of clastic influx can be inferred both by paleocurrent data (Fig. 4) and by the south-alpine affinity of the conglomerate composition (Massari *et al.*, 1974). Conglomerates consist of well-rounded clasts (granules to cobbles), mostly of limestones, dolostones, chert and cherty limestones, and are characterized by a matrix of fine- to medium-grained sandstone.

M. BALA MEMBER

This member consists of a muddy and marly sequence including sparse, thin-bedded turbidites and a few sandstone-conglomerate channelized bodies (Fig. 3). Foraminiferal assemblages suggest a shoaling trend from outer- to middle-neritic depths (Massari *et al.*, 1976).

Interchannel Deposits

Channelized bodies are enclosed in homogeneous, generally massive, light to dark-grey marly mudstone which becomes progressively siltier upwards and contains sparse sandstone interbeds. The mudstone is occasionally burrowed, and in places (mostly in the upper part of the member) displays a thin, quasi-varved lamination consisting of alternating silt and shale laminae. Sequences of thin-bedded sandstones usually display a sand/shale ratio well below unity, except near channels, where the ratio may approach unity. The sandstone beds, which are usually 3-12 cm thick, consist of fine to very fine sandstone, and show Bouma T_{b-e} or T_{c-e} sequences. They are sometimes laterally discontinuous and either scattered without a recognizable trend or organized into thickening- or thinning-upwards sequences 7-9 m thick. The former are in places capped by lenticular sandstone beds up to 50 cm thick, probably the fill of small, shallow channels. These sequences are interpreted in terms of overbank sedimentation, levee progradation and the fill of small crevasse channels cut into levees (Mutti, 1977). Zones of contorted stratification, particularly in proximal levee and crevasse deposits occur locally and point to occasional mobility of sediments. *Scolicia* and rare *Zoophycos* have been identified at sandstone-mudstone interfaces.

Channelized Bodies

Channelized bodies consisting of sandstone and conglomerate occur at two main levels in the M. Bala member. The lower body (C. Foda body) shows higher sandstone/conglomerate ratio, maximum thickness of 31 m, and can be traced laterally for more than 2.5 km in a section transverse to paleoflow. It consists of a number of channel-fills from 250 to 900 m wide, which are nested one into another (Fig. 5). Channel margins are either relatively steep (up to 60°) if incised into prodelta mudstones, or gently inclined if cut into previously formed channel fills.

Sole marks are common at the base of the bodies. Groove and impact casts predominate and show much higher variability in orientation than flute casts. The latter are rare and have unusually large dimensions: width may attain 40 cm and height 15 cm. Clast imbrication varies least in orientation (Fig. 4). Within individual channel-fills coarser deposits are generally concentrated in the axial part and in the lower part (Figs. 5 and 6): consequently a fining-thinning trend can be recognized both upwards and towards the channel margins (see also Johnson and Walker, 1979, Winn and Dott, 1979, Cazzola *et al.*, 1981, Hein and Walker, 1982).

The upper body has comparable geometry, with maximum thickness of 9.5 m and a minimum lateral extent of 1.5 km. Outcrop quality is too poor to allow recognition of internal organization.

The relative abundance of different units in the channelized bodies of the M. Bala member is shown in Figure 17. Abundance of 'traction' units (mostly R_{3a}: see section on Facies Model for the symbols) in the lowermost part of the channel fills suggests that 'traction' and perhaps 'lagging' effects were prominent in the stage during which the channel acted as a funnel for downslope movement, with little

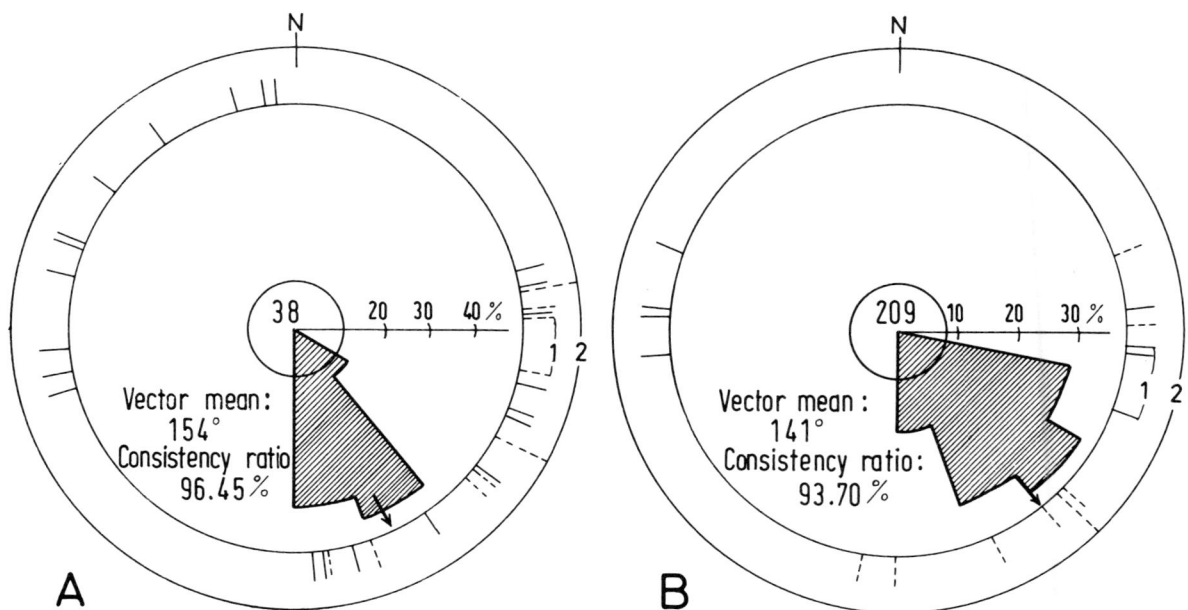

Fig. 4. Summary of paleocurrent data for M. Bala member (**A**) and for M. Altare and C. Posoccon members (**B**). Azimuths of imbrication and number of layers contributing to these observations are shown in the central part of the diagrams. Directions of groove and impact casts (the latter represented by dashed lines) are shown in the outer ring. The higher dispersion in the orientation of groove and impact casts compared to imbrication should be noted.

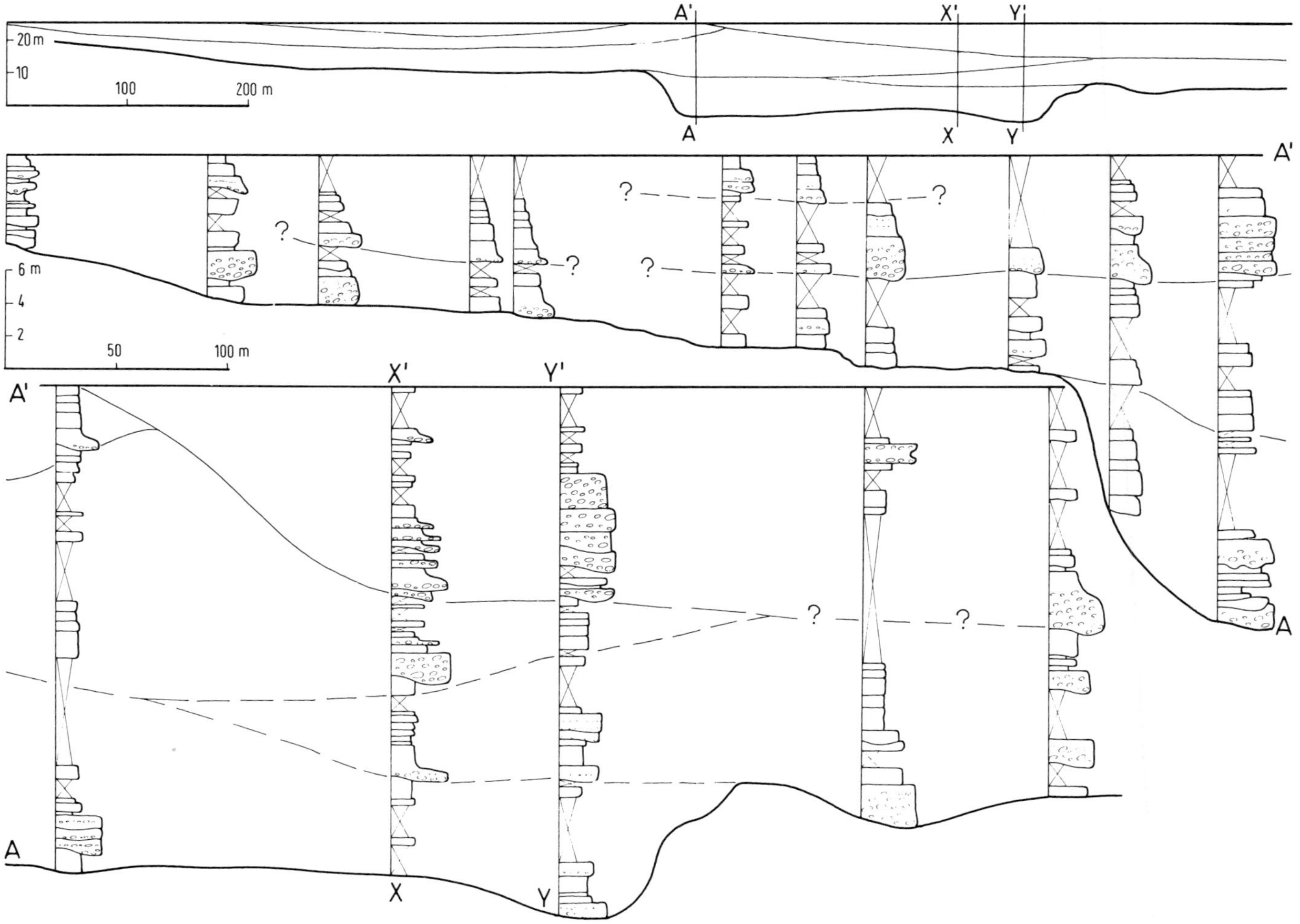

Fig. 5. The eastern portion of C. Foda body (M. Bala member) between C. Foda and the town of Vittorio Veneto, with measured sections and a suggestion of the internal organization. The section of the body is perpendicular to paleoflow. Vertical exaggeration in the lower figure is 7.5.

permanent sediment accumulation. Facies changes in a direction perpendicular to paleoflow are common, with spectacular transitions of massive into stratified units in a distance of a few metres.

Within the channel-fills contorted stratification was occasionally observed in some laminated silty lithologies suggesting episodic slumping of levee deposits into the channel. *Ophiomorpha* burrows occur in sandstones near the base of the body.

The depositional setting of the M. Bala member was probably a muddy slope located in an outer-neritic environment.

M. ALTARE MEMBER

This member consists of a number of composite channelized bodies, mostly made up of conglomerate and surrounded by interchannel deposits; these consist of an alternation of sandstone beds and silty mudstone or siltstone containing in places middle- to inner-neritic foraminifera.

Interchannel Deposits

The sandstone-mudstone ratio of interchannel deposits is usually much higher in this member than in the underlying one. It is in the order of 0.3 in the lower part of the member and of 0.8 in the upper. A shoaling trend can be inferred by the changes in the character of these deposits. In the lower part of the member the sandstone beds are very fine-grained and can be considered thin-bedded, T_{b-e} or T_{c-e} turbidites, sometimes rapidly wedging out laterally. The middle and upper parts of the member, on the other hand, are characterized by an increasing abundance of straight-crested and symmetrical, wave-generated ripples and occasional hummocky cross-strata. Laminae of finely macerated plant debris are common in the sandstone beds throughout this member. The interbedded mudstones show an upward increase in silt and fine sand sizes and a progressive reduction in diversity of the foraminiferal assemblage. This is dominated in the middle-upper part of the member by euryhaline forms such as *Ammonia beccarii* and *Elphidium* sp. The upward coarsening trend is accompanied by a change in the ichnofossil assemblage from a predominance of the *Phycodes-Teichichnus* group to abundance of *Ophiomorpha* and *Thalassinoides*.

Broadly lenticular sandstone beds 30-50 cm thick, interpreted as the filling of crevasse channels, increase in abundance upwards. Planar and low-angle cross-lamination, rows of clay chips and sometimes of pebbles are common features of these lenses.

Slump scars of variable extent can be found at several levels. They appear as smoothly curved and concave-upward surfaces of angular discordance. Sedimentary infills draped over them are wholly similar in lithology to the surrounding sediments and are often cut by other discordances. Similar features were described as typical of infilled slump scars by Laird (1968) and by Clari and Ghibaudo

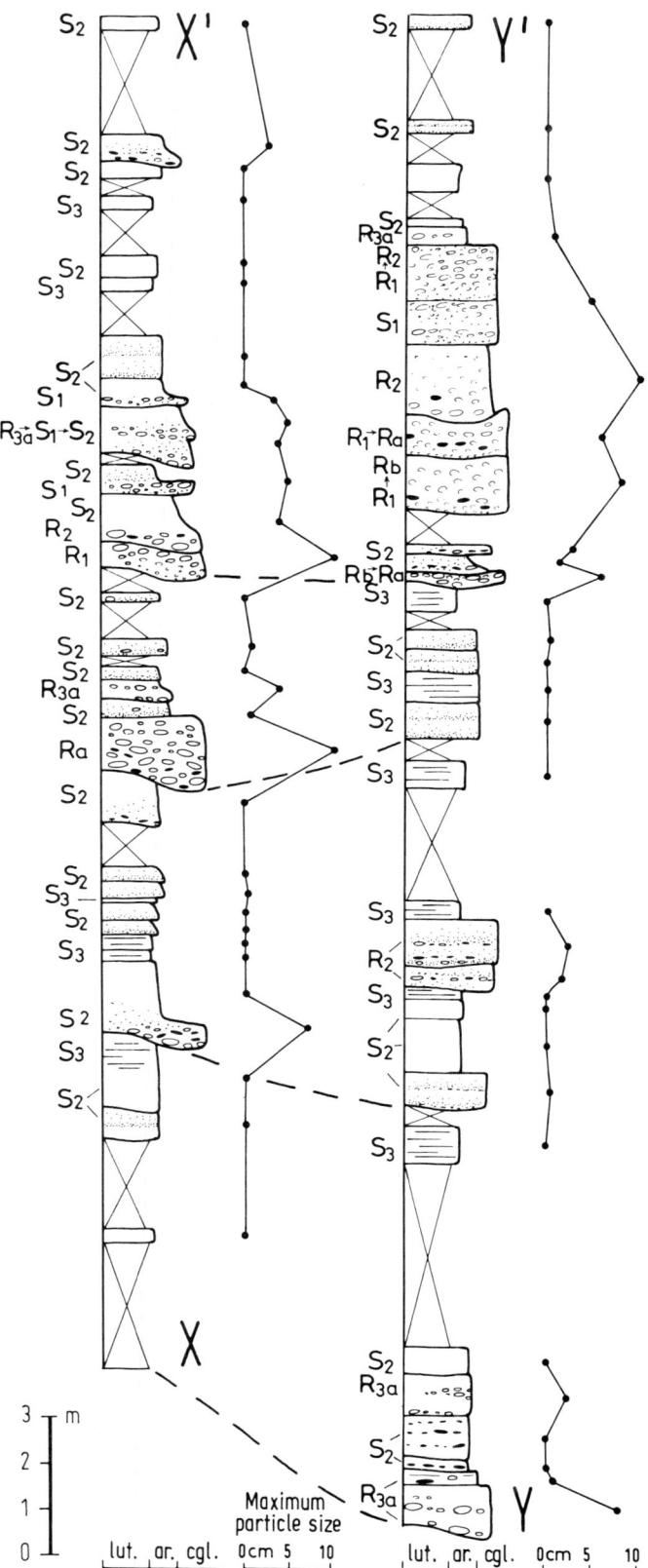

Fig. 6. Vertical log through sections XX' and YY' of the C. Foda body (see Figs. 2 and 5 for location) with trend of maximum particle size (mean of ten largest clasts). Division boundaries may exist between figured units. Unlabelled sandy layers are either S_2 or S_3 units. Arrow linking unit symbols indicates vertical transitions within sequences related to single depositional events.

(1979). In addition the interchannel lithologies occasionally show more or less contorted and folded stratification suggestive of repeated slumping.

Channelized Bodies

A number of composite channelized bodies, mostly consisting of conglomerates, can be identified in the M. Altare member and these usually show a concave-up, bowl-shaped base. The lower bodies cut down deeply into the mudstones of the M. Bala member and have on average higher relief (Fig. 7). An appreciable upward increase in width/thickness ratio of the bodies can be recognized. In addition a progressive eastward offsetting can be detected leading to a characteristic imbricate pattern. Three orders of conglomerate units can be identified. Major bodies may be as much as 100 m thick and 2.1 km wide and result from the coalescence of second-rank bodies (Figs. 2 and 3). The latter may occur also as isolated units and range from 20 to 70 m in thickness and from 350 to 1500 m in width. The second-order bodies are, in turn, composite as they consist of 2 to 9 amalgamated third-rank units (Fig. 3). The latter show pronounced upward changes in geometry and size. Those occurring at the base of the body are normally thickest (up to 20 m) and coarsest-grained and sometimes show a particular abundance of R-S units, whereas sequences occurring higher are characterized by a progressive increase in width/thickness ratio, decrease in average thickness (to 6-4 m), decline in size of conglomerates and increasing abundance of internally stratified layers. This results in an overall trend of thinning and fining upwards (Fig. 8) which probably reflects a gradual decrease in depositional gradients. In addition, individual sequences tend to develop a crude trend of fining and thinning upwards from massive to stratified units.

In some of the thickest bodies the bedding attitude of conglomerates gradually changes from conformable to the regional bedding at the base of the body to increasingly steep towards the top with a dip in the downflow direction. A steepening of up to 18° has been observed in the M. Altare body, where a slightly arcuate, concave upward attitude in inclined strata may also be observed in longitudinal section. This arrangement is very similar to the foreset bedding described by Postma (this volume) in the conglomeratic filling of arcuate slump scars of the Abrioja fan-delta.

In a section perpendicular to paleoflow the basal boundary of the M. Altare body appears complicated by a number of U-shaped troughs leading to a series of steps and terraces. Such troughs are 60-75 m wide and 6-30 m deep and are characterized by a flat bottom and steep flanks (up to 60°). These features seem consistent with the notion of a non-erosional origin for the troughs.

The relative abundance of different units in the M. Altare member is illustrated in Figure 17. Paleocurrent data are consistent in all conglomeratic bodies of the M. Altare member. However paleocurrent directions measured at different levels of third-rank sequences commonly show greater variability at higher levels than near the base.

Essentially three types of bedding have been identified in the conglomerates: 1) plane-parallel bedding; 2) cut-and-fill bedding; 3) large-scale, low-angle inclined bedding dipping either downstream or upstream. The third type of bedding is mostly found in the middle-upper part of the member. Larger exposures show that the inclined bedding is actually part of large symmetrical 'waves' (Figs. 9 A and B), mostly made up of R_a and S_1 units, but occasionally also of other units, such as R_1, R_2 and R_{3b} (see later for facies symbols). Such wavy features average 14 m in wavelength and 2.5 m in thickness, show a length/amplitude ratio of about 6, and generally consist of a bundle of beds (from 2 to 8). The largest ones generally extend beyond available outcrop. In larger examples the height may attain 5-6 m and the inferred wavelength may well be greater than 20-25 m. The base of the set may be either planar or erosional. These rather enigmatic features are the composite product of numerous events; they apparently resulted from vertical aggradation, and probably imply extreme conditions of flow.

C. POSOCCON MEMBER

This member consists of a network of closely-spaced lenticular channel-fill sequences and inter-channel sediments. The former appear as ribbon-like conglomeratic bodies showing a very poor lateral persistence, and surrounded by an irregularly stratified alternation of burrowed and commonly wave-rippled siltstone and sandstone beds; the latter commonly thin out and have lenticular cross-sections. the channel-fill sequences are from 1.5 to 9 m thick, and are dominated by massive, thick-bedded units of the R group, mostly R_a units (Fig. 17), which commonly show high to very high imbrication angles and the coarsest of any observed grain sizes. Clast-supported textures are dominant in the conglomerates. Internal bands in the stratified units as well as individual layers are thick and often have poorly defined boundaries leading to apparent

Fig. 7. Panoramic view of some channelized bodies of the M. Altare member west of highway. Top is at left.

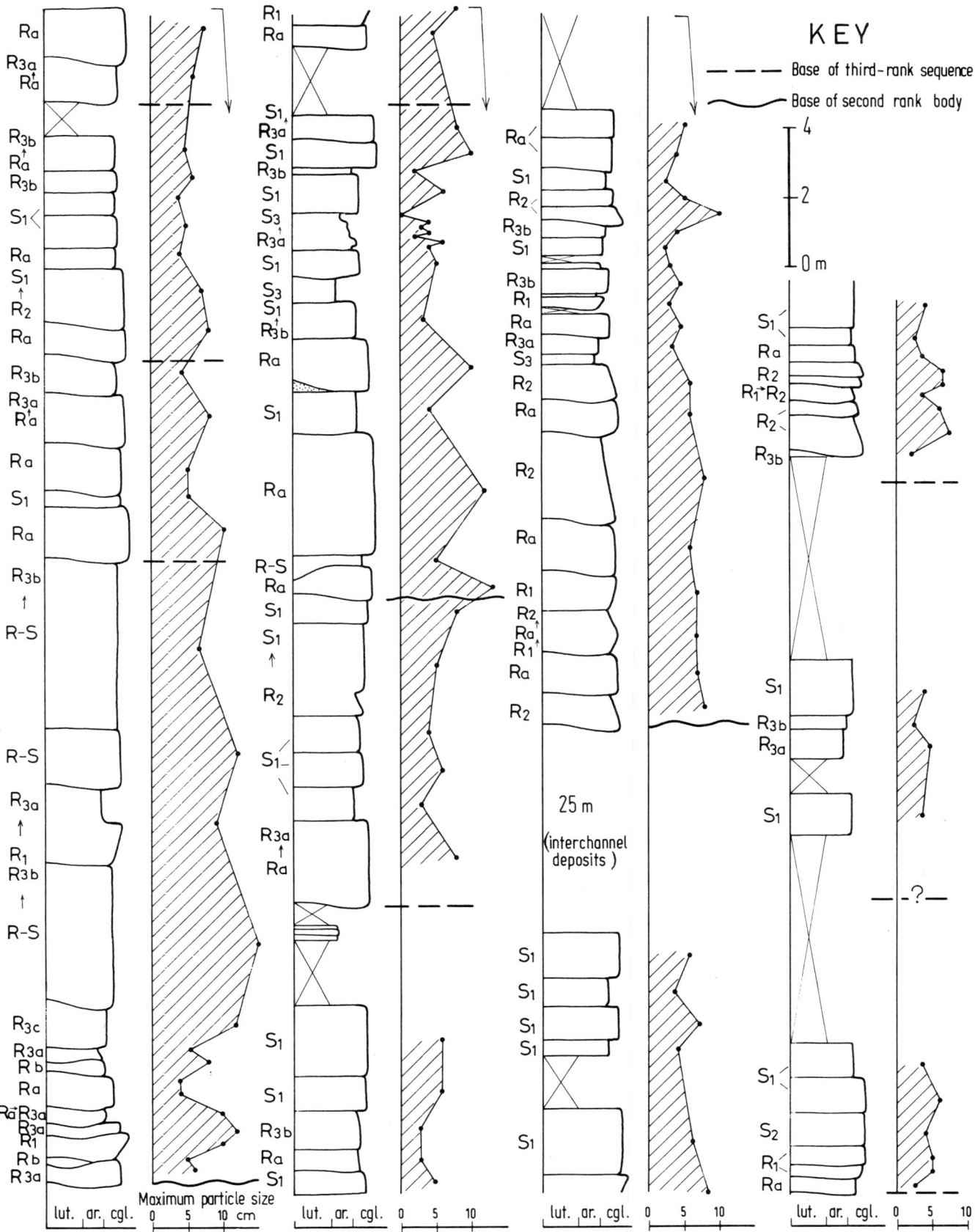

Fig. 8. Vertical log through a first-rank body of the M. Altare member (M. Altare section). Division boundaries may exist within figured units. The stacking of third-rank sequences and overall fining upwards trend within each second-rank body is suggested by the trend of maximum particle size and by upward changes in facies association. Arrows linking unit symbols indicate vertical transitions within sequences related to single depositional events.

amalgamation. No consistent vertical trends in grain size and bed thickness can be identified in the channel-fill sequences. Beds commonly show a pronounced lenticular geometry, with a markedly scoured base, mostly in the lower part of the channel fills; the resulting pattern of mutually interwoven cut-and-fill lenses resembles a braided channel facies.

VAL SIAN MEMBER

The uppermost deposits of the C. Posoccon member appear to be truncated at an angle of about 10° by an unconformity surface. Subsequent layers of Val Sian member show a gentler dip angle and lie in accordance with this surface. The Val Sian member consists of a series of coarsening and thickening upward cycles, each typified by bioturbated mudstone interstratified with beds of fine-grained sandstone, followed by thick sandstones showing amalgamated hummocky stratification (Dott and Bourgeois, 1982) and wave-megaripple cross-bedding and finally thick beds of conglomeratic sandstone and conglomerate; the latter sometimes include high-angle cross-bedded sandstone lenses, probably produced by landward migrating megaripples, as reported from modern shoreface deposits by Clifton et al. (1971). The sandstone interbeds occurring in the lower part of the sequence may show hummocky cross-bedding and wave-ripples at the top. Coarse wood debris is locally present, sometimes rounded. Trace fossils include *Ophiomorpha* and *Thalassinoides*. These sequences are interpreted as representing episodic progradation of a shoreline peripheral to a fan-delta complex.

FACIES MODEL

Mass-emplaced conglomerates and sandstones associated with neritic sediments have received little attention, and documented geological examples interpreted as neritic channels are rare (Walker, 1969; Jeletzky, 1975; Hyden, 1980; Lewis et al., 1980). However, as observed by Hyden (1980), the facies models proposed for sandstones and conglomerates of deep-water (turbidite) association which cannot be adequately described using the Bouma sequence (Mutti and Ricci Lucchi, 1972; Walker and Mutti, 1973; Aalto, 1976; Walker, 1978; Slaczka and Thompson, 1980; Lowe, 1982) can also be applied to resedimented neritic deposits (Fig. 10).

Figs. 9 A and B. Examples of large-scale, wave-like features in the upper part of M. Altare member. Paleoflow is in the downdip direction. Top is at left. Encircled person for scale.

The model and terminology proposed by Lowe (1982) have been adopted here with some modifications. As discussed by Lowe (1982), deposition from high-density turbidity currents can be treated in terms of several grain populations which are deposited during discrete sedimentation waves as the relative efficacy of different support mechanisms changes with flow deceleration. The pebble and cobble population produces a sequence of sedimentary divisions termed the R_{1-3} sequence; sand to small pebble-sized gravel produces the S_{1-3} sequence. According to Lowe, each sedimentation wave passes from an initial stage of traction sedimentation to one of mixed frictional freezing and suspension sedimentation within traction carpets, to a final stage of direct suspension sedimentation. It seems however more likely that within each sedimentation wave the highest concentration and highest shear stress are localized near the base of the flow and that this would lead to maximum efficacy of grain interaction in the basal grain carpet, at least in the case of relatively low flow viscosity (Bagnold, 1956). As noted by Carter (1975) the term traction carpet is not particularly appropriate, due to the absence of true traction movement.

The term basal inertia-flow carpet is probably more correct as it suggests that flow in this basal carpet is in the inertial regime of Bagnold (1956). On the other hand 'traction' effects could be expected to be mostly active in the final stage, owing to the shearing of the depositional interface by the more dilute current (Middleton and Hampton, 1973).

A modification of Lowe's model is therefore proposed, with the suggestion that each sedimentation wave passes from basal inertia-flow carpet sedimentation to suspension sedimentation, to a final stage of 'traction' sedimentation.

In most cases, as stressed by Lowe (1982), multiple support mechanisms probably act simultaneously, since different size populations of grains can be supported by different mechanisms. Moreover evidence of simultaneous traction and suspension sedimentation can be occasionally found. For instance, in some of the normally graded units examined in the present study long axes of larger clasts at the base of the bed are oriented transverse to flow: this suggests that, in addition to a suspended load maintained by one or more particle-support mechanisms, there also existed a bed load which was not fully supported

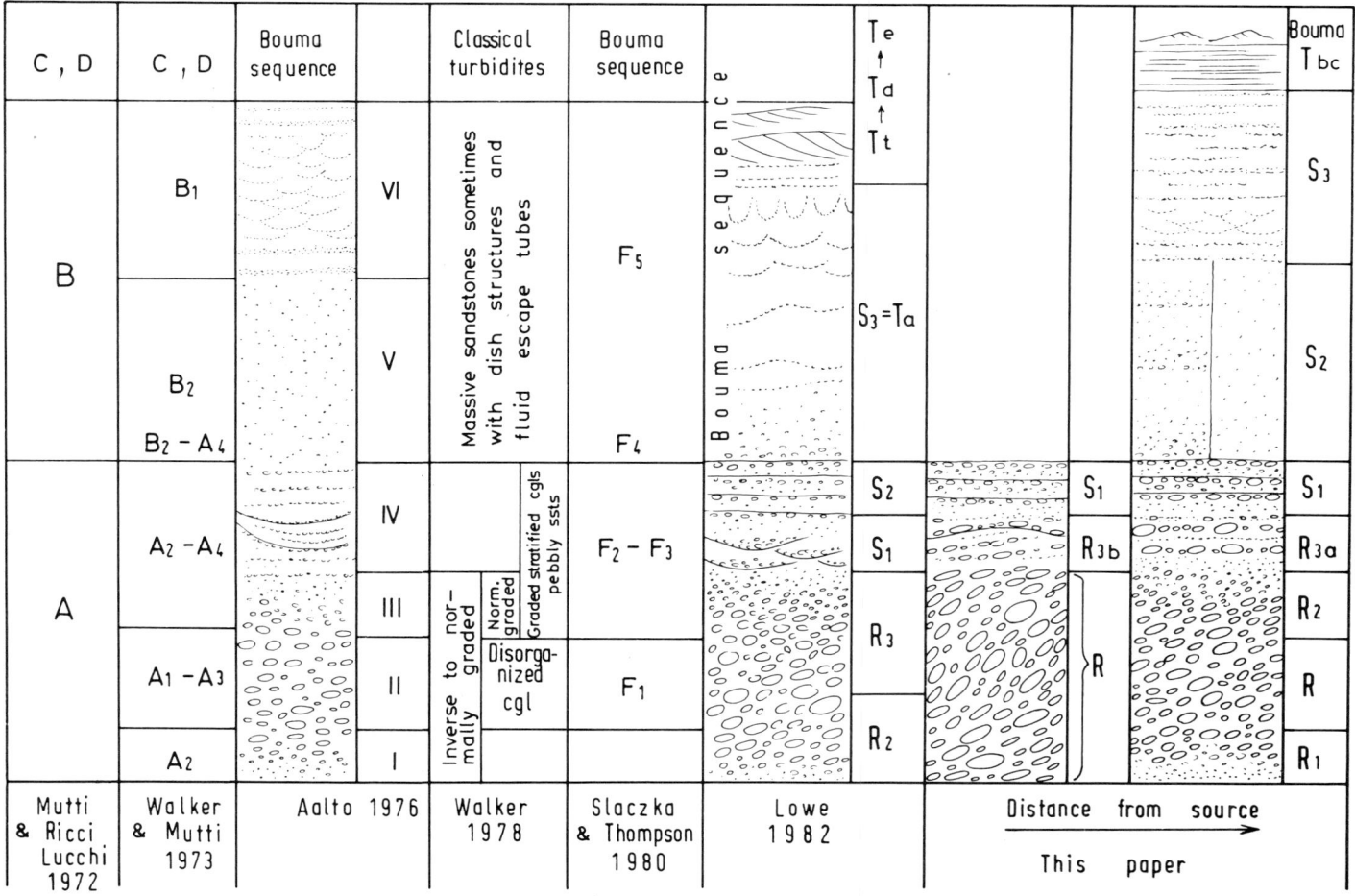

Fig. 10. A comparison of published sequence models for coarse deposits thought to be emplaced by highly concentrated turbidity flows, and suggestion of approximate equivalent facies.

or suspended. Within this traction load the clasts are probably dragged forward by the surrounding, more rapidly moving flow and are susceptible to rolling along the bed (Lowe, 1979; Hein, 1982).

Although the order of the units in the R + S sequence is generally maintained, the complete sequence has never been found in the study area, probably because much sand completely bypasses areas of gravel deposition. In most cases a depositional event is registered by only one division and in addition most sequences comprise no more than two or three divisions (Table 1). Individual divisions may be missing because of extreme flow unsteadiness and because of variations in the mean size of the suspended load. For instance flow composed largely of fine- and very fine-grained sand will not deposit layers from basal inertia-flow carpets, because of the negligible dispersive pressure between such fine grains (Bagnold, 1956; Lowe, 1982). Furthermore, if one of the support mechanisms strongly dominates, it will leave its characteristic imprint. For instance if the flow is fully turbulent and several grain populations are present, a depositional sequence exclusively consisting of suspension sediments and mixed 'traction' and suspension sediments will be formed.

Each of the R and S depositional sequences are composed of three elements indicated by numbers 1, 2 and 3 at the foot of the sequence symbol. These correspond to sedimentation from a basal inertia-flow carpet (1), suspension sedimentation (2) and 'traction' sedimentation (3).

The data of Table 1 have been tested for the presence of a Markov process using a log-linear model of quasi-independence but the results are not significant because of the presence of too many 'sampling' zero entries in the transition matrix below the diagonal of structural zeros. Some of the fabrics, textures and lateral facies associations are quite similar to those documented by Hein (1982) for the Cap Enragé Formation.

R SEQUENCE

The massive divisions of the R sequence are generally characterized by broad lensing of individual beds and tend to show a positive correlation between the bed thickness and maximum clast diameter. Three basic divisions have been distinguished, namely R_1, R_2 and R_3, respectively characterized by inverse grading, normal grading, and 'traction' structures. In addition, another unit is considered, corresponding to layer II of Aalto (1976), which is typically structureless and ungraded. This unit is indicated by the symbol R, and a clast-supported variety, R_a, is distinguished from a matrix-supported one, R_b. A central ungraded division occurs in some inverse-to-normally graded conglomerate beds.

A characteristic feature which may occur in the massive units of the R sequence is a vertical change of imbrication angles within the bed (Fig. 12). In some cases (pattern 1) the imbrication angles are highest at the top of the bed (with rare imbrication reversals at high levels in the layer) and flatten out gradually towards the base. In other cases (pattern 2) the imbrication angles are highest in the central part of the bed and flatten out gradually both towards the base and top of the bed. An estimate of the occurrence of vertical change of imbrication angle in different units/sequences (Table 2) illustrates that this particular fabric is typically developed in R_a units. It should also be noted that it is most abundant in proximal areas. Although patterns 1 and 2 are about equally common in R_a units, Table 2 suggests a differentiation between units/sequences including an inversely graded portion (and lacking normal grading) and those including a normally graded one (and lacking

Sequence	Number of examples	Sequence	Number of examples
R_1-R_1	2	R_2-R_2	1
R_1-R_a	18	R_2-R_2-R_2	1
R_1-R_b	1	R_2-R_3	1
R_1-R_a-R_2	7	R_2-R_{3a}	9
R_1-R_a-R_2-R_{3b}	2	R_2-S_1	3
R_1-R_a-R_2-S_2	4	R_2-S_1-S_2	1
R_1-R_a-R_2-S_2-S_3	1	R_2-S_2	1
R_1-R_a-R_{3a}	1	R_{3a}-R_a-S_2	1
R_1-R_a-S_1	3	R_{3a}-S_1	2
R_1-R_a-S_2	1	R_{3a}-S_2	2
R_1-R_2	4	R_{3a}-S_3	1
R_1-R_{3a}	1	R_{3b}-R_a	1
R_1-R_{3b}	1	R_{3b}-R_a-R_2	1
R_1-S_2	5	R_{3b}-S_1	3
R_a-R_a	1	S_1-R_{3b}	1
R_a-R_a-R_a	1	S_1-R_{3b}-S_1	1
R_a-R_2	5	S_1-S_2	8
R_a-R_2-S_2	1	S_1-S_2-S_3	1
R_a-R_{3a}	7	S_1-S_3	4
R_a-R_{3b}	9	S_1-S_3-S_2-S_3-S_2	1
R_a-R_{3c}	2	S_2-S_3	11
R_a-S_1	9	S_2-S_3-S_2	1
R_a-S_2	5	S_2-S_3-S_2-S_3-S_2-S_3-S_2	1

Table 1. Vertical sequences (related to single depositional events) observed within channelized deposits.

Unit/ sequence	Number of examined layers	Layers showing vertical change of imbrication angles	Pattern 1	Pattern 2
R_1	54	14	13	1
R_a	183	82	37	45
R_1-R_a	26	13	2	11
R_1(-R_a)-R_2	30	10	/	10
R_a-R_2	9	4	/	4
R_2	81	10	/	10
S_1	86	11	10	1
	469			

Table 2. Number of cases in which a vertical change of imbrication angles has been observed in different units/sequences, and number of layers showing respectively pattern 1 and 2 of imbrication.

inverse grading). In both cases steep imbrication is fairly rare but the former are dominated by pattern 1 and the latter by pattern 2.

Both imbrication patterns contrast with observations of Walker (1977) who found highest imbrication angles at the base of the beds and a progressive flattening out towards the top. Because the base of the flow is the zone of maximum shear stress and grain interaction, Walker concluded that the imbrication angle is a function of violence and number of clast collisions. According to some authors high imbrication angles are related to grain interaction (Rees, 1968; Hiscott and Middleton, 1980). It is stressed here however that in the conglomerates with patterns 1 or 2 there is a clear flattening out of imbrication angles toward

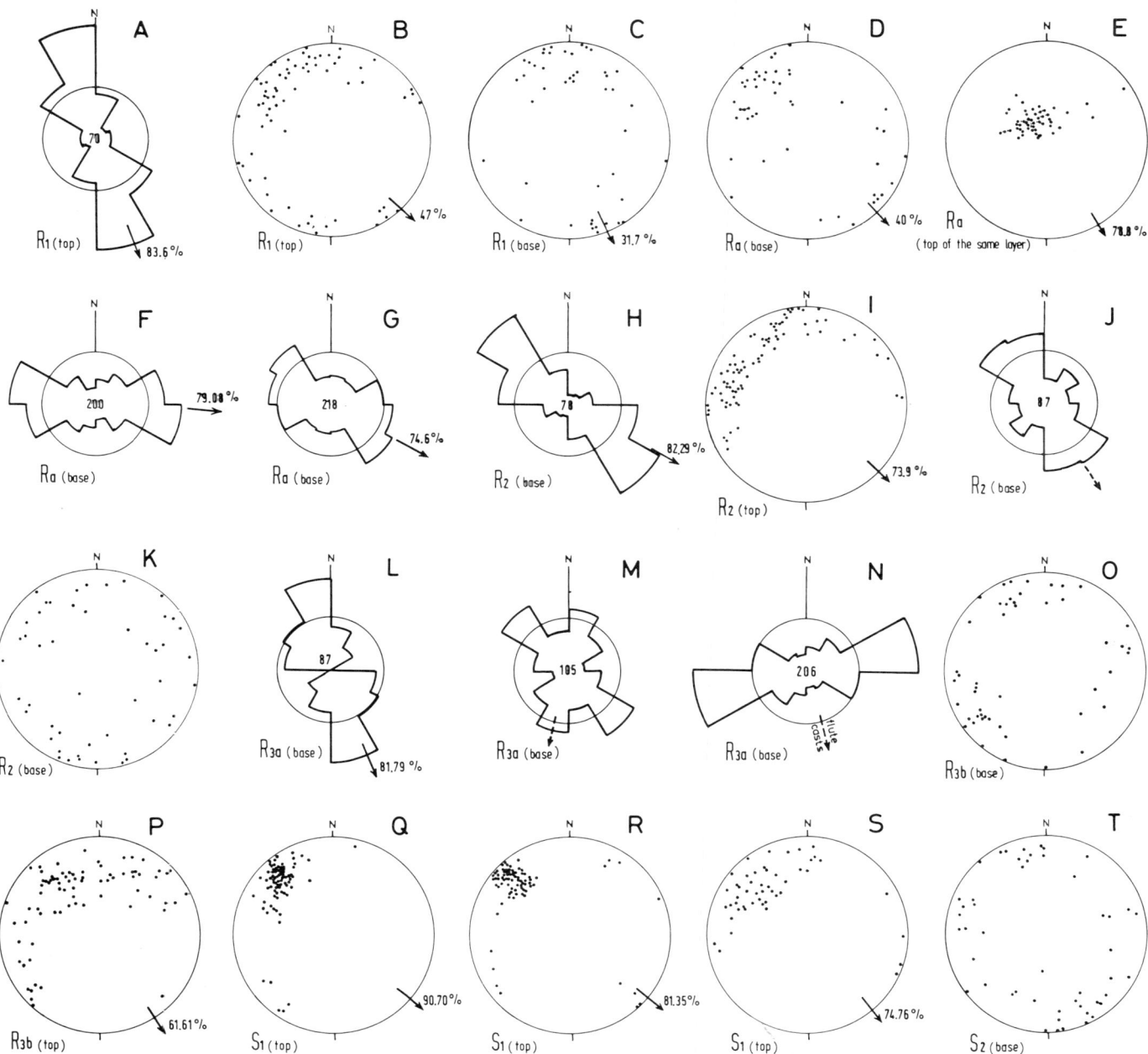

Fig. 11. Wulff stereoplots and rose diagrams of a-axis orientation in different conglomerate units. Vector mean and consistency ratio (in percent) in the stereonets have been computed for unimodal fabrics only. Arrow in rose diagrams indicates paleoflow direction determined in the field from imbrication; in the case of unimodal fabric parallel to flow this direction coincides with vector mean and is indicated by a solid arrow; in the case of bimodality or a-axis fabric transverse to flow, it is indicated by a dashed arrow. Number at the inside indicates the total observations contributing to each rose. Outer circle on each current rose encloses 20% of all observations for each 30° sector. Observations were made either at the base or at the upper boundary of the bed and refer to bases and tops of different layers, except for the couple D and E.

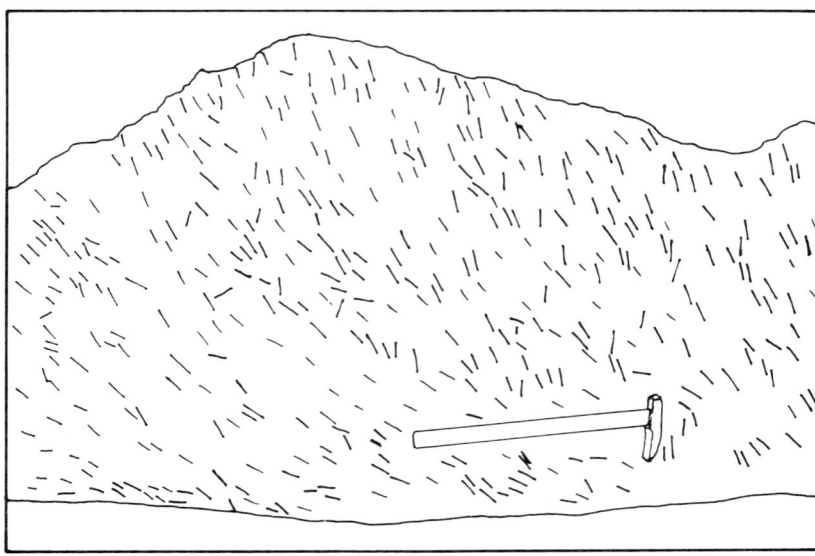

Fig. 12. (A) R_a unit showing progressive increase of imbrication angles from the base of the layer upwards. Top is at left. M. Altare member. (B) drawing from the photograph.

the base of the beds; in addition, inversely graded, repetitive bands interpreted as the deposits of inertia-flow carpets (S units) show mainly low imbrication angles (15° - 20°) in spite of an undoubted influence of grain interaction. It is likely therefore that in addition to grain interaction, some other conditions are necessary in order to produce the observed fabrics. Moreover it should be noted that some (rarely found) compound beds show lensing internal subdivisions (each showing pattern 1) separated by undulatory surfaces; this suggests that shear wavy surfaces developed between flows of different densities travelling at different velocities (Aalto, 1972; Hein, 1982).

It can be assumed that pattern 1 of imbrication reflects the development of internal shearing immediately before the freezing of a highly concentrated flow. On the other hand, the fact that pattern 2 is more common in units/sequences comprising an upper, normally graded portion, suggests that flattening-out of imbrication angles at the top of the layer may be related to the shearing action of a turbulent flow in rapid motion over the depositing bed.

R Unit

This unit has two variants, the one clast-supported (R_a) and the other matrix(sand)-supported (R_b). The occasional occurrence of lateral transitions between the two varieties suggests a similar emplacement process. Both are characterized by a pebble-cobble population and by lack of grading. In proximal areas R_a units commonly show a grain-size bimodality due to dispersion of cobble-sized clasts in a 'matrix' of pebbles. R_a beds average 90 cm thick, are bounded by an erosional base and commonly show the above-described patterns of vertically changing imbrication angles (Fig. 12; Figs. 11 D and E), with the steepest of any observed angles and even rare imbrication reversals at higher levels of the bed. The fabric at the base of the bed is either nearly isotropic or more or less anisotropic, with a-axes parallel to flow and imbricate upcurrent (Figs. 11 F and G) and sometimes also displaying a secondary mode dipping downcurrent (Fig. 11 D). A fan-like arrangement of clast long-axes, with the mean orientation parallel to flow, may appear in single areas at the base of the bed (Fig. 11 G). The imbrication within the bed is moderately to well developed; some beds, however, show a scatter of imbrication angles without a recognizable trend. Fabric deterioration may occur in the central part of R_a or R_a-R_2 or R_1-R_a-R_2 units, whereas well-developed imbrication occurs near bed boundaries.

A rarely occuring R_3 division at the base of R_a units (Table 1) might be interpreted as the result of deposition from turbulent flow which originated from the unstable head of a large highly concentrated gravelly mass-flow. As suggested by Postma (this volume) such a turbulent flow may outstrip the main mass and deposit its load before the arrival of the main gravelly mass flow.

The R_b variety is a rather uncommon unit in the study area and consists of structureless conglomeratic sandstone or matrix-supported sandy conglomerate (facies A_3 of Walker and Mutti, 1973), with poor fabric and imbrication ranging from highly variable (or unrecognizable) to moderately developed with upcurrent dip.

R_2 Unit

This unit (on average 90 cm thick) consists of normally graded, clast-supported, pebble-cobble conglomerate, bounded by an erosional base. The normal grading is generally of the distribution type. Outsize clasts are occasionally present. At the base of the bed the a-axis fabric is commonly almost isotropic (Fig. 11 K) or bimodal with a secondary mode transverse to flow (Fig. 11 J). In some cases however anisotropic fabrics with a-axes parallel to flow were observed (Figs. 11 H and I). The imbrication is poorly to well developed, with angles usually in the 15-20° range and generally quite consistent vertically. Pattern 1 of imbrication has never been observed in R_2 units and pattern 2 only in relatively rare cases (Table 3); in addition angles never exceed 40°.

Normal grading, scouring at the base, fabric inconsistency, bimodality or isotropism point to very rapid deposition from a highly concentrated turbulent suspension (Hiscott and Middleton, 1980). Occasional presence of a secondary mode transverse to flow in the clast long-axis fabric, mostly shown by larger clasts, may reflect tractional rolling along the bed. The occasional strong *a*-axis fabric parallel to flow and relatively steep imbrication suggest that, in addition to fluid turbulence, grain interaction may have had a secondary role as grain support, at least in stages immediately preceding the deposition.

R_1-(R_a)-R_2 Sequence

This sequence corresponds to the inverse-to-normally graded conglomerate facies of Walker (1975, 1977) and consists of clast-supported, pebble-cobble conglomerate with inverse-to-normal grading. The layer is on average 140 cm thick, shows an erosional base and lenticular geometry in section transverse to flow. Sometimes a thin sandstone skin with tool marks occurs at the base. Inverse-to-normal grading is sometimes very local and can be laterally replaced by simple normal or inverse grading. Grading, imbrication and anisotropic fabric occasionally occur only in the lowermost and uppermost parts of the bed. In some cases pattern 2 of imbrication occurs, and is usually accompanied by a fan-like orientation of clast long-axes in single areas at the base of the bed. In all cases, clast long-axes are predominantly parallel to flow and imbricate upcurrent at the base of the bed.

The presence of an erosional base suggests the existence of turbulent flow in stages preceding the formation and freezing of a basal inertia-flow carpet. According to Hiscott and Middleton (1980) the deterioration of the fabric in the middle of the bed may imply that this portion has moved as a rigid plug.

R_3 Unit

This unit is thought to have been deposited from 'traction' flow (fabric data do not always indicate pure tractional transport) during the last depositional stages of the R sequence and is characterized by internal stratification. There are two main varieties, respectively characterized by crude sub-horizontal stratification (R_{3a}) and by undular and low-angle cross-bedding (R_{3b}), and two minor varieties respectively characterized by backset bedding (R_{3c}) and trough cross-bedding (R_{3d}).

The R_{3a} variety, on average 55 cm thick, is characterized by a scoured basal contact and sub-horizontal stratification (Fig. 13). The latter is caused by rows of pebbles or cobbles in sandstone, by the alternation of sandstone bands (10-20 cm thick) with layers of conglomerate, or by varying grain size in conglomerate. Individual bands may be coarse-tail graded. The sandstone bands may show a crude and thick plane lamination. Imbrication is usually poorly to moderately developed.

Some beds show pronounced basal scouring with occasional large flute casts, rapid lateral changes in thickness and grain size, abundance of mudstone intraclasts and a-axis fabric bimodal (Fig. 11 M) or transverse to flow (Fig. 11 N). Such features suggest high turbulence and at least some bed-load traction. Other beds show groove and/or impact casts and anisotropic fabric (Fig. 11 L); in this case clast imbrication commonly deviates from sole markings, suggesting variable flow in the earliest stages of deposition.

The R_{3b} variety (Fig. 14) consists of sandstone and conglomerate showing low-amplitude symmetrical waves with crests transverse to flow. Layers are internally subdivided into bands of variable grain size, sometimes separated by scour surfaces. Such bands are either in phase with the wave profile or are horizontal to low-angle crossbedded; in the latter case they usually dip upcurrent at angles of 7-12°. The wavelength of undular pattern ranges

Fig. 13. R_{3a} unit (upper half of the photo). Scale bar 22 cm long. M. Altare member.

Fig. 14. R_{3b} unit. Paleoflow is in the downdip direction. Top is at left. M. Altare member.

from 0.8 to 6 m (on average 2.15 m), being 7-9 times the amplitude. The R_{3b} units are on average 65 cm thick and are bounded at the base by a plane or undular-erosional surface. It is difficult to determine whether the observed bedforms are bi- or tri-dimensional features because they are most often recognizable only in cross section. A crude thick lamination sometimes occurs in the sandstone bands and may be emphasized by granule or pebble stringers or by rows of rip-up mudstone clasts. Some bands occasionally show a normal grading. A bimodal long-axis fabric can often be recognized (Fig. 11 O); a mode displayed by larger clasts, mostly developed in the case of matrix support, is at right angles to the fine mode and shows b-axis imbrication. This would suggest that the former preserves a current-normal fabric due to rolling of clasts on the bed. Imbrication ranges from moderate to well-developed. R_{3b} units are often laterally gradational into S_1 units.

Davies and Walker (1974) recorded some wavy stratification with amplitude of 10 cm and wavelength of 1 m in the resedimented Cap Enragé conglomerates of Québec (Canada). Hiscott and Middleton (1979) described examples of internally stratified beds with wave-like geometry in the Ordovician 'base of slope' sandstones of Québec and thought that individual bands might represent stratification produced by migration and aggradation of long-wavelength antidunes beneath a supercritical flow. Colella (1980) described antidune-like bedforms with wavelength of 350-500 cm and low-angle inclined internal laminae in the Gorgoglione flysch of Southern Apennines.

The R_{3c} variety is quite rare and consists of a set of planar backset beds dipping upstream at angles of 15-25°. Set thickness ranges from 50-130 cm. The lithologies involved range from sandstone with scattered or aligned clasts in the pebble-cobble range to matrix- or framework-supported pebble- or pebble-cobble conglomerate. The sorting is usually very poor. In the sandier varieties there are thick laminae up to 4 cm thick, sometimes emphasized by pebble stringers. Normal grading occasionally occurs within some of the 'foresets'. Mudstone intraclasts and plant remains are present in places. This variety is interpreted as representing upstream migration of chute and pool bedforms.

The R_{3d} variety has been found only in two cases and consists of isolated sets, about 20-30 cm thick, of trough cross-bedding. The structure shows up because of the alternation of rows of matrix-supported small pebbles and granules, and sandstone bands with crude and thick lamination. Analogous medium-scale, high-angle cross-stratification is reported by some authors (Hendry, 1972; Mutti *et al.*, 1975; Walker, 1978; Surlyk, 1978; Colella, 1980; Hein, 1982) and may have resulted from migrating dunes.

S SEQUENCE

The units of the S sequence in the study area typically show internal stratification and are characteristically internally repetitive: in particular the S_1 units show repeated inverse grading and S_2 units almost always show multiple normal grading. These repetitions may be due to surging flows (Lowe, 1982). A depositional model can be established, as in the R sequence, with transition from inertia-flow carpet sedimentation to suspension sedimentation and finally 'traction' sedimentation. However, frequent irregularities and variations of the sequence have been found, probably because the three mechanisms of sedimentation may be active at almost any stage, depending on fluctuation in the rate of suspended-load fall-out (Lowe, 1982).

S_1 Unit

This unit (average 90 cm) is usually bounded by a planar, non-erosional contact and is characterized by near-horizontal stratification (average 24 cm thick bands). The most prominent feature of individual bands is an inverse grading, so that the whole bed is characterized by repetitive inverse grading (Figs. 15 and 16). No obvious overall grading occurs. Clasts are mostly in the small-pebble range, are well-sorted and supported by a medium-fine, sometimes granule-bearing sandstone matrix; less commonly they are framework supported. No obvious correlation between

Fig. 15. S_1 unit. Top is at right. Scale bar 22 cm long. M. Altare member.

Fig. 16. S_1 unit. Meter stick with 20 cm intervals for scale. M. Altare member.

bed thickness and maximum clast size occurs. Some inversely graded bands show a gradual increase in the sand content upwards, appearing clast-supported at the base and matrix-supported at the top. Fabric is strongly unimodal with clast long axes parallel to flow and consistently imbricate upcurrent, in most cases at angles of about 15-20° (Figs. 11 Q-S). It is stressed that S_1 units show the most strongly anisotropic fabric and the best imbrication of any of the examined units.

Sometimes the bands are not horizontal but form a low-angle (12-14°) cross-stratified set ranging in thickness from 25 cm to 2 m. The S_1 unit often gradationally overlies an R_a unit. Lateral gradations between S_1 and R_{3b} units are relatively common. Lateral transitions into massive, non-stratified conglomerates have occasionally been found.

Lack of erosion at the base of the beds may be explained by lack of erosive ability of the flow. Several workers have suggested that basal inertia-flow carpets would form a barrier between the current and the substrate, inhibiting its erosive capacity (Hsu, 1959; Walker, 1966). According to Hiscott and Middleton (1979 and 1980) stratification bands with inverse grading may have resulted from periodic 'freezing' of 'traction carpets' driven along by shear transmitted from an overlying turbidity current and upward migration of the sediment bed. Inverse grading and long-axis fabric parallel to flow and dipping upstream could have been produced by grain interaction resulting from bulk shearing of a high concentration of clasts. The periodic 'freezing' of inertia flow carpets at the base of the flow may have been caused by the surging character of the flow. Observed vertical transitions from unit R_{3b} or R_2 into S_1 demonstrate that it was possible to revert from a fluid turbulence to dispersive pressure grain-support mechanisms as the size population of available sediment became finer (see also Aalto, 1976). Moreover lateral transitions between R_{3b} and S_1 units suggest that grain interaction could alternate with 'traction' transport in supercritical regime (Aalto, 1976), probably due to the surging character of the flow.

S_2 Unit

This unit includes two varieties. The S_{2a} variety partly corresponds to facies A_4 of Walker and Mutti (1973). It is on average 90 cm thick and is characterized by repetitive normal grading in near-horizontal bands. Basal contacts are invariably sharp, commonly loaded and either nearly flat or broadly concave-up erosional; they occasionally show groove and impact casts, or (more rarely) large flute casts. Internal bands are on average 30 cm thick (range: a few cm - 70 cm) and are characterized by a 'dispersed' texture in which granules or small pebbles are scattered with coarse-tail grading throughout a fine- to medium-grained sandstone matrix (see also Hein and Walker, 1982). An overall normal grading is occasionally present throughout the bed. Mutual contacts between bands are either gradational or rather abrupt, but without textural break of the matrix. Sometimes banding is absent and normal grading occurs only once. The a-axis fabric is poor to nearly isotropic (Fig. 11 T) and imbrication is weakly developed. Sparse outsize clasts occasionally occur. Rip-up mudstone intraclasts and macerated plant debris are common. In places spectacular loading convolutions can be seen at the interface between some stratification bands and probably result from liquefaction due to very rapid sedimentation which trapped pore fluids and produced an unstable grain packing.

As discussed by Lowe (1982) this unit probably originated from surging flows showing an oscillatory decline, each surge characterized by an abrupt velocity increase followed by a gradual deceleration.

The S_{2b} variety consists of massive fine-grained sandstone and corresponds to layer V of Aalto (1976). It occurs only rarely in the study area and beds average 90 cm thick. Rarely, clusters of small pebbles or granules occur in the sandstone, and probably represent 'pocket'-like water-escape structures (Postma, 1983). This unit may be considered a structureless equivalent of variety S_{2a}. The massive nature of the bed and occasional presence of water-escape features suggest that it was affected by liquefaction during or immediately after deposition. Alternatively it may represent a deposit from liquefied flow, closely resembling that formed by settling of a laminar high-density suspension (Lowe, 1982).

S_3 Unit

This unit (on average 50 cm thick) corresponds to facies B_2 of Walker and Mutti (1973) and consists of fine- to medium-grained sandstone showing thick and crude subhorizontal laminations and rare dish structures. Laminae are between 2 mm and 4 cm thick and tend to display gradational contacts; they are defined by slight alternations in grain size. Lateral wedging of sets of laminae, in places accompanied by low-angle cross-bedding, has occasionally been observed. These features generally imply some internal scouring. In one case the horizontal banding was observed to pass gradationally downwards into crudely laminated and gently wavy features of low amplitude, with wavelength of about 95 cm and amplitude of 8 cm. Rhythmic alternations of S_2 and S_3 units have sometimes been found. In one case a 25 cm thick set showing planar cross-lamination and consisting of an alternation of fine- to medium-grained sandstone laminae and thick, medium-grained sandstone laminae with sparse granules and small pebbles was observed in association with crude horizontal lamination.

Deposition of the coarse-grained, high-density suspended load which formed the S_2 unit left a residual turbulent suspension which deposited 'traction' units. Low-amplitude wavy bedding may be related to migration and aggradation of antidunes beneath a supercritical flow. Medium-scale cross-bedding that is not part of the normal Bouma sequence may be generated by migration of dunes and should probably be included in this division of the S sequence. Similar structures have been reported by Mutti *et al.* (1975), Colella (1980) and Lowe (1982).

R-S UNITS

The designation R-S refers to a compound facies which cannot be included in the above-described units. It is defined as a complex and amalgamated bundle of pebble-cobble conglomerate layers interbedded with lenses or discontinuous layers of sandstone. The bundles average 150 cm thick and have a sharp, broadly concave-up base. Individual conglomerate beds in the bundle may show normal or inverse grading or absence of grading and are either matrix- or clast-supported. The sandstone bands sometimes contain outsize clasts and may show a planar base and convex-up top as well as a crude near-horizontal lamination (antidune-like bedforms ?); in other cases they show a concave-up erosional base and appear as scour-fills. They may grade laterally into conglomerates. A similar facies was described by Surlyk (1978) and interpreted in terms of progressive failure on a slope leading to a series of retrogressive flows. It is suggested that this mechanism would have caused surging flows, and if the transported sediment included a wide spectrum of sizes, a series of vaguely defined R and S sequences amalgamated together would have been deposited.

Lateral transitions	Number of examples	Lateral transitions	Number of examples
R_1-R_a	1	R_a-S_1	4
(R_1-R_a)-R_a	2	R_a-(R-S)	7
(R_1-R_a)-(R_1-R_a-R_2)	1	R_b-R_2	1
(R_1-R_a-R_2)-(R_1-R_a-R_{3a})	1	R_b-R_{3a}	1
(R_1-R_a-R_2)-(R_a-R_2)-R_2	1	R_b-S_1	1
R_1-R_2	3	R_b-S_2	3
(R_1-R_2)-R_2	2	R_2-(R_1-R_b)-R_{3a}	1
R_1(R_{3a}-R_1)	1	R_2-(R_2-R_{3a})	1
(R_1-R_a)-R_{3a}	2	(R_2-R_{3b})-(R_2-S_1)	1
R_1-R_{3a}	2	R_2-S_1	1
R_1-R_{3a}-R_{3d}	1	R_{3a}-R_{3c}	1
R_1-S_1	2	R_{3b}-R_{3a}-R_{3c}	1
R_a-R_b	2	(R_{3a}-R_a-S_2)-(R_{3a}-S_1-S_2)	1
R_a-R_2	1	R_{3b}-S_1	7
R_a-(R_a-S_1)	1	R_{3a}-S_2	5
R_a-R_3	1	R_{3a}-S_3	1
R_a-R_{3a}	9	S_1-S_2	2
R_a-(R_a-R_{3a})	1	S_2-S_3	1
R_a-R_{3b}	1		
(R_a-R_{3b})-R_{3a}	1		
R_a-R_{3a}-S_2	2		

Table 3. Lateral transitions observed within channelized deposits. In brackets: vertical sequences.

LATERAL TRANSITIONS

The identified units are linked by gradual transitions (Table 3). In general, lateral gradations are more rapid perpendicular to local paleoflow than parallel to it and in some cases spectacular lateral transitions can be identified within a few metres. Massive units, either graded or ungraded, can commonly be traced laterally into stratified units. Grading and preferred clast imbrication can appear and disappear when traced laterally; matrix content also varies greatly: a bed may show matrix-support in one part and clast-support in another. The base of some conglomerate beds may show a predominance of a-axis imbrication in some areas and of b-axis imbrication in others, even at very short distances. These conclusions have also been emphasized by others (e.g., Ghibaudo and Mutti, 1973; Aalto, 1976; Surlyk, 1978; Johnson and Walker, 1979; Winn and Dott, 1979; Cazzola et al., 1981; Hein and Walker, 1982). As observed by Surlyk (1978), the frequency of lateral facies transitions must imply a continuum in transporting mechanisms. This in turn suggests that the clast-supporting mechanism repeatedly changed during transport and deposition, or in different points of the flow.

DOWNSLOPE TRANSITIONS

The main changes of facies association and of type of sequence with distance from the source are summarized in Figs. 10 and 17. Proximal channel-fills (C. Posoccon member) are dominated by massive units of the R sequence, with the highest abundance of ungraded R_a units (see also Walker, 1975 and Hein, 1982) and of high imbrication angles. Downcurrent changes for R_a units imply segregation of a basal inertia-flow carpet and development of turbulence (Aalto, 1976), leading to inverse and normal grading respectively. This is suggested by 1) the significant downcurrent change in the relative abundance of graded and ungraded beds (Fig. 17) and 2) the fact that inverse-to-normally graded conglomerates commonly show an ungraded central portion which has either high imbrication angles or more isotropic fabric than near bed boundaries (see also Hiscott and Middleton, 1980). This conclusion for the R sequence contrasts with generalizations made by Walker (1975) on the downslope trend of grading. However, the downcurrent trend of the S sequence, which shows a downslope increase in percentage of S_2 units and a significant decrease in abundance of S_1 units, is consistent with Walker's views.

R_{3b}, R_{3c} and S_1 units have been almost exclusively found in the proximal facies associations. This means that conditions leading to the formation of multiple inertia-flow carpets and to supercritical flow regime are mostly realized in proximal areas and can seldom be reproduced downslope, even in the presence of a suitable size population.

It is further noted that stratified conglomerate units (R_3 and S_1) tend to become thinner-bedded and have better defined band-boundaries in the downslope direction. Finally,

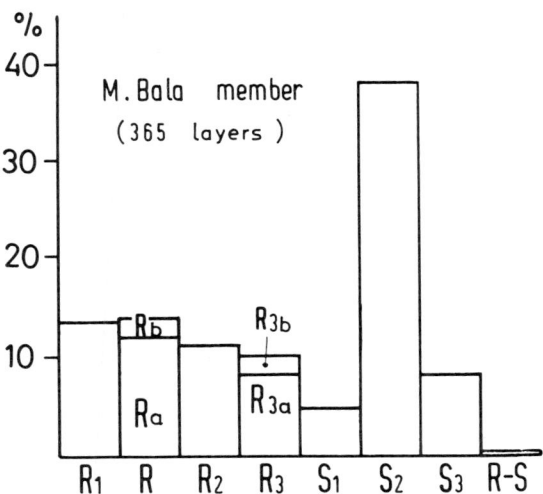

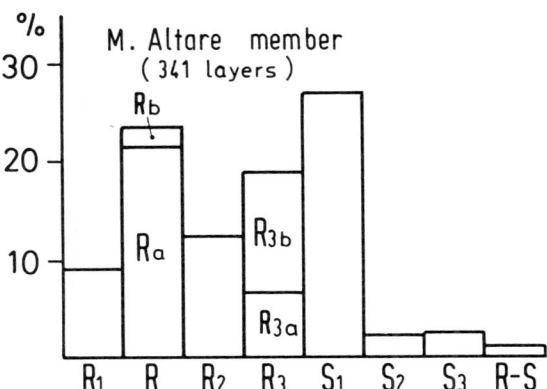

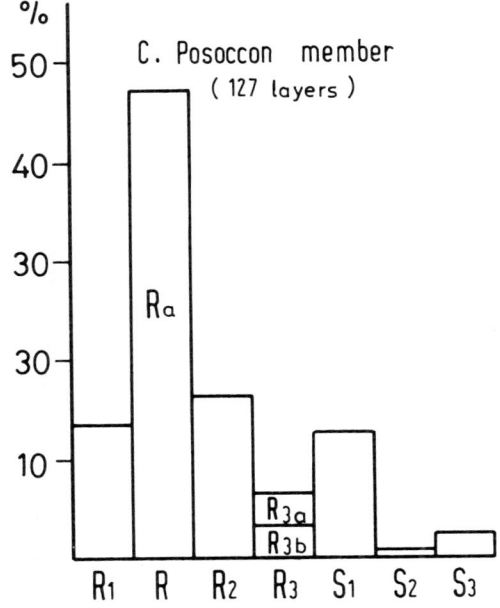

Fig. 17. Percent abundance of different units in M. Bala, M. Altare and C. Posoccon members. Proportions are for number of layers.

in the case of units including clast- and matrix-supported varieties, (such as R, R_2, R_{3a} and S_1) the former are generally more abundant nearer the source, pointing to a downcurrent increase in sand content (see also Nemec et al., 1980).

Similar downslope changes, though on a smaller scale, with proximal structureless deposits ('plugging' U-shaped troughs linked to slump scar zones) and distal dominantly stratified deposits (filling shallow troughs cutting each other) were recorded by Postma et al. (1983) in sandy-gravelly mass-flow deposits of an ice-marginal lake at Veluwe (The Netherlands). In particular, there is a striking similarity in the repetitive character of the stratified deposits which, they think, records flow pulsations possibly related to retrogressive flow slides, with scars progressively extending upslope.

Conclusions

The large bodies of the M. Altare member raise an important question concerning the maximum thickness of sequences of gravity-emplaced gravels that can be accumulated in a neritic environment. On the one hand it is necessary to assume the existence of definite slopes in a relatively shallow-water setting, in order to generate large-scale gravity flows, while on the other hand depressions are necessary to trap the gravity-displaced deposits at high levels on the slope, preventing a funnelling farther downslope into deeper waters.

Tectonically controlled slopes were probably present, considering both the general framework of sedimentation and the angular unconformity at the top of the C. Posoccon member. Tectonic gradients coupled with high sedimentation rates would certainly have led to gravity sliding. It is suggested that rotational slides combined with subsequent translational movements in a downslope direction could have taken place in such shallow-water areas, the process having been possibly aided by cyclic loading due to passage of storm waves and liquefaction (Marshall, 1978). Headward advance of such slumps, followed by spontaneous liquefaction (Andresen and Bjerrum, 1967), may generate elongate slide scars and gullies (Prior and Coleman, 1982), which may extend shorewards up to the surf zone (Reimnitz and Gutierrez-Estrada, 1970). Large volumes of sediment may be funnelled and partly trapped in such a system of gullies. The network of conglomerate lenses of the C. Posoccon member can perhaps be interpreted as the filling of small ephemeral gullies related to retrogressive headward slumping. They were probably subject to rapid filling and had low lateral mobility.

The bulk of sediment moving downslope might then have filled large, simple or composite slide scars or have been trapped on the downthrown side of synsedimentary faults. Such depressions may provide suitable traps for gravity-displaced sediments even at high levels on the slope. As an alternative possibility, such trap-depressions may have resulted from a complex interplay of synsedimentary faulting and slumping of large masses of sediment.

The coalescence of second-rank units of the M. Altare member to form major bodies suggests that formation and subsequent filling of closely adjacent depressions may have taken place in rapid succession.

No doubt the sedimentation of the Vittorio Veneto sequence was tectonically-controlled. However the additional influence of other external factors, such as a temporary lowering of sea level leading to a potential increase in gravitative mass-flow phenomena, cannot be excluded.

Wescott and Ethridge (1980) recognized two models of fan-deltas, those that prograde directly on to submarine slopes (Yallahs-type fan-deltas, localized along continental and island-arc collision coasts) and those that prograde on to continental or island shelves. Fan-deltas of the first type generally form thicker deposits, and are characterized by the importance of mass-gravity processes and the presence of numerous submarine channels and angular unconformities. The Vittorio Veneto complex probably represents an example of this type of fan-delta.

References

Aalto, K.R. 1972. Flysch pebble conglomerate of the Cap-des-Rosiers Formation (Ordovician), Gaspé peninsula, Québec. Journal of Sedimentary Petrology, v. 42, p. 922-926.

Aalto, K.R. 1976. Sedimentology of a mélange: Franciscan of Trinidad, California. Journal of Sedimentary Petrology, v. 46, p. 913-929.

Andresen, A. and Bjerrum, L. 1967. Slides in subaqueous slopes in loose sand and silt. In: Richards, A.F. (Ed.), Marine Geotechnique, Urbana, Chicago, p. 221-239.

Bagnold, R.A. 1956. The flow of cohesionless grains in fluids. Philosophical Transactions of the Royal Society of London, Series A, v. 265, p. 315-319.

Carter, R.M. 1975. A discussion and classification of subaqueous mass-transport with particular application to grain-flow, slurry-flow, and fluxoturbidites. Earth Science Reviews, v. 11, p. 145-177.

Cazzola, C., Fonnesu, F., Mutti, E., Rampone, G., Sonnino, M. and Vigna, B. 1981. Geometry and facies of small, fault-controlled deep-sea fan systems in a transgressive depositional setting (Tertiary Piedmont basin, Northwestern Italy). In: Ricci Lucchi, F. (Ed.), Excursion Guidebook, International Association of Sedimentologists, 2nd. European Regional Meeting, Bologna, p. 7-53.

Clari, P. and Ghibaudo, G. 1979. Multiple slump scars in the Tortonian type area (Piedmont Basin, northwestern Italy). Sedimentology, v. 26, p. 719-730.

Clifton, H.E., Hunter, R.E. and Phillips, R.L. 1971. Depositional structures and processes in the non-barred high-energy nearshore. Journal of Sedimentary Petrology, v. 41, p. 651-670.

Colella, A. 1980. Medium-scale tractive bedforms and structures in Gorgoglione Flysch (Lower Miocene, Southern Apennines, Italy). Bollettino della Società Geologica Italiana, v. 98, p. 483-494.

Davies, I.C. and Walker, R.G. 1974. Transport and deposition of resedimented conglomerates: the Cap Enragé Formation, Cambro-Ordovician, Gaspé, Québec. Journal of Sedimentary Petrology, v. 44, p. 1200-1216.

Dondi, L., Mostardini, F. and Rizzini, A. 1982. Evoluzione sedimentaria e paleogeografica nella Pianura Padana. In: Cremonini, G. and Ricci Lucchi, F. (Eds.), Società Geologica Italiana, Guide Geologiche regionali, p. 47-60.

Dott, R.J. jr. and Bourgeois, J. 1982. Hummocky stratification: significance of its variable bedding sequences. Geological Society of America Bulletin, v. 93, p. 663-680.

Ghibaudo, G. and Mutti, E. 1973. Facies ed interpretazione paleoambientale delle arenarie di Ranzano nei dintorni di Specchio (Val Pessola, Appennino Parmense). Memorie della Società Geologica Italiana, v. 12, p. 251-265.

Hein, F.J. 1982. Depositional mechanisms of deep-sea coarse clastic sediments, Cap Enragé Formation, Québec. Canadian Journal of Earth Sciences, v. 19, p. 267-287.

Hein, F.J. and Walker, R.G. 1982. The Cambro-Ordovician Cap Enragé Formation, Québec, Canada: conglomeratic deposits of a braided submarine channel with terraces. Sedimentology, v. 29, p. 309-329.

Hendry, H.E. 1972. Breccias deposited by mass flow in the Breccia Nappe of the French Pre-Alps. Sedimentology, v. 18, p. 277-292.

Hiscott, R.N. 1980. Depositional framework of sandy mid-fan complexes of Tourelle Formation, Ordovician, Québec. American Association of Petroleum Geologists Bulletin, v. 64, p. 1052-1077.

Hiscott, R.N. and Middleton, G.V. 1979. Depositional mechanics of thick-bedded sandstones at the base of a submarine slope, Tourelle Formation (Lower Ordovician), Québec, Canada. In: Doyle, L.J. and Pilkey, O.H. (Eds.), Geology of Continental Slopes. Society of Economic Paleontologists and Mineralogists, Special Publication 27, p. 307-326.

Hiscott, R.N. and Middleton, G.V. 1980. Fabric of coarse deep-water sandstones, Tourelle Formation, Québec, Canada. Journal of Sedimentary Petrology, v. 50, p. 703-722.

Hsu, K.J. 1959. Flute- and groove-casts in the Prealpine Flysch, Switzerland. American Journal of Science, v. 257, p. 529-536.

Hyden, F.M. 1980. Mass flow deposits on a mid-Tertiary carbonate shelf, southern New Zealand. Geological Magazine, v. 117, p. 409-424.

Jeletzky, J.A. 1975. Hesquiat Formation (new): a neritic channel and interchannel deposit of Oligocene age, western Vancouver Island, British Columbia. Geological Survey of Canada, Paper 75-32, 55p.

Johnson, B.A. and Walker, R.G. 1979. Paleocurrents and depositional environments of deep-water conglomerates in the Cambro-Ordovician Cap Enragé Formation, Québec Appalachians. Canadian Journal of Earth Sciences, v. 16, p. 1375-1387.

Laird, M.G. 1968. Rotational slumps and slump scars in Silurian rocks, Western Ireland. Sedimentology, v. 10, p. 111-120.

Lewis, D.W., Laird, M.G. and Powell, R.D. 1980. Debris flow deposits of Early Miocene age, Deadman Stream, Marlborough, New Zealand. Sedimentary Geology, v. 27, p. 83-118.

Lowe, D.R. 1979. Sediment gravity flows: their classification and some problems of application to natural flows and deposits. In: Doyle, L.J. and Pilkey, O.H. (Eds.), Geology of Continental Slopes. Society of Economic Paleontologists and Mineralogists, Special Publication 27, p. 75-82.

Lowe, D.R. 1982. Sediment gravity flows: II. Depositional models with special reference to the deposits of high-density turbidity currents. Journal of Sedimentary Petrology, v. 52, p. 279-297.

Marshall, N.F. 1978. Large storm-induced sediment slump reopens an unknown Scripps submarine canyon tributary. In: Stanley, D.J. and Kelling, G. (Eds.), Sedimentation in Submarine Canyons, Fans and Trenches. Stroudsburg, Pennsylvania: Dowden, Hutchinson and Ross, p. 73-84.

Massari, F., Rosso, A. and Radicchio, E. 1974. Paleocorrenti e composizione dei conglomerati tortoniano-messiniani compresi fra Bassano e Vittorio Veneto: Memorie degli Istituti di Geologia e Mineralogia dell'Università di Padova, v. 31, 20p.

Massari, F., Iaccarino, S. and Medizza, F. 1976. Depositional cycles in the Tortonian Messinian of the Southern Alps (Italy): transition from fan-delta to alluvial fan sedimentation. Messinian Seminar 2, Field trip Guide Book.

Middleton, G.V. and Hampton, M.A. 1973. Sediment gravity flows: mechanics of flow and deposition. In: Middleton, G.V. and Bouma, A.H. (Co-chairmen), Turbidites and Deep-Water Sedimentation, Pacific Section, Society of Economic Paleontologists and Mineralogists, Anaheim, Short Course Notes, p. 1-38.

Mutti, E. 1977. Distinctive thin-bedded turbidite facies and related depositional environments in the Eocene Hecho Group (South-central Pyrenees, Spain). Sedimentology, v. 24, p. 107-131.

Mutti, E. and Ricci Lucchi, F. 1972. Le torbiditi dell'Appennino settentrionale: introduzione all'analisi di facies. Memorie della Società Geologica Italiana, v. 11, p. 161-199.

Mutti, E., Parea, G.C., Ricci Lucchi, F., Sagri, M., Zanzucchi, G., Ghibaudo, G. and Iaccarino, S. 1975. Examples of turbidite facies and facies associations from selected formations of the Northern Apennines. 9th International Sedimentological Congress, Nice, France, Field Trip A11, 120p.

Nemec, W., Porebski, S.J. and Steel, R.J. 1980. Texture and structure of resedimented conglomerates: examples from Ksiaz Formation (Famennian-Tournaisian), southwestern Poland. Sedimentology, v. 27, p. 519-538.

Postma, G. 1983. Water escape structures in the context of a depositional model of a mass flow dominated conglomeratic fan-delta (Abrioja Formation, Pliocene, Almeria basin, SE Spain). Sedimentology, v. 30, p. 91-103.

Postma, G., Roep, T.R. and Ruegg, G.H.J. 1983. Sandy-gravelly massflow deposits in an ice-marginal lake (Saalian, Leuvenumsche Beck valley, Veluwe, The Netherlands), with emphasis on plug-flow deposits. Sedimentary Geology, v. 34, p. 59-82.

Prior, D.B. and Coleman, J.M. 1982. Active slides and flows in undercompacted marine sediments on the slopes of the Mississippi delta. In: Saxov, S. and Nieuwenhuis, J.K. (Eds.), Marine Slides and Other Mass Movements. New York: Plenum Press, p. 21-49.

Rees, A.I. 1968. The production of preferred orientation in a concentrated dispersion of elongated and flattened grains. Journal of Geology, v. 76, p. 457-465.

Reimnitz, E. and Gutierrez-Estrada, M. 1970. Rapid changes in the head of the Rio Balsas submarine canyon system, Mexico. Marine Geology, v. 8, p. 245-258.

Riba, O. 1976. Syntectonic unconformities of the Alto Cardener, Spanish Pyrenees: a genetic interpretation. Sedimentary Geology, v. 15, p. 213-233.

Ricci Lucchi, F. 1975. Miocene paleogeography and basin analysis in the Periadriatic Apennines. In: Squyres, C. (Ed.), Geology of Italy. Petroleum Exploration Society of Libya, p. 5-111.

Rizzini, A. and Dondi, L. 1978. Erosional surface of Messinian age in the subsurface of the Lombardian plain (Italy). Marine Geology, v. 27, p. 303-325.

Slaczka, A. and Thompson, S. III 1980. A revision of the fluxoturbidite concept based on type examples in the Polish Carpathian Flysch. Annales Societatis Geologorum Poloniae, v. 51, p. 3-44.

Surlyk, F. 1978. Submarine fan sedimentation along fault scarps on tilted fault blocks (Jurassic-Cretaceous boundary, East Greenland). Grønlands Geologiske Undersøgelse Bulletin 128, 108p.

Walker, R.G. 1966. Deep channels in turbidite-bearing formations. American Association of Petroleum Geologists Bulletin, v. 50, p. 1899-1917.

Walker, R.G. 1969. The juxtaposition of turbidite and shallow-water sediments. Study of a regressive sequence in the Pennsylvanian of North Devon, England. Journal of Geology, v. 77, p. 125-143.

Walker, R.G. 1975. Generalized facies models for resedimented conglomerates of turbidite association. Geological Society of America Bulletin, v. 86, p. 737-748.

Walker, R.G. 1977. Deposition of Upper Mesozoic resedimented conglomerates and associated turbidites in Southwestern Oregon. Geological Society of America Bulletin, v. 88, p. 273-285.

Walker, R.G. 1978. Deep-water sandstone facies and ancient submarine fans: models for exploration for stratigraphic traps. American Association of Petroleum Geologists Bulletin, v. 62, p. 932-966.

Walker, R.G. and Mutti, E. 1973. Turbidite facies and facies associations. In: Middleton, G.V. and Bouma, A.H. (Co-chairmen), Turbidites and Deep-Water Sedimentation. Pacific section, Society of Economic Paleontologists and Mineralogists, Anaheim, Short Course Notes, p. 119-157.

Wescott, W.A. and Ethridge, F.G. 1980. Fan-delta sedimentology and tectonic setting-Yallahs fan delta, Southeast Jamaica. American Association of Petroleum Geologists Bulletin, v. 64, p. 374-399.

Winn, R.D., Jr. and Dott, R.H., Jr. 1979. Deep-water fan-channel conglomerates of Late Cretaceous age, Southern Chile. Sedimentology, v. 26, p. 203-228.

CONGLOMERATIC FAN-DELTA SEQUENCES, LATE CARBONIFEROUS — EARLY PERMIAN, WESTERN SPITSBERGEN

K.L. KLEINSPEHN[1], R.J. STEEL[2], E. JOHANNESSEN[3] AND A. NETLAND[4]

Abstract

Nine conglomeratic fan-delta front sequences are used to describe the evolution of a Carboniferous-Permian fan-delta complex in southwestern Spitsbergen. Within these sequences, three conglomerate assemblages characterize the fan-delta front deposits. The fluvial assemblage consists of texturally immature, sandy conglomerates interbedded with tightly-packed, well-sorted and matrix-poor, normally graded conglomerates. Beds are typically flat or slightly inclined, although cross-stratified sets do occur. The barrier/spit assemblage also has flat or slightly inclined strata, but is characterized by thinner, more evenly developed lamination, by 'pebble-thick' conglomerates, and by greater amounts of sandstone. The channel mouth assemblage is dominated by beds in which granules to pebbles line shallow troughs and grade vertically into massive or ripple cross-laminated sandstone. These beds alternate with matrix-supported disorganized units, with sandy cross-stratified units, and with inversely-to-normally graded units. This channel mouth assemblage represents high-concentration gravity flows, ranging from high-density turbidity currents to cohesive subaqueous debris flows, which occur intercalated with traction deposits of both fluvial and marine origin. Deposits of mass flows, though still not widely recognized as a common fan-delta front feature, are abundant in the studied succession.

The internal organization of individual sequences is determined by an interplay between several intrabasinal and extrabasinal controls. Although tectonism and corresponding relative sea level changes do influence fan evolution, avulsion of fluvial channels, lateral migration of barriers or shoals, aggradation of mouth bars, and modification by wave and tidal currents exert primary control on the character of the fan-delta front deposits. In the studied succession, vertical changes in the basal three sequences are explained by these primary controls superimposed on a background of constant gradual subsidence. In contrast, abrupt sea level changes or punctuated subsidence is necessary to produce the offshore facies observed at the base of the upper sequences.

We support earlier suggestions that, as in conventional deltas, fan-deltas be classified according to processes dominating the delta front. Preservation of subaqueous mouth bars suggests a subsidence and burial rate sufficiently high to protect those deposits from reworking after channel abandonment. Abundant channel mouth deposits may record a fluvially-dominated fan-delta system, but in this case, it is preferable to interpret the channel mouth deposits as fluvial breaks in the wave reworked fan-delta front. The intercalation of abundant beach, barrier, and spit deposits with fluvial strata indicates that the fan-delta system was of the fluvial-wave interaction type.

Résumé

Neuf séquences frontales d'un cône de déjection conglomératique sont utilisées pour décrire l'évolution d'un complexe d'un cône de déjection du Carbonifère-Permien dans le sud-ouest du Spitsbergen. Dans ces séquences on distingue trois assemblages de conglomérat qui caractérisent les dépôts frontaux du cône de déjection. Un assemblage fluviatile qui comprend des conglomérates gréseux à texture immature, interstratifiés avec des conglomérats à granoclassement normal, compacts, bien triés et ayant peu de matrice. La stratification est typiquement horizontale ou légèrement inclinée, quoique des couches à stratification oblique soient présentes. Un assemblage de cordon littoral/flèche littorale exhibe également des strates à plat ou légèrement inclinées, mais il est caractérisé par des laminations plus minces et plus uniformes, par des conglomérats de l'épaisseur d'une 'monocouche de galets', et par de plus fortes quantités de grès. L'assemblage de l'embouchure du cône de déjection est caractérisé par des couches de granules et de galets qui garnissent les dépressions peu profondes et qui passent verticalement à des grès massifs ou avec rides à lamination oblique. Ces couches alternent avec des unités inorganisées à fragments non-jointifs, avec des unités gréseuses à stratification oblique et avec des unités à granoclassement allant d'inverse à normal. Cet assemblage de l'embouchure du chenal représente des coulées par gravité fortement chargées, variant de courants de turbidité de forte densité à des coulées de débris cohésives sous-aquatiques, qui apparaissent intercalées dans des dépôts d'écoulements en masse, bien que n'étant pas encore largement reconnus comme une particularité fréquente du front d'un cône de déjection, sont abondantes dans la succession étudiée.

L'organisation interne des séquences individuelles est déterminée par un effet réciproque des contrôles intrabassinaux et interbassinaux. Quoique le tectonisme et les changements relatifs du niveau de la mer associés influencent le développement du cône de déjection, l'avulsion des chenaux fluviaux, les barres latérales, et la modification résultant des courants des vagues et des marées, exercent un contrôle primordial sur les particularités des dépôts frontaux d'un cône de déjection. Pour la section étudiée, les changements verticaux dans les trois séquences basales sont expliqués par ces contrôles primordiaux surimposés aux effets d'une subsidence graduelle et constante. Au contraire, les variations brusques du niveau de la mer ou une subsidence ponctuée est nécessaire pour engendrer le faciès offshore observé à la base des séquences supérieures.

[continued]

---

[1] Geological Institute, (A), University of Bergen, 5014 Bergen, Norway.
 Current address: Department of Geology and Geophysics, University of Minnesota, 310 Pillsbury Drive S.E., Minneapolis, Minnesota 55455, U.S.A.
[2] Norsk Hydro Research Centre, P.O. Box 4313, 5013 Bergen, Norway.
[3] Statoil, Kanebogen, P.O. Box 40, 9401 Harstad, Norway.
[4] Statoil, P.O. Box 1212, 5001 Bergen, Norway.

Research for this paper is part of the University of Bergen's Svalbard Project and was undertaken as part of the 1982 Statoil Svalbard Expedition. Ole Aga and the Statoil staff are graciously thanked for their invitation to join the expedition and for permission to publish these results. Enlightening discussions with Mike Talbot and Gyrd Sundsbø greatly enhanced our understanding of the succession. Wojtek Nemec, Trevor Elliot, and George Postma provided helpful criticism of earlier drafts of the manuscript. Preliminary study was conducted by Netland as part of a Candidatus Realium degree at the University of Bergen. Kleinspehn gratefully acknowledges support from a Håkon Styri Fellowship of the American-Scandinavian Foundation, a Fulbright-Hays Travel Grant, and a NATO Postdoctoral Fellowship. The authors express their appreciation to Masaoki Adachi and Ellen Irgens for their assistance with technical drafting.

Copyright © 1984, Canadian Society of Petroleum Geologists

Nous sommes en accord avec les interprétations antérieures spécifiant que dans les deltas conventionnels, les côres de déjection soient classifiés selon les processus prédominants qui affectent le front du delta. La préservation des flèches barrantes de l'embouchure révèle une subsidence et un taux d'enfouissement suffisamment rapides pour protéger ces dépôts d'un remaniement après l'abandon du chenal. L'abondance des dépôts à l'embouchure du chenal peuvent témoigner de l'existence d'un système de cône déjection principalement fluviatile, mais dans ce cas, il est préférable d'envisager les dépôts de l'embouchure du chenal comme correspondant à des interruptions fluviales du remaniement par les vagues du front du cône de déjection. L'intercalation de dépôts abondants de plage, barrière, et flèche littorale avec des strates fluviatiles indiquent que le système du cône de déjection correspond au type d'une interaction d'un écoulement fluvial et des vagues.

INTRODUCTION

A growing number of modern and ancient fan-deltas and fan-deltaic successions have been documented in recent years. The accessibility of the subaerial part of recent fan-deltas has resulted in preferential documentation of the fan-delta plain deposits (Galloway, 1976; Wescott and Ethridge, 1980), whereas there is limited information on the facies and sequences developed on the fan-delta front (Ricci Lucchi et al., 1981; Wescott and Ethridge, 1982; Postma, 1984). Although accepting the *essentially alluvial fan nature* of fan-deltas (Sykes and Brand, 1976), we suggest, as with common river deltas, that fan-deltas should be classified according to the type of process dominating the delta front.

In the present study we describe delta-front deposits from nine fan-delta sequences and construct a composite model of the fan-delta type for the succession studied on Spitsbergen. In view of the abundance of conglomerates in the succession, an additional aim is to document conglomerate assemblages representative of fluvial, mouth bar, and beach/spit settings. The varying hydraulic conditions within these closely related subenvironments resulted in a spectrum of gravel facies.

TECTONIC AND STRATIGRAPHIC SETTING

The studied succession consists of some 150 m of conglomerates, sandstones, and sandy carbonates, apparently organized into nine vertical sequences with offshore to nearshore/subaerial character (Fig. 1). The rocks occur in the Bellsund area, at the western end of Van Keulenfjorden in southwestern Spitsbergen, and the main described section occurs on a ridge of Aldegondaberget at 77°30' North, 15° East (Fig. 2). The succession is part of the Reinodden Formation (Cutbill and Challinor, 1965) and is poorly dated. It is lithologically similar to the Treskelodden beds (Birkenmajer, 1964) and to the Drevbreen beds (Nysæther, 1977) in southern Spitsbergen. The latter succession has a proposed Late Carboniferous age on the basis of fusulinids (Nysæther, 1977) and brachiopods (Birkenmajer and Czarniecki, 1960), and an Early Permian age from the contained corals (Fedorowski, 1964; 1965). Immediately north of the study area, the coarse-grained succession interfingers with, and is eventually dominated by marine carbonates (Netland, 1981), referred to as the Nordenskiøldbreen Formation and dated as Late Carboniferous to Early Permian in age.

Mid-Carboniferous tectonism in southern Spitsbergen caused localized graben formation (Gjelberg and Steel, 1981) which was superseded by widespread development of a late Carboniferous to mid-Permian carbonate platform over most of Spitsbergen (Steel and Worsley, in press). Although the studied succession is generally coeval with this long-term transgression, it is immediately overlain by a tidal flat succession including stromatolitic dolomites (Tangen, 1981) and records a brief regressive interval prior to the widespread Permian transgression. We suggest that the studied conglomerate sequences in part reflect sporadic fault movements along the extensional margin of the Inner Hornsund Trough (Fig. 2), as well as intrabasinal processes.

At Aldegondaberget, the Reinodden Formation rests on Hecla Hoek metamorphic basement, while at nearby localities farther north there is a variable thickness of the Lower and Middle Carboniferous strata present above the Hecla Hoek. We suggest that Upper Carboniferous and Permian rocks overlap Middle Carboniferous units, which in turn overlap Lower Carboniferous strata across the northeastern edge of the Inner Hornsund Trough. This was probably due to repeated, but decreasing, instability of the graben margin into Permian time (Fig. 3).

REASONS FOR FAN-DELTA INTERPRETATION

A sedimentary body of inferred fan-delta origin contains alluvial fan deposits, usually in its upper portion, and delta front/slope deposits, usually in its lower part. Although signs of wave or tidal action in the lower or outer part of such a body would clearly demonstrate its deltaic nature, this should not be mandatory, because fluvially dominated fan-delta bodies, especially lacustrine ones, may well lack such facies. The delta front, in the latter case, will be dominated by channel mouth deposits.

The deltaic nature of the studied succession is quite obvious. That the conglomerates built into standing water is clear from the general stratigraphic relations, in that some conglomerates are fossiliferous, and that the conglomerates in general interfinger laterally with marine carbonates (Figs. 2, 3). In addition, the succession consists of a series of sequences, in which sandy delta-front deposits exhibiting evidence of wave and tidal processes are overlain by fluvial or shoreline conglomerates.

In the studied succession, the parts interpreted as subaerial fan deposits are conglomeratic, mainly flat or low-angle stratified, and thinly bedded. Their textures and

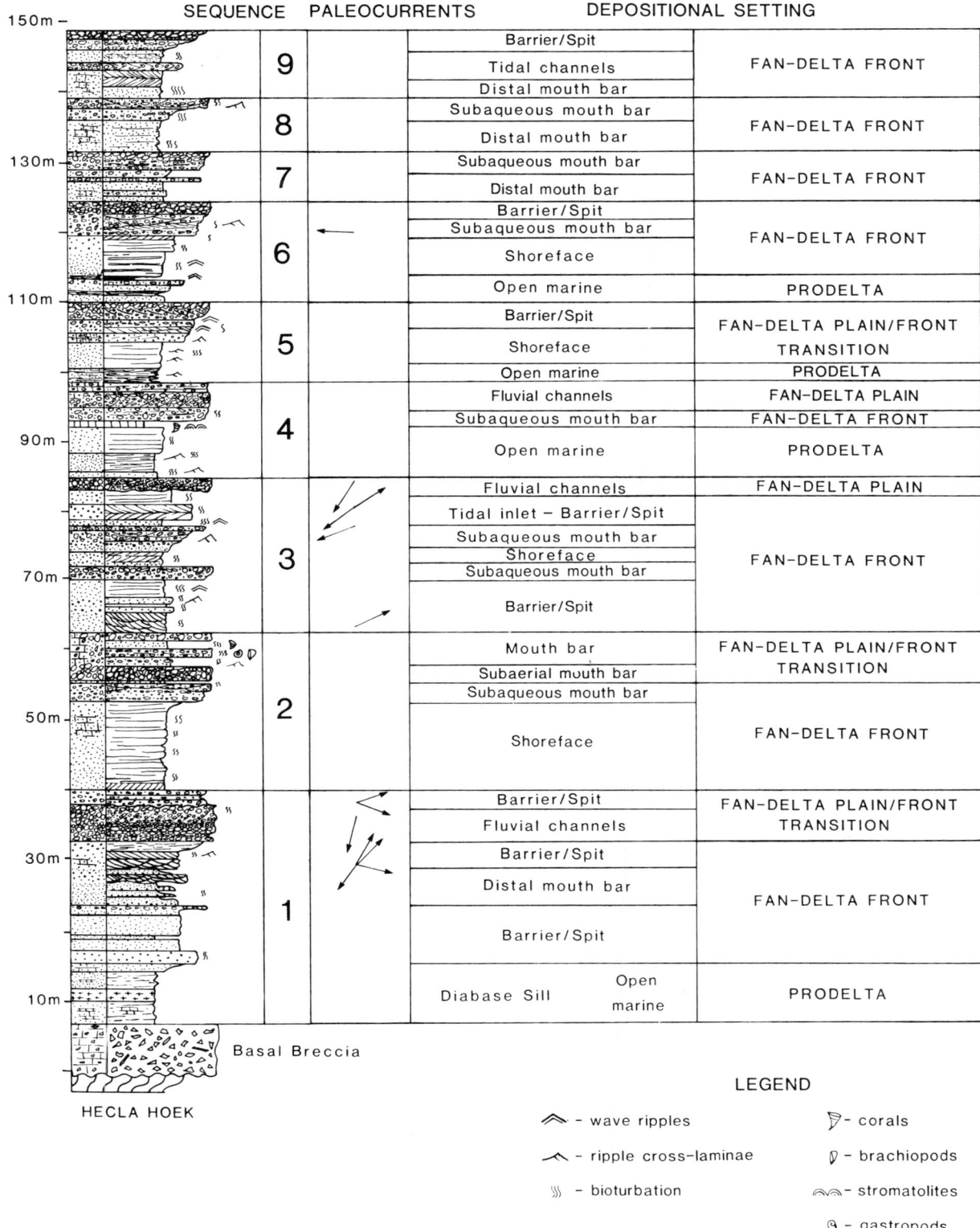

Fig. 1. A generalized stratigraphic profile of the Reinodden Formation at Aldegondaberget showing the organization of the succession into nine sequences.

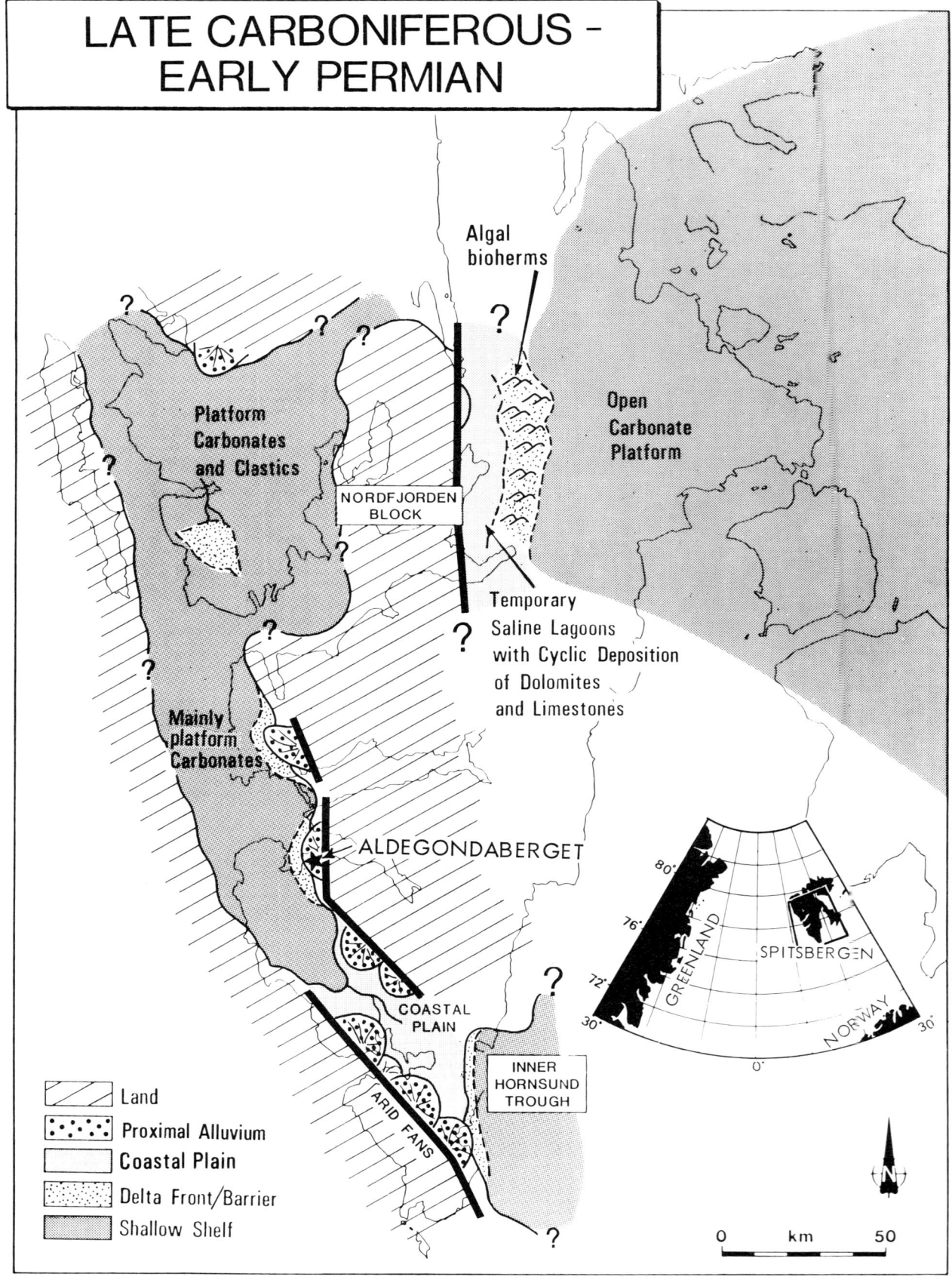

Fig. 2. Schematic paleogeographic reconstruction of southern Spitsbergen during Late Carboniferous-Early Permian time. The location of the larger map is outlined by a rectangle on the inset map. The present study locality is denoted by a star. Heavy dark lines demarcate fault zones that were presumably active during the designated time period.

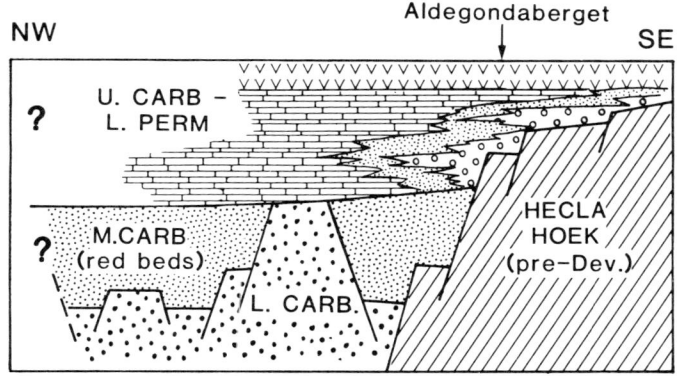

Fig. 3. Schematic palinspastic cross-section from the study area, showing Upper Carboniferous-Lower Permian strata onlapping basement. Present-day irregular distribution of Middle Carboniferous redbeds suggests complex Middle Carboniferous graben development.

characteristic sedimentation units, as discussed below, are typical of fluvial conglomerates, while their coarseness, structure, and thinly bedded nature favour braided rather than meandering stream deposits. In addition, the absence of epsilon cross-strata and of interbedded fine-grained sediments further supports a braided fluvial interpretation.

Conglomeratic Assemblages

FLUVIAL CONGLOMERATES

Conglomerate beds interpreted as being of fluvial origin are relatively rare in the studied section, occurring only in Sequences 1, 3 and 4 (Fig. 1). They occur in the upper parts of sequences and overlie marine sandstones and conglomerates of fan-delta front origin (Fig. 1). Facies have the following characteristics:

a) They typically show crudely or well-developed normal grading, though they are interbedded with less well-organized, ungraded beds.
b) They are typically clast-supported and well-packed, with a framework which is either open or sand-filled, though the ungraded beds can be sand matrix-supported.
c) Beds are dominantly flat or low-angle stratified, rarely exceeding 70 cm in thickness, but cross-stratified sets, up to 1.6 m thick, do occur.
d) Where sets are cross-stratified, the individual foresets have textural/structural characteristics similar to those noted for the flat-stratified conglomerates, but also display erosional channelized bases.

The texture and grading typical of these conglomerates are illustrated in Figure 4. The graded beds are generally thinner (mean 17 cm in Fig. 4) than the ungraded beds (mean 60 cm). The graded beds are usually well-packed and often lack matrix in their middle or lower parts (Fig. 4, Type a, Photo 2) though thicker varieties have a more poorly sorted, or sometimes bimodal (well-sorted pebbles in abundant sand matrix) and matrix-rich lower part (Fig. 4, Type b). The graded beds have a sand matrix-filled framework in their upper part and sometimes have a thin capping of well-packed granules or sandstone. Ungraded beds have a texture which resembles the lowermost portion (poorly sorted and bimodal) of the thicker graded beds (Fig. 4, Type c, Photo 1).

The coarseness and general stratigraphic position of these conglomeratic facies within individual sequences suggest a relatively proximal setting in the fan-delta system. The characteristic grading, winnowed textures, channeling, and bed thicknesses further suggest deposition in a fluvial setting. Graded gravel beds or units are fairly common in some fluvial systems (*e.g.* Smith, 1974; Bluck, 1976) and probably form during waning flood stage. The sorted and well-packed textures reflect both longer-term sorting on bar surfaces and winnowing attendant upon falling flow stage. Gravel units with a lower open and upper sand-filled framework indicate early winnowing and later sand infiltration during falling stage, and have been recorded both from recent (Smith, 1974) and ancient (Steel and Thompson, 1983) fluvial deposits. The thicker, disorganized beds probably were deposited relatively rapidly during early or peak flood stage, and so were 'protected' from the waning-stage, winnowing processes. The coarseness of the deposits, absence of interbedded fine sediments and of fossils, together with the absence of any epsilon cross-strata, suggest that the gravels formed on longitudinal bars in a braided fluvial system.

It is emphasized that it is the combination of criteria and dominance of certain features which allow the above interpretation. Flat or low-angle strata and sorted textures also typify some of our inferred beach conglomerates, but the beach strata are characteristically much thinner and the facies is richer in sand. Similarly, grading is common in many of our inferred subaqueous mouth bar gravels, but the type of grading and sedimentation units are different, as discussed below.

BARRIER/SPIT CONGLOMERATES

Conglomerate beds inferred to have been wave-generated or wave-reworked on beaches have been recognized in Sequences 1, 2, 6 and 9 (Fig. 1). They occur in the upper parts of sequences, either associated with conglomeratic mouth-bar deposits or, where the product of larger scale reworking of the fan-delta front, in the boundary zone between sequences (Fig. 5). This conglomerate assemblage is recognized by the following facies:

a) The conglomerates here are laterally persistent 'pebble thick' layers or discontinuous horizons of isolated pebbles, usually flat-lying in both cases. They are often merely a minor component in flat or low-angle stratified quartz arenites which contain *Skolithos* burrows and are interpreted as foreshore-upper shoreface deposits.
b) Thicker conglomerate units (up to 60 cm) also occur. These can be either clast-supported or sand matrix-supported, and in either case have a well-developed

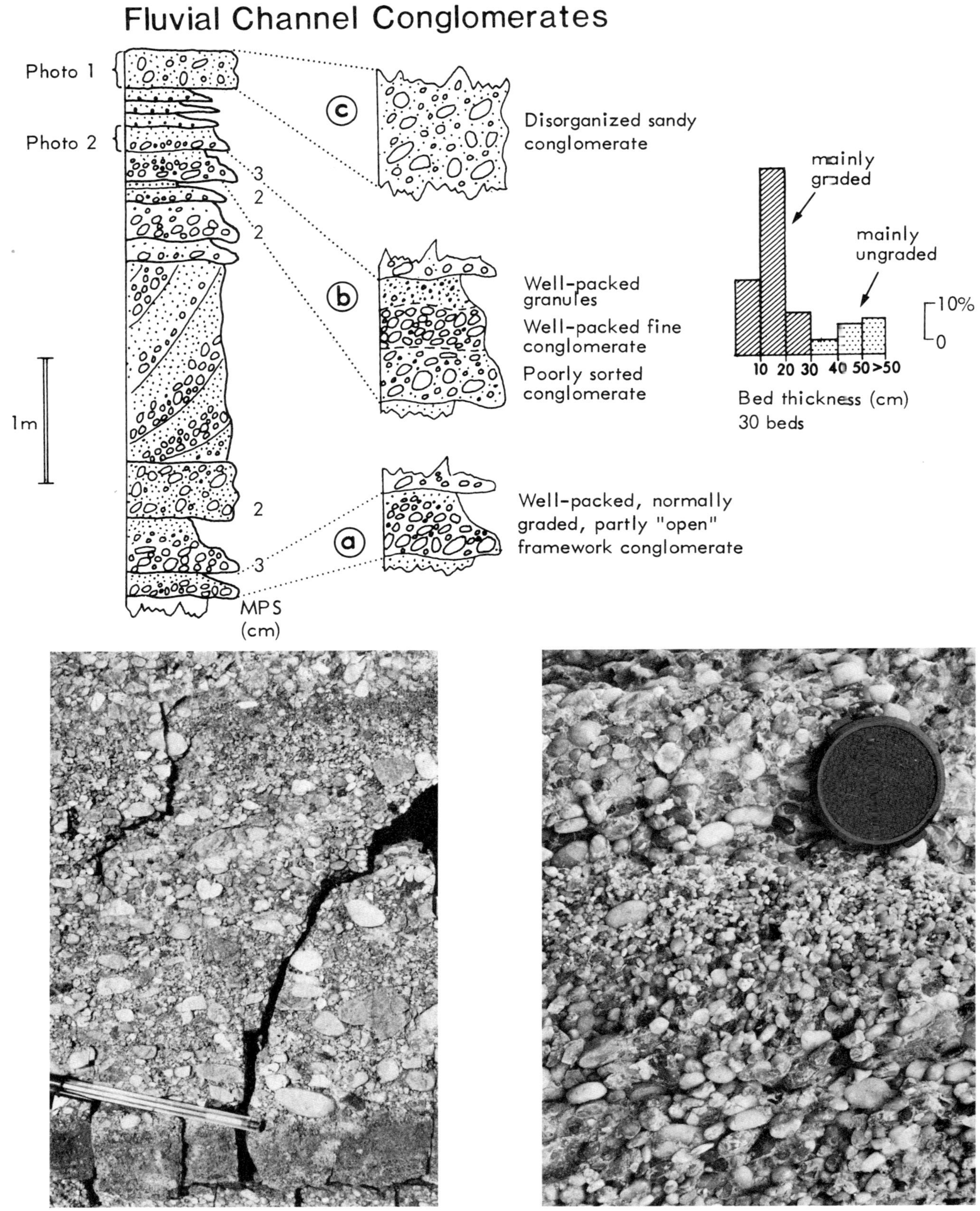

Fluvial Channel Conglomerates

(a) Well-packed, normally graded, partly "open" framework conglomerate

(b) Well-packed granules
Well-packed fine conglomerate
Poorly sorted conglomerate

(c) Disorganized sandy conglomerate

Bed thickness (cm)
30 beds

mainly graded
mainly ungraded

Fig. 4. Detailed sketch of fluvial conglomerates from Sequence 1 with two corresponding photographs. Pen in photo of 'disorganized' conglomerate is 12 cm long and lens cap in photo of well-packed, normally-graded, open-framework conglomerate is 5 cm in diameter. *MPS* is maximum particle size.

sub-horizontal, thin stratification (Fig. 5). The clast-supported units sometimes show alternating open and sand-filled framework layers, analogous to, but much thinner than, the 'falling stage' units in the fluvial conglomerates. The matrix-supported beds also contrast with analogous fluvial beds in having better sorted clast populations, i.e. they tend to be texturally bimodal, or have a significant shape sorting.

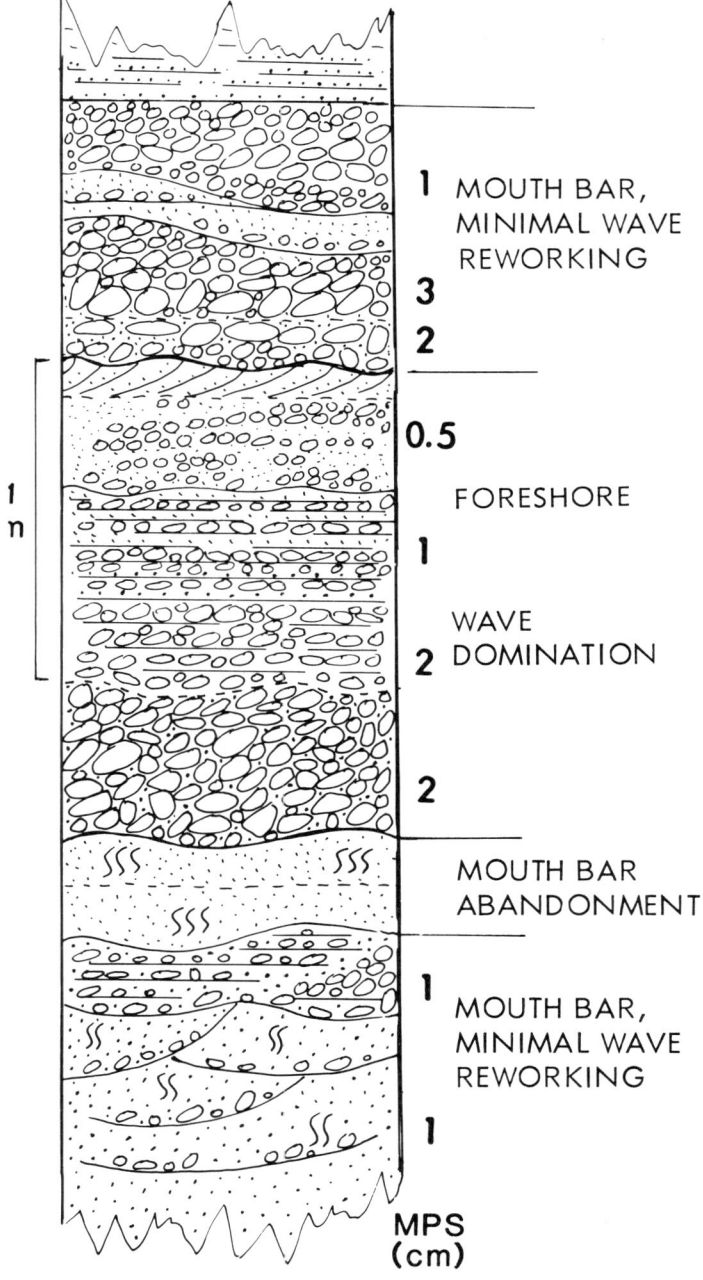

Fig. 5. Detailed sketch of wave-worked barrier/spit conglomerates from Sequence 2. Mouth bar conglomerates are shown in greater detail in Figure 6.

These types of conglomerates are not easily distinguished in the fan-delta sequences, but in general the inferred wave-reworked conglomerates have more flat-lying and more thinly developed stratification than in the two other assemblages. Our examples tend to confirm that Clifton's (1973) criterion of lenticularity is useful for distinguishing wave-reworked gravels from alluvial gravels. However, we find his pebble segregation criterion less useful. Our fluvial conglomerates frequently show better pebble segregation than our shoreline facies which are often sand-rich. This may be partly due to relatively continuous fluvial discharge, but probably more to the high-energy and high sand/gravel ratio in shoreline environments. Examination of some Recent gravelly beach-spits on Spitsbergen suggested that although waves are very efficient in segregating shoreline gravels, sand abundance often causes clast-supported frameworks to be re-filled. In addition, well-segregated gravel lenses are easily disrupted by wave or swash action, and piecemeal pebble incorporation into sand sheets produces bimodal textures. As well as creating texturally mature deposits, the high-energy of shorelines thus also tends to produce textural inversions.

CHANNEL MOUTH CONGLOMERATES

A conglomerate assemblage interpreted as a channel mouth bar complex occurs in eight of the nine sequences at Aldegondaberget. It consists of beds generated by both traction and mass flow processes (Figs. 6, 7). Post-depositional modification by organisms and basinal processes resulted in an assemblage which records an alternation of processes, but a fluvial dominance.

Although other conglomerate types are intercalated, this assemblage is generally recognized by the abundance of beds in which thin, normally-graded, granule to pebble conglomerates line shallow troughs. The conglomeratic part of the bed grades upward to heavily bioturbated sandstone in which lamination is locally preserved. These characteristics are displayed by 30 to 50% of the beds in this assemblage. Intimately associated beds occasionally contain a marine fauna including brachiopods, fusulinids, gastropods, and rugose corals.

The following conglomerate facies are used in this study as the basis for recognition of the channel mouth assemblage (terminology of Lowe (1982) is adopted for sediment-gravity flows):

A) Normally graded, pebble to granule, matrix-supported conglomerate; may be clast-supported at the very base of the bed; beds are lensoid and have a low thickness to width ratio; maximum thickness is 15 cm; base is a scoured surface; upper 50 to 80% of the bed is normally graded or ungraded pebbly to medium sandstone with plane-parallel laminae or ripple cross-laminae; upper portion of the bed may be extensively bioturbated (Fig. 7A). These beds are interpreted as flood-generated, high-density turbidites with a basal suspension-sedimentation unit overlain by traction-current structures.

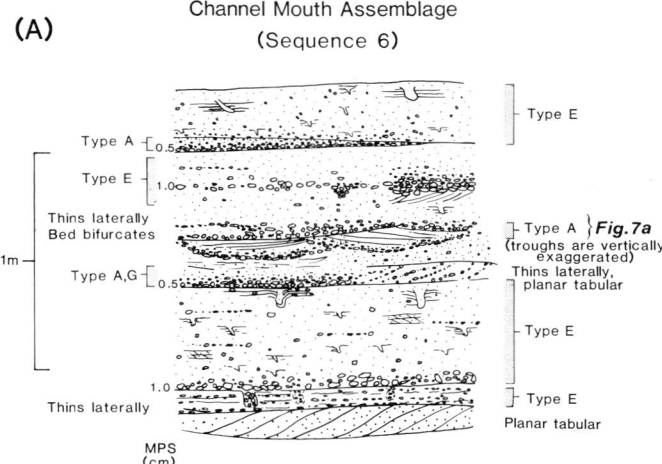
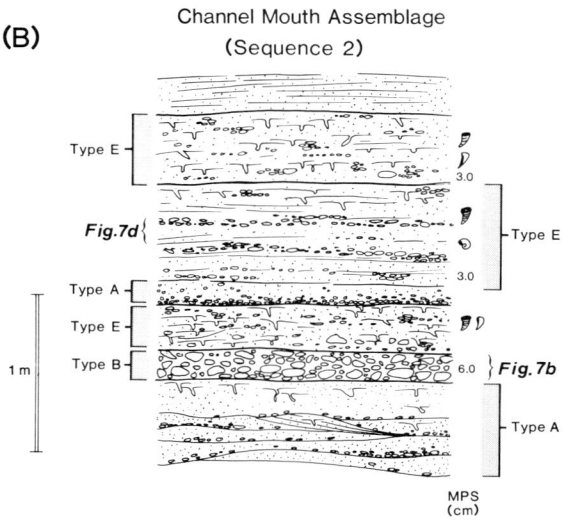
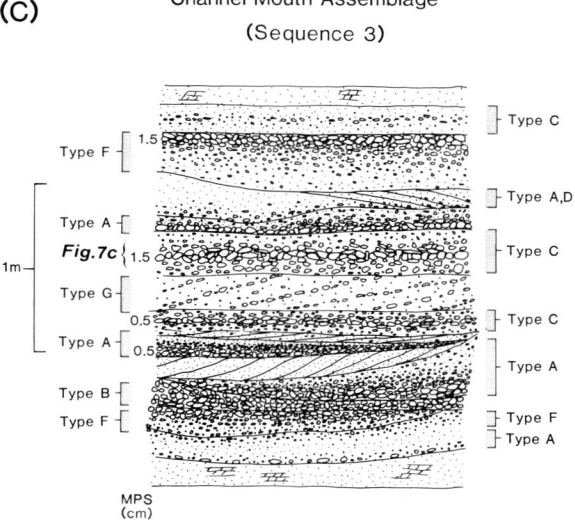

Fig. 6. Comparison of three detailed profiles of channel mouth deposits showing the intercalation of deposits of sediment-gravity flow and traction flow origin. *MPS* indicates maximum particle size.

B) Poorly sorted, matrix-supported conglomerate with clasts dispersed evenly throughout matrix of carbonate clay and sand; beds are ungraded, have diffuse internal laminae, and show no lateral change in thickness over 20 m (Fig. 7B). These beds are interpreted as subaqueous, cohesive debris flow deposits, where the diffuse laminae result from spasmodic downward migration of the resedimentation surface during freezing (*e.g.*, Hampton, 1975).

C) Inversely-to-normally graded conglomerate; beds are clast-supported only in their medial portion and are poorly-sorted; each bed maintains a constant thickness (Fig. 7C). These beds are interpreted as subaqueous debris flows with a turbulent upper part from which pebbles were able to settle.

D) Cross-stratified sandstone beds with granule to pebble clasts lining the base of the shallow troughs. These are interpreted as deposits of traction currents in shallow channels.

E) Pebbly sandstone with diffuse parallel laminae; clasts are both concentrated in isolated clumps and dispersed throughout; beds are extremely bioturbated and frequently bear an invertebrate marine fauna (Fig. 7D). These beds are interpreted as subaqueous sand-rich debris flows.

F) Inversely graded beds with pebbly sandstone at the base and clast-supported cobble conglomerate at the top; beds are poorly sorted and their matrix percent decreases upward; bed thickness is laterally persistent. These beds are interpreted as sediment gravity flows in which grains were supported largely by dispersive pressure.

G) Sandstone beds with planar-tabular cross-stratification in which metastably perched pebbles demarcate the foresets; beds are moderately well-sorted. These beds are thought to represent tractional deposition of rapidly migrating sand waves. These beds often change laterally to beds of type *(A)* with no sharp boundary.

No single conglomerate type permits recognition of the depositional setting, although as a package, the assemblage is interpreted as a channel mouth deposit. The presence of a marine fauna or faunal debris confines the deposit to the subaqueous realm. The assemblage frequently overlies sandy deposits with signs of wave influence, and its coarsening-upward trend is indicative of mouth bar progradation. Dispersal directions are concordant with fluvial paleocurrent directions (Fig. 1).

The channel mouth is a zone of interaction between wave, fluvial, and tidal processes with complex hydraulic conditions that produce a more varied sedimentary unit than would result from simple decrease in carrying capacity. A fluvial traction-current or debris flow may transform to a turbidity current as it moves basinward over the steeper gradient of the delta front. The resultant interbedding of conglomerate types is diagnostic of a channel mouth setting.

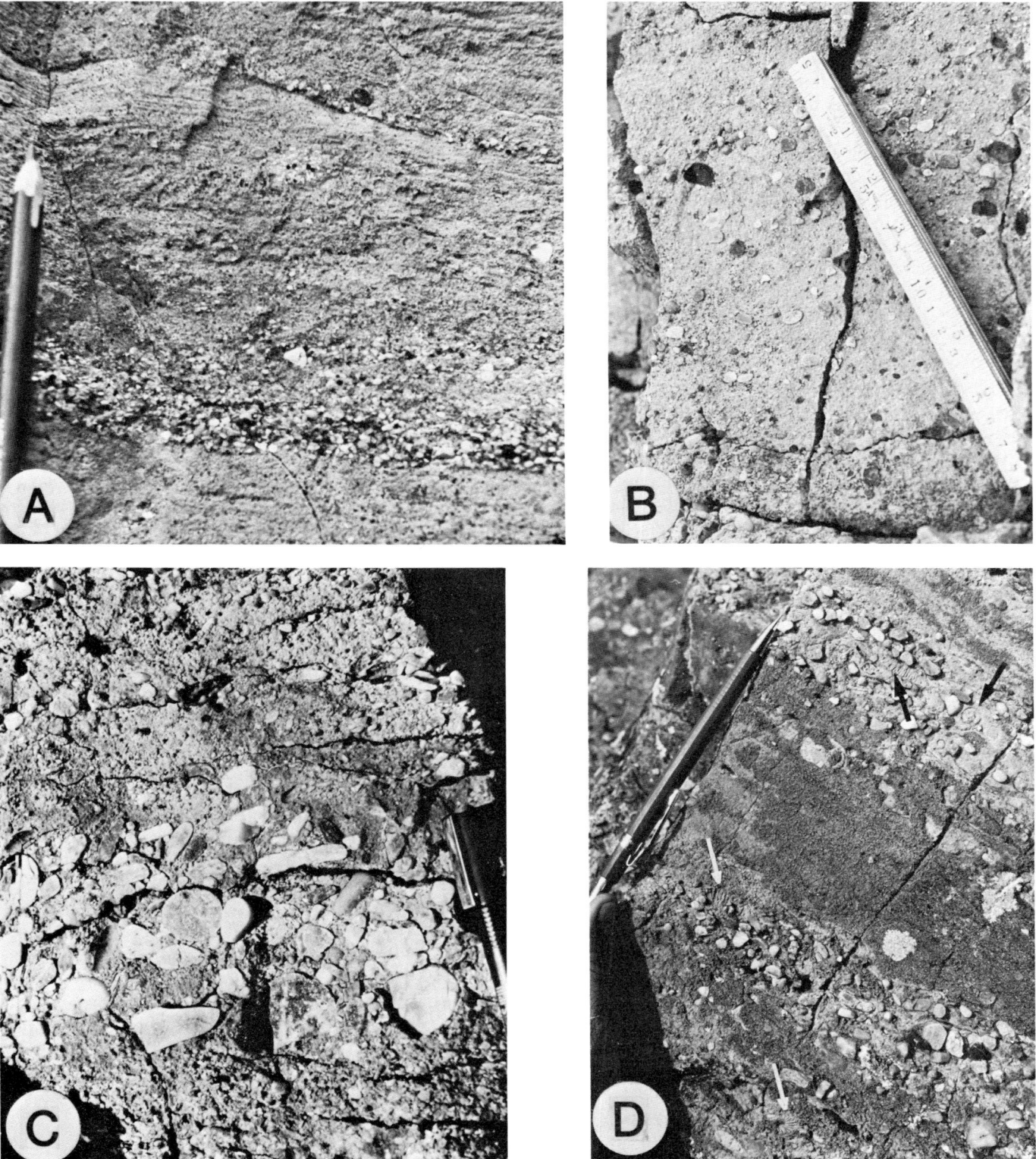

Fig. 7. Photographs of channel mouth conglomerates:
A) Conglomerate Type A showing clasts lining shallow troughs and grading upwards to bioturbated sandstones capped by parallel-laminated finer sandstone. Alternatively, the upper trough is cross-laminated in its upper portion.
B) Conglomerate Type B with sandy matrix-support and poor sorting.
C) Conglomerate Type C with inverse-to-normal grading.
D) Conglomerate Type E shows diffuse laminae, bioturbation, poor sorting, and preserved invertebrate debris (with arrows). Darker weathering in the central portion is due to a late dolomitic cement. Stratigraphic top is to the upper right.

Channel mouth deposits may accumulate until they are subaerially exposed. Beach conglomerates may overlie channel mouth deposits indicating the formation of a subaerial mouth bar or the lateral migration of a delta front spit after channel abandonment (Fig. 5). Alternatively, channel mouth conglomerates are capped by fluvial conglomerates indicating channel progradation.

High sedimentation rates on a fan-delta front occur where fluvial channels breach the delta front that has been modified by basin processes. Thus, the channel mouth deposits represent a volumetrically significant portion of the delta-front facies. Channel mouth deposits may record a local fluvial break through a barrier/spit or shoreline of a reworked fan-delta front. Alternatively, their presence may denote a fluvially-dominated system with little barrier development. Because fan-delta fronts experience abandonment phases, channel mouth deposits are commonly reworked and not preserved unless protected by rapid burial and subsidence.

Proximity to the channel mouth or lateral position on a mouth bar in part determines the dominant conglomerate facies in a channel mouth deposit. Fluvial discharge also controls conglomerate facies. Fluctuation in discharge influences the rate of sediment supply, the available grain size, the rheology of the sediment-fluid mixtures, the sediment-support mechanism, and the mixing of fresh and salt water wedges.

Because braided fluvial channels drain a fan-delta, most detritus is delivered to the fan-delta front as bed load. In such a system, friction-related processes should dominate over inertial or buoyancy effects. Although no data is available from modern fan-deltas, we speculate that the morphology of channel mouth bars in fan-deltas may be similar to that generated by frictional processes on a fluvially-dominated front of a common delta. Observed mass-flow deposits may be the product of oversteepening and failure on the slope of a mouth bar, but deposits indicative of both high (such as modified grain flows) and low gradients are intercalated within the channel mouth assemblage. Disturbed or slumped beds recording mass transport over short distances are also a component of the channel mouth assemblage.

Figure 6 compares representative channel mouth assemblages from three different Aldegondaberget sequences demonstrating the alternation between mass-flow and traction-flow deposits. In Figure 6A, gravity-flow conglomerates of Types *B* and *C* are conspicuously absent, while the assemblage is characterised by Types *A* and *E*. Although Types *A* and *E* dominate the profile in Figure 6B, conglomerate Type *B* is also present. The largest clasts in the succession occur in these cohesive debris-flow units (Type *B*). The beds in Figure 6B record a greater degree of biogenic reworking. The assemblage in Figure 6C is composed of a larger variety of conglomerate types, with an absence of Type *E*. The sand to pebble ratio is markedly lower relative to the other two profiles.

It is tempting to suggest that this arrangement of profiles (Fig. 6A-C) may typify a distal-to-proximal relation or a lateral variation along a time line within a single channel mouth. However, such an arrangement is speculative and a more thorough understanding of morphology and process segregation in a fan-delta channel mouth is necessary before lateral relationships can be predicted with certainty.

SANDSTONE ASSEMBLAGES

The sandstones of the studied succession, commented upon only briefly here, are clearly important in the overall interpretation of the fan-delta sequences. They provide further evidence for the relative roles played by flood, wave or tidal processes on the front of the fan-deltas. Sandstones normally occupy the lower parts of successive upward-coarsening sequences, and they either display a gradual vertical trend or are abruptly overlain by mouth-bar or shoreline conglomerates (Fig. 1).

Three sandstone facies associations are distinguished, representing sand deposition on (a) flood-dominated parts of the fan-delta front, (b) wave- or tide-influenced parts of the fan-delta front, and (c) the prodelta-shelf areas of the fan-deltas:

(a) Flood-Dominated (Mouth Bar) Fan-Delta Front

Sandstones of this association tend to form in large-scale (5-15 m) upward-coarsening sequences and smaller scale (3-4 m) upward-coarsening subsequences. Centimetre-thick, current-rippled siltstones and mudstones are overlain by massive or normally graded sandstone beds (25-50 cm) with sharp, medium-grained bases, and silty, bioturbated or convolute-laminated tops. In some sequences, the latter are further overlain by planar cross-stratified (up to 100 cm sets) sandstones with marine trace fossils. The multi-storey, upward-coarsening trends and the absence of wave-generated structures suggest mouth-bar progradational sequences. The normally graded beds with bioturbated and convolute-laminated tops may have been generated directly from flood events on the mouth bars. These normally graded beds resemble partial and complete Bouma (1962) T_{a-e} sequences, and may also represent low-density turbidites generated by storms. Storm-related density-flows can be produced on shallow gradient slopes by high suspension load (Hamblin and Walker, 1979; Lewis, 1980) and may provide a mechanism for delivering sediment offshore of a channel mouth. Type *(a)* assemblage is best developed in Sequence 1 (Fig. 1).

(b) Wave-Tide Influenced Fan-Delta Front

Sandstones of this assemblage may also form an overall upward coarsening sequence, but this tends to be crude and is internally more irregular than in *(a)*. Thin, fine-grained sandstone sets (<20 cm) display flat or slightly inclined cross-stratification (? hummocky strata), have siltstone caps, and are bioturbated to varying degrees.

Interbedded sets of wave ripple-laminae are common. Structureless beds with wave-rippled and/or bioturbated caps also occur.

These sequences are superficially similar to some of the thin bedded facies in assemblage *(a)*, but the abundance of low-angle cross-strata and wave ripples here suggests considerable wave influence on the delta front. Occasional sequences of coarser sandstone with high-angle cross-strata, often directed shoreward or sometimes with a clear bi-directional character, may have formed in tidal channels or inlets cutting the delta front. Facies assemblages of type (b) are best developed in Sequences 3 and 6 (Fig. 1).

(c) Prodelta-Shelf

Sandstones interpreted as having accumulated on the shelf area beyond the delta front also form crude coarsening-upward sequences. Rippled and bioturbated (notably *Thalassinoides, Asterosoma, Chondrites*) very fine sandstones and siltstones are common. The latter commonly grade upwards into very fine and fine sandstones which are massive or flat-laminated and thoroughly bioturbated. These sequences are associated with grey and black shales/siltstones and with open marine carbonates. Examples occur in Sequences 4, 5 and 6 (Fig. 1).

FAN-DELTA SEQUENCES

The fan-delta deposits at Aldegondaberget occur as nine sequences, each of which has a fine-grained basal part and a conglomeratic upper part (Fig. 1). Some sequences obviously coarsen upward gradually and have abrupt boundaries, while in others such a trend is less evident and sequence boundaries are more arbitrary.

Several intrabasinal and extrabasinal processes interact to generate such sequences. While relative sea level changes can produce cyclicity in coastal sediments, several fan-

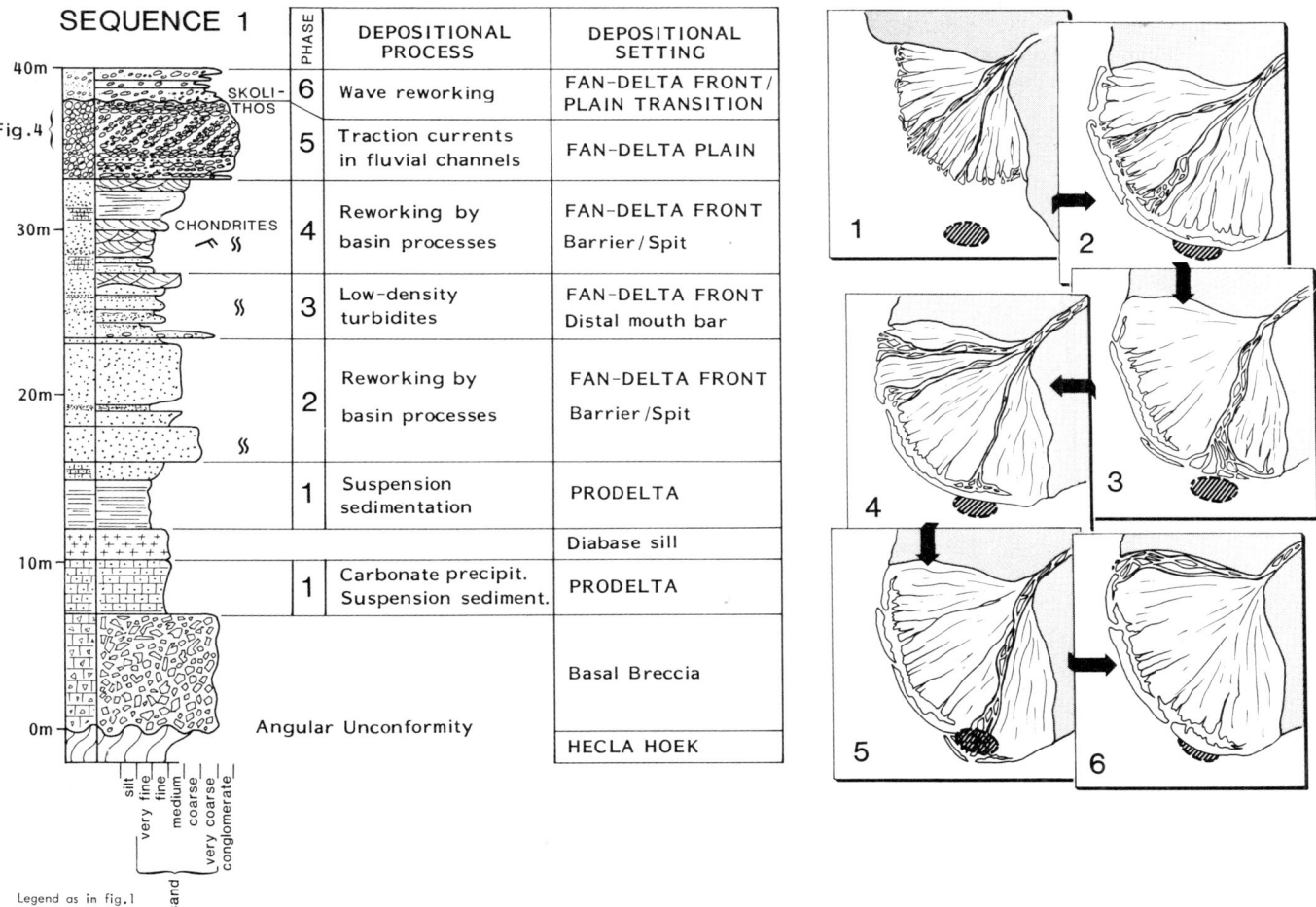

Fig. 8. Description of the fan-delta evolution during Sequence 1. Phases 1-6 in the stratigraphic column correspond to Phases 1-6 shown in plan view. In plan view, the Aldegondaberget succession is represented by a striped oval area that remains fixed in space while the fan-delta geometry changes relative to that area. The actual Aldegondaberget outcrop is not laterally extensive so the diameter of the oval in these illustrations is an exaggeration. It is important to note that these illustrations are hypothetical and represent only one of many possible fan-delta configurations for each phase.

related processes also can control sequence evolution. The internal organization of some Aldegondaberget sequences cannot be attributed to simple pulses of progradation nor to eustatic sea level changes. Avulsion of fluvial channels, lateral migration of spits and barriers, aggradation of mouth bars, and modification by wave and tidal currents has significantly influenced sedimentation on the Aldegondeberget fan-delta front.

Figures 8-11 describe the inferred evolution of representative Aldegondaberget sequences. No two sequences in the succession have a similar internal organization. It is important to emphasize that the illustrated depositional settings are suggested models and they represent one possible explanation for the observed stacking of rock types.

The base of the Aldegondaberget succession is marked by a regional transgression over the previously exposed surface of Late Precambrian-Paleozoic metamorphic units (Figs. 1, 8). This regional transgression was eventually reversed and the subsequent regression culminated in the deposition of the tidal-flat Gipshuken sediments which overlie the studied succession. Figure 8 schematically outlines a plausible fan history during the period represented by Sequence 1. In its first phase, vertical changes in the sequence reflect progradation of the fan. Wave energy is increasingly focused on the fan-delta as it becomes a protruding coastal landform around which waves refract (Phase 2). Avulsion of the fluvial channels to a new position shoreward of the Aldegondaberget section provided an influx of low-density turbidites (Phase 3). Subsequently reduced fluvial input on that part of the fan allows reworking of the channel mouth sediments and longshore drift to establish a barrier or spit complex (Phase 4). Because the slightly inclined and flat bedding in the quartz arenites may have originated fairly high up on a shoreface, we use the term 'barrier/spit', but point out that these may merely represent sandy shoals as no direct evidence of subaerial exposure was observed. Fluvial conglomerates overlying these sandy beds record avulsion and progradation of stream channels (Phase 5). Paleocurrent data from the fluvial conglomerates, although limited, suggest dispersal toward the southwest in contrast with those from the barrier/spit units where dominant transport was directed shoreward, or to the northeast (Fig. 1). This paleocurrent pattern appears consistent throughout the studied section. Washover fans, ridge and runnel systems, or flood-tidal deltas associated with a barrier/spit complex commonly

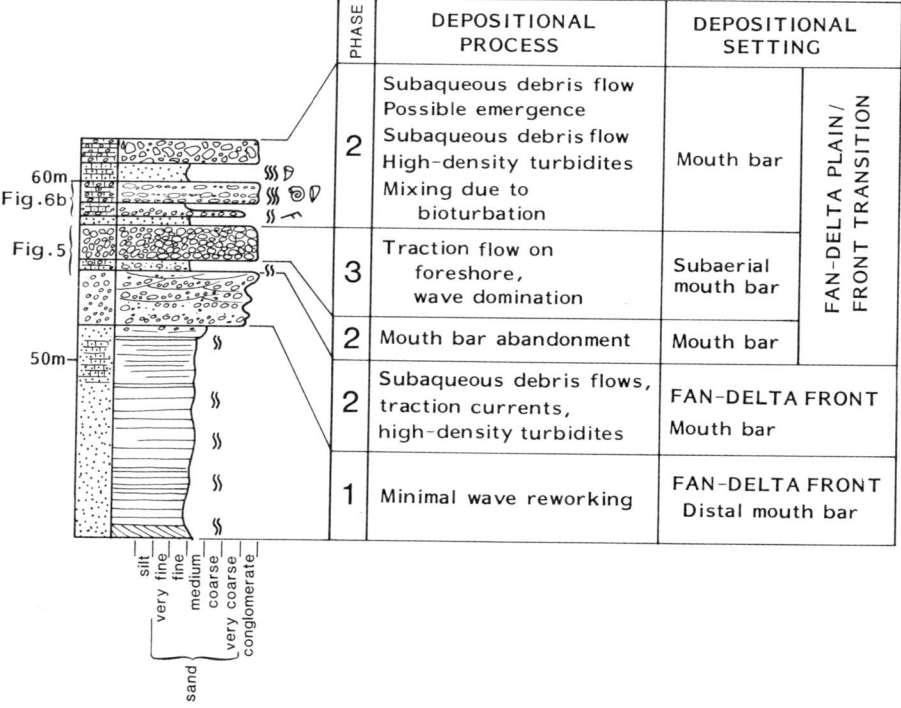
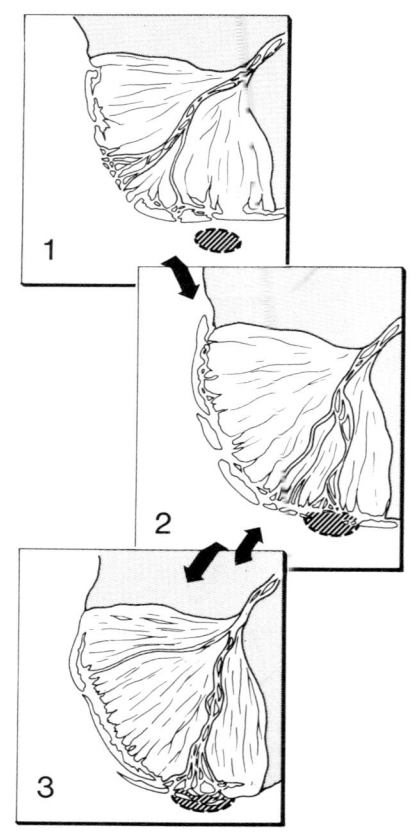

Fig. 9. Description of the fan-delta evolution during Sequence 2. Explanation is the same as in Figure 8.

contain paleocurrent indicators displaying a shoreward component of transport (McCubbin, 1982). The fluvial channels were subsequently abandoned (Phase 6) and the conglomeratic deposits were reworked into a barrier complex as the delta front was eroded. Matrix sand in the barrier complex may have been derived from underlying fluvial conglomerates or may have been winnowed from fluvial or beach sediment elsewhere and transported along the delta front by longshore currents.

The offshore facies at the base of Sequence 2 (Fig. 9) marks continued erosion of the delta front by wave action as the shoreline receded landward. Unless rejuvenated by return of the fluvial channels to that portion of the fan or unless receiving a longshore sediment supply, a fan-delta front may suffer extensive erosion, causing a local landward shift of the delta-front facies. A shift of the fluvial channels (Phase 2) led to the accumulation of well-developed, subaqueous channel-mouth bars. Although drawn as representing the subaerial tops of channel mouth bars, the beach conglomerate (Phase 3) may alternatively represent a barrier/spit shoreline migrating parallel to the delta front. Because the beach conglomerates occur between channel mouth deposits, we favour the former model. Note that the actual foreshore in a fan-delta is a very narrow belt compared to the total fan surface-area, and that beach deposits may therefore constitute a minor fraction of the preserved fan-delta complex.

Sequence 3 is characterized by an alternation of subaqueous mouth bars and laterally migrating barriers or shoals (Fig. 10). Inferred tidal sediments are preserved at only two intervals in the Aldegondaberget section and are recognized by the occurrence of herringbone cross-stratification (Phase 3) with a diverging azimuth of 173°. Although bi-directional planar cross-strata can be generated by a variety of processes, the orientation of those preserved in Sequence 3 is perpendicular to the inferred shoreline of the Aldegondaberget fan (Fig. 1), thus reducing the possibility that they originated from reversing longshore currents flowing parallel to the coastline. Breaks in a wave-modified fan-delta front are conduits through which tidal currents communicate with a back-barrier region. If fluvial channels actively drain through a break in the barrier, the flood current of the tidal cycle is masked by fluvial flow and preservable flood-tidal sediments are not deposited (as in Phase 2). However, if no fluvial current competes with the flood current, bi-directional flow may be recorded (Phase 3). Because wave-generated structures dominate the shoreface and barrier deposits at Aldegondaberget, it is assumed that wave reworking exceeded tidal reworking of the fan-delta front.

Sequence 5 displays a series of phases which record little or no fluvial influence and a coarsening-upward sediment supply (Fig. 11). The abrupt appearance of offshore sediments at the base of the sequence (Fig. 1) marks a

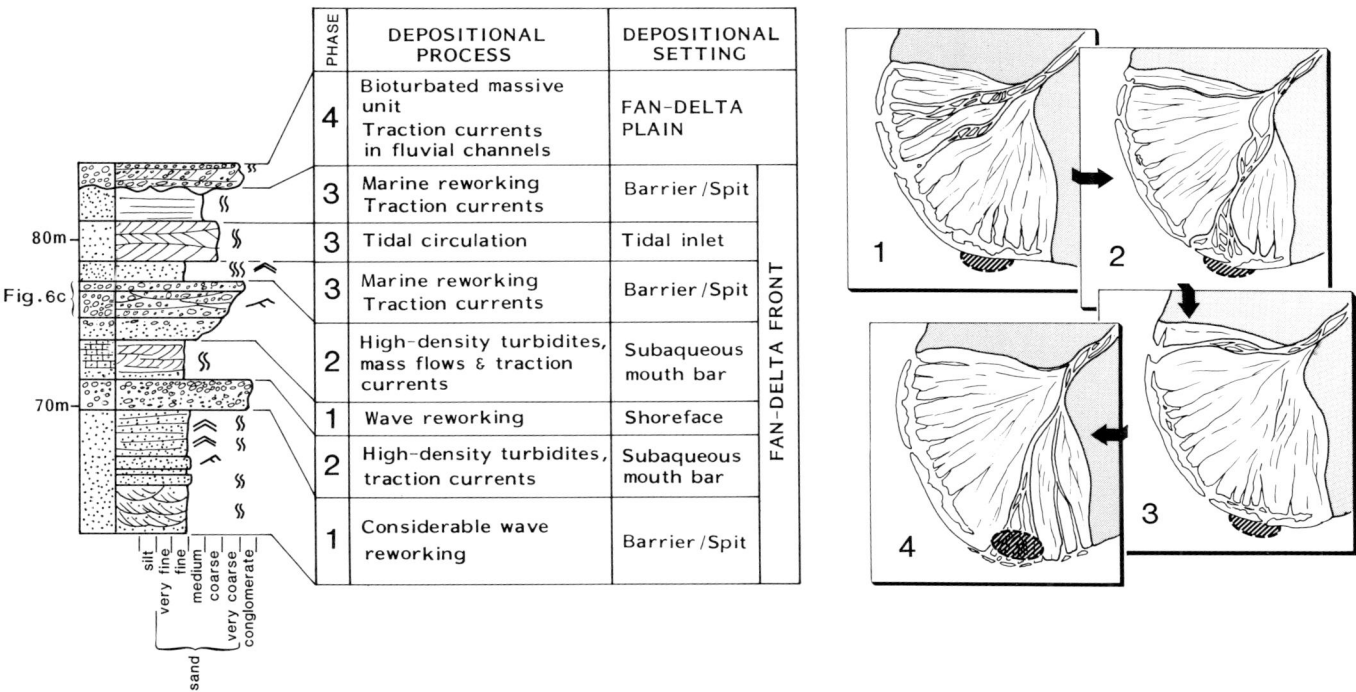

Fig. 10. Description of the fan-delta evolution during Sequence 3. Explanation is the same as in Figure 8.

relative rise in sea level, which was followed by progradation of the fan-delta front and progressive shoaling (Phase 2, Fig. 11) that culminated in the traction deposition of a beach conglomerate at the top of Sequence 5 (Phase 3). During this interval, it appears that fluvial channels were located elsewhere on the fan and that sediment was delivered to the studied sequence primarily by longshore currents.

Discussion and Conclusions

Although fluvial, beach, and channel mouth conglomerates can be intimately stacked within a fan-delta front succession, they are recognizable as distinct conglomerate types. We have documented such an association within the Aldegondaberget succession, and have pointed out that high sediment-concentration flows can be a major constituent of fan-delta front deposits.

With cursory examination, the nine observed conglomeratic sequences may be misinterpreted as simple progradational cycles following intermittent transgressions. It is preferable to interpret vertical changes in the basal three sequences as a record of avulsion of fluvial channels, local abandonment and erosion of the delta front, lateral migration of barriers or shoals, or aggradation of mouth bars overprinted on a background of constant gradual subsidence. Observed vertical changes requiring either a punctuated acceleration of subsidence or a rise in sea level are more common in the upper part of the succession. In the upper four sequences no delta plain sedimentation was documented.

We propose that fan-delta sequences can originate as a response to either extra-basinal or intra-basinal controls, and that while the two operate simultaneously, one type of control may dominate temporarily only to be masked by another in overlying strata. In the case of the Aldegondaberget section, extra-basinal controls dominated in the upper half of the succession. Because marine Reinodden units at Aldegondaberget overlie a basal angular unconformity and because the uppermost conglomerate beds are sharply overlain by tidal-flat sediments, it appears that extra-basinal controls effected abrupt changes. Given the regional stratigraphic relations (Fig. 3) and given that Reinodden successions at neighbouring localities cannot be regionally correlated, it is more likely that basin tectonics, rather than eustatic sea level changes, operated as the extra-basinal control on stratigraphic evolution of the succession.

A maximum sequence thickness of approximately 35 m occurs in the basal sequence (Fig. 1). Because this sequence is bounded by a subaerial surface at its base and other such

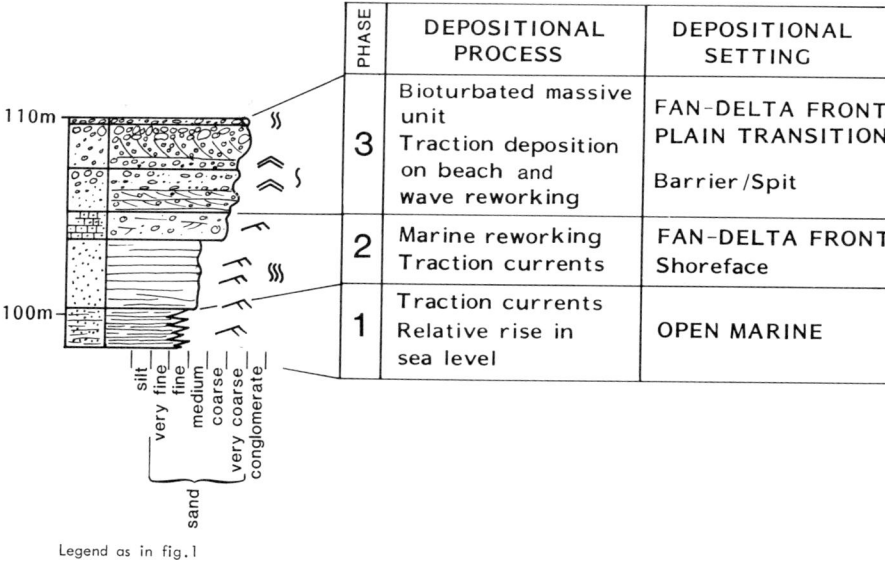

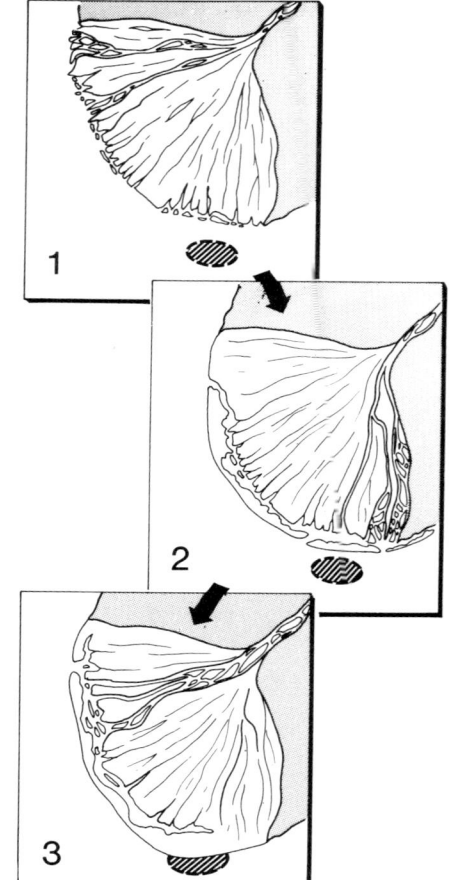

Fig. 11. Description of the fan-delta evolution during Sequence 5. Explanation is the same as in Figure 8.

surfaces within the fluvial conglomerates beds, we infer that the Aldegondaberget fan prograded into water that was no deeper than approximately 35 m. Sequence thickness in the upper part of the succession is reduced to 10-15 m reflecting a regressive trend. Although there are fewer subaerial surfaces to constrain water depth for the uppermost sequences, Sequences 5 could not have been deposited in more than 12 m of water, and an open-marine to foreshore transition is accomplished in less than 15 m of stratigraphic thickness in Sequence 6. These thicknesses represent a maximum possible water depth excluding consideration of syndepositional subsidence. Preservation of subaqueous mouth bars suggests a subsidence and burial rate sufficiently high to protect those deposits from reworking after channel abandonment. Therefore, subsidence rates must be considered significant in calculating water depth and it is likely that water depths were considerably less than the maximum figures discussed above. The fan-delta front sediments do not suggest deposition on a steep avalanche face. In contrast, their character is indicative of shallow gradients along which minor changes in relative sea level initiated major lateral shifts in the fan-delta front facies.

The stratigraphy of neighbouring Reinodden successions north of Van Keulenfjorden is not directly correlable to the Aldegondaberget succession (G. Sundsbø, M. Talbot, pers. comm., 1982). These differences suggest that neighbouring fan systems operated independently and limit the size of the Aldegondaberget fan-delta. We estimate that the subaerial Aldegondaberget fan was 1-5 km in diameter.

Because fan-delta plain sediments are minimally preserved in the Aldegondaberget succession, it is difficult, on the basis of direct evidence, to characterize the fan as 'humid' or 'semi-arid'. Paleogeographic reconstructions place Spitsbergen at approximately 35°N in Late Carboniferous-Early Permian time (Kanasewich et al., 1978; Irving, 1977) during which sedimentation in north-central Spitsbergen generated carbonate and evaporite successions (Fig. 2), indicating that the climate was at least locally semi-arid to arid. If the Aldengondaberget fan was veneered by ephemeral streams, some fan-delta front features herein attributed to stream avulsion may instead reflect intermittent fluvial discharge. However, the few observed fan-delta plain deposits reflect stream flow rather than subaerial debris-flow processes. Other indicators of arid conditions are also absent. The soil profile at the basal unconformity, although possibly relict, contains no carbonate (M. Talbot, pers. comm., 1982), and aeolian sediments are absent in the succession. Thus it may be appropriate to consider an Aldegondaberget fan developed in a warm climate but constructed by perennial streams.

The morphology of a marine fan-delta front is controlled by competing fluvial, tidal, and wave processes. Dominance of a single process or combination of processes will modify the preserved deposits, and we suggest that, as in common deltas, those deposits may be classified according to the predominant process. Probable tidal deposits are rare in the Aldegondaberget succession and occur in settings suggesting deposition in narrow, microtidal inlets. Because wave and fluvially influenced strata dominate the studied succession, we categorize the Aldegondaberget succession as a fluvial-wave interaction type of fan-delta.

REFERENCES

Birkenmajer, K. 1964. Devonian, Carboniferous and Permian Formations of Hornsund, Vestspitsbergen. Studia Geologica Polonica, v. 11, p. 47-123.

Birkenmajer, K. and Czarniecki, S. 1960. Stratigraphy of marine Carboniferous and Permian deposits in Hornsund (Vestspitsbergen), based on brachiopods. Bulletin de l'Academie Polonaise des Sciences, Serie des Sciences Geologiques et Geographiques, v. 8, p. 203-208.

Bluck, B.J. 1976. Sedimentation in some Scottish rivers of low sinuosity. Transactions of the Royal Society of Edinburgh, v. 69, p. 425-456.

Bouma, A.H. 1962. Sedimentology of Some Flysch Deposits. Amsterdam: Elsevier, 168p.

Clifton, H.E. 1973. Pebble segregation and bed lenticularity in wave-worked versus alluvial gravel. Sedimentology, v. 20, p. 173-187.

Cutbill, J.L. and Challinor, A. 1965. Revision of the stratigraphical scheme for Carboniferous and Permian rocks of Spitsbergen and Bjørnøya. Geological Magazine, v. 102, p. 418-439.

Fedorowski, J. 1964. On late Paleozoic Rugosa from Hornsund, Vestspitsbergen; Preliminary communication. Studia Geologica Polonica, v. 11, p. 139-146.

Fedorowski, J. 1965. Lower Permian Tetracoralla of Hornsund, Vestspitsbergen; Preliminary communication. Studia Geologica Polonica, v. 17, p. 7-173.

Galloway, W.E. 1976. Sediments and stratigraphic framework of the Copper River fan-delta, Alaska. Journal of Sedimentary Petrology, v. 46, p. 726-737.

Gjelberg, J. and Steel, R.J. 1981. An outline of Lower-Middle Carboniferous sedimentation on Svalbard: Effects of tectonic, climatic, and sea level changes in rift basin sequences. In: Kerr, J.W. and Ferguson, A.J. (Eds.), Geology of the North Atlantic Borderlands. Canadian Society of Petroleum Geologists, Memoir 7, p. 543-561.

Hamblin, A.P. and Walker, R.G. 1979. Storm-dominated shallow marine deposits: the Fernie-Kootenay (Jurassic) transition, southern Rocky Mountains. Canadian Journal of Earth Sciences, v. 16, p. 1673-1690.

Hampton, M.A. 1975. Competence of fine-grained debris flows. Journal of Sedimentary Petrology, v. 45, p. 834-844.

Irving, E. 1977. Drift of the major continental blocks since the Devonian. Nature, v. 270, p. 304-309.

Kanasewich, E.R., Havskov, J., and Evans, M.E. 1978. Plate tectonics in the Phanerozoic. Canadian Journal of Earth Sciences, v. 15, p. 919-955.

Lewis, D.W. 1980. Storm-generated graded beds and debris flow deposits with *Ophiomorpha* in a shallow offshore Oligocene sequence at Nelson, South Island, New Zealand. New Zealand Journal of Geology and Geophysics, v. 23, p. 353-369.

Lowe, D.R. 1982. Sediment gravity flows: II. Depositional models with special reference to the deposits of high-density turbidity currents. Journal of Sedimentary Petrology, v. 52, p. 279-297.

McCubbin, D.G. 1982. Barrier-island and strand plain facies. In: Scholle, P.A. and Spearing, D. (Eds.), Sandstone Depositional Environments. American Association of Petroleum Geologists, Memoir 31, p. 247-279.

Netland, A. 1981. Facies analyse av Drevbreen Bed og Nordskiøldbreen Formasjonen, Øvre Karbon til Undre Perm, Bellsund området, Svalbard. Unpublished Cand. Real. thesis, University of Bergen, Norway, 212p.

Nysæther, E. 1977. Investigations on the Carboniferous and Permian stratigraphy of the Torrell Land area, Spitsbergen. Norsk Polarinstitutt Årbok 1976, p. 21-42.

Postma, G. 1984. Slumps and their deposits in fan delta front and slope. Geology, v. 12, p. 27-30.

Ricci Lucchi, F., Colella, A., Ori, G.G., and Ogliani, F. 1981. Pliocene fan deltas of the Intra-Apenninic Basin, Bologna. *In:* Ricci Lucchi, F. (Ed.), Excursion Guidebook, 2nd European Regional Meeting. Bologna, International Association of Sedimentologists, p. 79-161.

Smith, N.D. 1974. Sedimentology and bar formation in the Upper Kicking Horse River, a braided outwash stream. Journal of Geology, v. 82, p. 205-224.

Steel, R.J. and Thompson, D.B. 1983. Texture and structure in Triassic (Bunter) braided stream gravels, northwest England. Sedimentology, v. 30, p. 341-368.

Steel, R.J. and Worsley, D. (in press). Svalbard's post-Caledonian strata — an atlas of sedimentational patterns and palaeogeographical evolution. *In:* Spencer, A. M. (Ed.), North European Marginal Basins. London: Graham & Trotman.

Sykes, R.M. and Brand, R.P. 1976. Fan-delta sedimentation: An example from the Late Jurassic-Early Cretaceous of Milne Land, Central East Greenland. Geologie en Mijnbouw v. 55, p. 195-203.

Tangen, O. 1981. A sedimentological and environmental interpretation of the Upper part of Nordenskioldbreen Formation and the Gipshuken Formation on Western Spitsbergen. Unpublished Cand. Real. thesis, University of Bergen, Norway, 256 pp.

Wescott, W.A. and Ethridge, F.G. 1980. Fan-delta sedimentology and tectonic setting — Yallahs fan-delta, Southeast Jamaica. American Association of Petroleum Geologists Bulletin, v. 64, p. 374-399.

Wescott, W.A. and Ethridge, F.G. 1982. Bathymetry and sediment dispersal dynamics along the Yallahs fan delta front, Jamaica. Marine Geology, v. 46, p. 245-260.

DOMBA CONGLOMERATE, DEVONIAN, NORWAY: PROCESS AND LATERAL VARIABILITY IN A MASS FLOW-DOMINATED, LACUSTRINE FAN-DELTA

W. Nemec[1], R.J. Steel[2], S.J. Porębski[3] and Å. Spinnangr[4]

Abstract

The Domba Conglomerate is a wedge-shaped (up to 10 m thick) conglomeratic body composed of mass flow deposits and enveloped by sandy fluvial sediments. It represents a small-radius (~ 2 km) alluvial fan which prograded into an inferred shallow body of water, probably a floodbasin-related lake. The fan body was generated during escarpment creation in response to local syndepositional faulting of the alluvial basin floor, and the inferred lake was probably formed by consequent damming of streams and change in water table.

The fan-delta wedge comprises clast-supported and subordinate matrix-rich conglomerates, while its marginal/frontal part consists of finer-grained, well-sorted, thinly layered deposits. *Clast-supported conglomerate beds* (Facies A) are sheet-like, ungraded or variously graded, often have clast *a*-axis imbrication, and represent density-modified grain flows. *Matrix-rich conglomerate and pebbly siltstone beds* (Facies B) have a disorganised clast fabric but commonly show coarse-tail grading, and are the product of mixing between fan gravels and lacustrine fines, mainly as cohesive debris flows. *Granule sandstones and fine conglomerates* (Facies C) form well-sorted, low-angle stratified, coarsening-up succession comprising thin (2-12 cm) sediment layers which have gently inclined erosional bases, distribution inverse grading, and clast alignment parallel to flow and imbrication. These formed from liquefied-flow grain flows (partly transitional to high-concentration turbidity currents) on the subaqueous fan slope, the steeper portions of which were probably subject to progressive liquefaction and 'gravitational winnowing' processes.

Within the main conglomerate body, both maximum particle size and bed thickness decrease downfan (respectively 2 cm and 6 cm per 100 m horizontal distance), and ungraded/inversely graded beds are transformed to normally graded ones. Vertical accretion rather than progradation was dominant on the fan. A rising lake level and progressive inundation of the fan slope are implied by the onlapping character of the delta front sequence.

Résumé

Le conglomérat de Domba constitue un dépôt conglomératique en forme de biseau (jusqu'à 10 m) composé de matériaux d'écoulements en masse et enveloppé de sédiments arénacés fluviatiles. Il représente un cône de déjection de faible rayon (~2 km) qui a avancé progressivement dans un bassin aquatique, possiblement une plaine d'inondation associée à un lac. Le corps du cône de déjection fut édifié lors de l'apparition d'un escarpement créé par un ajustement à un mouvement de faille synsédimentaire du plancher du bassin alluvial, et ce lac présumé doit possiblement son existence à un barrage des cours d'eau et à un changement du niveau de la nappe phréatique.

Le cône de déjection en biseau comprend un conglomérat à fragments jointifs et un conglomérat riche en matrice mais moins abondant, alors que la zone marginale/frontale est constituée de dépôts à grains plus fins, bien triés, et en couches minces. *Les couches de conglomérat à fragments jointifs* (Faciès A) se présentent en forme de nappe, elles sont massives ou granoclassées, souvent avec une imbrication selon l'axe *a*, et elles correspondent à des écoulements de matériaux grenus modifiés par des effets de densité. *Un conglomérat riche en matrice incluant des lits de 'siltstone' caillouteux* (Faciès B) possède une fabrique de fragments inorganisée, mais on observe fréquemment aux extrémités un granoclassement rudimentaire, et il résulte d'un mélange de graviers du cône de déjection et de fines particules lacustres, généralement sous forme de coulées de débris cohésives. *Des grès à granules et des conglomérats fins* (Faciès C) sont bien triés, à strates peu inclinées, la succession est de granulométrie croissante vers le haut et elle renferme des couches minces de sédiment (2-12 cm) à bases érodées légèrement inclinées, avec granoclassement inverse, et un alignement des fragments parallèle à l'écoulement et à l'imbrication. Ils dérivent d'écoulements granulaires liquéfiés (passage graduel à des courants de turbidité fortement chargés) sur la pente sous-aquatique du cône de déjection, dont les zones les plus inclinées furent probablement affectées par des processus de liquéfaction progressive et de 'vannage par gravité'.

Dans le corps principal du conglomérat, la granulométrie et l'épaisseur des couches diminuent le long de la pente du cône (2 à 6 cm par 100 m de distance horizontale, respectivement), et les couches non-granoclassées ou granoclassées inversement sont transformées en couches à granoclassement normal. L'accrétion verticale sur le cône de déjection prédomine sur la progradation. Une élévation du niveau du lac et une inondation progressive de la pente du cône de déjection sont suggérés par le caractère transgressif de la séquence frontale du delta.

Introduction

A fan-delta is an alluvial fan that has prograded into a standing body of water from an adjacent highland (Holmes, 1965; McGowen, 1970). There has been increased recent interest in stream-dominated fan-deltas, and a brief review of several published modern examples and ancient counterparts was recently given by Wescott and Ethridge (1980). Among the numerous other examples are those interpreted by Flores (1975), Ogliani (1981), Ricci Lucchi et al. (1981) and Fairchild (1982). In contrast, there is little information about the processes or deposits of mass flow-dominated fan-deltas (for interpreted examples see Larsen

---

[1] Institute of Geological Sciences, University of Wroclaw, 50-205 Wroclaw, Poland.
 Current address: Geological Institute (A), University of Bergen, 5014 Bergen, Norway.
[2] Norsk Hydro Research Centre, P.O. Box 4313, 5013 Bergen, Norway.
[3] Institute of Geological Sciences, Polish Academy of Sciences, 31-002 Kraków, Poland.
[4] Statoil, P.O. Box 40, 9401 Harstad, Norway.

Reviews by Karen L. Kleinspehn (Bergen), Dave Larue (Stanford), and George Postma (Amsterdam) considerably improved this manuscript. We appreciate their help, and we also thank Sveinulf Vågene (Bergen) for geological discussion and his logistical support of our field work in 1981.
The senior author (W.N.) wishes to thank Andrew D. Miall and his wife, Charlene, whose generous help and hospitality allowed him to present this paper at the I.A.S. Canadian Congress in 1982.

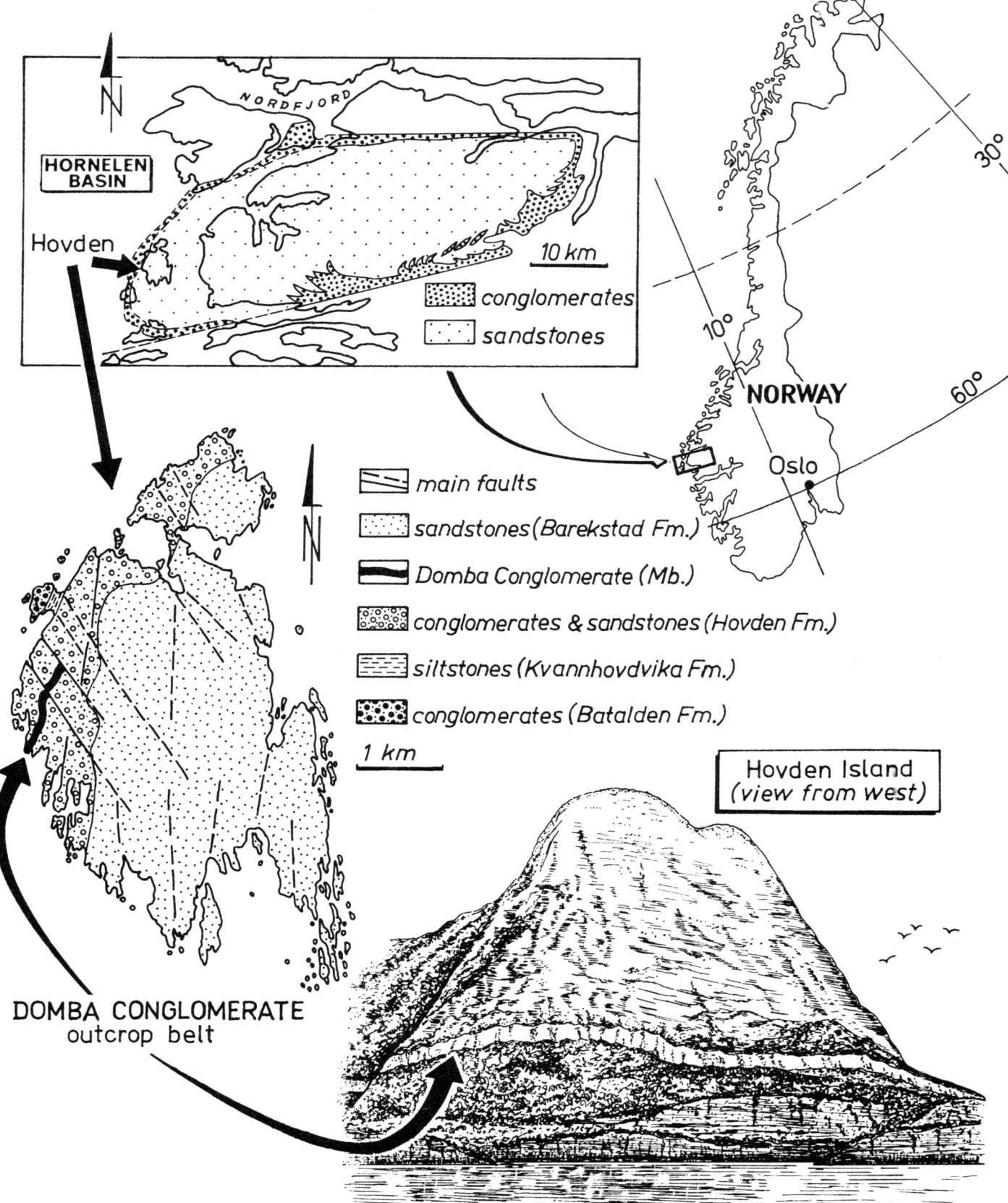

Fig. 1. Locality maps showing (**bottom**) the outcrop belt and lithostratigraphic context of the Domba Conglomerate unit in Hovden Island, and (**top**) the location of this island within the entire outcrop of the Devonian Hornelen Basin in southwestern Norway.

and Steel, 1978; Nemec et al., 1980; Gloppen and Steel, 1981; Porębski, 1981; Gnaccolini, 1982; Postma, 1984, and this volume). This is probably not because mass flow fan-delta are particularly rare in the sedimentary record, but because criteria for their recognition have not been available (e.g., see statement of Rust, 1979, p. 10).

The majority of published fan-delta examples come from various tectonically active coastlines, where the development of terrigenous clastic-wedge is primarily controlled by the shelf/slope settings of continental or island-arc margin, and where the fan-delta sediments are characteristically reworked and modified by a variety of marine processes (e.g., Galloway, 1976; Sykes and Brand, 1976; Surlyk, 1978; Wescott and Ethridge, 1980, 1982, 1983; Daily et al., 1980; Harbaugh and Dickinson, 1981; Ricci Lucchi et al., 1981; Kleinspehn et al., this volume). In contrast, relatively few studies have been devoted to lacustrine fan-deltas, especially to the gravelly ones (e.g., Gustavson et al., 1975; Larsen and Steel, 1978; Sneh, 1979; Clemmensen and Houmark-Nielsen, 1981). The lacustrine fan-deltas are themselves important because they probably serve as laboratory-like examples where we can attempt to understand the separation of subaerial and unreworked subaqueous deposits. The scarcity of such studies is somewhat surprising, particularly in view of the facts that alluvial fans are often associated with lakes, and that one of the classical descriptions of fan-deltas was from a lacustrine setting, by Gilbert (1890).

This paper describes in detail the sedimentological characteristics of an ancient, mass flow-dominated, lacustrine fan-delta that displays a subtle separation of subaerial and subaqueous parts, and is characterised by an unusual mode of delta front development. Particular attention has been paid to lateral variability in the characteristics of fan-wedge sediments, and to the nature of the processes responsible for their desposition. This study is also relevant to the problem of drawing distinction between ancient mass flow-dominated alluvial fans and fan-deltas.

Domba Conglomerate

Recent sedimentological studies of alluvial fan deposits in the Hornelen Basin (Devonian), southwestern Norway, have focused on a variety of fan bodies that built out from this basin's fault margins. Some of these marginal gravelly fans were constructed mainly by stream-flow processes, others entirely by mass flows or by a combination of both modes (Mæhle, 1975; Nilsen, 1975; Spinnangr, 1975; Larsen, 1977; Gloppen, 1978). This variation in the type of marginal alluvial sedimentation was itself related to varied tectonic setting of the basin margins (Steel, 1976; Steel and Gloppen, 1980). There was also an active interfingering of the marginal fan gravels with fine-grained floodbasin sands (Fig. 1, upper left) of a major alluvial-plain system occupying the central/axial part of the Hornelen Basin (Aasheim, 1977; Steel et al., 1977; Steel and Aasheim, 1978), and so the toes of many fans were often under shallow standing water (Larsen and Steel, 1978; Gloppen and Steel, 1981; Pollard et al., 1982).

In this paper, we document an unusual detail of the Hornelen Basin, namely an alluvial fan body which was itself not related to any of the prominent marginal fault-scarps, but which apparently originated in response to syndepositional faulting of the alluvial basin floor.

The stratigraphy of the oldest (western) part of the Hornelen Basin succession consists of the following formations overlying the Cambro-Silurian basement (Spinnangr, 1975; Vågene, 1982) (Fig. 1):

3. *Hovden Formation* — conglomerates fining upwards to sandstones, both fluvially derived from the east along the basin axis.

2. *Kvannhovdvika Formation* — reddish siltstones and mudstones of lacustrine origin.

1. *Batalden Formation* — various conglomerates of alluvial fan origin, derived from the basin margins in the northwest, west, and south.

The Hovden Formation, investigated in detail by Spinnangr (1975) and Vågene (1982), records the first period of expansion or migration of the eastern margin of Hornelen

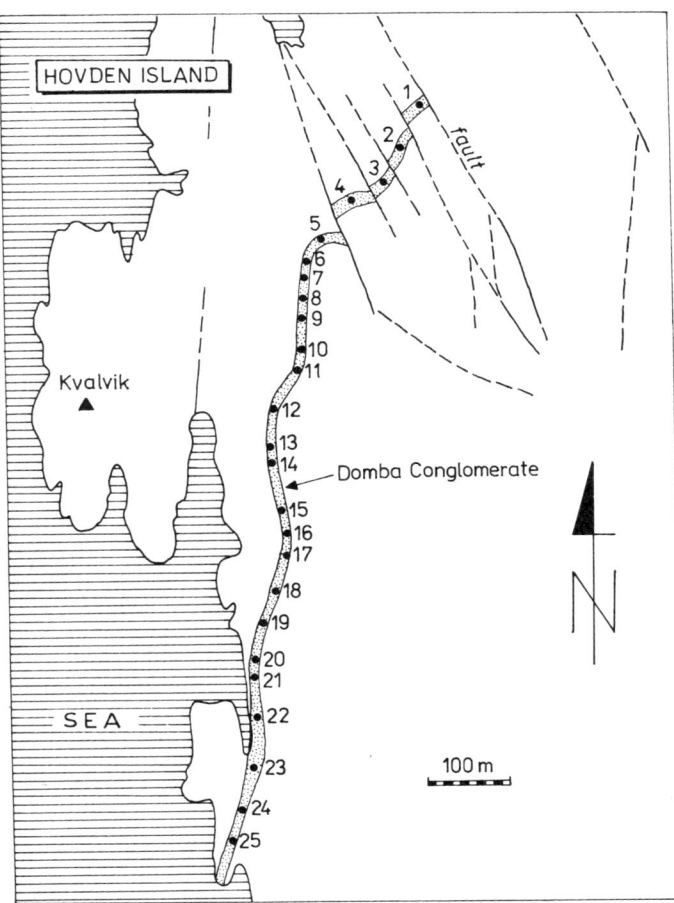

Fig. 2. Index map showing the location of data stations along the outcrop belt of Domba Conglomerate, western Hovden Island.

Basin. In the uppermost, dominantly sandy portion of this formation, there are two laterally equivalent conglomerate members with distinct clast compositions: one dominated by greenstone fragments and derived from the basin's northern margin, and the other, polymict, formed within the basin's central tract. This latter, the Domba Conglomerate, is the topic of this paper.

The Domba Conglomerate occurs as a 1 km long, almost continuous outcrop-belt in the western part of the Hovden Island (Figs. 1 and 2). This mappable, wedge-shaped conglomerate unit is relatively thin (up to 10 m), erosionally bounded, and both underlain and covered by thick sandstones and pebbly sandstones of fluvial origin (Figs. 3 and 4). On its northern end, the Domba outcrop is abruptly truncated by a fault plane, while on the southern end it becomes concealed offshore under the sea (Fig. 2). Palaeocurrent indicators show that the material was consistently dispersed towards the southwest (Fig. 5), and thus the present-day outcrop belt is itself oblique, at about 45°, to the inferred palaeoflow direction. Such an outcrop orientation allows us to study both lateral and downflow changes within this conglomeratic wedge (Fig. 6).

Vertical logging of texture and structure along the outcrop of this conglomeratic body (Figs. 2 and 7) has suggested that it represents an alluvial fan which prograded into a shallow body of water, probably an ephemeral lake. Among the important characteristics of this gravelly fan-delta is its relatively large radius (at least 1.5 - 2 km) in comparison to its thickness (<10m), a predominance of various mass-flow deposits throughout, and an unusual type of stratified 'delta-front' sequence (Fig. 6).

Three distinct lithofacies have been distinguished and mapped along the outcrop (Fig. 7), and these are first described and interpreted here. The development and lateral variability of the Domba fan will then be considered, and its subaqueous depositional aspects discussed in detail. Finally, a summary model for the textural evolution of sediments comprising the Domba fan-delta will be outlined briefly.

FACIES A: CLAST-SUPPORTED CONGLOMERATES

DESCRIPTION

Facies A deposits predominate volumetrically in the outcrop. They form the thickest, coarsest-grained ('proximal') part of the Domba wedge, though at the base of this unit they occur throughout the outcrop belt (Figs. 7 and 21). These conglomerates typically display a polymodal grain-size distribution, but are relatively well-sorted and have a clast-supported texture, with a matrix of mostly coarse sand and granules (Figs. 8, 9 and 10). Matrix content is generally less than 20%, though in most instances it does not exceed 10-15%. In the matrix, the amount of material finer than coarse silt is apparently no more than several per cent.

Despite their overall clast-supported nature, the tops of many beds reveal a strikingly bimodal texture, with closely packed pebbles surrounded by a reddish matrix of silt and very fine sand (Fig. 11). Moreover, in a few instances there is a lateral transition from clast-supported to silt-rich, matrix-supported texture within single beds, generally downflow (for details see Facies B below).

Clasts are mostly well-rounded to subrounded, being composed of quartz and various metamorphic/igneous rocks; intraformational or rip-up clasts of sedimentary rocks are generally lacking. Maximum particle size (MPS = mean of ten largest clasts per bed in station after omitting the two largest) is from 2 to 33 cm, averaging 8 cm. A highly significant linear correlation between the calculated bed MPS and the size of the largest clast in the bed (see Fig. 12) implies a relatively good sorting of the sediment fraction above the sand-size grade. Very few beds (thickest and coarsest) contain any markedly outsized clasts (Fig. 8).

Clast fabric has been measured in a number of randomly selected beds at various localities along the outcrop, and in each case there appears to be a strongly preferred clast orientation (Fig. 5; for detailed examples see Fig. 13). The clast longest (a-)axes are either parallel to flow, or are parallel to flow and imbricated, dipping upflow; this latter fabric type is usually coded as a(p)a(i) (Walker, 1975a, p. 134). It is worth noting that the same fabric is also displayed by the above mentioned bimodal bed tops; in only a very few instances, the bed top reveals a subordinate a(t)b(i) fabric mode (for diagram examples see Stations 1, 2 and 12 in Fig. 13).

Fig. 3. Typical appearance of the braided-stream/streamflood deposits underlying the Domba Conglomerate unit.

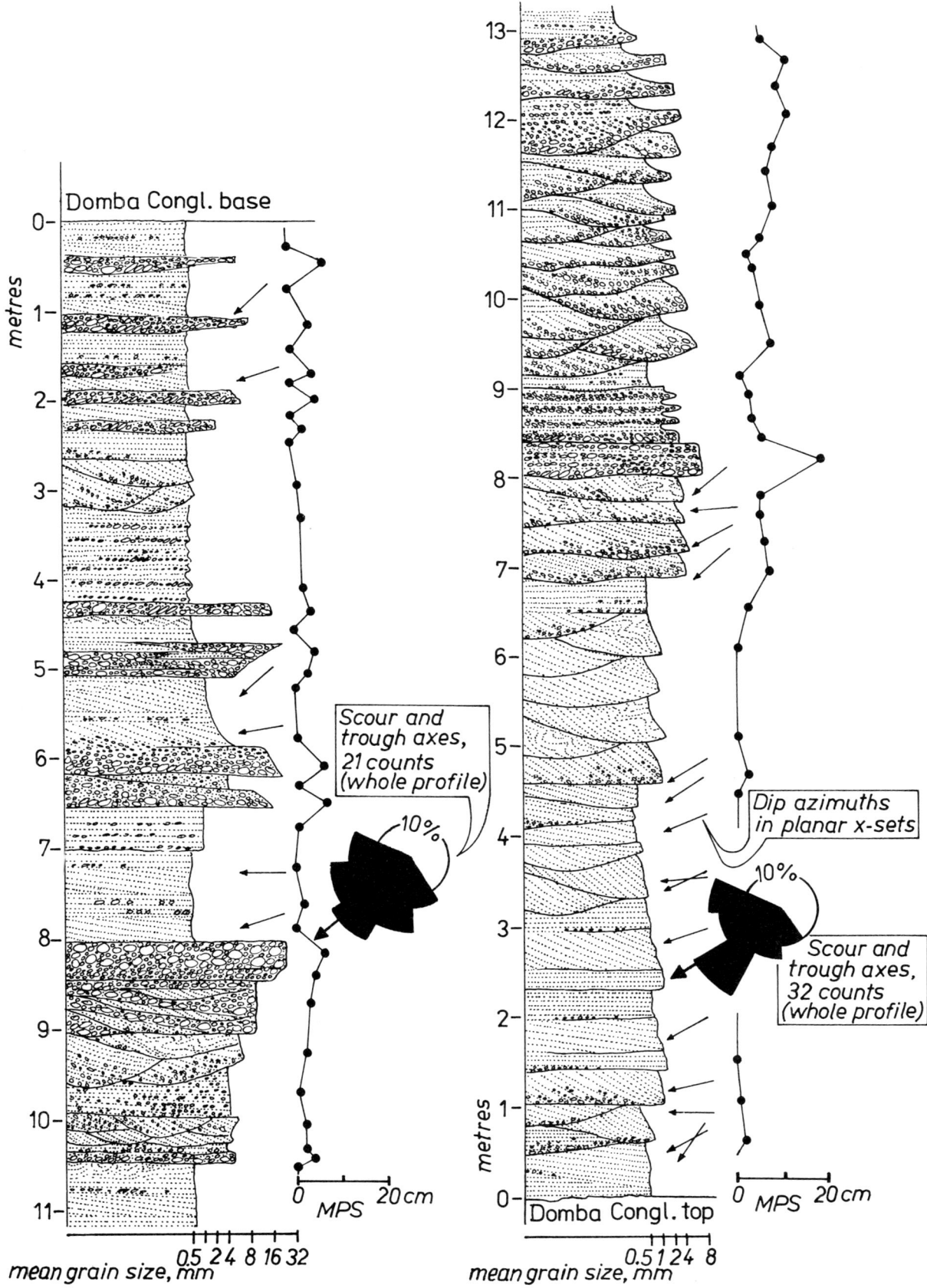

Fig. 4. Example of a detailed log of the fluvial deposits enveloping the Domba Conglomerate unit (data from the vicinity of Stations 4 and 5).

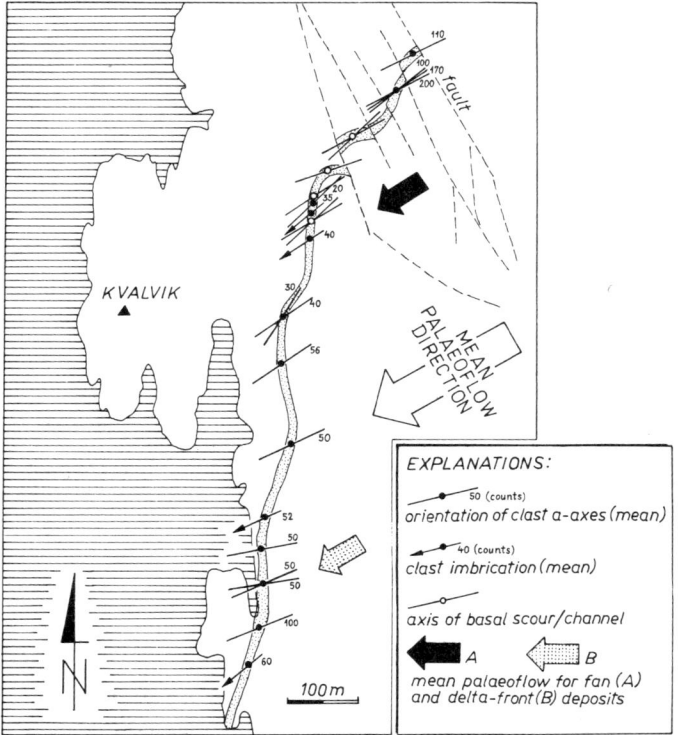

Fig. 5. Palaeoflow map showing inferred directions of sediment dispersal within the outcrop belt of Domba Conglomerate.

In the coarsest conglomerates, bedding is not always easy to identify (Figs. 8 and 9), and this is primarily due to bed amalgamation (as depicted in Fig. 7). However, there is often the helpful appearance of sandy interbeds (Fig. 10), the thickest of which are shown in Figure 7. These are lenticular in shape, a few to few tens of centimetres thick and up to a few tens of metres wide, and are often erosionally based. The sediment is mostly pebbly sandstone or granule sandstone, displaying moderate sorting and a frequent vague stratification (horizontal or inclined) (Fig. 10).

The bases of the conglomerate beds are generally flat or weakly undulatory, without any signs of significant erosion. However, this is not the case at the base of the Domba unit, where the conglomerates overlie an irregular erosion surface (see Stations 1 through 11 in Fig. 7). This basal channel-like feature (overall relief up to 3 m) comprises many local scours, with a relief of up to 0.6 m, which are often steep-sided and create a step-like morphology (Fig. 14). In other places, the basal erosion plane is flat within a distance of 10-50 m (Fig. 8).

Both this broad channel and its morphological details were probably formed prior to the deposition of the overlying conglomerates, because many of the local scours are veneered with a thin layer of fluvial deposits, consisting of vaguely cross-stratified pebbly sandstones up to 30-40 cm thick; however, the original channel morphology is likely to have been at least locally modified by the subsequent conglomerate emplacement.

Beds are generally sheet-like and their thicknesses range between 10 and 175 cm, averaging 60 cm. Strikingly, a frequency distribution of the bed thicknesses (Fig. 15) closely approximates a Gaussian, or normal, distribution. There is also a significant linear correlation between the thickness and maximum particle size (*MPS*) of the conglomerate beds as measured at the individual stations (Fig. 16). Both of these statistical facts are discussed and interpreted further below.

For every conglomerate bed at each station the presence or absence of vertical clast-size grading has been documented. Among all 94 bed occurrences measured, only 12 reveal no vertical grading (ungraded beds). The remaining occurrences show five general types of grading: weak inverse grading of only the basal part of the bed (14 occurrences); inverse grading throughout the bed (20); inverse-to-normal grading (15); normal grading of only the topmost part of the bed (11); and strong normal grading throughout the bed (22). The grading is invariably of distribution type, though, except for the normally-graded beds, is often not obvious within finer fractions and is more clearly revealed by the coarser fraction.

Lateral variation in the frequency of individual grading types is shown in Figure 17. When viewed from relatively proximal to relatively distal sites (Stations 1 through 22), there is a general tendency of ungraded and inversely graded beds to be replaced by their normally graded lateral equivalents. In order to distinguish any overall lateral trend, the discrete observations on all conglomerate beds at individual stations were compiled in the form of a lateral transition-count matrix and quantitatively evaluated with the use of Markov chain technique; the results are shown in Figure 18.

INTERPRETATION

Despite their general good sorting and clast-supported nature, Facies A conglomerates are likely to be mass flow deposits on account of their lack of stratification within beds, the common presence of inversely or inverse-to-normally graded beds (Figs. 10 and 11), the characteristic $a(p)a(i)$ clast fabric, the lack of obvious erosion surfaces between beds, and the significant relationship between the thickness of beds and their maximum particle size (Fig. 16). The latter relationship is a reflection of the dependence of both flow competence (*MPS*) and flow thickness (*BTh*) on the internal regime of the sediment gravity flow (Johnson, 1970, chapter 13; Lowe, 1976a; Gloppen and Steel, 1981, 1983; Shultz, 1983; Nemec and Steel, this volume). This *MPS/BTh* relationship has been documented for both subaerial (Bluck, 1967; Steel, 1974; Larsen and Steel, 1978; Nemec and Muszyński, 1982) and subaqueous (Nemec et al., 1980; Porębski, 1981) mass-flow deposits. The normality of the bed-thickness frequency distribution (Fig. 9) implies that these conglomerate beds themselves represent a single, coherent statistical population. Although this distribution is likely to have been slightly influenced

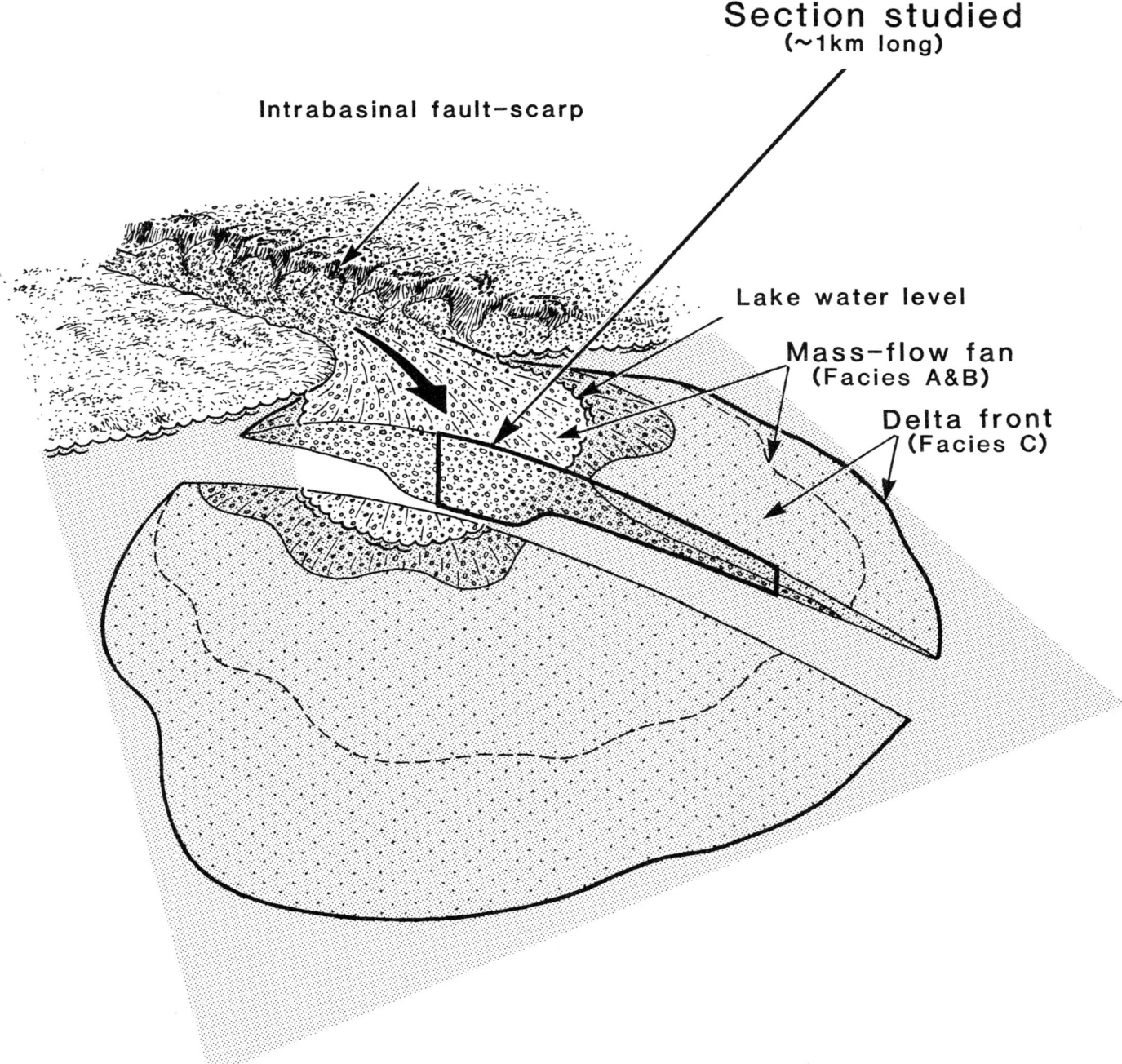

Fig. 6. Schematic, hypothetical palaeoreconstruction of the Domba fan-delta, showing the inferred position of the studied cross-section.

by collecting data along a fan cross-section, it also indicates that there was a certain preferred capacity of generated flows and that the flow capacities varied randomly about that preferred value. Thus, there is probably a suggestion that these conglomerate beds were emplaced by essentially a single type of flow mechanism. Some further inferences will be made below when interpreting the Domba fan development.

The clast-supported texture and polymodal coarse-sediment size distribution, together with the general scarcity of material finer than silt, all point to deposition from essentially cohesionless debris flows dominated by frictional/inertial effects (Fisher, 1971; Lowe, 1982); such flows are often termed combined grain flow/debris flows (Hampton, 1979, p. 757) or *density-modified grain flows* (Lowe, 1976a, 1982; Middleton and Southard, 1978). This type of emplacement mechanism is also strongly suggested by the conglomerate fabric and structure (*i.e.*, grading types).

The preferred *a*-axis orientation parallel to flow shows that the clasts were relatively free to move with respect to

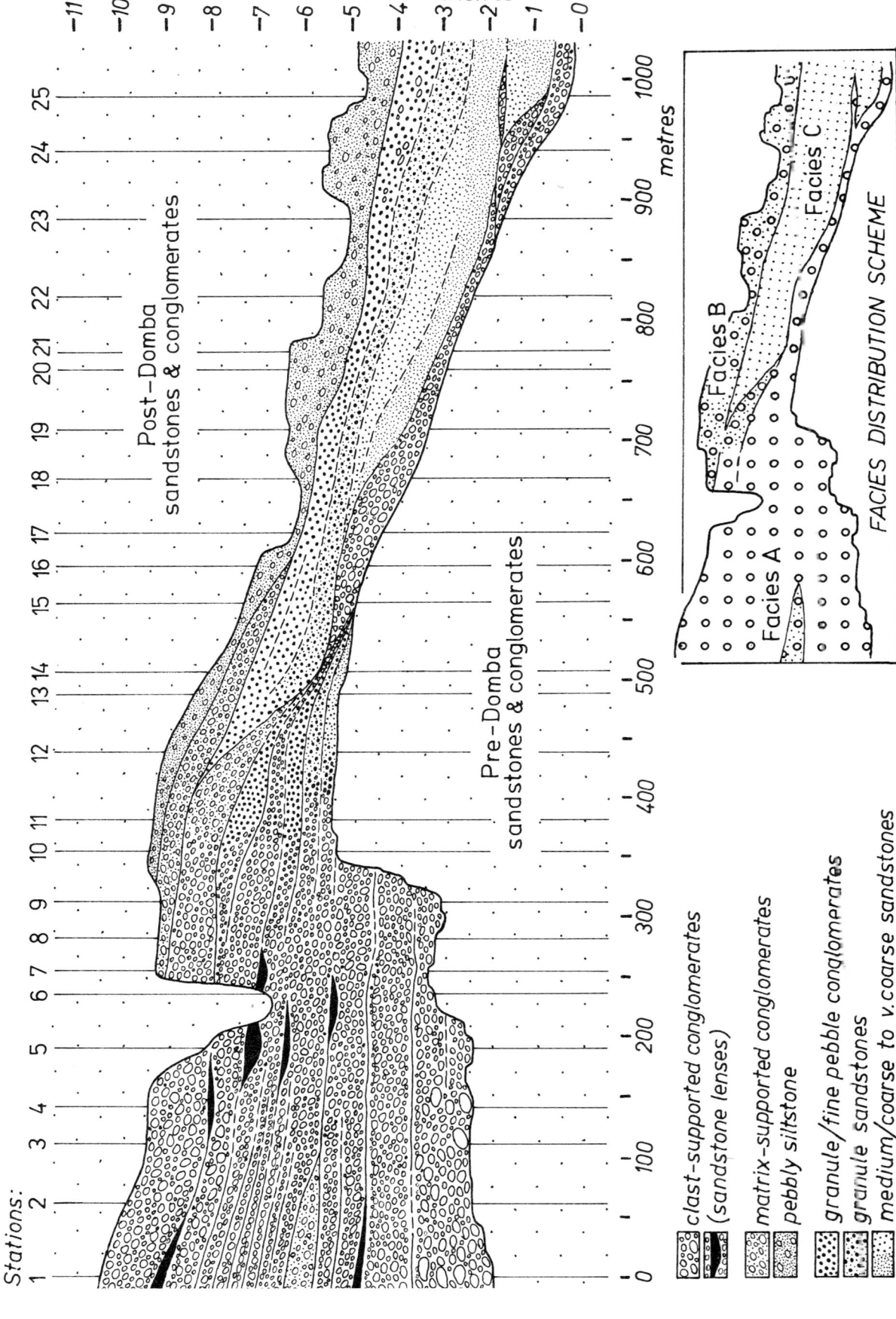

Fig. 7. Cross-section of the Domba Conglomerate outcrop belt (for interpreted palaeogeographic position see Fig. 6) and a schematic diagram (lower right) showing spatial relationships between the three main lithofacies. Note the considerable vertical exaggeration (37×).

each other, did not roll on the bed, and were supported above the bed (Walker, 1975b). According to Rees (1968), the *a*-axis imbrication is a result of tilting of the principal clast axes during collision of the clasts. In the present case, this type of fabric suggests that dispersive pressure, produced by clast collisions was probably the most important supporting mechanism during transport, though it must have been considerably aided by the matrix density (buoyancy), with a subsidiary role possibly played by some limited cohesiveness of the matrix. The density difference between the coarse sandy matrix and the clasts was probably only slight and the concentration of the coarse grains was high, thus strongly increasing the buoyancy effect. The conglomerates of Facies *A* are thus interpreted as density-modified, viscous to inertial grain flows; when significantly viscous, and containing slightly cohesive matrix, the flow would probably be also modified by the matrix strength and so would be transitional to a cohesive debris flow (Carter, 1975; Hampton, 1975; Lowe, 1982, Fig. 12).

For the ungraded conglomerate beds, the dispersive pressure was probably significantly aided by buoyancy with much control from high matrix-density (high viscosity), thus implying deposition from density-modified viscous grain flows (transitional to cohesive debris flows). The inverse-grading, whether near the base or throughout the bed, suggests that the dispersive pressure played a dominant role in both supporting the clasts and their size segregation, particularly in the basal zone of shearing. The inverse-grading effect is usually interpreted in terms of the larger clasts moving up to the regions of lesser shear stress (Bagnold, 1954), or of the smaller clasts sieving downwards between the larger ones (Middleton, 1970). The inversely graded beds were thus probably deposited from density-modified grain flows with a higher inertial component (higher shear-rates), and their close lateral relationship with the base-only inversely graded beds (Fig. 18) suggests that the latter were transitional to the former.

The normal grading, when restricted to the bed top, indicates only a slight density separation where the largest clasts have sunk down from the highest part of the flow. This may have been a result of downward migration of a

Fig. 8. Basal part of the Domba Conglomerate unit (Station 3), showing the clast-supported conglomerates of Facies *A*. Note the rare outsized clasts, and the sharp, flat contact with the stratified fluvial sandstone below. Scale is 1 m long.

Fig. 9. Basal part of the Domba Conglomerate unit (between Stations 6 and 7), showing three superimposed beds of the clast-supported conglomerates of Facies *A*. The long dimension of this photograph is 230 cm.

Fig. 10. Example of thin, lenticular beds of stratified fluvial sandstones between the beds of clast-supported (Facies A) conglomerates. Note the inverse-to-normal grading in the conglomerate bed shown in the middle. The open pocket-knife is 19 cm long.

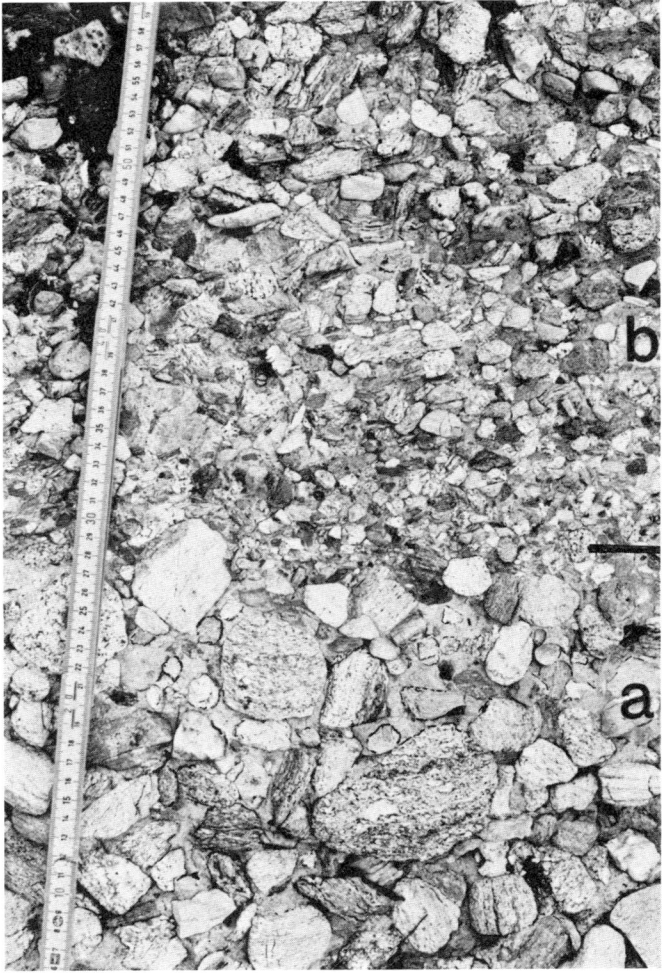

Fig. 11. Clast-supported conglomerates of Faices A: the lower bed has a strongly bimodal top (a), composed of pebbles/small cobbles and interstitial silty matrix, and the overlying bed (b) displays marked inverse grading. The plane of this photograph intersects the bedding surface at a low angle, thus only apparent thickness is revealed. Scale in centimetres.

cohesive 'rigid plug' within the flow (Johnson, 1970; Carter, 1975; Hampton, 1975), but more likely it was simply water entrained by the flow, at its top, that decreased the effective flow viscosity and so increased the possibility for vertical clast-size segregation (cf. Gloppen and Steel, 1981). The beds with normally-graded tops appear to display a close downflow relationship with the beds which are normally-graded throughout (Fig. 12), thus lending support to this latter interpretation. Flow dilution and the accompanying viscous effects are thought to have been particularly important, because with even a slight decrease in grain concentration, the viscosity may drop several orders of magnitude, whereas density decreases very little (Fisher, 1971, p. 918). Also the competence of fluid matrix is known to be proportional to its water content, and as postulated by Hampton (1975, p. 839), an increase of, say, 5% water to a slurry of low water content will produce a larger decrease in flow competence than will an addition of 5% water to a more dilute slurry.

The presence of normal grading implies that the clasts were able to move freely with respect to each other and that there was both vertical and lateral segregation of clast sizes within the flow (Walker, 1965). In the present case of matrix-poor normally-graded beds, a dominant role of the dispersive pressure is still implied by the clast $a(p)a(i)$ fabric, but these flows were probably least viscous and possibly even partly turbulent (Davies and Walker, 1974; Middleton and Hampton, 1976; Lowe, 1982, Fig. 12).

Accordingly, we suggest that the appearance of normal grading in our beds, whether at the bed top or throughout the bed profile, lends support to an interpretation of them as being the deposits of *subaqueous* mass-flows. This suggestion is consistent with the lateral/downflow changes of grading types in the conglomerate beds (Figs. 17 and 18). As theoretically deduced by Johnson (1970) and experimentally demonstrated by Hampton (1975), the upper zone of normal grading in debris flows results from shear stress

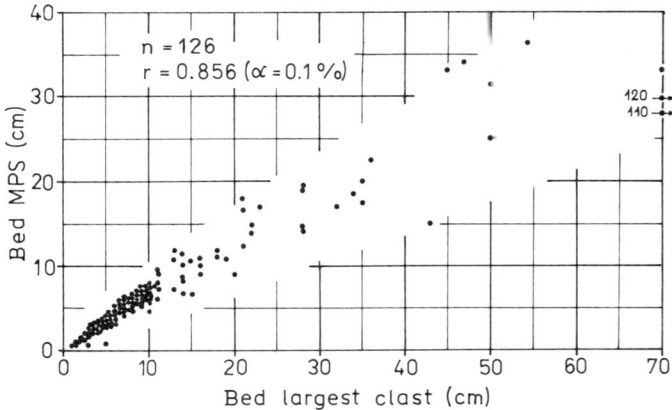

Fig. 12. Scatter plot of the largest clast diameter vs. calculated maximum particle size (MPS) for Facies A conglomerate beds. Data from individual beds from all stations (n = number of data; r = linear correlation coefficient; α = correlation significance level). Note that the largest clasts are outsized at only very few sites.

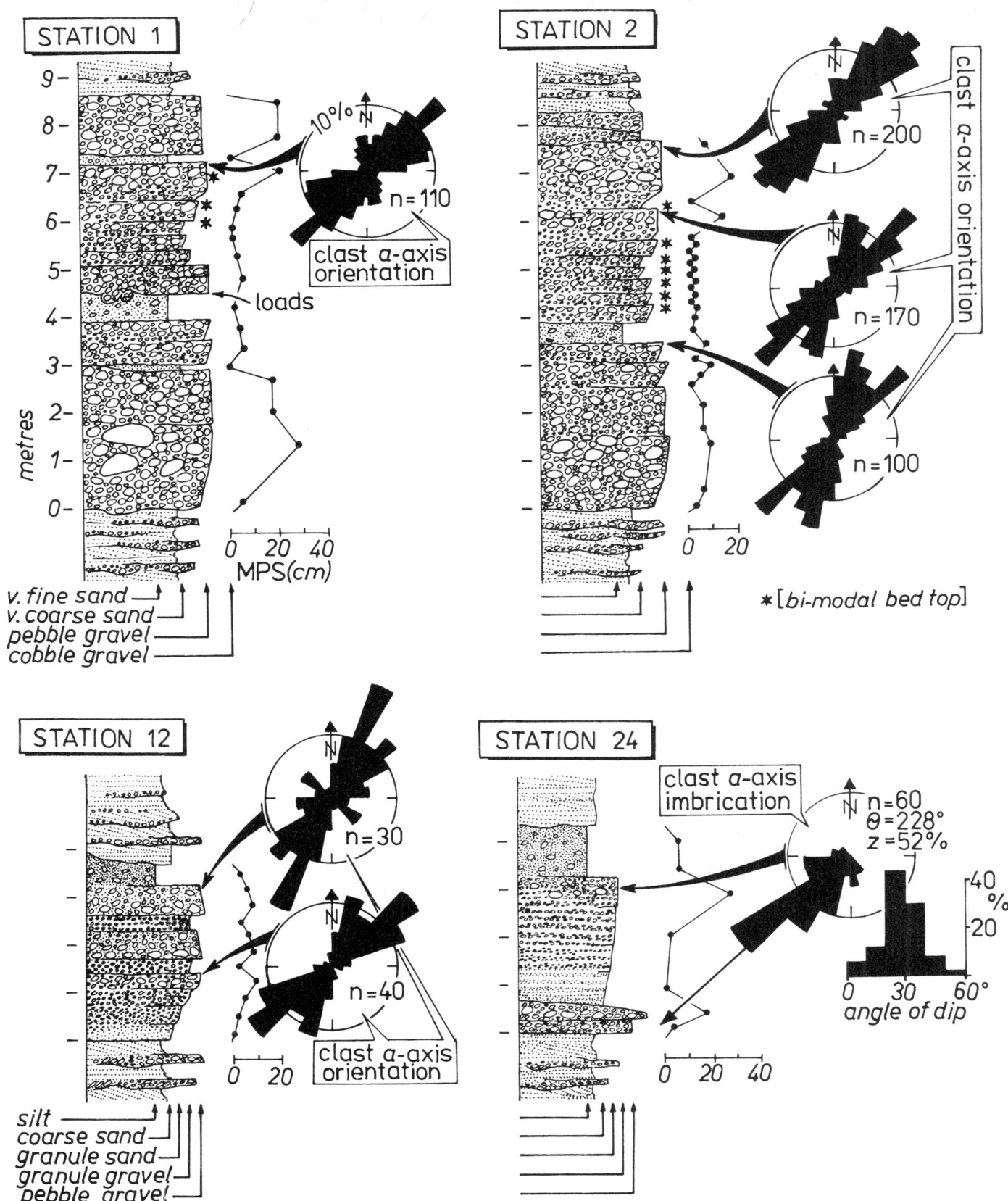

Fig. 13. Representative detailed logs of the Domba Conglomerate (vertical extent indicated by the maximum particle size, MPS, plots) and its fluvial envelope, illustrating lateral variability of the fan-delta body. (For the location of stations see Fig. 7).

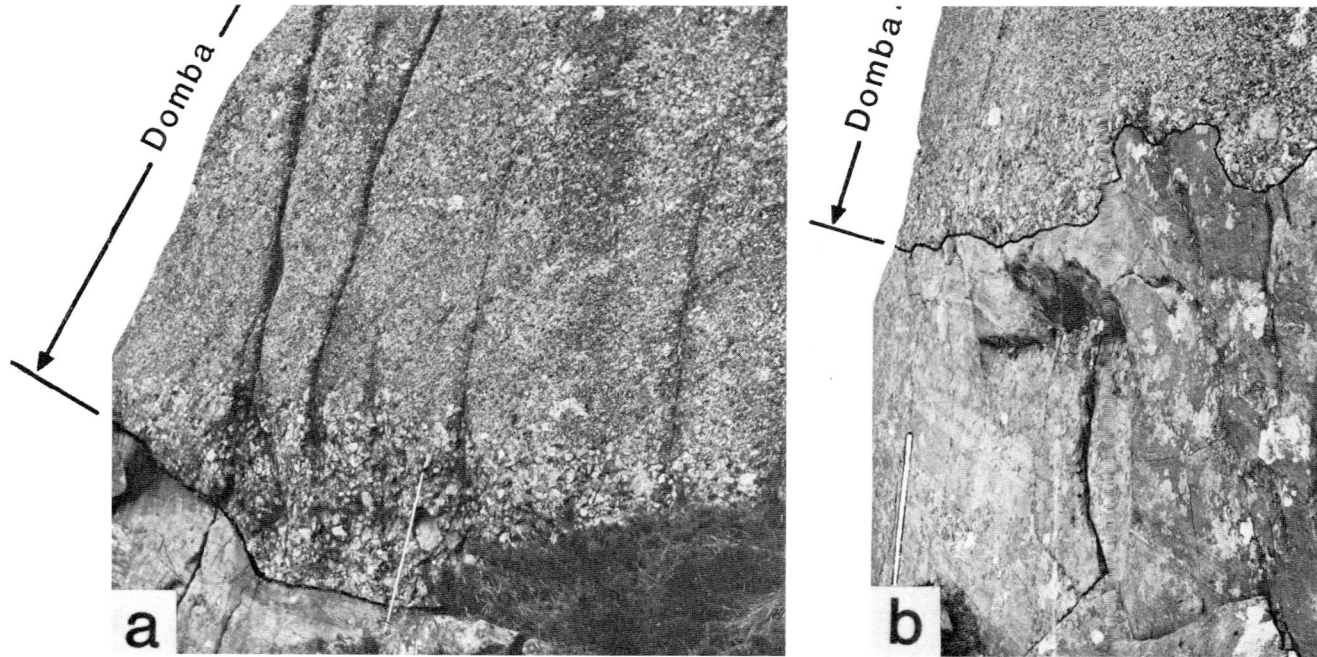

Fig. 14. Local marked irregularities (steep-sided scours) at the base of the Domba Conglomerate, as observed between Stations 5 and 9. The white ruler is 1 m long.

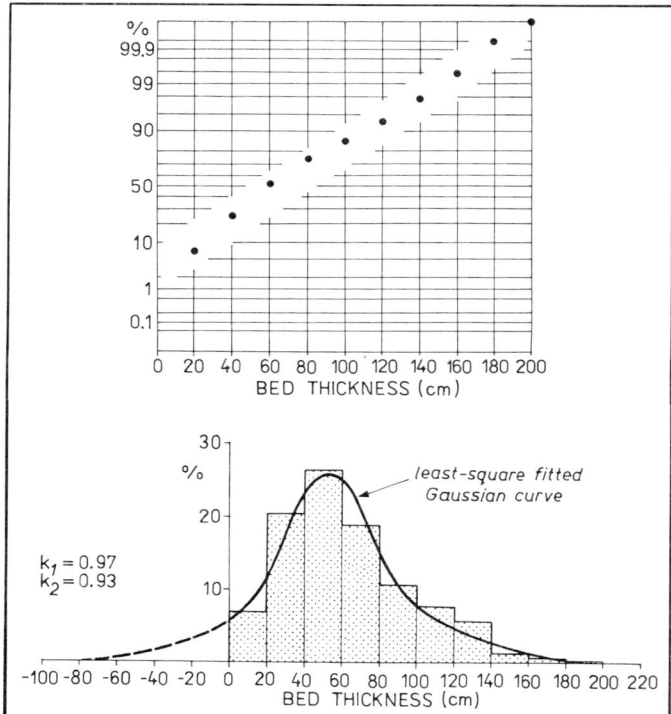

Fig. 15. Conglomerate bed-thickness data. **Upper diagram:** Cumulative number-frequency distribution of thickness for all conglomerate beds plotted on a probability scale (data from all stations, as in Fig. 16); note the close approximation to a normal distribution. **Lower diagram:** The same data presented as a number-frequency distribution with a corresponding Gaussian curve; note the left-side truncated nature of the fitted curve. The k_1 and k_2 values are Kolmogorov's coefficient and ratio, respectively, as calculated for the fitted normal curve (further details see text).

at the flow top, as generated by frictional resistance of overlying water, aided by accompanying water intake and flow dilution. As the flow dilution progresses with movement, the normal grading is readily developed throughout the flow thickness.

In addition to the abundance of normal grading, the marginal portions of many mass-flow beds reveal a drastic lateral decrease in both their thickness and grain size (see Stations 10 through 14 in Fig. 7). This is thought to be a reflection of the significantly decreased capacity and competence of mass-flows due to water being forced into the flows, possibly with some accompanying liquefaction (subaqueous portions of mass-flow beds). Thus, these bed margins were generated out of much thicker gravelly flows (Fig. 6) as the latter flowed down into an adjacent body of water and thinned markedly.

On the other hand, the sandy interbeds associated with the conglomerates of Facies *A* are interpreted in terms of subaerial sheetwash/rillwash processes operating on the alluvial fan surface. Some of these lenticular interbeds, when unstratified and lacking erosive bases, may possibly represent the deposits of thin sandy debris-flows. Because of the high permeability of surficial gravels, it is likely that a sediment-laden sheetflow could sometimes readily change into sandy debris-flow due to water percolation down into the bed.

The sheetwash processes of fluvial erosion are thought also to have been responsible for winnowing of finer sediment fractions from the tops of many conglomerate beds. The extremely bimodal tops of many clast-supported beds of Facies *A* imply infiltration of silt and very fine sand into

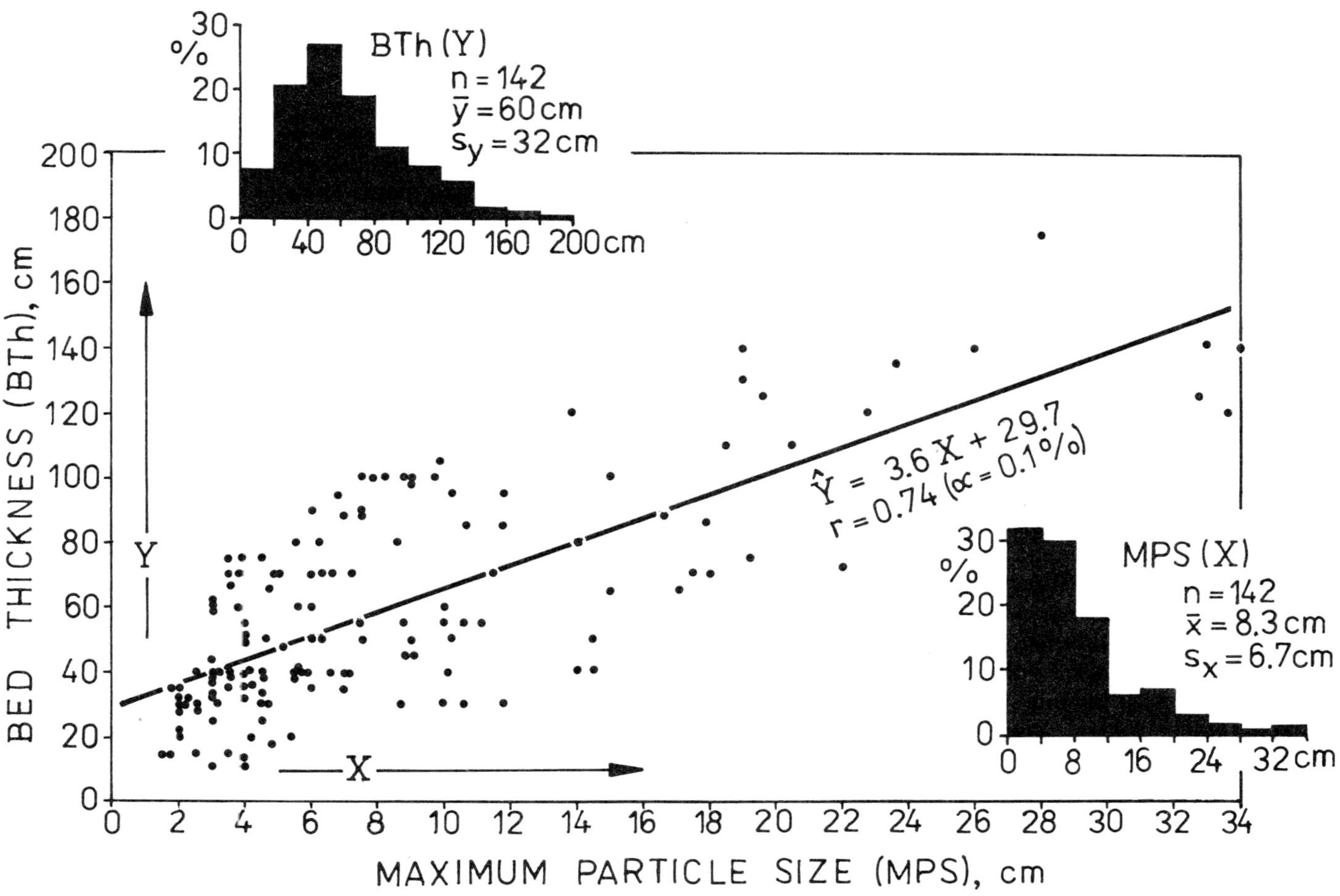

Fig. 16. Scatter plot of calculated maximum particle size *vs.* bed thickness for all conglomerate beds as measured at all stations. The histograms are number-frequency distributions of these two variables ($\bar{x}$, $\bar{y}$ and s_x, s_y are their mean values and standard deviations, respectively; n = number of data). The thick oblique line with corresponding equation is a least-squares regression line fitted to the data (r = linear correlation coefficient; α = correlation significance level).

earlier deposited gravelly mass-flow units whose tops had already been subject to local sorting by winnowing while lying on the fan surface (*cf.* Larsen and Steel, 1978). It is stressed that the marked scarcity of material intermediate between clast-supported framework gravel and silt/very fine sand within these bimodal bed tops is difficult to explain on a hypothesis that the debris in its entirety was derived *en masse* from the drainage area, or was supplied directly from the fan surface. As argued and discussed in more detail below, this fine material is inferred to have been deposited from an adjacent impinging lake.

Thus, the proximal part of the Domba wedge, composed mainly of Facies A (Stations 1 through 12 in Fig. 7), seems to provide evidence of a subtle alternation between subaerial and subaqueous deposition.

Facies B: Matrix-Rich Conglomerates and Pebbly Siltstones

DESCRIPTION

These deposits are volumetrically subordinate and are present mostly near the top of the Domba wedge (Fig. 7).

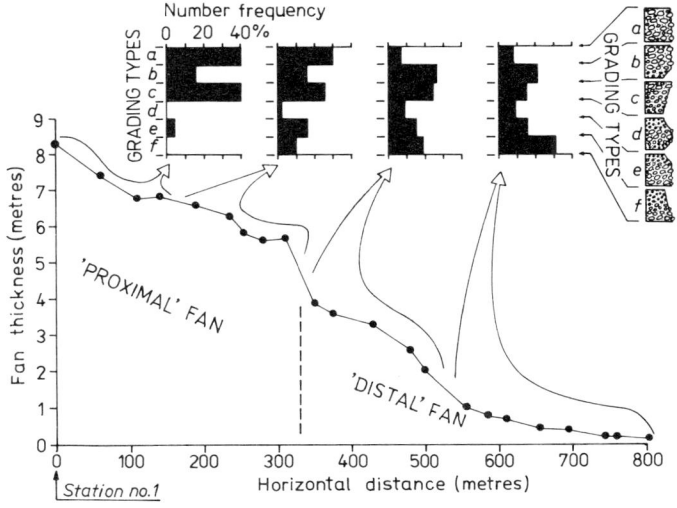

Fig. 17. Lateral variability in the number-frequency distribution of grading types for all conglomerate beds, presented against a background of the Domba fan thickness profile (thickness of both Facies C and overlying beds of Facies B not included).

As depicted in Figure 7, the few occurrences of this particular facies are, in ascending order: one pebbly sandstone bed which laterally becomes silt-enriched before wedging out (Stations 1 and 2); one matrix-rich conglomerate bed (matrix content up to 30-40%) whose texture becomes clast-supported and virtually silt-free over some short distance upflow (Stations 12 through 10); one matrix-supported conglomerate bed which becomes more matrix-rich laterally, generally downflow, and contains up to 60-70% of silty matrix (Stations 7 through 16); and an overlying, topmost bed which represents pebbly siltstone (Fig. 19) with some admixed coarse sand and scattered, 'floating' clasts of mostly granule and fine-pebble size grades (Stations 10 through 25). In the outcrop, there is locally some limited textural evidence that this latter bed may probably represent a composite depositional unit; moreover, there seems to be a good textural continuity between the 'coarse-tail' grain sizes of this pebbly siltstone and the underlying granule/fine pebble conglomerate itself.

In all these matrix-rich beds (Fig. 20), which are in fact matrix-enriched 'distal' segments of clast-supported beds, the matrix is reddish and consists mostly of a very fine sand/silt fraction with relatively little primary clay. Despite the considerable amount of fine matrix, there is otherwise a marked textural continuity in the coarse-fraction size distribution between the matrix-rich bed segment and its matrix-poor lateral (upflow) equivalent.

The beds of Facies B are 20 to 120 cm thick, and except for the topmost pebbly siltstone, they markedly wedge out over relatively short distances (50-250 m), generally downflow (Fig. 7). Their maximum particle size (MPS) ranges between 3.5 and 11 cm.

As measured in the individual stations (Fig. 7), the beds are often ungraded, though in many instances they display coarse-tail grading. This may be represented by an inversely graded base and/or a normally graded top of the bed, but more distally the beds have their coarse-tails normally graded throughout, often with a pronounced concentration of larger pebbles near the base. One type of grading may laterally change into another over a distance of a few tens of metres or less.

Clast composition is strictly the same as in the conglomerates of Facies A above, and no intraformational clasts have been identified here either. The clasts are randomly

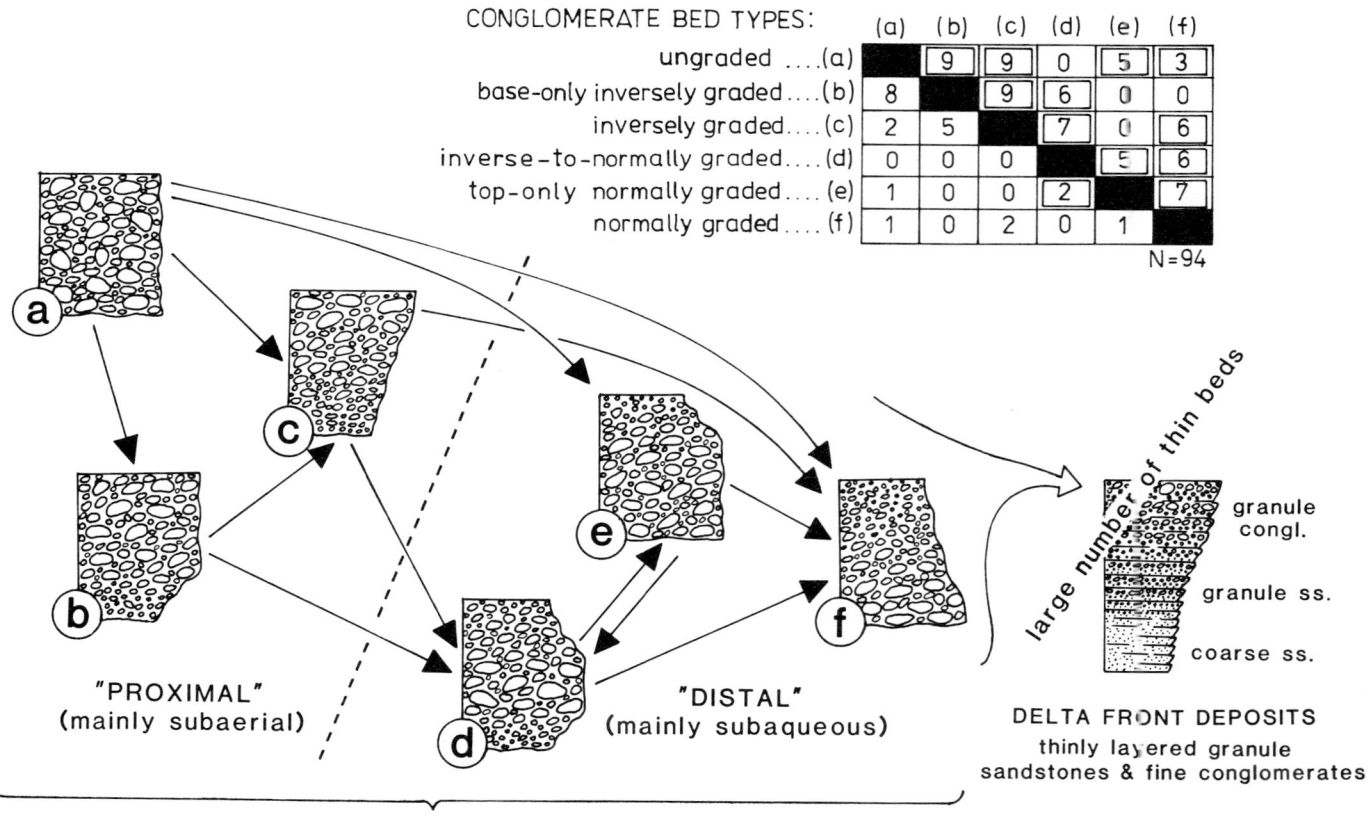

Fig. 18. Frequency data array of the observed downfan transitions between grading types in all conglomerate beds, as viewed by observations from Stations 1 through 25. The framed values in the transition-count matrix, and the corresponding flow-diagram below represent statistically preferred downfan transitions (with observed probabilities higher than if the transitions were random). Markov property not considered.

Fig. 19. Contact between matrix-supported conglomerate and the overlying, topmost pebbly siltstone (Facies B) in the vicinity of Station 12 (see Fig. 17). Note the steep to subvertical position of many elongate clasts in the conglomerate bed, and the floating clasts in the pebbly siltstone. The portion of ruler is 16 cm long.

Fig. 20. Matrix-rich conglomerate of Facies B. The plane of this photograph is parallel to the bedding surface and shows the top of the conglomerate bed. Note the alignment of elongate clasts parallel to palaeoflow direction (towards the upper left corner of the picture). The portion of ruler is 20 cm long.

oriented, and only occasionally display horizontal alignment of their longest (a-) axes parallel to an overall palaeoflow direction in the outcrop (Fig. 20). In the case of ungraded bed segments, many elongate clasts occur in a (sub-)vertical position in relation to the bed lower boundary (Fig. 19), and some of the large clasts are even slightly elevated above the bed top.

INTERPRETATION

The characteristics of Facies B suggest deposition by rapid 'freezing' of cohesive debris flows (Johnson, 1970; Fisher, 1971; Hampton, 1975; Middleton and Hampton, 1976). The textural bimodality of the deposits implies that the large amounts of fines were added to, or incorporated by the gravels either during or prior to the final stage of their deposition. The evidence of marked lateral changes in texture within the beds suggests that Facies B deposits originated as textural modifications of the flows that otherwise laid down the gravels of Facies A. This probably happened by the localised, distal passage of matrix-poor gravelly flows into layers of unconsolidated silt/very fine sand. We thus suggest that Facies B beds represent texturally inverted portions of Facies A flows, but we emphasise the difficulty in determining to what extent Facies A gravels actively 'invaded' soft bottom muds and to what degree some of the Facies A gravels, after being deposited and draped by mud, became unstable and were caused to resediment and mix. We believe, however, that this former possibility is more likely here as the reason for mixing and remobilisation of the gravel and inferred mud layers.

As already suggested above (see Facies A) and further discussed below, the fine silty material was itself added to the alluvial-fan surface from an impinging lake. Accordingly, we suggest here that the deposits of Facies B originated mostly, if not entirely, by *subaqueous* debris flows. Water-saturated mud is known to have a highly reduced strength (Johnson, 1970), and so is particularly susceptible to resedimentation. The natural instability of such a mud layer on the fan slope must have been critically increased when the water-saturated mud was suddenly overridden by a subsequent gravelly flow. This probably resulted in both loading of the gravel and liquefaction of the underlying mud, thus causing them to mix and to resediment together over some relatively short distance as a cohesive debris flow. For example, Van der Knaap and Eijpe (1968) experimentally induced slumping in inclined subaqueous sediments by raising the pore pressure at planes of permeability contrast, such as at the interface between a clay layer underlying a sand layer.

The processes postulated above seem to explain well both the occurence and internal features of Facies B deposits within the fan-delta wedge, though as further discussed below, the origin of the topmost pebbly mudstone unit (Fig. 19) may have been slightly more intricate. In this latter case, the coarse fractions before being mixed with mud and remobilised, are likely to have accumulated on the mud-layer surface in a more gradual manner. Their accumulation was probably due to 'freezing' of thin, essentially discrete flows which otherwise produced earlier the underlying sequence of Facies C sediments (see below), but which continued to originate farther upslope (somewhere beyond Station 10) after the intervening deposition of lacustrine mud. It was probably an additional rise in lake-water level and/or subsequent emplacement of the cohesionless debris flow(-s) of Facies A that caused a new, upslope portion of the fan surface to become unstable and so to release the above-mentioned, thin gravelly flows (see "gravity-winnowing" process below). Probably at least a few such depositional episodes, with an intervening accumulation of lacustrine mud, may have been involved in the origin of this pebbly siltstone unit discussed above.

Facies C: Assemblage of Thinly Layered Sandstones, Granule Sandstones and Granule-Rich Fine Conglomerates

DESCRIPTION

Facies C deposits occur distally, at the inferred toe of the Domba wedge (see Stations 13 through 24 in Fig. 7), and themselves constitute what we herein refer to as the fan-delta front (Fig. 6). They occur as three sequences (0.7-1.5 m thick) which are superimposed upon one-another, and which successively overlap each other as they onlap the gentle slope of the main conglomeratic wedge (Fig. 7). Despite this lateral overlapping relationship, the three units display a striking textural continuity with each other, and form what is essentially a single coarsening-upward sequence. The sequence starts with a medium/coarse, or coarse/very coarse sandstone at the base, and grades upwards into granule sandstone and then into granule-rich, very fine-pebble conglomerate (Fig. 7; for detailed examples see Figs. 21-23 and Station 24 in Fig. 13).

Coarse sand and granule particles are generally subangular to subrounded, but their size sorting is very good or excellent. Coarser clasts are subrounded to moderately well-rounded. Mica and any fragile rock fragments are absent. Maximum particle size (MPS) is from 0.2-0.5 cm in the granule sandstones, to 1-2 cm in the fine pebble conglomerate. Outsized pebbles are occasionally present, as described below.

In addition to the high degree of sediment sorting, there is a marked separation of grain size fractions from the sequence base upwards ('distribution type' coarsening on the sequence scale). Another notable textural feature is that the coarse granule/very fine pebble mode is abruptly cut-off above a certain maximum grain size, except for only occasional outsized clasts.

The granule-rich, fine pebble conglomerates, though generally clast supported, are usually bimodal and consist of small pebbles in a coarse sand/granule matrix. The matrix content varies from 15 to 25%, but abruptly increases near the sequence top, where both pebbles and the coarse granules together become more or less 'floating' in a reddish, muddy matrix rich in fine sand/silt fraction; this results in an almost gradational contact with the overlying, topmost pebbly siltstone of Facies B (Figs. 7 and 19) discussed above.

Fig. 21. Lower part of the Domba Conglomerate (Station 21), showing an isolated bed of Facies A conglomerate overlain by thinly layered, coarse sandstones of Facies C. The short dimension of this photograph is 130 cm.

An important characteristic of Facies C deposits is that they are low-angle stratified in relatively thick layers (Fig. 21). The layers are 1-12 cm thick (thickness partly proportional to the sediment grain size), and display a sheet-like to broadly lenticular geometry and distinct to diffuse boundaries (Figs. 21-23). They dip towards the southwest at an angle that ranges from 7-11° (most proximal sites) to 4-5° (distal sites) in relation to the Domba base. The layers are usually diffuse to moderately well-developed in the basal sandstones, are distinct and well-developed in the overlying granule sandstones, and again become diffuse or are almost invisible in the uppermost granule conglomerates. Basal surfaces of the strata often converge laterally and sometimes display very low-angle unconformities, although on the outcrop scale these surfaces appear as plane, low-angle inclined stratification (Fig. 21).

Fig. 22. Part of the Facies C sequence (Station 23), showing a gradual upward passage from coarse/very coarse sandstone to granule-rich sandstone with occasional small pebbles. The portion of the ruler is 96 cm long.

The stratification layers have a characteristic internal organisation (Fig. 24). They display sharp and roughly plane basal surface, a lower interval (1-4 cm) of distribution inverse grading, and an upper interval (2-8 cm) which is ungraded or only slightly inversely graded. Only when very thin or diffuse do the layers reveal a uniform (usually weak) inverse grading throughout. Some other thin sandy layers may instead be normally graded and display diffuse horizontal lamination. In addition, within the coarse sandstones and sometimes also in the granule sandstones, the 'upper interval' of many thicker layers is immediately overlain (distinct but non-erosive contact) by an additional interval: a finer grained, diffusely horizontally laminated sandstone (3-8 cm thick) which usually displays slight normal grading, particularly near its top (Fig. 24). It is emphasised that no sedimentary structures other than grading and the above-noted diffuse lamination occur in this facies.

Random measurements indicate that the individual stratification layers, at least in the granule sandstones and conglomerates, have strong and consistently organised grain fabric. The particle longest (a-)axes are parallel to flow and commonly imbricated upflow at an angle often in excess of 20-30° (see southern part of outcrop in Fig. 5; for a detailed example see Station 24 in Fig. 13). The recognition of this $a(p)a(i)$ fabric is based on evidence from the uppermost intervals of some well-exposed layers.

In places, Facies C deposits contain some markedly outsized pebbles which are either widely scattered, or occur in local lenticular concentrations. Pebble diameters range from 1-5 cm in the sandstones to 15-35 cm (occasionally up to 50-60 cm) in the granule conglomerates. These outsized clasts are present in sediment layers as thin as one pebble diameter or less, and so are often elevated above the layer top. The pebble-concentrates occur as thin, only one or two pebbles thick, erosionally-based lenses (up to 1 m wide) of well-sorted, packed clasts; in the sandstones, these clast 'pockets' are smaller but markedly scoured (3-5 cm deep and 10-15 cm wide), and are usually filled with very well-sorted coarse granules. All clasts are invariably of extraformational origin.

INTERPRETATION

The deposits of Facies C are thought to have been transported under the influence of gravity down the gentle subaqueous slope of the distal fan margin. It is suggested that a highly concentrated (high-density), highly fluid flow emplaced each individual layer in the sequence (cf. Fisher and Mattinson, 1968). The boundaries of the sediment layers are thought to represent a succession of sediment/water interfaces associated with sheet-like transport and deposition of repeated, discrete flows; we believe that they are not discontinuities imparted to the interior of a thicker sediment mass during movement (compare e.g., Carter, 1975, p. 159; Hiscott and Middleton, 1979; Lowe, 1982, p. 283-89).

The main mechanism of grain support in the flows was probably dispersive pressure due to grain collisions during movement, and so these sediment flows may have been similar to Bagnold's grain flows (Stauffer, 1967, p. 502; Lowe, 1976a). The distribution inverse grading and the type of fabric described here both are known to be characteristically produced by grain interaction during deposition from grain flow (Rees, 1968; Parkash and Middleton, 1970; Taira, 1976).

However, pure grain flows are probably not likely here because sand dispersions greater than a few centimetres thick cannot be maintained solely by dispersive pressure, and because slopes between 18 and 28° are necessary to maintain steady flows in which the interstitial fluid is water (Middleton and Hampton, 1976; Lowe, 1976a; Middleton and Southard, 1978, chapt. 8, p. 14). Accordingly, the inferred grain flows are thought to have been modified and aided by some other mechanism, most likely a liquefaction effect and the buoyant lift of a dense sand-water matrix. They would thus represent some sort of density-modified, liquefied-flow grain flows (Lowe, 1976b). The liquefaction factor has an extra appeal because: (a) it allows an explanation of the mechanism of grain flow triggering here (see below), and (b) it is consistent with the evidence that the flows moved over slopes of only several to few degrees (liquefied sediment has low resistance to shear and so its flow can be initiated on slopes as low as 3 or 4°; Lowe, 1976b).

The apparent lack of water-escape structures in the deposits of Facies C is thought to be due to their very good sorting and relatively coarse grain size (high effective porosity). The latter characteristics, although possibly aiding liquefaction initially (cf. Seed, 1958; Peacock and Seed, 1968), do also constrain the liquefaction to brief periods only.

Thus, the thinly layered sediments of Facies C are interpreted here in terms of essentially discrete-event deposition from a great number of thin, successive, density/liquefaction-modified grain flows. The deposits with crude, diffuse layering, or with only faint signs of the latter, are probably the result of a fusion of many rapidly deposited grain-flow layers, most likely due to a strongly surging character of the flows. The weakly inverse-graded tops of many thicker layers (Fig. 24) may possibly represent sediment that partly collapsed to a 'non-shearing plug' driven on top of the underlying dispersion (Lowe, 1976a, p. 195).

The sandy tops of many stratification layers, those displaying weak normal grading and diffuse horizontal lamination (Fig. 24), resemble the B-division of idealised turbidites. As such, they suggest deposition from a less concentrated, turbulent suspension entrained above the moving grain-flow sheet. These layers probably originated from sand-

Fig. 23. Middle part of the Facies C sequence (Station 22), showing coarsening-upward trend and broadly lenticular geometry of individual sediment layers (view almost perpendicular to palaeoflow direction). Note the low-angle erosional surfaces between some depositional layers. The ruler is 23 cm long.

Fig. 24. Details of some of the thickest depositional layers of Facies C sediment (overlay drawing from a photographic print). The lens cap is 5 cm.

rich flows in which the coarser, high-concentration dispersion was restricted to only a basal pseudo-viscous sublayer (cf. Hiscott and Middleton, 1979, p. 323), and when this basal traction-carpet collapsed and 'froze', an immediate deposition of the above-entrained turbulent suspension began. The deposition of the latter must have been rapid enough to prevent development of bedforms other than a diffuse plane bed, but not so rapid that vertical segregation of grain sizes (grading) was entirely prevented or that a 'quick' bed could result (Middleton and Hampton, 1976, p.204). Also the considerable sorting and narrow grain-size range of the sediment flow probably contributed to an abrupt cut-off of its turbulent suspension 'tail', thus preventing the development of higher turbidite divisions in the resulting deposit (a 'short-tailed' turbidity current).

These sediment layers with graded/laminated tops are thus interpreted as deposited from flows transitional between modified grain-flows and high-density turbidity currents. Lowe (1976a, 1976b, 1982) and Middleton and Southard (1978) suggested that there is probably a complete transition between grain-flows and liquefied flows, and between liquefied flows and turbidity currents. In the present case, the entrained turbulent tail-current was probably able also to travel some farther distance as a sole flow, thus giving rise to the solitary, thin sand layers with normal grading and diffuse lamination. Within the most 'distal' delta-front sequence, such laminated layers are often vertically stacked and amalgamated as thicker (up to 30-40 cm) intervals of this sequence.

This explanation is essentially consistent with the concept of Lowe (1982, p. 286-87), who postulated that deposition of a coarse-grained, high-concentration dispersion may leave a residual current containing, in turbulent suspension, finer sand grains that did not settle with the coarse material. These residual currents can move and possibly accelerate downslope as discrete turbidity currents of low to relatively high density. Although they may entirely bypass areas of the high-concentration 'traction carpet' deposition, they can have significant local effects; for example, they may partly erode or rework the underlying sediment, leaving thin sets of high-velocity plane lamination and producing low-angle erosional unconformities (as observed in Facies C deposits).

As an alternative to the turbidite explanation above, the diffuse plane laminations may have been produced by some minor surges active during cessation of the modified grain-flow or its main surge. This explanation is likely in the context of the overall mechanism of Facies C deposition postulated below, and seems also partly supported by the variable thicknesses of the laminated caps (comprising from 2-3 to more than 10 laminae) of otherwise identical depositional layers and by the thinning-up nature of the laminae (generally from 6 to 2 mm) in the individual sets; moreover, the thicker laminae are often inversely graded (Fig. 24).

The characteristics of Facies C sediments appear to resemble closely the depositional effect of the subaqueous 'gravity-winnowing' process described by Postma (1984, this volume) from his flume experiments on a low-plasticity sand-gravel mixture. In the present case, it was probably the frontal parts (or 'noses') of the cohesionless gravelly flows of Facies A which, after being subaqueously emplaced as a sediment pile (Fig. 7), directly became subject to the gravitational winnowing of both fine and progressively coarser grain-size fractions. An abrupt passage of the gravelly flow from subaerial to subaqueous environment, together with the polymodal grain-size of the sediment and its rapid mode of deposition, all probably caused an excessive pore-fluid pressure and consequently liquefaction/remobilisation of the finer fractions. This selectively remobilised sediment thus started partially to escape from the 'frozen' gravelly deposit as an almost continuous series of thin, surging grain-flows and associated turbulent flows (see interpretation above). Once the sand fractions began to escape, successively coarser fractions started to lose their support and thus became increasingly incorporated into the following flows of the same kind (cf. Postma, this volume). The overall coarsening-upward nature of the delta-front (Facies C) succession is thought to be the result of such progressive winnowing, involving grain fractions up to granule/fine-pebble size grade. Local slope failure due to winnowing and/or the emplacement of a new gravelly flow probably caused the smaller-scale irregularities observed in this coarsening-upward depositional trend.

Because of the gravitational winnowing of finer fractions from the gravelly sediment slope, many large pebbles and cobbles then also lost their support and fell freely downslope onto the surface of the finer sediment which was already moving as the thin, surging grain-flows discussed above. Initially, these large single clasts may have partly rolled, but then they probably began to 'glide' on the flow surface, with their freefall momentum providing the momentum to continue gliding downslope (Postma et al., 1984). When the clasts lost their momentum, they assumed the velocity of the flow below, sank down, and were eventually incorporated into the 'freezing' flow; some clasts were probably deposited on the static surfaces of already 'frozen' flows, or alternatively, may have actually by-passed the latter when maintaining a high enough momentum.

This seems to provide an explanation for the otherwise puzzling, scattered outsized clasts occasionally present within the delta-front (Facies C) sequence. This explanation is also consistent with the flow-parallel alignment or $a(p)a(i)$ fabric noted for many of these outsized clasts (cf. Postma et al., 1984). However, the scoured 'pockets' of outsized clasts reveal no such preferred fabric, and they are likely to have been emplaced either by small slumps derived through local failures of the winnowed and destabilised slope (cf. Andresen and Bjerrum, 1967), or by fluvial (rillflow) processes active on the subaerial fan surface, but occasionally extending subaqueously farther downslope.

As a whole, the deposits of Facies C provide the main evidence of a standing-water (lacustrine) region adjacent to the Domba fan toe. It is to be stressed that these deposits are difficult to explain on any hypothesis maintaining that the entire Domba wedge was constructed solely by subaerial processes.

DOMBA FAN-DELTA: DEVELOPMENT AND LATERAL VARIABILITY

SUBAERIAL ASPECTS

Palaeoflow data, the origin and distribution of sediment facies, the overall geometry of the conglomeratic wedge and its inferred palaeogeography (Figs. 5, 7 and 6 respectively), all suggest that the Domba Conglomerate represents an alluvial fan body which originated in response to syndepositional faulting of the alluvial basin floor. The Domba fan had a radius of about 2 km and grew southwestwards, and thus its development was towards the basin's western margin, rather than out from it (Fig. 1).

From the characteristics of the individual facies, there is enough compelling evidence to conclude that this alluvial fan prograded into a shallow, standing body of water, most probably a floodbasin-related lake. It is to be stressed, however, that the presumed contact between the fan-delta and lacustrine/floodbasin deposits is not exposed within the outcrop limits, and so the existence of a lacustrine region adjacent to the Domba fan toe has been inferred.

The inferred lake and fan-delta originated in a tectonically unstable, marginal area of the alluvial basin floor. They came into existence almost simultaneously, probably due to intrabasinal faulting and related escarpment creation, accompanied by locally rising water-table and damming of streams. The escarpment formation was probably rapid and reasonably the result of one or two large magnitude earthquakes. A large volume of sediment was suddenly made available for erosion, and so the construction of the fan prism was entirely by mass-flow processes. Fault-induced sliding of pre-existing basinal gravels was probably the major process responsible for supplying coarse, relatively well-sorted sediment onto the fan surface. The sediments of Facies A, forming the thickest and coarsest part of the fan body (Fig. 7), were most proximal and presumably banked directly against the intrabasinal fault scarp (Fig. 6).

At an early stage, the sediment dispersal was largely confined to a wide, pre-existing channel (Figs. 6 and 7). There is some evidence (see description of Facies A above) that this basal channel, itself about 3 m deep and a few hundred metres wide, was cut into the basin floor sediments by fluvial processes, rather than by the mass-flows themselves. The pre-Domba fluvial palaeocurrents (Fig. 4) and gently convex, basin-floor topography (Fig. 7) suggest that prior to the Domba fan development, the area was occupied by a braided stream system which constructed a broad, fan-like prism of mostly sandy deposits. Its aggradation was probably followed by stream entrenchment and the development of wide, axial channel (cf. McGowen, 1970). Because only thin, local veneers of fluvial deposits are now preserved in this flow conduit, there is a suggestion that the channel was entrenched across a slightly uplifted basin floor, so that almost all transported sediment actually bypassed the channel. When the escarpment was then formed and the construction of Domba fan began, this channel became the major conduit for gravelly mass-flows. As the channel was gradually filled with gravels, successive mass-flows started to overtop the channel margins and to construct what may be termed the channel 'levees' (see Stations 10 through 14 in Fig. 7). In physiographic terms, the slope of this latter feature presumably corresponded to the main, marginal slope of the fan itself (Fig. 6), and is therefore referred also to as 'fan toe' here.

The conglomerates of Facies A display a significant relationship between their maximum particle size and bed thickness (Fig. 16). Such a relationship in alluvial fan deposits (e.g., Bluck, 1967; Steel, 1974; Larsen and Steel, 1978, Nemec and Muszyński, 1982) emphasises the constant abundance of gravelly material available in the drainage area for entrainment by floods onto the fan surface, and thus partly explains also the unimportance of traction flow in the whole process of fan construction here.

As this MPS/BTh relationship (Fig. 16) reflects an interdependence between flow capacity (BTh) and flow competence (MPS), it is worth mentioning the difference noted in this particular respect between proximal and distal parts of the Domba fan (Fig. 25). A comparison of the respective regression coefficients (i.e., regression line gradients) suggests that, for the interpreted mass-flows, the downflow rate of flow-thickness reduction tends to be higher than the corresponding rate of flow-competence drop (see also Figs. 26 and 27). Although rather insignificant here in terms of an appropriate statistical test, the above downfan change is likely to reflect a respective change in the physical properties of the flows which terminated their emplacement subaqueously (Nemec and Steel, this volume, Fig. 23D).

Within the conglomeratic wedge, when viewed laterally from its relatively proximal to relatively distal sites, there is a marked general decrease in both particle size (Fig. 26) and bed thickness (Fig. 27). On the average, the maximum particle size decreases by 2 cm and the bed thickness by 6 cm per 100 m horizontal distance. In this same direction there is also a general tendency for ungraded and inversely graded beds to be replaced by their normally graded lateral equivalents (Figs. 17 and 18).

Another feature worth drawing attention to here is the specific nature of the Facies A bed-thickness distribution (Fig. 15). The frequency histogram apparently approaches a left-side truncated normal distribution (Fig. 15, lower diagram). This suggests that a Kolmogorov-type process of random-event deposition and erosion (Kolmogorov, 1951; Mizutani and Hattori, 1972) may have been responsible for the accumulation of Facies A sediments. If this assumption is true here, then the Gaussian curve fitted to the now incomplete (erosionally reduced) bed-thickness data may provide an estimate of the amount of sediment which was supplied to the fan, but which happened to bypass its individual sites. In other words, it seems possible to make an estimate of how strongly accumulative/erosive the depositional environment was. We have attempted to provide such an estimate for the Domba fan, and following the numerical procedure of Mizutani and Hattori (1972), the Kolmogorov's coefficient and ratio were calculated for the present data (Fig. 15, lower diagram).

Kolmogorov's coefficient deals with the number of beds, and so with the number of alternations of aggradation and erosion events. This value for Facies A beds ($k_1 = 0.97$)

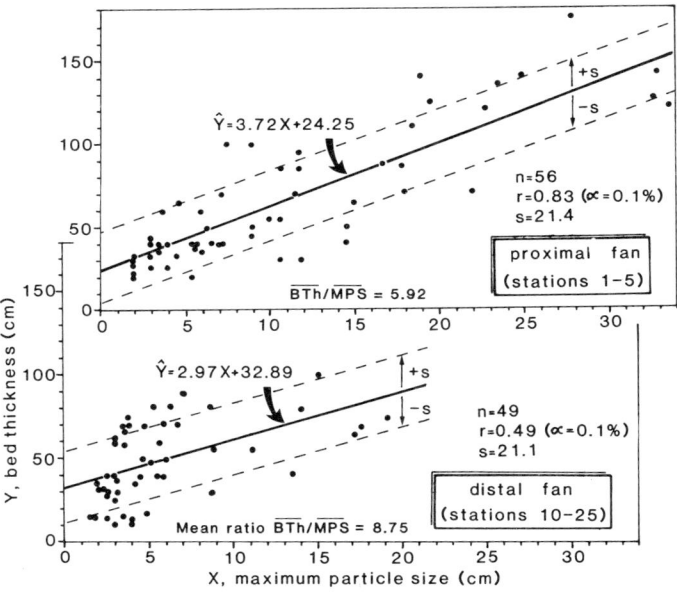

Fig. 25. Scatter plot of calculated maximum particle size (MPS) vs. bed thickness (BTh) for the conglomerates from proximal and distal parts of the Domba fan-delta. The thick oblique lines are least-square fitted regression lines; n = number of data; r = linear correlation coefficient (α = significance level); s = standard error of estimation.

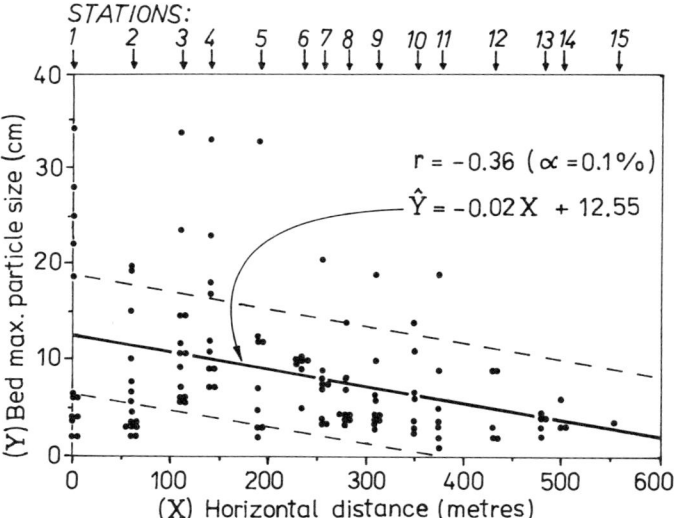

Fig. 26. Scatter plot of the calculated maximum particle size (*MPS*) of all conglomerate beds *vs.* horizontal distance along the outcrop belt (data from Stations 1 through 15). The solid line with corresponding equation is a least-squares regression line (± standard error of estimation) fitted to the data.

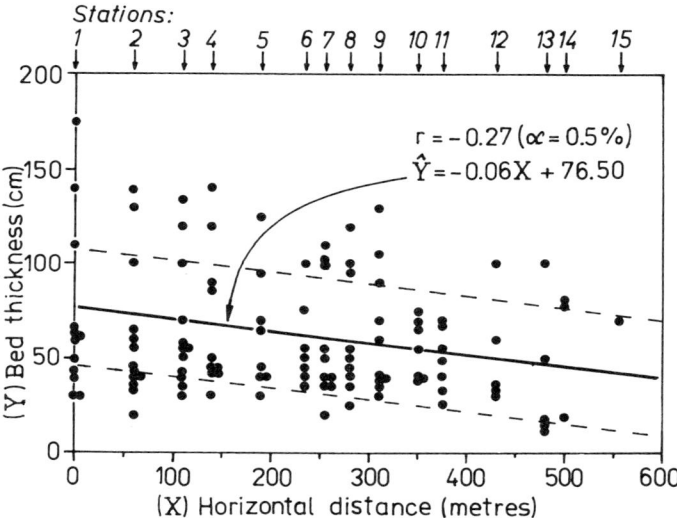

Fig. 27. Scatter plot of the thickness of all conglomerate beds *vs.* horizontal distance along the outcrop belt (data from Stations 1 through 15). Explanations as for Figure 26.

implies, in stochastic terms, that the overall preservation potential on this alluvial fan was such that in, say, 100 alternations or material accumulation and erosion, 97 beds were preserved and 3 were eliminated. Thus, it is rather improbable here that a conglomerate bed, once deposited on the fan surface, might have subsequently been entirely eroded. Kolmogorov's ratio deals with the total thickness of bed succession compared to the total thickness of sediment once transported and laid down at a given depositional site. In the present case, the ratio ($k_2 = 0.93$) suggests that about 93% of the total volume of material transported and deposited on the alluvial fan was preserved, while about 7% was eroded and removed during the course of sedimentation. Thus, in stochastic terms, the reduction of individual bed thickness by erosion was generally insignificant. These inferences clearly point to the highly accretionary nature of the Domba fan depositional system, and as such, are compatible with the inferred mass-flow domination and with the limited extent of erosion of the conglomerate beds by contemporaneous fluvial processes.

The scarcity of any thick stream-deposited units or channel-fills within the alluvial wedge indicates that the drainage system on the Domba fan was small. Water flows on the fan surface were shallow and relatively unconfined. The fan was modified by sheetwash-rillwash processes that served mostly to rework surface sediment by eroding only finer fractions (generally below coarse-pebble size grade). Such fluvial flow on recent alluvial fans has been described by Bull (1977, p. 230) as concentrated into "broad swales that are hardly noticable, except during times of flow".

SUBAQUEOUS ASPECTS

Although syndepositional faulting together with the Early Devonian climatic factor and limited vegetation primarily controlled the growth of Domba fan, the intrabasinal escarpment was probably also critical in causing the development of an ephemeral body of water adjacent to the fan toe. The rise in lake water-level was sufficient to inundate a considerable part of the fan slope. In effect, the mass-flow gravels of Facies A were, in their distal portions, emplaced largely as subaqueous flows. This is suggested by the downfan distribution of various grading types in the beds (Figs. 17 and 18), and possibly also by the inferred, respective change in flow behaviour (see discussion of Fig. 25 above, and Nemec and Steel, this volume, Fig. 23D).

Water level in the lake played an important role in the mode of fan margin development. The geometry and internal structure of the fan-toe sequence, or what has also been called the channel 'levee' (see Stations 10 to 14 in Fig. 7), point to a very limited lateral growth of the fan wedge and to its more significant vertical accretion. When water level in an adjacent reservoir is low fan-derived sediments are known to be distributed laterally over a considerable area and the fan-delta largely progrades; but when high, the lateral distribution is restricted and vertical accretion pronounced (*cf.* McGowen, 1970). In the present case, the 'levee' sequence consists of vertically stacked mass-flow units whose marginal portions are thin and finer-grained (Fig. 7). As already postulated above (see Facies A interpretation), this suite of marginal mass-flows provides independent evidence of subaqueous emplacement.

Despite these features above, however, there is no sharp demarcation between subaerial and subaqueous parts of the gravelly fan prism, and the main criterion for making a distinction between them here is probably the extent to which the gravelly mass-flows are mixed/infiltrated with fine sediment added to the fan surface from an impinging lake. This fine sediment, mostly reddish silt, is difficult to envisage on a hypothesis other than one involving lake/floodbasin suspended fines.

As on some other alluvial-fans in the Hornelen Basin (*e.g.*, Larsen and Steel, 1978), the fan-derived gravels here often mixed with lacustrine fines because they were accumulating in the same area. This occurred either by a passive, *in situ* infiltration of these former by the latter (thus producing the texturally bimodal tops of many Facies A beds), or by a vigorous passage of the gravelly mass-flows down into blankets of lacustrine mud when the latter had already accumulated on the fan slope (thus giving rise to the matrix-rich gravels of Facies B). The deposits of Facies B occur in the uppermost part of the Domba wedge, thus suggesting that it was mainly at a late stage that the depositional system eventually became conductive of this second type of mixing of the mass-flow gravels with lacustrine fines. This probably reflects an increasing time-delay between the emplacement of successive mass-flows, giving more and more time for an intervening accumulation of lacustrine sediment on the inundated fan slope.

On the other hand, there is a sharp textural/structural demarcation between the main conglomeratic wedge composed of Facies A and B, and the thinly layered deposits of Facies C (Figs. 6 and 7). The narrow grain-size range and strikingly good sorting of Facies C deposits suggest that they were derived from the fan surface by selective winnowing. As a whole, the characteristics of this facies point to the subaqueous process of 'gravitational winnowing', experimentally documented by Postma (this volume). Thus, there was otherwise an intimate relationship between the nature of Facies A gravels and the origin of Facies C sediments.

Once subaqueously emplaced, the fan-derived gravels of Facies A were mostly likely subject to partial liquefaction/fluidisation, restricted to finer grain-size fractions. Spontaneous liquefaction probably occurred because additional water was suddenly forced into the porous sediment and excessive pore-pressure was produced due to the sediment rapid deposition; possibly it was also cyclic stresses, resulting from the emplacement of successive mass-flows, that could contribute to liquefaction-triggering in the loosely packed sediment (Lowe, 1976b). The remobilised fine fractions, because of their lower frictional resistance, began then to escape from the gravelly fan slope as thin, surging grain-flows, sometimes transitional to high-density turbidity currents ('gravitational winnowing' process). The grain flows were probably modified by their relatively high matrix-density, and because the depositional slope was gentle (several to few degrees only), the gravity acting on sediment is thought to have additionally been aided by liquefaction.

The resultant, thinly-layered succession of the delta-front sediments (Facies C) itself provides the main, independent evidence of a lacustrine region adjacent to the Domba fan toe. Lake water energy was no higher than necessary to suspend silt and very fine sand, as there is no evidence of any erosion/reworking of the delta front deposits (except for the absence of any silt or clay veneers within the delta front sequence).

The delta front deposits apparently did not significantly prograde, but rather accreted to the fan-delta surface (Fig. 7). Chang (1967) produced similar depositional effects in a large, laboratory-model basin by progressively changing reservoir base level (rising water depth). Natural processes of this kind were also described by McGowen (1970) from the Gum Hollow fan-delta, Nueces Bay, Texas. In the present case, in addition to the evidence already discussed above, a rising lake-level and progressive inundation of the fan slope are also implied by the geometry of the delta-front (Facies C) prism. This depositional prism consists of three units which apparently overlapped one another and onlapped the fan slope (Fig. 7), thus suggesting that the delta-front deposits were probably formed during a period of rapidly rising lake-level (*cf.* Van Straaten, 1960).

A rising lake level and subtle interplay between subaerial and subaqueous deposition have also been inferred earlier, from independent evidence, for the Domba fan gravels (Facies A and B). Moreover, the subaerial rillflows were occasionally strong enough to penetrate onto the gentle subaqueous slope of the delta front, thus producing the narrowly scoured 'pockets' of well-sorted, outsized gravel within the delta-front sequence.

SUMMARY OF SEDIMENT TEXTURAL EVOLUTION ON THE DOMBA FAN-DELTA

From the data discussed above, a summary model for the textural evolution of sediments on the Domba fan-delta is proposed (Fig. 28). The grain size and sorting of the fan-derived sediment (Facies A) were primarily controlled by the textural characteristics of the pre-existing fluvial deposits being eroded. The main stages of further textural evolution of the sediment are summarised as follows:

(1) The texturally polymodal, fan-derived gravels (Facies A) were laid down from cohesionless mass-flows, interpreted as density-modified grain flows. These deposits were often slightly reworked by surficial erosion of finer fractions, due to shallow water flows (fluvial sheetwash/rillwash) following the emplacement of many individual mass-flows. In effect, the tops of many beds became developed as coarse, openwork gravel 'lag', essentially devoid of the finer fractions. Elsewhere on the fan surface, the fluvially derived sediment was deposited in broad, shallow scours or other morphological depressions.

(2) The fan-derived gravels often mixed with fine sediments deposited from an adjacent, impinging lake. This happened either passively, by infiltration (or 'sieving') of the lacustrine mud into the openwork tops of earlier depos-

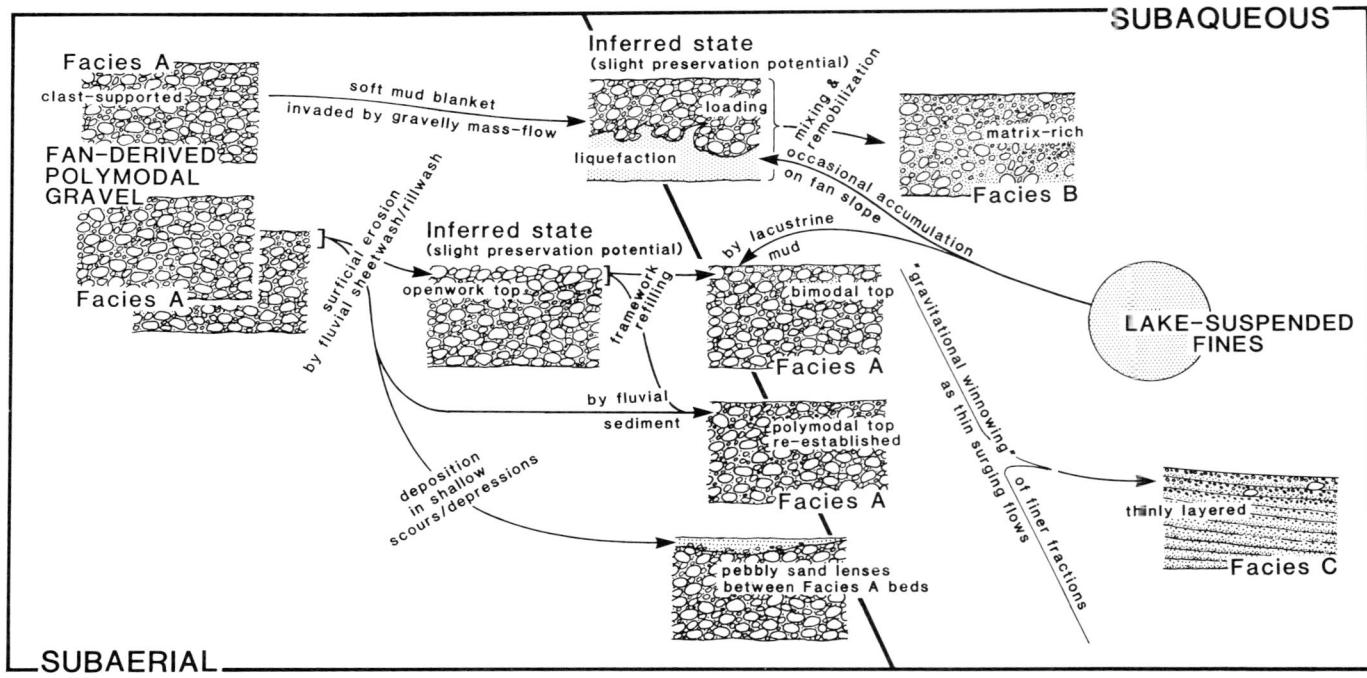

Fig. 28. Graphical summary model for sediment textural evolution on the Domba fan-delta.

ited mass flow beds, or actively as the fan gravels vigorously invaded soft blankets of lacustrine mud which had occasionally accumulated on the fan slope and became remobilised by the invading flows. It was only at a late depositional stage that the fan slope became conducive to this latter type of sediment mixing. The resultant bimodal bed-tops (Facies A) and matrix-rich deposits of Facies B are thus typically the product of 'textural inversion' (Larsen and Steel, 1978).

(3) Frontal, subaqueous portions of the fan-derived gravelly beds were subject to the process of 'gravitational winnowing' (Postma, this volume). Here, an intergranular remobilisation released sediment fractions up to fine-pebble size grade, and these accumulated as the thinly-layered deposits of Facies C.

Thus, the Domba fan-delta encompasses a number of textural varieties most of which were due to the subaqueous depositional setting of the alluvial fan toe. Together with several other depositional characteristics, this textural evolution clearly points to the ability of a standing-water reservoir to considerably modify the distal deposits of a mass flow-dominated, gravelly alluvial wedge.

References

Aasheim, S. 1977. Devonian sediments along a segment of the northern margin of the Hornelen Basin, with emphasis on flood-generated cyclic sedimentation in the Aalfoten area, western Norway. Cand. Real. Thesis, University of Bergen, 542p.

Andresen, A. and Bjerrum, L. 1967. Slides in subaqueous slopes in loose sands and silts. In: Richards, A.F. (Ed.), Marine Geotechnique. Urbana: University of Illinois Press, p. 221-239.

Bagnold, R.A. 1954. Experiments on a gravity-free dispersion of large solid spheres in a Newtonian fluid under shear. Proceedings of the Royal Society of London, Series A, v. 225, p. 49-63.

Bluck, B.J. 1967. Deposition of some Upper Old Red Sandstone conglomerates in the Clyde area: a study in the significance of bedding. Scottish Journal of Geology, v. 3, p. 139-167.

Bull, W.B. 1977. The alluvial-fan environment. Progress in Physical Geography, v. 1, p. 222-270.

Carter, R.M. 1975. A discussion and classifiction of subaqueous mass-transport with particular application to grain-flow, slurry-flow and fluxoturbidites. Earth Science Reviews, v. 11, p. 145-177.

Chang, H.Y. 1967. Hydraulics of rivers and delta. Ph.D. thesis, Colorado State University, Fort Collins.

Clemmensen, L.B. and Houmark-Nielsen, M. 1981. Sedimentary features of a Weichselian glaciolacustrine delta. Boreas, v. 3, p. 229-245.

Daily, B., Moore, P.S. and Rust, B.R. 1980. Terrestrial-marine transition in the Cambrian rocks of Kangaroo Island, South Australia. Sedimentology, v. 27, p. 379-399.

Davies, I.C. and Walker, R.G. 1974. Transport and deposition of resedimented conglomerates: the Cap Enragé Formation, Cambro-Ordovician, Gaspé, Québec. Journal of Sedimentary Petrology, v. 44, p. 1200-1216.

Fairchild, I.J. 1982. The Orustdalen Formation of Brøggerhalvøya, Svalbard: A fan delta complex of Dinantian/Namurian age. Polar Research, no. 1, p. 17-34.

Fisher, R.V. 1971. Features of coarse-grained, high-concentration fluids and their deposits. Journal of Sedimentary Petrology, v. 41, p. 916-927.

Fisher, R.V. and Mattinson, J.M. 1968. Wheeler Gorge turbidite-conglomerate series, California; inverse grading. Journal of Sedimentary Petrology, v. 38, p. 1013-1023.

Flores, R.M. 1975. Short-headed stream delta: model for Pennsylvanian Haymond Formation, West Texas. American Association of Petroleum Geologists Bulletin, v. 59, p. 2288-2301.

Galloway, W.E. 1976. Sediments and stratigraphic framework of the Copper River Fan Delta, Alaska. Journal of Sedimentary Petrology, v. 46, p. 726-737.

Gilbert, G.K. 1890. Lake Bonneville. Monograph of United States Geological Survey, v. 1, 438p.

Gloppen, T.G. 1978. Hornelen basin (Devonian), Western Norway: A study of various fan bodies and their deposits. Cand. Real. Thesis,

University of Bergen, 186p.

Gloppen, T.G. and Steel, R.J. 1981. The deposits, internal structure and geometry of six alluvial fan - fan delta bodies (Devonian, Norway) — A study in the significance of bedding sequences in conglomerates. *In:* Ethridge, F.G. and Flores, R.M. (Eds.), Recent and Ancient Nonmarine Depositional Environments: Models for Exploration. Society of Economic Paleontologists and Mineralogists, Special Publication 31, p. 49-69.

Gloppen, T.G. and Steel, R. 1983. The deposits, internal structure and geometry of six alluvial fan - fan delta bodies (Devonian, Norway) — A study in the significance of bedding sequence in conglomerates: Reply. Journal of Sedimentary Petrology, v. 53, p. 328-329.

Gnaccolini, M. 1982. Oligocene fan-delta deposits in northern Italy: a summary. Rivista Italiana Paleontologia e Stratigrafia, v. 87, p. 627-636.

Gustavson, T.C., Ashley, G.M. and Boothroyd, J.C. 1975. Depositional sequences in glaciolacustrine deltas. *In:* Jopling, A.V. and McDonald, B.C. (Eds.), Glaciofluvial and Glaciolacustrine Sedimentation. Society of Economic Paleontologists and Mineralogists, Special Publication 23, p. 264-280.

Hampton, M.A. 1975. Competence of fine-grained debris flows. Journal of Sedimentary Petrology, v. 45, p. 834-844.

Hampton, M.A. 1979. Buoyancy in debris flows. Journal of Sedimentary Petrology, v. 49, p. 753-758.

Harbaugh, D.W. and Dickinson, W.R. 1981. Depositional facies of Mississippian clastics, Antler foreland basin, central Diamond Mountains, Nevada. Journal of Sedimentary Petrology, v. 51, p. 1223-1234.

Hiscott, R.N. and Middleton, G.V. 1979. Depositional mechanics of thick-bedded sandstones at the base of a submarine slope, Tourelle Formation (Lower Ordovician), Québec, Canada. *In:* Doyle, L.J. and Pilkey, O.H. (Eds.), Geology of Continental Slopes. Society of Economic Paleontologists and Mineralogists, Special Publication 27, p. 307-326.

Holmes, A. 1965. Principles of Physical Geology (2nd edition). London and Edinburgh: Nelson, 1288p.

Johnson, A.M. 1970. Physical Processes in Geology. San Francisco: Freeman, Cooper and Co., 577p.

Kolmogorov, A.N. 1951. Solution of a problem in probability theory connected with the problem of mechanism of stratification. Transactions of the American Mathematical Society, v. 53, p. 171-177.

Larsen, V. 1977. Aspects of the sedimentology and paleogeography along a segment of the northern margin of Hornelen Basin (Devonian), western Norway, with emphasis on alluvial fan bodies and their deposits between Myklebustsaetra and Karlskaret. Cand. Real. Thesis, University of Bergen, 246p.

Larsen, V. and Steel, R.J. 1978. The sedimentary history of a debris flow-dominated, Devonian alluvial fan — a study of textural inversion. Sedimentology, v. 25, p. 37-59.

Lowe, D.R. 1976a. Grain flow and grain flow deposits. Journal of Sedimentary Petrology, v. 46, p. 188-199.

Lowe, D.R. 1976b. Subaqueous liquefied and fluidized sediment flows and their deposits. Sedimentology, v. 23, p. 285-308.

Lowe, D.R. 1982. Sediment gravity flows: II. Depositional models with special reference to the deposits of high-density turbidity currents. Journal of Sedimentary Petrology, v. 52, p. 279-297.

Mæhle, S. 1975. Devonian conglomerate-sandstone facies relationships and their palaeogeographic importance along the margin of the Honelen Basin, between Storevann and Grondalen, Sunnfjord, Norway. Cand. Real. Thesis, University of Bergen, 163p.

McGowen, J.H. 1970. Gum Hollow Fan Delta, Nueces Bay, Texas. Austin: University of Texas, Bureau of Economic Geology, Report of Investigations 69, 91p.

Middleton, G.V. 1970. Experimental studies related to flysch sedimentation. *In:* Lajoie, J. (Ed.), Flysch Sedimentology in North America. Geological Association of Canada, Special Paper 7, p. 253-272.

Middleton, G.V. and Hampton, M.A. 1976. Subaqueous sediment transport and deposition by sediment gravity flows. *In:* Stanley, D.G. and Swift, D.J.P. (Eds.), Marine Sediment Transport and Environmental Management. New York: John Wiley, p. 197-218.

Middleton, G.V. and Southard, J.B. 1978. Mechanics of Sediment Movement (2nd printing). Binghamton, New York: Society of Economic Paleontologists and Mineralogists, Short Course No. 3, 242p.

Mizutani, S. and Hattori, I. 1972. Stochastic analysis of bed-thickness distribution of sediments. Journal of the International Association for Mathematical Geology, v. 4, p. 123-146.

Nemec, W. and Muszyński, A. 1982 Volcaniclastic alluvial aprons in the Tertiary of Sofia district (Bulgaria). Annales Societatis Geologorum Poloniae, v. 52, p. 239-303.

Nemec, W., Porębski, S.J. and Steel, R.J. 1980. Texture and structure of resedimented conglomerates — examples from Książ Formation (Famennian-Tournaisian), southwestern Poland. Sedimentology, v. 27, p. 519-538.

Nilsen, H.R. 1975. Sedimentological studies along the central part of the southern margin (Haukå-Storevann) of Hornelen Devonian Basin, western Norway. Cand. Real. Thesis, University of Bergen, 223p.

Ogliani, F. 1981. Transgressive-regressive phases in the proximal part of a fan delta system, Early Pliocene, Intra-Apenninic Basin, Bologna. Bologna: International Association of Sedimentologists, 2nd European Regional Meeting, Abstracts, p. 126-129.

Parkash, B. and Middleton, G.V. 1970. Downcurrent textural changes in Ordovician turbidite graywackes. Sedimentology, v. 14, p. 259-293.

Peacock, W.H. and Seed, H.B. 1968. Sand liquefaction under cyclic loading and simple shear conditions. American Society of Civil Engineers, Journal of Soil Mechanics, v. 94, p. 689-708.

Pollard, J.E., Steel, R.J. and Undersrud, E. 1982. Facies sequences and trace fossils in lacustrine/fan delta deposits, Hornelen Basin (M. Devonian), western Norway. Sedimentary Geology, v. 32, p. 63-87.

Porębski, S.J. 1981. Świebodzice succession (Upper Devonian-lowest Carboniferous, Western Sudetes): a prograding, mass-flow dominated fan-delta complex. Geologia Sudetica, v. 16, p. 101-192.

Postma, G. 1984. Slumps and their deposits in fan delta front and slope. Geology, v. 12, p. 27-30.

Postma, G., Kleinspehn, K.L. and Nemec, W. 1984. Outsized clasts in high-density turbidity currents: a mechanism for their transport. Marseille: International Association of Sedimentologists, 5th European Regional Meeting, Abstracts of Papers, p. 366-367.

Rees, A.I. 1968. The production of preferred orientation in a concentrated dispersion of elongated and flattened grains. Journal of Geology, v. 76, p. 457-465.

Ricci Lucchi, F., Colella, A., Ori, G.G. and Ogliani, F. 1981. Pliocene fan deltas of the Intra-Apenninic Basin, Bologna. Bologna: International Association of Sedimentologists, 2nd European Regional Meeting Excursion Guidebook, p. 79-161.

Rust, B.R. 1979. Coarse alluvial deposits. *In:* Walker, R.G. (Ed.), Facies Models. Geoscience Canada, Reprint Series 1, p. 9-21.

Seed, H.B. 1968. Landslides during earthquakes due to liquefaction. American Society of Civil Engineers, Journal of Soil Mechanics, v. 94, p. 1053-1122.

Shultz, A.W. 1983. The deposits, internal structure and geometry of six alluvial fan - fan delta bodies (Devonian, Norway) — a study in the significance of bedding sequence in conglomerates: Discussion. Journal of Sedimentary Petrology, v. 53, p. 325-327.

Sneh, A. 1979. Late Pleistocene fan-deltas along the Dead Sea Rift. Journal of Sedimentary Petrology, v. 49, p. 541-552.

Spinnangr, Å. 1975. Some sedimentary and stratigraphic studies of the Devonian strata across the western part of the Hornelen Basin, western Norway. Cand. Real. Thesis, University of Bergen, 247p.

Stauffer, P.H. 1967. Grain flow deposits and their implications, Santa Ynez Mountains, California. Journal of Sedimentary Petrology, v. 37, p. 487-508.

Steel, R.J. 1974. New Red Sandstone floodplain and piedmont sedimentation in the Hebridean Province, Scotland. Journal of Sedimentary Petrology, v. 44, p. 336-357.

Steel, R.J. 1976. Devonian basins of western Norway — sedimentary response to tectonism and to varying tectonic context. Tectonophysics, v. 36, p. 207-224.

Steel, R.J. and Aasheim, S.M. 1978. Alluvial sand deposition in a rapidly subsiding basin (Devonian, Norway). *In:* Miall, A.D. (Ed.), Fluvial

Sedimentology. Canadian Society of Petroleum Geologists, Memoir 5, p. 385-412.

Steel, R. and Gloppen, T.G. 1980. Late Caledonian (Devonian) basin formation, western Norway: signs of strike-slip tectonics during infilling. *In:* Reading, H.G. and Ballance, P.F. (Eds.), Sedimentation in Oblique-Slip Mobile Zones. International Association of Sedimentologists, Special Publication 4, p. 79-103.

Steel, R.J., Mæhle, S., Nilsen, H., Røe, S.L. and Spinnangr, Å. 1977. Coarsening-upward cycles in the alluvium of Hornelen Basin (Devonian), Norway: sedimentary response to tectonic events. Geological Society of America Bulletin, v. 88, p. 1124-1134.

Surlyk, F. 1978. Submarine fan sedimentation along fault scarps on tilted fault blocks (Jurassic-Cretaceous boundary, East Greenland). Grønlands Geologiske Undersøgelse Bulletin, v. 126, p. 1-108.

Sykes, R.M. and Brand, R.P. 1976. Fan-delta sedimentation: an example from the late Jurassic-early Cretaceous of Milne Land, Central East Greenland. Geologie en Mijnbouw, v. 55, p. 195-203.

Taira, A. 1976. Grain orientation and depositional processes — fabric analyses of modern and laboratory flume deposits. Ph.D. thesis, Part 1, University of Texas, Dallas.

Van der Knapp, W. and Eijpe, R. 1968. Some experiments on the genesis of turbidity currents. Sedimentology, v. 11, p. 115-124.

Van Straaten, L.M.J.U. 1960. Some recent advance in the study of deltaic sedimentation. Liverpool and Manchester Geological Journal, v. 2, p. 411-442.

Vågene, S. 1982. Aspects of the sedimentology and geometry of the lower part of the Hovden Formation (Devonian), western Norway. Cand. Real. Thesis, University of Bergen, 270p.

Walker, R.G. 1965. The origin and significance of the internal sedimentary structures of turbidites. Proceedings of the Yorkshire Geological Society, v. 35, p. 1-32.

Walker, R.G. 1975a. Conglomerates: Sedimentary structures and facies models. *In:* Depositional Environments as Interpreted From Primary Sedimentary Structures and Stratification Sequences. Dallas: Society of Economic Paleontologists and Mineralogists, Short Course No. 12, p. 133-161.

Walker, R.G. 1975b. Generalized facies models for resedimented conglomerates of turbidite association. Geological Society of America Bulletin, v. 86, p. 737-748.

Wescott, W.A. and Ethridge, F.G. 1980. Fan-delta sedimentology and tectonic setting — Yallahs fan delta, southeast Jamaica. American Association of Petroleum Geologists Bulletin, v. 64, p. 347-399.

Wescott, W.A. and Ethridge, F.G. 1982. Bathymetry and sediment dispersal dynamics along the Yallahs fan delta front, Jamaica. Marine Geology, v. 46, p. 245-260.

Wescott, W.A. and Ethridge, F.G. 1983. Eocene fan delta-submarine fan deposition in the Wagwater Trough, east-central Jamaica. Sedimentology, v. 30, p. 235-247.

DAGBREEK FAN-DELTA: AN ALLUVIAL PLACER TO PRODELTA SEQUENCE IN THE PROTEROZOIC WELKOM GOLDFIELD, WITWATERSRAND, SOUTH AFRICA

C.S. Kingsley[1]

Abstract

The Dagbreek Formation represents a fan-delta which covers an area of approximately 400 km^2. The fan is wedge-shaped in section and thickens from 25 m in the west to over 200 m some 30 km further east. Its base, marked by the Leader placer, rests unconformably on a debris flow arenite of the Harmony Formation. The Leader placer consists of two conglomerate placers, namely the lower oligomictic, mature, Alma placer and the overlying polymictic, less mature, Bedelia placer. Their combined thickness varies between 30 cm to more than 3 m. Intercalated conglomerate and quartzite beds occur within the lower 14 m of the formation, and these fine upwards into khaki-coloured, coarse-grained protoquartzites with rare pebble lags. Small scale trough cross-bedding is ubiquitous. The fining-upward megasequence continues through medium- to fine-grained subgraywackes and thin khaki to black shale beds near the top. The subgraywackes in the top 20 m display medium to large scale cross-bedding and form fining-upward units which contain shale intraclasts at their bases. Palaeocurrent measurements and pebble size distribution patterns indicate a unimodal fan-like distribution trending eastward. The sandy and silty shale on top is plane-bedded, ripple cross-laminated, and graded bedded, these facies being arranged in coarsening-upward sequences.

The placers are interpreted as braided channel-fill sands and gravels. The overlying sands represent the more distal braided equivalents of the alluvial fan whereas facies sequences in the topmost part suggest single channels on the lower fan area. The shaly sequence on top represents shallow subaqueous sediments deposited in a lacustrine prodelta area.

The model presented here is that of a fan-delta produced by rejuvenation of the source followed by lowering of relief and recession of the source area.

Résumé

La formation de Dagbreek représente un cône de déjection qui couvre une superficie approximative de 400 km^2. La coupe verticale du cône de déjection est en forme de biseau et l'épaisseur passe de 25 m à l'ouest à plus de 200 m à quelques 30 km plus à l'est. Sa base est repérée par le placer Leader, lequel repose en discordance sur une coulée de débris arénacés de la formation de Harmony. Le placer Leader inclut deux placers conglomératiques, soit à la base le placer Alma, oligogénique et mature, et au-dessus le placer Bedelia, polygénique et moins mature. L'épaisseur dex deux placers réunis varie de 30 cm à plus de 3 m. Des couches intercalées de conglomérat et de quartzite sont présents dans les 14 derniers mètres du bas de la formation, et en remontant leur texture devient plus fine, la couleur kaki, et appraissent des protoquartzites à gros grains accompagnés de quelques rares galets résiduels. Des dépressions de petitie échelle et une stratification oblique se rencontrent partout. La mégaséquence positive (granoclassement normal) se prolonge au travers les subgrauwackes à grains moyen et fin et les lits de shale kaki à noir près du sommet. Les subgrauwackes des 20 premiers mètres de la partie supérieure exhibent une stratification oblique de grande échelle et constituent des unités positives qui contiennent à leurs bases des intraclastes de shale. Des mesures des paléocourants et les motifs de distribution de la grosseur des galets révèlent une distribution en éventail unimodale pointant vers l'est. Le shale gréseux et silteux du sommet se présente en couches planes, il est caractérisé par des rides à lamination oblique et il est granclaseé, ces faciès sont disposés en séquences à granoclassement négatif (inverse).

Les placers sont considérés comme des remplissages de sables et de graviers dans des chenaux anastomosés. Les sables sus-jacents correspondent aux équivalents anastomosés les plus distaux du cône de déjection tandis que les séquences des faciès de la partie la plus élevée suggèrent le developpement de chenaux simples dans la zone inferieure du cône de déjection. La séquence de shale du sommet représente des sédiments sous-aquatiques accumulés dans une zone de prodelta lacustre.

Le modèle présenté ici est celui d'un cône de déjection engendré par un rajeunissement de la région nourricière suivie d'une atténuation du relief et d'une régression de la région nourricière.

Introduction

The Dagbreek Formation, which is 700 m stratigraphically beneath the top of the Witwatersrand Supergroup (Fig. 1), thickens from 25 m in the west to over 200 m some 30 km further east; the Leader reef lies at its base. This formation represents a fan-delta which covers an area of approximately 400 km^2. The Witwatersrand rocks in the study area are covered patchily by Ventersdorp lava and sediments and are covered completely by a Carboniferous to Permian blanket of fluvial and lacustrine sandstones and

---

[1]Geology Department, Anglo American Corporation of South Africa Limited, P.O. Box 20, Welkom 9460, South Africa.

I gratefully acknowledge the permission to publish this paper and the support granted by the management of Anglo American Corporation of South Africa for presenting it at the 11th International Congress on Sedimentology at Hamilton, Canada, in August 1982. I wish to thank W.E.L. Minter and W.M. Stear for critically reviewing an early draft of the paper. I also wish to thank the following people for their assistance: R. King for typing the manuscript; C Carlile, M. Hanson and J. Tatalias for drafting the diagrams; J. Reeves for preparing the photos; and the assistance given by the geological staff and management of Welkom and Saaiplaas Divisions.

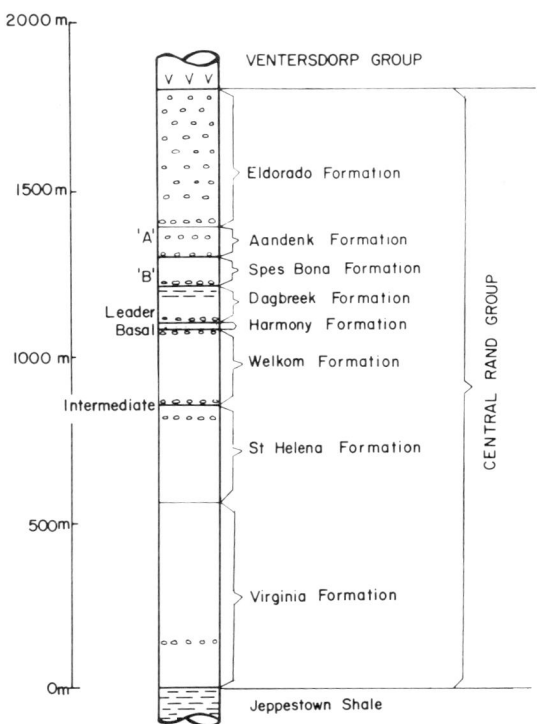

Fig. 1. Stratigraphic column of the Central Rand Group in the Welkom area, indicating the principal 'reefs' developed in the area.

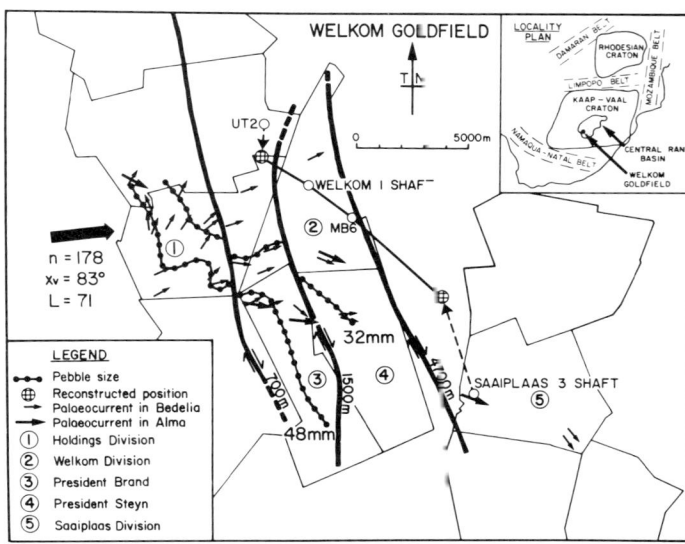

Fig. 2. Locality plan of the Welkom Goldfield, depicting the pebble size and palaeocurrent distribution of the Leader reef. Note that the positions of two of the sections have been restored to compensate for postdepositional dextral shifts.

mudstones. The Witwatersrand subcrops 500 to 1000 m below surface and generally dips 10 to 30° to the east.

A conceptual model of sedimentation of Witwatersrand-type placer deposits was published by Pretorius (1976a), in which each placer was considered to be the result of cyclic sedimentation caused by the progradation and recession of an alluvial fan. Pretorius (1976b) also postulated that longshore currents diverted the outer parts of fluvial fans in a clockwise direction. Minter (1978) has interpreted most of the Witwatersrand rudaceous and arenaceous succession as representing fluvial braided stream deposits. He has described the lateral facies relationships in four different placers in the Witwatersrand. He concluded that braided channels on alluvial fans played a major role in the deposition of most of these placers in which proximal-distal relationships are well defined by the size of clasts and heavy minerals and by mineral ratio changes. In a more specific study on Western Holdings Limited (Welkom Division) Smith and Minter (1980) concluded that the Leader reef is a well channelised midfan deposit situated on an alluvial fan. They furthermore described in detail some heavy mineral concentration mechanisms and their relationship to the various bedforms.

In a synthesis of a depositional model for the Witwatersrand Supergroup, Vos (1975) described the vertical megasequences in this supergroup and concluded that the main depositional environment contained wet alluvial fans which he interpreted as being analogous either to the Kosi River fan of India, or to glacial outwash fans. He postulated a braided alluvial plain with coarse gravels grading downstream into braided channel sands and finally into lacustrine silts and muds.

In studying mineralogical variations in the Leader reef, Callow (1982) concluded that the average gold content of the conglomerate facies decreases from 13.5 ppm proximally to 4.6 ppm distally, whilst uranium remains consistent at 200 ppm.

It is the purpose of this paper to document the first fan-delta facies sequence in the Witwatersrand and the detailed environmental changes that took place during deposition of its fining-upward megasequence. This sequence, the Dagbreek Formation, forms part of the Central Rand Group with an age of approximately 2.5 to 2.6 b.y. (Fig. 1). It occurs near the southern extremity of the Witwatersrand Basin in the Welkom Goldfield (Fig 2) where formations are generally thinner than in the Central Rand type area at Johannesburg.

Several ancient fan-deltas have been documented which display fining-upward sequences (Ricci Lucchi et al., 1981; Gnaccolini, 1981; Larsen and Steel, 1978; Gloppen and Steel, 1981; Gjelberg and Steel, 1983) and which show some similarities to the depositional character and tectonic setting of the Dagbreek fan-delta. A direct comparison is, however, very difficult because the Dagbreek fanhead was removed during subsequent tectonic movement and erosion. However, enough of the midfan and outer fan is preserved to derive a general tectono-environmental model.

Few vertical sequence analyses have been applied to ancient fan-deltas (Scholle and Spearing, 1982). Although the Witwatersrand rocks, and especially the conglomerate reef deposits, have been studied extensively by many authors, only a few studies (Vos, 1975; Eriksson et al., 1981) were directed towards a generalised vertical sequence analysis. Thus relatively little is known of the arenaceous and argillaceous units of the sequence which make up more than 95% of the succession. Likewise, little is known about the environmental changes that took place through the depositional history of the Witwatersrand Supergroup. This paper describes a complete sedimentational sequence which ranges from a subaerial gravel placer deposit at the base through a sandy unit into a subaqueous shale on top.

This study was suggested to me by Minter in 1981 when, during a routine investigation of borehole core, the present author realised that fluvial gravel and sand deposits in the Dagbreek Formation are overlain sharply but conformably by deltaic silts and muds. This led to a more extensive study in which three boreholes, namely UT 2, MB 5 and MB 6, were logged in detail on a vertical scale of 1 : 100. In addition, two carefully chosen underground sections at Welkom Division No 1 Shaft and Saaiplaas Division No 3 Shaft were mapped to supplement the borehole observations. Because the rocks in the underground exposures are fresh, sedimentary structures are well preserved and the various lateral and vertical changes in lithology can be identified quite easily. The localities of four of these observations are detailed in Figure 2.

Vertical sequence analysis was found to be the best method of analysing the Dagbreek Formation for two reasons. Firstly, as the formation changes through a continuous transgressive sequence from braided to sinuous channel deposits and ultimately to prodelta deposits, the method is suitable for tectonoenvironmental analysis. The second reason is that many boreholes and a few good underground exposures are easily accessible for sedimentological analysis. Thus, by applying Walther's facies rule in the sense used by Visher (1965), it was possible to reconstruct a sedimentological model of the Dagbreek Formation.

Description of Sequence

The Proterozoic Dagbreek Formation, a clastic wedge-shaped deposit (Fig. 3) overlying a prominent unconformity, is itself, unconformably truncated by the younger Proterozoic Spes Bona Formation. The Dagbreek Formation, in general, is a fining-upward sequence which is subdivided into three members namely the Leader Reef Zone, the Dagbreek Quartzite and the Upper Shale Marker (Fig. 4). The topmost member, the Upper Shale Marker, is truncated to the west and south so that it is absent in the area where the Dagbreek Formation is thinner than 80 m. To some extent the thickening of the formation eastward is also a primary feature.

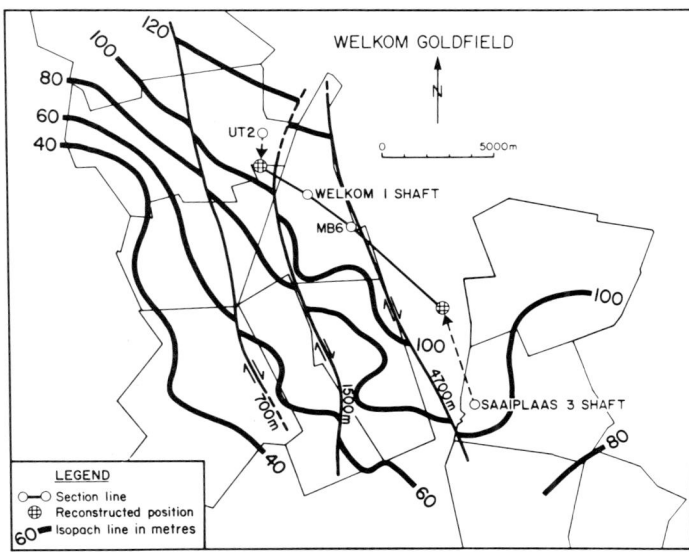

Fig. 3. Isopach plan of the Dagbreek Formation. Note that the position of two of the sections has been restored to compensate for post-depositional dextral shifts.

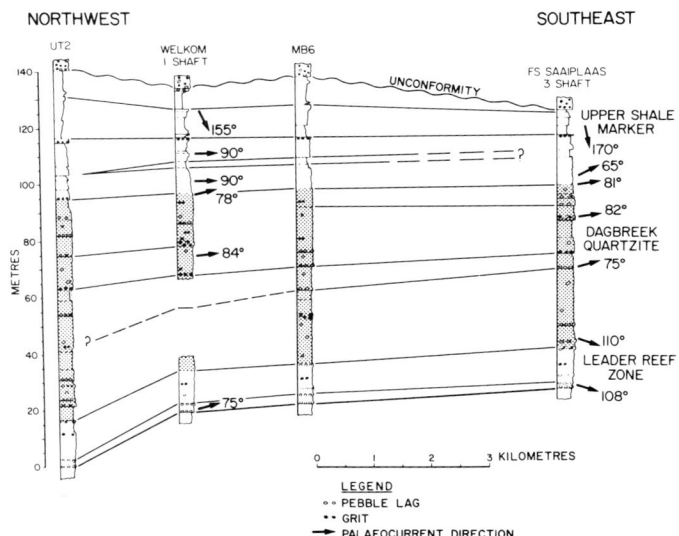

Fig. 4. Correlation diagram of the four investigated sections showing the fining-upward megacycle and the three members of the Dagbreek Formation.

THE LEADER REEF ZONE

The Leader reef is clearly composed of two different deposits. The older oligomictic, texturally mature Alma placer, occurs as a remnant beneath the polymictic Bedelia placer. The Alma placer is a light grey, very pyritic pebbly quartzite which often shows kerogen veneers at the base and, in some cases, also within the deposit. Thin conglomerate lags are occasionally developed. The pyrite occurs as small to medium-sized detrital grains which are arranged in a stringer-like fashion, probably because of the presence of very thin, poorly defined cross-beds. The pebble population consists mainly of vein quartz and black chert. The gold mineralisation is low except where kerogen occurs, in which case fairly high values are recorded.

The Bedelia placer is channelised into the Alma placer and its footwall, and channels may be individual and a few metres in width or multiple and up to 200 m wide. Its thickness usually varies between 1 and 3 m within the channelised areas and 1 m in the inter-channel areas. The highest gold content is confined to conglomerate in the channelised areas and the best gold and other heavy mineral concentration generally occur in clast-supported conglomerate. Quartz, quartzite, a variety of cherts, yellow silicified shale fragments and silicic acid lava dominate the pebble assemblage in the conglomerate. The matrix of the conglomerate and the intercalated reef quartzite varies from being muddy to sandy in the proximal area but the maturity increases downslope so that a very mature deposit, devoid of muddy material, is located at Western Holdings Mine (Saaiplaas Division).

Mud cracks have been identified in muddy silt in the reef horizon at one locality, whereas current ripple marks and poorly preserved oscillation ripples were observed at many localities.

Overlying the Leader reef is a sequence of muddy, gritty, yellowish quartzite (protoquartzite), irregularly interbedded with thin orthoquartzite lenses and polymictic pebble stringers which may be poorly mineralised in places.

The combined thickness of the Leader reef - Leader Reef Zone varies between 12 and 16 m. The top is fairly well defined at the uppermost orthoquartzite (siliceous) bed in the sequence.

THE DAGBREEK QUARTZITE

The Lower Part

The Dagbreek Quartzite is a protoquartzite which has the same appearance as the underlying muddy quartzite, but it is devoid of siliceous beds or gravel. Rare polymictic pebble lags do occur, however, at a few horizons high in the sequence, and the pebble size does not vary much between the investigated sections. The protoquartzite is occasionally gritty with scattered pebbles, and trough cross-bedding, although poorly defined, is very common; sets are up to 20 cm thick and set boundaries are erosive. Planar cross-bed sets are less frequent but where present they are up to 60 cm thick.

Coset units are well-defined and commonly range between 50 cm and 90 cm thick. These units are commonly separated by lenticular siltstone and yellowish shale drapes up to 15 cm thick. It is likely that the original sedimentation units were thin fining-upward sequences some 30 to 70 cm thick but subsequent erosion has cut off the upper parts in most cases.

The Upper Part

The topmost part of the Dagbreek Quartzite is different in several respects from the lower part and represents a transition zone (Fig. 5).

A fairly abrupt textural change takes place vertically within the uppermost 18 to 22 m of the Dagbreek Quartzite in all the investigated sections. The arenite here generally grades vertically from protoquartzite to khaki-colored subgraywacke which is texturally more immature as regards the amount of clay matrix than the underlying sequence. The subgraywacke is generally medium- to coarse-grained while gritty material is scarce and scattered pebbles are rare. Some well-defined, fining-upward units 40 cm to 8 m thick, with black shale intraclasts, occur in this transition zone. These fining-upward units are well displayed in the UT 2 borehole core and to a lesser extent in the section at Saaiplaas Division No 3 Shaft (Fig. 6). Cross-bedding in the subgraywacke is difficult to detect due to the large amount of clay matrix. The gritty quartzite 25 cm thick just underneath the top contact of this member at Saaiplaas Division No 3 Shaft is texturally mature and can almost be termed an orthoquartzite; some pyrite foresets, common at this horizon, occur in it, but are quite exceptional within the Dagbreek Quartzite sequence.

The detailed correlation diagram (Fig. 5) shows that the uppermost 10 m at UT 2, which consists of a fining-upward arenite unit, correlates with two coarsening-upward shale-siltstone units at Welkom Division No 1 Shaft 3 km to the southeast. At the latter locality several load cast structures and slump structures are present, resembling similar structures in the overlying Upper Shale Marker. Asymptotic trough cross-bedding with sets up to several metres wide and 50 cm thick is well developed in the top 5 m of

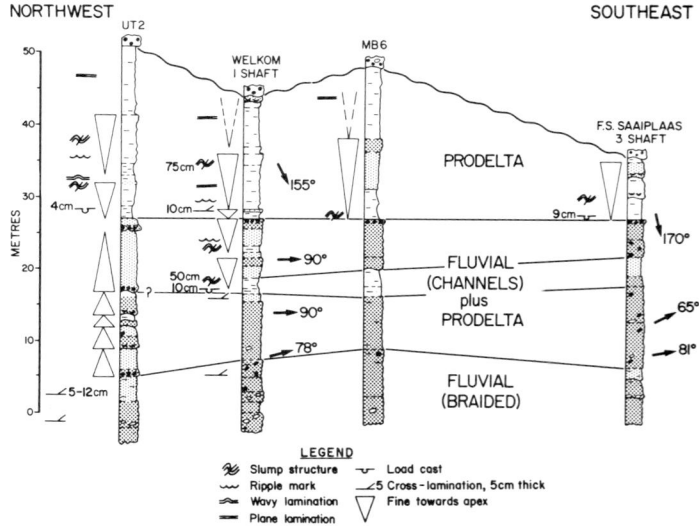

Fig. 5. Detailed lateral variation in the uppermost Dagbreek Quartzite and the Upper Shale Marker. Note the environmental change within the topmost 20 m of the Dagbreek Quartzite.

this member at Saaiplaas Division No 3 Shaft (Fig. 7). The scale and orientation of individual sets are completely different from sets in the underlying cross-bedded arenites, also indicating a significant environmental change.

THE UPPER SHALE MARKER (USM)

Lithofacies and Sedimentary Structures

The transition from the Dagbreek Quartzite to the Upper Shale Marker is sharp and distinctive (Fig. 8). The contact between the arenite and overlying silty yellowish shale is either a smooth and slightly undulating surface capped with a yellow mud veneer of a few millimetres thick over a distance of at least 4 m, or it is a ripple-marked surface.

The outstanding characteristic of the USM is its textural immaturity and finer grain size compared with the sandy facies of the Dagbreek Quartzite. Three main facies are recognised, here called Facies *A*, *B* and *C*.

Black shale laminae and beds of Facies *A* are up to 8 cm thick and separate the silty and sandy facies at a few places (Fig. 8). Yellowish-grey and dark greenish-grey silty shale (Facies *B1*) and siltstone (Facies *B2*) constitute the major part of this member. Both Facies *B1* and *B2* are faintly plane laminated and ripple laminated (Fig. 9). Beds of Facies *B1* vary between a few centimetres and several metres in thickness, while individual beds of Facies *B2* are characteristically 5 to 15 cm thick but may reach 50 cm in some cases. Facies *C* consists of greenish-grey and khaki-colored fine-grained graywacke with a high matrix content; thin coarse-grained laminae, a few millimetres thick, occur sporadically in borehole MB 6. Beds of Facies *C* are usually less than 30 cm thick in UT 2 but reach 80 cm at Welkom Division No 1 Shaft and Saaiplaas Division No 3 Shaft. Although the thicker beds of Facies *C* have a massive appearance they are plane-bedded; the thin beds (*ca.* 10 cm) are dominantly ripple laminated.

Slump structures abound in many beds of Facies *B1* and *B2*. In both underground sections small slump structures, up to 110 cm thick, are common (Fig. 9). Closely associated with these slump structures are load casts varying from 1 to 15 cm in thickness and developed mainly at the bases of Facies *B2*. In some cases small load casts occur

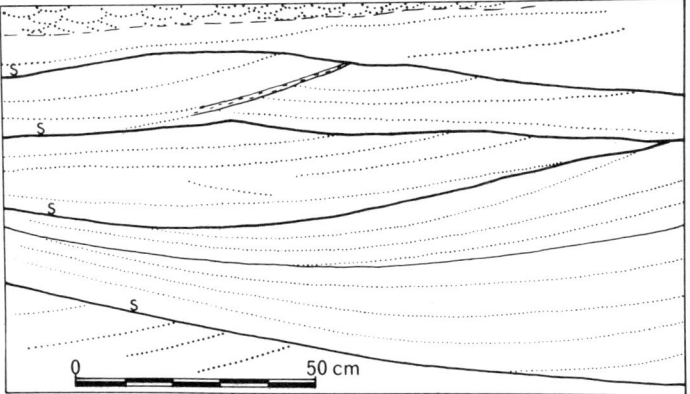

Fig. 7. Sets of low-angle, cross-bedding showing asymptotic transition into bottomsets. Note the thinning of some of the foreset beds (stippled) towards their toes. Tops of individual sets are cut off by scours (S). Note also the mud drape on the minor scour near the top. Small-scale cross-bedding occurs on top.

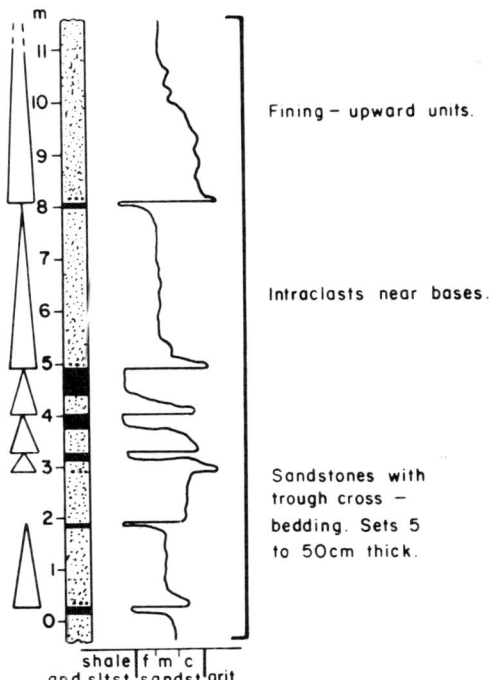

Fig. 6. Fining-upward channel sands capped by yellow silty shale in borehole UT 2. Bottom of section is 17 m underneath the USM.

Fig. 8. Base of USM at white mark. Note the rippled bedding plane. Topmost 30 cm is a subgraywacke and bottom part of photo is an orthoquartzitic grit. Note scour surface and cross-bedding in the topmost sand. Pyrite foresets are faintly visible in lower left corner next to the cm scale. RL denotes ripple lamination and PL plane lamination in facies *B2*.

above a graded bed and on several horizons within the USM thin, normal graded beds with sharp bases are displayed (Fig. 10). Their thicknesses vary between 2 and 5 cm but may reach 20 cm in rare cases. Two to four of these graded beds often occur together and some of them are directly overlain by convolute lamination or poorly developed ripple lamination, in which case they resemble distal turbidites of the Bouma A, C and D beds. In a few cases small load casts occur within the top part of the graded bed.

Facies Sequences

Most typically Facies A, B and C are arranged in crude coarsening-upward sequences 4 to 11 m thick. An individual coarsening-upward unit may have transitional contacts between the facies as at UT 2 or abrupt contacts with shale intraclasts. There are two main coarsening-upward units which are correlatives while a third incomplete cycle is developed on top in UT 2 (Fig. 5).

Facies A occurs predominantly at or near the base of a cycle and forms less than 2% of the sequence. Facies B1 constitutes up to 25% of the lower part of a specific cycle where it is commonly interbedded with Facies B2. The latter is the most common, about 60%, and mainly comprises the upper two-thirds of a specific cycle. Facies C completes the coarsening-upward cycle at the top but thin beds of it are also interbedded with Facies B2 within the upper half of a specific cycle. This facies forms up to 15% of both cycles in UT 2, 44% of the lower cycle at Welkom Division No 1 Shaft and 50% of the lower cycle at MB 6. At the same time this lower cycle increases from 5.2 m at UT 2 to 9 m at Welkom Division No 1 Shaft, 11.4 m at borehole MB 6 and decreases to 8.3 m at Saaiplaas Division No 3 Shaft. It is quite clear that Facies C is responsible for the thickening of this cycle between UT 2 and MB 6.

The unconformity at the top of this marker transgresses the Upper Shale Marker southward so that at Saaiplaas Division only one coarsening-upward unit is preserved. At both underground sections directly beneath the Spes Bona Formation a yellow argillaceous sandstone, with spectacular intraclasts of up to 30 cm across, was deposited in a channel about 1 m deep (Fig. 11).

PALAEOCURRENTS

Palaeoslope indicators in the Leader reef are the orientation of channels and trough cross-bedding within intercalated arenites. In general, the long axes of troughs at any particular locality are roughly parallel to the orientation of channel edges so that both were used in combination to determine the palaeoflow direction.

Fig. 9. Ripple laminated silty shale of Facies *B1* ('RL') overlies plane laminated siltstone of Facies *B2*. A thin slump unit (5 cm thick) overrides the rippled unit. Photo taken a few metres above base of USM.

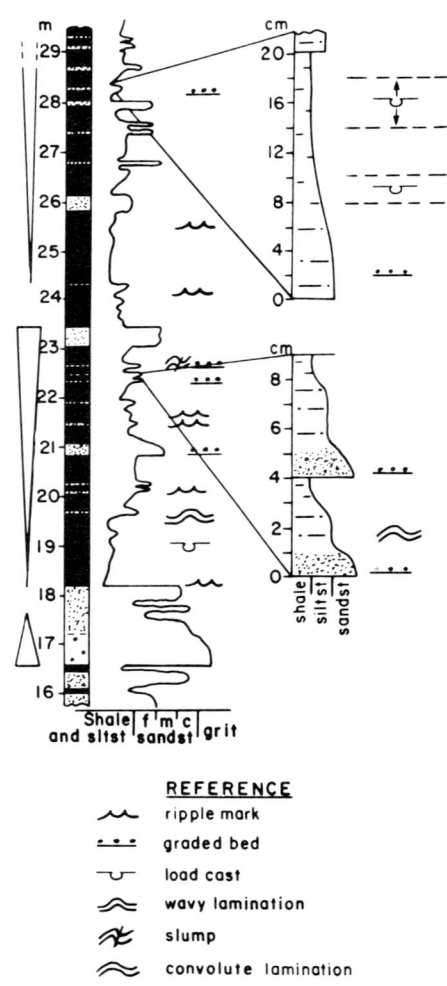

Fig. 10. Base of USM is at 18.2 m. Details of lithofacies and sedimentary structures in lower part of the USM in borehole UT 2. There is about 5 m of section between the bottom in this figure and the top of the section in Figure 6.

Fig. 11. Ripped-up black shale clasts in plane-laminated subgraywacke. Note the sharp top contact (unconformity) and overlying pebbly grit of Spes Bona Formation, Welkom Division, No 1 Shaft.

A trough cross-bed axis is the most reliable feature for measuring a palaeocurrent direction and this feature was measured wherever possible. About one third of the measurements were taken on foresets and the rest were measured from the orientation of these trough axes. Figure 2 depicts 178 palaeocurrent measurements taken at 31 localities on the Bedelia Leader reef. The vector mean for the Bedelia reef is 083° indicating a source to the west. Only 25 measurements taken at 5 localities on the Alma reef gave 107° for the vector mean, possibly indicating a different source area to the north of the Bedelia source.

A second, most reliable palaeoslope indicator regionally is pebble size. The long axes of the ten largest quartz pebbles, measured within 1 m^2 of Leader placer on a stope face, were recorded; the pebble size isopleths on Figure 2 are based on 56 stations, spread mostly over Holdings Division and President Brand Mine. This parameter indicates a northeasterly palaeoslope thus agreeing very well with the general palaeocurrent direction and with the isopach plan.

Stratigraphically higher in the succession, palaeocurrent measurements could only be done on certain horizons. Although only 18 measurements were taken, all of them are reliable. It is quite clear that the palaeocurrent flow remained consistently eastward and east-northeastward throughout the deposition of the Dagbreek Quartzite, but a significant change to the south and southeast occurred at the top of the unit within the gritty quartzite bed (Fig. 4). Low-angle cross stratification in the overlying siltstone at Welkom Division No 1 Shaft dips southward and confirms this trend.

Environmental Interpretation

PROXIMAL AND MID FAN-DELTA PLAIN

As a result of the mining operations the Leader reef provided much more detailed sedimentological data for interpreting its environmental setting than the overlying sequence. The contrast between the Alma and Bedelia reefs is a consequence of different sources. The abundant silicified yellow shale fragments which are characteristic of the Bedelia source are not present in the Alma placer. Although only a few palaeocurrent measurements have been taken in the Alma placer these are sufficient to indicate that it was probably fed from an area situated north of the Bedelia placer's source. Furthermore, the textural maturity (ca. 95% of quartz and chert) of the Alma placer indicates a more prolonged reworking of sediment before final deposition. The thin conglomerate beds, indicating minimal gravel supply, and stringer-like pyrite laminae possibly support this conclusion and suggest long periods of perennial flow, sorting and reworking.

The Bedelia placer, on the other hand, is texturally a more immature deposit which was mainly deposited by rapidly flowing currents in broad channelised areas. Episodes of ephemeral floods with intermittent perennial flow conditions introduced immature or clayey sand, gravel and heavy minerals to the large Leader reef alluvial fan which covered at least 400 km^2. In the channels, the local migration of gravel and sand bars gave rise to the main channelised areas being filled with a complex arrangement of conglomeratic bars and interbedded sands. Several mechanisms were responsible for the heavy mineral concentrations (amongst others gold and uranium) in the gravel; one of them being the 'bank hugger' setting which is well documented in the Leader reef by Smith and Minter (1980). Other aspects of modern braided stream processes which were probably operative during Bedelia deposition include the confluence mechanism of channels (Read, 1982) and entrapment of sand and heavy minerals in open framework gravels (Smith, 1974; Steel and Thompson, 1983).

The overlying immature sands and thin lenticular pebble beds, known as the Leader Reef Zone, are spread as thin sheet-like deposits over the Leader reef, during a time of less confined, mostly perennial flow, similar to the conditions described by Minter (1978). At times some gritty sands were reworked sufficiently to produce lenticular well sorted deposits within the lower 16 m of the Dagbreek succession.

The 70 to 90 m of overlying muddy grits, sands and rare pebble lags were derived from the same source area to the west and accumulated relatively rapidly in shallow unconfined braided channels. The vertical maintenance of uniform grain size and other features suggest a fine balance between aggradation and subsidence. The abundance of trough cross-bedding, the occasional planar cross-beds and the absence of major erosional surfaces indicate continuous stacking of braided stream sands over a long period of time.

LOWER FAN-DELTA PLAIN

A sharp change in environmental conditions took place during sedimentation of the interval located about 10 to

20 m below the USM. Here there are typical fining-upward units interpreted as channel fill deposits. In addition, large, low-angle trough cross-bedding, with basal intraclasts and silty and muddy cappings with ripple lamination all support more confined channel sands than the stacked braided sands underneath. They are interpreted as fluvial sands which formed further down the depositional dip within distinct channels while the associated muds and silts are interpreted as overbank and abandoned channel-fill deposits. These lower fan-delta fluvial sediments locally grade into subaqueous deposits identical to those of the USM (compare sections of UT 2 and Welkom Division No 1 Shaft, Fig. 5). This facies association suggests that fluvial channels discharged directly into a lake with very little reworking by waves.

This uppermost sequence signifies a retreat of the alluvial fan with a concommitent transgression of the fan-delta shoreline. This event is strongly suggested by the presence of fluvial deposits at UT 2, while at the same time shallow deltaic deposition took place at Welkom Division No 1 Shaft. This first stage of the transgression took place either as a result of a relatively rapid subsidence or of a marked decrease in source sediment supply. Environmentally, this transition in the sequence may be interpreted in terms of sedimentation close to the fan-delta shoreline.

LOWER SUBAQUEOUS FAN-DELTA

The texturally mature grit at the top of the Dagbreek Quartzite is interpreted as the product of low-energy, wave action along the lacustrine shoreline on the fan. The very thin and patchy development of this facies indicates impersistent or insignificant wave activity. Reworking by these low-energy waves on the shoreline of the Dagbreek Fan Delta took place probably only during a short period before increased subsidence or decreased sediment supply caused a sharp transgression. In this uppermost sandy unit one would expect to find some evidence of tidal deposits but due to limited exposures this could not be established. Eastward-flowing fluvial currents on the lower fan-delta plain were diverted southward possibly as longshore currents. This change in direction is already noticeable in the mature grit and, therefore, the uppermost metre or two of the sandy unit represents a transgressive deposit.

The most striking differences between the arenaceous Dagbreek Quartzite and the muddy Upper Shale Marker are the coarsening-upward sequences, slump structures, load casts, graded bedding, plane bedding and ripple lamination present in the latter but absent in the former.

The sandier coarsening-upward units at MB 6 and Welkom Division No 1 Shaft, combined with their greater thickness compared to the less sandier unit to the northwest, signify a higher deposition rate than at UT 2.

The abundance of slump structures points to unstable slope conditions, while the common occurrence of load casts is indicative of rapid deposition. Interfingering of small ripple-laminated siltstone and thin graded beds points to subaqueous sedimentation. The ripple laminated beds were the result of wave action and the graded beds may be attributed to small density flows which could have originated in shallow water as the result of storm events. All these features suggest prodelta deposits rapidly accreting on unstable slopes. The maximum water depth during deposition was probably of the order of 5 to 10 m, as derived from the thickness of individual cycles. The recurrence of cycles might have been inherent in the depositional system with successive cycles produced by switching of major distributaries on the lower fan-delta, during progradational episodes. At least three such progradations occurred. The youngest progradation was probably the most active as suggested by the large black shale clasts on Saaiplaas Division No 3 Shaft and Welkom Division No 1 Shaft indicating high erosive energy. This event may have marked the first pulse of renewed uplift in the source area.

It is more difficult to interpret the plane-laminated sands of Facies C occurring towards the top of the coarsening-upward cycles. These sand beds may represent flood events characterised by high sediment discharge. They were probably fluvially dominated sands deposited near the subaqueous delta front.

Discussion and Conclusions

Bull (1972) recognised three main types of fans based on radial or longitudinal cross-section. His third type is wedge-shaped, thin and coarse-grained adjacent to the mountain front and thicker away from it, a type which seems to fit the tectonic setting of the Dagbreek fan-delta (Fig. 12). This type reflects a lengthy interval of source erosion and retreat combined with retrenchment and redistribution of proximal fan deposits farther downfan.

The broad fining-upward megacycle of the Dagbreek Formation may be attributed to a single tectonic uplift followed by a long period of erosion, denudation, sedimentation and associated subsidence of the clastic wedge. Fining-upward megacycles may result from waning sediment supply due to the wearing down of a source area without tectonic rejuvenation. Although the influence of climate is difficult to determine, this factor could also have played a role in reducing sediment supply. If alluvial fans are dependent solely on initial erosive or tectonic topography, only relatively thin fining-upward cycles accumulate (Selley, 1965; Laming, 1966; Williams, 1969; Steel, 1974); this is the case with the Dagbreek fan. The gradual reduction of the highland and continuous retreat of the source area supplied a decreasing amount of debris and lowering of the slope (Heward, 1978, Fig. 6b). In the case of the Dagbreek fan the retreat of the source area was accompanied by the sourceward migration of the locus of sedimentation, causing transgression and lacustrine deposition on top of the fluvial sequence. This was the result of a faster subsidence and/or decrease in sediment supply. The best ancient analogue for the Dagbreek fan-delta is probably the Perry

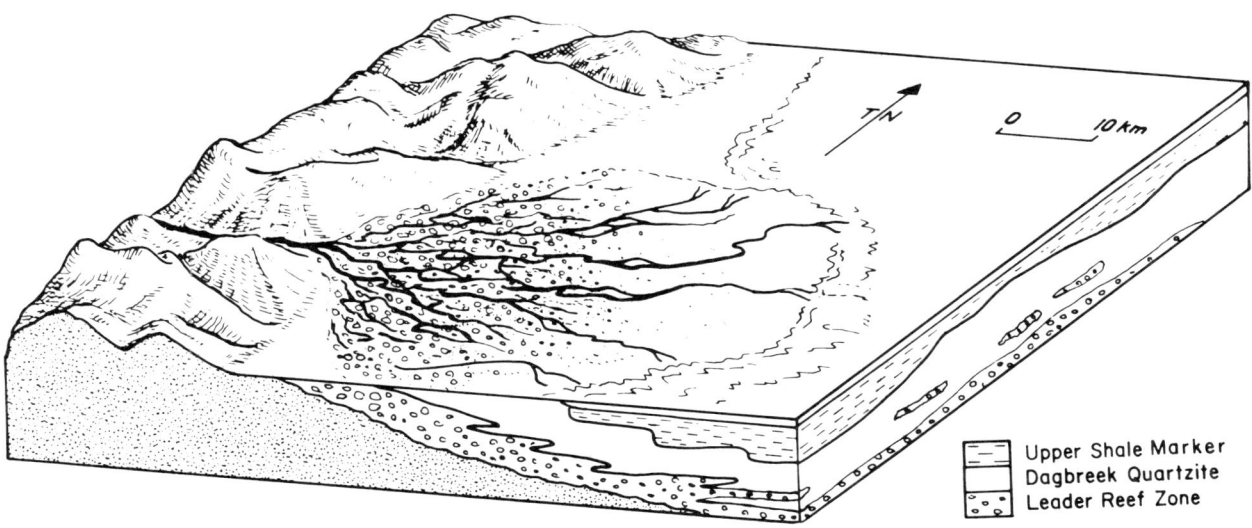

Fig. 12. Environmental model of the Dagbreek fan-delta. Note the single fluvial channels in the lower fan area.

Formation, New Brunswick, described by Schluger (1973). In Heward's terminology the Dagbreek sequence can therefore be explained as the product of 'a geomorphological cycle of erosion' between two rejuvenation events.

The conceptual model proposed by Pretorius (1975) for Witwatersrand placer formations has, as a principal component, a border fault, suggesting a fault scarp which was rejuvenated periodically to produce the megacycles Unlike many other fan-deltas described by Heward (1978), there is no evidence for an active fault scarp in the Dagbreek fan setting. Therefore, Pretorius's model does not fit the tectono-environmental model for the Dagbreek fan. The explanation by Snowden (pers. comm., June 1983) seems more feasible. He proposed compression from the west over an extended period resulting in progressive folding, and thrusting, to the east which gave rise to periodic rejuvenation of the source area on a stable craton.

On the Dagbreek fan the braided, low-sinuosity channel pattern changed downfan possibly to less braided and more single channels in the lower fan area. In this sense it is analogous to the Reno Fan (Ori, 1982). The lower fan graded directly into the subaqueous toe slopes which were locally unstable (slumping) beneath only 5 to 10 m of water. The thin sedimentation cycles in this area probably reflected avulsion of channels in the immediate upstream position from the prodelta area of the lake. Longshore currents probably played a role in diverting the eastward palaeoflow in a southerly direction. The shoreline of the fan-delta was well defined but wave action was weak while microtides could have been present but not indicated in all exposures. The fan was thus fluvially dominated.

It is difficult to point to any particular modern analogue to the Dagbreek fan-delta because most modern fan-deltas are associated with faults scarps (Tanner, 1976).

One reasonable comparison can be made with the Gulf of Aquaba fan-deltas (National Geographic, April 1982) on the east coast of the Sinai Peninsula. Although the size of this fan is much smaller than the Dagbreek fan, the braided pattern and low wave energy of the shoreline are consistent features.

References

Bull, W.B. 1972. Recognition of alluvial fan deposits in the stratigraphic record. *In*: Rigby, J.K. and Hamblin J. (Eds.), Recognition of Ancient Sedimentary Environments. Society of Economic Palaeontologists and Mineralogists, Special Publication 16, p. 63-83.

Callow, M. 1982. Facies variation and heavy mineral distribution within an Upper Witwatersrand auriferous placer deposit. 11th International Sedimentological Congress, Hamilton, Canada. Abstracts of Papers, p. 20.

Eriksson, K.A., Turner, B. R. and Vos, R. G. 1981. Evidence of tidal processes from the lower part of the Witwatersrand Supergroup, South Africa. Sedimentary Geology, v. 29, p. 309-325.

Gjelberg, J. and Steel, R.J. 1983. Middle Carboniferous marine transgression, Bjornoya, Svalbard: facies sequences from an interplay of sea level changes and tectonics. Geological Journal, v. 18, p. 1-19.

Gloppen, T. G. and Steel, R.J. 1981. The deposits, internal structure and geometry in six alluvial fan - fan delta bodies (Devonian-Norway) - a study in the significance of bedding sequence in conglomerates. *In*: Ethridge, F.G. and Flores, R.M. (Eds.), Non-Marine Depositional Environments: Models for Exploration. Society of Economic Paleontologists and Mineralogists, Special Publication 31, p. 49-69.

Gnaccolini, M. 1981. Oligocene fan-delta deposits in northern Italy: a summary. Rivista di Italiana Paleontologia e Stratigrafia, v. 87, p. 627-636.

Heward, A.P. 1978. Alluvial fan sequence and megasequence models, with examples from Westphalian D - Stephanian B coalfields, northern Spain. *In*: Miall, A.D. (Ed.), Fluvial Sedimentology. Canadian Society of Petroleum Geologists, Memoir 5, p. 669-702.

Laming, D.J.C. 1966. Imbrication, paleocurrents and other sedimentary features in the Lower New Red Sandstone, Devonshire, England. Journal of Sedimentary Petrology, v. 36, p. 940-959.

Larsen, V. and Steel, R.J. 1978. The sedimentary history of a debris flow-dominated alluvial fan - a study of textural inversion. Sedimentology, v. 25, p. 37-59.

Minter, W.E.L. 1978. A sedimentological synthesis of placer gold, uranium and pyrite concentrations in Proterozoic Witwatersrand sediments. In: Miall, A.D. (Ed.), Fluvial Sedimentology. Canadian Society of Petroleum Geologists, Memoir 5, p 801-829.

Ori, G.G. 1982. Braided to meandering channel patterns in humid-region alluvial fan deposits, River Reno, Po Plain (Northern Italy). Sedimentary Geology, v. 31, p. 231-248.

Pretorius, D.A. 1975. The depositional environment of the Witwatersrand goldfield: a chronological review of speculations and observations. Minerals Science and Engineering, v. 7, p. 18-47.

Pretorius, D.A. 1976a. Gold in Proterozoic sediments of South Africa: systems, paradigms and models. In: Wolf, K.H. (Ed.), Handbook of Strata-Bound and Stratiform Ore Deposits. New York: Elsevier, v. 7, p. 1-27.

Pretorius, D.A. 1976b. The nature of the Witwatersrand gold-uranium deposits. In: Wolf, K.H. (Ed.), Handbook of Strata-Bound and Stratiform Ore Deposits. New York: Elsevier v. 7, p. 29-88.

Read, S.E. 1982. Concentration of heavy minerals in braided channels: the effect of convergent flow. M.Sc. Thesis, University of Illinois at Chicago Circle, Illinois, 79p.

Ricci Lucchi, F., Colella, A., Ori, G. G., Ogliani, F. and Colalongo, M.L. 1981. Pliocene fan deltas in the Intra-apenninic Basin, Bologna. 2nd European Regional Meeting, International Association of Sedimentologists, Excursion Guidebook, p. 79-162.

Scholle, P.A. and Spearing, D. (Eds.), 1982. Sandstone Depositional Environments. American Association of Petroleum Geologists. 410p.

Schluger, P.R. 1973. Stratigraphy and sedimentary environments of the Devonian Perry Formation, New Brunswick, Canada, and Maine, U.S.A. Geological Society of America Bulletin, v. 84, p. 2533-2548.

Selley, R.C. 1965. Diagnostic characteristics of fluviatile sediments of the Torridonian Formation (Precambrian) of northwest Scotland. Journal of Sedimentary Petrology, v. 35, p. 366-380.

Smith, N.D. 1974. Sedimentology and bar formation in the Upper Kicking Horse River, a braided outwash stream. Journal of Geology, v. 82, p. 205-223.

Smith, N.D. and Minter, W.E.L. 1980. Sedimentological controls of gold and uranium in two Witwatersrand paleoplacers. Economic Geology, v. 75, p. 1-14.

Steel, R.J. 1974. New Red Sandstone floodplain and piedmont sedimentation in the Hebridean province, Scotland. Journal of Sedimentary Petrology, v. 44, p. 336-357.

Steel, R.J. and Thompson, D.B. 1983. Structures and textures in Triassic braided stream conglomerates ('Bunter' Pebble Beds) in the Sherwood Sandstone Group, North Staffordshire, England. Sedimentology, v. 30, p. 341-367.

Tanner, W.F. 1976. Tectonically significant pebble types: sheared, pocked, and second-cycle examples. Sedimentary Geology, v. 16, p. 69-83.

Visher, G.S. 1965. Use of vertical profile in environmental reconstruction. American Association of Petroleum Geologists Bulletin, v. 49, p. 41-61.

Vos, R.G. 1975. An alluvial plain and lacustrine model for the Precambrian Witwatersrand deposits of South Africa. Journal of Sedimentary Petrology, v. 45, p. 480-493.

Williams, G.E. 1969. Characteristics and origin of a Precambrian pediment. Journal of Geology, v. 77, p. 183-207.

WAVE/TIDE-DOMINATED SYSTEMS

WAVE-WORKED CONGLOMERATES — DEPOSITIONAL PROCESSES AND CRITERIA FOR RECOGNITION

JOANNE BOURGEOIS AND ELANA L. LEITHOLD[1]

ABSTRACT

Wave-worked conglomerates are preserved in progradational and transgressive sequences typically associated with tectonically active coastlines and high-energy wave climates. These settings are commonly subjected to sporadic, intense storm activity, leaving a record that may either eliminate fair-weather deposits or may juxtapose sediments deposited in the shifting dynamic zones of the nearshore.

Biogenic features associated with wave-worked conglomerates not only help to distinguish marine from non-marine conglomerates, but they also help elucidate the processes that operate in nearshore zones. The study of fossil hard parts, including their taphonomy, can contribute to detailed paleoecologic analysis of conglomeratic sequences. Trace fossils are commonly a good indicator of the depositional and erosional intensity of the nearshore zone.

Beach gravels and conglomerates, deposited in the swash zone, are typified by well-defined layers of well-sorted, imbricated, disc-shaped pebbles. Detailed knowledge of sorting by size and shape of beach clasts, both across shore and along shore, has yet to be applied to (ancient) beach conglomerates.

Upper shoreface conglomerates are typified by crudely graded, tabular beds and trough cross-bedded pebbly sandstone. Lower shoreface conglomerates are characterized by low-angle-stratified pebbly sandstone and lenticular conglomerates. Breaking waves, surf, rip currents and longshore currents are the major processes that control their deposition, but the interaction and effect of these processes, as well as of tides, river mouths, and biologic activity, on nearshore gravels, are not yet well understood.

RÉSUMÉ

Des conglomérats maniés par les vagues sont préservés dans des séquences progradationnelles et transgressives typiquement associés avec les lignes côtières tectoniquement actives et les climats qui produisent des vagues de grande énergie. Ces contextes sont fréquemment le lieu de fortes tempêtes sporadiques, lesquelles peuvent effacer du registre géologique les dépôts de climats plus cléments ou mener à une juxtaposition des sédiments accumulés dans les zones dynamiques mouvantes du littoral.

Les particularités biogéniques associées avec les conglomérats maniés par les vagues n'aident pas seulement à distinguer entre les conglomérats marins et non-marins, mais elles facilitent la compréhension des processus actifs dans les zones littorales. L'étude des parties durcies des fossiles, incluant leur taphonomie, peut fournir des enseignements concernant les analyses paléoécologiques des séquences conglomératiques. Les empreintes fossiles sont en général un bon indicateur du taux de sédimentation et d'érosion dans la zone littorale.

Les graviers des plages et les conglomérats, déposés dans la zone de clapotement des vagues, sont caractérisés par des couches bien définies de galets bien triés, imbriqués et discoïdes. La connaissance détaillée du triage selon la grosseur et la forme des fragments, aussi bien au travers de la rive que le long de la rive, est encore à faire sur les conglomérants (anciens) de plage.

Les conglomérats de la zone infratidale supérieure sont caractérisés par un granoclassement rudimentaire, des couches tabulaires, et des creux remplis de grès cailloutex à stratification oblique. Les conglomérats de la zone infratidale inférieure sont caractérisés par des grès cailloutex en couches faiblement inclinées et des conglomérats lenticulaires. Les vagues déferlantes, le ressac, les courants d'arrachement et les courants de dérive littorale constituent les principaux processus qui contrôlent l'accumulation des sédiments, mais l'interaction et l'effet de ces processus, tout autant que ceux des marées, des embouchures de rivière, et de l'activité biologique, sur les graviers littoraux, ne sont pas encore bien compris.

INTRODUCTION

Conglomerates deposited in nearshore, high-energy, wave-dominated environments have received only limited attention, whereas models for fluvial and deep-sea deposition of conglomerates are relatively well established. With the growing interest in fan-delta sequences and in shelf deposition in wave-dominated settings, we expect significant advances to be made in developing models for gravel deposition in the nearshore. A major stumbling block, however, is the extreme difficulty in observing and sampling modern coastal environments where coarse debris is being reworked and deposited beneath waves. Furthermore, our knowledge of the complex interaction of shoaling and breaking waves, tides, river-mouth processes, and biologic activity in such settings is in its infancy.

In this paper we outline what is known of conglomerate deposits in beach and nearshore settings, with illustrations

---

[1]Department of Geological Sciences, University of Washington, Seattle, Washington 98195, U.S.A.

Funding for parts of this study were provided by the donors to the American Chemical Society Petroleum Research Fund (JB), the University of Washington Graduate School Research Fund (JB), a Geological Society of America Research Grant (ELL), a Sigma-Xi Grant-in-Aid of Research (ELL), and the University of Washington Department of Geological Sciences Corporation Fund (ELL). We would like to thank J.D. Smith and R.J. Steel for advice in the field and for valuable discussions afterward, and R.J. Steel and A. Nøttvedt for critically reviewing the manuscript. Others who have offered advice on aspects of this study and to whom we are grateful include W.O. Addicott, W.T. Fox, R.E. Hunter, R.D. Kreisa and V.S. Mallory.

Copyright © 1984, Canadian Society of Petroleum Geologists

from the literature and from our own studies of coarse clastic sequences, principally in California and Oregon (Bourgeois and Leithold, 1983). Our most detailed examples and illustrations will come from the Sandstone of Floras Lake (Miocene, SW Oregon - Leithold and Bourgeois, 1983; Fig. 1) and the lower Cape Sebastian Sandstone (Campanian, SW Oregon - Bourgeois, 1980, Fig. 2), as well as from Pleistocene terraces (see also Clifton, 1973; Hunter, 1980). Excluded from this review are fluvial-dominated and tide-dominated settings: for the latter see Phillips (this volume) - for the former, see papers in this volume on ancient fan-delta systems, especially Kleinspehn *et al.*

During the preparation of this review, we have noted that much more has been published on modern beach gravels than on ancient beach conglomerates, and that the opposite is true of shoreface deposits. There are two major reasons for this discrepancy. First, modern beach gravels are much more easily observed than shoreface gravels; second, beach deposits are less likely to be preserved in the geologic record than shoreface deposits. Our review, as a result, is skewed toward modern beach gravels and ancient shoreface conglomerates.

TERMINOLOGY

Beach, foreshore, beachface, shoreface, nearshore—the terminology for shallow-marine environments is complex and overlapping (see Fig. 3). In this paper we will use the following terms, all with reference to surfaces underlain by unconsolidated, cohesionless sediment. The *backbeach* is the coastal zone above mean high high water (defined as the average of the higher of the two daily high tide levels), affected by swash washover and storm surges; eolian and fluvial processes operate at other times. The *berm crest* is a positive-relief, constructional feature separating the backbeach from the beachface. It is built up by deposition of sediment when swash washes up over the beachface onto the backbeach, decelerates, and sinks into the unsaturated sediment. The *beachface* extends from mean high high water to mean low low water; it is affected by swash and backwash, surf, rip currents, and longshore currents, as well as by periodic subaerial exposure during low tide. Shore-parallel ridges within this zone are called *ridge-and-runnel*. The *shoreface* extends from mean low low water to the point where waves first break; it is critical to recognize that this area, which includes the breaker and (part of the)

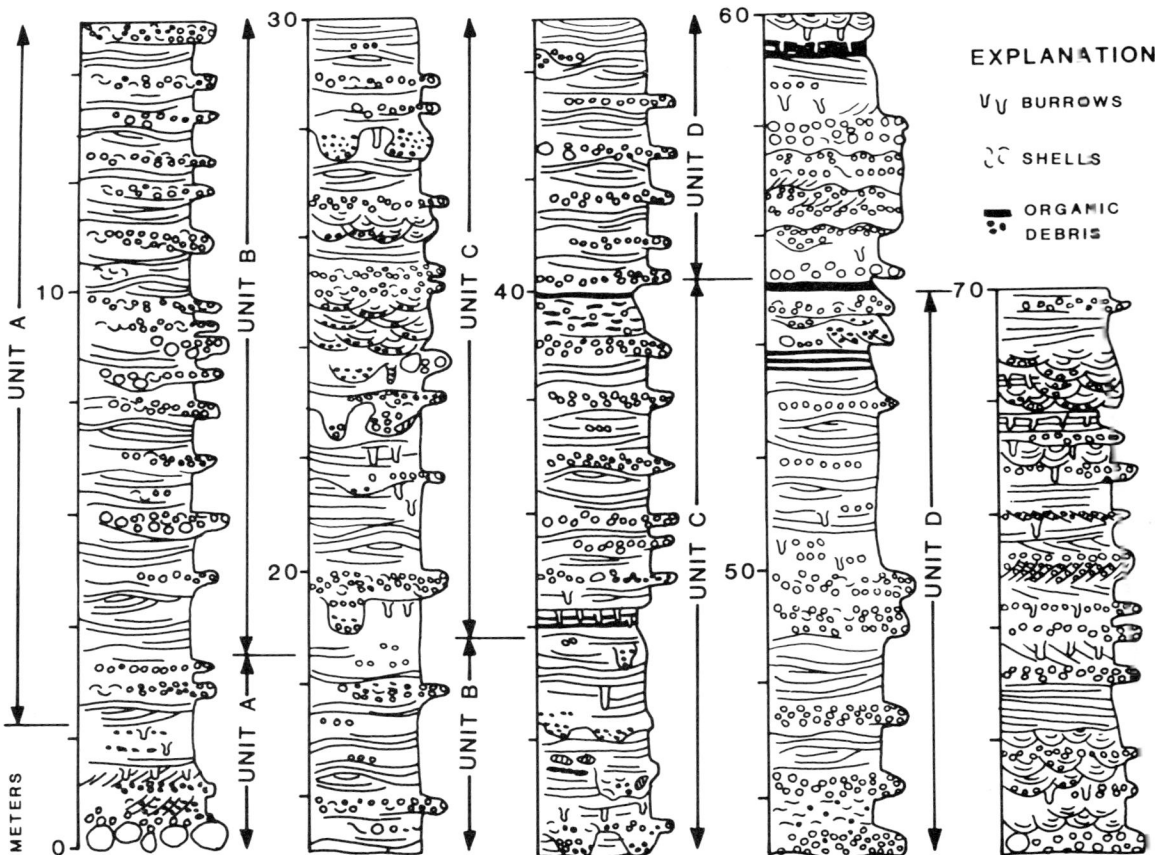

Fig. 1. Generalized measured stratigraphic section of the conglomeratic sandstone (lower 70 m) of the Sandstone of Floras Lake (Leithold and Bourgeois, 1983 and in press), Miocene, SW. Oregon. Units A and C are interpreted as lower shoreface and transition-zone deposits; Units B and D are principally upper shoreface deposits.

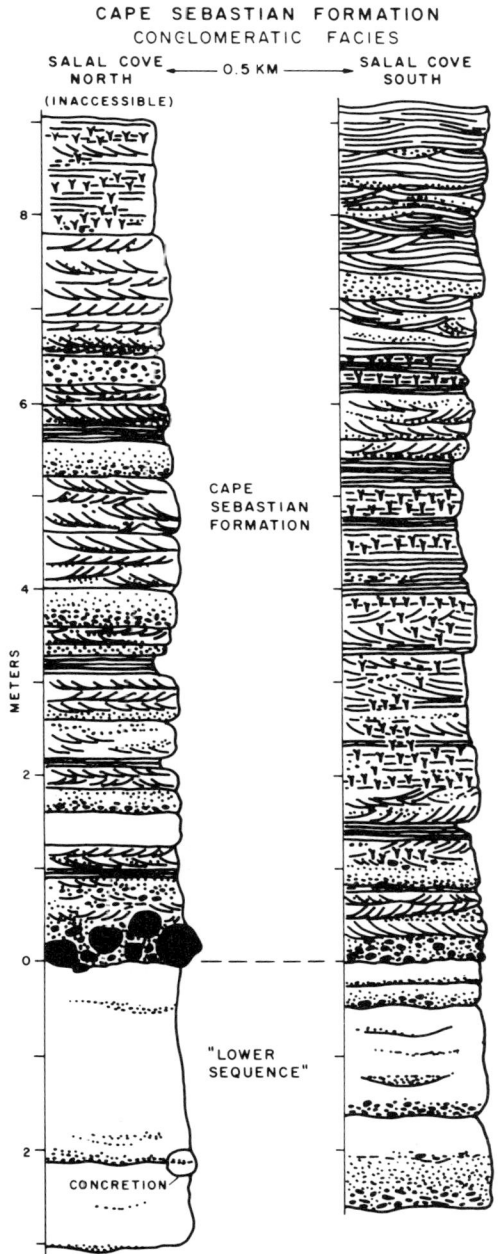

Fig. 2. Two measured sections of the lower 9 m of the Cape Sebastian Sandstone (Bourgeois, 1980), Campanian, SW. Oregon. This transgressive sequence represents a progression from upper shoreface to (southern section) transition-zone deposits.

surf zones, is much wider during storms than during fair-weather conditions, and thus the shoreface has no specific lower depth limit. Breaking waves, bore, and longshore and rip currents are active in the shoreface zone. The shoreface/offshore transition is the zone affected by wave build-up; it has no distinct boundaries. Biologic processes become important in this zone. *Bars* are coast-parallel, low-relief, linear ridges in the shoreface and transition zones. The *offshore* is the zone characterized by oscillatory wave motion; coarse sediments are rarely transported to this zone on the shelf, which will not be included in the following discussion.

Some authors (*e.g.*, Clifton *et al.*, 1971; Hunter *et al.*, 1979; Dupre *et al.*, 1980) in describing modern coastal processes and their deposits have chosen to use dynamic-zone terminology (*e.g.*, surf zone) rather than geomorphic terminology (*e.g.*, shoreface). Although their usage has advantages, in that 'shoreface' may suggest a narrow, topographically restricted zone, shoreface-type deposits in the geologic record will typically respresent a mix of several zones and processes (*e.g.*, surf, rip currents, longshore currents), and thus the term 'surf-zone deposits' may also be misleading. We have, therefore, chosen to use the more general terms.

One of the major controls on nearshore, high-energy deposition of conglomerates is storm activity and associated large waves. During storms, waves will break much farther offshore than during fair weather; unusually large waves may also approach the nearshore during fair weather. Storms and associated waves vary in intensity as well, so the zone of breaking waves will shift in position and change in width from storm to storm, with intervening periods of fair weather. Thus the depositional features associated with the lower beachface, with the upper and lower shoreface, and with the transition zone may be complexly interfingered in the stratigraphic record.

THE STRATIGRAPHIC RECORD

Relatively thick accumulations of shallow-marine conglomerate deposited in wave-dominated environments are preserved in both progradational and transgressive sequences. Repetitive coarsening-upward, conglomeratic, progradational sequences punctuated by omission surfaces of thin lag deposits representing transgression and erosion of the beach and upper shoreface have been recently

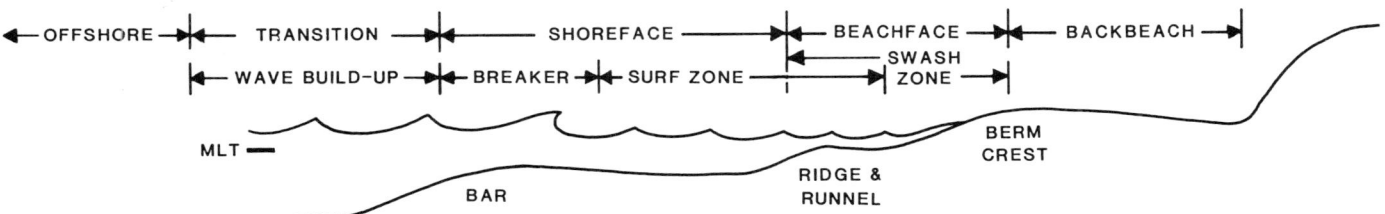

Fig. 3. Terminology of nearshore zones and features used in this paper.

described by Nilsen (1973), Clifton (1981), and Wright and Walker (1981). Leckie and Walker (1982) describe similar repetitive sequences, which are overlain by fluvial and beach conglomerates. Thick transgressive sequences, in some cases with pulses of progradation, that contain wave-worked conglomerates have been described by Bourgeois (1980), Bourgeois and Leithold (1983), and Leithold and Bourgeois (in press). In most of these examples, backbeach and beachface conglomerates are rare, thin, or absent, destroyed by active wave erosion of the shore during transgression. Beach conglomerates are rare, overall, in the geologic record, but there are some good examples preserved in emerged Pleistocene terraces (Clifton, 1973; Kumar and Sanders, 1976; Hunter, 1980; Dupre et al., 1980).

In progradational sequences, several workers have interpreted paleodepth as the stratigraphic thickness below a datum in strata inferred to have been deposited at sea level (e.g., Dupre et al., 1980; Clifton, 1981). The use of this technique is possible in simple prograding sequences, although subsidence and eustatic sea-level change may produce a sequence thicker than the total relief represented by various facies, and even in simple cases, nearshore facies may be interfingered (as discussed above). In transgressive sequences, as well as in prograding sequences in rapidly subsiding basins (e.g., the Scotia Bluffs Sandstone; Bourgeois and Leithold, 1983), these paleodepth reconstructions are not possible.

Deposits of similar appearance from subenvironments of the nearshore (e.g., inner and outer rough facies, see Dupre et al., 1980) have sometimes been distinguished by their relative position in a regressive ('shoaling upward') sequence. Again, use of this technique is possible only in very simple progradational packages, and care must be taken to identify vertical facies changes that are due to secular variations in storm intensity as deposition occurred.

MARINE OR NON-MARINE?

Significant accumulations of shallow-marine conglomerates are typically (almost invariably) associated with fluvial conglomerates because active fluvial systems are necessary to supply an abundance of coarse sediment to the coastline. (Rarely, glaciers or coastal erosion could be the supplier of coarse sediment.) Thus it is critical to be able to distinguish fluvial from shallow-marine conglomerates, before one can examine the characteristics of subenvironments of the coastal zone. Clifton (1973) reviewed some of the criteria used to distinguish wave-worked versus alluvial gravel and proposed that pebbles in wave-worked gravels are better segregated into discrete beds, and bedding in these gravels is laterally more regular (less lenticular) than in alluvial gravel. Other physical characteristics that have been used include clast imbrication, pebble shape and other textural parameters, and relative preservation of resistant clasts (see Clifton, 1973, for review; also Stratten, 1974; Clifton, 1981; Leckie and Walker, 1982; Leithold and Bourgeois, 1983). However, where available, the most diagnostic criterion is faunal evidence.

BIOGENIC FEATURES ASSOCIATED WITH WAVE-WORKED CONGLOMERATES

In shallow-marine conglomeratic sequences, macro-fossils are most commonly preserved within conglomeratic beds and lenses or along scoured horizons below sandstone beds. These accumulations are the result of concentration and reworking of faunal debris by high-energy events, and they typically comprise mixed assemblages. Faunal diversity is generally low in high-energy, nearshore environments where conglomerates accumulate, reflecting a limited number of benthic species adapted to living on an unstable substrate, but population density may be quite high. Bivalve and gastropod fossils typically have thick, robust shells (Fig. 4). Shells and shell fragments from the few thin-

Fig. 4. Conglomeratic nearshore deposits in the Cretaceous Hornbrook Formation, northern California, with numerous, redeposited, thick-shelled fossil molluscs. Scale bar in each photograph is 15 cm long.

shelled species living in the mobile substrate are less commonly preserved, but might be present in fossil accumulations deposited rapidly during high-energy events and not subsequently reworked.

Accumulations of fossil hard parts in beach deposits are rare. Where present they typically represent a mixture of fauna from several environments; shells from many parts of the nearshore may be thrown onshore during storms. These shells may, like lithic clasts, be sorted (by size and shape) by swash and backwash processes (see below).

A shoreface fossil assemblage is illustrated in the lower, most conglomeratic portion of the Sandstone of Floras Lake (Fig. 2; Addicott, 1980; Leithold and Bourgeois, 1983, in press). The association of *Mytilus*, *Balanus* and *Nucella* in the unit indicates deposition at depths less than 10 m in proximity to a rocky shoreline. *Mytilus* and *Balanus* are epifaunal, suspension-feeding organisms that have very efficient means of attachment and are therefore well-adapted to life on a rocky substrate in turbulent water. *Nucella* is a common intertidal gastropod on modern rocky coastlines. It is distinguished by its thick, spired shell, which is capable of resistance to surf. *Nucella* is a very efficient predator that most commonly feeds on barnacles (*Balanus*). Sand-dwelling organisms, most of which are highly efficient burrowers, are also common in the lower portion of the Sandstone of Floras Lake, suggesting a mix of rock- and sand-bottom communities as on an irregular coastline with rocky headlands and sandy coves.

Biogenic structures are relatively rare in shallow-marine, conglomeratic sequences. This scarcity may be attributed to inhospitability of the environment to burrowing organisms, to rapid sedimentation rates, and/or to lack of preservation of traces due to scouring and amalgamation of storm-deposited beds. Additionally, as pointed out by Howard (1978), bioturbation by very small organisms, such as amphipods, may be pervasive yet not apparent in coarse-grained sediments because the minute biogenic structures produced do not significantly alter the primary physical record. Relatively large organisms may live in gravel and not leave an identifiable record. Those burrows that are present and apparent in shallow-marine conglomeratic sequences are typically vertical, cylindrical or U-shaped dwelling structures of the *Skolithos* ichnofacies (Seilacher, 1967; Frey and Seilacher, 1980). Escape structures, characterized internally by protrusive or retrusive *spreiten* are common, reflecting the response of benthic organisms to rapid aggradation or degradation of the substrate in an unstable environment.

Trace fossils are uncommon in beach deposits, particularly conglomeratic ones. In sandstones associated with conglomeratic beach deposits, traces may be present and even common. *Macaronichnus segregatus* (Clifton and Thompson, 1978) is characteristic of sandstone deposited at the base of the beachface (see below). On the beachface itself, animals adapted to moving up and down the beachface with change in tide are successful. For example, *Donax* is a modern clam that lives in the sandy swash zone; it emerges with each change in tide and re-establishes itself approximately in the middle of the swash zone. The sand crab *Emerita* also lives in the swash zone and can bury itself quickly or emerge from the substrate in response to rapid changes in the beachface. In backbeach sediments, insect, crustacean, vertebrate, and plant traces may be present.

Diversity of biogenic structures in shoreface conglomeratic sequences is generally low, but traces may be locally abundant, particularly in sandstone interbedded with the conglomerate. Vertical, passively filled burrows of the ichnogenus *Ophiomorpha*, which have been compared to structures made by modern Callianassid shrimp (Weimer and Hoyt, 1964; Frey *et al.*, 1978), are particularly common. Small (average 4 mm in diameter), sinuous, circular (in cross-section) burrows termed *Macaronichnus segregatus* (Clifton and Thompson, 1978) have been described from sandstone in a number of conglomeratic nearshore units (Hunter, 1980; Clifton, 1981; Leckie and Walker, 1982). These burrows are lined with biotite, presumably as the result of segregation by a deposit-feeding organism. They have been compared to burrows made by the modern polychaete *Ophelia limacina*, which lives in intertidal and shallow subtidal sands (Clifton and Thompson, 1978). Large, funnel-shaped burrows, which are typically filled with pebbles and sand and are ten to a few tens of centimeters deep and wide, may also be common (Fig. 5). The burrows typically taper downward to a pointed bottom and internally have a chaotic, swirled structure indicative of backfilling. These burrows, which have been described from ancient nearshore conglomeratic sequences by Clifton (1981) and Leithold and Bourgeois (in press), are similar to ray-feeding structures described by Howard *et al.* (1977). They are also similar to structures interpreted by Postma (1983) as fluid-escape structures in fan-delta foresets.

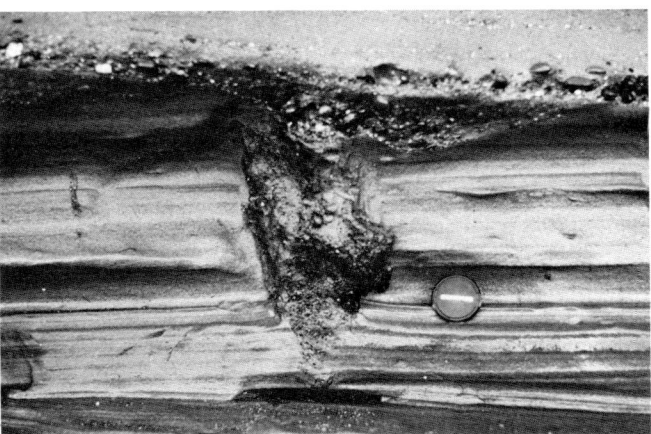

Fig. 5. Large pebble- and coarse sand-filled burrow with chaotic, swirled fill indicative of active filling. Note scour at interface of burrow top with overlying gravel bed. Camera lens cap for scale.

BEACH CONGLOMERATES

BEACH PROCESSES

The beachface is the coastal zone affected by swash and backwash. As water surges up the beachface, carrying sediment as bedload and suspended load, it decelerates and percolates into the underlying sand and gravel. This spatial deceleration and decrease in discharge produces a gradient of decreasing grain size and sediment load from the base of the beachface to the top. (The largest clasts, however, are moved only during violent storm events, and may end up highest on the beach.) Water that does not overtop the berm crest or percolate into the beach returns downslope as backwash, which is principally sheetflow. At equilibrium, onshore sediment transport by wave-driven swash is balanced by offshore sediment transport by gravity-driven backwash. The processes and energy distribution during swash and backwash are discussed by Komar (1976), Dupre et al. (1980), and Kirk (1980).

The beachface dips at a steeper angle than any other part of the nearshore zone, typically 4-16°, with slopes of up to 25° reported from cobble beaches; the slope depends on the asymmetry in intensity of swash and backwash and resulting asymmetry of sediment transport (Bascom, 1951; Komar, 1976; Kirk, 1980). More specifically, the percent of swash that percolates into the beach governs the slope angle; the more percolation, the greater the asymmetry between swash and backwash, and the steeper the beachface. Because percolation rates are primarily a direct function of grain size, beachface steepness increases with increasing grain size; coarser particles also maintain a steeper angle of repose. Prograding beaches have steeper faces than retreating beaches (Kirk, 1980); Komar (1976) and Kirk (1980) discuss other factors that influence beachface slope.

The berm crest separates the beachface from the backbeach, which is built up over time by seaward (lateral) accretion of the beachface. These laterally accreted sediments may be overlain by (berm) washover deposits, by fluvial sediments reworked and deposited in streams crossing the beach, by berm-top-pond deposits, and by eolian sediments. Berm height is a function of the upper limit of the swash, which is a function of wave height (the highest during high tide) and percolation rates. Even as a beach is eroded, a high, narrow, storm-berm crest may be deposited at the back of a beach (Bascom, 1954); Kirk (1980) documents gravel storm berms as high as 14 m above sea level. Accretionary beaches frequently affected by storms may contain several of these storm ridges (Davis, 1978); Leckie and Walker (1982) described a possible example of accreted storm berms in the Cretaceous Gates Formation.

BEACH GRAVELS

The most detailed studies of modern beach gravels have considered sorting by particle shape and size (e.g., Bluck, 1967, 1969; Carr, 1969, 1971; Carr et al., 1970; Dobkins and Folk, 1970; Humbert, 1968; Matthews, 1980; Orford, 1975; Orford and Carter, 1982) and attrition of particles under wave action (for reviews see Zenkovich, 1978; Kirk, 1980; Matthews, 1983). In particular, many authors have noted a shape selection where disc-shaped particles are characteristic of the upper beachface, and spherical particles are more common on the lower beachface. Most authors attribute this shape sorting to the greater suspension potential of discs (important in swash) and the greater pivotability of spheres (important in backwash); others have failed to confirm these effects (for review, see Orford, 1975). Orford pointed out that few authors have considered the effects of the nature of the pebble surface on clast transport.

Bluck (1967) provided a detailed description of downbeach distribution of particle size and shape on modern beaches in South Wales. He described four zones seaward from the berm crest (Fig. 6): 1) a large disc zone, typified by cobble-sized discs; 2) an imbricate zone composed primarily of imbricate, disc-shaped pebbles; 3) an infill zone where spherical and rod-shaped pebbles fill in a framework of spherical cobbles; and 4) the outer frame, made up of spherical cobbles. Bluck noted, as others have, that "particles are not so much made as used on the beaches," i.e., that the *sorting* is characteristic of beaches, but the shapes are not produced there. On the other hand, Dobkins and Folk (1970) studied beaches where only basalt was available for clasts and correlated development of discoidal clasts with wave energy and character of the beach surface. Orford (1975) tested Bluck's (1967) hypothesis that downbeach zonal associations were dependent on wave energy received by the beach, confirming a correlation of maximum pebble zonation with swell (wave) action; wave phase and breaker type were also found to be important. Kirk (1980) concluded that both source and process are important in determining particle-shape distribution on gravelly beaches.

BEACH CONGLOMERATES

Beach deposits of any type are not common in the geologic record; conglomeratic beaches, in particular, are typically associated with eroding coastlines and thus are rarely preserved in any significant thickness. In prograding depositional systems supplied with coarse sediment, beach conglomerates may be preserved, whereas during transgression, beach deposits are commonly reworked in the nearshore zone. On fan-deltas and in other settings where coarse sediment is supplied to a coastline, beach conglomerates may accumulate and be preserved. Conglomerates interpreted as beach deposits recently have been described by Clifton (1973), Nilsen (1973, partly reinterpreted by Kumar and Sanders, 1976), Hunter (1980), Dupre et al. (1980), Wescott and Ethridge (1980), Leckie and Walker (1982), Bourgeois and Leithold (1983), and Kleinspehn et al. (this volume).

Conglomeratic beach deposits may be identified by good size and shape sorting (Bluck, 1967; Clifton, 1973). The presence of imbricated, disc-shaped pebbles is probably the single, best criterion for identifying beach conglomerates. Imbrication is generally seaward-dipping; some clasts may dip landward on the seaward side of the berm crest (Bluck, 1967; Dupre et al., 1980). Associated sandstone with planar lamination rich in heavy minerals would also be diagnostic (Thompson, 1937; Clifton, 1969); bed continuity and position of the units between shoreface and fluvial or eolian deposits have also been used as criteria for distinguishing beach deposits. The detailed work on modern beaches on sorting by particle size and shape, and its correlation with wave energy and type, have yet to be applied to ancient beach conglomerates.

Our illustrations of beach conglomerates (Fig. 7) are from Pleistocene terraces in southwest Oregon, also described by Clifton (1973) and Hunter (1980). Figure 7A illustrates upper beachface deposits above lower beachface deposits. The upper beachface is characterized by planar beds of well-segregated, imbricated, disc-shaped clasts. The lower beachface conglomerate comprises less well-sorted, more spherical pebbles and sand. Figure 7B illustrates a conglomeratic sandstone interpreted as a beach deposit. The pebbles are distinctively disc-shaped and occur in thin lenses and along channel floors, which Hunter (1980) interpreted as stream channels crossing the beach.

Shoreface Conglomerates

SHOREFACE PROCESSES AND PRODUCTS

Shoreward of the transition zone, the steepening of waves as they shoal culminates in their breaking. From the point of breaking, which defines the lower limit of the shoreface, the wave progresses as a bore through the surf zone. The breaking waves and bores transport water shoreward and pile it against the shoreline. The hydraulic head thus created drives unidirectional flow parallel to the shoreline, as longshore currents, and offshore, as rip currents (Komar, 1976; Hunter et al., 1979; Dupre et al., 1980).

Features of conglomeratic shoreface deposits would be expected to reflect dominantly onshore-directed transport on the lower shoreface and the increasing importance of longshore and offshore transport on the upper shoreface. Features of lower shoreface sequences may include cross-bedded conglomerate and pebbly sandstone deposited by the shoreward migration of bars and megaripples. In upper shoreface deposits longshore- and offshore-dipping trough cross-bedded pebbly sandstones and gravel-filled scours associated with unidirectional currents may be common. Sheet conglomerates may be present in both lower and upper shoreface sequences, representing lag pavements created by storm-wave reworking of coarse sediments supplied to the nearshore. These conglomerate beds and other conglomeratic shoreface deposits may be molded into gravel ripples, reflecting reworking by (offshore) oscillatory currents when the nearshore zone contracts during periods of lower energy.

Nearshore conglomeratic sequences with features suggestive of deposition on the shoreface have been described by several workers. Kumar and Sanders (1976) described crudely graded gravel layers in cores of shoreface sediment off Long Island, New York and interpreted them as lags formed during storms. They compared these beds to similar features in Quaternary and Tertiary sequences in New York, Virginia, and California, and to some older deposits. Dupre et al. (1980) described outer-surf-zone (lower shoreface) deposits in Pleistocene terraces of the Santa Cruz region, California, characterized by sand and fine gravel with landward- and seaward-dipping cross-bedding. Inner-surf-zone (upper shoreface) deposits in the same sequence were characterized by parallel-laminated

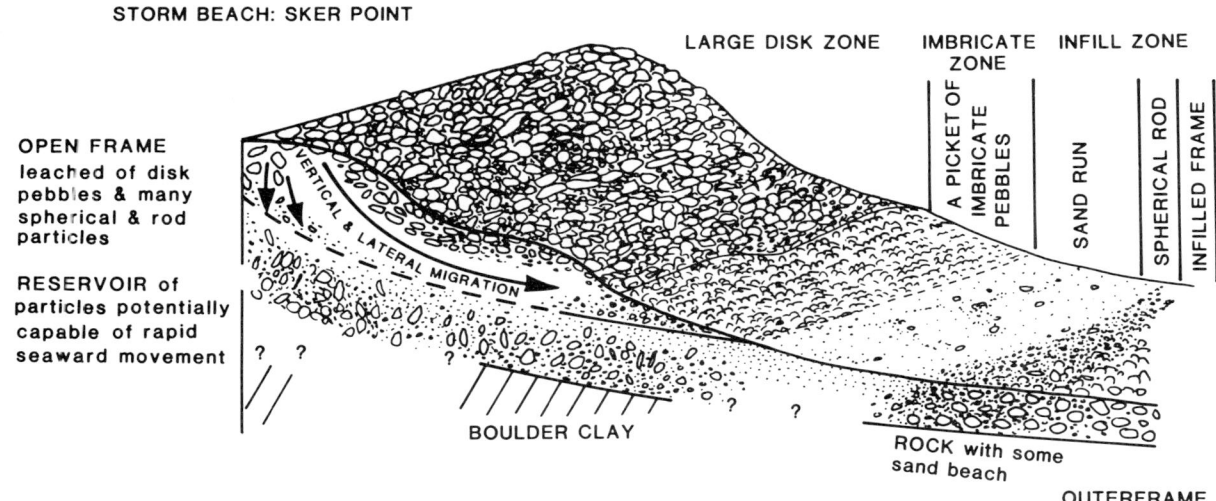

Fig. 6. Textural zonation of beach gravels at Sker Point, Wales, from Bluck (1967).

and cross-bedded pebbly sand and by structureless gravel beds. Clifton (1981) interpreted rippled pebbly sandstone in Miocene strata of the Caliente Range, California, to have been deposited in depths of about 8 m (?lower shoreface). He suggested that trough cross-bedded pebbly sandstones and thin, laterally extensive gravel beds in these rocks were deposited in the surf zone of a high-energy, nearshore system. Leckie and Walker (1982) described 'offshore' gravel bars covered with symmetrical ripples and graded, sharply based conglomerate layers from the Cretaceous Moosebar-Lower Gates interval of western Canada. These features were diagrammatically suggested to have formed in depths ranging from 20-100 m, but they may represent shoreface deposits.

UPPER SHOREFACE CONGLOMERATES — EXAMPLES FROM SOUTH-WEST OREGON

Upper shoreface conglomerates are characterized by tabular conglomerate beds (Fig. 8), trough cross-stratified pebbly sandstone, and, in the Sandstone of Floras Lake, by high-angle, gravel-filled scours. Tabular conglomerate beds are sharply based, sheet-form units of cobbles, pebbles and coarse sand, which show no evidence of pinching or thinning (for over 30 m laterally in the Sandstone of Floras Lake, for example). The beds are typically 10-50 cm thick and alternate with beds of coarse sandstone. The conglomerates are clast-supported and tend to be the most tightly packed of all shoreface conglomerate types. In the Sandstone of Floras Lake clasts average 3-5 cm in diameter and cobbles up to 10 cm in diameter are common; shell debris is present in some beds as a minor component. Internal stratification is not generally apparent in these tabular conglomerate beds; crude normal grading is common. In the Cape Sebastian Sandstone (Fig. 2), tabular conglomerates are typically graded from pebbles, up to 5 cm in diameter, to coarse sand.

In the Sandstone of Floras Lake, at least three tabular conglomerate beds are molded into large, symmetrical

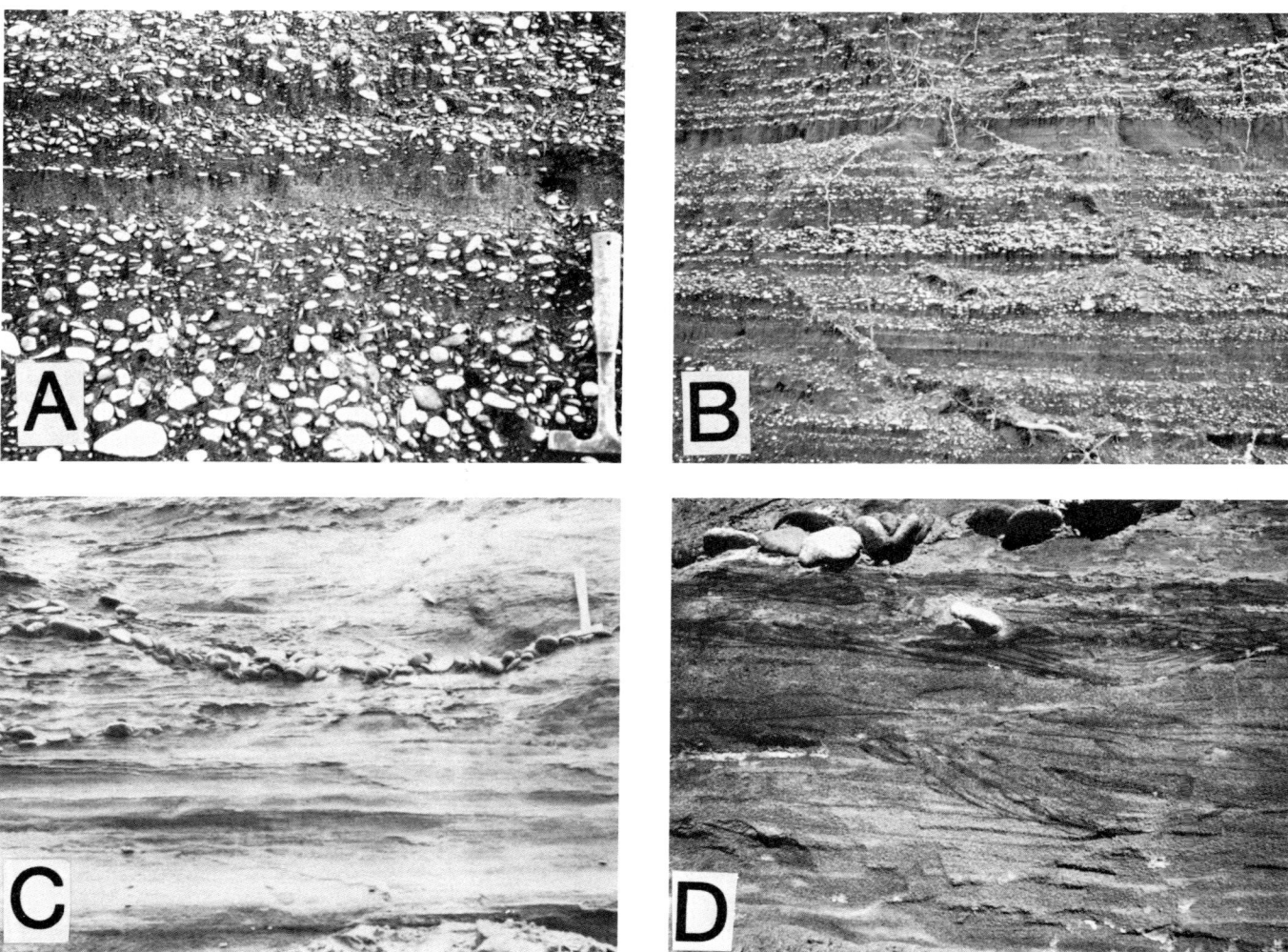

Fig. 7. Beachface conglomerates from Pleistocene terraces located at **(A) (B)** Cape Blanco and **(C) (D)** Otter Point, SW Oregon. **(A)** shows detail of lower portion of **(B)**. The disk-lined channel **(C)** and cross-laminated heavy minerals **(D)** are interpreted as the product of streams crossing the beach (Hunter, 1980).

gravel ripples (Fig. 9). The ripples are sinusoidal in outline and have wavelengths averaging 1 m and heights averaging 10 cm. Clasts within the ripples are somewhat smaller than in most of the tabular conglomerates, averaging 1 cm in diameter. The conglomerate beds retain a thickness of at least a few centimeters in the troughs, suggesting remolding of a bed initially more uniform in thickness. Each ripple set is draped by a layer of medium-grained sandstone, which tends to thicken in troughs and thin over crests of the ripples. The drapes have an average thickness of about 2 cm; they are as thick as 15 cm in the troughs and are entirely absent over some ripple crests. The gravel ripples are associated both above and below with unrippled, tabular conglomerate beds.

The horizontal persistence of tabular conglomerate beds indicates deposition by an agent that acted uniformly over a broad area. These beds are interpreted to represent the reworking and widespread distribution of fluvially supplied gravels by long-period swell (as documented by Kumar and Sanders, 1976; Dupre et al., 1980; Clifton, 1981). The presence of large gravel ripples at the tops of some of the tabular conglomerate beds provides further evidence of reworking. Such ripples have been described by other workers in ancient conglomeratic storm deposits (Clifton, 1981; Leckie and Walker, 1982; Wright and Walker, 1981) and have been observed in modern nearshore systems (Clifton et al., 1971; Cacchione et al., 1983).

Trough cross-bedded, medium- to coarse-grained pebbly sandstone beds occur in nearshore sequences, typically in beds on the order of one-half meter thick (Figs. 1, 2). Small pebbles and rounded wood fragments are concentrated near the bases of troughs, creating a 'pseudo-layering' of conglomerate and sand beds where laterally extensive cross-sets are exposed in cross-section (Fig. 10). Paleocurrent

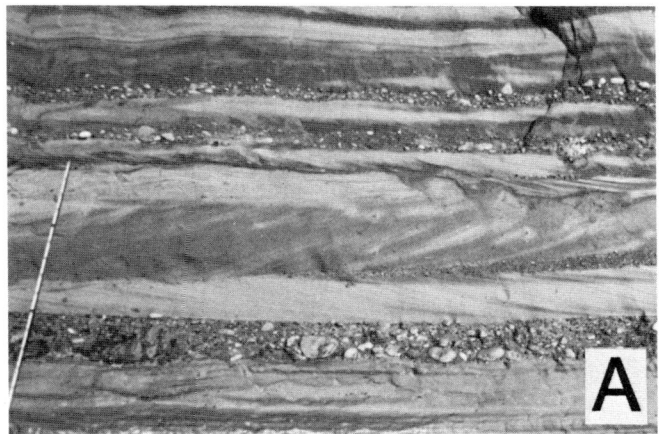

Fig. 8. Upper shoreface deposits with crudely graded tabular conglomerate beds. Photo **A**, which is from the Sandstone of Floras Lake (see Fig. 1), has a 1.4-meter-long rod for scale. Photo **B** is from the Cape Sebastian Sandstone (see Fig. 2); scale bar is 15 cm long.

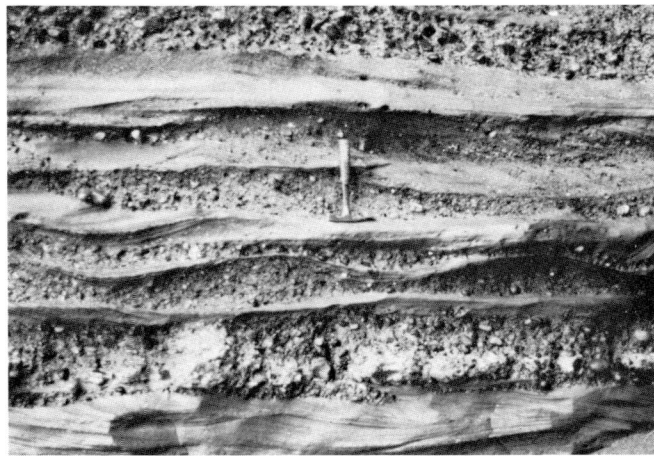

Fig. 9. Symmetrical gravel ripples (middle of photo) in the Sandstone of Floras Lake (in Unit D in Fig. 1).

Fig. 10. Pseudo-layering of conglomerate and sandstone (middle of photo) produced by migration of lunate megaripples; tabular conglomerate in lower portion of photo. From the Sandstone of Floras Lake (Unit D of Fig. 1); rod is 95 cm long in photo.

analysis of these trough cross-beds may indicate onshore, longshore, and offshore directions of bed-form migration (Bourgeois, 1980; Leithold and Bourgeois, in press).

Trough cross-bedded pebbly sandstone is interpreted as having been deposited by unidirectional migration of megaripples in a zone above wave base. In the high-energy nearshore environments of the modern Oregon coast, Clifton et al. (1971) observed shoreward-migrating, lunate megaripples in medium- to coarse-grained sand. These bed forms were interpreted to have formed as the result of the onshore-directed velocity asymmetry associated with long-period swell. Longshore- and offshore-plunging trough axes probably represent the migration of lunate megaripples within longshore troughs, rip channels, or at the mouths of small coastal streams (as documented by Cook, 1970; Clifton et al., 1973; Davidson-Arnott and Greenwood, 1976; Hunter et al., 1979; Clifton, 1981).

High-angle scours are common in upper shoreface deposits of the Sandstone of Floras Lake. In cross-section these scours typically are asymmetric, with one wall at an angle close to vertical and the other less steeply inclined, at about 30-45° (Fig. 11). The high-angle scours are filled with varying amounts of gravel and wood fragments, and shell fragments in medium- to coarse-grained sandstone. Many of the scour-fills show an asymmetric distribution of clasts, with the largest and greatest abundance of cobbles, pebbles, and round wood fragments, and/or shell fragments concentrated against the steeper wall of the scour. The concentration of coarser clastic debris against the steeper walls suggests that unidirectional currents were largely responsible for emplacement of the scour-fills, and the graded nature of the fill indicates deposition under waning current flow. Although no modern analogue to these high-angle scours has been described, the association of many large burrow structures with scoured horizons in the Sandstone of Floras Lake (Fig. 11) suggests that these burrows may have been the initial sites of some type of secondary flow (eddy motion) that enlarged them (see Leithold and Bourgeois, in press).

LOWER SHOREFACE CONGLOMERATES — EXAMPLES FROM SOUTH-WEST OREGON

Conglomeratic lower shoreface deposits are typically adjacent to or interstratified with beds of amalgamated hummocky-stratified sandstone, interpreted as transition-zone deposits (see below). The most common conglomerate beds exhibit low-angle cross-stratification and are interpreted to represent migrating, gravelly bars. Lenticular conglomerate units are also part of the lower shoreface association. In the Sandstone of Floras Lake (Fig. 1), the low-angle cross-stratified conglomerate beds are thin (10-20 cm) and laterally discontinuous, and are vertically spaced from a few tens of centimeters to one meter apart. The pebbles (average 1-3 cm diameter) and shell fragments in the majority of conglomerate beds are aligned along low-angle, sandy cross-laminae comprising cross-sets up to 1 m thick but averaging less than 50 cm thick (Fig. 12). Laminae commonly have gently convex-upward outlines and with few exceptions are oriented onshore with respect to an inferred paleoshoreline (Leithold and Bourgeois, in press). The conglomerates typically comprise pebbles within a sandy matrix; locally pebbles are more tightly packed. Shell debris, especially fragments of barnacle plates, is commonly an important constituent of these conglomerates and is typically oriented parallel to bedding. Similar low-angle cross-stratified conglomerate beds are present but not common in the lower Cape Sebastian Sandstone (Fig. 2).

Low-angle cross-stratified conglomerates in nearshore sequences are probably the product of deposition by

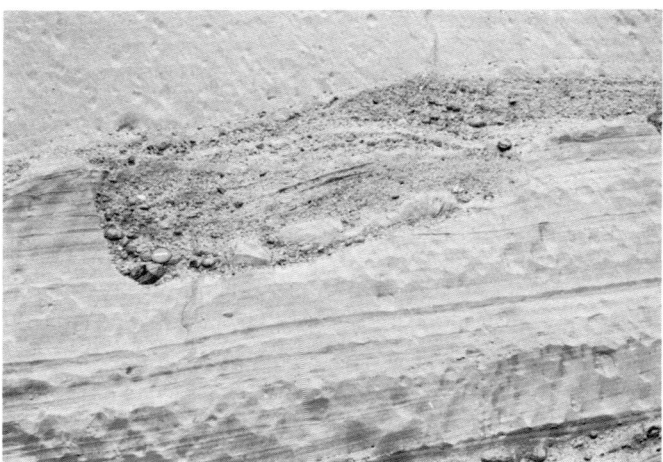

Fig. 11. High-angle, asymmetric, gravel-filled scour in the Sandstone of Floras Lake (Unit B of Fig. 1). Note large, elongate intraclast at bottom of scour and conglomerate with low-angle cross-stratification (bar) at top of scour. Camera lens cap for scale.

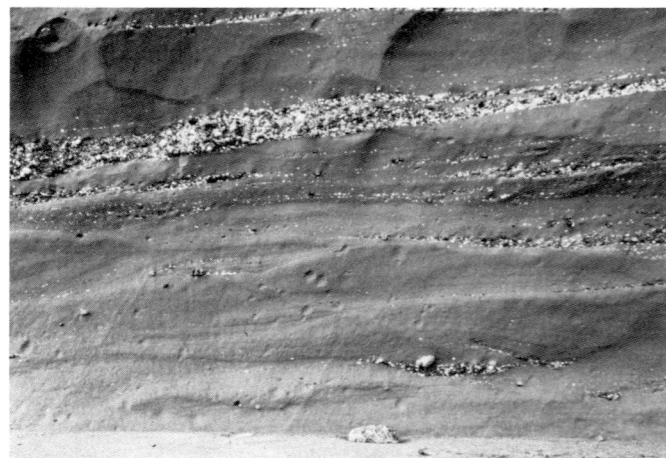

Fig. 12. Low-angle cross-stratified conglomeratic sandstone with numerous barnacle plates in the Sandstone of Floras Lake (Unit A, Fig. 1). Exposure is about 3 m high.

migrating, low-amplitude bars. Pebbles and shell fragments are concentrated along the bases of sandy cross-laminae because they rolled faster and farther than did the finer-grained sediments. Possible modern analogues for these bars are the nearshore bars described by Shepard (1950) and Greenwood and Davidson-Arnott (1979). These bars, which are associated with vortices under high, plunging waves have heights that are small compared to wave heights. They are asymmetric in form, with the steeper face on the landward side of the bar, migrate rapidly, and are considered to be highly unstable features.

Lenticular conglomerates in lower shoreface sequences are characterized by laterally discontinuous, channel-form lenses of pebbles and cobbles which may or may not show crude normal grading and poorly defined, low-angle cross-stratification. In the Sandstone of Floras Lake (Fig. 2), the lenses generally have sharp bases and width:depth ratios averaging 3:1 (Fig. 13). The conglomerates are clast-supported and commonly contain abundant shell fragments. Clasts average 3-5 cm in diameter, and are up to 15 cm in diameter. These clasts, which are larger than those within the low-angle cross-stratified conglomerates, may represent the coarsest fraction of sediments transported by rip currents to, and reworked on, the lower shoreface. They were probably concentrated as lags within shallow scours, possibly in the troughs of landward-migrating bars.

COARSE SEDIMENTS IN THE TRANSITION ZONE

The transition zone, as defined herein, is the zone of wave build-up seaward of the shoreface and of well-developed offshore bars. During fair weather, this environment will be relatively hospitable for organisms; for example, Howard and Reineck (1981) described dense populations of sand dollars in this zone. During storms, however, the transition zone will be disrupted by random, intense scour by waves, and by rapid deposition of sediment eroded by waves breaking on the shoreface or offshore bars. Thus physical processes obliterate most evidence of biologic activity, and hummocky stratification of the amalgamated type (Bourgeois, 1980; Dott and Bourgeois, 1982) is deposited in the inner part of the transition zone.

The transition zone is a sandy, not a gravelly zone. There is no common, physical mechanism to transport coarse bed load seaward of the shoreface and offshore bar. Unusually strong rip currents may transport gravel to the lower shoreface, but shoaling waves will transport this material landward again. Only rarely may a pebble be temporarily suspended and travel offshore beyond the shoreface.

Although the mechanism for transporting coarse debris seaward of the shoreface has not been documented, in amalgamated hummocky stratified sandstone such as the lower Cape Sebastian Sandstone (Fig. 2), scattered pebbles occur along hummocky laminae, and thin, lenticular conglomerate beds, typically one or two clasts thick, may be preserved as concentrated lags of scattered pebbles that were exhumed by scour (Fig. 14). One possible mechanism for transport of the scattered pebbles is that they may originally have been dropped from plant debris (tree roots, kelp, *etc.*) carried seaward in surface rip currents. If dropped seaward of the offshore bar, these pebbles could be moved around only locally; during intense storms they would be concentrated as scour lags. The presence of scattered pebbles along hummocky laminae does not indicate very high values of boundary shear stress (or shear velocity), as suggested by Hunter and Clifton (1982). The boundary shear stress necessary to move a pebble on a sand bed is much lower than that necessary to initiate pebble transport in a pebble bed because a pebble on a sand bed protrudes into higher-velocity flow and may even move before the sand bed itself is put in motion.

CONCLUSIONS

Knowledge of wave-worked gravels and conglomerates and models for their deposition are still in their infancy. Modern beach gravels and ancient shoreface conglomerates have been described in some detail, but their ancient and modern counterparts, respectively, are not at all well-known, nor are the complex mechanisms that operate in the nearshore zone. The use of biologic features, both hard parts and burrows, where present, can advance our knowledge of gravelly nearshore zones. Advances can be made in applying what is known of modern beach gravels to ancient beach conglomerates. Characteristics of ancient shoreface and 'transition-zone' conglomerates present problems yet to be solved with respect to mechanisms for gravel transport and deposition in the nearshore. The study of wave-worked gravels and conglomerates is a challenging frontier.

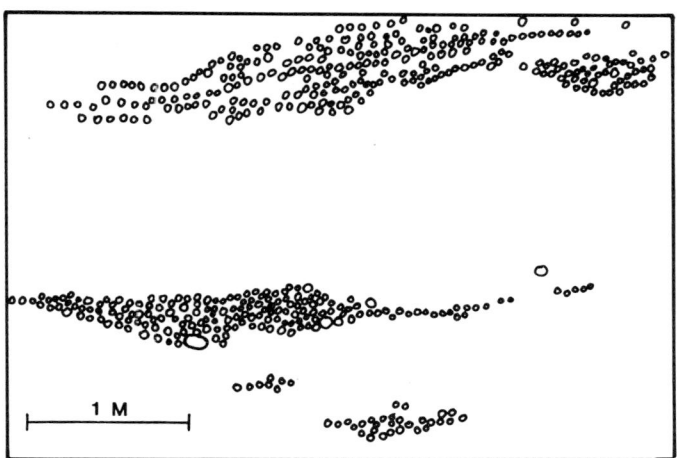

Fig. 13. Sketch from photo of lenticular conglomerate bed in the Sandstone of Floras Lake (Unit A, Fig. 1). Low-angle cross-stratified conglomerate bed at top of photo.

Fig. 14. A: Amalgamated hummocky-stratified sandstone with pebbles concentrated along bases of scours, from the Cape Sebastian Sandstone (see Fig. 2). **B:** detail of hummocky lamination with a single pebble (dropstone?) in middle of laminaset.

References

Addicott, W. O. 1980. Miocene stratigraphy and fossils, Cape Blanco, Oregon. Oregon Geology, v. 42, p. 87-98.

Bascom, W. 1951. The relationship between sand size and beachface slope. Transactions of the American Geophysical Union, v. 32, p. 866-874.

Bascom, W. 1954. Characteristics of natural beaches. Proceedings of the 4th Conference on Coastal Engineering, p. 163-180.

Bluck, B. J. 1967. Sedimentation of beach gravels: examples from South Wales. Journal of Sedimentary Petrology, v. 37, p. 128-156.

Bluck, B. J. 1969. Particle rounding in beach gravels. Geological Magazine, v. 106, p. 1-14.

Bourgeois, J. 1980. A transgressive shelf sequence exhibiting hummocky stratification: the Cape Sebastian Sandstone (Upper Cretaceous), southwestern Oregon. Journal of Sedimentary Petrology, v. 50, p. 681-702.

Bourgeois, J. and Leithold, E. L. 1983. Sedimentation, tectonics and sealevel change as reflected in four wave-dominated shelf sequences in Oregon and California, in: Larue, D. K. and Steel, R. J., (Eds.), Cenozoic Marine Sedimentation, Pacific Margin, U.S.A. Pacific Section, Society of Economic Paleontologists and Mineralogists, p. 1-16.

Cacchione, D. A., Drake, D. E., Grant, W. D., Williams, A. J. III and Tate, G. B. 1983. Variability of sea-floor roughness within the Coastal Ocean Dynamics Experiment (CODE) region. Woods Hole Oceanographic Institute Technical Report WHOI-83-25, 48p.

Carr, A. P. 1969. Size grading along a pebble beach: Chesil Beach, England. Journal of Sedimentary Petrology, v. 39, p. 297-311.

Carr, A. P. 1971. Experiments on longshore transport and sorting of pebbles: Chesil Beach, England. Journal of Sedimentary Petrology, v. 41, p. 1084-1104.

Carr, A. P., Gleason, R. and King, A. 1970. Significance of pebble size and shape sorting by waves. Sedimentary Geology, v. 4, p. 89-101.

Clifton, H. E. 1969. Beach lamination: nature and origin. Marine Geology, v. 7, p. 553-559.

Clifton, H. E. 1973. Pebble segregation and bed lenticularity in wave-worked versus alluvial gravel. Sedimentary Geology, v. 20, p. 173-187.

Clifton, H. E. 1981. Progradational sequences in Miocene shoreline deposits, southeastern Caliente Range, California. Journal of Sedimentary Petrology, v. 51, p. 165-184.

Clifton, H. E., Hunter, R. E. and Phillips, R. L. 1971. Depositional structures and processes in the non-barred high-energy nearshore. Journal of Sedimentary Petrology, v. 41, p. 651-670.

Clifton, H. E., Phillips, R. L. and Hunter, R. E. 1973. Depositional structures and processes in the mouths of small coastal streams, southwestern Oregon, In: Coates, D. R. (Ed.), Coastal Geomorphology. New York: State University of New York at Binghamton, p. 115-140.

Clifton, H. E. and Thompson, J. K. 1978. *Macaronichnus segregatus*: a feeding structure of shallow marine polychaetes. Journal of Sedimentary Petrology, v. 48, p. 1293-1302.

Cook, D. 0. 1970. The occurrence and geologic work of rip currents off southern California. Marine Geology, v. 9, p. 173-186.

Davidson-Arnott, R. G. D. and Greenwood, B. 1976. Facies relationships on a barred coast, Kouchibouguac Bay, New Brunswick, Canada. Society of Economic Paleontologists and Mineralogists, Special Publication 24, p. 149-168.

Davis, R. A., Jr. 1978. Beach and nearshore zone, In: Davis, R.A., Jr. (Ed.), Coastal Sedimentary Environments. New York: Springer-Verlag, p. 237-285.

Dobkins, J. E. and Folk, R. L. 1970. Shape development on Tahiti-Nui. Journal of Sedimentary Petrology, v. 40, p. 1167-1203

Dott, R. H. Jr. and Bourgeois, J. 1982. Hummocky stratification: significance of its variable bedding sequences. Geological Society of America Bulletin, v. 93, p. 663-680.

Dupre, W. R., Clifton, H. E. and Hunter, R. E. 1980. Modern sedimentary facies of the open Pacific coast and Pleistocene analogs from Monterey Bay, California, In: Field, M. E. et al. (Eds.) Quaternary Depositional Environments of the Pacific Coast. Pacific Section, Society of Economic Paleontologists and Mineralogists, Pacific Coast Paleogeography, v. 4, p. 105-120.

Frey, R. W., Howard J. D. and Pryor, W. A. 1978. *Ophiomorpha*: its morphologic, taxonomic, and environmental significance. Palaeogeography, Palaeoclimatology, Palaeoecology, v. 23, p. 199-229.

Frey, R. W. and Seilacher, A. 1980. Uniformity in marine invertebrate ichnology. Lethaia, v. 13, p. 183-278.

Greenwood, B. and Davidson-Arnott, R. G. D. 1979. Sedimentation and equilibrium in wave-formed bars: a review and case study. Canadian Journal of Earth Sciences, v. 11, p. 312-332.

Howard, J. D. 1978. Sedimentology and trace fossils. Society of Economic Paleontologists and Mineralogists, Short Course 5, p. 11-42.

Howard, J. D., Mayou, T. V. and Heard, R. W. 1977. Biogenic sedimentary structures formed by rays. Journal of Sedimentary Petrology, v. 47, p. 339-346.

Howard, J. D. and Reineck, H.-E. 1981. Depositional facies of high-energy beach-to-offshore sequences: comparison with low-energy sequence. American Association of Petroleum Geologists Bulletin, v. 65, p. 807-830.

Humbert, F. L. 1968. Selection and wear of pebbles on a gravel beach. Geologic Institute of Gröningen, Report 190, 144p.

Hunter, R. E. 1980. Depositional environments of some Pleistocene coastal terrace deposits, southwestern Oregon—case history of a progradational beach and dune sequence. Sedimentary Geology, v. 27, p. 241-262.

Hunter, R. E. and Clifton, H. E. 1982. Cyclic deposits and hummocky cross-stratification of probable storm origin in Upper Cretaceous rocks of the Cape Sebastian area, southwestern Oregon. Journal of Sedimentary Petrology, v. 52, p. 127-143.

Hunter, R. E., Clifton, H. E. and Phillips, R. L. 1979. Depositional processes, sedimentary structures, and predicted vertical sequences in barred nearshore systems, southern Oregon coast. Journal of Sedimentary Petrology, v. 49, p. 711-726.

Kirk, R. M. 1980. Mixed sand and gravel beaches: morphology, processes and sediments. Progress in Physical Geography, v. 4, p. 189-210.

Komar, P. D. 1976. Beach Processes and Sedimentation. New Jersey: Prentice-Hall, 429p.

Kumar, N. and Sanders, J. E. 1976. Characteristics of shoreface storm deposits: modern and ancient examples. Journal of Sedimentary Petrology, v. 46, p. 145-162.

Leckie, D. A. and Walker, R. G. 1982. Storm-and tide-dominated shorelines in Cretaceous Moosebar-Lower Gates interval—outcrop equivalents of deep-basin gas traps in western Canada. American Association of Petroleum Geologists Bulletin, v. 65, p. 138-157.

Leithold, E. L. and Bourgeois, J. 1983. Sedimentology of the Sandstone of Floras Lake (Miocene)—transgressive, high-energy shelf deposition, southwestern Oregon, In: Larue, D. K. and Steel, R. J. (Eds.), Cenozoic Marine Sedimentation, Pacific Margin, U.S.A. Pacific Section, Society of Economic Paleontologists and Mineralogists, p. 17-28.

Leithold, E. L. and Bourgeois, J. (in press). Characteristics of coarse-grained sequences deposited in nearshore, wave-dominated environments — examples from the Miocene of southwest Oregon. Sedimentology.

Matthews, E. R. 1980. Observations of beach gravel transport, Wellington Harbour entrance, New Zealand. New Zealand Journal of Geology and Geophysics, v. 23, p. 209-222.

Matthews, E. R. 1983. Measurements of beach pebble attrition in Palliser Bay, southern North Island, New Zealand. Sedimentology, v. 30, p. 787-799.

Nilsen, T. H. 1973. Facies relations in the Eocene Tejon Formation of the San Emigdio and western Tehachapi mountains, California. Annual Meeting, Society of Economic Paleontologists and Mineralogists Anaheim, Guidebook Trip 2, p. 7-23.

Orford, J. D. 1975. Discrimination of particle zonation on a pebble beach. Sedimentology, v. 22, p. 441-463.

Orford, J. D. and Carter, R. W. G. 1982. Coastal overtop and washover sedimentation on a fringing sandy gravel barrier coast, Carnsore Point, southwest Ireland. Journal of Sedimentary Petrology, v. 52, p. 265-278.

Postma, G. 1983. Water escape structures in the context of a depositional model of mass flow dominated conglomeratic fan-delta (Abrioja Formation, Pliocene, Almeria Basin, SE Spain). Sedimentology, v. 30. p. 91-103.

Seilacher, A. 1967. Bathymetry of trace fossils. Marine Geology, v. 5, p. 413-428.

Shepard, F. P. 1950. Longshore-bars and longshore-troughs. United States Army Corps of Engineers, Beach Erosion Board, Technical Memo 15, 30p.

Stratten, T. 1974. Notes on the applicability of shape parameters to differentiate between beach and river deposits in southern Africa. Geological Society of South Africa Transactions, v. 77, p. 59-64.

Thompson, W. 0. 1937. Original structures of beaches, bars, and dunes. Geological Society of American Bulletin, v. 48, p. 723-752.

Weimer, T. J. and Hoyt, J. H. 1964. Burrows of *Callianassa major* Say, geologic indicators of littoral and shallow neritic environments. Journal of Sedimentary Petrology, v. 38, p. 761-767.

Wescott, W. A. and Ethridge, F. G. 1980. Fan-delta sedimentology and tectonic setting—Yallahs fan delta, southeast Jamaica. American Association of Petroleum Geologists Bulletin, v. 64, p. 374-399.

Wright, M. C. and Walker, T. G. 1981. Cardium Formation (Upper Cretaceous) at Seebe, Alberta—storm-transported sandstones and conglomerates in shallow-marine depositional environments below fair-weather wave base. Canadian Journal of Earth Sciences, v. 18, p. 795-809.

Zenkovich, V. P. 1978. Attrition. In: Fairbridge, R. W. and Bourgeois, J. (Eds.), Encyclopedia of Sedimentology. Pennsylvania: Dowden, Hutchinson and Ross, p. 18-20.

DEPOSITIONAL FEATURES OF LATE MIOCENE, MARINE CROSS-BEDDED CONGLOMERATES, CALIFORNIA

R. Lawrence Phillips[1]

Abstract

Cross-bedded, sandy conglomerate and pebbly coarse sandstone form the dominant facies of the basal part of the Santa Margarita Sandstone in the Santa Cruz Mountains of central California. Fossils in this unit of late middle and early late Miocene age, as well as its geometry, indicate deposition in a shallow, northeast-trending seaway connecting the Pacific Ocean with an interior sea. Cross-stratified sets, up to 8 m thick, are interbedded with conglomeratic lag deposits up to 1.3 m thick. Foresets are defined by variations in conglomerate concentration or by alternating conglomerate and sandstone layers. Most of the pebbles are matrix-supported. The conglomeratic foresets typically dip at angles of less than 20°. Some conglomerate layers overlie reactivation surfaces that separate sandy foresets that dip at angle-of-repose (31°). Intense bioturbation disrupts many of the pebbly foresets.

The lithological character of the basal part of the Santa Margarita Sandstone changes vertically. The coarseness and abundance of conglomerate diminish upwards and the geometry of the cross-bedded units changes from trough to tabular in style. Concomitant with these changes, the foreset inclination direction reverses from northeasterly in the lower, trough units to southwesterly in the overlying tabular units. The lateral array of paleocurrent directions also shows a consistent trend: to the northeast along the southeastern flank of the seaway and to the southwest in the central and northwestern parts. The size and direction of cross-stratification and the abundance of reactivation surfaces indicate that the seaway was swept by strong tidal flow. The trough cross-beds reflect the passage of lunate or sinuous-crested dunes; the tabular units, straight- to sinuous-crested sand waves. The bedform migration was sufficiently slow or infrequent to permit intensive infaunal reworking of the substrata. The vertical sequence observed in the basal part of the Santa Margarita Sandstone appears to record a relative rise in sea level during deposition.

Résumé

Un conglomérat arénacé à stratification oblique et un grès à grain grossier et caillouteux composent le faciès dominant de la partie basale du grès de Santa Margarita dans les montagnes de Santa Cruz du centre de la Californie. Les fossiles de cette unité d'âge Miocène moyen tardif et début supérieur, ainsi que sa géométrie indiquent une accumulation des sédiments dans un détroit orienté nord-est et reliant l'océan Pacifique à une mer intérieure. Des ensembles de couches à stratification oblique, jusqu'à 8 m d'épaisseur, sont interstratifiées avec des dépôts conglomératiques résiduels qui peuvent atteindre une épaisseur de 1,3 m. La présence des lits frontaux est révélée par les variations des proportions de conglomérat ou par l'alternance de couches de conglomérat et de grès. La majorité des galets sont non-jointifs dans la matrice. Les lits frontaux conglomératiques sont typiquement inclinés à des angles inférieurs à 20°. Certaines couches de conglomérat chevauchent des surfaces de réactivation qui séparent les lits frontaux arénacés dont le pendage correspond à l'angle de repos (31°). Une bioturbation intense a détruit plusieurs lits frontaux caillouteux.

La nature lithologique de la partie basale du grès de Santa Margarita varie verticalement. La grosseur des particules et l'abondance du conglomérat diminuent de bas en haut et la géométrie des unités à stratification oblique passe du style d'une dépression à tabulaire. Associé à ces variations, on observe un renversement de la direction de l'inclinaison des lits frontaux, passant de nord-est pour les sédiments dans les dépressions constituant les unités de la base à sud-ouest pour les unités tabulaires sus-jacentes. La répartition latérale des directions des paléocourants indiquent également une tendance uniforme: vers le nord-est le long du flanc sud-est du détroit, et vers le sud-ouest dans les parties centrales et nord-est. La dimension et la direction de la stratification oblique et l'abondance des surfaces de réactivation indiquent que le détroit était balayé par les flots des marées. Les couches à stratification oblique des dépressions reflètent le passage de dunes en forme de croissant ou à crêtes sinueuses; unités tabulaires, barres sableuses de crêtes rectilignes à sinueuses. La migration des formes de litage était suffisamment lente ou peu fréquente pour pouvoir permettre un remaniement important par la faune interne des sous-strates. La séquence verticale observée dans la partie basale du grès de Sante Margarita semble indiquer une élévation relative du niveau de la mer durant l'épisode de sédimentation.

Introduction

Shallow marine conglomerates have been less well investigated then have their associated finer clastic deposits. Most studies have focused on conglomerates as storm deposits, either as planar conglomeratic lag deposits (Dott, 1974; Anderton, 1976; Dupre *et al.*, 1980; Graham, 1982) or within cross-beds associated with hummocky cross-stratification (Wright and Walker, 1981; Leckie and Walker,

---

[1]United States Geological Survey, 345 Middlefield Road, Menlo Park, California 94025, U.S.A.

I wish to thank Jody Bourgeois, Ron Kreisa and Hugh McLean for their critical reading and constructive criticisms on an earlier version of this manuscript and Ed Clifton for the final review. I am indebted to Ed Clifton for numerous invaluable comments and helpful suggestions as well as continuing discussions which have aided substantially in writing this paper.

The following quarry companies allowed access to their lands during this study for which I am extremely grateful: Granite Rock, Kaiser Sand and Gravel, Lone Star Industries, Inc., and Santa Cruz Aggregate. I wish to thank Mr. Martin, Granite Rock, and Mr. Ross, Lone Star Industries, Inc. for their generous help and numerous discussions during investigations within their quarries. Mr. Lee Herford, present owner of the Santa Cruz Aggregate quarry land (this study), has also been extremely helpful during investigation of the gravel deposits.

1982; Bourgeois and Leithold, 1983; Leithold and Bourgeois, 1983) or deposited in a nearshore environment (Dupre *et al.*, 1980; Wright and Walker, 1981; Leckie and Walker, 1982). Occurrences interpreted as non-storm include planar lag deposits produced by migrating bedforms or winnowing by non-storm currents (Levell, 1980). The few examples of large-scale sets of cross-stratified conglomerate have been interpreted as offshore or shoreface gravel bars (Leckie and Walker, 1982; Leithold and Bourgeois, 1983).

Extensive surficial gravel deposits presently exist in a variety of modern marine shelf environments: the North Sea (Veenstra, 1969), the English Channel (Stride *et al.*, 1982; Larsonneur *et al.*, 1982), the northeast Atlantic continental shelf off the United States (Schlee, 1973), and the North Pacific continental shelf (Venkatarathnam and McManus, 1973). Belderson and Stride (1966), Belderson *et al.* (1972), and Kumar and Sanders (1976) have investigated contemporary erosional gravel lag deposits. Reports of gravel bedforms on the continental shelfs are limited; examples exist in the North Sea (Flemming and Stride, 1976), off the southern coast of England (Dyer, 1971, 1972; Langhorne, 1982), from a variety of shelf areas in water depths ranging from 8 - 59 m (Belderson *et al.*, 1972) and at depths between 85 - 105 m on the northeast Pacific continental shelf off Vancouver Island (Yorath *et al.*, 1979). The maximum reported height of marine gravel bed forms is 2 m (Dyer, 1972).

This report describes the occurrence and character of large-scale (0.1 - 8 m thick), marine cross-stratified conglomerate beds within the Santa Margarita Sandstone in the Santa Cruz Mountains of central California and discusses their origin and environmental setting (Fig. 1). The Santa Margarita Sandstone of late middle and early late Miocene age comprises a marine onlap sequence. Most of the observations reported here are derived from two adjacent (1.3 km) sand and gravel quarries cut into the basal part of the Santa Margarita Sandstone, the Santa Cruz Aggregate Quarry and the Lone Star Quarry (Fig. 1). Lateral tracing of the conglomeratic lag deposits indicates that the strata exposed in the two quarries are nearly equivalent. Adjacent quarries and some road-cut exposures containing basal strata of the Santa Margarita Sandstone are included to help define lateral relationships. The quarries provide excellent three-dimensional exposures from which the geometry of the cross-stratified units could be readily defined.

Geologic Setting

The Santa Margarita Sandstone rests unconformably on the middle Miocene Monterey Formation and older Tertiary strata, and nonconformably on Cretaceous granodiorite and associated Paleozoic or Mesozoic metamorphic rocks. Erosion of the granitic rocks supplied much of the sediment to the arkosic sandstone in the Santa Margarita. The Santa Cruz Mudstone, of late Miocene age, conformably overlies the Santa Margarita Sandstone.

Marine transgression over irregular bedrock within the Santa Cruz Mountains of central California, during late middle Miocene time, created a connection with the interior marine San Joaquin Basin to the east. The seaway, bounded and partly controlled by topographic bedrock highs, was subject to strong tidal current flow. Nondeposition and pinchout of the Santa Margarita Sandstone as well as deposition of the Santa Cruz Mudstone on bedrock defines the paleotopographic highs (Fig. 2). Eventual deepening of the marine environment due to transgression or subsidence, resulted in widening of the seaway and reduced current velocities. The Santa Cruz Mudstone was then deposited over the Santa Margarita Sandstone.

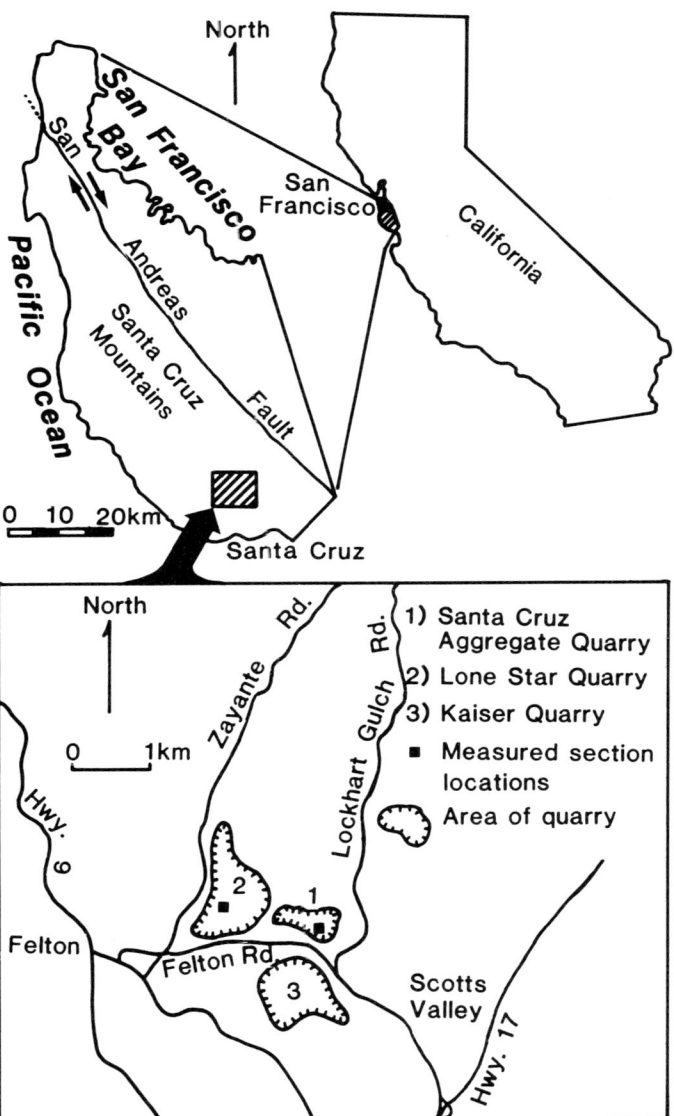

Fig. 1. Study area and location of quarries within the Santa Margarita Sandstone, Santa Cruz Mountains, California.

The Santa Margarita Sandstone varies in thickness from zero (in areas of pinchout) up to 130 m. To the northwest of the study area the sandstone is discontinuously exposed for 45 km eventually thinning and becoming interbedded with mudstone; to the southwest toward the Pacific Ocean the strata dip beneath the overlying mudstone; to the northeast (toward the San Andreas fault) and to the southeast most of the sandstone has been eroded. The sandstone is continously exposed for 20 km in an east-west direction parallel to the seaway.

The formation is gently folded. A syncline, dipping less than 10°, forms the major structure within this study area, whereas to the west a southwest dipping homocline characterizes the coastal region. Normal faults, with displacements less than 5 m, generally trend to the northwest, parallel to the San Andreas fault. Exposures of the Santa

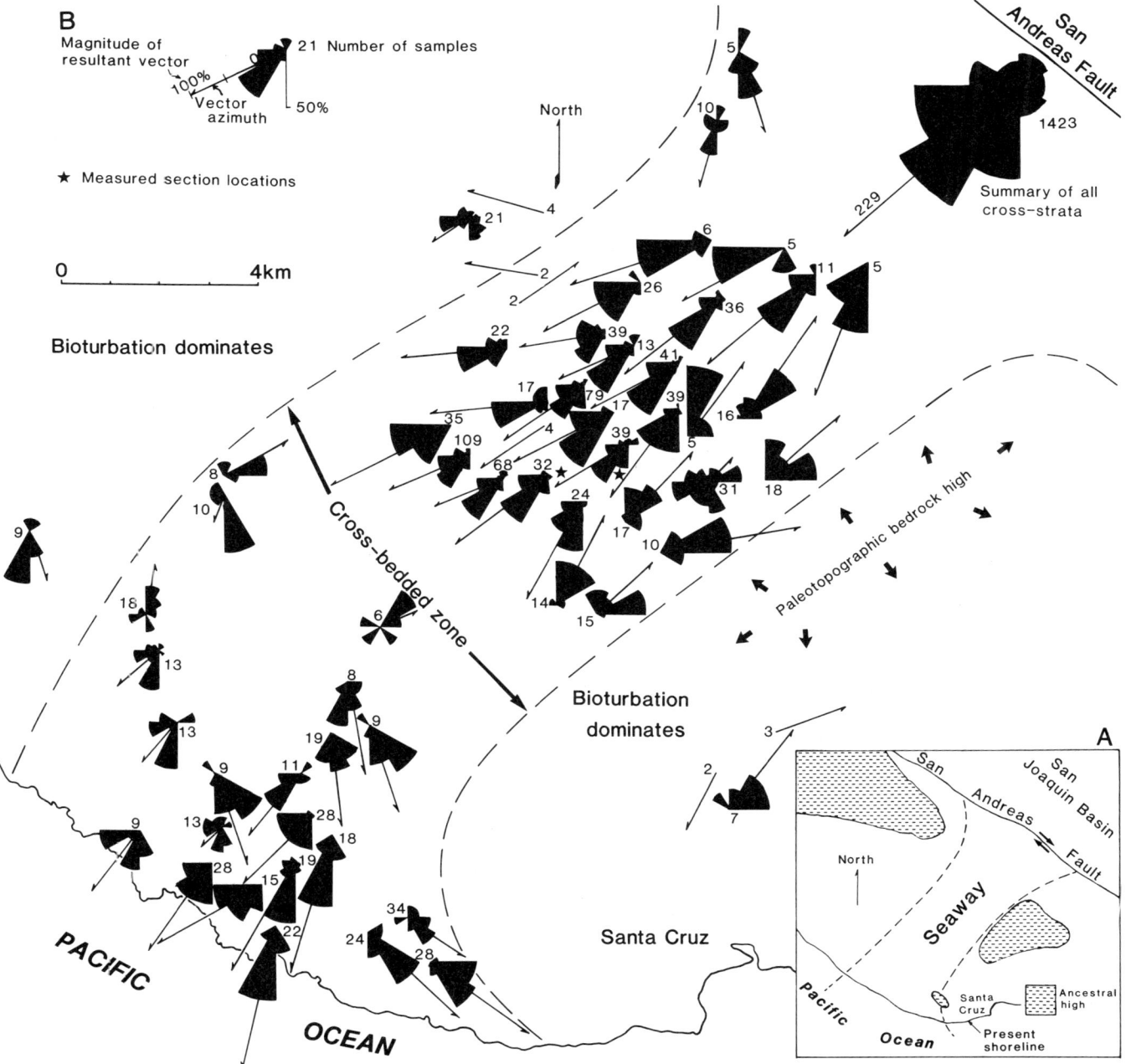

Fig. 2. A) Late Miocene paleogeographic setting. A seaway, bounded by ancestral highs, connected the interior San Joaquin Basin with the Pacific Ocean. The Santa Margarita Sandstone has been palinspastically displaced 224 km to the south along the San Andreas Fault. **B)** Paleocurrents within part of the Santa Margarita Sandstone, Santa Cruz Mountains, California. The seaway deposits, approximately 8 to 10 km in width, contain abundant unidirectional large-scale cross-bedded sets and cosets. A southwest-directed current pattern dominates; an opposed northeast-directed current pattern is preserved on the southeast flank of the tidal seaway.

Margarita Sandstone range from excellent within 25 quarries and in cliff exposures within stream valleys to poor in brush-covered terrane.

General Character of the Santa Margarita Sandstone

The Santa Margarita Sandstone is a friable arkosic sandstone with abundant cross-strata, local conglomerate, extensive invertebrate remains, abundant bioturbation and local tar accumulations. An overall fining-upward grain size characterizes most of the deposit. The sandstone varies from well-bedded to indistinctly bedded where bioturbation dominates. Conglomerate is concentrated in the basal part of the unit (this study) but also occurs as up to 6 m sets of trough and tabular cross-strata near the top of the formation to the southwest. Abundant sets of cross-bedding, up to 17 m in thickness, associated with invertebrate fossils and bioturbated sediment are characteristic of the studied seaway tract, which is some 8 - 10 km wide (Phillips, 1983). Bioturbated sandstone interfingers with, and bounds the cross-bedded strata to the northwest and to the southeast on the flanks of the seaway. Within the thickest vertical section (130 m), the cross-bedded sets change from trough to tabular geometry and eventually to bioturbated sandstone. Laterally the cross-beds may change to repeated beds of invertebrate remains or to bioturbated sandstone. Across the depositional trend of the seaway deposits the strata can be traced and correlated laterally by erosional surfaces containing gravel. In the direction of sediment transport, to the southwest, only local correlation is possible due to facies changes and erosion, as well as to thickening and thinning of the sandstone on ancestral highs.

The abundant fossils in the Santa Margarita Sandstone indicate a marine environment. The marine invertebrate fauna includes pectens, gastropods, barnacles and echinoids; the vertebrate fauna includes remains of cetaceans, pinnepeds, sirenians and teleosts, as well as teeth of rays and sharks (Repenning and Tedford, 1977; and Clark et al., 1979; Clark, 1981). Terrestrial vertebrate remains, found only in the basal gravels of the Santa Margarita Sandstone, include desmostylians, a mastodont, a camel and horses (*Archeohippus*, *Hipparion*, and *Pliohippus*), (Mitchel and Repenning, 1963; Repenning and Tedford, 1977; Clark et al., 1979). The terrestrial vertebrate remains are disarticulated and lie in association with articulated cetaceans and marine fish remains, thereby implying that the land mammal remains are reworked (Phillips, 1981). Trace fossils are also abundant throughout the Santa Margarita Sandstone and many beds are thoroughly bioturbated.

The abundant cross-bedding in the Santa Margarita Sandstone provides an excellent basis for reconstructing paleocurrent patterns in detail. A summary of paleocurrent measurements (Fig. 2) indicates that flow was predominantly parallel to the axis of the inferred seaway. At most exposures the paleocurrents are unidirectional although some cosets bounded by widespread gravel lag deposits exhibit 180° reversals in direction. Paleocurrents to the southwest dominate in most of the exposures. However, a persistent northeasterly trend occurs along the southeastern margin of the seaway (Fig. 2). Such a pattern is consistent with the coriolis right-hand deflection of large-scale, fluid flow in the northern hemisphere.

Conglomerates in the Basal Part of the Santa Margarita Sandstone

The pebbly deposits exposed in the quarries range from sandy pebble conglomerates to pebbly coarse sandstones (classification of Folk, 1974). A bimodal texture is evident from size analysis (Fig. 3). The largest pebbles (10 long-axis measurements per bed) average slightly less than 10 cm (range from 5 - 20 cm), whereas the average mean pebble diameter is about 1 cm. The pebbles are rounded to well-rounded and composed of siliceous volcanics, granite, quartzite, meta-arkose, schist, siltstone, mudstone, and arkosic sandstone in decreasing order of abundance (Clark, 1981). The sandstone is coarse- to medium-grained and arkosic.

Large-scale sets of trough and tabular cross-bedding and tabular conglomeratic lag deposits dominate within the quarries (Figs. 3 and 4). Most of the pebbly trough sets range in thickness from 3 - 8 m (Figs. 5 and 6). Trough sets lacking gravel tend to be less than 1 m thick. The largest tabular set is 4.4 m thick (Fig. 7). All cross-stratified sets have sharp, erosional basal contacts. Foresets in the trough cross-bedded units may be either planar or tangential; those in the tabular units are tangential only. The tabular, conglomeratic lag deposits vary in thickness from 0.4 - 1.3 m and are gently inclined, lie on undulatory surfaces and can be discontinous (Fig. 8).

Foresets in the cross-bedded units are defined by concentrations of pebbles and by textural variations in the sandstone (Figs. 8 and 9). Although some foresets consist of clast- to matrix-supported pebble layers several centimeters thick (Fig. 10A), most of the pebbles along the foresets are isolated (Fig. 10B). Many of the pebbly foresets are wedge-shaped and pinchout up-dip (Figs. 8 and 10C). Conglomerate is typically concentrated in the axial portion of the trough sets, and where concentrated in thicker sets, the pebbles may be imbricated. Generally, however, preferred fabric is poorly developed, and sometimes appears to be completely disorganized (Fig. 11), apparently as a result of intensive bioturbation. Pebbly foresets in the trough units may be either normally or inversely graded; in the tabular sets, where pebbles are generally less abundant, the pebble-bearing foresets grade upward into coarse sandstone. An upward decrease in clast size occurs toward the top of the foresets in some trough units (Fig. 12).

Reactivation surfaces are abundant throughout the Santa Margarita Sandstone both in trough and tabular cross-stratified sets. Conglomerate overlying reactivation surfaces is common within some of the trough sets in the

Lone Star Quarry. Spacing between the reactivation surfaces within the tabular sets ranges from 0.2 - 3.4 m. The foreset bundles between successive erosional surfaces commonly show cyclic increasing-decreasing distances and these are interpreted as representing bedform movement and deposition during spring-neap tidal conditions. Superimposed on the cycles, in some sandy foreset beds only, are repeated increasing-decreasing distances between adjacent bundles, possibly due to diurnal tides, similar to the pattern of foreset bundles identified by Visser and de Boer (1982). Mud-draped reactivation surfaces have been found only within the basal part of the Santa Margarita Sandstone. Within the Santa Cruz Aggregate Quarry large-scale, tabular cross-beds contain mud-layer couplets which also suggest a diurnal depositional pattern (Visser and de Boer, 1982). The mud-draped reactivation surfaces also show increasing-decreasing distances attributed to deposition during neap-spring tidal cycles. Mud drapes within trough cross-stratified sets have not been observed.

The dip angle of the foresets ranges from near angle-of-repose (31°) to less than 12°. Most of the foresets in the pebbly trough cross-beds dip at angles less than 20°, and the foresets composed of only clast- to matrix-supported conglomerate (Fig. 10A) are inclined typically at angles less than 12°. Some trough units at Lone Star Quarry contain reactivation surfaces that are overlain by gravel (Figs. 8 and 12). These reactivation surfaces dip at angles of less than 20°, in contrast to the more steeply dipping intervening sandy foresets.

Bioturbation is common throughout the section. Vertically oriented pebbles, discontinous pebble beds and V-shaped pebble-filled burrows document faunal mixing within the cross-bedded deposits (Fig. 13). Vertically oriented, sand-filled burrows, 3 - 4 cm wide and at least 20 cm in depth, disrupt many of the conglomerate cross-strata at Lone Star Quarry (Fig. 14). Bioturbation increases toward the axial portion of many of the trough sets. Some of the larger burrow structures containing disrupted V-shaped

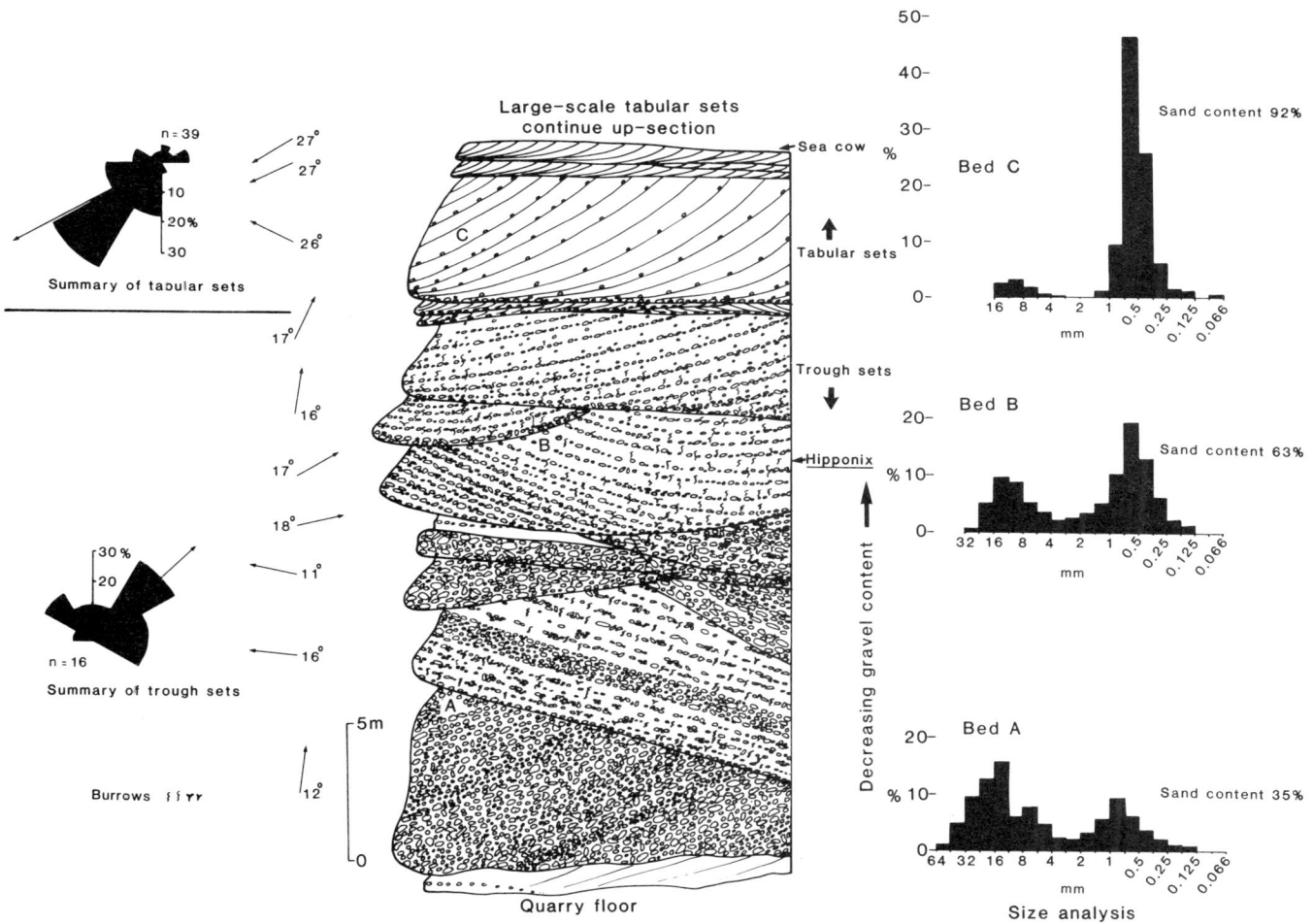

Fig. 3. Measured section, paleocurrent trends and texture of the Santa Margarita Sandstone in the Santa Cruz Aggregate Quarry. The number beside the individual vectors represents the dip-angle of the cross-bed. The sample for size analysis was obtained from approximately a 1 m vertical section by 40 cm wide and 20 cm deep.

pebble beds (Fig. 8) resemble ray feeding traces (Howard et al., 1977; Gregory et al., 1979) and the presence of ray teeth within the trough sets supports such an interpretation. Pelecypod molds found within the conglomerates suggest that some of the vertical burrows may represent pelecypod escape structures or feeding traces of rays and fish. Conglomerate-lined burrows, over 1 m long in vertical section, containing a sand-filled central core up to 2 cm in diameter may represent decapod dwelling structures (Fig. 8). The intense bioturbation results in the complete disruption of the coarse-grained sediment in some of the trough sets. Bioturbation within the large-scale tabular sets is generally restricted to the bottomset beds or to fine-grained laminations within the foreset beds

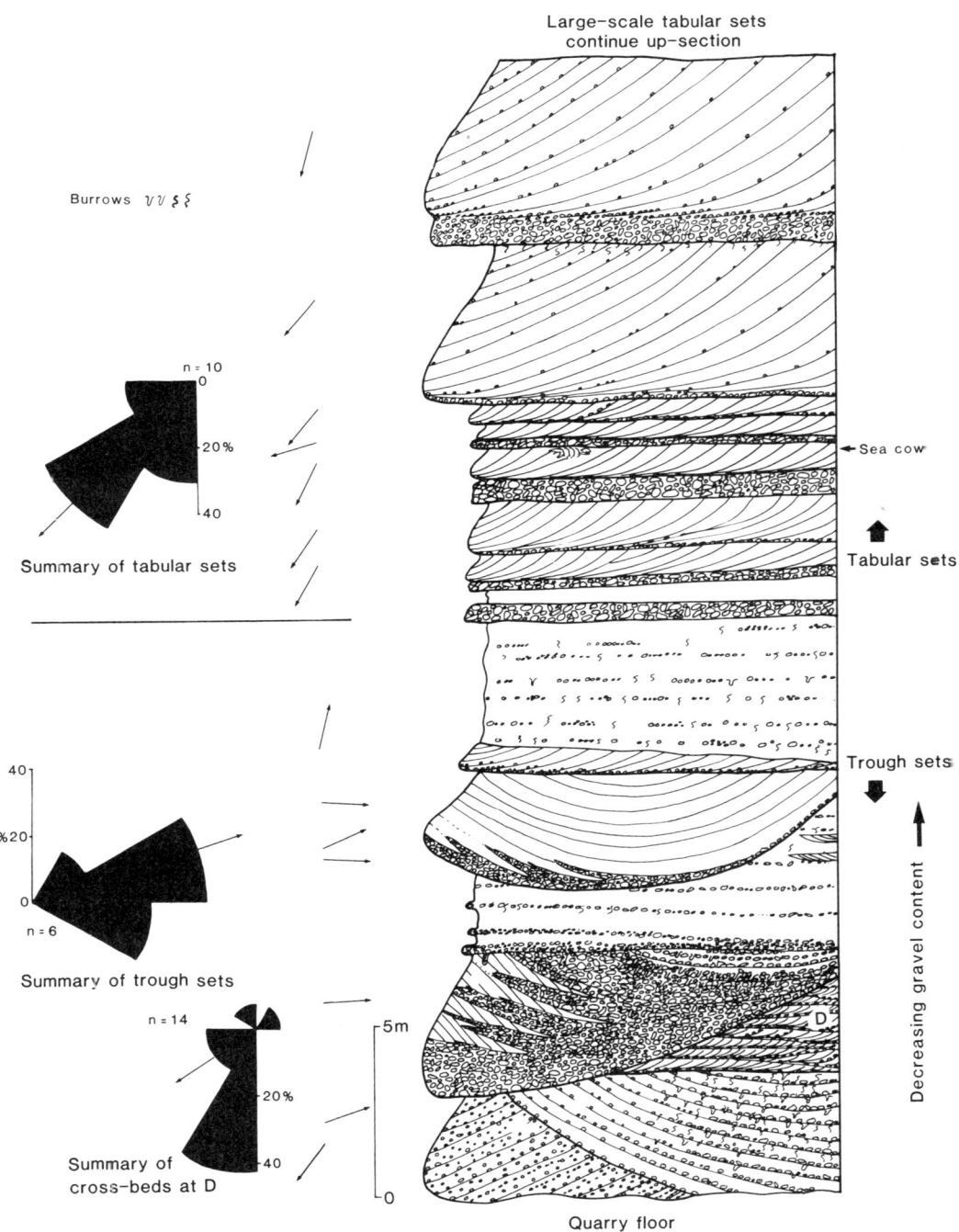

Fig. 4. Measured section and paleocurrent trends within the Santa Margarita Sandstone in Lone Star Quarry. The cross-strata change from conglomeratic trough cross-beds to large-scale tabular cross-beds in the vertical sequence.

Fig. 5. Part of a large-scale, 8 m thick, conglomerate trough cross-bed in the Santa Margarita Sandstone in the Santa Cruz Aggregate Quarry (bed A, Fig. 3). The pick is 50 cm.

In both quarries there is a consistent vertical trend of diminishing upward abundance and coarseness of conglomerate (Figs. 3 and 4). Superimposed on this upwards fining is a change in the geometry of the cross-bedded units from trough to tabular as well as a paleocurrent reversal. Paleocurrent trends as indicated by the orientation of the trough cross-strata are variable but have a strong northeasterly-directed mode. In contrast, the overlying large-scale tabular cross-beds record mostly southwesterly-directed paleocurrents (Figs. 3 and 4).

The nature of the vertical sequence in the basal Santa Margarita Sandstone changes laterally away from the two quarries described here. Thick, northeasterly-directed trough cross-beds persist in sand conglomerate south and east of the Santa Cruz Aggregate Quarry (in the Kaiser Quarry and along Lockhart Gulch road, see Fig. 1). However, to the northwest of Lone Star Quarry, north of a local bedrock high of the Monterey Formation, changes occur.

Fig. 6. Sets of conglomeratic trough cross-beds within bioturbated sediment in the Santa Margarita Sandstone in Lone Star Quarry. The basal set of alternating sandstone and conglomerate cross-beds is cut obliquely. The upper trough set is 70 m wide.

Fig. 7. Large-scale tabular cross-bed within the Santa Margarita Sandstone in the Santa Cruz Aggregate Quarry. The cross-stratified set in the center of the photograph is 4.4 m thick (bed C, Fig. 3). The cross-beds are composed of medium- to coarse-grained sand with pebbles scattered along the foreset beds.

Fig. 8. Tabular to wedge-shaped bioturbated conglomerate lag deposits in the Santa Margarita Sandstone within Lone Star Quarry. The lag deposits range in thickness from a clast to 80 cm here. The biogenetic structures represent ray feeding traces (V-shaped beds upper left) and possible decapod dwelling structures (vertical pebble-filled burrows). The underlying alternating conglomerate-sandstone cross-beds represent part of a trough set approximately 4 m thick. The conglomerate cross-beds rest on reactivation surfaces.

Fig. 9. Alternating conglomerate-sandstone cross-beds in the Santa Margarita Sandstone in the Santa Cruz Aggregate Quarry (bed B, Fig. 3). The pick is 50 cm.

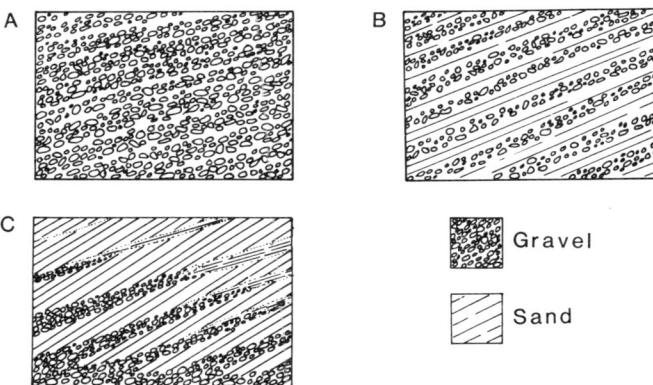

Fig. 10. Three cross-bed types containing conglomerate are identified in large-scale, cross-stratified sets in the Santa Margarita Sandstone. **A)** Repeated thick cross-beds of inversely or normally graded clast- to matrix-supported conglomerate which usually dip at angles less than 12°. **B)** Alternating matrix-supported conglomerate and sandstone cross-beds. **C)** Alternating wedge-shaped conglomerate cross-beds and sandstone cross-beds. The wedge-shaped conglomerate cross-beds rest on reactivation surfaces, become progressively thinner and finer-grained up-slope and dip at angles less than 20°. The sand cross-beds overlying the conglomerate dip at angles up to 31°.

Large trough cross-beds continue to occupy the lower part of the section, but the sediment here is predominantly medium-grained sandstone and the foresets dip to the southwest. The trough cross-bedded sandstone is overlain, not by tabular cross-beds, but by bioturbated sand and a series of tabular lag deposits composed predominantly of bioclastic debris, mostly echinoids, pectens and barnacles (Phillips, 1983).

ORIGIN OF THE CONGLOMERATIC DEPOSITS

Much evidence within the Santa Margarita Sandstone points to the presence of tidal influences:

1) The overall paleocurrent pattern (Fig. 2) shows opposed bipolar flow within the seaway.

2) Local cosets showing current reversals (herringbone cross-strata).

3) Abundant reactivation surfaces within the foreset units.

4) The local presence of mud drapes and mud layer couplets within the foresets.

5) A systematic lateral increase and decrease in foreset thickness within the tabular sets.

The paleocurrent pattern whereby flow indicators at a single locality show a unidirectional trend but the overall pattern shows areas of reversal, is typical of tidal currents (Anderton, 1976; Klein, 1977; Johnson, 1978; Clifton and Phillips, 1980). Herringbone cross-strata and reactivation surfaces may be produced by mechanisms other than tidal flow (Collinson, 1970; Clifton et al., 1971; Allen, 1973) but are commonly considered good criteria for a tidal environment (Klein, 1977). Mud drapes are common in tidal deposits (Clifton, 1982) and mud couplets on foresets are thought to be diagnostic of tidal conditions (Visser, 1980). Alternating rhythmic increases and decreases in foreset spacing reflect neap-spring tidal cycles (Visser, 1980; Allen, 1981) and the observed increasing distance between mud-draped foresets is useful for establishing a tidal origin (Visser and de Boer, 1982).

The geometry of the cross-bedded units suggests the nature of the bedforms that produced them. The large-scale, essentially unidirectional trough cross-stratification represents dune bedforms fronted by well-developed scour pits (Harms, 1975) migrating on the sea floor. The large-scale tabular cross-stratified sets that overlie the trough sets represent straight- to sinuous-crested sand waves (Harms, 1975) migrating on the sea floor. The change from dunes to sand waves suggests a reduction in the flow velocity (Southard, 1975), perhaps due to increasing water depth during the transgression. The tabular gravel lag deposits resulted from current scouring of the sea floor or avalanching and concentration of gravel in the bottomset beds as the bedforms swept by.

Three-dimensional analysis of the trough sets in both quarries and measurements of cross-bed orientation on

Fig. 11. Abundant bioturbation produced the disorganized conglomerate fabric within the gentle dipping strata within the Santa Margarita Sandstone in the Santa Cruz Aggregate Quarry. The scale is 20 cm.

Fig. 12. Alternating conglomerate-sandstone cross-beds within the Santa Margarita Sandstone in Lone Star Quarry. The clast-supported cross-beds rest on erosional surfaces and dip less than 20°. The clasts decrease in size toward the top of the cross-beds. The pick is 50 cm.

Fig. 13. Part of a trough cross-bed set in the Santa Margarita Sandstone containing disorganized and bioturbated cross-beds. Vertically oriented pebbles, sand- or gravel-filled burrows and pebble nests identify bioturbation of the sediment. The scale is 20 cm.

both flanks of individual sets suggest the presence of two distinct dune morphologies. The trough sets in the Lone Star Quarry represent strongly curved lunate dunes (Fig. 15), up to 70 m wide, at least 4 - 5 m high, on which flank deposition produced nearly opposing cross-beds. The trough sets within the Santa Cruz Aggregate Quarry represent lunate dunes that, based on cross-bedding direction on the flanks, were not as strongly curved. These trough sets represent either solitary lunate dunes or catenary gravel waves (Fig. 15), at least 85 m wide and some at least 8 m high. The structures resemble solitary trough sets (140 m wide and 10 m high) in other ancient fluvial gravel-rich systems (Steel and Thompson, 1983).

The intensity of bioturbation within the trough foresets suggests that the dunes either moved slowly or infrequently. In this respect they resemble the large (up to 7 m high) gravel and sand waves off southern England (Dyer, 1971) or the sand waves (greater than 7 m high) presently in the axis of lower Cook Inlet, Alaska (Rappeport, 1980). The lower and therefore more preservable parts of these bedforms in Cook Inlet are characterized by a high degree of faunal activity that diminishes toward the crests where the effects

Fig. 14. Sand-filled burrows cut through the conglomeratic foreset beds in the Santa Margarita Sandstone in Lone Star Quarry. The origin of these burrow structures is uncertain; they may represent ray or fish feeding traces or decapod or pelecypod escape structures.

of fluid flow are more pronounced. Studies by Whitney *et al.* (1979) suggest that the structures have not migrated significantly in the past seven years. The rate or infrequency of movement of the Santa Margarita dunes may have been comparable, although this cannot be verified.

The conglomerate in the cross-beds, whether clast-supported or matrix-supported, was undoubtedly transported as bed load that rolled, slid or avalanched down the lee side of the bedforms probably in a similar manner to the overpassing of gravel on bars (Allen, 1983). The sand in the foresets may have been emplaced by avalanching or by falling into place from suspension. The latter mechanism would account for the matrix-supported fabric of most gravel foresets. Suspended load transport of much of the sand is likely. Most is less than 1 mm in diameter (Fig. 3), a size likely to be placed in suspension by currents capable of moving the associated coarser gravel clasts (Walker, 1975). Flow separation, which would allow the sand to fall on the lee surface, will start on lee surface slopes of approximately 10° (Allen, 1982), well below the dip inclination of most of the sand-gravel foresets.

The textural differentiation and formation of alternating conglomerate-sand cross-beds observed within many trough sets could result from a combination of factors including fluctuating current strength, opposing tidal flow or migration of superimposed bedforms. Fluctuations in the current strength could have been random short-lived events, systematic alternations in flow induced by the tides, or a combination of the two. Superimposed bedforms migrating on the stoss flank of large-scale bedforms can

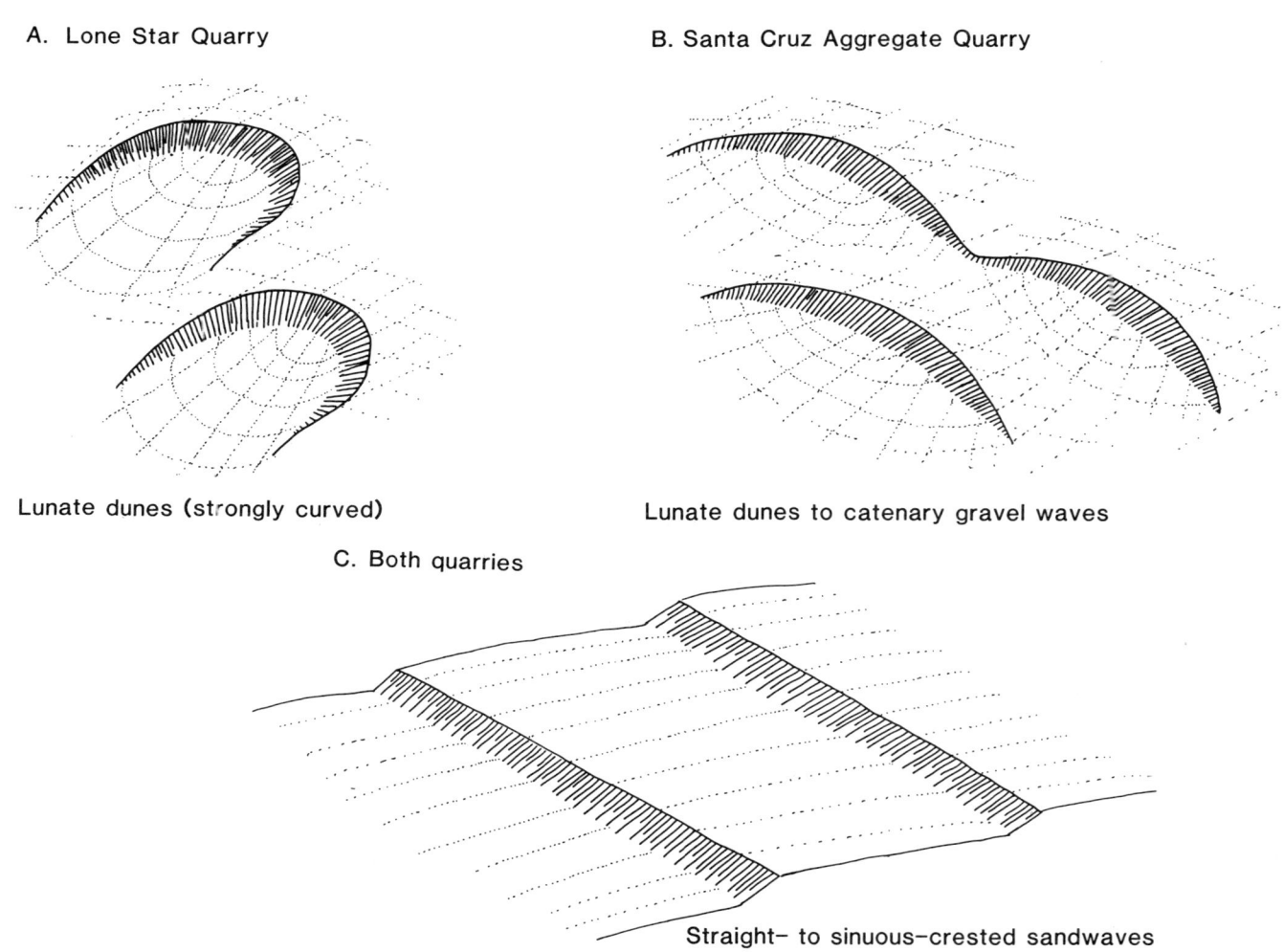

Fig. 15. Bedform morphologies in the Santa Margarita Sandstone determined from cross-stratified sets in the Lone Star and Santa Cruz Aggregate Quarries. **A)** Strongly curved lunate dunes composed of sandy conglomerate. **B)** Lunate dunes to catenary gravel waves composed of sandy pebble gravel. In vertical section within both quarries the gravel content decreases and straight- to sinuous-crested sand waves **C)** composed of pebbly coarse sand form the dominant bedforms.

also produce gravel-sand segregation when they reach the crest and avalanche down the slip face (Smith, 1972; Baker, 1973). The lateral persistence of the conglomerate and sandstone cross-beds across the breadth of the trough set however, suggests continuous bed load transport and deposition on the slip face rather than intermittent avalanching of smaller bedforms.

The wedge-shaped conglomerate beds that lie on reactivation surfaces and the more steeply inclined sandstone cross-beds (Figs. 8 and 10C) may reflect the effects of opposing tidal currents. The persistent concentration of conglomerate in the lower part of the units suggests that it resided preferentially in the troughs of the bedforms. Tidal currents flowing opposite to the direction of bedform migration could erode the slip face, forming a reactivation surface, upon which gravel was carried from the trough. The smaller pebbles would be carried further up the slope, thereby producing the observed up-slope fining of the conglomerate. Upon reversal of the current, the large bedform would again migrate, burying the reactivation surface and the gravel with sand that avalanched down the renewed slip face (Fig. 16).

It is unlikely that the conglomeratic wedges resulted from sand-gravel segregation during avalanching on the slip face of the dune, though this is another mechanism of producing up-slope fining (the reversal tangential graded bedding of Allen, 1983) in cross-bedded deposits (Allen and Nayaran, 1964; Boersma, 1967). It does not explain the difference in dip angle that exists between the conglomeratic and the overlying sandstone foresets, nor does it explain the absence of pebbles from within the sandstone cross-beds (sand foresets dipping at near angle-of-repose in the tabular sets commonly contain scattered pebbles).

The origin and significance of the low dip-angle of the conglomerate cross-bed within all the trough sets (less than 20°) is uncertain. It is unlikely that these cross-beds represent only the bottomset beds of still larger bedforms because the associated sand laminae dip at angles up to 31°. Moreover, even where topset, foreset and bottomset beds can be discerned in a set, the foresets dip at angles between 18 - 20°.

The low-angle of cross-bed dip may in fact be the result of high current velocity. Experimental investigations show that as the velocity increases the slip face angle decreases (Jopling, 1965). A similar relationship of decreasing slip face angle was identified in fluvial sediments by Smith (1972). Modern marine gravel waves contain lee slope-angles less than 20° (Dyer, 1971). The origin of the low dip angle of the conglomerate foresets in the Santa Margarita Sandstone may be due to texture, opposing currents or current strength.

The flow conditions capable of moving gravel and processes producing large-scale dunes to 8 m height are poorly understood. The minimum velocity 1 m above the bed require to move a pebble of 1.2 cm is approximately 175 cm/sec (quartz at 20°C., Miller et al., 1977). The shear velocity for the same size pebbles is approximately 11.0 cm/sec. The high velocities required to move the gravel and turbulence generated by the dune bedforms would indicate an extremely dynamic environmental setting.

The high velocities required to move the gravels and produce large-scale bedforms suggest that transport occurred mainly during spring tides when maximum velocity was generated. Periodic movement of the dune bedforms is also suggested by the abundant bioturbation of the trough cross-strata.

The transitions from gravel dunes to straight- to sinuous-crested sand waves may indicate a reduction in flow velocity due to increasing water depth during the transgression or subsidence or may only reflect a grain-size change. Changes in grain size, velocity and water depth within deep flow conditions can produce different bedform morphologies (Rubin and McCulloch, 1980).

The change from northeast migrating dunes to southwest migrating sand waves within the vertical sequence apparently reflects the local tidal current dominance. There is a lateral shift to the southeast flank of the seaway of the northeast-directed currents, as reflected in cross-bed orientation in vertical sections, with an apparent increase in water depth during the transgression (Phillips, 1983).

The depth at which these deposits formed is uncertain. The thickness of the stratigraphic section to the first laterally continuous erosional lag deposit above the basal trough cross-beds is 25 m, suggesting a possible minimum depth. No wave-generated structures (hummocky cross-beds) or nearshore depositional sequences have been identified within the Santa Margarita Sandstone to date. This suggests that much of the Santa Margarita Sandstone was deposited below wave base at depths greater than 20 m. Islands, shoals or the strong tidal current flow, however, may have dampened the effects of the surface waves, so a shallower depth could be possible.

Summary

The Santa Margarita Sandstone was deposited in a marine, tidally-dominated seaway within which there were currents of sufficient strength to transport gravel as large-scale dunes and sand waves. Erosion of the sea bed by currents or migration of bedforms also resulted in the formation of gravel lag deposits. Conglomeratic trough cross-beds, to 8 m height, contain at least three types of foresets consisting of: 1) clast- to matrix-supported conglomerate, 2) alternating conglomerate and sandstone and 3) conglomerate resting on reactivation surfaces. The conglomerate foresets dip at angles less than 20° whereas sandy foresets overlying conglomeratic cross-beds on reactivation surfaces dip at angles up to 31°. The sets of conglomeratic cross-beds contain abundant bioturbation, suggesting that the giant bedforms moved infrequently, possibly only during spring tides.

Depositional cycle

1) Erosion

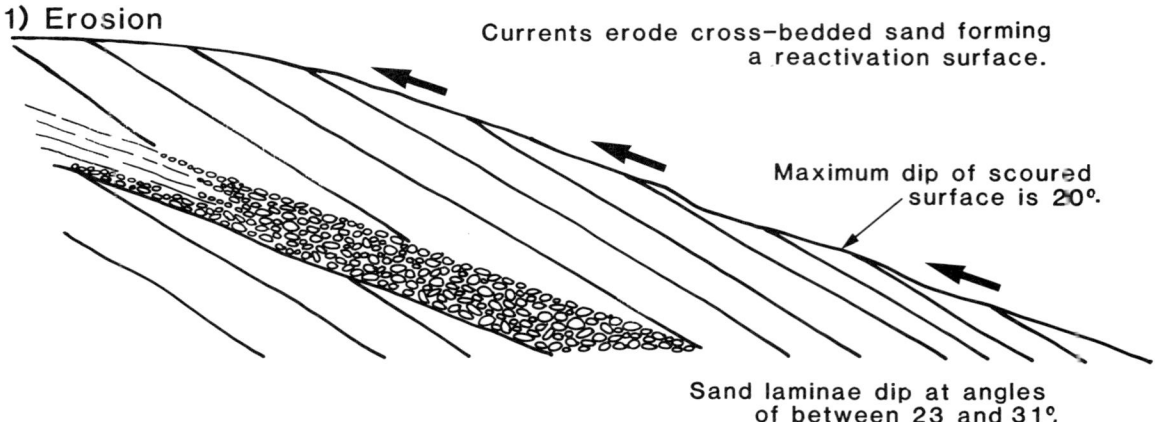

2) Gravel deposition (Maximum flow velocity)

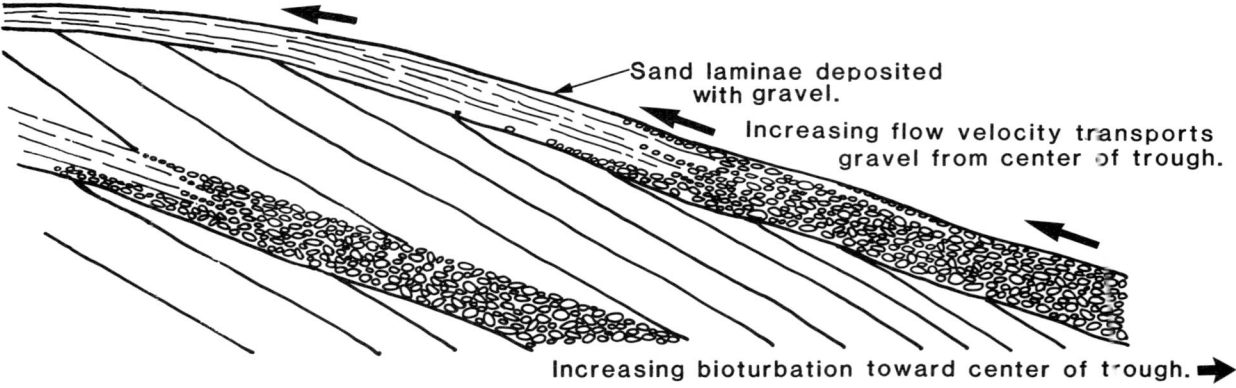

3) Sand deposition

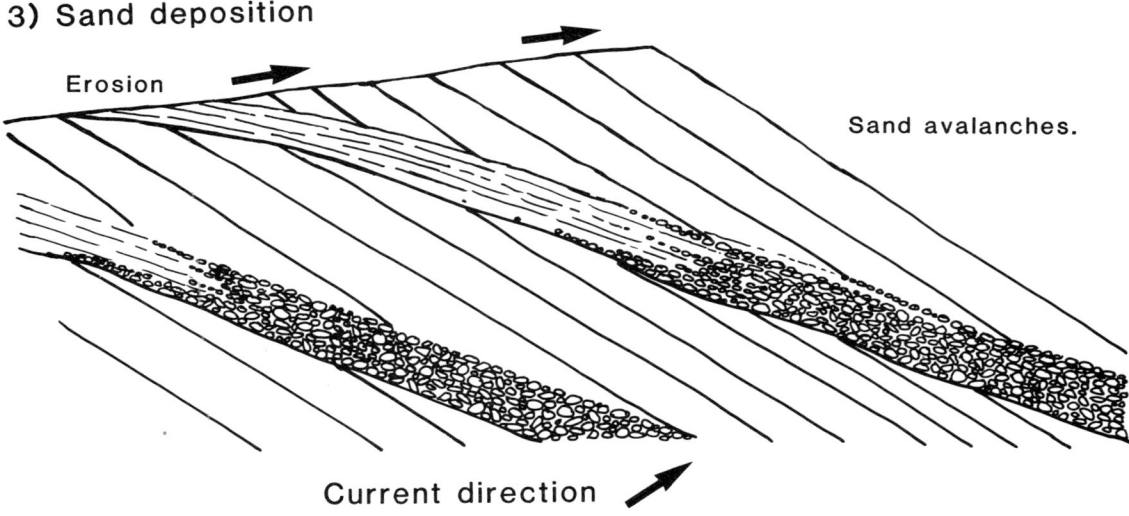

Fig. 16. Suggested depositional sequences in the formation of wedge-shaped conglomerate cross-beds resting on reactivaction surfaces. 1) Currents flowing opposite to the direction of bedform migration initially erode substrate. 2) Increasing tidal velocity transports gravel from center of trough resulting in deposition of gravel with an up-slope decrease in clast size. 3) The current reverses and bedform migrates with sand avalanching down the renewed slip face of the bedform.

In vertical section the conglomeratic trough cross-beds (dune bed forms) change to tabular cross-beds (straight- to sinuous-crested sand waves) of pebbly coarse sandstone, along with a change from essentially northeast-directed to southwest-directed paleocurrent trends. The change in cross-strata geometry may reflect a reduction in the flow velocity, possibly due to increasing water depth, whereas the change in cross-strata orientation may be due to local tidal current dominance.

REFERENCES

Allen, J.R.L. 1973. Features of cross-stratified units due to random and other changes. Sedimentology, v. 20, p. 189-202.

Allen, J.R.L. 1981. Lower Cretaceous tides revealed by cross-bedding with mud drapes. Nature, v. 29, p. 579-581.

Allen, J.R.L. 1982. Sedimentary structures their character and physical basis. Developments in Sedimentology 30B, v. 2, New York: Elsevier Scientific Publishing Company, 663p.

Allen, J.R.L. 1983. Gravel overpassing on humpback bars supplied with mixed sediment: examples from the Lower Old Red Sandstone, southern Britain. Sedimentology, v. 30, p. 285-294.

Allen, J.R.L. and Narayan, J. 1964. Cross-stratified units, some with silt bands, in the Folkstone Beds (Lower Greensands) of southeast England, Geologie en Mijnbouw, v. 10, p. 451-461.

Anderton, R. 1976. Tidal shelf sedimentation: an example from the Scottish Dalradian, Sedimentology, v. 23, p. 429-458.

Baker, V.R. 1973. Paleohydrology and sedimentology of Lake Missoula flooding of eastern Washington. Geological Society of America, Special Paper 144, 69p.

Belderson, R.H. and Stride, A.H. 1966. Tidal current fashioning of a basal bed. Marine Geology, v. 4, p. 237-257.

Belderson, R.H., Kenyon, N.H., Stride, A.H. and Stubbs, A.R. 1972. Sonographs of the Sea Floor. Amsterdam: Elsevier, 185p.

Boersma, R.J. 1967. Remarkable types of mega cross-stratification in the fluviatile sequence of a subrecent distibuary of the Rhine Amergongey: the Netherlands. Geologie en Mijnbouw, v. 6, p. 217-235.

Bourgeois, J. and Leithold, E.L. 1983. Sedimentation, tectonics and sea-level change as reflected in four wave-dominated shelf sequences in Oregon and California. *In:* Larue, D.K. and Steel, R.J. (Eds.), Cenozoic Margin Sedimentation, Pacific Margin, U.S.A. Society of Economic Paleontologists and Mineralogists, Pacific Section, p. 1-16.

Clark, J.C. 1981. Stratigraphy, paleontology, and geology of the central Santa Cruz Mountains, California Coast Ranges. United States Geological Survey Professional Paper 1168, 51p.

Clark, J.C., Brabb, E.E. and Addicott, W.O. 1979. Tertiary paleontology and stratigraphy of the central Santa Cruz Mountains, California Coast Ranges. Field trip guide book, Geological Society of America, Cordilleran Section meeting, San Jose, California, 23p.

Clifton, H.E. 1982. Estuarine deposits. *In:* Horne, M.K. (Ed.), Sandstone Depositional Environments. American Association of Petroleum Geologists Memoir 31, p. 179-189.

Clifton, H.E., Hunter, R.E. and Phillips, R.L. 1971. Depositional structures and processes in the non-barred high energy nearshore. Journal of Sedimentary Petrology, v. 41, p. 651-670.

Clifton, H.E. and Phillips, R.L. 1980. Lateral trends and vertical sequences in estuarine sediments, Willapa Bay, Washington. *In:* Field, M.E., Bouma, A.H., Colbourn, I.P., Douglas, R.G., and Ingle, J.C., (Eds.), Quaternary Depositional Environments of the Pacific Coast. Pacific Coast Paleogeography Symposium 4, Society of Economic Paleontologists and Mineralogists, Pacific Section, p. 55-71.

Collinson, J.D. 1970. Bedforms of the Tana River, Norway. Geografiska Annaler, v. 52, p. 31-56.

Dott, R.H. 1974. Cambrian tropical storm waves in Wisconsin. Geology, v. 2, p. 243-246.

Dupre, W.R., Clifton, H.E. and Hunter, R.E. 1980. Modern sedimentary facies of the open Pacific Coast and Pleistocene analogs from Monterey Bary, California. *In:* Field, M.E., Bouma, A.H., Colburne, I.P., Douglas, R.G., and Ingle, J.C. (Eds.), Quaternary Depositional Environments of the Pacific Coast. Pacific Coast Paleogeography Symposium 4. Society of Economic Paleontologists and Mineralogists, Pacific Section, p. 105-120.

Dyer, K.A. 1971. The distribution and movement of sediment in the Solent, southern England. Marine Geology, v. 11, p. 175-187.

Dyer, K.A. 1972. Bed shear stress and the sedimentation of sandy gravels. Marine Geology, v. 13, p. 31-36.

Flemming, N.C. and Stride, A.H. 1967. Basal sand and gravel-patches with separate indications of tidal current and storm-wave path near Plymouth. Journal of Marine Biology Association, United Kingdom, v. 47, p. 433-444.

Folk, R.L. 1974. Petrology of Sedimentary Rocks. Austin, Texas: Hemphill Publishing Company, 182p.

Graham, J.R. 1982. Wave-dominated shallow-marine sediments in the lower Carboniferous of Morocco. Journal of Sedimentary Petrology, v. 52, p. 1271-1276.

Gregory, M.R., Ballance, P.F., Gibson, G.W. and Ayling, A.M. 1979. On how some rays (Elasmobranchia) excavate feeding depressions by jetting water. Journal of Sedimentary Petrology, v. 49, p. 1125-1130.

Harms, J.C. 1975. Stratification produced by migrating bed forms. *In:* Harms, J.C., Southard, J.B., Spearing, D.R. and Walker, R.G. (Eds.), Depositional Environments as Interpreted from Primary Sedimentary Structures and Stratification Sequences. Society of Economic Paleontologists and Mineralogists, Short Course Notes No. 2, p. 45-61.

Howard, J.D., Mayou, T.V. and Head, R.W. 1977. Biogenic sedimentary structures formed by rays. Journal of Sedimentary Petrology, v. 47, p. 339-346.

Johnson, H.D. 1978. Shallow siliciclastic seas. *In:* Reading, H.G. (Ed.), Sedimentary Environments and Facies. New York: Elsevier, p. 207-258.

Jopling, A.V. 1965. Hydraulic factors controlling the shape of laminae in laboratory deltas. Journal of Sedimentary Petrology, v. 35, p. 777-791.

Klein, G. de V. 1977. Clastic Tidal Facies. Champaign, Illinois: Continuing Education Publication Company, 149p.

Kumar, N. and Sanders, J.E. 1976. Characteristics of shoreface storm deposits: modern and ancient examples. Journal of Sedimentary Petrology, v. 46, p. 145-162.

Langhorne, D.N. 1982. The characteristics and entrainment of gravel in the tidal marine environment. 11th International Sedimentological Congress, Hamilton, Canada, Abstracts of Papers, p. 94.

Larsonneur, C., Bouysse, P. and Auffret, J.P. 1982. The superficial sediments of the English Channel and its western approaches. Sedimentology, v. 29, p. 851-864.

Leckie, D.A. and Walker, R.G. 1982. Storm- and tide-dominated shorelines in Cretaceous Moosebar - Lower Gates interval-outcrop equivalents of deep basin gas traps in western Canada. American Association of Petroleum Geologists Bulletin, v. 66, p. 138-157.

Leithold, E.L. and Bourgeois, J. 1983. Sedimentology of the sandstone of Floras Lake (Miocene) transgression, high-energy shelf deposition, SW Oregon. *In:* Larue, D.K. and Steel, R.J. (Eds.), Cenozoic Marine Sedimentation, Pacific Margin U.S.A. Society of Economic Paleontologists and Mineralogists, Pacific Section, p. 17-28.

Levell, B.K. 1980. Late Precambrian tidal shelf deposit, the Lower Sandfjord Formation, Finnmark, north Norway. Sedimentology, v. 27, p. 539-557.

Miller, M.C., McCave, I.N. and Komar, P.D. 1977. Threshold of sediment motion under unidirectional currents. Sedimentology, v. 24, p. 507-527.

Mitchell, E.D. Jr. and Repenning, C.A., 1963. The chronologic and geographic range of desmostylians. Los Angeles County Museum Contributions in Science, no. 38, 20p.

Phillips, R.L. 1981. Depositional environments of the Santa Margarita Formation in the Santa Cruz Mountains, California. Thesis, University of California, Santa Cruz, 358p.

Phillips, R.L. 1983. Late Miocene tidal shelf sedimentation Santa Cruz Mountains, California. *In:* Larue, D.K. and Steel, R.J. (Eds.), Cenozoic Marine Sedimentation, Pacific Margin, U.S.A. Society of Economic Paleontologists and Mineralogists, Pacific Section, p. 45-61.

Rappeport, M.L. 1980. Depositional environments and Quaternary sedimentary units within lower Cook Inlet, Alaska. *In:* Field, M.E., Bouma, A.H., Colbourn, I.P., Douglas, R.G., and Ingle, J.G. (Eds.), Quaternary Depositional Environments of the Pacific Coast. Pacific Coast Paleogeography Symposium 4, Society of Economic Paleontologists and Mineralogists, Pacific Section, p. 73-88.

Repenning, C.A. and Tedford, R.H. 1977. Otarioid seals of the Neogene. United States Geological Survey Professional Paper 992, 93p.

Rubin, D.M. and McCulloch, D.S. 1980. Single and superimposed bedforms: a synthesis of San Francisco Bay and flume observations. Sedimentary Geology, v. 26, p. 207-231.

Schlee, J. 1973. Atlantic continental shelf and slope of the United States — sediment texture of the northeast part. United States Geological Survey Professional Paper 529-L, p. 1-64.

Smith, N.D. 1972. Some sedimentological aspects of planar cross-stratification in a sandy braided river. Journal of Sedimentary Petrology, v. 42, p. 624-634.

Southard, J.B. 1975. Bed configurations. *In:* Harms, J.C., Southard, J.B., Spearing, D.R. and Walker, R.G. (Eds.), Depositional Environments as Interpreted from Primary Sedimentary Structures and Stratification Sequences. Society of Economic Paleontologists and Mineralogists, Short Course Notes No. 2, p. 5-43.

Stride, A.H., Belderson, R.H., Kenyon, N.H. and Johnson, M.A. 1982. Offshore tidal deposits: sand sheet and sand bank facies. *In:* Stride, A.H. (Ed.), Offshore Tidal Sands, Processes and Deposits. Andover, England: Chapman and Hall Association, p. 95-125.

Steel, R.G. and Thompson, D.B. 1983. Structures and textures in Triassic braided stream conglomerates ('Bunter' Pebble Beds) in the Sherwood Sandstone Group, North Staffordshire, England. Sedimentology, v. 30, p. 341-367.

Veenstra, H.J. 1969. Gravels of the southern North Sea. Marine Geology, v. 7, p. 449-464.

Venkatarathnam, K. and McManus, D.A. 1973. Origin and distribution of sands and gravels on the northern continental shelf off Washington. Journal of Sedimentary Petrology, v. 43, p. 799-811.

Visser, M.J. 1980. Neap-spring cycles reflected in Holocene subtidal large-scale bedform deposits: a preliminary note. Geology, v. 8, p. 543-546.

Visser, M.J. and de Boer, P.L. 1982. The effect of the diurnal inequality on tidal sediments: a tool for the recognition of tidal influences. 11th International Sedimentological Congress, Hamilton, Canada, Abstracts of Papers, p. 162.

Walker, R.G. 1975. Conglomerate: sedimentary structures and facies models. *In:* Harms, J.C., Southard, J.B., Spearing, D.R. and Walker, R.G. (Eds.), Depositional Environments as Interpreted from Primary Sedimentary Structures and Stratification Sequences. Society of Economic Paleontologists and Mineralogists, Short Course Notes No. 2, p. 133-161.

Whitney, J.W., Noonan, W.G., Thruston, D., Bouma, A.H. and Hampton, M.A. 1979. Lower Cook Inlet, Alaska: Do those large sand waves migrate? 11th Offshore Technology Conference, Paper 3484, p. 1071-1082.

Wright, M.E. and Walker, R.G. 1981. Cardium Formation (Upper Cretaceous) at Seebe Alberta - storm-dominated sandstones and conglomerates in shallow marine depositional environments below fair-weather wave base. Canadian Journal of Earth Sciences, v. 18, p. 795-809.

Yorath, C.J., Burnold, B.D. and Thompson, R.E. 1979. Oscillation ripples on the northeast Pacific continental shelf. Marine Geology, v. 31, p. 45-58.

ANCIENT SUBMARINE
SLOPE-FAN SYSTEMS

FAN-DELTA TO SUBMARINE FAN CONGLOMERATES OF THE VOLGIAN-VALANGINIAN WOLLASTON FORLAND GROUP, EAST GREENLAND

Finn Surlyk[1]

Abstract

A major episode of block faulting and tilting occurred in East Greenland roughly at the Jurassic-Cretaceous boundary. In northern East Greenland an up to 3 km thick syntectonic wedge constituting the Wollaston Forland Group was deposited during Middle Volgian to Valanginian time. The group was formed as a coalescent fringe of fan-deltas - submarine fans along a major scarp on tilted blocks. Water depth ranged from zero at the scarp to about 1 km at the basin axis, 15 km east of and parallel to the scarp. The fan sediments are coarse conglomerates, pebbly and coarse sandstones deposited from sediment gravity flows. The flow types include rock-fall avalanches, retrogressive flow slides, sandy debris flows, density modified grain flows, gravelly high-density turbidity currents sometimes transitional to liquefied flows, and low-density turbidity currents. Over 15 km from the scarp to the basin axis there is no downslope change in conglomerate facies types or in the relative proportion. These observations are incompatible with the predicted downslope changes in published conglomerate models. It is shown that there is no unequivocal correlation between proximality (defined as distance from source or fan apex) and travel distance of flows for the Wollaston Forland conglomerates. Furthermore, several of the features used to define the models appear to reflect mechanisms active during deposition rather than transportation. This means that the models only to some extent reflect a succession of stages in flow maturity. The observed contrast between theory and observation can thus easily be accounted for and the Wollaston Forland conglomerates may represent the rule rather than the exception. In this case the use of the published conglomerate models in basin analysis should be discontinued.

Résumé

Un épisode important de formation des blocs faillés et de basculement est apparu dans le Groënland oriental approximativement à la limite séparant le Jurassique du Crétacé. Dans le nord du Groënland oriental un prisme syntectonique jusqu'à 3 km de puissance forme le groupe Wollaston Forland dont l'accumulation des matériaux s'est accompli entre le Volgien moyen et le Valanginien. Le groupe résulte d'une combinaison des zones limitrophes de cônes de déjection-deltas de canyon sous-marin le long d'un escarpement majeur sur des blocs basculés. La hauteur de l'eau variant de zéro à environ 1 km à l'axe du bassin, soit 15 km à l'est de et parallèle à l'escarpement. Les sédiments du cône de déjection comprennent des conglomérats grossiers, des grès grossiers et cailouteux; le dépôt provien de sédiments d'écoulements par gravité. Les types d'écoulements incluent des éboulements d'avalanche, des glissements de terrain rétrogressifs, des coulées de débris de sable, des coulées de grains modifiées par densité, des courants de turbidité graveleux de forte charge parfois transitionnels avec des coulées liquéfiés, et des courants de turbidité de faible charge. Au-delà de 15 km de l'escarpement vers le bassin, aucun changement n'est observé de haut en bas de la pente dans les types de faciès conglomératiques ou de leur proportion relative. Ces observations contredisent les changements de haut en bas de la pente anticipés par les modèles publiés sur les conglomérats. Il est démontré qu'aucune corrélation franche n'existe entre la proximalité (définie comme la distance de la source ou du point le plus élevé du cône de déjection) et la distance de parcours des coulées pour les conglomérats de Wollaston Forland. En plus, plusieurs des particularités utilisées pour définir les modèles semblent refléter les mécanismes qui furent actifs durant la sédimentation plutôt qu'au cours du transport. Ceci signifie que les modèles reflètent seulement dans une certaine limite une succession de stades vers une maturité des coulées. On peut donc rendre compte assez facilement des disparités entre la théorie et l'observation pour les conglomérats de Wollaston Forland et elles peuvent être la règle plutôt que l'exception. Dans ce cas l'utilisation des modèles publiés sur les conglomérats dans l'analyse d'un bassin doit être abandonnée.

Introduction

In the last ten years several authors have attempted to formulate facies models for the varieties of resedimented conglomerates. In the Cambro-Ordovician of Quebec, Davies and Walker (1974) grouped conglomerate beds into five facies using grain size as the main distinguishing feature. They then condensed the five facies into two depositional models, one dominated by inverse and normal grading and the other dominated by stratification. Characteristically, both models have a preferred grain fabric. A third group, the so-called 'disorganized' conglomerates, were excluded from the models following Walker and Mutti (1973).

Walker (1975) combined the facies into a classification scheme, which also relied on information from an extensive literature review. Although the 'disorganized' con-

---

[1]Grønlands Geologiske Undersøgelse, Øster Voldgade 10, DK-1350 København K, Denmark.

The main field work forming the basis for this study was carried out in 1974 in connection with the Geological Survey of Greenland. I had the opportunity to revisit the key sections for a week in 1983 while demonstrating the sequence to a group of geologists from Marathon Oil Co. Ltd. I am very grateful to Karen Kleinspehn for a penetrating and constructive review of the manuscript. I thank John M. Hurst for critically reading the revised version. Bodil Sikker Hansen and Nina Turner are thanked for artwork and typing respectively. The paper is published with permission of the Director of the Geological Survey of Greenland.

Copyright © 1984, Canadian Society of Petroleum Geologists

glomerates were now included in the scheme, Walker (1975) stressed that the classification only applied to clast-supported conglomerates. Thus while the original classification of Walker and Mutti (1973) was based on the presence/absence of grading and organized fabric, Walker's (1975) classification only included clast-supported conglomerates, irrespective of their possessing grading and/or organized fabric.

Both classifications have the advantage that they are purely descriptive but have a weakness in that important classes of conglomerates are excluded. Walker (1975) further suggested that the three facies models could be used in basin analysis, particularly within the framework of the submarine fan model. Thus it was suggested that the disorganized conglomerates were characteristic of the main feeder channel (canyon), whilst the inverse-to-normally graded, the graded, and the graded-stratified conglomerates occupied successive down-channel positions in the main upper fan channels.

Other coarse-grained resedimented facies, which can be generally described as pebbly sandstones have also proved difficult to classify. Walker (1975) noted that the graded-stratified conglomerates which include some pebbly sandstones were superficially similar to the Bouma sequence. However, he stressed that there was probably no direct gradation of facies between the conglomerate models and the Bouma sequence. Aalto (1976) presented a model for coarse sediment gravity-flow deposits, which could not be accommodated with the Bouma model or any other facies model. He recognized a larger sequence of six different layer types which were interpreted as representing transitions from grain support by dispersive pressure to fluid turbulence. As in the case of Walker's (1975) conglomerate models, Aalto (1976) suggested that the sequence can be used to determine proximality in specific submarine fan environments. Recently Lowe (1982) proposed a unified scheme containing Walker's (1975) conglomerate models, Aalto's (1976) pebbly sandstone model, Bouma's (1962) classical fine-grained turbidite model, as well as the somewhat neglected deposits from cohesive flows.

The Volgian-Valanginian conglomerates described in the present study were deposited on a fringe of coalescent fan-delta to submarine fan systems located along a fault scarp on a tilted block. The main field work was done in 1974 and a large number of sections were measured with special emphasis on bed boundaries, grain sizes and sorting, fabric, and grading (Surlyk, 1975a, b, 1978a). The conglomerates can be followed downslope from the scarp, for at least 15 km. There are no observable changes in facies over this distance. More precisely, the ratio of clast versus matrix-supported beds does not change, nor does the relative proportion of non-graded, inverse-to-normally graded, and normally graded beds. Even the clasts do not seem to show any marked size decrease. Furthermore, different conglomerate facies pass rapidly into each other within individual beds and the facies alternate rapidly both vertically and laterally. The only downslope changes over 10-15 km are an overall decrease in thickness of all lithostratigraphic units which is governed by the wedge-shaped geometry of the basin, and a relative volumetric increase in mudstone and sandstone compared to conglomerate.

These observed features are incompatible with the downslope changes hypothesized on the basis of theoretical considerations by Davies and Walker (1974), Walker (1975, 1977), Aalto (1976), Porębski (1981), and Lowe (1982) for resedimented conglomerates. This discrepancy may have two implications. First, the inferred downslope changes in conglomeratic gravity flow do not occur. Second, the conglomerate facies models (Walker, 1975; Aalto 1976; Lowe, 1982) cannot be used in basin analysis and certainly not in interpreting specific fan subenvironments.

I consider the first possibility highly unlikely as the theoretical considerations underlying the concept of downslope changes in conglomeratic flows appears to be basically sound. In particular, Lowe's (1982) outstanding synthesis has wide applicability. In contrast, the Wollaston Forland data clearly show that the environmental importance of the conglomerate models must be considered very doubtful.

It may be argued that the Wollaston Forland Group presents a special case. However, it should be remembered that the published facies models are based on limited field work almost exclusively undertaken in extremely complex tectonic settings where the depositional system cannot be mapped out in three dimensions. In contrast, the Wollaston Forland Group is an up to 3 km thick undisturbed unit which is exposed for 10-20 km across strike and for tens of kilometres along strike (Figs. 1, 2). The conglomerate facies are of the same type as those forming the basis for most of the published models.

In this paper I describe the Wollaston Forland conglomerates specifically in an attempt to explain the apparent discrepancy with the environmental implications of all published conglomerate facies models.

Geological Setting

Post-Devonian East Greenland experienced a series of tectonic episodes characterized by rifting. Jurassic sedimentation was controlled by the interplay of rifting and eustatic sea-level changes (Surlyk, 1978b; Surlyk et al., 1981; Surlyk and Clemmensen, 1983). The tectonic activity reached a climax during Middle Volgian to Valanginian time (Jurassic-Cretaceous) when faulting was associated with pronounced block tilting (Maync, 1947, 1949; Vischer, 1943; Surlyk, 1978a). This resulted in a very irregular topography characterized by cliff-forming fault scarps, narrow deep basins, and elongate islands and shores localized over block crests. The main scarps became fringed by fan-deltas which passed into submarine fans (Fig. 3). The

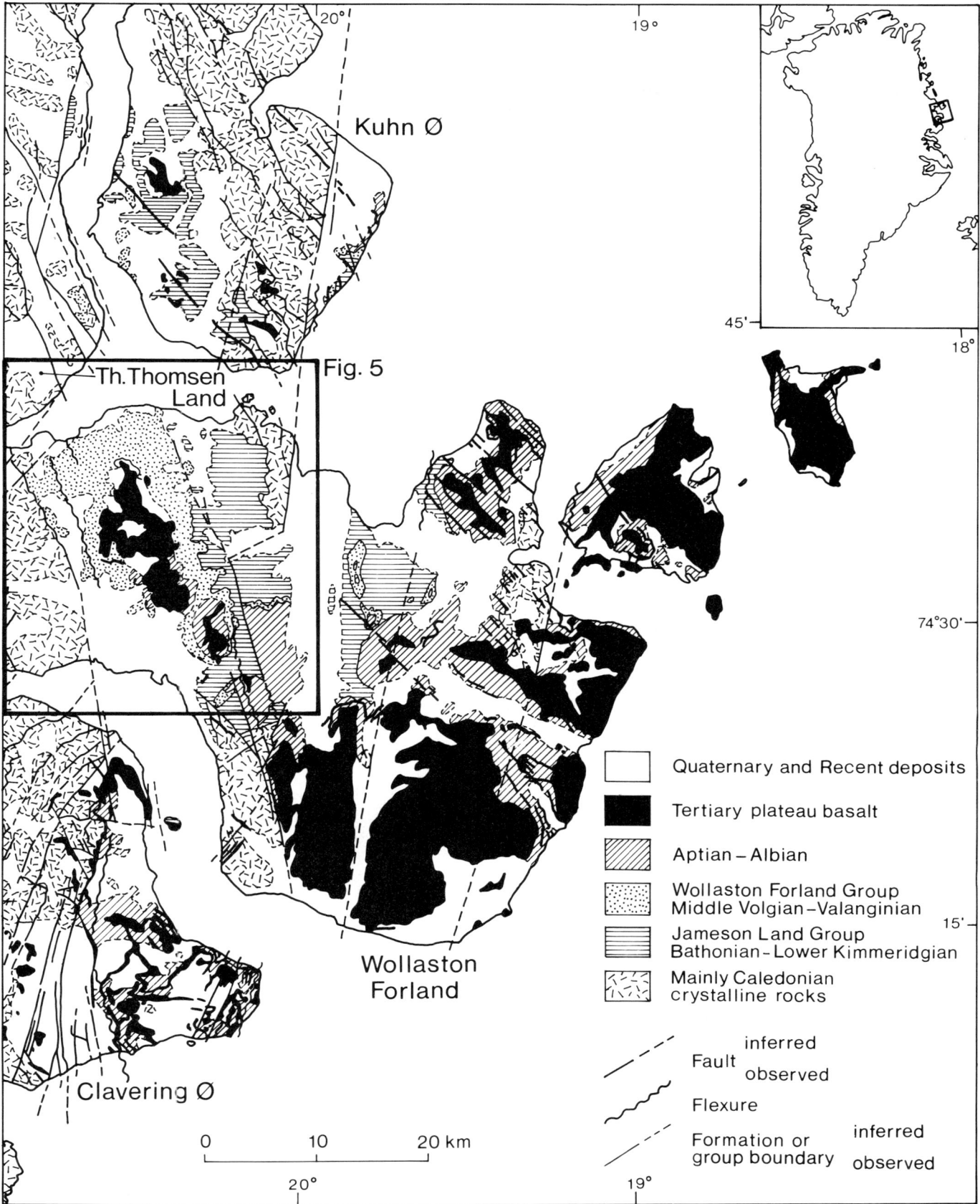

Fig. 1. Geological sketch map of the Wollaston Forland region, northern East Greenland. The main area of study is located in northwestern Wollaston Forland immediately below the name Th. Thomsen Land on the map.

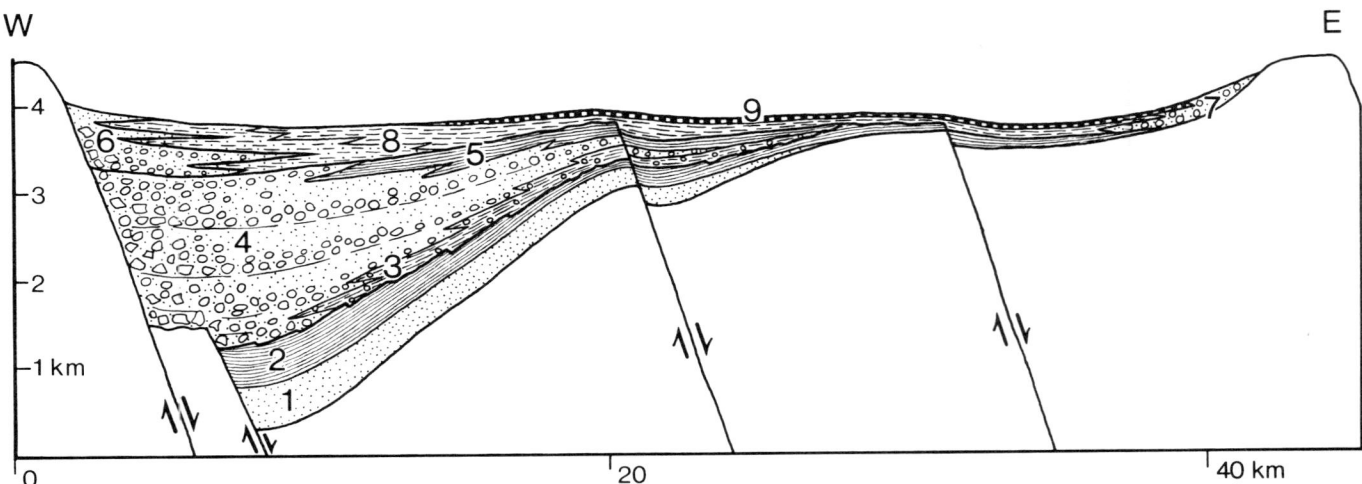

Fig. 2. W-E palinspastic cross-section through northern Wollaston Forland representing Early Cretaceous time (end Valanginian). 1: Vardekløft Formation (Bathonian?-Lower Oxfordian), 2: Bernbjerg Formation (Upper Oxfordian-Lower Kimmeridgian). 3-5: Lindemans Bugt Formation (Middle Volgian-Ryazanian), 3: Laugeites Ravine Member, 4: Rigi Member, 5: Niesen Member. 6-9: Palnatokes Bjerg Formation (Ryazanian-Valanginian); 6: Young Sund Member, 7: Falskebugt Member, 8: Albrechts Bugt Member, 9: Rødryggen Member.

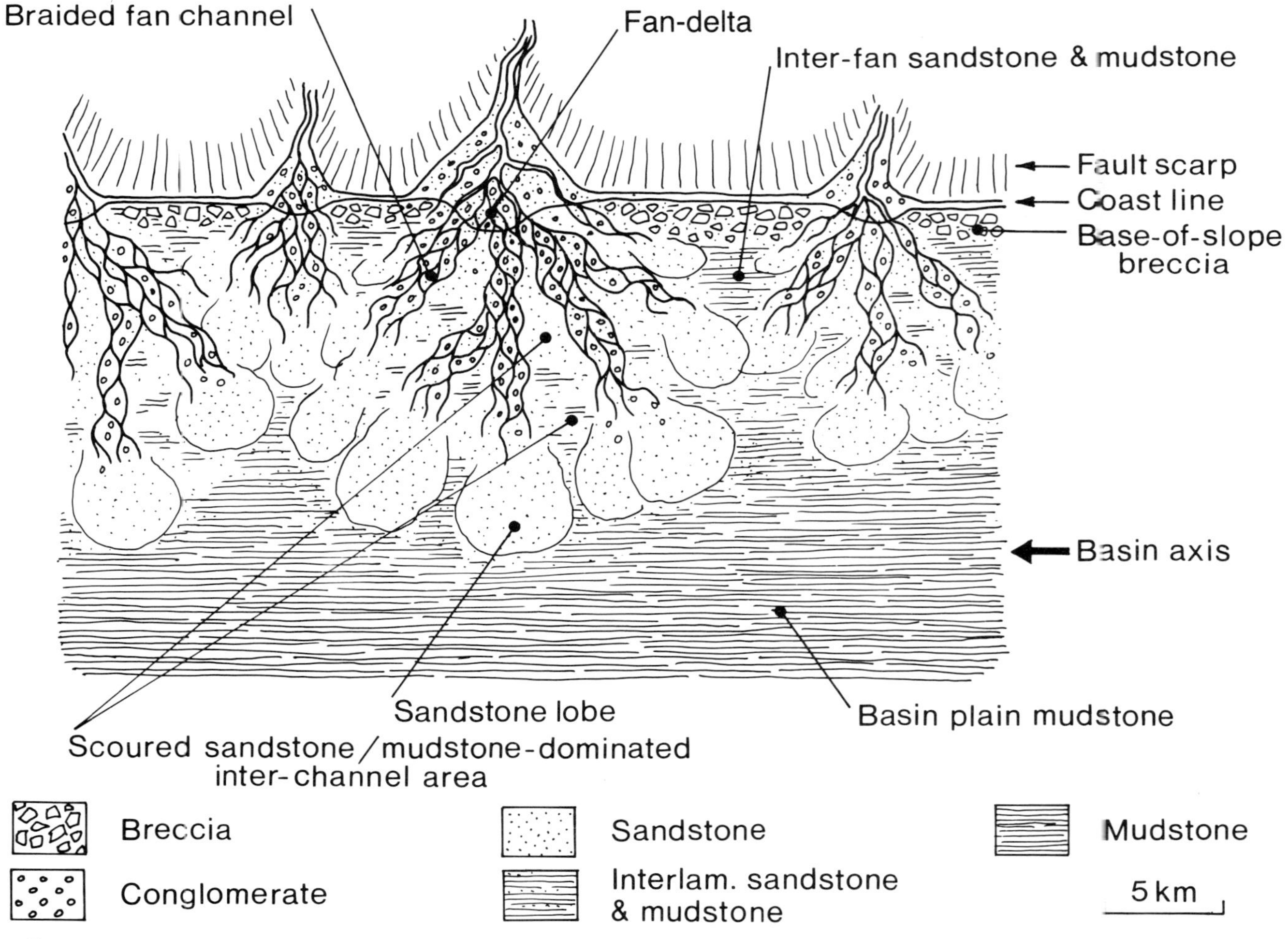

Fig. 3. Plan view of model of a coarse-grained fan-delta to submarine fan based on the Wollaston Forland Group. The model relates facies, fan morphology, depositional environment and tectonic style. A three-dimensional version of the model is shown in Surlyk (1978a, Fig. 41). The differences between this model and the 1978a model are mainly semantic and reflect the standing controversy concerning the use of the terms midfan, suprafan, outer fan, and adjacent basin plain (see Nilsen, 1980). A real difference is the recognition of the distal sandstone lobes. In the 1978a version these fan segments were grouped as the channelled midfan environment reflecting the common occurrence of small-scale fining-upwards cycles. Many of these cycles are here re-interpreted as deposited from surging flows and are thus not indicative of gradual channel filling and abandonment.

water of the depositional basin was deepest immediately adjacent to scarps due to the tilting of the underlying fault blocks (Fig. 4). Fan deposits are thus almost exclusively marine. The ubiquitous occurrence of marine fossils in all facies from the deepest part of the basin to the scarp supports this conclusion.

Fan sediments of the Wollaston Forland Group (Surlyk, 1978a) are dominantly conglomerate with associated pebbly sandstone, sandstone and mudstone. All coarse-grained facies were deposited from various types of sediment gravity flow. The sedimentary sequence displays a pronounced vertical cyclicity (Fig. 4). Small-scale, fining-upward cycles reflect sedimentary processes on the fan surface, while large scale, fining-upward cycles result from major phases of block faulting and tilting (Surlyk, 1978a). A number of facies associations can be recognized, which together with the paleocurrent pattern (Fig. 5), and the small-scale paleotopography leads to recognition of a number of fan subenvironments (Fig. 3).

The Conglomerates

A number of conglomerate facies can be recognized on the basis of clast size, shape, segregation, and fabric, the nature of bed boundaries and geometry, and absence or presence and nature of grading. The conglomeratic facies are described and interpreted below. In this connection it is important to stress that the facies gradually merge into several of the other facies. The facies are described in a systematic way including all possible combinations of grading, segregation and fabric. In practice several of the combinations will of course be combined to facilitate overlook and communication. To complete the facies spectrum the finer grained facies are also included in the scheme of Figure 6, and some of them are briefly described.

1. Non-graded, matrix-supported, conglomerate with random fabric (Figs. 6, 7).

The clasts of this facies vary from pebbles to coarse boulders. Sorting of clast population is poor to good, but is generally better in pebble and cobble conglomerates, while coarse boulder conglomerates may be very poorly sorted. The matrix is poorly sorted coarse, pebbly sandstone with a clay content generally below 1%. Ammonites, belemnites and *Buchia* commonly occur. In most cases bed thickness is between 1 and 8 m. Base of beds may be erosive or conformable.

The lack of grading, absence of sedimentary structures and of an organized fabric indicate deposition from sediment gravity flows related to debris flows. The very low content of clay and the coarse poorly sorted nature of the matrix suggest that deposition was from inertial, non-cohesive sandy debris flows (*cf.* Hampton, 1975; Lowe, 1976, 1982).

2. Non-graded, matrix-supported, imbricated conglomerate (Fig. 6).

This facies is identical to facies 1, except for the *a*-axis clast imbrication. This feature clearly shows that clast interaction played an important role in generating dispersive pressure (Rees, 1968; Walker, 1975).

The lack of grading, the high content of poorly sorted almost clay-free matrix, and the imbrication suggest deposition from sandy debris flows transitional to density modified grain flows (Lowe, 1976, 1982; Rodine and Johnson, 1976).

3. Non-graded, clast-supported conglomerate with random fabric (Fig. 6).

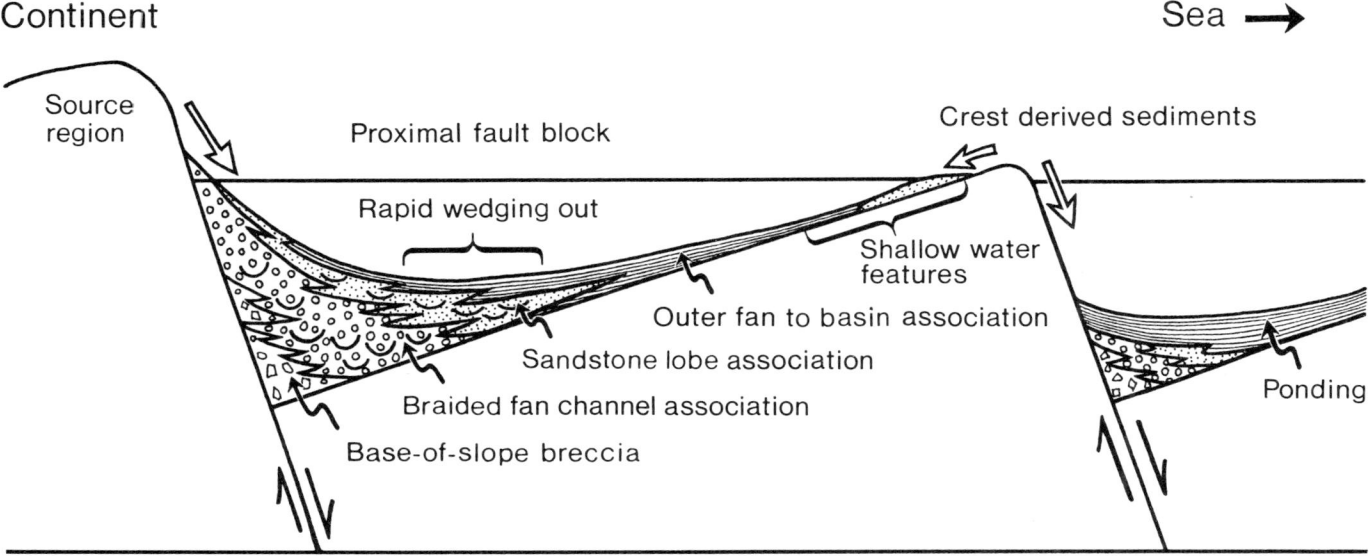

Fig. 4. Model for fan-delta to submarine fan sedimentation along scarps in a tilted fault block situation. The upwards-fining basin fill corresponds to one phase of down-faulting and tilting (modified from Surlyk, 1978a).

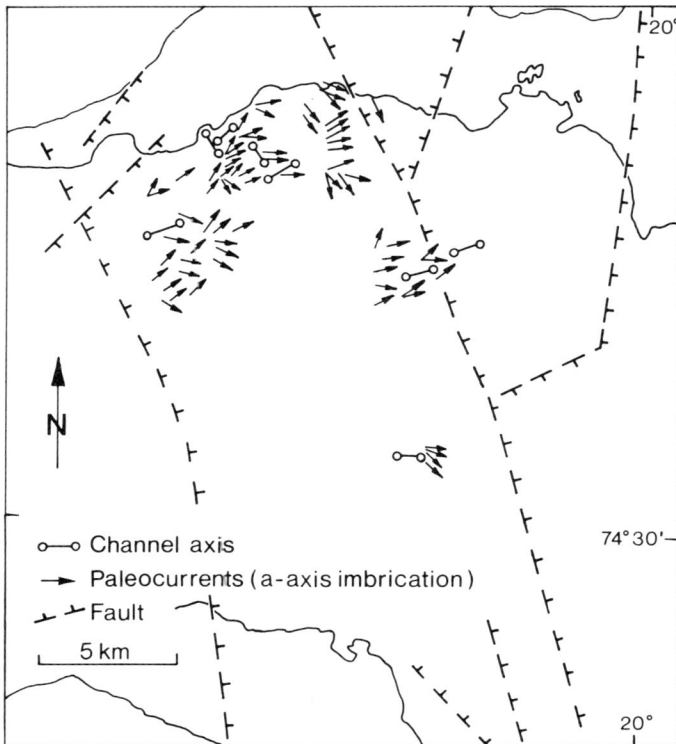

Fig. 5. Paleocurrent map of the western part of Wollaston Forland (inset map in Fig. 1). Each arrow indicates the paleocurrent direction for one major conglomerate bed. Note that the current system was fan shaped with mean direction towards the ENE away from the western border fault.

This facies is identical to facies 1, except for the relatively higher clast content. The absence of imbrication suggests that clast collisions and by implication dispersive pressure played only a minor role in supporting the clasts during the latest stages of the flow. The facies is accordingly interpreted as a result of relatively high-density, non-cohesive sandy debris flows.

Coarse-grained breccias displaying indistinct bedding and almost without a matrix occur at some localities close to the fault scarp. Clast size varies from cm-sized pebbles up to 20-30 m, the most common size being 0.5-1 m. Clasts are usually angular, but all transitions to well rounded clasts may be found, even within the same bed (see Fig. 10). The best section occurs in south-eastern Thomsen Land where a strongly tilted breccia sequence about 100 m in thickness rests on Caledonian basement. The facies invariably contains marine fossils, but the fossils are always rare.

The breccias are interpreted as having been emplaced by catastrophic submarine rock fall avalanches in the scarp zone (Surlyk, 1978a).

4. Non-graded, clast-supported, imbricated conglomerate (Fig. 6).

This facies differs from facies 2 only in its relatively higher clast content. This suggests that dispersive pressure in the present facies played a somewhat greater role in maintaining the clasts in a dispersed state. Furthermore, the lack of grading and stratification suggests that fluid turbulence was relatively unimportant as a clast supporting process.

The facies is thus interpreted as the deposits from sandy debris flows transitional to density modified grain flows with the latter process being of greater importance than in facies 2.

5. Inverse-to-normally graded, matrix-supported conglomerate with random fabric (Fig. 6).

This facies and the succeeding three facies (6-8) are compositionally identical to the preceding facies. They also include some inverse-to-non-graded beds where normal grading has not developed in the upper part of the beds. The facies can be classified as composed of divisions R_2-R_3 of Lowe (1982) although the inverse-to-non-graded beds classify as pure R_2 divisions.

The presence of basal inverse grading suggests deposition from traction carpets followed by suspension sedimentation from high-density turbidity currents.

6. Inverse-to-normally graded, matrix-supported, imbricated conglomerate (Fig. 6).

The a-axis imbricated clast fabric suggests that dispersive pressure caused by clast interactions played a more important role than in facies 5. The facies can be described in terms of Lowe's (1982) R_2-R_3 divisions. Deposition was from frictional freezing of a basal traction carpet followed by suspension sedimentation from high-density turbidity currents (Lowe, 1982) which during the latest stage of flow was transitional to a density modified grain flow.

7. Inverse-to-normally graded, clast-supported conglomerate with random fabric (Fig. 6).

The flows from which this facies was derived only differ from the flows of facies 5 in having a higher clast concentration. The facies is composed of divisions R_2-R_3 of Lowe (1982) and was deposited from below by freezing of a basal traction carpet followed by suspension sedimentation from a high-density turbidity current.

8. Inverse-to-normally graded, clast-supported, imbricated conglomerate (Figs. 6, 8, 9 and 10A).

The basal inverse grading, clast support and c-axis imbrication suggest that clast interactions followed by frictional freezing played a major role also during sedimentation of the normally graded R_3 division. Thus deposition was from a basal traction carpet followed by suspension sedimentation from a high-density turbidity current transitional to a density modified grain flow.

9. Graded, matrix-supported conglomerate with random fabric (Fig. 6).

This facies and the following three facies (10-12) can all be classified in terms of division R_3 of Lowe (1982). Petrographically the graded facies are identical to the preceding conglomerate facies.

The facies results from direct suspension sedimentation from a high-density turbidity current without traction carpet development. Dispersive pressure played little or no

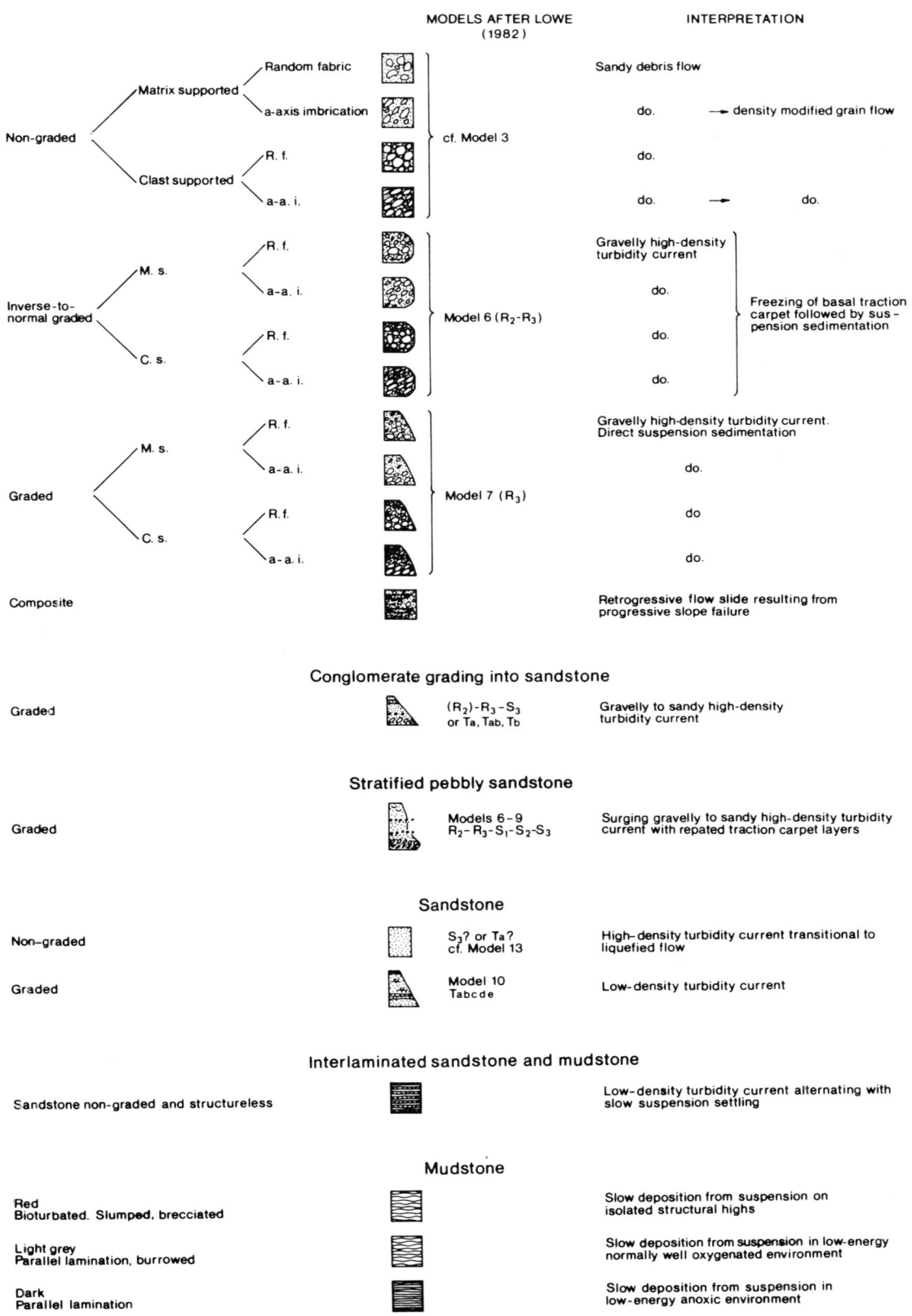

Fig. 6. List of conglomerate facies, with emphasis on their descriptive parameters.

role as a clast supporting factor as evidenced from the lack of an organized fabric and the high matrix content.

10. Graded, matrix-supported, imbricated conglomerate (Fig. 6).

This facies can also be described as the R_3 division of Lowe (1982). Deposition was directly from suspension from a high-density turbidity current. However, the a-axis imbrication shows that dispersive pressure was a dominant factor during the late stage of the flow.

11. Graded, clast-supported conglomerate with random fabric (Figs. 6 and 8A).

This facies can also be included in Lowe's (1982) R_3 division and was deposited directly from suspension from high-density turbidity currents. The high clast concentration suggests that dispersive pressure developed in the flows, but suspension deposition was probably too fast to allow the formation of an a-axis imbricated fabric.

12. Graded, clast-supported conglomerate with imbricated fabric (Fig. 6).

This facies can also be classified in terms of the R_3 division of Lowe (1982), but the high clast concentration and a-axis imbrication suggest that direct suspension sedimentation was immediately followed by a final stage of movement as a density modified grain flow.

13. Composite conglomerate (*sensu* Hendry, 1973) (Fig. 6).

This facies differs from the preceding relatively simple facies in that the beds are often structured in a highly complex manner (Fig. 11, section to the right, 353-360 m). The conglomerates contain more or less well-defined internal, discontinuous sandstone layers. These layers usually have a relatively flat upper surface whereas the lower surface is irregular and always convex downward. Where they wedge out, they thin from the base upwards. The interbedding of

Fig. 7. Matrix-supported conglomerates of sandy debris flow origin from the Rigi Member. **A.** The matrix is coarse sandstone. To the left (out of picture) the bed becomes richer in clasts and an a-axis imbrication is developed. A weak imbrication is seen in the top part of the bed above the 14 cm long pencil. The bed is overlain by an inverse-to-normally graded, a-axis imbricated, pebble conglomerate. **B.** Non-graded, poorly sorted pebble conglomerate. Note that almost all grain sizes up to fine cobbles are present. The clast fabric is random. The conglomerate is jointed from upper right to lower left. Silva compass (encircled) is 10.5 cm long.

Fig. 8. Inverse-to-normally graded clast-supported conglomerates, Rigi Member. **A.** Pebbly sandstone bed showing ripples draped with coaly detritus, overlain by inverse-to-normally graded pebble-cobble conglomerate with random fabric. **B.** Conglomerate showing rather flat a-axis imbrication and inverse-to-normal grading. Base of bed is poorly defined and diffuse. Deposition from gravelly high-density turbidity current with basal traction carpet layer. Pencil in both A and B is 14 cm long.

sandstone and conglomerate varies from isolated thin sandstone layers in conglomerate units many metres thick to extremely complex interbedding.

This facies strongly resembles the composite conglomerates described from Lower Ordovician rocks of Quebec (Hendry, 1973). According to Hendry (1973, p. 135), the deposition of composite beds occurred as a series of events which were closely related in space and time. The development of events is related to the mechanism of initiation of a slide or flow rather than to the transport or depositional mechanism. The composite beds are thus thought to be the products of successive pulses of coarse-grained mass flows resulting from progressive headward failure on a slope (Hendry, 1973). As an alternative to the hypothesis proposed by Hendry (1973), it could be suggested that segregation of different grain size population during flows which when stacked during deposition may produce the complex interbedding. Indeed, this process seems to have been very important in the Wollaston Forland Group, and it is used to interpret facies 15, stratified pebbly sandstone (see Figs. 17, 18, 19 and 20). The present facies 13 is, however, much more complex and it cannot be described by the Lowe (1982) divisions. Thus Hendry's (1973) interpretation involving retrogressive slides and flows on a slope, is followed here.

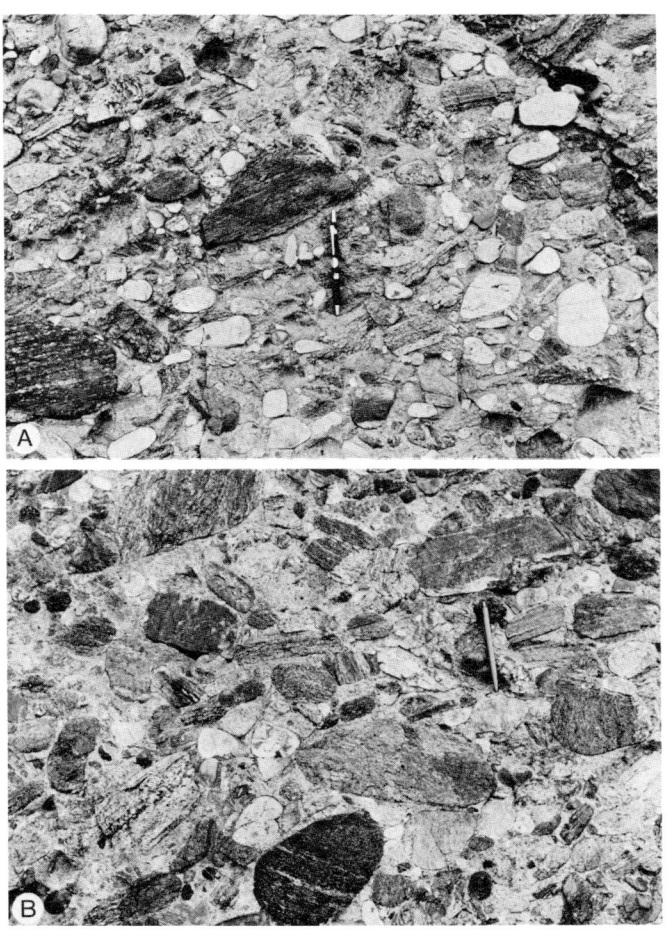

Fig. 9. Clast-supported, a-axis imbricated cobble conglomerate, Rigi Member. A and B show the same bed. Photo A shows a view perpendicular to flow which was from left to right, while B is viewed parallel to flow which was towards the observer. Note that, because of its orientation the section in B shows an apparent subhorizontal clast fabric. Deposition was from a gravelly high-density turbidity current. Pencils (14 cm) for scale.

Fig. 10. A. Inverse-to-normally graded, a-axis imbricated, clast-supported, pebble-cobble conglomerate — Rigi Member; flow was from left to right. Note that the imbricated cobbles in the right-hand side of the photo appear reminiscent of cross stratification. This effect is commonly seen in imbricated conglomerates and can easily be mistaken for foresets formed by currents running in the opposite direction. Deposition was from a gravelly high-density turbidity current with a basal traction carpet. The conglomerate overlies a structureless graded sandstone deposited as Ta division from a sandy high-density turbidity current. B. Chaotic breccia-conglomerate — Rigi Member. Note the presence of outsize clasts comprising extremely angular to rounded boulders of 1-2 m in diameter. Deposition was from a distal rock-fall avalanche transitional to a sandy debris flow. A person stands at the extreme right for scale.

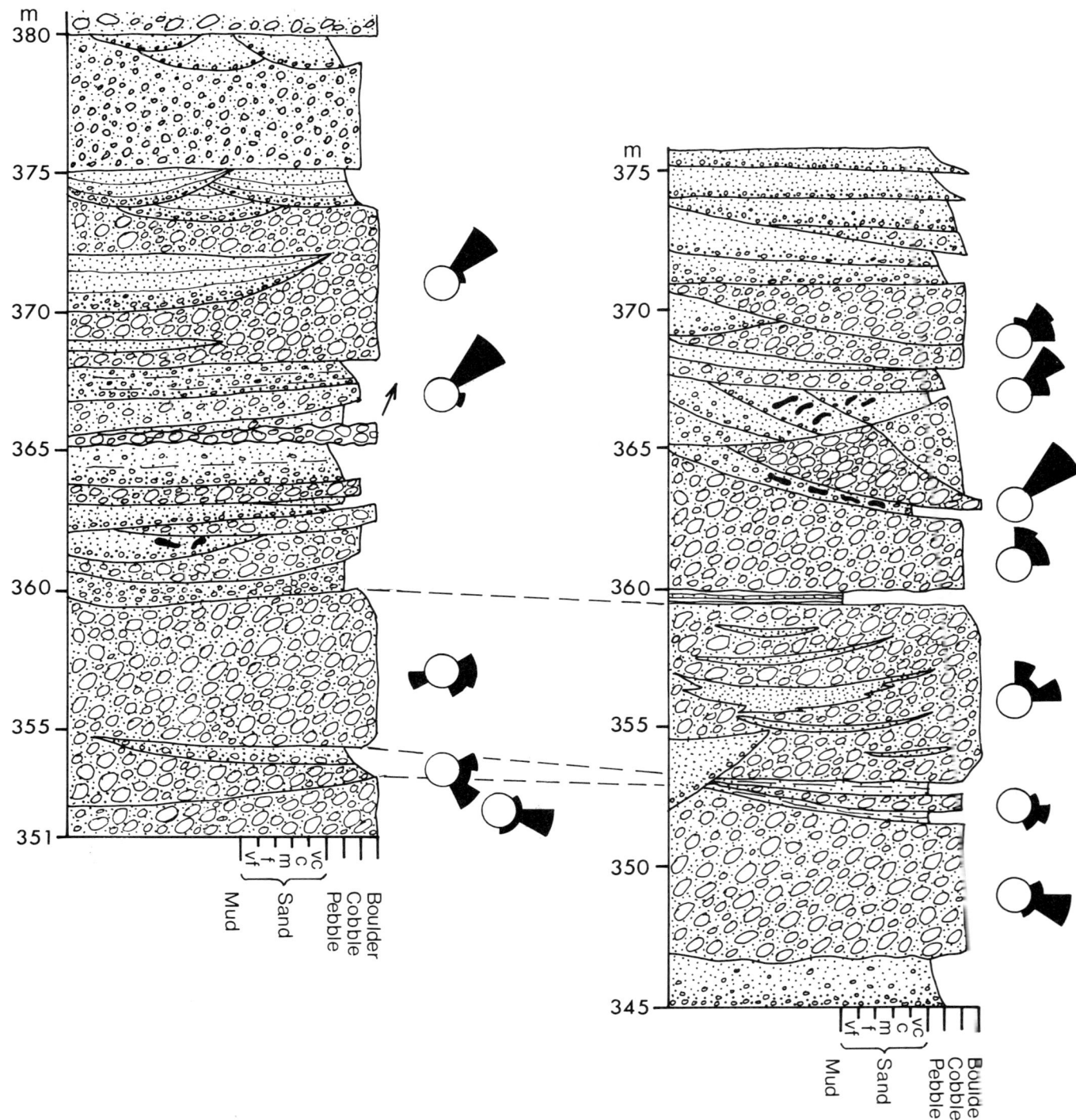

Fig. 11. Strongly channelled and scoured conglomerates and pebbly sandstones, Rigi Member. Section to the left located about 100 m down-current of section to the right. The altitudes indicate the same stratigraphical level in both sections. Only one rather complex bed (ca. 355-360 m) can be traced between sections. Note the systematic change from E to NE orientated paleocurrent directions at about 360 m. The section to the right shows an example of facies 13, composite conglomerate. Modified from Surlyk (1978a).

Fig. 12. Large symmetrical channel filled by five successively thinner and finer grained graded beds, Rigi Member. In bed 1, the grading is from boulder conglomerate at the base upwards to coarse pebbly sandstone. Beds 1 to 5 are interpreted as essentially one complex sedimentation unit deposited by a surging gravelly to sandy high density turbidity current. Flow into the picture away from the observer. Person to the right indicates scale.

Fig. 13. Detail of graded beds filling channel shown in Figure 12. The beds are numbered 3 and 4 in Figure 12. The lowest bed grades from coarse cobble conglomerate to coarse sandstone. The conglomerate (R_3) is clast supported in its lower part and becomes matrix supported in its highest part. The clasts are a-axis imbricated. The oblique lines cutting both beds are joints. R_3 and S_3 are Lowe (1982) symbols for divisions of gravelly and sandy high density turbidities.

14. Conglomerate grading into sandstone (Figs. 6 and 12-16).

Most of the beds referred to this facies (Subfacies 7a and 7b of Surlyk, 1978a) were deposited in channels or scours and are highly lenticular in shape. Lateral and vertical grading within beds occurs and varies from coarse-grained, often pebbly sandstone or pebble/cobble conglomerate at the base to fine-grained sandstone or mudstone at the top. Inverse grading of the basal few centimetres has been noted in a few beds. Distribution grading is most common (Fig. 13), but the coarse-tail type also occurs. The beds are rarely laminated parallel to the channel floor. The clasts characteristically show a well developed a-axis imbrication. The beds are defined in terms of the classical turbidite model (Bouma, 1962) as Ta, Tae, Tab, $Tabe$, Tb or Tbe turbidites. The large grain size, imbrication, dominance of the A-division, and absence of the C- and D-division suggest deposition from high-density turbidity currents. Following the classification of Lowe (1982), they can also be described as R_2-R_3-S_3 turbidites which in some cases are capped by a clay layer representing passive suspension deposition (Te).

Sequences of upwardly thinner and finer-grained beds are common. The example shown in Figures 12 and 13 where five successive beds fill a large channel is interpreted as caused by surging flows. Each surge was characterized by an abrupt velocity increase followed by a gradual deceleration (cf. Fig. 13 and Lowe, 1982, Fig. 11B). The ubiquitous occurrence of small-scale, up to about 10-20 metres, fining-upward cycles in this and comparable facies, both in Wollaston Forland (Fig. 11, section to the right, top 5 m) (Surlyk, 1978a) and elsewhere, are to a large extent the results of surging flows. This type of fining- and thinning-upward cycles occurs to a much larger extent than hitherto recognized.

15. Stratified pebbly sandstone (Figs. 6, 15 and 17-19).

This facies is closely related to the preceding facies, cobble/pebble conglomerate grading into sandstone. It comprises fine to coarse pebble conglomerate layers in coarse sandstone with floating larger clasts scattered throughout (Figs. 17-20). The basal part of the stratified beds is commonly coarser grained than the higher parts and may reach cobble grade (Figs. 17, 18B). The stratification is outlined by layers of pebbles in a matrix of coarse sandstone. The pebble layers are up to about 10 cm thick, and non-graded, inverse-to-non-graded, inverse-to-normally graded, or normally graded. Layers with inversely graded base are most common (Fig. 17). The stratification is in some cases parallel and horizontal, but it may also be curved, apparently outlining bar-like forms (Figs. 19, 20). In the latter case the stratification is superficially reminiscent of large-scale cross-bedding formed by lee side migration of bed forms. However, the pebbles show a distinct a-axis imbrication, and were probably transported near the bed within a highly concentrated traction carpet. Deposition took place by freezing of the traction carpet followed by direct suspension deposition of slightly finer grained material. This depositional mechanism results in inverse-to-normally graded gravel layers. The origin and occurrence of repeated traction carpet layers has been described in detail by Lowe (1982), who classified them as S_2 divisions of his high-density turbidity current model.

The curved pebbly layers in some beds probably reflect lateral accretion during deposition by a single high-density turbidity current over an extremely irregular, scoured surface. The numbers on Figures 19 and 20 indicate beds interpreted as deposited by single flows, while the letter code (following Lowe, 1982) indicates the individual divisions within each flow unit.

The stratified part of the beds are sometimes overlain by weakly graded sandstone (Fig. 17) which rarely contains vague dish structures and is interpreted as a suspension deposit. The paucity of well-preserved dish structures is due to the coarse grain size which allows rapid escape of pore water even at high suspended-load fallout rates.

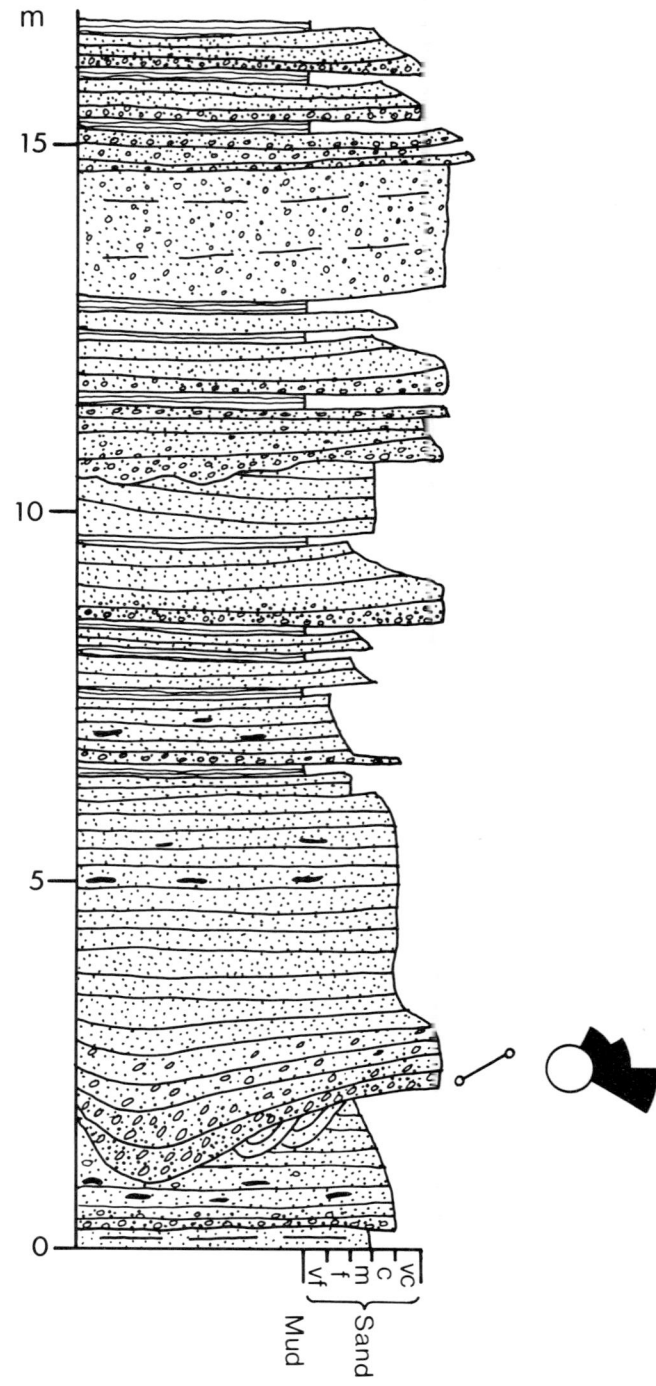

Fig. 15. Lenticular pebbly sandstones, Rigi Member. Note the asymmetric filling of basal scour as seen both from lamination and divergence between scour axis and a-axis pebble imbrication (shown by rose diagram). Modified from Surlyk (1978a).

Fig. 14. Three successive conglomerate and sandstone sedimentation units (1-3) with scoured boundaries, Rigi Member. R_2, R_3 and S_2, S_3 indicate Lowe (1982) divisions deposited by single gravelly and sandy high-density turbidity currents corresponding to units 1, 2 and 3. All conglomerates are clast supported and a-axis imbricated.

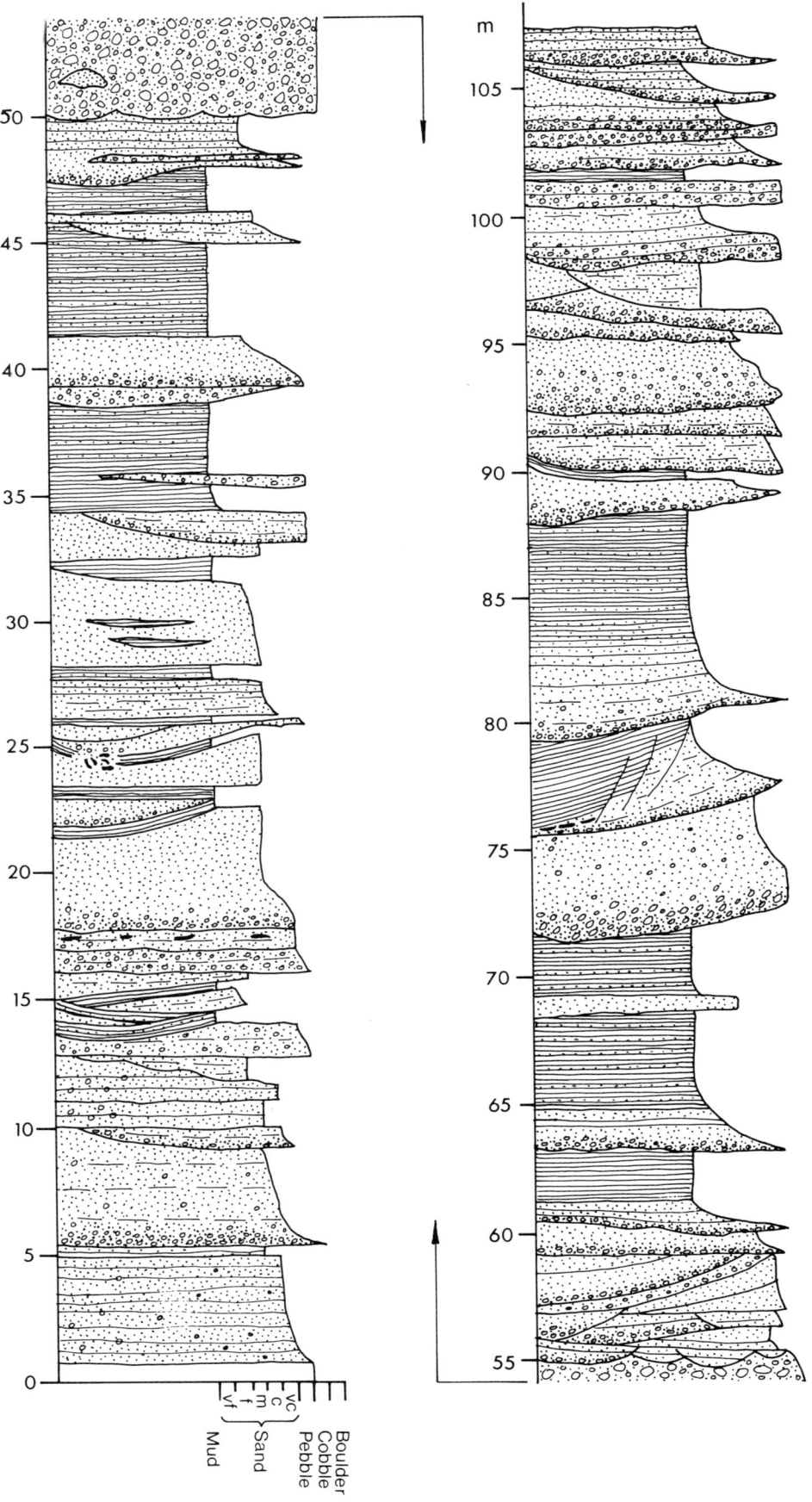

Fig. 16. Channelled and scoured conglomerates, pebbly sandstones, sandstones and mudstones, Rigi Member. Note mud filled channels, mud drapes of channel flows, slumping of channel wall (75-80 m) and dominance of fining-upward cycles. The section is located at the same stratigraphic level as sections on Figure 24 about 1 km north along depositional strike in an inter-channel area (Fig. 3).

Complex, upwards-fining sedimentation units deposited by surging high-density turbidity currents are very common in this facies although not as clearly seen as in the preceding facies.

16. Non-graded sandstone (Figs. 6 and 15).

This facies and the remaining finer grained non-conglomeratic facies are only briefly listed for the sake of completeness.

Fig. 17. Sequence deposited by a single high-density turbidity current during gravelly and sandy sedimentation waves (*cf.* Lowe, 1982, Fig. 11A). R_2-R_3 is an *a*-axis imbricated, inverse-to-normally graded, clast-supported pebble conglomerate deposited by freezing of a traction carpet (R_2) followed by deposition direct from suspension (R_3). S_1 and S_2 are not easy to distinguish in the present case. S_1 reflects deposition from traction as indicated by plane lamination and very low-angle cross stratification. S_1 is characterized by thin horizontal layers of fine, inversely graded gravel representing traction carpet deposits. S_3 was deposited by suspension sedimentation. It is devoid of primary sedimentary structures but contains diffuse dish-structures. Rigi Member. Pocket knife for scale.

Fig. 18. Complex sedimentation unit from a single gravelly to sandy high-density turbidity current, Rigi Member. **A** shows a detailed view of **B** (see inset). The unit starts with a coarse *a*-axis imbricated, graded, clast-supported pebble conglomerate (R_3), followed by a coarse sandstone grading inversely into a coarse pebbly sandstone (A) which can be classified as a combined S_2-R_2 Lowe (1982) division.

The non-graded sandstones are coarse-grained, poorly sorted and in some cases contain scattered pebbles. Sedimentary structures are absent except for rare dish structures. Bed thickness varies from a few centimetres to several metres. They are interpreted as deposited from high-density turbidity currents which during deposition became transitional to liquefied flows (Surlyk, 1978a). The rarity of water escape structures is probably due to the coarse grain size.

17. Graded sandstone (Figs. 6, 15, 16 and 21).

This facies can be classified after the Bouma (1962) model for turbidites as *Ta, Tae, Tab, Tabe*, and in rare cases *Tabcde, Tbcde* and *Tcde*. Deposition was from sandy low density turbidity currents, with the basal *Ta* division which dominates in the present case deposited by direct suspension from high-density flows (Middleton, 1967).

18. Interlaminated sandstone and mudstone (Figs. 6, 16 and 22).

The mudstones are dark, poorly sorted, coarse-grained and well laminated. The sandstones vary in thickness from a few millimetres up to 10 cm but are most commonly about 1 m thick. They are normally parallel laminated but small-scale cross-lamination occurs in some layers. They are non-graded or weakly graded and can be described as Bouma (1962) *Tbc, Tc*, and *Td* divisions. The interbedding does not show any pronounced cyclicity. Deposition of this facies was by quiet mud sedimentation out of suspension under anoxic conditions commonly interrupted by deposition of thin sand layers from small-scale low-density rather immature turbidity currents.

19. Red, bioturbated mudstone (Fig. 4).

The mudstones are massive or weakly laminated, commonly brecciated and contain abundant ammonites, belemnites and *Buchia*. Bioturbation is strong but well preserved trace fossils such as *Zoophycos* are characteristic. Intraformational breccias and slump structures are common. Deposition took place from suspension under low-energy, well-oxygenated conditions.

The facies is restricted to isolated structural highs such as submerged block crests (Surlyk, 1978a p. 40).

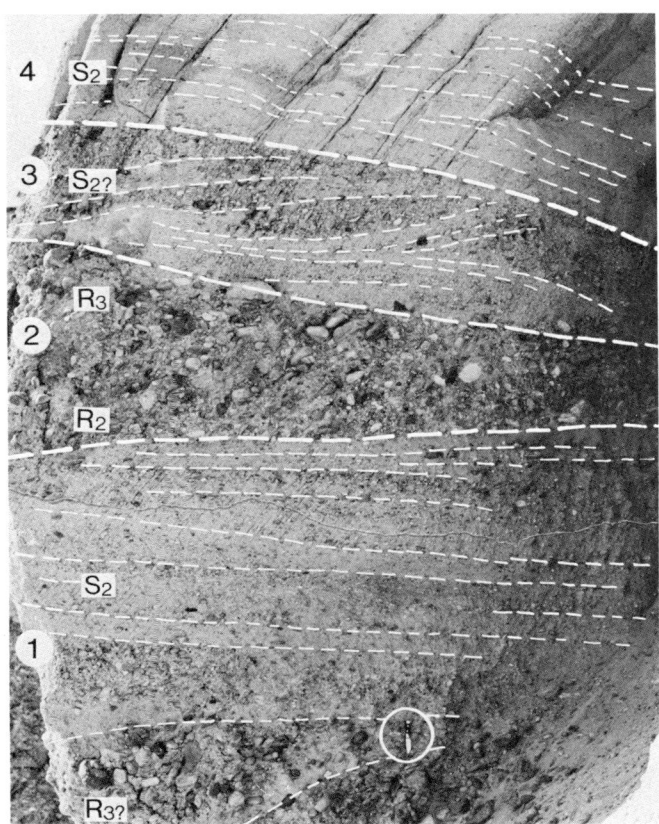

Fig. 19. Extremely complex conglomerate - pebbly sandstone sequence, Rigi Member, same locality as sections on Figure 11. 1-4 indicate individual sedimentation units. Note the irregular, scoured boundaries. R_2, R_3 and S_2 indicate Lowe (1982) divisions in gravelly and sandy high-density turbidity currents. The pebble layers (S_2 and S_2?) are a-axis imbricated and inverse-to-normally graded. Flow was outwards towards right. Note also the curved nature of the S_2 traction carpet layers in bed 3 reflecting formation of bar-like forms during depositon of the bed. This type of oblique stratification is characteristic for highly scoured sequences and should not be confused with real cross-bedding caused by lee-side migration of bed forms. Encircled knife for scale.

Fig. 20. Extremely complex conglomerate - pebbly sandstone sequence, Rigi Member, same locality as Figure 19. 1-9 show successive sedimentation units, while R_2, R_3, and S_2 indicate Lowe (1982) divisions for gravelly and sandy high-density turbidity current divisions, and *Te* is the Bouma *E*-division. Note the curved nature of the S_2 traction carpet layers indicating oblique, lateral accretion of bar-like forms during deposition of the bed. Flow was towards the observer. Encircled knife for scale.

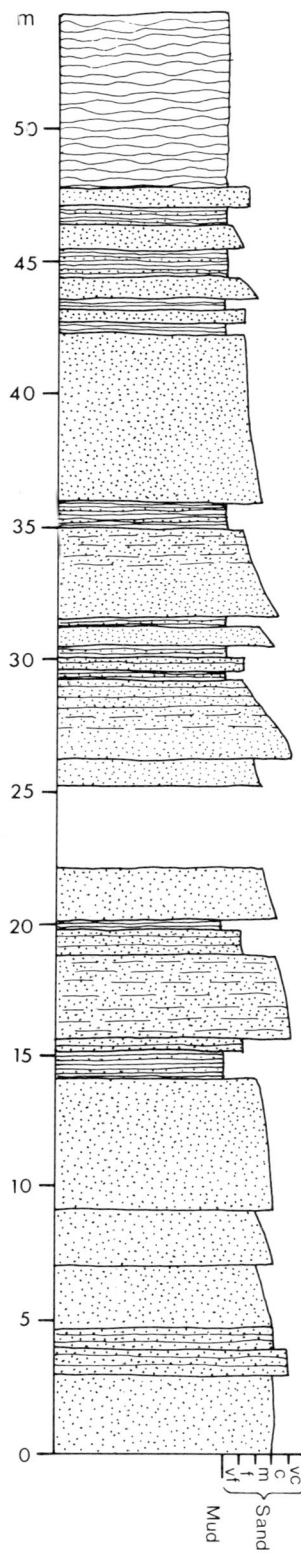

Fig. 22. Interlaminated sandstone and mudstone. **A.** Distal facies, about 13 km from fault-scarp. Deposition of sand from low-density turbidity currents. Laugeites Ravine Member. **B.** About 1 m thick fine-grained unit draping channelled graded sandy turbidite, about 8 km from fault-scarp, Rigi Member.

Fig. 21. Thick graded parallel sided sandstones and thin mudstones, Young Sund member. Note the prominent fining-upward cycles. The sandstone beds can be classified as *Ta*, *Tb* and *Tab* Bouma divisions and were deposited by sandy high-density turbidity currents.

Fig. 23. Channelled and scoured conglomerates and pebbly sandstones, Rigi Member. 1-8 indicate successive sedimentation units. R_2, R_3, and S_2, S_3 indicate divisions deposited by a single gravelly respectively sandy high-density turbidity current (Symbols after Lowe, 1982). Arrow points at boulder which is about 2 m long.

20. Light grey mudstone (Figs. 6 and 21, top 8 m).

The mudstone is bioturbated, laminated or massive and contains *Buchia*, ammonites and belemnites. *Zoophycos* is common. Deposition took place from suspension in a low-energy well oxygenated environment (Surlyk 1978a p. 39-40).

21. Dark, laminated mudstone (Fig. 6).

This facies contains ammonites and *Buchia* at some levels, but trace fossils and bioturbation are generally lacking suggesting deposition under low-energy, poorly oxygenated or anoxic conditions (Surlyk, 1978a, p. 39).

Spatial Distribution of Conglomerate Facies

An important feature characterizing all the conglomerates is the marked lenticularity of the beds (Figs. 11, 14, 19, 20, 23 and 24). Most beds wedge out within a few metres and even the thickest beds can only be traced for at maximum a few tens of metres (Fig. 24). The lenticularity is equally important in slope parallel and downslope directions.

Another very characteristic feature is the rapid lateral and vertical changes in facies (as in the models of Walker, 1975) within the conglomeratic units. The changes between beds possessing different combinations of grading types fabric and matrix and clast support takes place in a totally non-systematic way. The conglomerate facies thus occur in the same relative proportion immediately adjacent to the scarp and for over 10-15 km in a downslope direction. This is clearly revealed by close inspection of a large number of conglomerate sections measured along a 15 km long downslope transect at a right angle to the fault scarp in northern Wollaston Forland (see Surlyk, 1978a — Appendix, sections 14, 15, 17, 20-23, 28, 30-33, 45, 46, 49 and 50).

Rapid lateral facies changes not only take place between beds but also within individual beds. This is particularly the case with the fabric which often shifts gradually from random to *a*-axis imbricated within a few metres. Likewise a thin basal inversely graded zone is commonly observed over a few metres in beds which are otherwise normally graded.

Lateral changes in the clast/matrix ratio occur in many beds and in extreme cases a bed is clast supported in one part and matrix supported in another part. Many thick beds which are generally non-graded show a weak coarse-tail or content grading of the top 0.5 m (Fig. 24, section to the right).

Based on an analysis of the relations between different types of grading, fabric, and clast or matrix support in all measured beds, Surlyk (1978a) concluded:

1. That there is no correlation between type of grading and a specific clast fabric. However, subsequent field

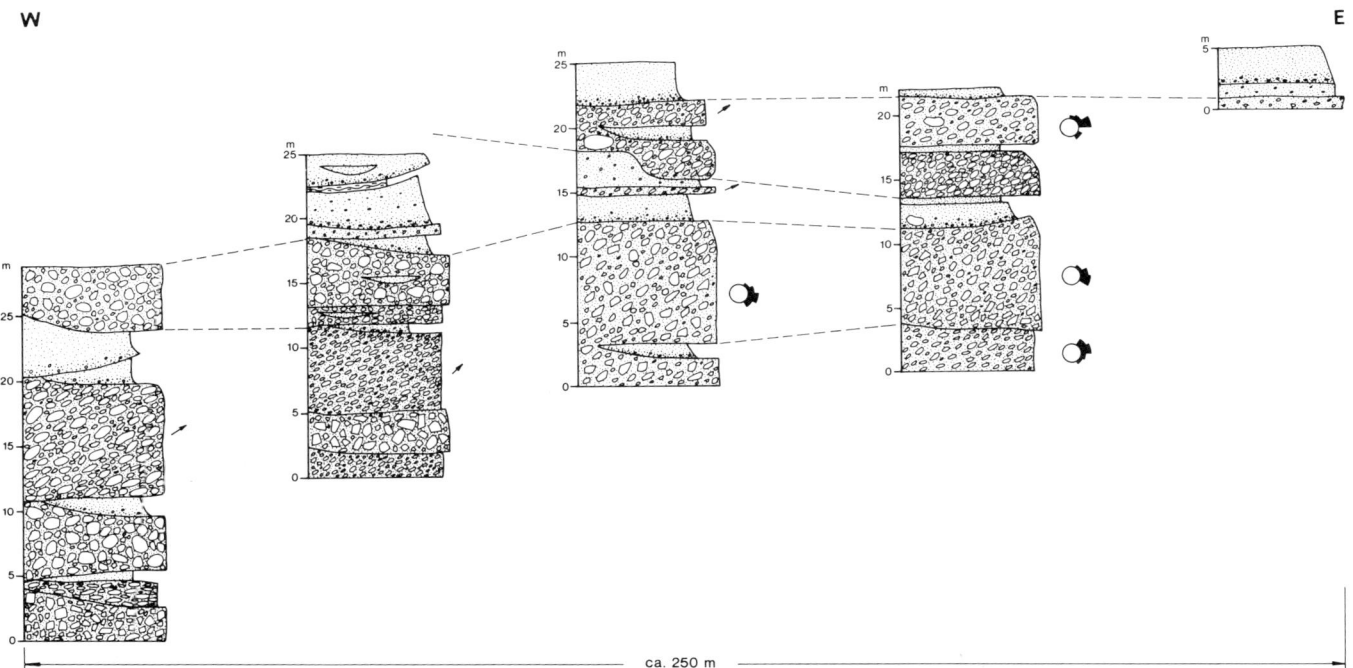

Fig. 24. Correlation of five sections from the Rigi Member. The distance between each section is *ca*. 50-80 m. The sections are distributed on a straight line parallel to paleocurrents which were from left (W) to right (E) as shown by arrows (estimates for whole beds) and roses (based on measurements of a number of imbricated clasts). Note the extreme lenticularity of all beds. None of the beds can be followed between two sections, and only some major conglomeratic or sandy units can be correlated (stippled lines). There are no systematic downslope changes in clast size, fabric, grading type or clast/matrix ratio over the *ca*. 250 m covered by the section line. Modified from Surlyk (1978a).

work in 1983 seems to suggest that inverse-to-normally graded beds tend to show *a*-axis imbrication;

2. That there is no significant correlation between matrix percentage and clast fabric; and

3. That there is no correlation between bed thickness and maximum clast size.

These aspects of the conglomerates are illustrated in Figures 25-27.

The three conclusions correlate well with the observed lack of downslope changes in conglomerate facies.

FACIES MODELS

The facies models for conglomerate and pebbly sandstone proposed by Walker (1975), Aalto (1976) and Lowe (1982) are all descriptive and are based on studies of actual field examples rather than theoretical considerations. After the recognition of the conglomerate models the next logical step in their analysis was to investigate if they occurred in a systematic downslope sequence within individual flows, thus perhaps representing stages in an evolutionary continuum.

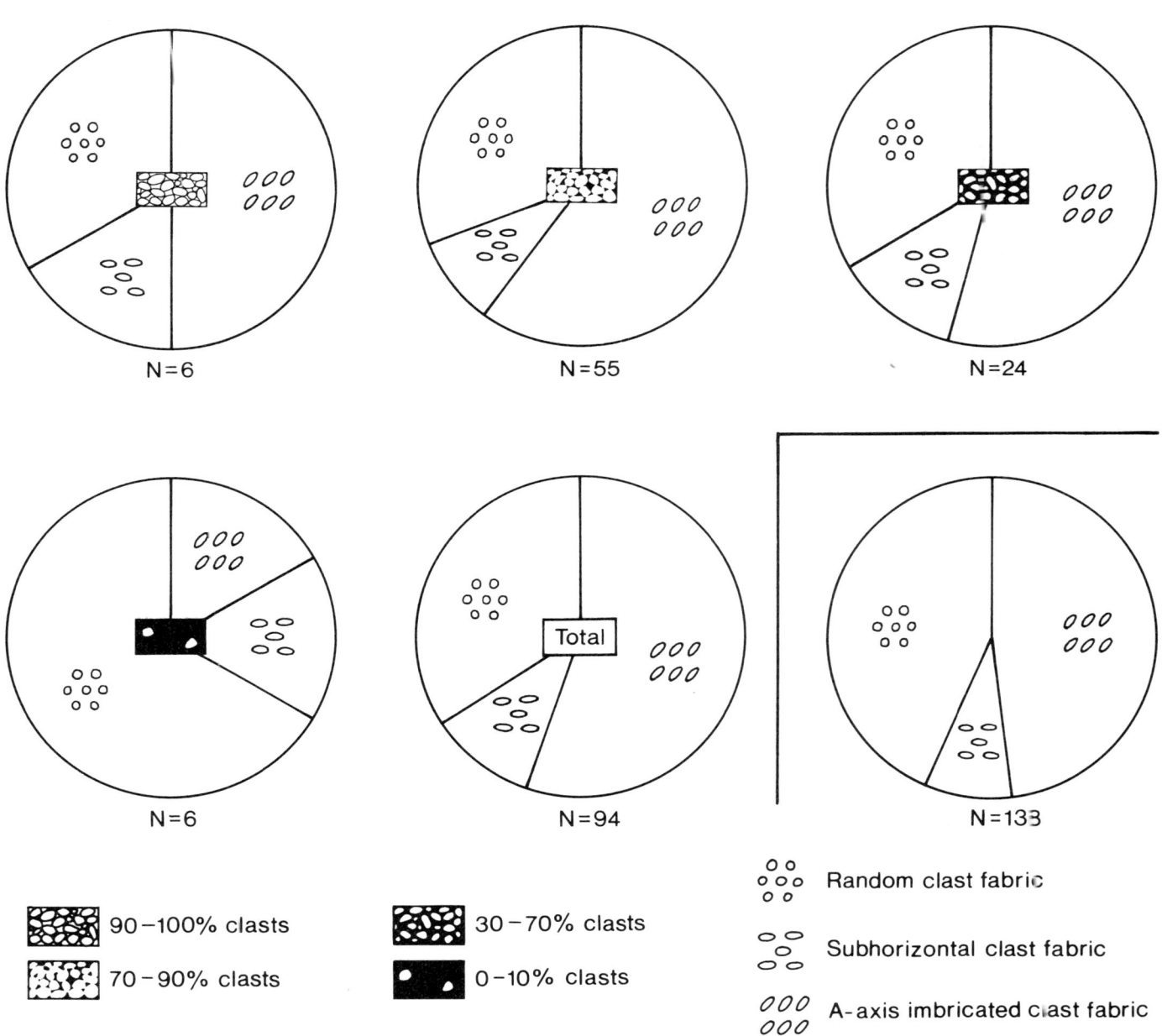

Fig. 25. Circle diagrams showing the relative proportion of the three different types of fabric in clast-supported and matrix-supported conglomerates. The lower right circle includes 44 beds not measured for matrix content. Note that the relative proportion of fabric types is roughly similar for the clast-supported and the matrix-supported conglomerates (compare $N = 55$ with $N = 24$). This suggests that the matrix content played a minor role in the development of a specific fabric.

Walker (1975) suggested that his disorganized-bed model (includes facies 1 and 3 of the present paper) is characteristic of the feeder channels or canyons. The inverse-to-normally graded model (includes facies 5-8 of the present paper) was thought to reflect flow on a relatively steep slope, although downstream from the disorganized-bed model. Finally he suggested that the inverse-to-normally graded model passes into the graded-stratified model where the slope flattens out in midfan areas. Classical turbidites occur distal to the conglomeratic facies, but Walker (1975, p. 746) stated that there is probably no gradation between the conglomerate models and the Bouma model. Aalto (1976) and Lowe (1982) both stressed that none of their layers or models have been *observed* to occur in a down-current succession reflecting the changes in size population and mechanics of transport and sedimentation which occur during a single flow event.

It is thus important to note that the inferences of Walker (1975), Aalto (1976) and Lowe (1982) about downslope changes in facies or models are based on theoretical flow mechanical considerations and not actual tracing of beds in the field or statistical analysis of large numbers of beds investigated along a slope transect. The data of previous studies are in contrast to the field data collected in the relatively well exposed and undisturbed Wollaston Forland Group (Surlyk, 1978a). It should also be mentioned that

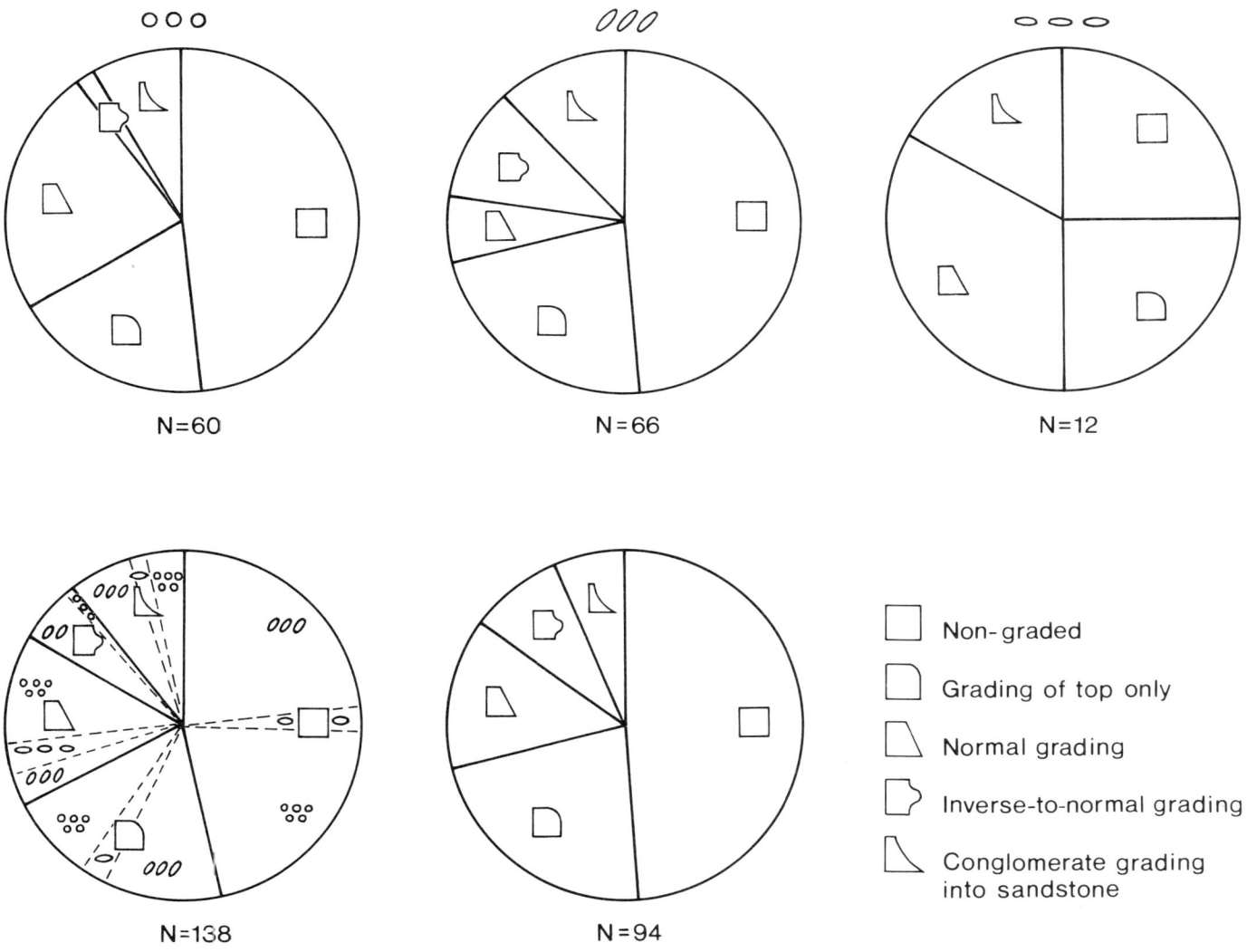

Fig. 26. The upper three circles show the relative proportion of non-graded, weakly graded, strongly graded, inverse-to-normally graded conglomerates, and conglomerates grading into sandstone. The three circles correspond to conglomerates with random, a-axis imbricated, and subhorizontal clast fabric. Note that the different types of grading occur in the same relative abundance in both imbricated and non-imbricated conglomerates. The subhorizontal fabric appears different, however, but this may be due to the small sample number. The circle in the lower right corner includes 94 measurements which were not investigated for relative matrix content. The circle in the lower left shows the sum of the three first circles. Each sector is subdivided to show the proportion of imbricated to horizontal to random fabric. This proportion is practically identical for each type of grading. Consequently there is no correlation between type of fabric and type of grading.

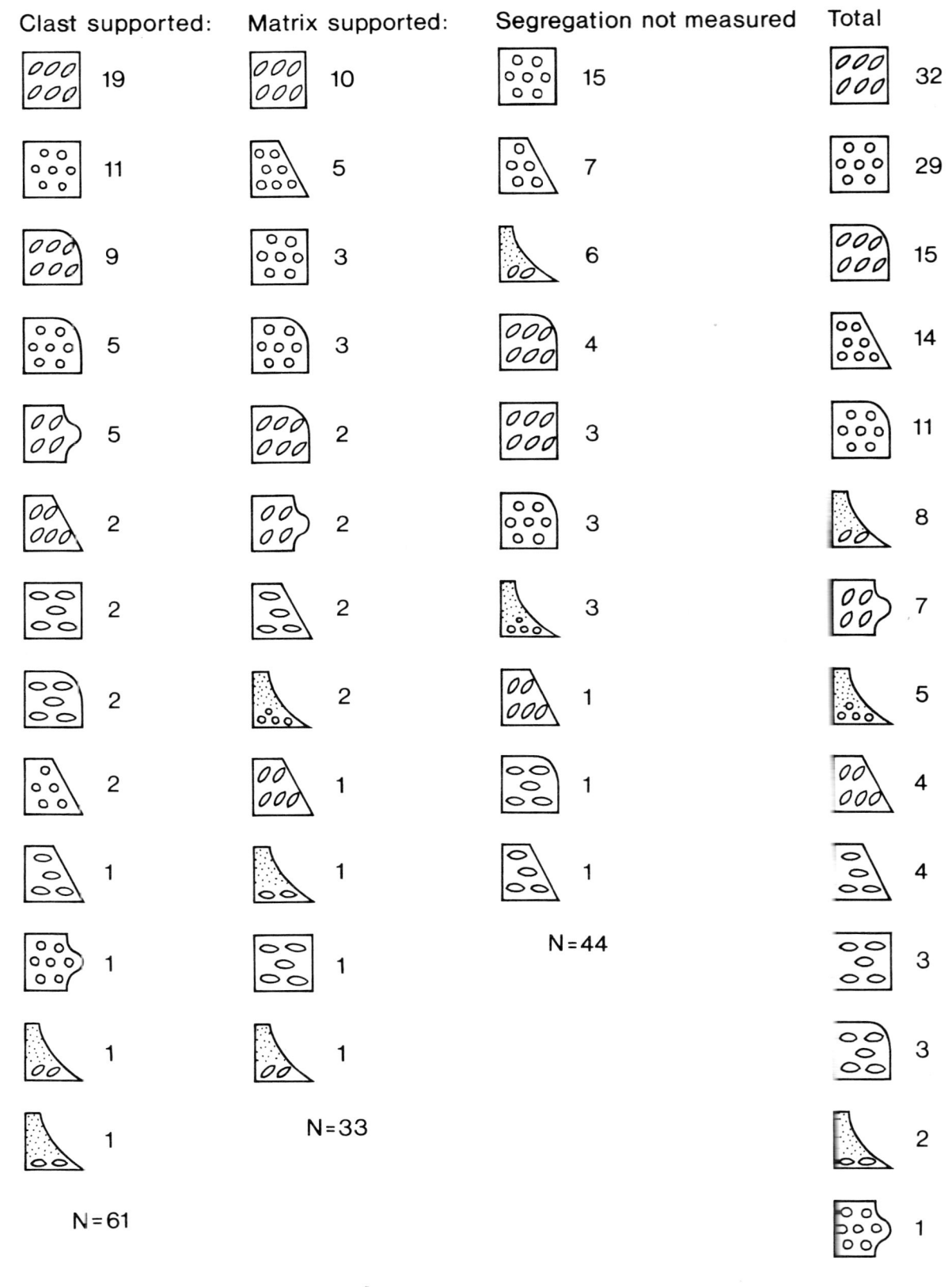

Fig. 27. Ranking of all possible fabric/grading type combinations for clast- and matrix-supported conglomerates respectively. Note that the rank order is roughly the same for the two groups.

this field work was done in 1974 before publication of any conglomerate models, and the field data are thus not prejudiced or influenced by the published models.

As shown above the Wollaston Forland conglomerates do not show any downslope changes in facies types, and are thus contrary to the predictions of Walker (1975), Aalto (1976) and Lowe (1982).

Assuming that the basic premises in their theoretical considerations concerning down slope changes within flows (*i.e.*, flow maturity) are correct, there are several possible explanations for this marked conflict between theory and empirical observations.

A major weakness in the use of conglomerate models in environmental and basin analysis is the mixing of the proximal/distal concept for individual flows and for whole depositional systems. It is apparent that to some extent conglomerate students have repeated the introduction of this mixed concept for turbidites by Walker (1967) (see discussion in Nilsen (1980)).

It is thus implicit in the use of the conglomerate models in inferences of proximality and basin analysis that all or at least most of the flows originated at the same point.

With little justification most authors seem to have taken it for granted that the depositional system of major units of resedimented conglomerates was a submarine fan. In this case the point of origin for the flows would be the fan apex. There is no reason to believe that this is always the case, and it is strongly contradicted by the field evidence from the Wollaston Forland Group. The ubiquitous scouring (Figs. 14-6 and 23), the rarity of real channels, the extreme bed lenticularity (Figs. 14, 19, 20, 23 and 24), the abundance of composite conglomerates (facies 13), the common occurrence of sequences deposited from surging flows probably triggered by retrogressive slumping (Figs. 12 and 13), together with the steep-slope immature coalescing fringe-like nature of the fans (Fig. 3) clearly indicate that flows originated on all parts of the fan surface.

This implies that there is no correlation between distance from the fan apex, *i.e.*, proximality, and actual travel distance of the individual flows. This again shows that the conglomerate models cannot automatically be used in environmental analysis.

Another major premise for using the conglomerate models in basin analysis is that fabric, structure, and grading reflect conditions during transport. If this is true the observational features give information on the maturity or travel distance of the flow. Most conglomerate students seem, however, to agree that several of these features were developed in their final form during the latest stage of flow and deposition. Normal grading indicates that lateral size sorting has taken place within the flow. Walker (1975) suggested on the basis of observations by Middleton (1966) that the degree of development of grading to some extent relates to decreasing slopes and increasing distance of flow. Grading thus seems to reflect the maturity of flow and processes active during transport rather than during deposition. It is thus a feature with an obvious potential in basin analysis. Inverse grading, on the other hand, results from the development and subsequent freezing of a traction carpet at the base of a turbulent flow. It thus reflects concentration effects at the base of the flow, and the occurrence of a succession in one sedimentation unit of inversely graded layers (see Fig. 17 and Lowe, 1982 Figs. 5-8) clearly shows that new carpets have reformed at the rising bed surface after collapsing and freezing of previous traction carpets. This indicates that basal inverse grading mainly reflects processes active during deposition, and its presence is not of any obvious value in basin analysis. Within a particular flow it is, however, very likely that further downslope from the inversely graded deposit extreme flow unsteadines results in direct suspension sedimentation of a graded bed as suggested by Walker (1975, 1977) and Lowe (1982).

Fabric such as the a-axis imbrication, which is so characteristic of many resedimented conglomerates, seems to be the least important feature in basin analysis. As shown above, fabric can change from random to imbricated over a few metres within one bed. Furthermore, imbrication is equally common in clast- and matrix-supported beds (Fig. 25, Surlyk, 1978a). Imbrication is very common in graded beds, and normal grading suggests direct suspension sedimentation from a relatively mature turbulent flow. A-axis imbrication is related to the statistically preferred orientation of moving and colliding clasts (Rees, 1968), and was developed during bulk shearing of high concentration of clasts (Walker, 1977). A fully turbulent flow will not show any clast imbrication but perhaps a clast alignment parallel to flow. Consequently it is suggested that the imbrication is developed during rapid deposition from a highly concentrated suspension. There is no clear correlation between the travel distance of the flow, and the presence or absence of inverse grading and imbricated fabric as these features to a large extent reflect mechanisms active during deposition.

It is apparent that the conglomerate models of Walker (1975, 1977), Aalto (1976) and Lowe (1982) are of little or no use in basin analysis of the Wollaston Forland Group. They are, however, very useful as descriptive models and in detailed analyses of transport and sedimentation mechanics of individual beds. Most of the resedimented conglomerate sequences described in the literature are very similar to the Wollaston Forland conglomerates in terms of facies characteristics, but their tectonic-sedimentologic setting and the nature of the depositional system are unknown, in most if not all cases, and are only indirectly inferred from facies. I therefore conclude that the conglomerate models should not be used to directly infer depositional environments in basin analysis.

Although this conclusion contrasts with much of the published literature on resedimented conglomerates it is simply analogous 1) to the much earlier realisation that specific sedimentary structures in shallow marine or fluvial sandstones are not diagnostic of specific environments, and 2) to the abandonment of Walker's (1967) proximality index for turbidites in *environmental* reconstruction.

SUMMARY AND CONCLUSIONS

The Middle Volgian-Valanginian Wollaston Forland Group was deposited as a coalescent fringe of fan-deltas - submarine fans along a major scarp on tilted blocks (Figs. 2 and 3). The subaerial part of the fan fringe was very narrow (max. 1 km) and rarely preserved.

Water depth ranged from zero at the scarp to about 1 km at the basin axis, 15 km east of and parallel to the scarp.

The dominant fan sediments are coarse conglomerates, and pebbly and coarse sandstones. All sediments were deposited from sediment gravity flows.

The flow types include rock-fall avalanches, retrogressive flow slides, sandy debris flows, density modified grain flows, gravelly high-density turbidity currents with basal traction carpets, sandy high-density turbidity currents sometimes transitional to liquefied flows, and low-density turbidity currents. Transition between flow types, especially during deposition, seems to have been common.

The deposits can be described in terms of the classification of Lowe (1982). Two types of cyclicity are recognized: small-scale fining-upwards cycles a few metres thick, and large-scale fining-upwards cycles. Coarsening-upwards cycles have not been observed.

Most of the small-scale cycles result from surging flows. This is considered a far more satisfactory explanation than the generally accepted one; that is gradual filling and abandonment of channels. However, the latter process appears to have been responsible for successions of conglomerate, pebbly sandstone and sandstone beds separated by mud drapes which fill some channels or scours. In the latter case the surging flow interpretation cannot be used.

The large-scale cycles are interpreted as reflecting major tectonic episodes of block faulting and tilting (Fig. 26; Surlyk, 1978a). Tilting caused deepening of the basin immediately adjacent to the scarp and thus steepening of fan slopes. Simultaneously the subaerial scarp area was uplifted and rejuvenated. The gradual filling of the basin and progressive erosion of scarp area and immediate hinterland resulted in gradual flattening of fan slope and deposition of the large-scale upwards-fining basin fill cycles (Fig. 4).

Fan progradation, resulting in large-scale coarsening-upwards cycle, did not take place in this particular tectonic setting because the steepest slopes and deepest water occurred immediately adjacent to the source area. Water depth gradually decreased east of the basin axis and the slope of the sea-bottom reflects the dip-slope of the underlying fault block (Fig. 4). The source area was the Caledonian gneisses, granites and related rocks exposed west of the fault scarp.

The individual conglomeratic facies have been examined with respect to clast-matrix proportion, maximum and average clast size, matrix, grain size, petrography, grading, fabric, stratification, bed boundaries and fossil content.

The conglomerate clasts vary from pebbles to boulders with dominance of coarse cobble and boulder conglomerates.

The matrix is coarse-grained extremely poorly sorted sandstone. The dominant clast fabric is a-axis imbrication, but random and subhorizontal fabric also occurs. A b-axis imbrication has only been observed in a few cases immediately adjacent to the scarp. This is interpreted as reflecting fluvial deposition in the narrow subaerial part of the fan-delta. Marine fossils, particularly ammonites and the bivalve *Buchia* occur in all facies. There is no correlation between clast percentage and type of fabric.

That is, clast and matrix-supported beds show the same proportion of random, subhorizontal or a-axis imbricated fabric.

Furthermore, there is, no correlation between type of fabric and type of grading. However, field work in 1983 suggests, that inverse-to-normally graded beds are generally imbricated. This was not so clearly reflected in the sections illustrated by Surlyk (1978a).

Finally, there is no correlation between maximum clast size and bed thickness (Surlyk, 1978a). The lack of correlation between fabric, clast percentage, and grading are somewhat unexpected but it is supported by quantitative data (Figs. 25-27; Surlyk, 1978a) and was carefully checked in a large number of beds in 1983. The positive correlation between bed thickness and maximum clast size commonly reported in the literature (Bluck, 1967; Steel, 1974) seems to mainly apply to deposits from viscous debris flows. The Wollaston Forland sediments were all deposited from non-viscous, inertial flows. This difference may be of importance and may deserve further investigation.

Over 15 km from the scarp to the basin axis there is no downslope change in conglomerate facies types or in their relative proportion. These observations are incompatible with the predicted downslope changes in resedimented conglomerates suggested by Walker (1975, 1977), Aalto (1976) and Lowe (1982).

This contrast between theory and empirical observations is intriguing and important to understand, not least because, the use of the published conglomerate models in basin analysis has to be abandoned if the Wollaston Forland conglomerates represent a universally applicable case.

Three major premises are implicit in the use of the models in basin analyses:

1) The depositional environment is in most cases a submarine fan.

2) All flows originated from one point, *i.e.*, the fan apex. In this way there is correlation between travel distance of flow and position of the bed on the fan.

3) The main features of the models *i.e.*, clast percentage, grading, stratification and fabric reflect processes active during transport.

The first point has only been independently demonstrated in very few cases (*e.g.*, Surlyk, 1978a; Cazzola *et al.*, 1981). Coarse resedimented conglomerates are normally found in strongly disturbed often exotic terranes, and the environmental interpretation is based solely on facies analysis.

The second point is invalidated in the case of the Wollaston Forland Group, where it can be demonstrated that flows originated from all parts of the fan surface. This is most probably also the case for a large part of other submarine conglomeratic fans and related systems described in the literature as their constituent facies are highly reminiscent of those described here (see examples in Davies and Walker, 1974; Hendry, 1976, 1978; Johnson and Walker, 1979; Porębski, 1981; Walker, 1975, 1977).

There is thus no unequivocal correlation between proximality (defined as distance from source or fan apex) and travel distance of flows. This again implies that published conglomerate models cannot be used to define proximality, but only to give inferences about flow maturity.

Concerning the third point it is well known that several of the features used to define the models reflect mechanisms active during deposition rather than transportation. This means that the models only to some extent reflect a succession of stages in flow maturity.

The observed contrast between theory and observation can thus easily be accounted for and the Wollaston Forland conglomerates may represent the rule rather than the exception. In this case the use of the published conglomerate models in basin analysis should be discontinued.

It is accordingly emphasized that sequences dominated by resedimented conglomerates should not be interpreted only on the basis of the facies types present. Interpretation of depositional environment has to be based on the sum of tectonic setting, overall geometry of the deposit, spatial distribution of facies associations, large- and small-scale cyclicity, small-scale paleotopography, paleocurrents and fossil content.

References

Aalto, K.R. 1976. Sedimentology of a mélange: Franciscan of Trinidad, California. Journal of Sedimentary Petrology, v. 46, p. 913-929.

Bluck, B.J. 1967. Deposition of some Upper Old Red Sandstone conglomerates in the Clyde area: a study in the significance of bedding. Scottish Journal of Geology, v. 3, p. 139-167.

Bouma, A.H. 1962. Sedimentology of some flysch deposits. A graphic approach to facies interpretation. Amsterdam: Elsevier Publishing Company, 168p.

Cazzola, C., Fonnesu, F., Mutti, E., Rampone, G., Sonnino, M. and Vigna, B. 1981. Geometry and facies of small, fault-controlled deep-sea fan systems in a transgressive depositional setting (Tertiary piedmont basin, northwestern Italy). *In:* Ricci Lucchi, F. (Ed.), International Association of Sedimentologists, 2nd European Regional Meeting Excursion Guidebook, Bologna, p. 5-53.

Davies, J.C. and Walker, R.G. 1974. Transport and deposition of resedimented conglomerates: the Cap Enragé Formation, Cambro-Ordovician, Gaspé, Québec. Journal of Sedimentary Petrology, v. 44, p. 1200-1216.

Hampton, M.A. 1975. Competence of fine-grained debris flows. Journal of Sedimentary Petrology, v. 45, p. 838-844.

Hendry, H.E. 1973. Sedimentation of deep water conglomerates in Lower Ordovician rocks of Québec — composite bedding produced by progressive liquefaction of sediment? Journal of Sedimentary Petrology, v. 43, p. 125-136.

Hendry, H.E. 1976. The orientation of discoidal clasts in resedimented conglomerates, Camro-Ordovician, Gaspé, Eastern Québec. Journal of Sedimentary Petrology, v. 46, p. 48-55.

Hendry, H.E. 1978. Cap des Rosiers Formation at Grosses Roches, Québec — deposits of the mid-fan region on an Ordovician submarine fan. Canadian Journal of Earth Sciences, v. 15, p. 1472-1488.

Johnson, B.A. and Walker, R.G. 1979. Paleocurrents and depositional environments of deep water conglomerates in the Cambro-Ordovician Cap Enragé Formation, Quebec Appalachians. Canadian Journal of Earth Sciences, v. 16, p. 1375-1387.

Lowe, D.R. 1976. Grain flow and grain flow deposits. Journal of Sedimentary Petrology, v. 46, p. 188-189.

Lowe, D.R. 1982. Sediment gravity flows: II. Depositional models with special reference to the deposits of high-density turbidity currents. Journal of Sedimentary Petrology, v. 52, p. 279-297.

Maync, W. 1947. Stratigraphie der Jurabildungen Ostgrönlands zwischen Hochstetterbugten (75° N.) und dem Kejser Franz Joseph Fjord (73° N.). Meddr Grønland, v. 132, 223p.

Maync, W. 1949. The Cretaceous beds between Kuhn Island and Cape Franklin (Gauss Peninsula), northern East Greenland. Meddr Grønland, v. 133, 291p.

Middleton, G.V. 1966. Experiments on density and turbidity currents. I. Motion of the head. Canadian Journal of Earth Sciences, v. 3, p. 523-546.

Middleton, G.V. 1967. Experiments on density and turbidity currents. III. Deposition of sediment. Canadian Journal of Earth Sciences, v. 4, p. 474-505.

Nilsen, T. 1980. Modern and ancient submarine fans; discussion of papers by R.G. Walker and W.R. Normark. American Association of Petroleum Geologists Bulletin, v. 64, p. 1094-1112.

Porębski, S.J. 1981. Swiebodzice succession (Upper Devonian-lowest Carboniferous; Western Sudetes): a prograding, mass-flow dominated fan-delta complex. (English summary). Geologia Sudetica, v. 16, p. 101-192.

Rees, A.I. 1968. The production of preferred orientation in a concentrated dispersion of elongated and flattened grains. Journal of Geology, v. 76, p. 457-465.

Rodine, J.D. and Johnson, A.M. 1976. The ability of debris, heavily freighted with coarse clastic materials, to flow on gentle slopes. Sedimentology, v. 23, p. 213-234.

Steel, R.J. 1974. New Red Sandstone floodplain and piedmont sedimentation in the Hebridean Province, Scotland. Journal of Sedimentary Petrology, v. 44, p. 336-357.

Surlyk, F. 1975a. Fault controlled marine fan-delta sedimentation at the Jurassic-Cretaceous boundary, East Greenland. 9th International Sedimentological Congress, Nice, France, Theme 4, p. 305-312.

Surlyk, F. 1975b. Block faulting and associated marine sedimentation at the Jurassic-Cretaceous boundary, East Greenland. Norwegian Petroleum Society, Oslo, Jurassic Northern North Sea Symposium, Paper 7, p. 1-31.

Surlyk, F. 1978a. Submarine fan sedimentation along fault scarps on tilted fault-blocks (Jurassic-Cretaceous boundary, East Greenland). Grønlands Geologiske Undersøgelse Bulletin 128, 108p.

Surlyk, F. 1978b. Jurassic basin evolution of East Greenland. Nature, v. 274, p. 130-133.

Surlyk, F., Clemmensen, L.B. and Larsen, H.C. 1981. Post-Paleozoic evolution of the East Greenland continental margin. *In:* Kerr, J.W. and Ferguson, A.J. (Eds.), Geology of the North Atlantic Borderlands. Canadian Society of Petroleum Geologists, Memoir 7, p. 611-645.

Surlyk, F. and Clemmensen, L.B. 1983. Rift propagation and eustacy as controlling factors during Jurassic inshore and shelf sedimentation in northern East Greenland. Sedimentary Geology, v. 34, p. 119-143.

Vischer, A. 1943. Die postdevonische Tektonik von Ostgrönland zwischen 74° and 75° N. Br., Kuhn Ø, Wollaston Forland, Clavering Ø und angrenzende gebiete. Meddr Grønland, v. 133, 195p.

Walker, R.G. 1967. Turbidite sedimentary structures and their relationship to proximal and distal depositional environments. Journal of Sedimentary Petrology, v. 37, p. 25-43.

Walker, R.G. 1975. Generalized facies models for resedimented conglomerates of turbidite association. Geological Society of America Bulletin, v. 86, p. 737-748.

Walker, R.G. 1977. Deposition of upper Mesozoic resedimented conglomerates and associated turbidites in southwestern Oregon. Geological Society of America Bulletin, v. 88, p. 273-285.

Walker, R.G. and Mutti, E. 1973. Turbidite facies and facies associations. *In*: Middleton, G.V. and Bouma, A.H. (Co-chairmen), Turbidites and Deep-Water Sedimentation. Pacific Section, Society of Economic Paleontologists and Mineralogists, Anaheim, Short Course Notes, p. 119-157.

DEPOSITIONAL PROCESSES AND FLUID MECHANICS OF UPPER JURASSIC CONGLOMERATE ACCUMULATIONS, BRITISH NORTH SEA

L.G. Kessler, II[1] and Kit Moorhouse[2]

Abstract

Upper Jurassic fault-controlled fan-deltas and submarine fans which have been drilled and cored along the western edge of the South Viking Graben in the British North Sea contain three major lithofacies: (1) interbedded sandstone and shale; (2) massive to bedded sandstone; and (3) pebble to boulder conglomerate. These lithofacies are indicative of depositional setting and processes within individual fan systems. Depending on a subaerial or submarine setting, lithofacies 1 represents fan-delta front and submarine channel overbank deposition; lithofacies 2 represents tractive deposition in a fan-delta channel and high-density turbidity current activity in a submarine fan channel. Lithofacies 3, the pebble to boulder conglomerate, was largely deposited as debris flows in subaerial fan-delta and submarine fan channels. Subaerial debris flows were differentiated from subaqueous flows by examination of fabric order and degree of correlation between maximum clast size and flow thickness. Better ordered fabric, matrix support, and poor correlation between maximum clast size and flow thickness were indicative of subaqueous debris flows with the converse true for subaerial flows.

Assuming debris flows to be Bingham substances, yield strengths were calculated for 14 selected flows from various cores in the study area. All of the calculated values using both largest boulder and flow unit parameters fell in 10^3-10^4 dynes/cm^2 range, and correlated well with yield strengths computed by other workers for modern debris flows and experimental laboratory slurries. Calculation of rigid plug flow velocities gave results which also compared favorably with velocity measurements in natural and experimental debris flows.

Résumé

Des cônes de déjection et des deltas de chenaux sous-marins d'âge Jurassique supérieur associés à des failles ont été forés et échantillonnés le long de la bordure occidentale de la fosse tectonique South Viking dans la Mer du Nord britannique et ils comportent trois lithofaciès dominants: (1) un schiste argileux et un grès interstratifiés; (2) un grès massif et lité; et (3) un conglomérat à cailloux et à blocs. Ces lithofaciès renseignent sur le contexte de sédimentation et sur les processus actifs dans les systèmes individuels de cônes de déjection et de deltas sous-marins. Etant en relation directe avec le contexte subaérien ou sous-marin, le lithofaciès 1 représente les dépôts du front des cônes de déjection et de ceux produits par débordement audessus des rives des chenaux sous-marins; le lithofaciès 2 correspond aux matériaux transportés par traction au fond du chenal du cône de déjection et de courants de turbidité de forte densité à l'intérieur d'un chenal sous-marin. Le lithofaciès 3, le conglomérat à cailloux et à blocs, est un dépôt de coulées de débris dans les chenaux des cônes de déjection subaériens et des deltas sous-marins. L'examen de l'ordre de la fabrique et le degré de corrélation entre la grosseur maximale des fragments et l'épaisseur de la coulée a permis de différencier les coulées de débris subaériennes d'avec les coulées sous-aquatiques. Les fabriques les mieux ordonnées, les fragments non-jointifs dans la matrice, et une faible corrélation entre la grosseur maximale des fragments et l'épaisseur de la coulée, étaient des révélateurs de coulées de débris sous-aquatiques contrairement à des coulées subaériennes.

En supposant que les matériaux de Bingham appartiennent à des coulées de débris, nous avons calculé la limite élastique de 14 coulées sélectionnées au moyen de diverses carottes prélevées dans la région étidiée. Toutes les valeurs calculées en utilisant les paramètres des plus gros blocs et des unités formées de coulées sont comprises entre 10^3-10^4 dynes/cm^2, et il existe une bonne corrélation avec les indices élastiques calculés par d'autres chercheurs pour des coulées de débris récentes et pour des suspensions de particules solides mesurées en laboratoire. Le calcul des vitesses d'écoulement avec effet de bouchon rigide fournit des résultats qui se comparent favorablement avec les mesures des vitesses de coulées de débris dans la nature et en laboratoire.

Introduction

During the course of petroleum exploration in the British North Sea, significant hydrocarbon discoveries have been made in Upper Jurassic conglomerates and sandstones shed into South Viking Graben off the Fladen Ground Spur, and from related and reactivated Caledonian fault trends (Fig. 1). These conglomerates, sandstones, and associated interlaminated black shales and fine-grained rippled sandstones have been continuously cored in numerous wells in the Brae area (U.K. Block 16/7A - Pan Ocean -Marathon) and T-Block (U.K. Block 16/17 - Phillips). These cores afford an excellent opportunity to examine the vertical succession of lithologic types, and comment on the variation in depositional processes associated with the emplacement of conglomerates in this area.

Using core examples from both the Brae and T-Block areas, we will discuss the variability in gravel, cobble, and boulder distribution within and between individual sediment gravity flow units and the fluid mechanical reasons for such differences. Where possible, and within the limits of using ancient rocks, calculations have been made of yield strength (ability to support boulders of a particular

---
[1] Marathon Oil Company, Denver Research Center, Littleton, Colorado 80160-0269, U.S.A.
[2] Phillips Petroleum Company, The Adelphi, John Adam Street, London WC2, England.

Copyright © 1984, Canadian Society of Petroleum Geologists

size at the top of or within a given flow) and rigid plug flow velocity (using the methods of Johnson, 1970). In addition, the method of Gloppen and Steel (1981) for differentiating subaerial (high viscosity) debris flows from subaqueous (low viscosity) debris flows by examining the ratio of maximum particle size to flow thickness has been applied to selected flow units.

TECTONICS AND DEPOSITIONAL SETTING

Brae and T-Block are situated astride a complex, generally north-south trending fault zone which marks the eastern edge of the Fladen Ground Spur and the western margins of the Triassic to Lower Cretaceous South Viking Trough (Fig. 1). The principal extensional phase in the development of this half-graben took place in late Oxfordian to Volgian, when the western margin collapsed along a roughly north-south fault zone; detailed mapping reveals many Caledonian north-northeast/south-southwest elements and northwest-southeast elements possibly Precambrian in age (Johnson and Dingwall, 1981; Fig. 2). During this tectonic phase, pre-Upper Jurassic and intraformational Upper Jurassic sediments on the Fladen Ground Spur and the upthrown sides of smaller-scale reactivated Caledonian features were eroded and redeposited in a variety of fan settings. These fans range from partially subaerial to subaqueous fan-deltas in the southern Brae area (Harms et al., 1981) to crude proximal submarine fans and debris tongues in T-Block.

Clast types observed in individual sandstone and conglomerate flow units within the various Brae and T-Block fans range widely in lithology and age; a majority are Devonian, Permo-Triassic, Triassic, Jurassic, or intraformational Upper Jurassic in age. The principle lithologies are sandstone, shale, and orthoquartzite with rare occurrences of volcanic and plutonic rock fragments. The variation of clast type both with depth and laterally over the fan complexes appears to be random, suggesting the exposure of many pre-Upper Jurassic (and perhaps even Upper Jurassic) sequences at any one time in the sediment

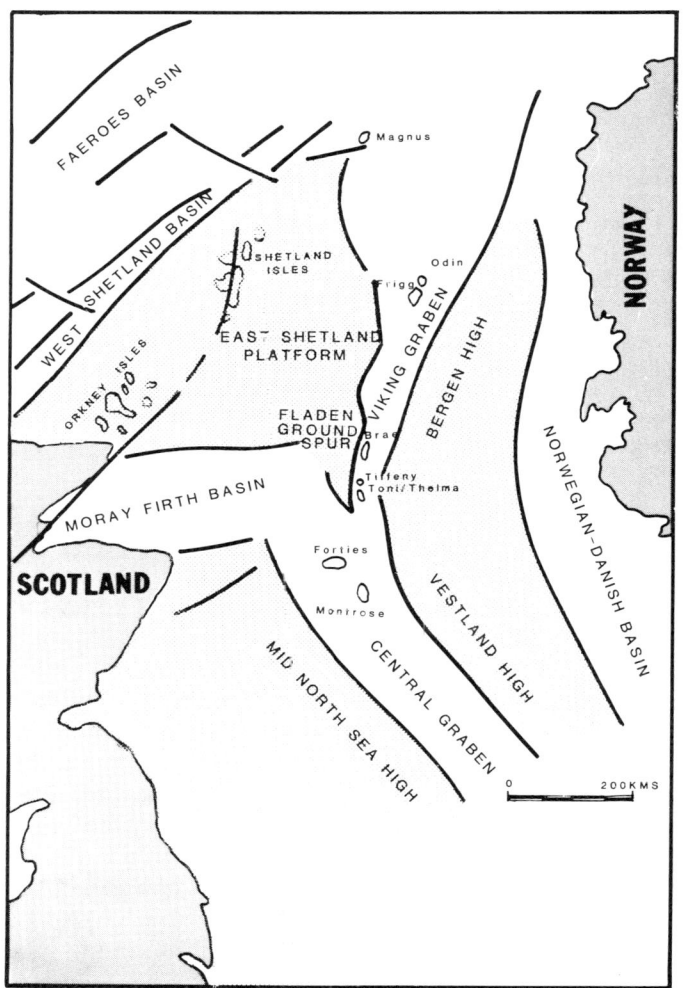

Fig. 1. Central and northern North Sea; tectonic framework and field locations.

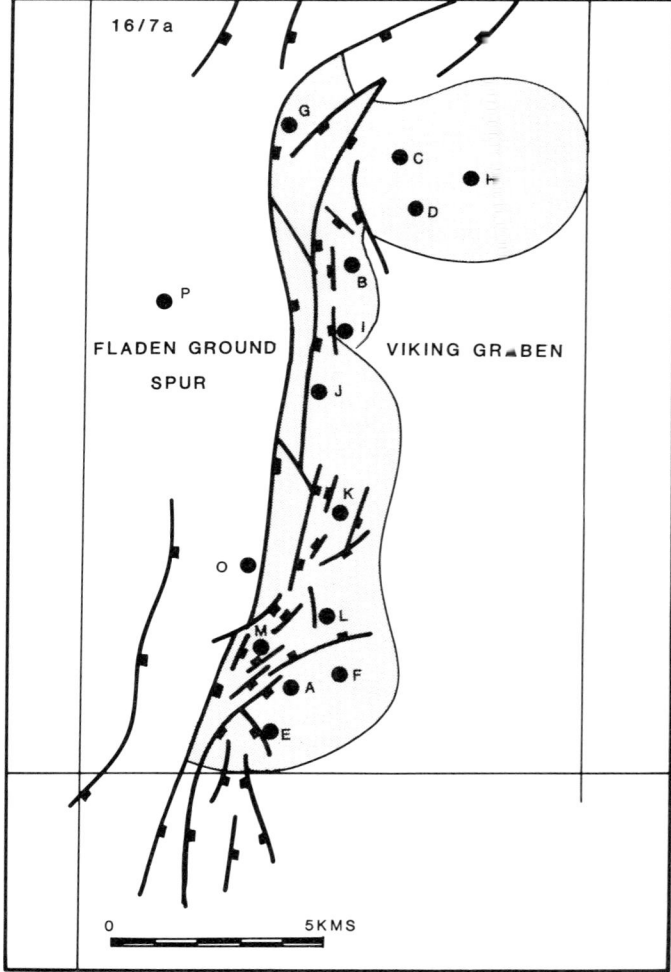

Fig. 2. Structural framework, thick sediment accumulation, and well locations; Brae area, British North Sea (modified after Stow et al., 1982).

source area(s) in a complex system of fault blocks at the margin of the trough.

Individual fans and fan lobes and their internal details will not be discussed beyond the generalizations of Figure 2 and comments on several individual flow units. Other than generalized outlines of fans, the subsurface data (well logs, cores, and seismic lines) from both Brae and T-Block do not allow us to make the detailed correlations necessary for delineation of individual small fan lobes and channels. About thirty wells and seismic reflections highly affected by complex faulting do not supply the quality, amount, and distribution of data necessary for such detailed correlations with any reasonable degree of certainty.

Lithofacies Classification

Three major lithofacies types (interbedded sandstone and shale, sandstone, and pebble-boulder conglomerate) predominate in the cored intervals from the Brae and T-Block areas. Each lithofacies represents a different depositional site and set of processes in the larger scheme of the Brae and T-Block fan systems.

INTERBEDDED SANDSTONE AND SHALE

Intervals of fine- to medium-grained sandstone interbedded with dark grey-black, slightly micaceous silty shales have been encountered in all wells. These intervals represent both lateral and distal stratigraphic equivalents of coarse sandstone and pebble to boulder conglomerate sequences.

Individual sandstone beds which are usually either plane bedded or rippled, vary in thickness from 0.5 to 100 mm with rare beds up to 250 mm thick (Fig. 3A-B). Individual shale beds rarely exceed 50 mm in thickness. In addition to ripples and planar beds, rare dune cross-bedding and normal grading from medium to fine sand are occasionally observed. Rip-up clasts of shale or organic material are present in a few intervals. Soft sediment deformational features, mainly load casts, slumps and flame structures are observed in many of the sandstone intervals.

The detrital mineralogy of the sandstones is dominated by quartz, with less than 5% feldspar and rare metamorphic and igneous rock fragments.

This lithofacies was deposited primarily by low-velocity tractive currents reworking older clastic material and/or by very dilute, thin turbidity currents (Stow et al., 1982).

SANDSTONES

Medium- to coarse-grained sandstone intervals encountered in all the wells are thought to represent bodies intimately related to coarser clastic deposits and independent and multiple (amalgamated) flow events. Such sandstone intervals are usually on the order of a few meters thick and it cannot be determined if they actually consist of thinner amalgamated flow units.

Most sandstone intervals are composed of massive or vague, normally graded flow units between 0.2 m and 1.0 m thick, with coarse-tail grading frequently developed in the basal 0.1 m of many units (Fig. 4). Rare and very faint dish structures are occasionally observed, and some units contain faint planar laminations.

The detrital mineralogy of this lithofacies is similar to that observed in the interbedded sandstone and shale lithofacies with complete domination by quartz and less than 5% feldspar. Rare igneous and metamorphic rock fragments are observed.

The sandstone units of this lithofacies were probably deposited by at least two different types of sediment gravity flow. As indicated by the massive bedding (with occasional normal and coarse-tail normal grading) in most flow units, the most common depositional agent was probably low density turbidity currents (Lowe, 1982). The presence of a few flow units with water release structures and others with rapid normal grading in the top few centimeters suggests that fluidized flows and density modified grain flows may also have been operational in this lithofacies (Lowe, 1976, 1982).

In the south Brae area some sandstone units assigned to this lithofacies contain cross-stratified beds of tractive current origin. Harms et al. (1981) believe that sandstone units of this sort were deposited in subaerial fan-delta channels.

CONGLOMERATES

Pebble to boulder conglomerates are a very common lithofacies type, and a very important hydrocarbon reservoir in the T-Block and Brae areas. Several wells in the T-Block area have encountered coarse clastic sequences greater than 500 m thick, thought to be composed predominantly of this lithofacies. Similar sequences close to 200 m thick have been observed in some Brae wells.

The clasts are generally pebble-to cobble-sized, although boulders up to 2.5 m in diameter have been recognized in core. In some coarse clastic bodies, a general upward decrease in maximum clast size occurs. As mentioned, a wide range of clast types of varying ages has been observed. Most of the clasts in this lithofacies are sandstone with some quartz pebbles and shale clasts. Clasts vary in shape from angular to well rounded with some brittle and softer shale and sandstone clasts displaying extensive fracturing or plastic deformation.

The matrix in conglomerate flow units is a fine to medium (occasionally coarse) sand very similar to the sandstones described above, generally with a slightly higher feldspar content. Less than 2% clay/silt fraction is observed in the matrix. All variations in matrix/clast ratio are present, with matrix-supported conglomerates tending to be predominant.

Clast-supported conglomerates show virtually no imbrication (on the scale of a core) and only rare organization. Matrix-supported conglomerates only show imbrication (and then only rarely) in south Brae and occasionally show the following forms of organization:

Fig. 3A. Interbedded siltstone and fine sandstone lithofacies, 16/7A-F well, British North Sea.

Fig. 3B. Same as 3A with development of current ripple about 1.5 cm thick, 16/7A-F well, British North Sea.

a. Normal coarse-tail grading, and very rare inverse to normal coarse-tail grading. This is restricted to pebble and small cobble conglomeratic, flow units with thicknesses of 0.2–0.6 m.

b. Vague inverse coarse-tail grading near the base with larger clasts floating on the top of a flow unit. Definition of amalgamation surfaces presents a problem in poorly organized lithologies such as these, but flow units ranging from 0.2 to 0.8 m in thickness can often be differentiated on the basis of the occurrence of larger cobbles and boulders floating at or near the top of individual flow units. Flow units ranging from 0.2 to 0.8 m in thickness are capable of carrying clasts of 20 cm or slightly greater diameter.

In addition to the conglomerate associations mentioned in (a) and (b), solitary boulders (0.3–0.7 m in diameter) surrounded by sand are also observed. These boulders, which are sometimes coated by a thin patina of very coarse sand and gravel, were probably deposited by rockfall, avalanching, and considerable rolling (Cook et al., 1972). The rolling probably accounts for the armored surface. A typical conglomeratic flow unit is shown in Figure 5.

The principal agent for the deposition of the cobble-boulder conglomerates is debris flow where larger clasts may have been supported by a variety of mechanisms including cohesive strength of clay-sand-water matrix, buoyancy, and dispersive pressure (Johnson, 1970; Hampton, 1972, 1975, 1979; Middleton and Hampton, 1973; Rodine and Johnson, 1976; Lowe, 1979, 1982; and Pierson, 1981). Other mechanisms of particle support in debris flows such as static grain-grain contact, turbulence, and maintenance of pore pressure have been discussed by various workers (Middleton and Hampton, 1973; Rodine and Johnson, 1976; Enos, 1977; Lowe, 1979; Lewis *et al.*, 1980; and Pierson, 1981). Debris flows are restricted to both grain and matrix-supported flow units with occasional inverse or inverse-to-normal grading, or rare imbrication where the largest cobbles are supported at or near the top of the flow. Intervals of the conglomerate lithofacies, which consist of cobbles and gravel, grade upward into medium-coarse sandstone (Fig. 6B) and are inversely or inversely-to-normally graded. They may have been deposited by density modified grain flows or high density turbidity currents (Lowe, 1976, 1982). Large cobbles or boulders supported at the top of flow units were not observed in these flows.

Examples of variability in the conglomerate lithofacies are shown in Figures 7-11. Cobble and boulder debris flows will be discussed further in the fluid mechanics section.

ASSOCIATION OF CONGLOMERATES WITH OTHER LITHOFACIES

Cobble and boulder conglomerates in the Brae and T-Block areas most commonly represent deposition in proximal (relative to deposition of the other two lithofacies) parts of fan channels and sandy-bouldery debris tongues not constrained by channel sides. Commonly, cobble and boulder conglomerate flow units abruptly overlie the interbedded sandstone and shale lithofacies with non-erosive contacts (Fig. 6A). Usually these contacts appear to be conformable, but occasionally some evidence of erosion, *i.e.*, rip-up clasts or a flow unit base cutting into bedding, is observed. This sort of abrupt contact is suggestive of sudden debris flow deposition perhaps after a catastrophic storm in an upland or shallow marine sediment source area (Pierson, 1981; Kessler, 1983), or submarine avalanching after renewed tectonism (Cook *et al.*, 1972).

Less abrupt occurrences of cobble and boulder conglomerates are associated with the sandstone lithofacies. In

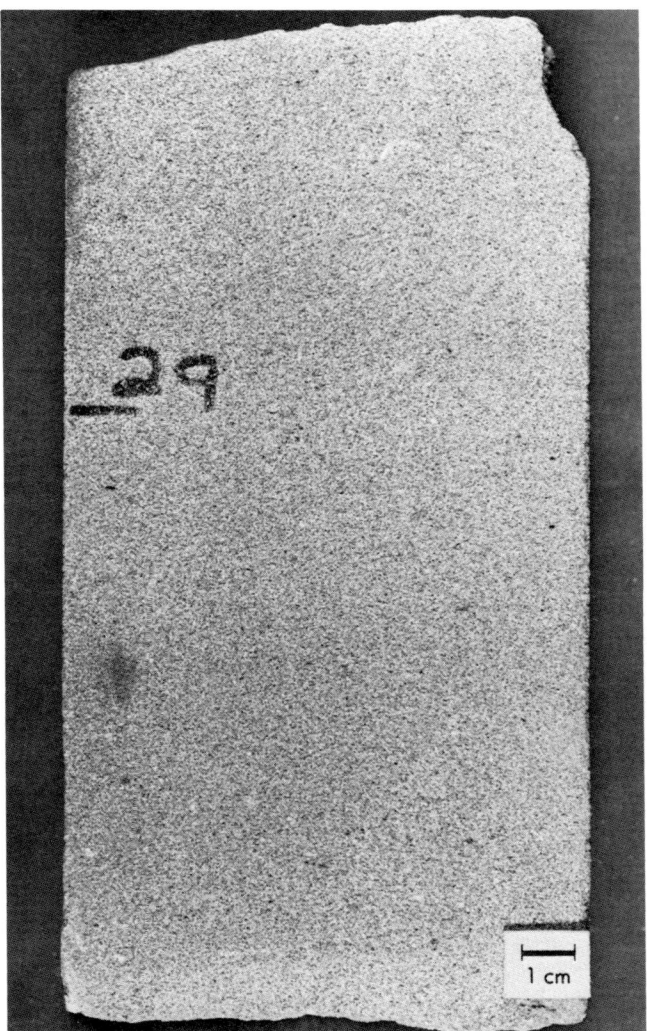

Fig. 4. Massive sandstone flow unit with faint grading near top, possibly of dilute turbidity current origin, 16/7A-C well, British North Sea.

Fig. 5. Typical South Brae area debris flow unit, note large sandstone clast at top supported by yield strength of flow, 16/7A-E well, British North Sea. This debris flow is Example 1 in fluid mechanical calculations, Tables 2, 3, and 4.

this association, conglomeratic flow units (including those which grade upward into medium-coarse sandstone) are interbedded with massive or graded medium-coarse sandstone flow units (Fig. 6B). Based on our experience in T-Block and Brae, this association most likely occurs in channels and represents reasonably continuous sediment gravity current deposition where individual flow units are episodic but in rapid succession. Lower and upper boundaries of these sandstone/conglomerate packages are gradational with the under- and overlying thinly interbedded black shale and sandstone lithofacies. Although the subsurface data are as yet inconclusive, we believe that this lithofacies association represents gradual lateral fan channel migration or the gradual reoccupation and eventual abandonment of an earlier channel.

Although we have stated that the cobble and boulder conglomerate lithofacies occurs in a rather proximal setting in most of the examples cited in this paper, a warning is in order. Surlyk (1978) observed multiple disorganized boulder conglomerates with clasts >0.7 m in diameter approximately 15 km from their fault scarp in the Wollaston Forland, East Greenland. Ricci-Lucchi and Valmori (1980) and Wain and Kessler (1982) mention the occurrence of axial plain turbidites which represent transportation of sandy and coarser material far out into narrow basins at a great distance from their original source. Wain (1980, pers. comm.) observed cobble and boulder conglomerates of probable marine debris flow origin which were deposited along the axial plain of the South Viking Graben. Thus, conglomerates of the sort discussed in this paper do not necessarily represent a highly proximal depositional site.

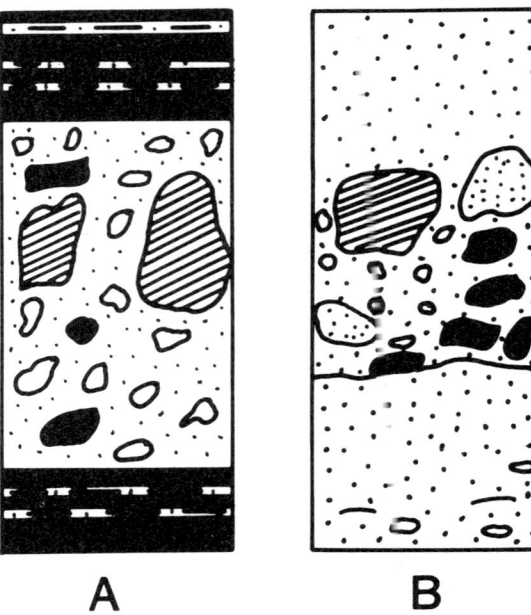

Fig. 6. (A) Cobble and boulder conglomerate flow unit showing abrupt upper and lower contacts with interbedded grey-black shale and sandstone deposition; (B) Cobble and boulder conglomerate flow unit showing amalgamated and/or gradational upper and lower contacts with massive or crudely graded medium-coarse sandstone lithofacies, indicative of nearly continuous but episodic debris flow and sandy high density turbidity current (Lowe, 1983) deposition.

Depositional Setting, Processes, and Fluid Mechanics of Conglomeratic Flow Units

Selected conglomeratic debris flow units in the 16/7A-A, B, C, D, E, and F wells have been examined in detail in an effort to learn more about their depositional setting and mode of emplacement. All of the debris flow units in the 16/7A-A, B, and C cores have been analyzed using the method of Gloppen and Steel (1981) for differentiation of subaerial from subaqueous flows by computation of the ratio of maximum clast size to flow thickness. Calculations of yield strength and rigid plug flow velocity in the 16/7A-A, B, C, D, and F cores and an outcrop example from the Ordovician of Québec have been made using equations discussed in Johnson (1970).

Maximum Particle Size Versus Flow Thickness

Gloppen and Steel (1981) observed a significant correlation between maximum particle size (*MPS*) and bed thickness (*BTh*) in Devonian debris flow deposits from the Hornelen Basin in Norway. They observed that subaerial (high viscosity) flows which were very coarse and mostly

Fig. 7. Debris flow unit with largest clast submerged in flow, 16/7A-E well, British North Sea. This debris flow is Example 2 in fluid mechanical calculations, Tables 2, 3, and 4.

clast supported had a *BTh/MPS* ratio of 3 or less, a general unordered fabric, and virtual lack of grading. They state that the "lack of imbrication and grading in beds is taken as an indication of lack of clast-to-clast movement during flow" (Gloppen and Steel, 1981, p. 51). The low *BTh/MPS* ratio is indicative of high yield strength for these debris flows. This description is consistent with the presence of a 'rigid plug' in modern subaerial debris flows as described by Johnson (1970). Their subaqueous (low viscosity) flows have a higher percentage of sandstone matrix (though not always matrix supported), are generally finer grained, are often imbricated, and can be inverse or inverse-to-normally graded. For these flows higher ratios (often 5 or 6) of *BTh/MPS* are often observed. When both types of debris flows are examined with regression analysis, lower but still significant correlation between *BTh/MPS* is observed in subaqueous flows.

Previously, the *BTh/MPS* ratio has been used to describe the nature of debris flows observed in outcrop (Larsen and Steel, 1978; Gloppen and Steel, 1981). Though obviously more information can be gained about individual debris flows in outcrop, we were able to see significant correlations between maximum particle size and bed thickness in debris flow units identified in cores from the 16/7A-A, B, and C wells.

In the 16/7A-A cored interval the maximum clast size and bed thickness were measured in 51 debris flow units (Table 1) The mean value of the *BTh/MPS* ratios was 3.0 with a range of 1.9-6.2. Thirty-five of the ratios fell in the 2.5-3.5 range. Regression analysis of these data gave a correlation coefficient of 0.90 and, as Figure 12 shows, there is tight packing and a strong linear relationship between maximum particle size and flow unit thickness in this well. Deposition of the debris flows observed in the 16/7A-A

Fig. 8. Debris flow unit with largest clast mostly submerged in flow, 16/7A-C well, British North Sea. This debris flow is Example 8 in fluid mechanical calculations, Tables 2, 3, and 4.

Fig. 9. Debris flow unit with very large clast totally supported at top of flow, 16/7A-B well, British North Sea. This debris flow is Example 3 in fluid mechanical calculations, Tables 2, 3, and 4.

Fig. 10. Debris flow unit with large boulder partially submerged at top, flow is matrix supported, note imbricated angular-subangular pebble-size clasts near base, 16/7A-D well, British North Sea. This debris flow is Example 13 in fluid mechanical calculations, Tables 2, 3, and 4.

Fig. 11. Solitary large boulder in sandy matrix with associated gravel size clasts; boulder is of possible rockfall origin, 16/7A-D well, British North Sea.

cores was in the largely subaerial part of a fan-delta (Harms et al., 1981).

Measurement and regression analysis of maximum particle size and bed thickness data in the 16/7A-B and C yielded some slightly different results. The 40 debris flow units measured in the 16/7A-B well had a mean *BTh/MPS* ratio of 3.6 with a range of 1.5-9.0. Only 11 of the ratios fell in the 2.5-3.5 range. Regression analysis of these data gave a correlation coefficient of 0.74 with a wide dispersion of data points in a virtually non-linear relationship (Fig. 13). Despite a *BTh/MPS* ratio of 2.9, the debris flow units in the 16/7A-C well show a similar nonlinear dispersion pattern and a very low correlation coefficient = 0.57 (Table 1, Fig. 14). Only 20 out of 42 *BTh/MPS* ratios fell in the 2.5-3.5 range.

The *BTh/MPS* ratios (3.6 and 2.9) computed for debris flows in the 16/7A-B and C wells are initially suggestive of a subaerial depositional setting. However, several lines of evidence suggest alternative interpretations. The common occurrence of marine macrofossils (including ammonites, belemnites, bivalves, and sponge spicules), a strong presence of marine dinoflagellates and virtually all regional subsurface facies mapping indicate that deposition throughout the cored intervals in the 16/7A-B and C wells was in a submarine setting (Harms et al., 1981; Stow et al., 1982; Harms and McMichael, 1983). As mentioned, in both wells a wide range of *BTh/MPS* ratios were observed with less

	16/7A-A	16/7A-B	16/7A-C
No. of debris flows examined	51	40	42
MPS	12.5 cm	14.9 cm	13.7 cm
BTh	35.7 cm	46.6 cm	37.8 cm
BTh/MPS	3.0	3.6	2.0
Range of BTh/MPS Values	1.9-6.2	1.5-9.0	1.2-5.5
Number of BTh/MPS ratios in the 2.5-3.5 range	35	11	20

Table 1. Summary of maximum particle (*MPS*) and bed thickness (*BTh*) data for 17/7A-A, B, and C cores.

than half of the ratios falling in the 2.5-3.5 range which Gloppen and Steel (1981) considered as clearly subaerial. Consequently, regression analysis for debris flows in both wells revealed a non-linear dispersion of points and, relative to the 16/7A-A well, low correlation coefficients. Examination of textural and fabric characteristics of the debris flows in the 16/7A-B and C wells showed wide variability with slightly more than half the flows being matrix supported and ordered with some semblance of inverse or crude normal grading and/or imbrication. The remainder of the debris flows in these two wells were very similar to those observed in the 16/7A-A well with less matrix, more clast support, and a lack of grading or imbrication. This presence of both high and low-viscosity debris flows (*sensu* Gloppen and Steel, 1981) in a strongly marine part of Brae is suggestive of a complex sediment input history. We are probably seeing an interbedded sequence of high-viscosity, clast-supported debris flows of subaerial origin which have continued offshore to a marine shelf setting, and low viscosity matrix-supported debris flows of subaqueous origin. These low viscosity flows may be resedimented conglomerates of submarine fault origin or may be derived from, and the distal equivalents of, high viscosity debris flows similar to those with which they are interbedded.

Despite the limitations of core data (only 10 cm wide views instead of outcrops) used in this study enough can be seen in individual debris flow units to describe bedding fabric and to use the techniques of Larson and Steel (1978) and Gloppen and Steel (1981) for comparison of bed thickness (*BTh*) and maximum particle size (*MPS*). By analogy to their results in the Devonian of Norway, the debris flows in the 16/7A-A well cored section can be clearly recognized as mostly high viscosity debris flows deposited in a subaerial setting. This interpretation fits with the

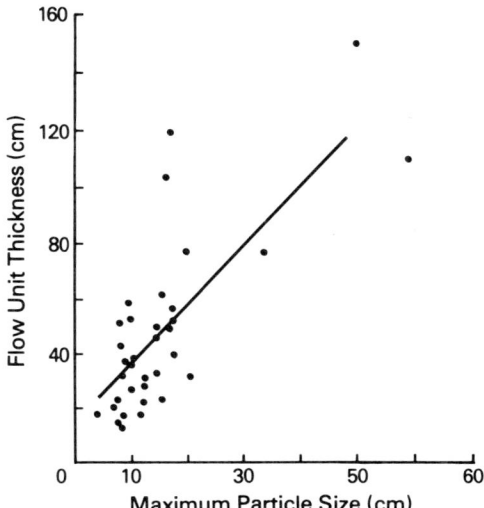

Fig. 13. Bed thickness (*BTh*) and maximum particle size (*MPS*) characteristics of subaqueous debris flows, 16/7A-B well, British North Sea, *BTh/MPS* = 3.6, r = 0.74.

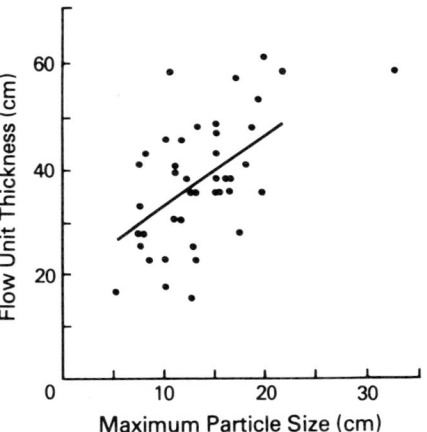

Fig. 14. Bed thickness (*BTh*) and maximum particle size (*MPS*) characteristics of subaqueous debris flows, 16/7A-C well, British North Sea, *BTh/MPS* = 2.9, r = 0.57.

views of Harms *et al.* (1981) and Harms and McMichael (1983) for the south Brae area. Despite low mean values for *BTh/MPS* ratios in debris flows from the 16/7A-B and C wells, the wide range and poor correlation of these ratios is suggestive of deposition by both high and low viscosity debris flows. Based on strong paleontologic evidence which indicates a marine depositional site, both high and low viscosity debris flows are being deposited in a subaqueous depositional setting in the northern part of Brae (Harms and McMichael, 1983).

FLUID MECHANICS OF CONGLOMERATIC DEBRIS FLOW UNITS

An interesting feature of many debris flows is their ability to transport extremely large boulders and blocks on relatively gentle slopes for great distances, even up to tens of kilometers. These boulders and blocks are often trans-

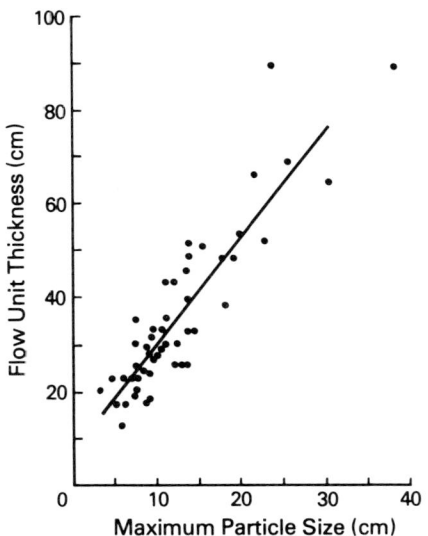

Fig. 12. Bed thickness (*BTh*) and maximum particle size (*MPS*) characteristics of subaerial debris flows, 16/7A-A well, British North Sea, *BTh/MPS* = 3.0, r = 0.90.

ported while virtually floating at the top or partially submerged in the front or middle part of a flow which is made up of finer granular material with varying amounts of entrained water. How then can a flow support and transport these boulders and blocks? Based on field and experimental observations, Johnson (1970) proposes in detail a mechanism of debris flow transport that is a "combination of high density and high strength of debris" which prevents large clasts from sinking to the bottom of a flow. More specifically, he ascribes the ability of a debris flow to support large clasts to the strengths of a non-Newtonian fluid matrix composed of clay minerals, fine granular material and water, plus the buoyancy of this matrix. From this, Johnson (1970) describes the competence of the flow as the weight of the largest grain being equated to the sum of the strength and buoyancy of the flow. This is best expressed in equation form by Hampton (1975):

$$\rho_s g \left(\frac{1}{6} \pi D^3\right) = \rho_f \left(\frac{1}{6} \pi D^3\right) + 1.4 \, k \, \pi D^2 \quad (1)$$

weight = buoyancy + strength

ρ_s = density of the clast
g = acceleration of gravity
D = diameter of grain
ρ_f = density of fluid matrix
k = yield strength of matrix

Drawing on Johnson (1970), Hampton (1975) points out that the weight and buoyancy terms in this equation are determined for spherical grains and that the strength term describes the support which a flat 'punch' (i.e., penetrating body) receives on a plastic surface. According to Johnson (1970, p. 464-487), this is a reasonable approximation of the support a boulder might receive in a clay-water mixture. This equation can be solved for D (maximum grain diameter) and ultimately for k or the yield strength of the debris flow. The application and implications of the solution for k or yield strength will be discussed after we review other mechanisms of clast support in debris flows.

Other workers have suggested a variety of clast support mechanisms in debris flows. In addition to stressing the importance of coarse grain support by the strength and buoyancy of debris flow matrix, Hampton (1975) pointed out that the collision of grains in natural debris flows generates dispersive pressure which makes a major contribution to grain support (Bagnold, 1954). Hampton (1975) also suggested that in boulder-rich, carbonate debris flows, the yield strength may be controlled by the electrical properties of the matrix particles. To our knowledge there has been no further research into this latter concept. Rodine and Johnson (1976) suggested that this buoyant force acting on a boulder in a debris flow is equal to the weight of all the displaced material (including sand, gravel etc.) not just a clay-water fluid matrix. They combine strength and buoyancy in a hypothesis where larger and larger clasts essentially pyramid upon the clay-water slurry until all grains are supported. This final support of the largest clasts is thought to occur as a result of a small density difference between these clasts and the debris, in addition to the cohesive strengths of the clay-water part of the debris flow slurry. Hampton (1979) further stressed the importance of buoyancy in the support of granular solids in debris flows. Based on experimental results, he found that the competence of debris increases with the concentration of coarse grains which are supported by the clay-water matrix. This increase appears to be the result of excess pore pressure which would increase buoyant forces. Hampton proposed that the weight of the clay and coarse grains is transferred to the fluid and this causes an increase in pore pressure. In application of these theories to natural flows, however, he points out that often the flow competence is much less than the size of the largest boulders supported in the flow. He concluded that these boulders were largely supported by other other forces in addition to strength and buoyancy such as dispersive pressure which was generated as a result of grain collisions during flow (Bagnold, 1954).

Lewis (1976) and Lewis et al. (1980) in studies of Pleistocene and Miocene sandy and muddy matrix debris flows in New Zealand stressed the importance of buoyancy, excessive pore pressure, and dispersive pressure in large boulder support. More specifically, Lewis et al. (1980) in Miocene clast-supported coarse sand-gravel matrix conglomerates advocated transportation by high density debris flow deposits similar to those hypothesized by Rodine and Johnson (1976). Low proportions of fine sand and mud in conjunction with water do not possess the strength to support the coarse clasts in these flows, so Lewis et al. (1980) envisaged a support mechanism based on the pyramid effect of the increased unit weight of water and all debris sizes plus excess pore pressures. The excessive pore pressure resulted from the inability of fluids to escape from highly concentrated, poorly sorted debris. In addition to these buoyant effects and excess pore pressures they infer that dispersive grain pressure during flow (larger clasts being impacted by smaller disoriented clasts) served to orient larger clasts in these flows due to low continuous phase (water plus clay and fine sand) strength.

Pierson (1981) found that in modern debris flows forming at an alluvial fan head near Mount Thomas, New Zealand, he could not account for the suspension of small boulders in these flows by any combinations of strength, buoyancy and dispersive pressure. Debris flow surge samples contain only enough clay to support medium or finer grained sand (Kuenen 1951; Hampton, 1972). In an effort to understand the dominant mechanisms in these flows, Pierson made laboratory measurements of pore pressure on whole debris samples and slurries with varying amounts of the coarse fraction removed. He observed that when high concentrations (0.58-0.66) of sand or sand and gravel were added to silt-clay-water slurries, settling slowed and the suspension competence of the mixture increased greatly. 1-2 cm diameter pebbles were supported by such mixtures.

As grain concentrations were increased, pore pressure and buoyancy increased. Pierson found buoyancy provided by the fluid phase in surges supported ⅔ of the coarse sediment load. Grain to grain contact and cohesive strength were thought to provide the remainder (in lieu of kinetic support mechanisms) of the forces necessary to support large particles. Examination of pore pressure gradient data from his experiments indicated to Pierson that very little of large clast support is taken up by cohesive strength (only 2% as compared to 79% by buoyancy in a 1 cm sandstone cube). This support decreased with increasing large clast size. Pierson concludes that the life and movement of debris flows with strong buoyant support due to excess pore pressure depends on partical size distribution. The poor sorting of the debris creates narrow and tortuous pore necks which can be readily plugged by even a small amount of clay or fine silt, thus trapping fluid under pressure.

Based on the various concepts of clast support in debris flows suggested above the calculation of yield strength (k) using the methods of Johnson (1970) takes on a more complex meaning.

Since equation (1) from which the equations (4 and 5) for yield strength (k) are derived contains a buoyancy factor, it appears that these equations provide a method of calculating the total strength necessary to support boulders in the Upper Jurassic and Lower Ordovician debris flows examined in this study. If one accepts the studies on pore pressure in sandy debris flows culminating in the work of Pierson (1981), about ⅔ of the value of k will be accounted for by buoyancy caused by excess pore pressure and the remainder by cohesive strength and grain to grain contact. Because of the large size of boulders in our flows, cohesive strength probably supplies less than 1.0% of the yield strength.

Johnson (1970) described two methods for the calculation of yield strength using either (1) the properties of the largest clast transported or (2) the thickness and surface slope of the debris deposit where the clast is resting:

(1) Assuming the boulder is sub-spherical, the debris flow is at rest and possesses just the strength to support the boulder, and that the upward forces exerted by the flow (acting as a rigid plastic substance) equal the downward force of the boulder penetrating the flow, the equilibrium equation can be written:

$$-\sigma yy \pi (ab) = [v_b(\frac{4}{3}) \pi (abc) - v_d \pi (\frac{4}{3})(\frac{abc}{n})] \quad (2)$$

where σyy = upward pressure of plastic flow
a = height of boulder
b = width of boulder
c = length of boulder
v_b = density of boulder
v_d = density of flow
n = fraction of boulder penetrating flow

Johnson (1970) indicated that if the thickness of the plastic material (the flow) is approximately equal to the radius of the punch (the penetrating large boulder) then the upward pressure of the flow (σyy) has the following approximate relationship with the yield strength (k) of the matrix:

$$\sigma yy = -5.5 k \quad (3)$$

By substituting this into equation (2) a relationship for approximating yield strength (k) is derived:

$$k \approx \frac{C}{4} [v_b - v_{d/n}] \quad (4)$$

(2) Yield strength of debris material in a flow can also be determined by consideration of the angle of the surface slope, flow thickness, and density of the flow:

$$k \approx T_c \, v_d \sin \delta \quad (5)$$

where δ denotes the angle of surface slope and T_c flow thickness.

Because the ancient debris flows which were examined in this study were either wholly or in part subaqueous and upon hand sample and petrographic examination were found to possess very little (1% or less) clay in the flow matrix, slightly lower flow density (v_d) values than those of Johnson (1970) were used in our calculations of yield strengths. Our values ranged from 2.01-2.28 whereas Johnson used values ranging from 2.10 to 2.50. Because of the scarcity of clay and abundance of coarse sand and gravel in our flows, we used high water volume percentage (29-46%) in order to create further buoyancy through increased pore pressure in the flow.

Table 2 lists all the data used in the calculations of yield strength for our 15 selected debris flows. With the exception of flow and boulder densities, all of the parameters are from cores and outcrop. The assumptions concerning determination of flow density have already been discussed. With the exception of the debris flow from Grosses Roches, Québec, a density of 2.65 (i.e., quartz) was used as a boulder density for the largest sandstone boulder in each flow. A density of 2.30 was used for highly porous limestone boulders at the top of the Grosses Roches example. Yield strength for each of the fifteen examples using both methods of calculation are shown in Table 3.

Using Methods 1 and 2, Johnson (1970, p. 486-489) calculated the yield strength of an ancient debris flow at Surprise Canyon, California, which supported an almost immersed boulder, 165 cm high. The flow, which was 160 cm thick, had a calculated yield strength (k) 3×10^4 dynes/cm^2 using Method 1 and 4×10^4 dynes/cm^2 range. From examination of highest computed yield strength values boulder size (volume) appears to be the major factor in the high values (Tables 2 and 3). For instance, the three highest yield strengths determined from Method 1 (Examples 5, 12 and 15) correspond to the three largest boulders observed in the study. The large size of the boulder from Example 15, the Grosses Roches outcrop, debris flow is shown in Figure 15. Smaller boulders with yield strength ranging from 1.76 − 4.55 × 10³ dynes/cm² are shown at or near the tops of the debris flow units in Figures 5 and 7-10. Examples 3 and 13 (Figs. 9 and 10) illustrate another major contributory factor to high yield strength values. The largest boulders in these two flow units are virtually floating on top of the flows, being only one-sixth and one-fourth submerged, respectively (Table 2). Thus, both boulder size and amount of boulder submergence in the flow exert the strongest influences on the yield strength as computed by Method 1.

Yield strength values calculated by Method 2 (Equation 5) which considers only flow related parameters do not vary greatly from those calculated by Method 1 (Equation 4). Johnson (1970) seems to intimate that differences of up to 50% between the two methods indicate a good comparison. Only three (8, 10 and 11) of our examples fall outside this limit. In all three examples the largest boulder was quite small and almost completely submerged in the flow which was relatively thick compared to the largest boulder size. This leads to a very high computed yield strength from flow parameters. This indicates that the flow is probably competent to carry larger boulders than observed in the core pieces.

The computations of yield strengths of fifteen example debris flows from the 16/7A cores and the Grosses Roches outcrop indicate that seemingly reasonable values for important debris flow parameters like yield strength can be computed for ancient rocks without making outrageous assumptions. The values obtained for yield strength from both boulder (Method 1) and flow (Method 2) characteristics compare reasonably to those computed by Johnson (1970) for recently active modern debris flows and experimental slurries. Thus, it appears that equations for yield strength which have been applied to modern subaerial debris flows can be accurately applied to their ancient subaerial and subaqueous counterparts. The only major problem with this application of equations and concepts is the earlier-mentioned lack of clay in the matrix in our debris flow units. Whereas Johnson (1970) and Hampton (1979) observed and required a minimum of 3.5% and often have 10+% of clay in their modern debris flows or experimental slurries, we observe a maximum of 1.5% and often less in our example flows. Though we use these low percentages of clay to compute flow density for the yield strength equations, we still obtain k values which are consistent with modern debris flow data. This suggests that in addition to cohesive strengths, other processes are serving as the major supporting mechanisms in the sandy matrix debris flows considered in this study. As suggested earlier, it appears that static grain contact, buoyancy as created by excess pore pressure (Pierson, 1981), and dispersive pressure as suggested by Hampton (1979) supply large parts of the yield strength of the debris flows considered in our study.

An interesting oddity observed in the 16/7A-D cores was the occurrence of solitary lithified sandstone boulders contained at various levels in medium-coarse massive sandstone flow units (Fig. 11). The flows with which these boulders are associated do not possess the yield strengths to carry clasts of this size. For instance, the boulder shown in Figure 11 would have required a yield strength of 7.75×10^3 dynes/cm², whereas the non-conglomeratic flow unit in which it is contained has a maximum yield strength of 2.10×10^3 dynes/cm². A boulder of this size

	Flow density (v_d)	Boulder density (v_b)	Flow thickness (T_c)	Flow surface slope (δ)	Amount of boulder submerged (n)	Height of boulder (a)	Length of boulder (b)	Width of boulder (c)
16/7A-E								
1	2.01	2.65	18.0 cm	5°	³⁄₈	7.7 cm	10.5 cm	9.6 cm
2	2.21	2.65	14.5 cm	5°	entire	9.5 cm	10.8 cm	11.2 cm
16/7A-B								
3	2.18	2.65	27.4 cm	5°	¹⁄₆	8.9 cm	22.0 cm	19.0 cm
4	2.20	2.65	28.6 cm	4°	⁴⁄₉	10.9 cm	19.1 cm	18.0 cm
5	2.19	2.65	51.3 cm	5°	¹⁄₅	36.6 cm	29.8 cm	30.5 cm
16/7A-C								
6	2.28	2.65	21.5 cm	4°	²⁄₇	9.9 cm	17.0 cm	15.5 cm
7	2.21	2.65	35.8 cm	4°	⁴⁄₉	11.1 cm	13.9 cm	13.1 cm
8	2.05	2.65	26.4 cm	6°	⁹⁄₁₀	9.2 cm	8.3 cm	7.9 cm
16/7A-A								
9	2.18	2.65	23.9 cm	5°	3.1⁄10	6.6 cm	19.8 cm	15.5 cm
10	2.03	2.65	47.5 cm	4°	⁷⁄₈	7.9 cm	9.4 cm	8.6 cm
11	2.01	2.65	21.7 cm	4°	¹⁰⁄₁₁	6.2 cm	5.6 cm	5.9 cm
16/7A-D								
12	2.18	2.65	34.7 cm	5°	¹⁄₅	16.9 cm	20.8 cm	18.5 cm
13	2.18	2.65	28.3 cm	5°	¹⁄₄	9.6 cm	14.7 cm	10.5 cm
14	2.05	2.65	67.9 cm	5°	¹⁄₅	10.5 cm	16.9 cm	12.1 cm
Grosses Roches								
15	2.03	2.30	91.6 cm	5°	⁶⁄₇	46.8 cm	87.7 cm	59.4 cm

Table 2. Debris flow parameters from selected flow units in 16/7A cores and Grosses Roches outcrop.

could not have been transported by this flow unit. Movement of these solitary boulders was probably by avalanching, rockfall, and considerable rolling.

RIGID PLUG FLOW VELOCITY

In a further effort to understand the nature of large clast transport in debris flows we have attempted to calculate the velocity of rigid plug movement in our ancient debris flow examples. Based on field and laboratory examples Johnson (1970) observed a discrepancy between the behavior of debris flows and theoretical plastic substances which was suggestive of viscous resistance of debris. He overcame this discrepancy by adding Newtonian viscosity to the plastic rheological model, so making debris into a Bingham substance:

$$[\sigma_s] = k + \eta_b \epsilon_s, \text{ where } [\sigma_s] \geq k. \quad (6)$$

This relationship states if the shear stress (σ_s) is greater than the yield strength (k) of a material, the resistance is equal to the sum of the plastic and viscous resistances. η_b is the coefficient of Bingham viscosity and ϵ_s is the rate of shear strain.

With the knowledge that debris flows are Bingham substances and Johnson's (1970) calculations of Bingham viscosities (η_b) for modern debris flows, we can closely estimate the velocity of rigid plug movement in our ancient debris flow examples. In a theoretical analysis of the Bingham substance flow in semi-circular and infinitely wide channels, Johnson verified the existence of the rigid plug and showed that it is possible to calculate the Bingham viscosity (η_b) of a substance by determining the radius and velocity of the plug with the following equation:

$$\eta_b = \frac{\omega^* T_c^2 \, v \sin \delta}{\omega} \quad (7)$$

where ω = velocity of the rigid plug
  v = unit weight of the debris (debris density $\times$ gravity)
  ω^* = velocity variable (from Johnson, 1970, p. 507, graph 14-7B)

The use of this equation requires the calculation of channel radius (a) and the plug radius (X_0). This then leads to the determination of X_0/a (plug width variable) and the reading of a value for k_* (strength variable) off the x-axis along the $a/b = 2$ line from Johnson's (1970, p. 507) Figure 14.7A graph. For all of our example flows, $a/b = 2$. With these results Johnson's (1970, p. 507) Figure 14.7B graph can be entered for the computed value of k^* along the curve for $a/b = 2$. The resultant value of the velocity variable ω^* can then be read from the y-axis of the graph.

We now have all the variables to solve Equation 6 for the velocity of the plug (ω) save one, the Bingham viscosity (η_b). In order to solve for ω we have to estimate this value. Here lies the most potentially dangerous assumption of our fluid mechanical considerations. Johnson (1970,

p. 511-513) calculated a value of 7.6×10^2 poise for the Bingham viscosity of the Wrightwood, California debris flow. This flow, which was 90 cm thick and had a rigid plug width of approximately 100 cm, is about the same scale as some of our larger example flows, i.e., Grosses Roches. Though ω is a major variable in the calculations of η_b, so is T_c or the thickness of the flow. Using flow thickness as the principal variable (it varies from 14.5 to 91.6 cm in our examples), we have estimated the value of using the 90 cm thickness and 7.6×10^2 poise calculations as our base values. Because of flow density differences (our flow densities being slightly lower), the estimated values of (Bingham viscosity) are about 15% lower than a straight line interpolation from the Wrightwood data. These estimated Bingham

	Method 1	Method 2
16/7A-E		
1	3.60×10^3 dynes/cm^2	3.09×10^3 dynes/cm^2
2	2.47×10^3 dynes/cm^2	2.78×10^3 dynes/cm^2
16/7A-B		
3	5.00×10^3 dynes/cm^2	5.10×10^3 dynes/cm^2
4	4.47×10^3 dynes/cm^2	4.31×10^3 dynes/cm^2
5	1.99×10^4 dynes/cm^2	9.60×10^3 dynes/cm^2
16/7A-C		
6	4.86×10^3 dynes/cm^2	3.35×10^3 dynes/cm^2
7	4.55×10^3 dynes/cm^2	5.41×10^3 dynes/cm^2
8	1.82×10^3 dynes/cm^2	5.54×10^3 dynes/cm^2
16/7A-A		
9	3.30×10^3 dynes/cm^2	4.45×10^3 dynes/cm^2
10	1.76×10^3 dynes/cm^2	6.46×10^3 dynes/cm^2
11	1.26×10^3 dynes/cm^2	2.98×10^3 dynes/cm^2
16/7A-D		
12	9.24×10^3 dynes/cm^2	6.46×10^3 dynes/cm^2
13	4.96×10^3 dynes/cm^2	5.27×10^3 dynes/cm^2
14	5.77×10^3 dynes/cm^2	1.19×10^3 dynes/cm^2
Grosses Roches		
15	1.04×10^4 dynes/cm^2	1.73×10^4 dynes/cm^2

Table 3. Debris flow yield strengths from selected flow units in 16/7A cores and Grosses Roches outcrop.

Fig. 15. Debris flow unit with large rounded carbonate boulder partially submerged at top (next to handle of geologist's hammer), 55 m level, measured section at Grosses Roches, Quebec. This debris flow is Example 15 in fluid mechanical calculations, Tables 2, 3, and 4.

viscosities and other data necessary for solving Equation 6 for plug velocity (ω) are listed in Table 4.

The computed velocities (ω) or rigid plug movement in our debris flow examples are listed in Table 4. They range from 134.0 cm/sec for Sample 2, which was only 14.5 cm thick, to 396.5 cm/sec for Sample 14, which was 67.9 cm thick. The Ordovician outcrop sample for Grosses Roches had a computed plug velocity of 698.7 cm/sec for a 5° slope. These computed velocities seem reasonable when compared to measured velocities in modern debris flows and experimental slurries studied by Johnson (1970) and Hampton (1975). For example, Johnson measured a rigid plug velocity of 110 cm/sec in a major natural debris flow at Wrightwood, California. Hampton (1975) created highly competent slurry flows in the laboratory which performed best at velocities of 101.5-284.0 cm/sec. Based on evidence of this sort, it would appear that our velocity calculations are fairly accurate. Considering what can be measured in and reasonably assumed about ancient debris flows, our adaptation of Johnson's (1979) method of velocity calculation is a useful means to learn more about the dynamics of such flows with a comparison to modern and experimental analogs.

Summary and Conclusions

Examination of well cores through the Upper Jurassic clastic interval in the Brae and T-Block areas of the British North Sea reveals the presence of three interrelated lithofacies types, thinly interbedded sandstone and shale, massive and bedded sandstone, and pebble-boulder conglomerate. These lithofacies each represent a different depositional part of subaerial-submarine fan systems derived from the Fladen Ground Spur and the upthrown sides of smaller scale reactivated Caledonian features. The interbedded sandstone and shale lithofacies was deposited by low velocity bottom-hugging tractive currents and/or by very dilute, thin turbidity currents in either a submarine channel overbank or fan-delta front setting. The sandstone lithofacies, depending on bedding characteristics (massive or graded versus planar or cross-bedded) was either deposited by high-density turbidity currents in a submarine channel or fluid gravity flow currents in shallow subaqueous or subaerial fan-delta channels. The pebble-boulder conglomerate lithofacies suggests deposition by debris flow movement in subaerial and submarine channels in a generally (though not always) proximal depositional setting. Individual debris flow units with larger boulders supported at or near the surface of the flow can be recognized in cores throughout the Brae and T-Block areas.

The conglomeratic debris flow units recognized in cores from the Brae and T-Block fan systems have afforded an excellent opportunity to study this sort of deposition using an ancient subsurface example. Debris flows in the 16/7A-A, B and C wells were examined as to textural characteristics and compared by regression analyses of maximum particle size versus flow thickness. The results of these comparisons allowed us to differentiate debris flows in a more subaerial setting (16/7A-A) from those in a subaqueous setting (16/7A-B and C). These results agreed with previous interpretations of the Brae area.

Cursory examination of T-Block conglomerates suggests that they would fit in a subaqueous setting using this method.

Using the fluid mechanical assumptions and derivations of Johnson (1970) the yield strengths (the ability of the flow to support heavy objects) and rigid plug velocities of 14 selected debris flows from cores throughout the Brae

	Thickness of flow (T_c)	Unit Weight (v) density × gravity)	Bingham Viscosity (η_b) (estimated)	Velocity variable (ω^*)	Computed plug velocity (ω)
16/7A-E					
1	18.0 cm	1969.8 dn/cm^3	0.96 × 10^2 poise	0.24	139.1 cm/sec
2	14.5 cm	2165.8 dn/cm^3	0.80 × 10^2 poise	0.27	134.0 cm/sec
16/7A-B					
3	27.4 cm	2136.4 dn/cm^3	1.61 × 10^2 poise	0.27	236.0 cm/sec
4	28.6 cm	2156.0 dn/cm^3	1.71 × 10^2 poise	0.30	227.9 cm/sec
5	51.3 cm	2146.2 dn/cm^3	3.60 × 10^2 poise	0.29	396.8 cm/sec
16/7A-C					
6	21.5 cm	2234.2 dn/cm^3	1.18 × 10^2 poise	0.28	171.1 cm/sec
7	35.8 cm	2165.8 dn/cm^3	2.40 × 10^2 poise	0.27	226.1 cm/sec
8	26.4 cm	2009.0 dn/cm^3	1.53 × 10^2 poise	0.28	267.8 cm/sec
16/7A-A					
9	23.9 cm	2136.4 dn/cm^3	1.34 × 10^2 poise	0.30	238.2 cm/sec
10	47.5 cm	1989.4 dn/cm^3	3.30 × 10^2 poise	0.31	294.3 cm/sec
11	21.7 cm	1969.8 dn/cm^3	1.20 × 10^2 poise	0.28	151.1 cm/sec
16/7A-D					
12	34.7 cm	2136.4 dn/cm^3	2.29 × 10^2 poise	0.27	264.5 cm/sec
13	28.3 cm	2136.4 dn/cm^3	1.69 × 10^2 poise	0.30	264.9 cm/sec
14	67.9 cm	2009.0 dn/cm^3	5.50 × 10^2 poise	0.27	396.5 cm/sec
Grosses Roches					
15	91.6 cm	1989.4 dn/cm^3	6.25 dn/cm^3	0.30	698.7 cm/sec

Table 4. Rigid plug velocities and related parameters from selected flow units in 16/7A cores and Grosses Roches outcrop.

area, and one Ordovician outcrop example from Québec for comparison, were calculated. The yield strength results fell in the 10^3-10^4 dynes/cm^2 range and matched closely with computations by Johnson (1970) for modern subaerial debris flows. These yield strengths represent the total strength necessary to support the largest boulders in our debris flows. Based on work by Lewis *et al.* (1980) and Pierson (1981), about ⅔ of this strength should be accounted for by buoyancy resulting from excess pore pressure and the remainder by grain-to-grain contact and cohesive strength. Based on the large size of boulders in the studied flows and the virtual lack of clay in the continuous phase of our flows, cohesive strength could supply as little as 1% of the boulder support. Using a formula for Bingham viscosity (with this parameter estimated), rigid plug velocities in our flows were calculated with a resulting range (strongly dependent on flow thickness) from 134 to 396.5 cm/sec. These results compare favorably with measured velocities in modern debris flows and experimental slurries. Though our numerical results from calculations of yield strength and rigid plug velocity do not help us to differentiate between subaerial and subaqueous debris flows (no discernible pattern was observable from our results), they do offer a fairly accurate estimate of the magnitude of total strength necessary for large clast support and possible rigid plug velocities in sandy matrix debris flows. More importantly, this study is an example of the application to ancient rocks of concepts developed for the study of modern sediments and experimental sedimentology.

References

Bagnold, R.A. 1954. Experiments on a gravity-free dispersion of large solid spheres in a Newtonian fluid under shear. Proceedings of the Royal Society of London, Series A, v. 225, p. 49-63.

Cook, H.E., McDaniel, P.N., Mountjoy, E.W. and Pray, L.C. 1972. Allochthonous carbonate debris flows at Devonian bank reef margins, Alberta, Canada. Bulletin of Canadian Petroleum Geology, v. 20, p. 439-497.

Enos, P. 1977. Flow regime in debris flows. Sedimentology, v. 24, p. 133-142.

Gloppen, R.G. and Steel, R.J. 1981. The deposits, internal structure and geometry in six alluvial fan - fan delta bodies (Devonian, Norway) — A study in the significance of bedding sequences in conglomerates. *In:* Ethridge, F.G. and Flores, R.M. (Eds.), Recent and Ancient Nonmarine Depositional Environments: Models for Exploration. Society of Economic Paleontologists and Mineralogists, Special Publication 31, p. 49-69.

Hampton, M.A. 1972. The role of subaqueous debris flow in generating turbidity currents. Journal of Sedimentary Petrology, v. 42, p. 775-793.

Hampton, M.A. 1975. Competence of fine-grained debris flows. Journal of Sedimentary Petrology, v. 45, p. 834-844.

Hampton, M.A. 1979. Buoyancy in debris flows. Journal of Sedimentary Petrology, v. 49, p. 753-758.

Harms, J.C. and McMichael, W.J. 1983. Sedimentology of the Brae oilfield area, North Sea. Journal of Petroleum Geology, v. 5, p. 437-439.

Harms, J.C., Tackenberg, P., Pollock, R.E. and Pickles, E. 1981. The Brae oilfield area. *In:* Illing, L.V. and Hobson, G.D. (Eds.), Petroleum Geology of the Continental Shelf of Northwest Europe. Heyden, p. 352-357.

Johnson, A.M. 1965. A model for debris flow. Ph.D. Thesis, Pennsylvania State University, State College, 232p.

Johnson, A.M. 1970. Physical Processes in Geology. San Fransico: Freeman, Cooper and Company, 577p.

Johnson, R.J. and Dingwall, R.G. 1981. The Caledonides: their influence on the stratigraphy of the Northwest European continental shelf. *In:* Illing, L.V. and Hobson, G.D. (Eds.), Petroleum Geology of the Continental Shelf of Northwest Europe, Heyden, p. 85-97.

Kessler, L.G. 1983. Origin of large rounded boulders in subaqueous debris flows. Geological Society of America, 96th Annual Meeting, Abstract, p. 611.

Kuenen, Ph.H. 1951. Properties of turbidity currents of high density. *In:* Hough, J.L. (Ed.), Turbidity Currents and the Transportation of Coarse Sediments to Deep Water. Society of Economic Paleontologists and Mineralogists, Special Publication 2, p. 14-33.

Larsen, V. and Steel, R.J. 1978. The sedimentary history of a debris flow-dominated alluvial fan — a study of textural inversion. Sedimentology, v. 25, p. 37-59.

Lewis, D.W. 1976. Subaqueous debris flows of Early Pleistocene age at Montunau, North Canterbury, New Zealand. New Zealand Journal of Geology and Geophysics, v. 19, p. 535-567.

Lewis, D.W., Laird, M.G. and Powell, R.D. 1980. Debris flow deposits of Early Miocene age, Deadman stream, Marlborough, New Zealand. Sedimentary Geology, v. 27, p. 83-118.

Lowe, D.R. 1976. Grain flow and grain flow deposits. Journal of Sedimentary Petrology, v. 46, p. 188-199.

Lowe, D.R. 1979. Sediment gravity flows, their classifications and some problems of applications to natural flows and deposits. *In:* Doyle, L.J. and Pilkey, O.H. (Eds.), Geology of Continental Slopes. Society of Economic Paleontologists and Mineralogists, Special Publication 27, p. 75-82.

Lowe, D.R. 1982. Sediment gravity flows: II. Depositional models with special reference to the deposits of high-density turbidity currents. Journal of Sedimentary Petrology, v. 52, p. 279-297.

Middleton, G.V. and Hampton, M.A. 1973. Sediment gravity flows: mechanics of flow and deposition. *In:* Middleton, G.V. and Bouma, A.H. (Co-chairmen), Turbidites and Deep-Water Sedimentation. Pacific Section, Society of Economic Paleontologists and Mineralogists, Anaheim, Short Course Notes, p. 1-38.

Pierson, T.C. 1981. Dominant particle support mechanisms in debris flows at Mt. Thomas, New Zealand, and implications for flow mobility. Sedimentology, v. 28, p. 49-60.

Ricci Lucchi, F. and Valmori, E. 1980. Basin-wide turbidites in a Miocene, over-supplied deep-sea plain: a geometrical analysis. Sedimentology, v. 27, p. 241-270.

Rodine, J.D. and Johnson, A.M. 1976. The ability of debris, heavily freighted with coarse clastic materials to flow on gentle slopes. Sedimentology, v. 23, p. 213-234.

Stow, D.A.V., Bishop, C.D. and Mills, S.J. 1982. Sedimentology of the Brae oilfield, North Sea: fan models and controls. Journal of Petroleum Geology, v. 5, p. 129-148.

Surlyk, F. 1978. Submarine fan sedimentation along fault scarps on tilted fault-blocks, (Jurassic-Cretaceous boundary, East Greenland). Gronlands Geologiske Undersøgelse Bulletin 128, 108p.

Wain, A.S. and Kessler, L.G. 1982. Coarsening/thickening upward infill of an active tectonic trough, Carboniferous, Jackfork Group, Arkansas, U.S.A. 11th International Sedimentological Congress, Hamilton, Canada, Abstracts of Papers, p. 188.

CLAST SIZE AND BED THICKNESS TRENDS IN RESEDIMENTED CONGLOMERATES: EXAMPLE FROM A DEVONIAN FAN-DELTA SUCCESSION, SOUTHWEST POLAND

SZCZEPAN J. PORĘBSKI[1]

ABSTRACT

The Upper Devonian-lower Tournaisian Świebodzice succession (ca. 4000 m thick) is interpreted as a submerged fan-delta complex of resedimented conglomerates, sandstone turbidites and fossiliferous mudstones. The mud-poor, dominantly clast-supported conglomerates are divided into five main facies, including: inversely graded beds (IG), inverse-to-normally graded beds (ING), normally graded beds (NG), normally-graded-stratified beds (GS), and ungraded beds (UG). All of these bed types show a positive relationship between bed thickness (BTh) and maximum clast size (MCS). The five MCS distributions are interrelated in having the same coefficient of variation. The same sort of an intrinsic relationship links the BTh distributions. The variability in MCS is larger than in BTh. There is a general rapid decrease in mean MCS associated with a less rapid decrease in mean BTh, and an increase in occurrence of sandstone cappings, in the order: IG-UG-ING-NG-GS. This order possibly indicates maturing flow if MCS is a measure of transport distance. A comparison between sections suggests a downfan decline in the amount of IG and UG beds and an increase in the proportion of NG and GS beds, and of sandstone cappings within conglomerate bodies. This is accompanied by a tendency for the BTh/MCS gradient and BTh/MCS ratio to increase basinwards. These trends are interpreted in terms of a distally decreasing slope and concentration of clast dispersion.

RÉSUMÉ

La succession de Świebodzice du Dévonien supérieur — Tournaisien inférieur (environ 4 000 m d'épaisseur) est interprétée comme un complexe d'un cône de déjection submergé formé de conglomérats resédimentés, de turbidites arénacées et de 'mudstones' fossilifères. Les conglomérats à dominance de fragments jointifs, pauvres en boue, sont divisés en cinq faciès, incluant: des couches à granoclassement inverse (GI) des couches à granoclassement inverse-à-normal (GIN), des couches à granoclassement normal (GN), des couches normales-granoclassées-stratifiées (GS), et des couches non-granoclassées (NG). Tous ces types de couches présentent une corrélation positive entre l'épaisseur des couches (EpC) et la grosseur maximale des fragments (GMF). Les cinq distributions de GMF sont interreliées par un même coefficient de variation. Les distributions des EpC affichent le même type de corrélation intrinsèque. La variabilité de la GMF est plus grande que celle de l'EpC. Il y a une diminution rapide générale de la GMF moyenne associée avec une diminution moins rapide de l'EpC moyenne, et une apparition plus fréquente des grès de recouvrement, dans l'ordre: GI-NG-GIN-GN-GS. Cet ordre reflète possiblement une maturation de la coulée en autant que la GMF soit une mesure de la distance du transport. Une comparaison des sections indique de haut en bas du cône de déjection une diminution de l'abondance des couches GI et NG et une augmentation de la proportion des couches GN et GS, et des grès de recouvrement au sein des corps conglomératiques. Ceci s'accompagne d'une tendance à des valeurs plus élevées du gradient EpC/GMF et du rapport EpC/GMF en direction du bassin. On interprète ces tendances comme une diminution de la pente en fonction de la distance à la source nourricière et une concentration dans la dispersion des fragments.

INTRODUCTION

Features of gravel-rich submarine-slope sequences are difficult to recognise in modern settings and attempts at construction of facies models for coarse, resedimented deposits have been derived mainly from the ancient record. Documentation of the downcurrent and lateral trends in bed properties is seldom possible due to poor or discontinuous exposure. Existing facies models for resedimented conglomerates have therefore been based largely on theoretical and experimental understanding of the origin of texture and stratification from sediment-gravity flows, as well as an overall facies sequences (e.g., Middleton and Hampton, 1976; Davies and Walker, 1974; Walker, 1975, 1977; Carter and Norris, 1977; Lowe, 1976, 1982; Hein, 1982). It is suggested here that insight into the problem of lateral variability in resedimented conglomerates can be gained from the analysis of bedding and maximum clast size (cf. Bluck, 1967). An example of such a semi-quantitative analysis is provided here for dominantly clast-supported, mud-poor conglomerates which were deposited in a proximal submarine-slope environment.

REGIONAL SUMMARY

The rocks under consideration form a folded, coarsening upward clastic succession (min. 3800 m thick) which fills a small (ca. 100 sq. km), fault-bounded basin in central Sudety Mountains (Teisseyre, 1968; Fig. 1). The Świebodzice Basin originated due to thrust-faulting and subsidence along the northern periphery of a Precambrian high-metamorphic terrain (Sowie Góry Uplift). During Late Devonian - early Tournaisian times (Gunia, 1968) the basin

---

[1]Polska Akademia Nauk, Instytut Nauk Geologicznych, Senacka 3, PL-31-002 Kraków, Poland.

Funding for the fieldwork was provided by the Polish Academy of Sciences Research Project MR.1.16. Presentation of the first version of this paper at XI IAS Congress in Hamilton, Ontario, was made possible through a travel grant from the IAS Congress Organizing Committee and through generous help from A. D. Miall, C. Miall, R. J. Steel, and K. Kleinspehn. The manuscript was critically commented upon by R. J. Steel, H. E. Clifton, R. Gradziński and W. Nemec. The author is solely responsible for any errors and misconceptions.

Copyright © 1984, Canadian Society of Petroleum Geologists

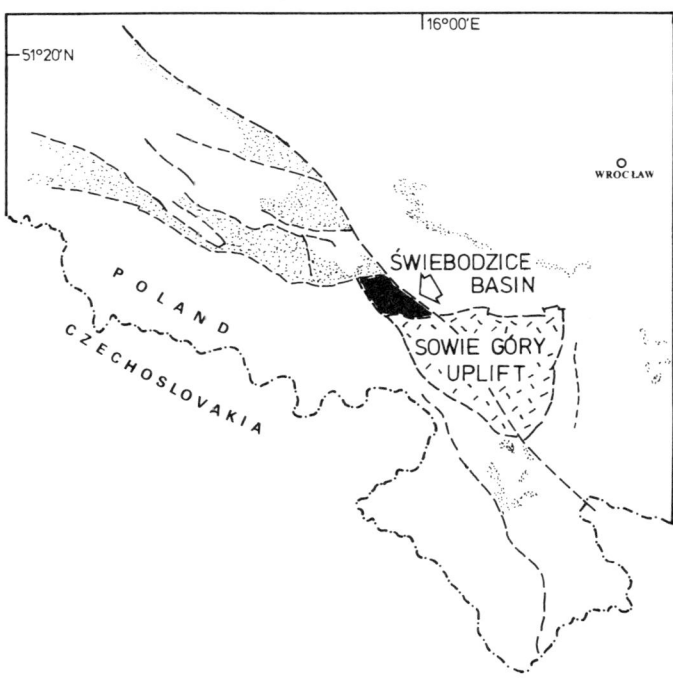

Fig. 1. Location of the Świebodzice Basin in the Sudety Mountains. Generalised distribution of Eocambrian through Lower Carboniferous mainly pelagic sediments and associated basic volcanics is also shown (stippled pattern).

was the site of conglomerate accumulation marking the first incursion of coarse land-derived clastics into a pelagic facies realm, the Sudetic segment of the Variscan geosyncline. The Świebodzice succession (Fig. 2) is typified by:

(a) a complex intertonguing of coarse conglomerate-sandstone bodies with fossiliferous mudstones and flysch-like sequences within a lower, 1300 m thick part (Pogorzała and Pełcznica Formations; upper Frasnian-Famennian) and mostly conglomeratic upper 2500 m (Chwaliszów and Książ Formations; upper Famennian-lower Tournaisian),

(b) a predominance of immature conglomerates of consistently great coarseness and thickness and of sediment gravity flow character,

(c) the occurrence and stacked nature of coarsening-up sequences (at least 150 m thick), and

(d) a fan-shaped regional pattern of palaeocurrents with the dominant sediment dispersal towards the north and north-east.

The succession records a progressive uplift of the nearby southern source and is interpreted as having originated from the progradation of an alluvial fan system directly onto a submarine, structurally controlled slope.

Facies Associations and Sedimentary Environment

Because a detailed description and interpretation of the facies and facies associations have been given elsewhere (Porębski, 1981), the depositional system is only briefly reviewed here.

PROXIMAL-FAN ASSOCIATION

This association forms large-scale coarsening-upward sequences (80-150 m thick), vertically stacked to form basin-wide bodies (300-2500 m thick) which consist mainly of repetitions of massive, pebbly to bouldery conglomerates interstratified with coarse- to medium-grained, graded, structureless or flat-laminated sandstones. Finer interbeds are scarce as are diamictites and associated slumped units. These deposits are thought to have been formed entirely subaqueously (Nemec et al., 1980) from flood- and slump-generated gravity flows which dropped their load in a proximal, probably largely non-channelised slope environment. The landward equivalents of this association, including shoreline and alluvial-fan facies, are unknown, probably due to removal during the basin-margin faulting.

DISTAL-FAN ASSOCIATION

The distal-fan association consists of fine pebble to cobble conglomerates which are interbedded with pebbly sandstones, fine sandstone turbidites and poorly fossiliferous mudstones. These lithologies are commonly organised into a number of small-scale (a few metres thick), coarsening- or fining-up cycles which form broadly coarsening-upward units (20-120 m thick). The latter either grade upwards into the proximal-fan association or occur as lens-shaped bodies (1-3 km along depositional dip) within fine-grained facies (cf. Fig. 2C). This association is believed to have been deposited as basinward, lobate extensions of the proximal gravels or within channel-distal lobe systems which prograded onto temporarily abandoned fan segments. The channel fills (up to 15 m thick, or more) consist mainly of mud-rich, matrix-supported conglomerates, amalgamated pebbly sandstone units and mudstone plugs.

FAN-FRINGE ASSOCIATION

The fan-fringe sediment comprises flysch-like sequences (up to 50 m thick, or more) consisting of regular repetitions of sharp-based, thin- to medium-bedded, fine- to medium-grained sandstones with dark, lenticular- to wavy-laminated silty mudstones and few fine pebble conglomerates. The sandstones display graded bedding and a variety of structures indicative of deposition from waning turbulent suspensions. Redeposited fauna, oriented plant remains and soft-sediment deformation, including mud dikes, are particularly common.

BASIN-AND-INTERLOBE ASSOCIATION

The basin plain, interlobe areas and abandoned-fan segments were dominated by a slow settling of suspended load. This resulted in accumulations of dark, homogeneous mudstones (up to ?700 m thick) with variable admixture of thin, laminated siltstones and fine, graded and/or ripple-laminated sandstones. The mudstones contain mainly infau-

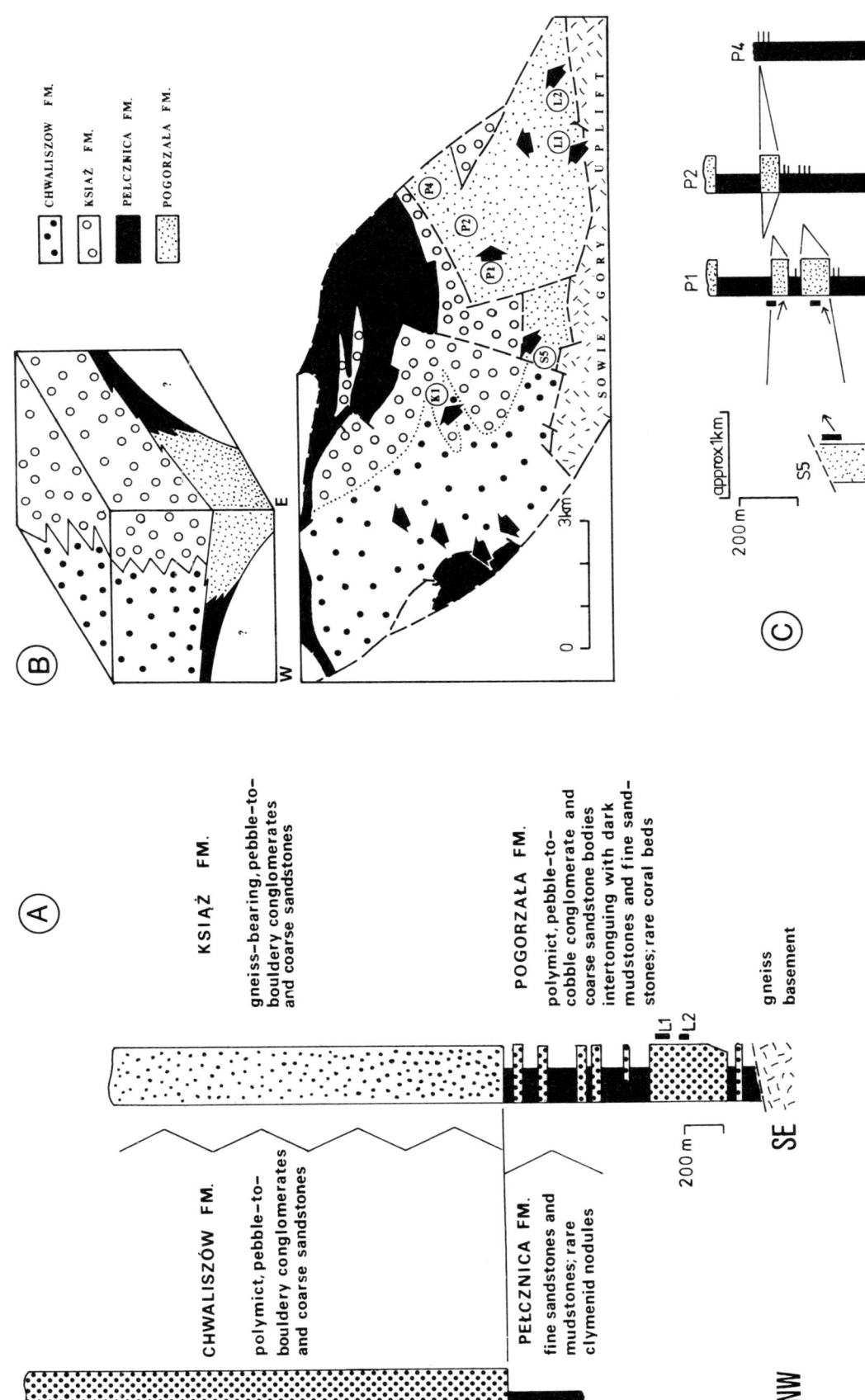

Fig. 2. (A) Generalised stratigraphic columns from northwestern and southeastern portions of Świebodzice Basin. (B) Simplified geological map of study area with positions of sections mentioned in text and with palaeocurrent directions. Inset diagram shows suggested spatial relationships between main lithostratigraphic units. (C) Example of lateral and downslope lithologic variability in Pogorzała Formation. Position of exposures marked by black rectangles.

nal molluscs, brachiopods, tentaculites, orthoceratids and rare goniatites. Unbioturbated finely laminated horizons with disseminated pyrite and coalified plant matter are frequent. Those areas characterised by low clastic input and low-subsidence rates were sporadically the site of small, biogenic carbonate accumulations. From these sites the carbonate debris, mixed with siliciclastics, was occasionally swept down into deeper areas.

SEDIMENTARY MODEL

The sedimentary model for the Świebodzice succession, as schematically depicted in Figure 3, involves:

(1) the existence of steep, tectonically rejuvenated, marine slopes backed by a coastal mountain range,

(2) a narrow or absent shelf and a supply of coarse, river-borne clastics by flood-generated flows directly onto the marine slope,

(3) the scarcity of large-scale channelling in the proximal-fan environment, and

(4) the fan-like spatial arrangement of the prograding, coarse bodies separated by the abandonment-phase fine sediment (cf. Stanley, 1980).

If one accepts Holmes' (1965) classical definition of fan-deltas, then within a spectrum of possible fan-delta systems the model postulated here would represent an end member typified primarily by a steep delta-front slope. Most of the published examples of recent and ancient gravelly fan-deltas of humid regions reveal proximal gradients not exceeding 50 m/km (Ricci Lucchi et al., 1981, Table 1). This may explain why the gravels described in such a setting have been trapped mainly within distributary-channel networks of the delta plain or, rarely, in the subaqueous delta-slope channels (e.g., Flores, 1975; Massari, 1978; Gnaccolini, 1981), whereas finer sediment swept further offshore was redistributed by wave and tidal processes. With the proximal slopes at least twice as steep, a magnitude not unlikely in mobile belts, deposition of the gravel would easily extend far on into deeper areas and the wave- and tidal-reworking would be limited to narrow, nearshore zones (cf. Stanley, 1980). Whatever label is attached to such a depositional system it is concluded that in the absence of a well-developed shelf the association of a steep, marine slope bounded landwards by an active coastal range, high clastic input, moderate- to high-subsidence rate, and low-energy environments (microtidal regime and weak wave activity) can generate thick, coarse clastic successions which exhibit features common to some alluvial fans (e.g., Heward, 1978; Gloppen and Steel, 1981) and some deep-sea fans (e.g., Marschalko, 1975; Rupke, 1977; Stanley and Hall, 1978).

As suggested here, the Świebodzice succession represents primarily basin-plain and steep, fan-delta-front slope facies which were presumably attached landwards to a thin zone of marine-worked clastics and alluvial-fan facies (Fig. 3). Another possible interpretation of the succession could be one of a small, 'low-efficiency' deep-sea fan (cf. Mutti, 1979) similar to the Rocchetta fans described from the Tertiary Piedmont Basin, NW Italy (Cazzola et al., 1981). It was shown that the Rocchetta fans are separated landwards by fine-grained slope facies from the coeval, wave-reworked, fluvially-introduced gravels (fan-deltas).

The proposed solution (Fig. 3) is highly speculative as it is based upon little field evidence.

DESCRIPTION OF THE CONGLOMERATES

Conglomerates, which form much of the basin-infill, typically have a mud-poor sandstone matrix (up to 30%, on average 4% of mainly phyllosilicate matrix) and fall into two general categories: 1) clast- to matrix-supported, dominantly polymict and 2) matrix-rich, dominantly gneiss-bearing. The latter category tends to prevail in the Książ Formation (Nemec et al., 1980); the former includes up to 90% of beds in the Pogorzała and Chwaliszów Formations and is a main theme in the ensuing discussion.

TEXTURE AND BEDDING

Beds range from a few cm to 4 m in thickness; thicker units are rare and probably composite in origin. The beds appear to be sheet-like to broadly lenticular with flat bases, or with rare basal scours up to 50 cm in depth. The conglomerates consist of poorly to moderately-well sorted mixtures of mostly pebble to cobble grade clasts set in a matrix of medium to granule sandstone that comprises between 15 and 50% of the rock volume. The maximum clast diameter seldom exceeds 60 cm, although boulders of gneiss, up to 200 cm in size, were occasionally noted. Pebbles and cobbles vary in roundness from subangular to well-rounded, depending on composition. A clast-supported framework tends to dominate though it is often seen to

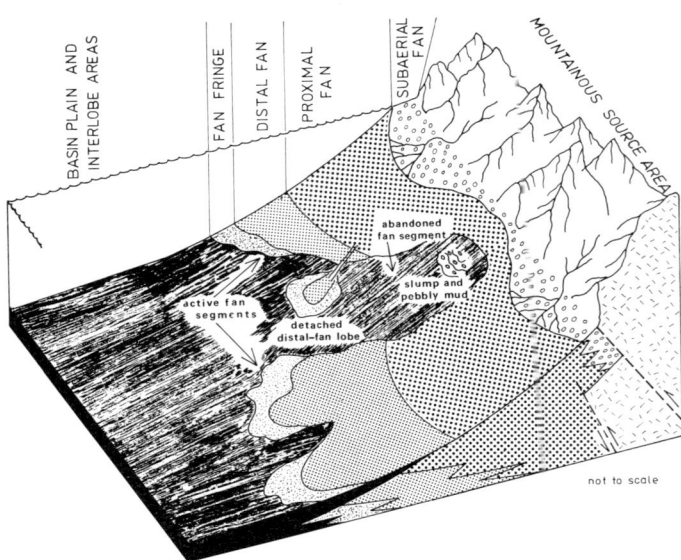

Fig. 3. Inferred depositional setting of Świebodzice succession.

pass (usually upwards) into a disrupted framework (Fig. 4A, B); tightly fitted clast fabrics are remarkably rare and never occur at the top of beds. An upcurrent long-axis imbrication is commonly evident (Fig. 5A); however, the systematic study of clast-fabric was precluded by the nature of the outcrops.

GRADING AND CONGLOMERATE FACIES

Most beds show either inverse or normal grading (75% of 344 beds); and therefore the presence and type of grading were used as the main variable to classify the conglomerate beds into five facies which generally correspond to Walker's (1975, 1977) resedimented conglomerate models. The inversely graded facies (coded here *IG*) consists of units in which inverse grading extends about two-thirds of the way up the bed, or is less commonly present throughout the whole bed (Fig. 4A, B). The inverse-to-normally graded facies (*ING*) consists of beds showing a zone of inverse grading in the lower one-third or one-half of the bed thickness, followed upwards by a normally graded portion (Fig. 4C). The normally graded facies (*NG*) comprises beds showing either coarse-tail grading (especially in coarse deposits; Fig. 5A) or distribution grading that is often associated with an upward transition into a granule to medium-grained sandstone (Fig. 5B). The graded-stratified facies (*GS*) includes units of conglomerate and pebbly sandstone, and is typified by the association of normal grading and flat or low-angle, crude lamination (Fig. 5C). The ungraded facies (*UG*) comprises beds of massive, very poorly to moderately-well sorted conglomerate lacking any discernible trends in vertical clast-size segregation. Thin zones of inverse grading at the base and of normal grading at the top may be locally developed, and clast imbrication is not uncommon in this facies.

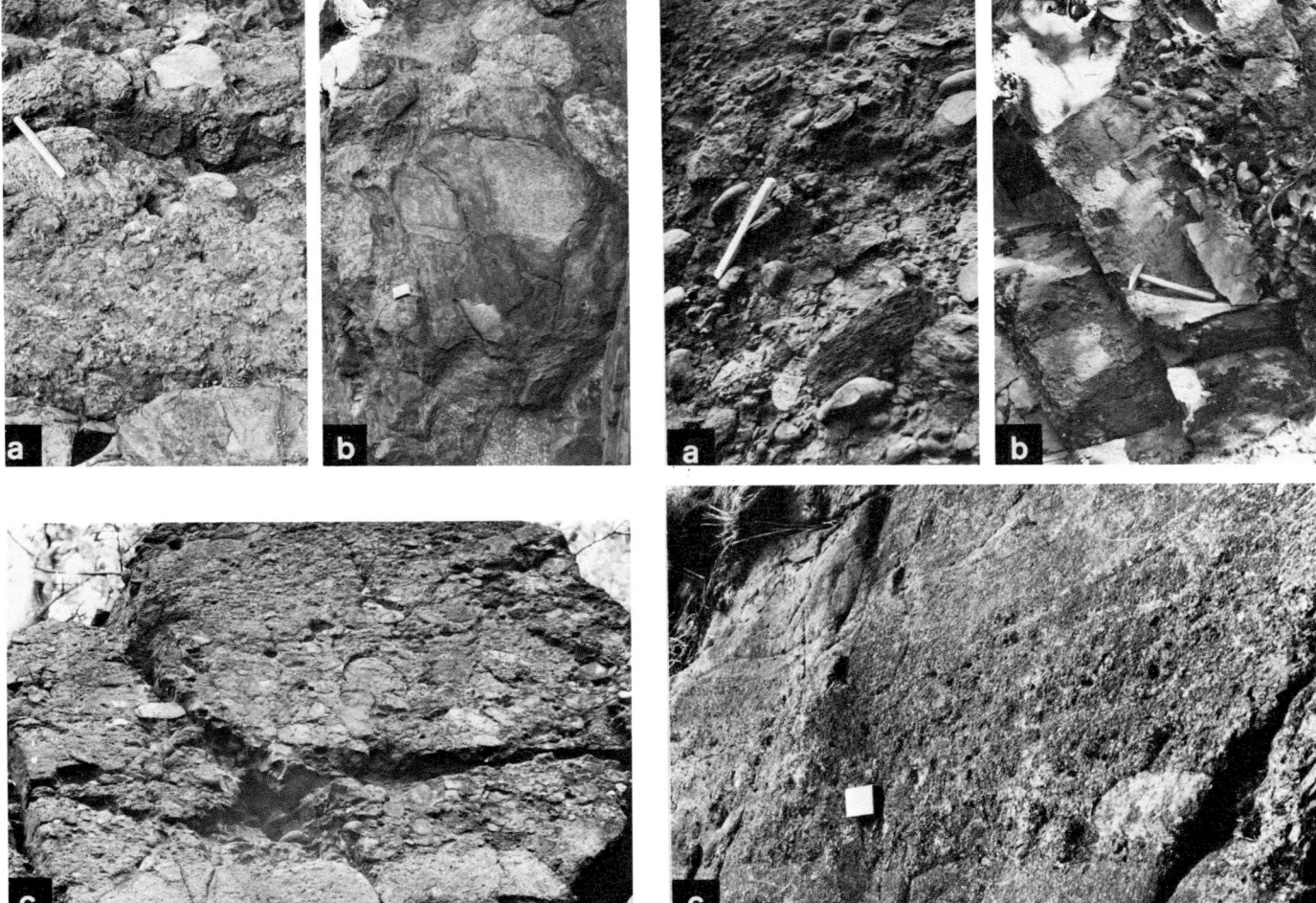

Fig. 4. (A) Lower portion of 160 cm thick inversely graded bed. Ruler is 20 cm long. (B) Uppermost portion of the same bed showing bimodal, disrupted framework and rounded boulders up to 64 cm in diameter. (C) Inverse-to-normally graded bed, 90 cm thick, with thin sandstone capping.

Fig. 5. (A) Coarse-clast variety of normally graded bed with well developed upcurrent long-axis clast imbrication. Ruler is 20 cm long. (B) Normally graded beds with sandstone tops. Hammer is 35 cm long. (C) Conglomeratic sandstone with normal grading and crude, flat lamination. Bed is 60 cm thick.

ASSOCIATED LITHOLOGIES

Most of the conglomerates are interbedded with coarse sandstones, although some intercalate with matrix-rich conglomerates and diamictites, and a few are associated with fine sandstone turbidites and lenticularly laminated silty mudstones. Beds of granule to medium sandstone as much as 250 cm thick (31 cm on average) form tabular sheets or thin impersistent lenses. The sheet-like beds usually show a normal grading commonly coupled with a flat lamination in the uppermost level, whereas the lenticular variety is essentially ungraded, structureless and often contains small floating pebbles. As a result, upper contacts of the sandstone layers with associated conglomerates are sharper than lower boundaries. All possible transitional stages exist between the graded sandstones and graded-stratified conglomerates, although no lateral transition within a single bed could be demonstrated in the field.

QUANTITATIVE COMPARISONS

The five conglomerate facies distinguished here are compared in terms of their maximum clast size (*MCS* — mean of ten largest clasts locally per bed), bed thickness (*BTh*), bed thickness/maximum clast size relationship (method of Bluck, 1967) and the frequency of sandstone layers above the conglomerate. Only those beds (259) were considered in this analysis for which the bed boundaries were determined with a reasonable certainty. A 5% significance level is used in statistical testing.

MAXIMUM CLAST SIZE DISTRIBUTIONS

The five *MCS* distributions show conspicuous differences in range, mean and standard deviation (Fig. 6A). When the facies are arranged according to decreasing magnitude of the above parameters the following trend emerges: *IG-UG-ING-NG-GS* (Fig. 6A). All the distributions tend to be positively skewed and show a high, positive correlation between the means and corresponding standard deviations. The plot of Figure 7A suggests that there is a direct relationship between these statistics, which implies a constant coefficient of variation ($V = 0.57$). In order to set up hypotheses about the means, the raw data

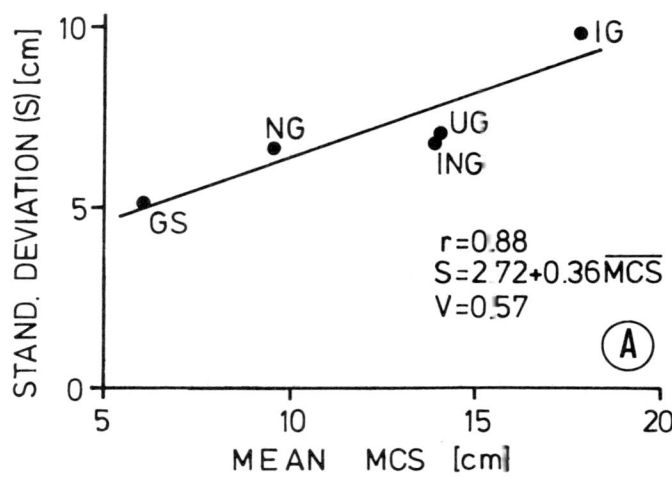

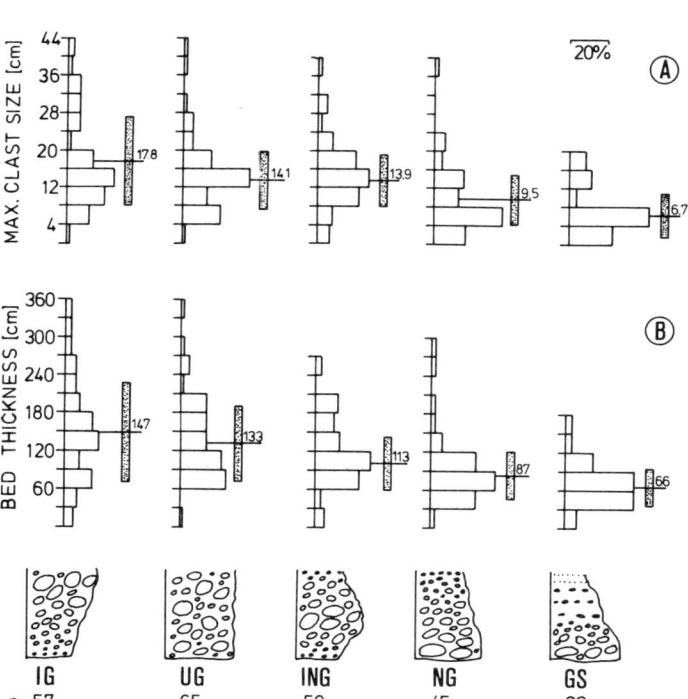

Fig. 6. Maximum clast size and bed thickness distributions of five conglomerate facies. Means and standard deviations (stippled rectangles) are also indicated; n = number of beds.

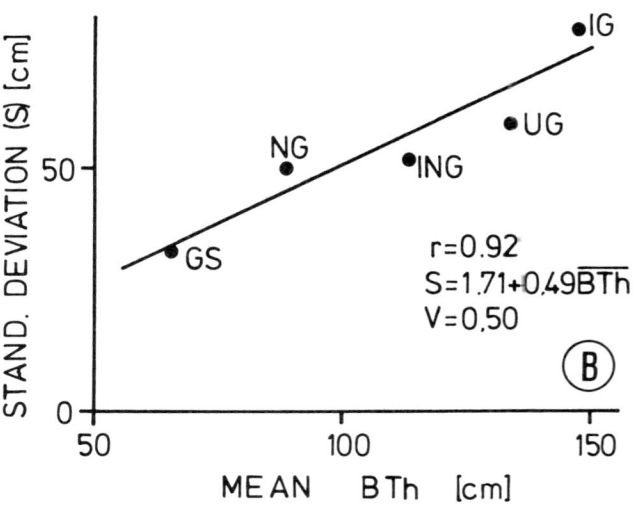

Fig. 7. Standard deviation against mean for maximum clast size (A) and bed thickness (B) in five conglomerate facies. Diagrams indicate that standard deviations are proportional to means, suggesting common coefficients of variations (V).

were subjected to a logarithmic transformation ($\log_{10} MCS$ cm). Although such a transformation only resulted in partly normalised distributions, it considerably smoothed the variances (Fig. 8A). The transformed data were next analysed in terms of a single factor analysis of variance and of Student-Newman-Keuls test (an equivalent of Student's t-test for multisample comparisons). Both these tests are known to be unaffected by considerable departure from normality, especially in cases where homogeneity of variance exists. The result of testing suggests that the *IG*, *UG*, and *ING* samples came from populations having the same *MCS* which is larger than the mean of the parent population for the *NG* beds, itself coarser than the mean of the *GS* population.

BED THICKNESS DISTRIBUTIONS

The bed thickness mean and standard deviation, when arranged according to decreasing magnitude, provide the same order of the conglomerate facies as that shown above for the maximum clast size (Fig. 6B). Similarly, there is a significant correlation between the means and the corresponding standard deviations (Fig. 7B), suggesting that the bed thickness distributions are interrelated in having the same variation coefficient ($V = 0.50$). The logarithmic transformation of the raw data strongly homogenised variances (Fig. 8B), and the parametric testing yielded a conclusion that the order of mean thicknesses is: $IG = UG = ING > NG > GS$.

THICKNESS/SIZE RELATIONSHIPS

There is a high, positive correlation between the thickness of the conglomerate beds and the size of the largest clasts (Fig. 9). The best correlation is shown by the *IG*, *ING*, and *NG* facies; the correlations for the *UG* and *GS* beds are poorer, but still significant at the 0.001 level. The data, when fitted to a straight line by least squares, taking the *MCS* as the independent variable, reveal some differences with respect to the regression equation. For all graded facies, there is a slight decrease in the regression coefficient and an increase in the elevation for finer and thinner facies (Fig. 9). However, the F-testing for coincidental regression concludes that this trend may be due to chance. Consequently, the $\overline{BTh}/\overline{MCS}$ relationship for the four graded facies were estimated by a common regression and the observed growth in the $\overline{BTh}/\overline{MCS}$ ratio, accompanying decreasing $\overline{MCS}$ and $\overline{BTh}$, takes place along the common regression slope (Fig. 11B).

SANDSTONE CAPPINGS

Some 46% of the documented beds pass upwards into a sandstone. Gradational and abrupt contacts between the conglomerate and the overlying sandstone are equally common. The most common contact involves an upward transition without any discernible break from a clast-supported zone to a matrix-rich portion to a purely sandstone top (*cf.* Clifton, 1981). This is displayed by all facies, except the *GS* beds. The latter usually show a distribution grading from a lower, dominantly pebble-size zone to an upper, sand-grade portion (Fig. 5C).

The frequency of a sandstone capping above particular conglomerate facies varies considerably (Fig. 10). As shown by chi-square goodness-of-fit testing, the ratio of sandstone-capped to non-capped conglomerate beds increases from 1:3 for the *IG* and *UG* facies, through 1:1 for the *ING* and *NG* beds, up to 8:1 in the case of the *GS* beds. This increase is accompanied by an insignificant variation in the mean sandstone thickness (between 27 and 37 cm). In contrast, the conglomerate beds which have the sandstone capping tend to be thinner than those which lack the sandy caps (70 vs 110 cm; t-test), and such a relationship is independently held by the *UG*, *IG*, and *ING* facies. This may imply a sort of inverse correlation between the thickness of the conglomerate and the thickness of the overlying sandstone (*cf.* Steel, 1974). However, no such significant correlation was found for the whole set of data and within conglomerate facies.

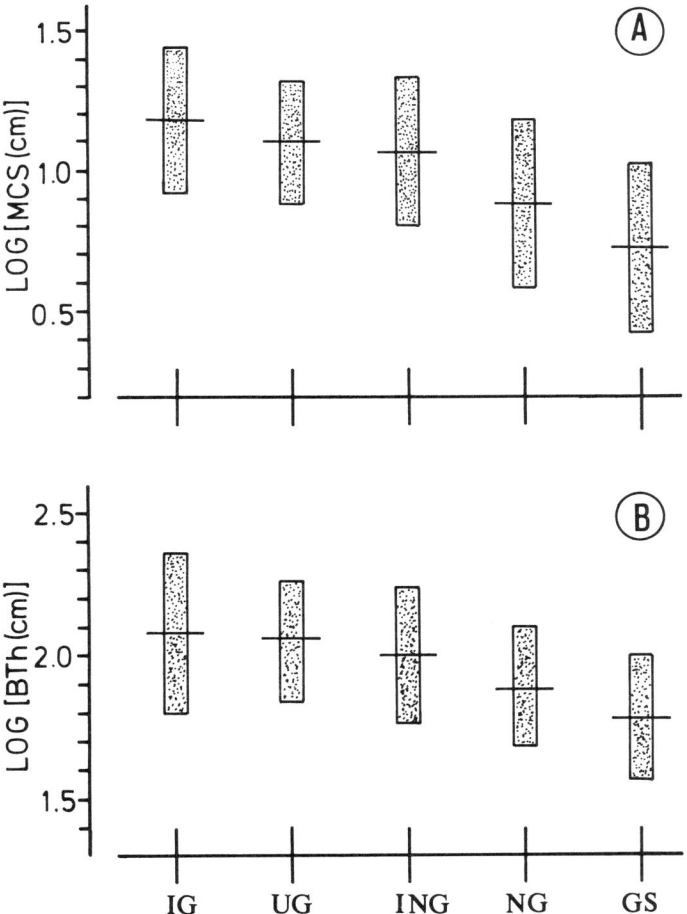

Fig. 8. Comparison of conglomerate facies in terms of mean and standard deviation for logarithmically transformed maximum clast size and bed thickness data.

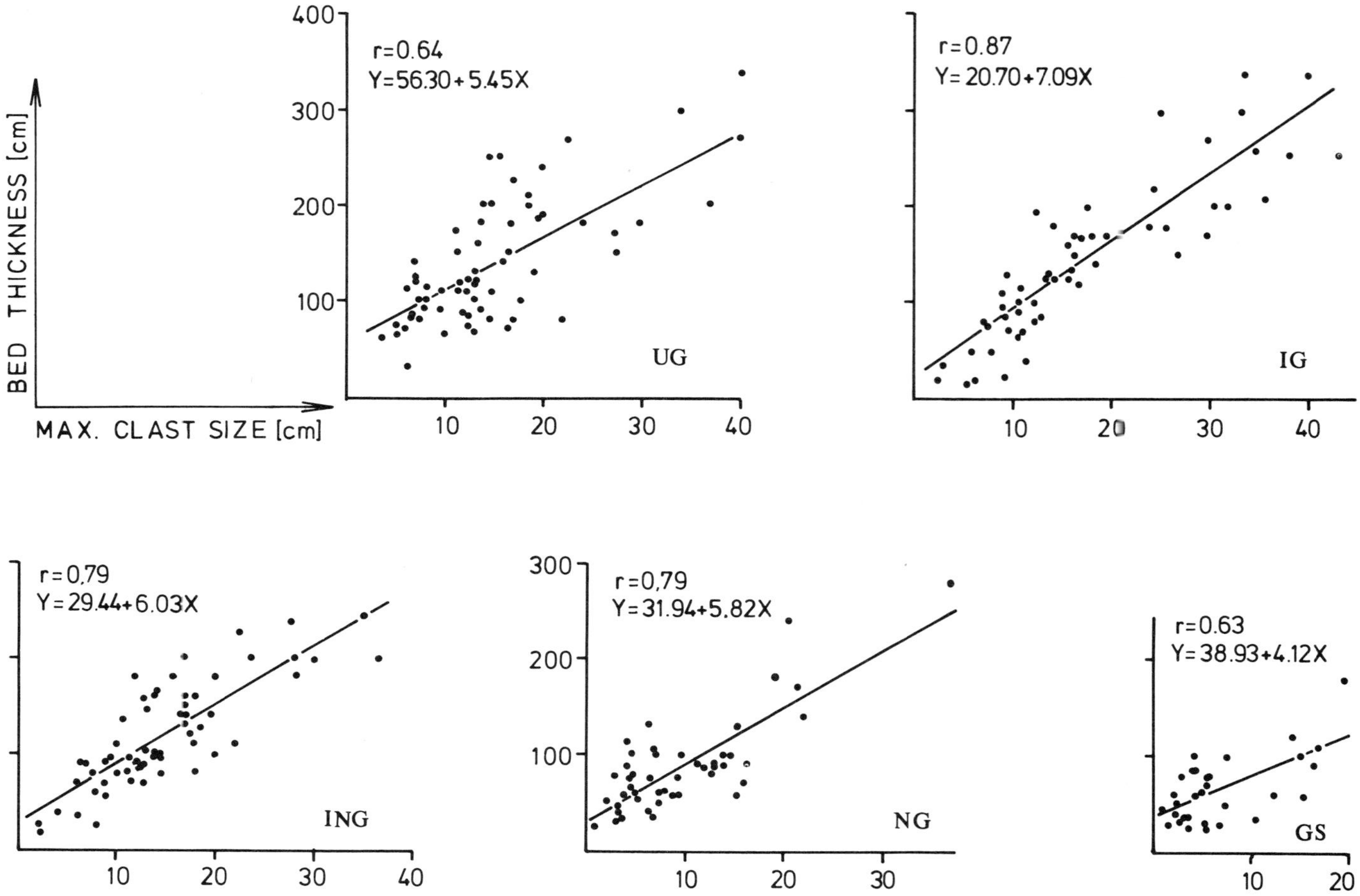

Fig. 9. Plots of bed thickness against maximum clast size for each conglomerate facies. All correlation coefficients and regression slopes are significant at 0.001 level.

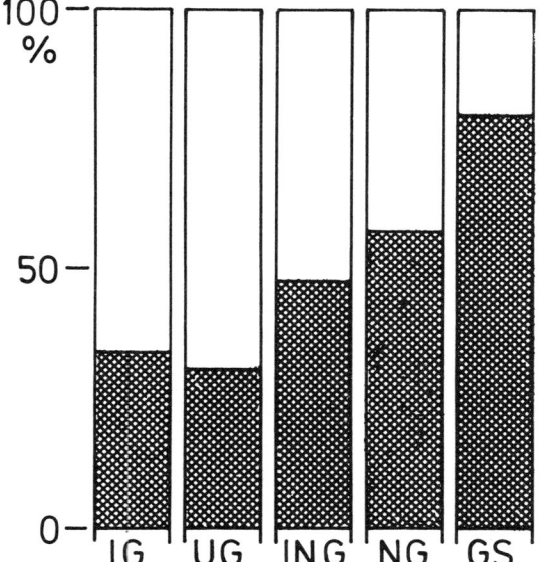

Fig. 10. Frequency of occurrence of sandstone layer above different conglomerate facies (shaded area).

PROXIMAL — DISTAL TRENDS

Poor exposure, strong tectonics and lack of correlative horizons did not permit a lateral tracing of facies changes within individual conglomerate bodies. Some indirect inferences were gained through mapping, analysis of the composite sections and from palaeocurrent data.

The southern conglomerate bodies in the Pogorzala Formation (sections L1, L2, S5 in Fig. 2B, C) are interpreted as having been deposited generally within a proximal fan, on account of their relative proximity to the source area and consistent coarseness and thickness. They reveal a predominance of the *UG* and *IG* facies which comprise up to 61% of the conglomerate beds; the proportion of the *ING* and *NG* facies varies between 34 and 58%, whereas that of the *GS* facies is low, ranging between 5 and 10% (Fig. 12). The thickness/size plots exhibit a largely constant, moderately steep slope, and the $\overline{BTh}/\overline{MCS}$ ratios vary from 7.4 to 8.7. Data from northern, distal-fan bodies are few (sections P1 and P2 in Fig. 2B, C), but they indicate a conspicuous growth in the amount of the *NG* and *GS* facies, totalling up to 78–91% of the documented beds (Fig. 12). The thick-

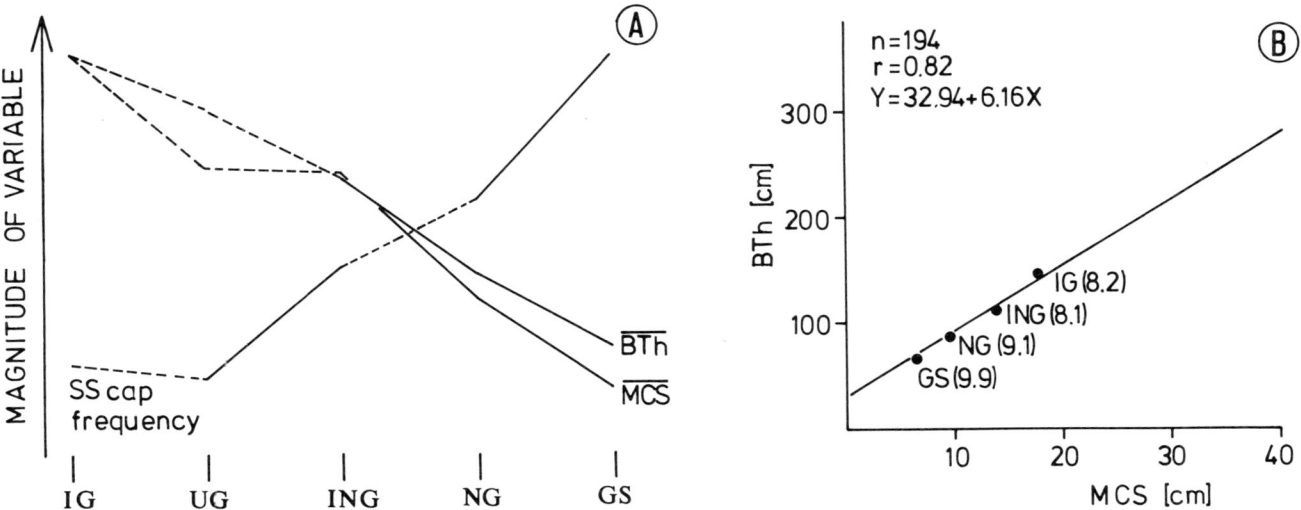

Fig. 11. (A) Comparison of five conglomerate facies in terms of rate of change of mean maximum clast size, bed thickness, and sandstone capping frequency. Heavy lines denote significant transitions. (B) Common regression for four graded conglomerate facies. Note that $\overline{BTh}/\overline{MCS}$ ratios, given in parentheses, change along common regression slope.

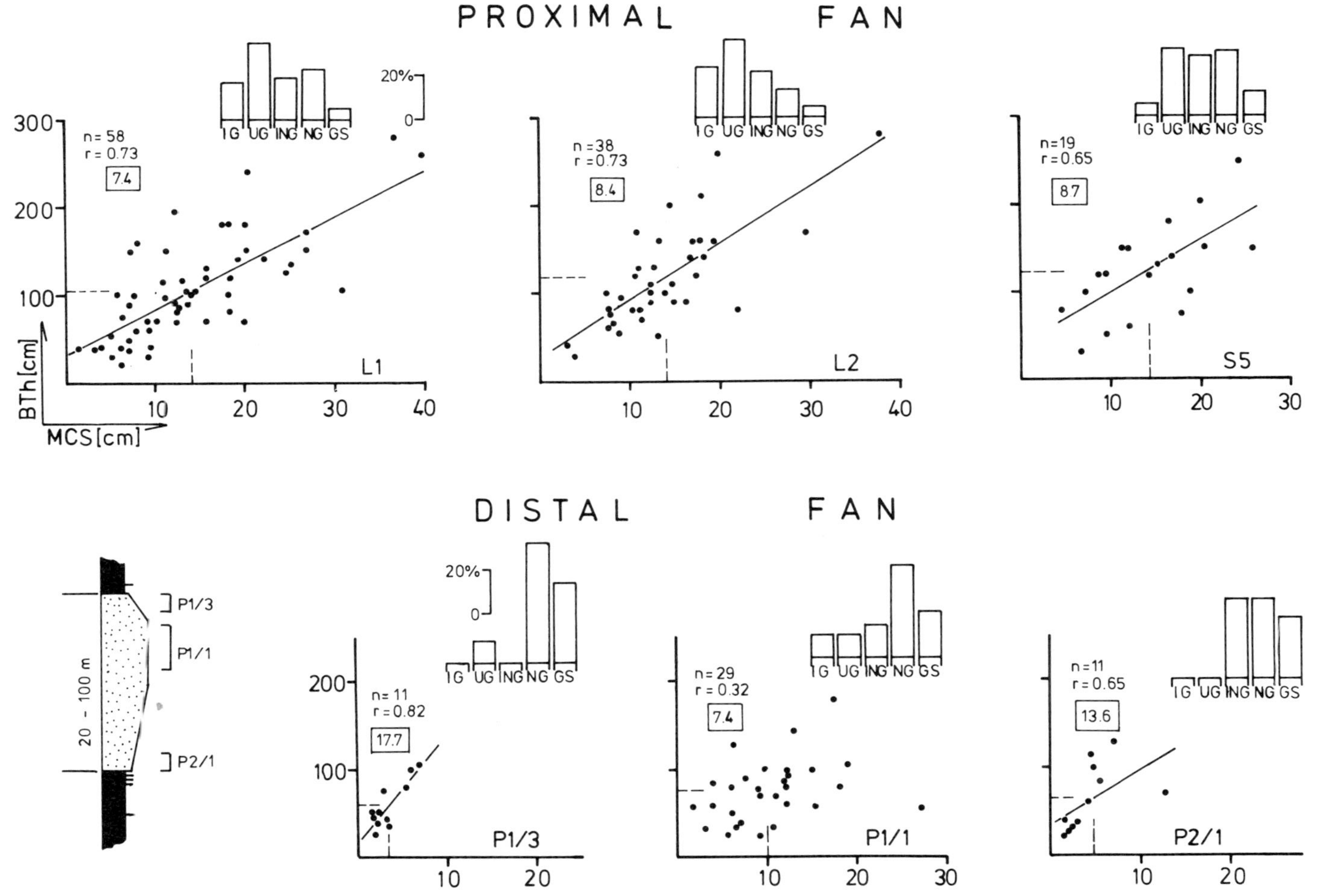

Fig. 12. Comparison between proximal and distal-fan associations in terms of conglomerate-facies distribution, thickness/size relationship, mean bed thickness and maximum clast size (dashed lines), and of $\overline{BTh}/\overline{MCS}$ ratio (in rectangles). Locations of sections given in Figure 2.

ness/size relationships may vary from insignificant to highly significant with either moderately steep or steep regression slope and the corresponding BTh/MCS ratios 13.6 and 17.7 (Fig. 12). The proportion of sandstone interbeds increases from 26-28% in the proximal bodies up to 75-95% in their basinward counterparts, and this is accompanied by an insignificant growth in the corresponding mean sandstone thickness from 25 to 32 cm.

The upper 2500 m of coarse conglomerates and sandstones in Świebodzice succession (Figs. 1 and 2) record an intense progradation of the fan system in response to a renewed faulting and enhanced erosion in the source area. This upper part of the basin-fill can be subdivided into a number of large-scale (100-150 m thick), coarsening- and thickening-upward sequences with a thinner, fining- and thinning-up capping. They show the following features (Nemec et al., 1980; Fig. 13): (1) a tendency for pebbly sandstones and matrix-supported conglomerates to give way upwards to clast-supported conglomerates, (2) a constant high proportion of the UG beds (42-54%) and a tendency for the NG beds to give way upwards to the IG facies, and (3) an upward decrease in the $\overline{BTh}/\overline{MCS}$ regression slope and in the $\overline{BTh}/\overline{MCS}$ ratio. On the basic assumption that there is a general decrease of clast size and bed thickness downfan, these large-scale sequences are thought to represent individual prograding (and abandoned) proximal-fan bodies and, consequently, the above textural and grading trends may reflect lateral (downfan) changes within the proximal-fan environment (Nemec et al., 1980; see also therein for discussion of extrabasinal controls upon textural trends).

INTERPRETATION

THE CONGLOMERATE BEDS

The common presence of inverse grading and little or no mud in the conglomerate matrix point to a high dispersive pressure arising from clast interactions during flow (Davies

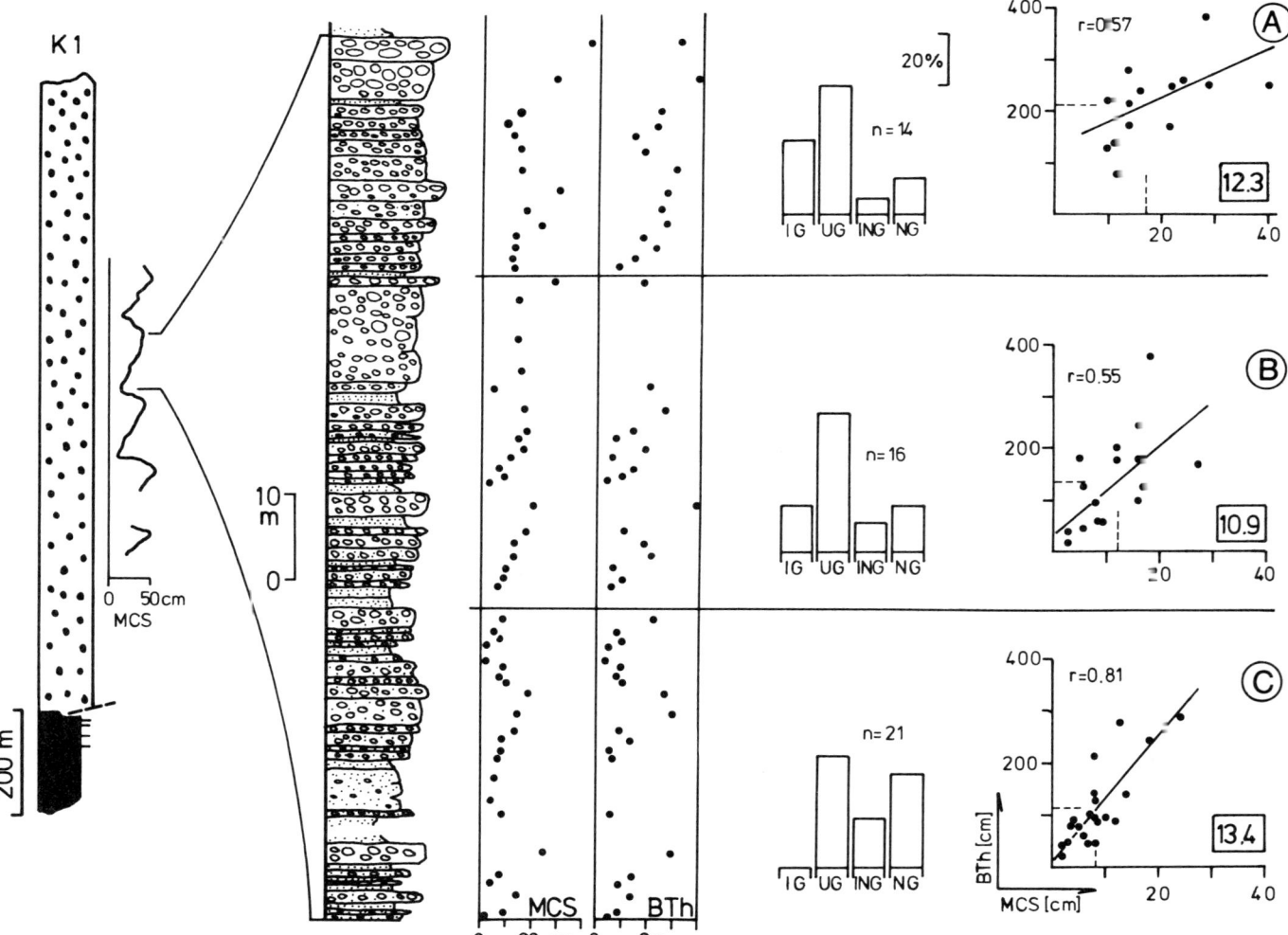

Fig. 13. Large-scale coarsening and thickening upward sequence together with distribution of conglomerate facies, thickness/size plots, mean bed thickness and maximum clast size (dashed lines), and $\overline{BTh}/\overline{MCS}$ ratios (in rectangles).

and Walker, 1974). The abundance and poor sorting of the sandy phase, usually associated with disrupted frameworks, indicate that buoyant lift has significantly contributed to the overall support mechanism. As shown by Hampton (1979), buoyancy forces due to high clast concentration in the matrix may account for the maintenance of highly mobile and competent debris flows with surprisingly low contents of clay (see also Rodine and Johnson, 1976). The association of normal grading and stratification is thought to reflect deposition from decelerating turbulent dispersions, followed by tractional flow conditions (Rocheleau and Lajoie, 1974; Walker, 1975; Aalto, 1976; Hein, 1982). The flows which deposited the conglomerate beds in question were probably combined grain flow/debris flows (Hamptom, 1979) and density-modified grain flows (Lowe, 1976), which tended to evolve toward high-density turbidity currents (Walker, 1975, 1977; Lowe, 1982).

THE SANDSTONE BEDS

The sandstone interbeds are much thinner and do not show the complexity of some sandstone fluxoturbidites. The latter usually grade upslope into and interfinger laterally with resedimented conglomerates in flysch basins (*e.g.,* Leszczyński, 1981; Ślączka and Thompson, 1981).

The graded to flat-laminated sandstones are interpreted as the deposits from fully turbulent flows which either were residual high-density, sandy turbidity currents in the sense of Lowe (1982), or induced through a dilution of gravel flows in the manner described by Hampton (1972). These two mechanisms of a downcurrent separation of sand-sized particles from a gravel population could be responsible for the observed tendency of sand to be concentrated in distal parts of the fan. The massive, structureless sandstones do not provide features easily attributable to any of the end members of the sediment gravity-flow family (Middleton and Hampton, 1976) and they may have originated from some intermediate processes (*cf.* Clifton, 1981).

CLAST SIZE, BED THICKNESS AND GRADING TRENDS

The *MCS* distributions of the five conglomerate facies are closely interrelated in that they have the same coefficient of variation; the same is also true for the *BTh* distributions (Fig. 7). This suggests that these distributions reflect various stages of basically one depositional process. It is believed that this fundamental process was deposition from high-concentration dispersions that left different though probably intergradational products, depending on the maturity achieved by the flows.

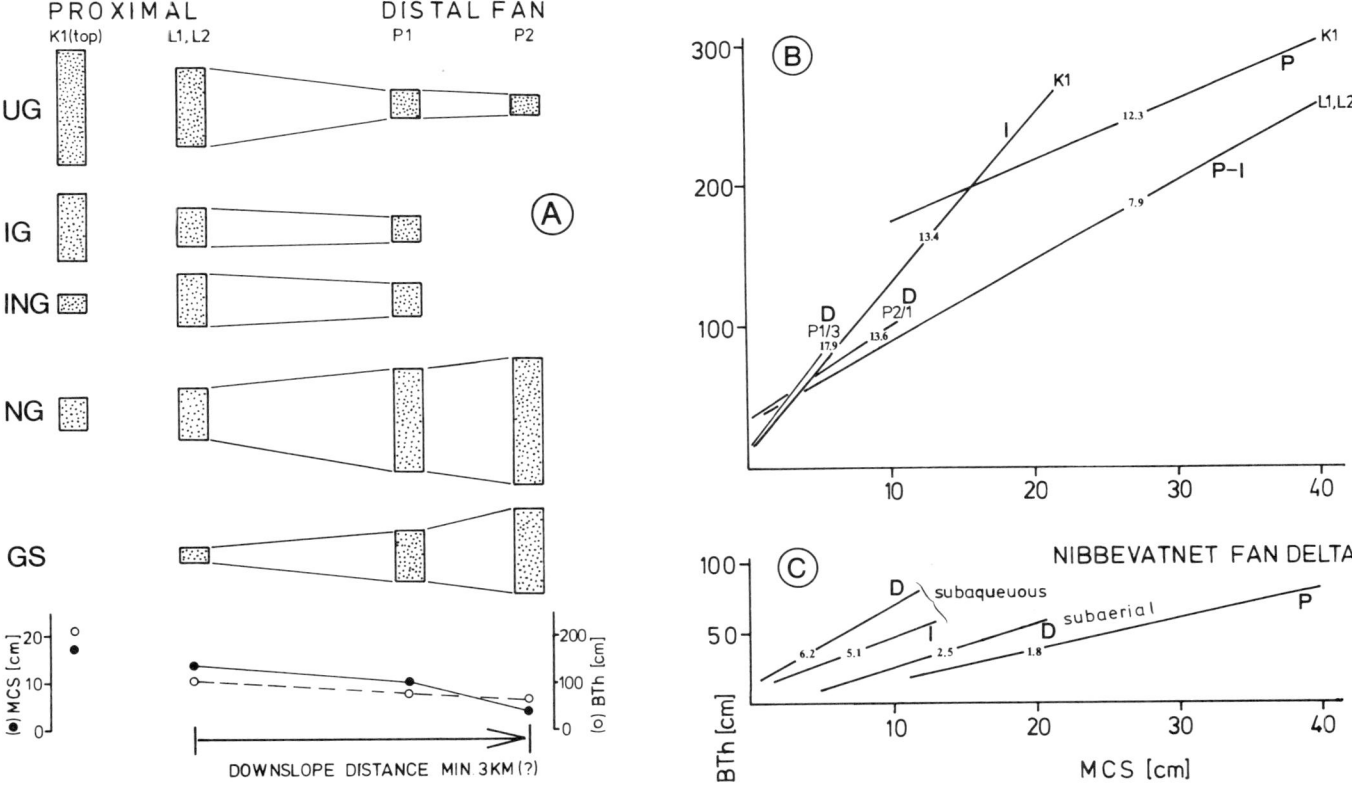

Fig. 14. (A) Lateral (downslope) distribution of different conglomerate facies within the fan body. (B) Summary of contrasts in bed thickness/maximum clast size relationship. $\overline{BTh}/\overline{MCS}$ ratios are indicated for each line; P — proximal, I — intermediate, D — distal fan. (C) Thickness/size characteristics for Nibbevatnet fan delta (Devonian, Norway) from Gloppen and Steel (1981); regression lines for subaqueous portion were fitted by eye to data in Figure 6 of Gloppen and Steel.

The facies sequence suggested here from the decreasing maximum clast size and bed thickness and from the increasing number of sandstone cappings (Fig. 11A) closely resembles the downcurrent grading pattern in deep-sea fan resedimented conglomerates, predicted by Walker (1975, 1977) on theoretical grounds and from experimental results of Middleton (1967, 1970). Field evidence shows, though not quite independently, that there is a downfan reduction in the amount of the ungraded and inversely graded beds associated with an increasing content of the normally graded and graded-stratified beds (Fig. 14A). In terms of Walker's model, such a facies distribution reflects a distally decreasing slope and greater segregation of clast sizes with increased distance of transport. If MCS and BTh can be taken as reflecting competence and capacity of depositing flows (as implied from the positive thickness/size correlations — Fig. 9; cf. Bluck, 1967; Steel, 1974), and assuming similar volumes of sediment set in motion, it is likely that finer and thinner beds have been deposited from less competent, less concentrated, and hence more distal flows.

THICKNESS/SIZE RELATIONSHIPS

The present data indicate greater intrinsic variability in the maximum clast size than in the bed thickness (Fig. 7) and it is suggested that a downfan decrease in BTh took place at a slower rate than in MCS (Figs. 11A and 14A). This in turn suggests that the flows suffered a more rapid decrease in competence than in capacity (Nemec et al., 1980).

A comparison between the facies implies the same thickness/size gradient over the whole range of MCS (Fig. 11B), suggesting a straight-linear relationship between flow capacity and competence. By comparing the sections it is evident, however, that the distal conglomerate bodies tend to yield steeper regression lines (Fig. 14B), which may indicate that, for a unit drop in competence, the distal flows decreased in capacity at a more rapid rate than their proximal counterparts. A similar tendency in thickness/size proximal-distal contrasts was noted by Kelling and Holroyd (1978) for Lower Palaeozoic canyon-fan rudites of Scotland and Wales, and interpreted in terms of a distally increasing role of tractional processes in deposition from sediment gravity flows. Gloppen and Steel (1981) demonstrated for Devonian fanglomerates of Norway that downfan thinning and fining trends accompany an increase of both the $\overline{BTh}/\overline{MCS}$ ratio and regression gradient (Fig. 14C). In a discussion to that paper, Shultz (1983) pointed out that these trends can be explained in terms of debris-flow theory. He predicted that in subaerial debris flows the competence/critical thickness ratio should be proportional to depositional slope and/or debris-flow density, and that slope is the dominant control for small ratios. It is therefore not unlikely that the rapid growth in the $\overline{BTh}/\overline{MCS}$ gradient, accompanying the supposed proximal — intermediate fan transition (Fig. 13 and K1 in Fig. 14B), may reflect a slope break. A further growth in the gradient is only slight for the most distal bodies though the $\overline{BTh}/\overline{MCS}$ ratio increases markedly; P1/3 in Figs. 12 and 14B) and may perhaps be accounted for by both decreasing slope and diminished clast concentration with increased distance of travel.

However, the thickness/size patterns are by no means consistent for all the distal sections examined. Occasionally there appear coarse, relatively thin beds (usually NG and GS facies — Fig. 5B) which result in either lowering of regression slopes (P2/1 in Fig. 12) or even in local insignificant correlations (P1/1 in Fig. 12). Moreover, data from the distal-fan association, when considered together, provide a low-gradient regression line associated with a poor though significant correlation. These facts suggest that several different processes could have been involved in producing the thickness/size patterns observed in the distal-fan bodies. Occasional deposition of basinwide layers from megaflows (cf. Ricci Lucchi and Valmori, 1980), deposition in channelled-to-unconfined flow transitions, and tractional reworking could be most critical in this respect. Certainly more work is needed, in better exposed areas, to evaluate fully the potential value of thickness-size patterns in distinguishing between various depositional agencies and in facies analysis.

REFERENCES

Aalto, K.R. 1976. Sedimentology of a mélange: Franciscan of Trinidad, California. Journal of Sedimentary Petrology, v. 46, p. 913-929.

Bluck, B.J. 1967. Deposition of some Upper Red Sandstone conglomerates in the Clyde area; a study in the significance of bedding. Scottish Journal of Geology, v. 3, p. 139-167.

Carter, R.M. and Norris, R.J. 1977. Redeposited conglomerates in a Miocene flysch sequence at Blackmount, Western Southland, New Zealand. Sedimentary Geology, v. 18, p. 289-319.

Cazzola, C., Fonnesu, F., Mutti, E., Rampone, G., Sonnino, M. and Vigna, B. 1981. Geometry and facies of small, fault-controlled deep-sea fan systems in a transgressive depositional setting (Tertiary Piedmont Basin, Northwestern Italy). In: Ricci Lucchi F. (Ed.), Excursion Guidebook with Contributions on Sedimentology of Some Italian Basins. International Association of Sedimentologists, 2nd European Regional Meeting, Bologna, p. 7-53.

Clifton, H.E. 1981. Submarine canyon deposits, Point Lobos, California. In: Frizzel, V. (Ed.), Upper Cretaceous and Paleocene Turbidites, Central California Coast. Society of Economic Paleontologists and Mineralogists, Pacific Section, Field Trip Guide Book No. 6, p. 79-92.

Davies, I.C. and Walker, R.G. 1974. Transport and deposition of resedimented conglomerates: the Cap Enragé Formation, Cambro-Ordovician, Gaspé, Québec. Journal of Sedimentary Petrology, v. 44, p. 1200-1216.

Flores, R.M. 1975. Short-headed stream delta: model for Pennsylvanian Haymond Formation, west Texas. American Association of Petroleum Geologists Bulletin, v. 59, p. 2288-2301.

Gloppen, T.G. and Steel, R.J. 1981. The deposits, internal structure and geometry in six alluvial fan - fan delta bodies (Devonian - Norway) a study in the significance of bedding sequences in conglomerates. In: Ethridge, R.G. and Flores, R.M. (Eds.), Recent and Ancient Nonmarine Depositional Environments: Models for Exploration. Society of Economic Paleontologists and Mineralogists, Special Publication 31, p. 49-69.

Gnaccolini, M. 1981. Oligocene fan-delta deposits in northern Italy: a summary. Rivista Italiana di Paleontologia e Stratigrafia, v. 87, p. 627-636.

Gunia, T. 1968. On the fauna, stratigraphy and conditions of sedimentation of the Upper Devonian in the Świebodzice Depression. Geologia Sudetica, v. 4, p. 115-220 (English summary).

Hampton, M.A. 1972. The role of subaqueous debris flow in generating turbidity currents. Journal of Sedimentary Petrology, v. 42, p. 775-793.

Hampton, M.A. 1979. Buoyancy in debris flows. Journal of Sedimentary Petrology, v. 49, p. 753-758.

Hein, F.J. 1982. Depositional mechanisms of deep-sea coarse clastic sediments, Cap Enragé Formation, Québec. Canadian Journal of Earth Sciences, v. 19, p. 267-287.

Heward, A.P. 1978. Alluvial fan and lacustrine sediments from the Stephanian A and B (La Magdalena, Ciñera-Matallana and Sabero) coalfields, northern Spain. Sedimentology, v. 25, p. 451-488.

Holmes, A. 1965. Principles of Physical Geology. London: Nelson and Sons, 1288p.

Kelling, G. and Holroyd, J. 1978. Clast size, shape, and composition in some ancient and modern fan gravels. *In:* Stanley, D.J. and Kelling, G. (Eds.), Sedimentation in Submarine Canyons, Fans, and Trenches. Stroudsburg: Dowden, Hutchinson and Ross, p. 138-159.

Leszczyński, St. 1981. Ciężkowice Sandstones of the Silesian Unit in Polish Carpathians: a study of coarse-clastic sedimentation in deepwater. Annales Societatis Geologorum Poloniae, v. 51, p. 435-502 (English summary).

Lowe, D.R. 1976. Grain flow and grain flow deposits. Journal of Sedimentary Petrology, v. 46, p. 188-199.

Lowe, D.R. 1982. Sediment gravity flows: II. Depositional models with special reference to the deposits of high-density turbidity currents. Journal of Sedimentary Petrology, v. 52, p. 279-297.

Marschalko, R. 1975. Depositional environment of conglomerate as interpreted from sedimentological studies (Paleogene of Klippen Belt and adjacent tectonic units in East Slovakia). Nauka o Zemi, Geologica 10, 109p. (English summary).

Massari, F. 1978. High-constructive coarse-textured delta systems, Tortonian, southern Alps. Evidence of lateral deposits in delta slope channels. Memorie della Societa Geologica Italiana, v. 18, p. 93-124.

Middleton, G.V. 1967. Experiments on density and turbidity currents. III. Deposition of sediment. Canadian Journal of Earth Sciences, v. 4, p. 475-505.

Middleton, G.V. 1970. Experimental studies related to problems of flysch sedimentation. *In:* Lajoie, J. (Ed.), Flysch Sedimentology in North America. Geological Association of Canada, Special Paper 7, p. 253-272.

Middleton G.V. and Hampton, M.A. 1976. Subaqueous sediment transport and deposition by sediment gravity flows. *In:* Stanley, D.J. and Swift, D.J.P. (Eds.), Marine Sediment Transport and Environmental Management. New York: John Wiley and Sons, p. 197-218.

Mutti, E. 1979. Turbidites et cônes sous-marins profonds. *In:* Homewood, P. (Ed.), Sédimentation détritique (fluviatile, littorale et marine). Institute Géologie, Université Fribourg, p. 353-419.

Nemec, W., Porębski, S. J. and Steel, R. J. 1980. Texture and structure of resedimented conglomerates: examples from Książ Formation (Famennian-Tournaisian), southwestern Poland. Sedimentology, v. 27, p. 519-538.

Porębski, S.J. 1981. Świebodzice succession (Upper Devonian-lowest Carboniferous; western Sudetes): a prograding, mass-flow dominated fan-delta complex. Geologia Sudetica, v. 16, p. 101-102 (English summary).

Ricci Lucchi, F., Colella, A., Ori, G.G., Ogliani, F., Colalongo, M.L., Padovani, A., Pasini, G., Raffi, S. and Venturi, L. 1981. Pliocene fan deltas of the Intra-Apenninic Basin, Bologna. *In:* Ricci Lucchi F. (Ed.), Excursion Guidebook, International Association of Sedimentologists, 2nd European Regional Meeting, p. 81-138.

Ricci Lucchi, F. and Valmori, E. 1980. Basin-wide turbidites in a Miocene, over-supplied deep-sea plain: a geometrical analysis. Sedimentology, v. 127, p. 241-270.

Rocheleau, M. and Lajoie, J. 1974. Sedimentary structures in resedimented conglomerate of the Cambrian flysch L'Islet, Quebec Appalachians. Journal of Sedimentary Petrology, v. 44, p. 826-836.

Rodine, J.D. and Johnson, A.M. 1976. The ability of debris, heavily freighted with coarse clastic material, to flow on gentle slopes. Sedimentology, v. 23, p. 213-234.

Rupke, N.A. 1977. Growth of an ancient deep-sea fan. Journal of Geology, v. 85, p. 725-744.

Shultz, A.W. 1983. The deposits, internal structure and geometry in six alluvial fan - fan delta bodies (Devonian-Norway) — a study in the significance of bedding sequence in conglomerates — Discussion. Journal of Sedimentary Petrology, v. 53, p. 325-327.

Stanley, D.J. 1980. The Saint-Antonin Conglomerate in the Maritime Alps: a model for coarse sedimentation on a submarine slope. Smithsonian Contributions to the Marine Sciences, No. 5, 25p.

Stanley, D.J. and Hall, B.A. 1978. The Bucegi Conglomerate: a Romanian Carpathian submarine slope deposit. Nature, v. 276, p. 60-64.

Steel, R.J. 1974. New Red Sandstone floodplain and piedmont sedimentation in the Hebridean Province, Scotland. Journal of Sedimentary Petrology, v. 44, p. 336-357.

Ślączka, A. and Thompson III, S. 1981. A revision of the fluxoturbidite concept based on type examples in the Polish Carpathian Flysch. Annales Societatis Geologorum Poloniae, v. 51, p. 3-44.

Teisseyre, H. 1968. Stratigraphy and tectonics of the Świebodzice Depression. Instytut Geologiczny, Biuletyn 222, p. 77-106.

Walker, R.G. 1975. Generalized facies models for resedimented conglomerates of turbidite association. Geological Society of America Bulletin, v. 86, p. 737-748.

Walker, R.G. 1977. Deposition of upper Mesozoic resedimented conglomerates and associated turbidites in southwestern Oregon. Geological Society of America Bulletin, v. 88, p. 273-285.

RESEDIMENTED CONGLOMERATES IN A MIOCENE COLLISION SUTURE, HOKKAIDO, JAPAN

H. Okada[1] and S. K. Tandon[2]

Abstract

Resedimented conglomerates and associated turbidites occur in the Miocene late orogenic basin of northwestern Hokkaido, Japan. The Kotanbetsu Formation, which mainly contains these deposits, occurs at the front of the Hidaka uplift which resulted from a major collision in Miocene time. The sediments below and above the Kotanbetsu Formation represent shallow water deposition. From the facies changes and micropalaeontological data, an abrupt subsidence of the order of 1000 m is inferred for the beginning of the Kotanbetsu deposition. Due to this rapid subsidence of the basin floor, at least 3000 m of resedimented conglomerates and turbidites were deposited in the basin. The conglomerates comprise the following facies: (a) massive (i.e., unstratified) conglomerate; (b) massive conglomerate with major zones of disrupted bedding; (c) conglomerate with clast imbrication, associated with cross-bedded sandstone; (d) inverse- to normally-graded conglomerate; (e) graded-stratified conglomerate; and (f) pebbly mudstone.

These sediments were deposited by a variety of processes, including gravity flows in proximal reaches, slumps, channelised mass flows and debris flows and turbidity currents. Compositional, textural and fabric data from the conglomerates correlate well with the directional data obtained from the associated turbidites.

Résumé

Des conglomérats resédimentés et des turbidites associées apparaissent dans les bassin orogénique, d'âge Miocène inférieur, dans le secteur nord-ouest de Hokkaido, Japan. La formation de Kotanbetsu, composée principalement de ces dépôts, est observée sur le front du soulèvement de Hidaka, ce dernier fut engendré par une collision majeure datant du Miocène. Les sédiments sous-jacents et sus-jacents à la formation de Kotanbetsu représentent des matériaux déposés en eau peu profonde. Les changements de faciès et les données micropaléontologiques suggèrent la création d'une subsidence abrupte de l'ordre de 1 000 m, laquelle a provoqué le commencement de l'accumulation des matériaux sédimentaires de la formation de Kotanbetsu. A cause de cette subsidence rapide du plancher du bassin, au moins 3 000 m de conglomérats resédimentés et de turbidites se sont accumulés dans le bassin. Les conglomérats incluent les faciès suivants: (a) un conglomérat massif (i.e., nonstratifié); (b) un conglomérat massif avec des zones importantes marquées par une rupture du litage; (c) un conglomérat avec imbrication des fragments, associé à un grès à stratification oblique; (d) un conglomérat à granoclassement inverse à normal; (e) un conglomérat granoclassé - stratifié et; (f) un 'mudstone' caillouteux.

Une variété de processus sont responsables du dépôt de ces sédiments tels les écoulements par gravité dans des biefs proximaux, les glissements, les écoulements en masse chenalisés et les coulées de débris ainsi que les courants de turbidité. Les données de la composition, la texture et la fabrique obtenues sur les conglomérats montrent une bonne corrélation avec les données des directions obtenues sur les turbidites associées.

Introduction

Resedimented conglomerates have been the subject of considerable interest during the last decade, and much progress has been made towards an understanding of their various characteristics and depositional mechanisms (Aalto, 1976; Hendry, 1973; Middleton and Hampton, 1973; Davies and Walker, 1974; Carter, 1975; Chapter 7 in Harms et al., 1975; Walker, 1975; Carter and Norris, 1977; Long, 1977; Surlyk, 1978; Nemec et al., 1980; Lowe, 1982). This study analyses the resedimented conglomerates from the Miocene basin of northwestern Hokkaido. They originated within the framework of a collision-type orogenesis related to a major arc-trench system. We follow the original definition of 'resedimented' (Migliorini, 1946) to refer to deposits that originally accumulated as an unstable pile in shallow water, and were later transported down into deeper water largely due to gravity. The study is concerned primarily with the description of sedimentary features of the resedimented facies, including textural details and clast fabric.

---

[1] Institute of Geosciences, Shizuoka University, Shizuoka 422, Japan.
[2] Department of Geology, University of Delhi, Delhi - 11007, India.

We are grateful to Dr. Yutaka Ikebe of the Japan Petroleum Exploration Company for his encouraging support. We are also indebted to the Indian National Science Academy and the Japan Society for Promotion of Science for a grant enabling one of us (SKT) to visit Japan during March to November, 1981.

Special thanks are due to the Tokyo and Sapporo Offices of the Japan Petroleum Exploration Company (JAPEX). The study was also partly supported by a Grant-in-Aid for Scientific Researchers given by the Japan Ministry of Education, Science and Culture (Monbusho).

We thank Prof. M. Kato, Dr. S. Sato and our several other colleagues of the Hokkaido University (Sapporo), and Dr. K. Takahashi of the Geological Survey of Hokkaido (Sapporo) for discussing various aspects of our data and interpretations.

Our sincere thanks are also due to Dr. Ron Steel for his encouraging interest in our research and for inviting us to contribute to this volume. Earlier drafts of the manuscript were critically reviewed and considerably improved by Dr. Ron Steel, Dr. Karen Kleinspehn, Dr. Wojtek Nemec and Dr. Emlyn Koster.

Copyright © 1984, Canadian Society of Petroleum Geologists

The conglomerate fabrics have been compared with the directional data on sole marks, ripple cross-lamination, and trough cross-stratification from the associated turbidite facies.

The tectonic setting and evolution of the Miocene Hokkaido basin were previously considered by Nagao (1938) and subsequent workers in terms of a molasse tectono-sedimentary realm. More recently, however, it was recognized that the sedimentary fill of this basin is entirely of deep marine origin.

TECTONO-SEDIMENTARY CONTEXT OF THE BASIN

The geological evolution of the Hokkaido region is best explained in terms of a multi-phase collision event marked by the tectonic suture of the western Hidaka marginal sub-belt and by the Abashiri Tectonic Line (Fig. 1A) (Okada, 1983). The former represents the first phase of collision during the Oligocene to Middle Miocene and the latter, the second phase in the latest Miocene. The development of the first collision phase between the Okhotsk Palaeoland to the east and the Yezo arc-trench system to the west is outlined in Figure 2. According to this model (Okada, 1982, 1983), conglomerates and gravitational slide deposits were formed in the Ishikari and Kamuikotan belts. Due to collision and related westerly overthrusting, the Hidaka belt became uplifted while the pre-existing Yezo forearc area subsided as deep basins that received extensive amounts of gravitationally-derived sediments. These tectonic movements, themselves defining the Miocene Hidaka orogeny, were responsible for the deposition of the coarse basin-fill clastics of the Kotanbetsu Formation to the north and Kawabata Formation to the south.

It is noteworthy from the standpoint of palaeogeography that some well defined sedimentary basins have been recognised by Ishiwada and Ogawa (1976) between the eastern coast of Tohoku Arc (northern Honshu) and the Japan Trench on the basis of seismic data (Fig. 1B). One of these basins — the Hidaka Trough — occurs behind this trench, wedges out towards Hokkaido Island and extends farther northwards along the western margin of the Hidaka belt. This structural depression corresponds to the Ishikari belt shown in Figure 1A.

THE KOTANBETSU FORMATION

The Kotanbetsu Formation (*ca.* 3000 m) is an early to middle Miocene unit of the Ishikari belt (Fig. 1A) in northwestern Hokkaido. It is underlain unconformably by the Chikubetsu Formation (*ca.* 450 m) consisting of shallow marine and brackish water deposits (Fig. 3). The Kotanbetsu Formation consists of resedimented conglomerates with

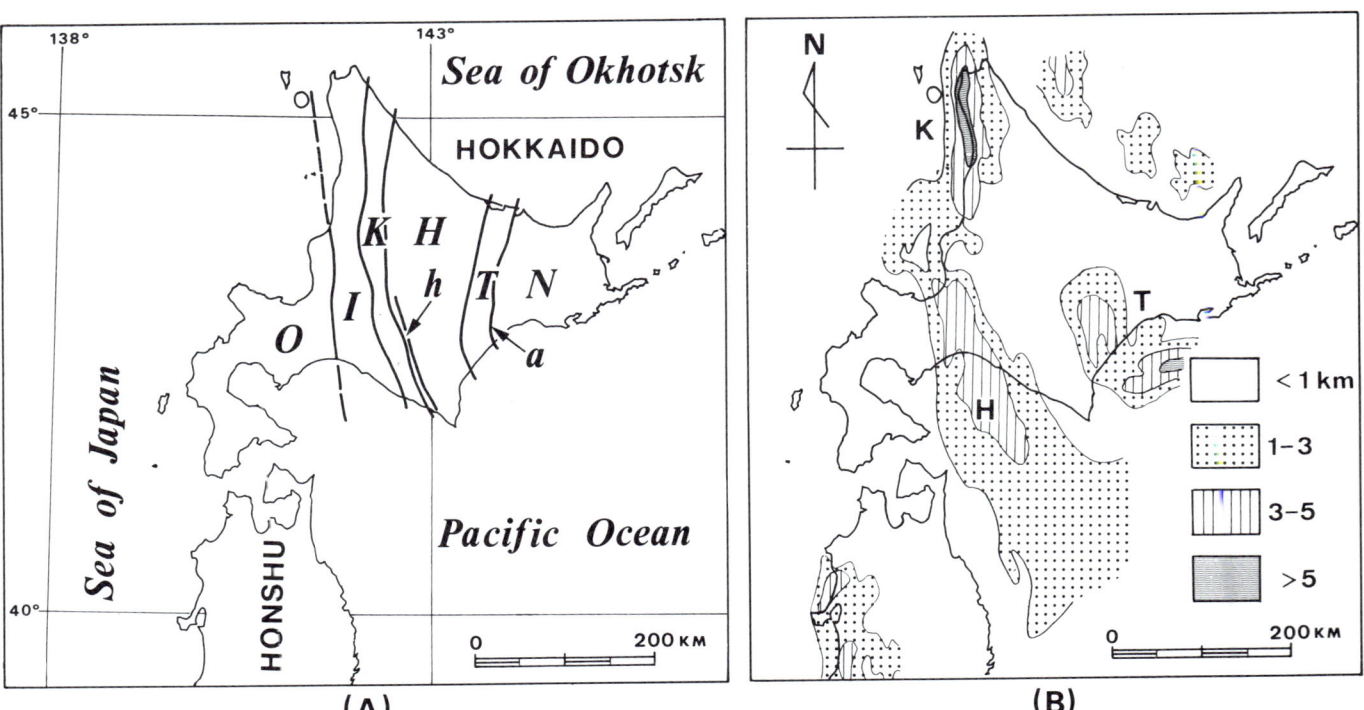

Fig. 1. **A.** Tectonic divisions of Hokkaido (from Okada, 1982). Symbols are as follows — O: Oshima-Rebun Belt, I: Ishikari Belt, K: Kamuikotan Belt, H. Hidaka Belt (h: Western Hidaka Margin Tectonic Subbelt), T: Tokoro Belt, N: Nemuro Belt, a: Abashiri Tectonic Line. **B.** Isopach map of Neogene basins in Hokkaido (from Ishiwada and Ogawa, 1976). Symbols are as follows — K: Kotanbetsu Basin, H: Hidaka Trough, T: Tokachi Basin. The Kotanbetsu Basin and Hidaka Trough are related to the first phase of collision, the Tokachi Basin to the second phase.

large olistostromal masses associated with turbidite sequences several thousand metres thick. The Formation occurs mainly within two sub-basins in northwestern Hokkaido (Fig. 4). In their present structural situation, they generally trend N-S and are separated by a NW-SE fault system.

In the Kotanbetsu Formation, the following three major facies have been distinguished (Okada, 1978, 1980): 1) chaotic conglomerate facies; 2) graded sandstone facies; and 3) ripple-laminated sandstone facies (Fig. 5). The graded bed facies consists of graded or graded-stratified sandstone corresponding to the Ta-b interval of the Bouma sequence. Graded sandstone beds average 3-4 m, but also occur on a scale of 5-10 cm. The ripple-laminated facies consists of frequent intercalations of rippled thin beds of fine- to medium-grained sandstones within pelitic intervals. The rippled sandstone units have sharp contacts and good sorting. The graded bed facies represents turbidites and the ripple-bed facies represents contourites (Okada, 1978).

The massive conglomerate facies comprises resedimented conglomerates that are described in detail in the following sections. This facies is well developed near the southeastern margin of the Embetsu sub-basin and along the western and eastern sides of the Haboro dome in the Haboro sub-basin. These areas probably represent relatively proximal depositional sites, while the graded-bed facies predominates in more distal areas. The ripple-laminated bed facies characterises the most distal parts of the basins.

The Chikubetsu succession implies deposition in a shallow water domain (*ca.* 200 m deep), whereas the overlying Kotanbetsu Formation clearly indicates deposition well below wave base. Also on the basis of micropalaeontological data, Dr. Maiya (1981, pers. comm.) inferred a water depth of 1000 m for the Kotanbetsu Formation. On the other hand, the low-angle unconformity that separates the Chikubetsu and Kotanbetsu Formations does not appear

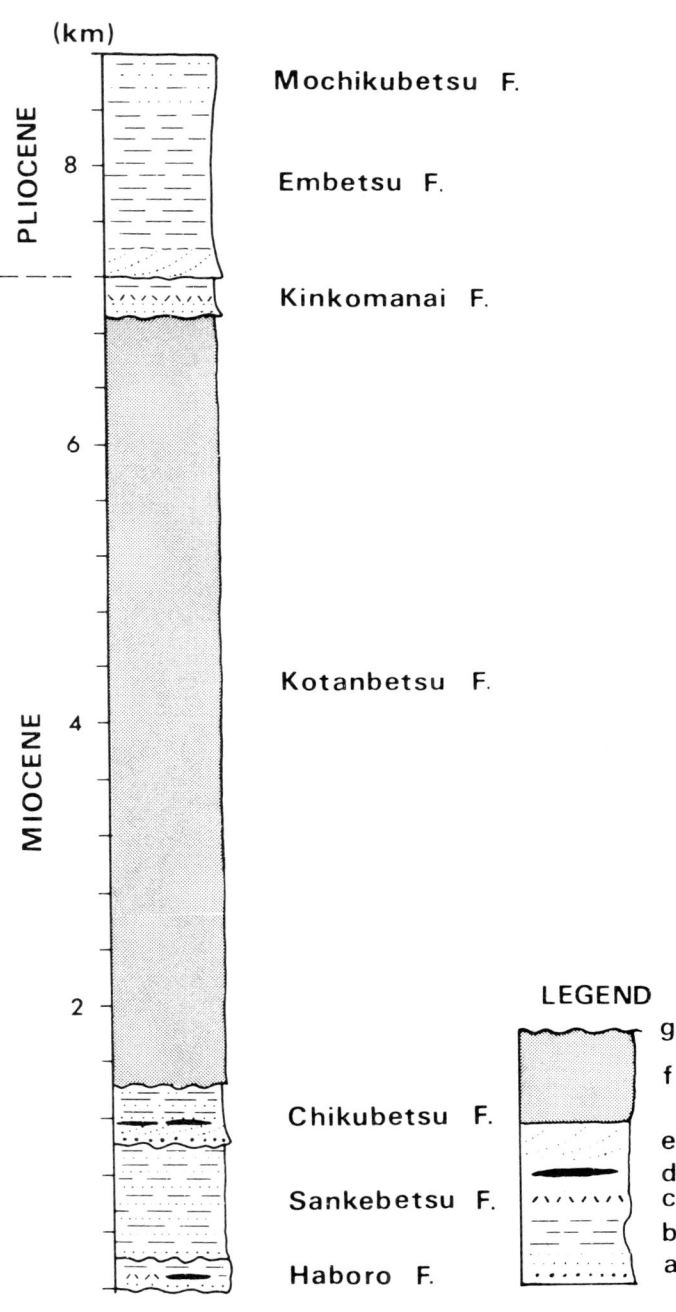

Fig. 3. Stratigraphy of the Neogene sequence in the Haboro Sub-basin (adapted from Okada, 1980). a: coarse clastics, b: mudstone, c: acidic tuff, d: coal seams, e: cross-stratification, f: sediment gravity-flow deposits, g: unconformity.

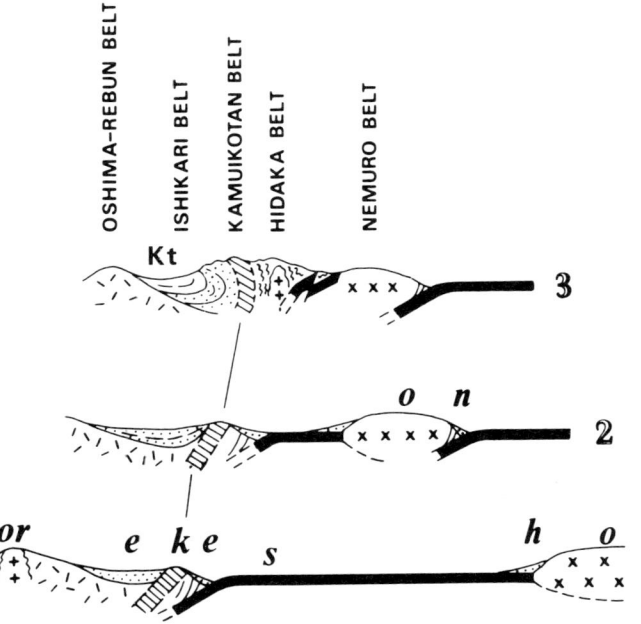

Fig. 2. Schematic diagrams showing the stage-wise collision history (first place) of the Okhotsk paleoland in the east and the Yezo Arc-Trench System in the west (Okada, 1982). Symbols are as follows— 1: early Cretaceous 2: latest Cretaceous, 3: middle Miocene. or: Oshima-Rebun Belt, e: Yezo Supergroup, k: Kamuikotan Belt, s: Sorachi Paleo-ocean, h: Hidaka Supergroup, o: Okhotsk Paleoland, n: Nemuro Group, Kt: Kotanbetsu Basin.

to indicate any dramatic bathymetric change in response to subsidence. Nevertheless, the accumulation of a thick succession of resedimented conglomerates and associated turbidites seems to suggest an abrupt subsidence and a well-defined depositional episode in the Neogene history of Hokkaido.

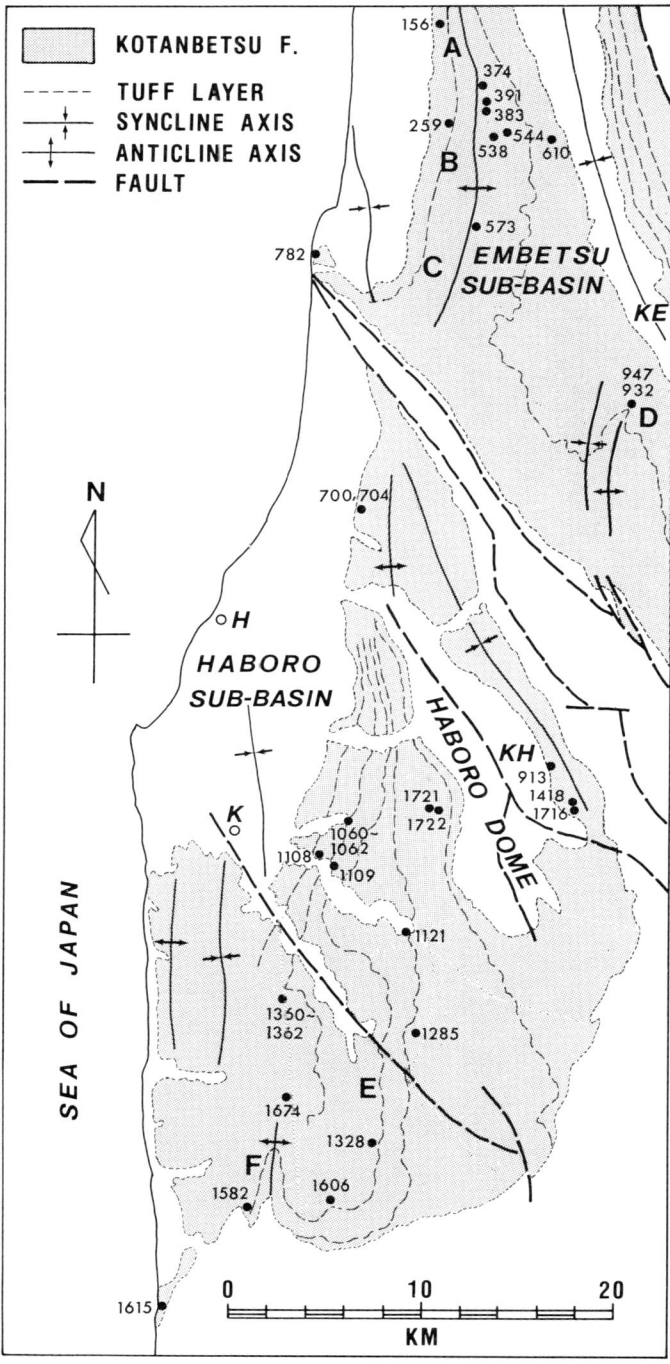

Fig. 4. Schematic geologic map of the Kotanbetsu Formation showing the two depocenters of the Embetsu sub-basin and Haboro sub-basin. Numerals indicate the locality number of the measurement sites. A to F indicate stratigraphic sections shown in Figure 5. KE: Kami-Embetsu, H: Haboro, KH: Kami-Haboro, K: Kotanbetsu.

METHODS OF STUDY

At each outcrop, and for the purpose of fabric measurements, the attitude of conglomerate beds was measured in relation to that of the adjacent turbidite beds. In the case of unexposed bed boundaries, the conglomerate bedding was estimated by data extrapolation from the nearest exposure of the associated turbidites. Detailed fabric measurements were performed on all individual conglomerate beds in which the first 25 measured clasts revealed a preferred orientation of fabric. The bed area covered by measurements was generally kept to 1 m². Extracted clasts larger than 1 cm with a minimum A to C axis ratio of 1:1.5 were selected. In more indurated outcrops, we measured only those clasts whose projection was good enough to identify their AB planes. The number of measurements per bed ranged from 25 to 100 (the latter consisting of two data sets, 50 each).

Data collected on individual clasts included the following properties: lithology, triaxial dimensions, a visual estimate of roundness, surface features, orientation of AB plane, and type of imbrication. Also included were bed geometry, the nature of adjacent sediments, percent matrix, clast/matrix relationships, maximum particle size and a visual estimate of the clast size distribution. Data on clast composition in the individual outcrops or beds were collected by counting about 200 clasts per square metre.

In outcrops containing abundant intraclasts, their fabric properties were also measured for comparison with those of the extrabasinal clasts. All fabric data were corrected for tectonic tilt.

RESEDIMENTED CONGLOMERATE FACIES

The resedimented conglomerates of the Kotanbetsu Formation have been grouped into the following six major facies:

FACIES $C1$

This facies comprises disorganized conglomerates (Fig. 6) with no observable bedding, and occurs on thickness scales of the order 5-20 m. The maximum clast size approaches 1 m, although the majority are between 5-10 cm. These internally chaotic units are often seen superimposed, forming complex conglomeratic sequences at the scale of outcrop (i.e., 5-15 m). Such sequences display a markedly contrasting basal portion which may represent either a channel slide or partly channelised debris flow deposits. The overlying conglomerate probably represents emplacement of mostly unchannelised mass-flows on submarine fan surfaces.

A distinctive feature of this facies is that the margins of the individual units are commonly rimmed by outsized blocks or rock fragments mostly of extrabasinal origin. At Kami-Embetsu, however, the base of conglomeratic units contain huge blocks of turbidite sandstone (up to 3 x 1.5

m). These blocks were apparently derived from the underlying turbidite sequence during emplacement of the gravelly mass-flow and were incorporated by the flowing mass probably due to its turbulent behavior. Basal parts of the facies *C1* units display preferred clast fabric with *A* axes generally parallel to the basal channel axis. Variations in the percentage of muddy sand to clay matrix seem to suggest depositional mechanisms ranging from slumping and debris flows to probably some sort of density-modified grain flows.

FACIES *C2*

This facies comprises conglomerates that are in many respects similar to the facies *C1*, inasmuch as they both display a chaotic internal character. Here, however, there

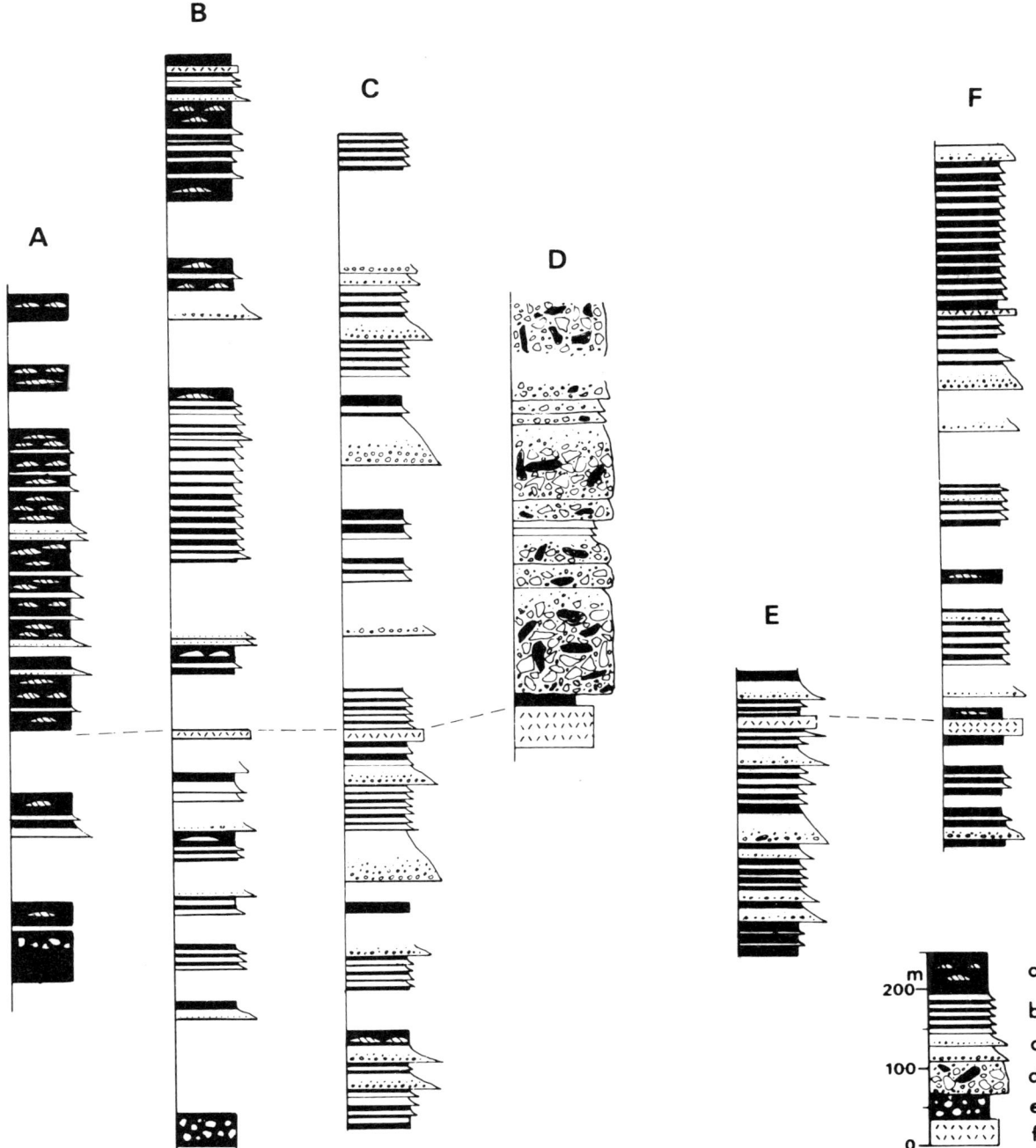

Fig. 5. Columnar sections showing lithofacies changes in the Kotanbetsu Formation at sections A to F indicated in Figure 4 (Okada, 1980). a: ripple-laminated bed facies, b-c: graded bed facies (b: sandstone, c: granule to small pebble conglomerate), d-e: massive (unstratified) conglomerate — pebbly mudstone facies, f: rhyolitic tuff layer.

Fig. 6. Large redeposited blocks of Facies *C1* at Locality No. 700. Note that individual blocks are rimmed by outsized clasts of extrabasinal origin.

Fig. 7. Disorganised conglomerates with individually recognizable zones of disrupted bedding (Facies *C2*). Locality No. 1419 (Kami-Haboro).

Fig. 8. Channel-fill conglomerates (Facies *C3*). Locality No. 1418 (Kami-Haboro). Hammer is 30 cm long.

are distinct zones of disrupted bedding and complex sediment mixing along the contacts between the conglomerates and associated turbidite sandstone units (Fig. 7). At Kami-Haboro, for example, the section commences as a sequence of thin turbidite/ripple-laminated beds and culminates with a 2 m thick, graded sandstone bed representing a turbidity current deposit. Above is a sharp, erosive contact (up to 70 cm relief) with a conglomeratic body whose boundaries parallel the bedding planes of the underlying turbidite sequence. Large rafted sediment blocks of penecontemporaneous origin occur about 1 m above the base, and their partly diffuse outline suggests emplacement as a sheared rip-up mass. The succession above consists of superimposed units of conglomerates and turbiditic sandstones that display complicated interrelationships. Fragments of disrupted sandstone beds commonly occur in a sub-vertical position relative to the low dip of the underlying strata. This package of deposits extends over the 50 m width of outcrop, and was probably deposited as gravelly mass-flows and turbidites, with a contemporaneous disruption of the latter. At the top, there occurs a conglomerate unit (3 m thick and at least 50 m wide) whose base is markedly scoured and which represents a channelised mass-flow deposit. In this unit, larger clasts tend to occur well above the base and there is also a 20 cm thick zone of inverse grading. The latter two features suggest 'rigid plug' transport within a debris flow and probably a basal shearing zone, respectively. An inversely graded basal zone and normally graded tops of many other beds seem to be a consequence of their subaqueous emplacement (Johnson, 1970).

The thickness estimates of individual units of facies *C2* are usually difficult to determine because of their disrupted geometry and later tectonic deformation.

FACIES *C3*

This facies comprises conglomerates (*ca.* 3 m thick) which lack stratification and grading and consist of subangular to subrounded, imbricated clasts (Fig. 8) set in a muddy sand matrix of 10-15%. Maximum clast size reaches 30 cm though is normally 3-10 cm. Because of their close association with cross-stratified sandstone, these conglomerates are interpreted as channel-fill deposits. Tractive transport of the associated sandstone beds (30-50 cm thick) is indicated by the common presence of planar or trough cross-stratification. This facies also includes some irregular, thinner (10 cm) conglomerate layers and discontinuous pebble stringers. These erosive layers are interpreted as having formed due to local channeling (or scouring) at the base of certain turbidity currents.

FACIES *C4*

This facies consists of unstratified conglomerates which characteristically display inverse-to-normal grading; the inversely graded lower zone is seldom thicker than 20 cm. The conglomeratic beds occur as well-defined layers (*ca*. 2 m thick) within a thicker succession of sandy and shaley turbidites. As shown in Figure 9, the conglomerate sequence starts with a thick inverse-to-normally graded bed whose basal contact is planar and sharp upon either a pelitic or a graded sandy interval of the underlying turbidite unit (Figs. 9, 10). The uppermost conglomeratic units are sharply overlain by claystone of a thicker pelitic unit (Fig. 11). In the latter, herein referred to as heterolithic facies, the sandstone interlayers are rippled and probably represent contourites as deduced from palaeoflow relationships. In the Kami-Haboro section there are at least two units of conglomeratic turbidites that are overlain by such contourites.

In the conglomeratic units of this facies, intraclasts are common, especially toward the tops, suggesting clast segregation during flow. The clasts in facies *C4* are generally relatively well-sorted (mainly 1-3 cm in diameter; *MPS* = 15 cm). Inverse-to-normally graded conglomerate units have been inferred (Chapter 7 in Harms *et al.*, 1975) to represent deposition from gravity flows on relatively steep slopes in proximal settings.

Fig. 9. Inverse-to-normally graded conglomerates (Facies *C4*) overlying the thinly bedded pelitic interval of the associated turbidite facies. Locality No. 947 (Kami-Embetsu). Vertical scale bar is 100 cm long.

FACIES *C5*

This facies consists of sharp-based conglomerate-sandstone beds generally *ca*. 5 m thick, which are graded near the base (coarse-tail grading) and horizontally stratified in the upper part (Fig. 12). Grain size varies from fine pebble fractions in the lower graded part through granules to very fine sand in the stratified upper part; the latter grades upwards into an overlying pelitic deposit. Layers and concentrates of intraclasts are common in the upper part of the individual units, and are parallel or sub-parallel to the stratification planes.

Fig. 10. Inverse-to-normally graded conglomerates (Facies *C4*) resting on a graded sandstone bed with a sharp, planar contact. Locality No. 1716 (Kami-Haboro). Hammer at the base of the conglomerate at the extreme right of the photo is 30 cm long.

FACIES *C6*

These are unstratified pebbly mudstones, with a *ca*. 5 m bed thickness, that contain 'floating' rounded to subrounded clasts, usually concentrated in pockets. Large (*ca*. 1 m) sandstone blocks and smaller intraclasts (Fig. 13) have also been identified, and these suggest contemporaneous sediment derivation by slumping/erosion processes. Sediment intervals containing granule to cobble-sized clasts are common, but boulders are rare. Facies *C6* is interpreted in terms of high viscosity debris-flows. Associated with the pebbly mudstone facies are occasional scours filled with coarse sandstone rich in intraclasts.

Conglomerate Texture

Textural data on the extrabasinal clasts from the various facies of resedimented conglomerates are summarised in

Fig. 11. Inverse-to-normally graded conglomerates (Facies *C4*) showing a sharp contact with an overlying ripple-laminated bed sequence. Locality No. 1721 (east of Kotanbetsu).

Fig. 12. Graded-stratified conglomerates (Facies C5). Locality No. 704 (northeast of Haboro). Intraclasts and calcareous concretions are common in the upper part. Hammer is 30 cm long.

Fig. 13. Pebbly mudstone (Facies C6) in which rounded clasts show distinct staining. Locality No. 1108 (east of Kotanbetsu). Hammer is 30 cm long.

Table 1. In the channel-fill conglomerates of facies $C3$, the predominant matrix is a muddy sand. On the other hand, facies $C1$, $C2$, $C4$ and $C6$ show a predominant sandy clay matrix. Highly angular clasts are noticeably absent in all conglomerate facies, though the matrix-rich variety usually contain a higher proportion of unbroken clasts. Subrounded clasts predominate in most facies although rounded to well-rounded clasts are common (36%) in certain localities of pebbly mudstone facies (e.g., Kotanbetsu locality). Maximum clast size commonly ranges from 4 cm to 1 m (for example, the channel-fill deposits of facies $C3$), but with an average in the 2-5 cm range.

It is interesting to note that intraclasts are usually of larger size than the extrabasinal clasts, commonly being larger than 1 m in diameter (Fig. 7). Intrabasinal sedimentary blocks greater than 10 m are not uncommon in the proximal parts of the basin (Fig. 6; Okada, 1980, Fig. 10). In general, the maximum intraclast sizes rapidly decrease from proximal to distal reaches of the basin (cf. Okada, 1980, Fig. 10).

CLAST COMPOSITION

As already mentioned, the resedimented conglomerate facies consists of both extra- and intrabasinal clasts. The former comprises mainly fragments of older sandstones, slates, hornfelses, granite, and basic tuff as well as occasional mudstone, limestone, chert, gabbro, rhyolite and acidic pyroclastics. In general, the ratio of sandstone to slate clasts increases westwards across the Embetsu sub-basin (Fig. 14). This spatial trend corresponds to the inferred proximal-distal direction within the sub-basin; however, a comparable trend could not be discerned in the Haboro sub-basin. Although there is a relative increase of the slate clast content at certain localities to the south, this may be related to sediment transport by higher viscosity flows as indicated by the presence of sandy clay matrix. The local abundance of intraclasts is illustrated by Figure 7.

The trends in clast composition together with the palaeocurrent data (Fig. 14, see also Figs. 18, 19) suggest a source area in the Hidaka and Kamuikotan belts (Fig. 1A). Clasts of sandstone, slate and granite were probably derived from the Hidaka terrain (cf. Takahashi, 1974), but compositional data does not indicate a provenance in the Kamuikotan area for the detritus in the Kotanbetsu Formation. It therefore appears unlikely that the Kamuikotan belt was subject to uplift.

CLAST FABRIC

Although clast fabric of resedimented conglomerates has often been considered as being of limited value to the interpretations of transport directions of internally chaotic flows, Walker (1977) and Winn and Dott (1979) have demonstrated their utility. These and other workers have cross-checked their results with palaeoflow data obtained from other indicators in the associated turbidites.

The resedimented conglomerates in Hokkaido are intimately associated with turbidites, so it has been possible to compare their clast fabric data with the palaeoflow data obtained from sole marks, ripple cross-lamination and other features of the turbidites. The data from individual localities are presented in Figures 15-17, and a palaeocurrent map is given in Figure 18.

Based on the pattern of major lithofacies distribution in the Embetsu sub-basin, a proximal-to-distal depositional trend has been inferred to have been from southeast to northwest. However, the palaeocurrent data as deduced from flute casts in the turbidite sandstones show a considerable degree of variability. This suggests a polymodal flow pattern, probably related to sediment input from different parts of the basin margin. Alternatively, a variable topography on the basin floor may have been the primary influence. This second view receives support from the polymodal directional data of the ripple cross-lamination of the heterolithic facies (Fig. 19). The multi-directional character of the palaeocurrents is suggestive of tractive transport from rapid and relatively short-lived episodes of

Locality No. (Sample Size) Local Name	Facies	Matrix	Roundness					Textures		Size Distribution		Maximum Clast Size
			A	SA	SR	R	WR	U:B	P:T			
1285 (n=50) Sankebetsu	C1	Sandy clay (10-15%)	0	34	48	16	2	4.0:1	3.3:1	<3 cm 3-10 cm >10 cm	20% 70% 10%	29 cm
1674 (n=50) Onne River	C1	Sandy clay	0	40	46	12	2	3.5:1	2:1	<2 cm 2-5 cm >5 cm	30% 60% 10%	15 cm
1582 (n=50) Otodokko	C1	Sandy clay (40%)	8	32	46	14	0	3.5:1	3:1	<2 cm 2-5 cm >5 cm	35% 50% 15%	30 cm
1360A (n=50) Sankei	C1	Sandy clay (10-15%)	6	26	64	10	0	5.3:1	6:1	<1 cm 1-3 cm <5 cm	35% 60% 5%	10 cm
1360B (n=50)	C1	Sandy clay	2	38	56	4	0	16:1	2.2:1	<1 cm 1-3 cm <5 cm	35% 60% 5%	8 cm
932 (n=50) Kami-Haboro	C3	Muddy sand (15-20%)	2	12	66	20	0	2.1:1	2.6:1	<5 cm 5-10 cm >10 cm	80% 15% 5%	100 cm
1716D (n=50) Kami-Haboro	C3	Muddy coarse sand (15%)	0	20	60	10	10	2.1:1	1.1:1	<2 cm 2-5 cm <5 cm	70% 20% 10%	10 cm
700 (n=50) Ariake	C3	Clay (15%)	0	12	50	34	4	3.5:1	4.3:1	<2 cm 2-5 cm >5 cm	20% 50% 30%	15 cm
544 (n=23) Utakoshi	C4	Mud (15-20%)	4	52	26	18	0	3.6:1	2.3:1	<5 cm 5-10 cm	75% 25%	10 cm
538 (n=50) Utakoshi	C4	(deficient)	0	32	62	6	0	6.1:1	3.2:1	<5 cm 5-10 cm >10 cm	50% 35% 15%	25 cm
1716 (n=50) Kami-Haboro	C4	Muddy sand (10-15%)	0	36	40	14	10	1.0:1	11.5:1	<5 cm 5-10 cm	70% 30%	16 cm
782 (n=30) Shosanbetsu	C4	Fine sand (15%)	0	47	37	10	6	5.0:1	3.3:1	<2 cm 2-5 cm	90% 10%	20 cm
1721 (n=50) Kami-Haboro	C4	Sandy clay (15%)	2	34	46	18	0	4.5:1	3.6:1	<2 cm 2-5 cm >5 cm	30% 60% 10%	25 cm
1722 (n=50)	C4	Sandy clay	2	30	44	20	4	5.2:1	3:1	<2 cm 2-5 cm >5 cm	35% 60% 5%	6 cm
1606 (n=25) Tappu	C4	Sandy clay (60-70%)	0	16	44	40	0	11.5:1		<2 cm >20 cm	70% 30%	4 cm
1328 (n=50) Sankei	C4	Sandy clay (25%)	0	14	62	22	2	9:1	3.6:1	<0.5 cm 0.5-2cm >2 cm	15% 80% 5%	2.7 cm
573 (n=50) Utakioshi	C5	Medium sand (70%)	0	30	52	18	0	1.9:1	8.0:1	<5 cm 5-15 cm	80% 20%	
1108 (n=50) Kotanbetsu	C6	Sandy clay (40%)	2	2	48	36	12	5.4:1	4.8:1	<2 cm 2-4 cm >4 cm	20% 70% 10%	10 cm
1060 (n=100)	C6	Sandy clay	2	51	33	14	0	9:1	1:1.3	<2 cm 2-5 cm >5 cm	40% 50% 10%	10 cm
1615 (n=50) Hanaoka		Coarse sand (15%)	0	0	14	32	54	24.0:1	1:5.5	<2 cm 2-5 cm >5 cm	35% 60% 5%	15 cm

N.B. A: angular, SA: subangular, SR: subrounded, R: rounded, WR: well-rounded.
U:B = Unbroken : broken clasts.
P:T = Parallel : transverse fabric.

Table 1. Representative textural properties of the resedimented conglomerate facies.

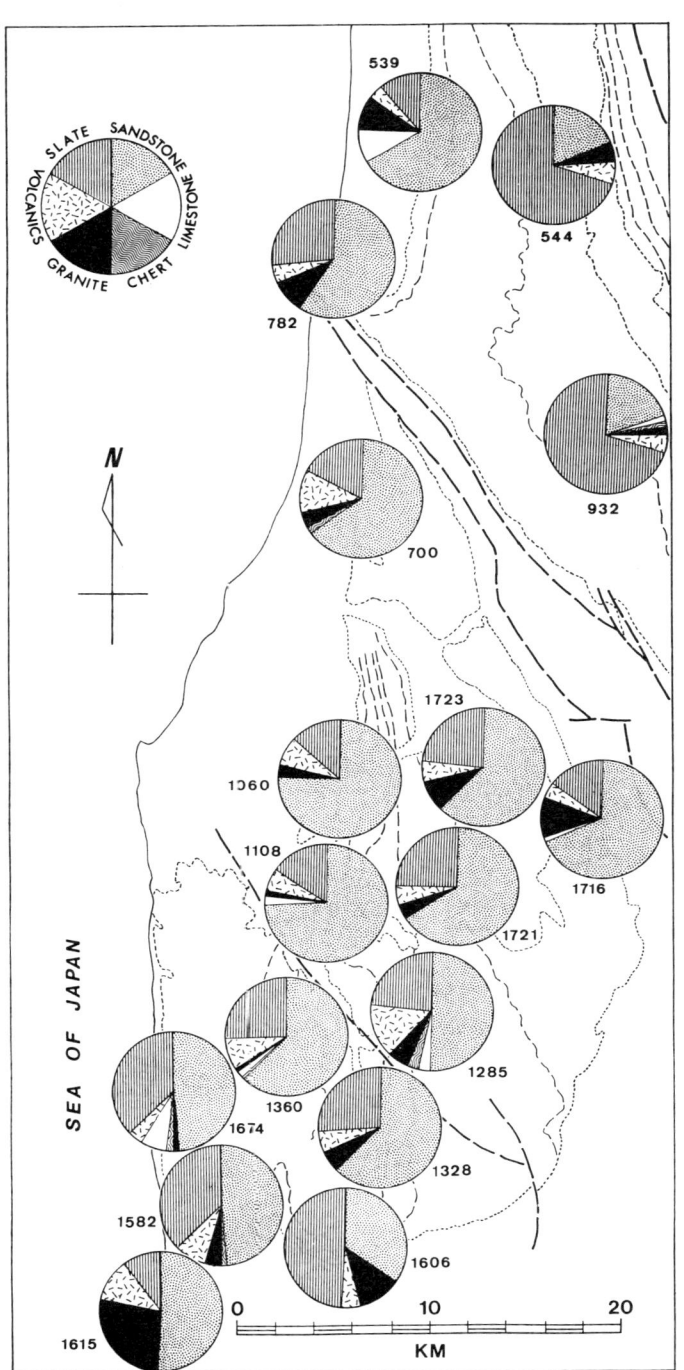

Fig. 14. Clast composition trends in the resedimented conglomerates. Numerals indicate the locality number.

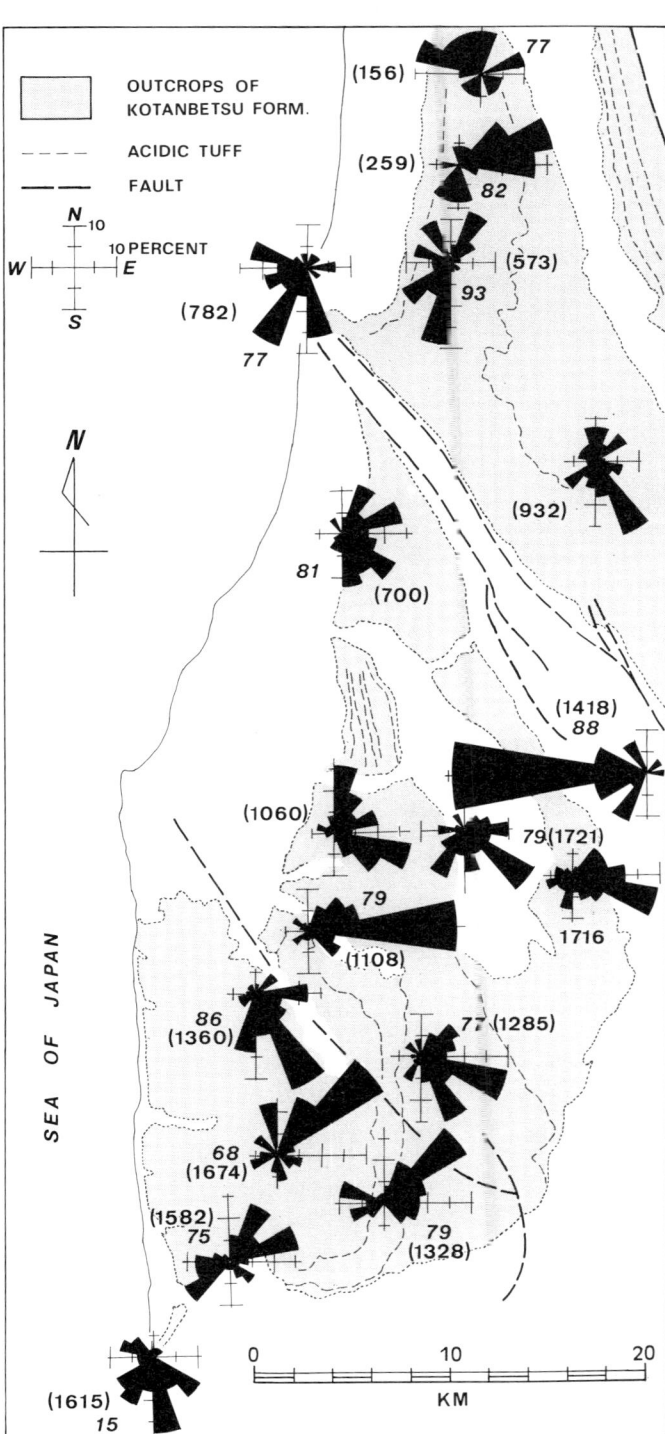

Fig. 15. Rose diagrams showing the azimuth of the *AB* plane for clasts of the resedimented conglomerates. Numeral in italics indicates the percentage of clasts for which the azimuth of the *AB* plane and that of the *A* axis are identical. Bracketed numerals indicate the locality number.

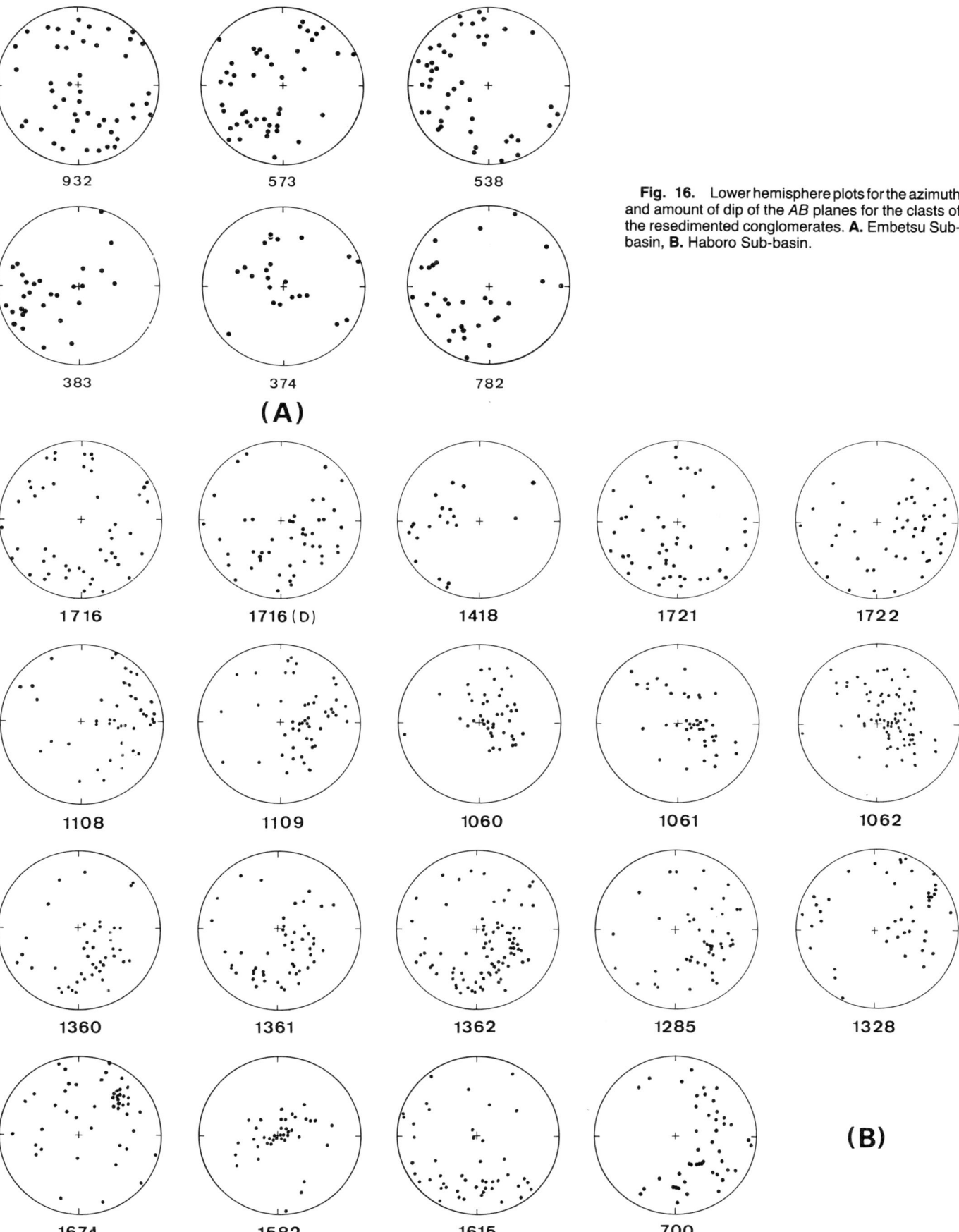

Fig. 16. Lower hemisphere plots for the azimuth and amount of dip of the AB planes for the clasts of the resedimented conglomerates. **A.** Embetsu Sub-basin, **B.** Haboro Sub-basin.

current activity. Similar situations are commonly reported from modern deep-sea environments (*e.g.*, Takano and Hara, 1970; Stow and Lovell, 1979; Okada and Ohta, 1983).

In the Haboro sub-basin, in contrast, the directional data deduced from the turbidite sole marks and cross-lamination, and from the conglomerate fabric all show a considerable degree of consistency. The prevailing palaeo-flow direction is towards the west and southwest (Figs. 16 and 19). Local discrepancies, for example near the eastern margin of the basin, are interpreted in terms of locally irregular basin-floor topography, based on the occurrence of some prominent, possibly syn-depositional, faults in this area.

Most of the resedimented conglomerates contain a preferred *A*(p)*A*(i) fabric (Fig. 15) probably due to gravity-flow transporting mechanisms that maintained clasts in suspension above the bed (Chapter 7 in Harms *et al.*, 1975). The same type of fabric has been reported also from subaerial and shallow-water subaqueous (*e.g.*, lacustrine) gravity flows (for example see Nemec *et al.*, this volume: Porębski, 1981). Our data are, however, consistent with Harms *et al.*'s (Chapter 7, 1975) interpretation.

It is interesting to note, however, that at Hanaoka the conglomerates display an *A*(t)*B*(i) fabric (Table 1). This type of fabric is known to be indicative of clast transport by rolling along the bed, as for example due to traction processes in fluvial channels or on a beach. The conglomerates of this locality are composed of well-sorted and well-rounded clasts. Despite their conformable basal contact with the underlying turbidites, the conglomerates represent shallow-marine sedimentation and, as such, reflect a marked bathymetric change in the Haboro sub-basin. There is also an accompanying major change in the palaeo-flow pattern — from a southerly transport of the underlying deposits to a northerly transport of the above conglomerates. Contrary to previous authors (Tsushima *et al.*, 1956), we suggest that these uppermost conglomerates be considered as a basal part of the overlying Embetsu Formation rather than the top of the Kotanbetsu Formation.

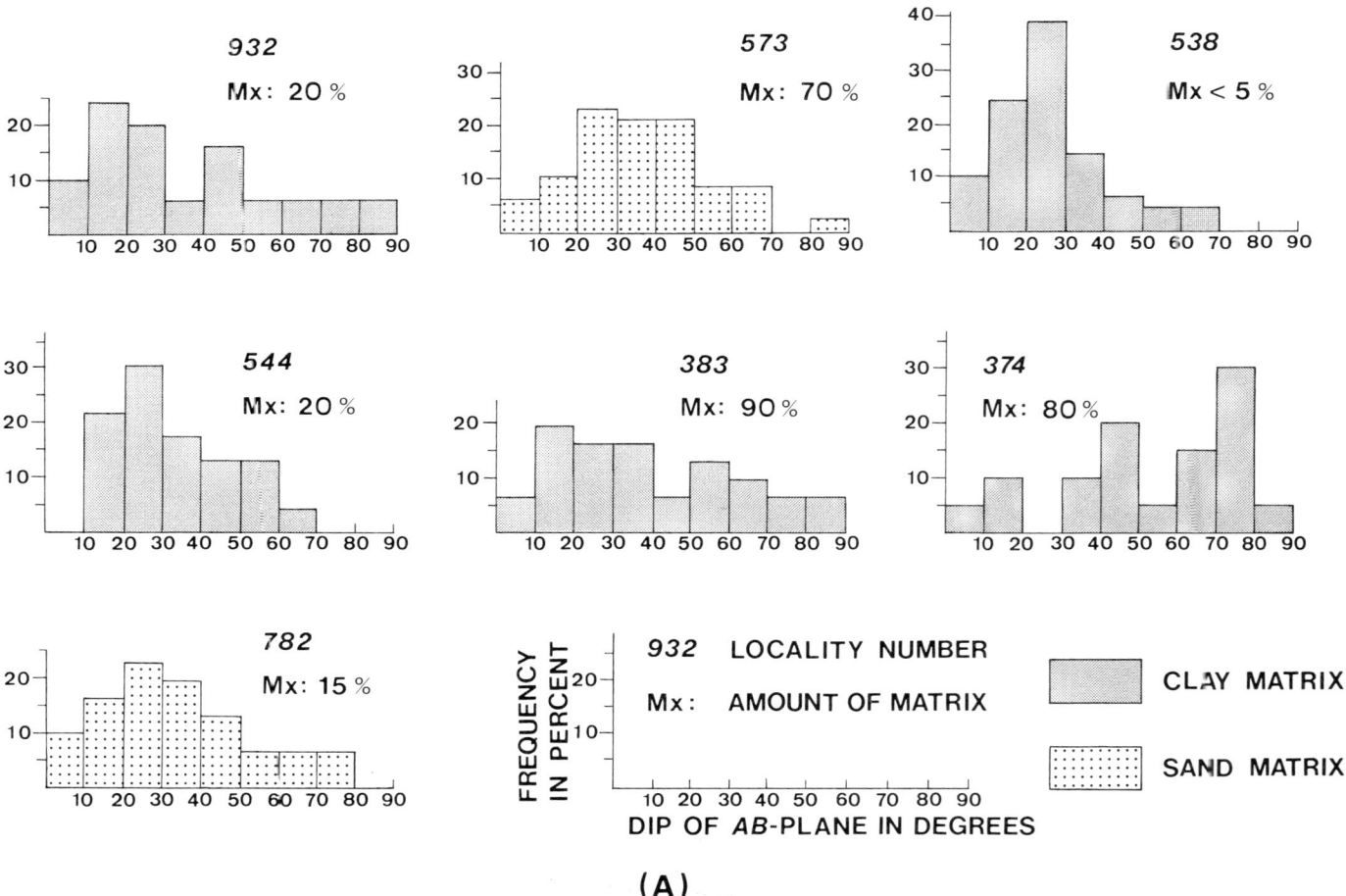

Fig. 17. Histograms showing the frequency distribution of the dip of the *AB* planes for the clasts of the resedimented conglomerates. **A.** Embetsu Sub-basin, **B.** (opposite): Haboro Sub-basin.

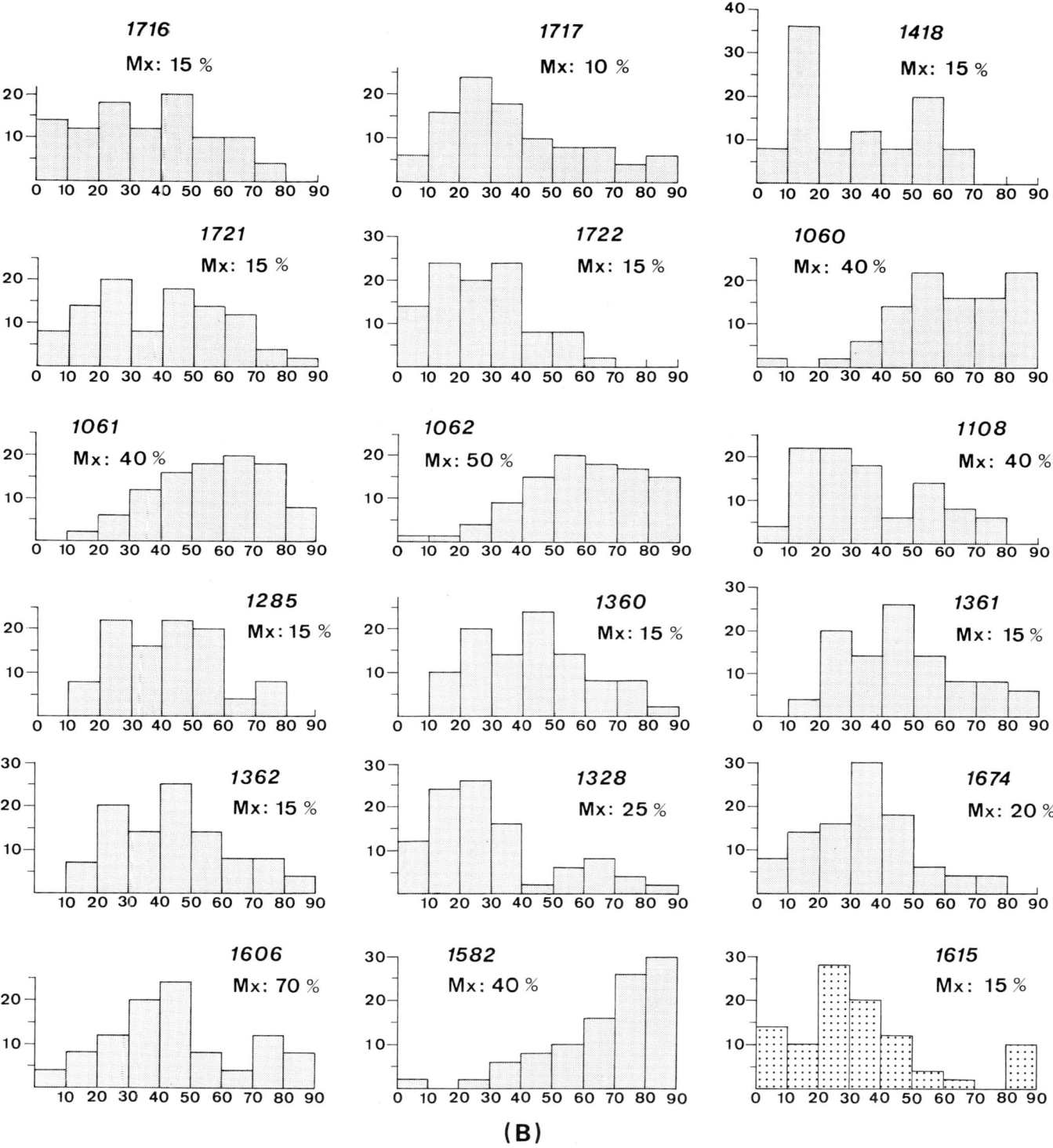

(B)

Conclusions

The Miocene basin in northwestern Hokkaido resulted from major subsidence (*ca.* 1000 m) following the Hidaka Orogeny. This basin records change from shallow-water to deep-water facies, the latter consisting of a great amount of resedimented conglomerates. The newly uplifted terrain supplied sandstone and slate clasts, but a significant quantity of intrabasinal sandstone and mudstone clasts are also present. In zones of collision orogenesis, late orogenic sedimentation is largely related to the syntectonic deformation (reactivation) of the basin margin and floor. Palaeoflow data of this Miocene basin suggest that the basement topography was complicated by syntectonic

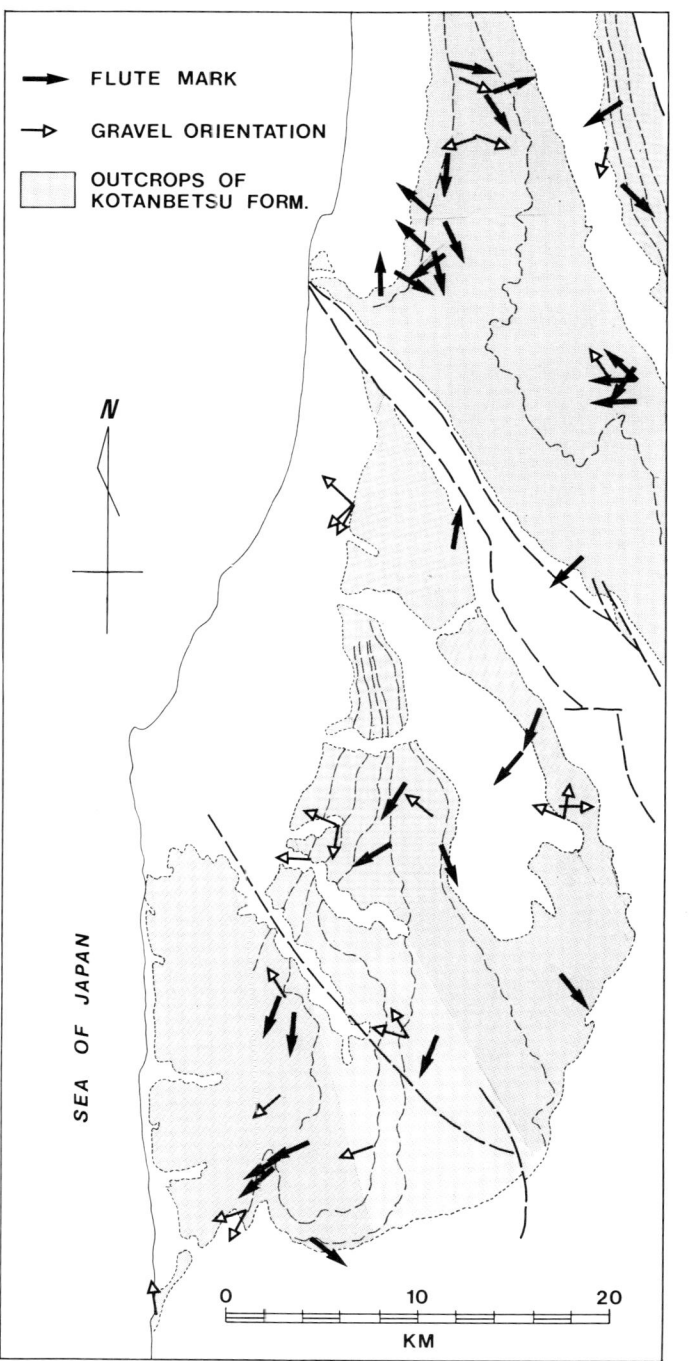

Fig. 18. Clast fabric of the resedimented conglomerates and directions of flute marks in the associated turbidites.

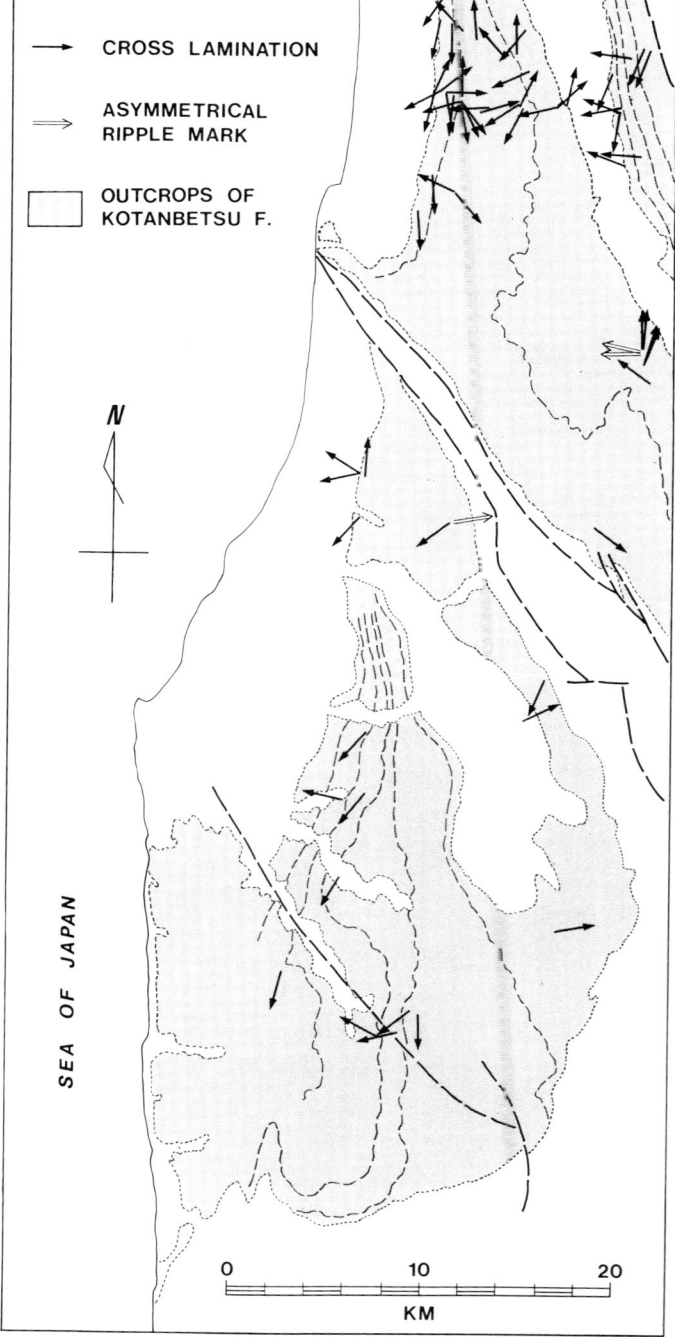

Fig. 19. Palaeocurrent map of the Kotanbetsu Formation based on small-scale cross-laminations and asymmetrical ripple marks.

deformation, but not to the point of disrupting the overall distribution of resedimented conglomerates in relation to the gross proximal-distal trends of the basin. Unlike the late orogenic deposits of continent-to-continent (*e.g.*, Himalaya) collision sutures, complex sediment mixing and disrupted zones of bedding appear to be common in this basin-fill succession.

Clast fabric data from the resedimented conglomerate of Kotanbetsu Formation are generally consistent with other flow-direction indicators, such as sole marks and cross-lamination in the associated turbidites. The typical clast fabric of $A(p)A(i)$ type accords with that suggested by Harms *et al.* (Chapter 7, 1975) for deep-marine gravity-flow conglomerates.

References

Aalto, K.R. 1976. Sedimentology of a melange: Franciscan of Trinidad, California. Journal of Sedimentary Petrology, v. 46, p. 913-927.

Carter, R.M. 1975. A discussion and classification of subaqueous mass-transport with particular application to grain-flow, slurry-flow and fluxoturbidite. Earth Science Reviews, v. 11. p. 145-177.

Carter, R.M. and Norris R.J. 1977. Redeposited conglomerates in a Miocene flysch sequence at Blackmount, western Southland, New Zealand. Sedimentary Geology, v. 18, p. 289-319.

Davies, I.C. and Walker, R.G. 1974. Transport and deposition of resedimented conglomerates: the Cap Enragé Formation, Cambro-Ordovician, Gaspé, Québec. Journal of Sedimentary Petrology, v. 44, p. 1200-1216.

Harms, J.C., Southard, J.B., Spearing, D.B. and Walker, R.G. 1975. Depositional environments as interpreted from primary sedimentary structures and stratification sequences. Society of Economic Paleontologists and Mineralogists, Short Course No. 2, 161p.

Hendry, H.E. 1973. Sedimentation of deep water conglomerates, in Lower Ordovician rocks of Quebec — composite bedding produced by progressive liquefaction of sediment? Journal of Sedimentary Petrology, v. 43, p. 125-136.

Ishiwada, Y. and Ogawa, K. 1976. Petroleum geology of offshore areas around the Japanese Islands, United Nations Economic and Social Commission for Asia and the Pacific, Committee for Co-ordination of Joint Prospecting for Mineral Resources in South Pacific Offshore Areas, Technical Bulletin, v. 10, p. 23-34.

Johnson, A.M. 1970. Physical Processes in Geology. San Francisco: Freeman, Cooper and Company, 577p.

Long, D.G.F. 1977. Resedimented conglomerates of Huronian (Lower Aphebian) age from the north shore of Lake Huron, Ontario, Canada. Canadian Journal of Earth Sciences, v. 14, p. 2495-2509.

Lowe, D.R. 1982. Sediment gravity flows: II. Depositional models with special reference to the deposits of high density turbidity currents. Journal of Sedimentary Petrology, v. 52, p. 279-297.

Middleton, G.V. and Hampton, M.A. 1973. Sediment gravity flows: mechanism of flow and deposition. *In:* Middleton, G.V. and Bouma, A.H. (Co-chairmen), Turbidites and Deep-Water Sedimentation. Pacific Section, Society of Economic Paleontologists and Mineralogists, Anaheim, Short Course Notes, p. 1-38.

Migliorini, C.I. 1946. Leta del macigno dell'Appennino Sulla Sinistra del Serchio e considerazione sul rimanggiamento del macroformaminiferi. Societa Geologica Italiana Bollettino, v. 63, p. 75-90.

Nagao, T. 1938. Tertiary orogeny in Hokkaido. Journal of the Faculty of Science, Hokkaido Imperial University, Series 4, v. 4, p. 23-30.

Nemec, W., Porębski, S.J. and Steel, R.J. 1980. Texture and structure of resedimented conglomerates: examples from Ksiaz Formation (Famennian — Tournaisian), southwestern Poland. Sedimentology, v. 27, p. 519-538.

Okada, H. 1978. Sedimentary patterns in apparent back-arc basins: a case study of the Neogene sequence in northwestern Hokkaido, Japan. Journal of Earth Physics, v. 26 (Suppl.), p. 477-490.

Okada, H. 1980. Sedimentary environments on and around island arcs: an example of the Japan Trench area. Precambrian Research, v. 12, p. 115-139.

Okada, H. 1982. Geological evolution of Hokkaido, Japan: an example of collision orogenesis. Proceedings of the Geologists' Association, v. 93, p. 201-212.

Okada, H. 1983. Collision orogenesis and sedimentation in Hokkaido, Japan. *In:* Hashimoto, M. and Uyeda, S. (Eds.), Accretion tectonics in the Circum-Pacific regions. Tokyo: Terra Scientific Publishing Company, p. 91-105.

Okada, H. and Ohta, S. 1983. Benthic biological activity. Proceedings of International Symposium on Sedimentation on the Continental Shelf with Special Reference to the East China Sea, Hangzhou, China, v. 1, p. 102-116.

Porębski, S.J. 1981. Swiebodzice succession (Upper Devonian-lowest Carboniferous; western Sudetes): a prograding, mass—flow dominated fan—delta complex (English summary). Geologia Sudetica, v. 16, p. 101-192.

Stow, D.A.W. and Lovell, J.P.B. 1979. Contourites: their recognition in modern and ancient sediments. Earth Science Reviews, v. 14, p. 251-291.

Surlyk, F.J. 1978. Submarine fan sedimentation along fault scarps on tilted fault blocks (Jurassic — Cretaceous boundary, East Greenland). Grønlands Geologiske Undersøgelse Bulletin, v. 126, p. 1-108.

Takahashi, K. 1974. Composition of Tertiary conglomerates in northern Hokkaido. Reports of the Geological Survey, Hokkaido, v. 46, p. 17-43.

Takano, K. and Hara, H. 1970. A preliminary analysis of current meter records. La mer (Bulletin of Franco-Japan Society on Oceanography), v. 8, p. 205-228.

Tsushima, K., Matsuno, K. and Yamaguchi, S. 1956. "Onishika'. Explanatory text of the geological map of Japan. Scale 1:50,000. Geological Survey of Japan, p. 1-17.

Walker, R.G. 1975. Generalized facies models for resedimented conglomerates of turbidite association. Geological Society of America Bulletin, v. 86, p. 737-748.

Walker, R.G. 1977. Deposition of upper Mesozoic resedimented conglomerates and associated turbidites in southwestern Oregon. Geological Society of America Bulletin, v. 88, p. 273-285.

Winn, R.D., Jr. and Dott, R.H., Jr. 1979. Deep-water fan-channel conglomerates of Late Cretaceous age, southern Chile. Sedimentology, v. 26, p. 203-228.

SEDIMENTATION UNITS IN STRATIFIED RESEDIMENTED CONGLOMERATE, PALEOCENE SUBMARINE CANYON FILL, POINT LOBOS, CALIFORNIA

H. Edward Clifton[1]

Abstract

Excellent exposures at Point Lobos, California provide a three-dimensional view of coarse-grained submarine canyon fill of Paleocene age. Although most of the conglomerate fill is stratified on a large scale (decimeters to several meters), individual sedimentation units can be clearly delineated only where sandstone interbeds isolate individual beds of conglomerate. Analysis of sandstone-conglomerate interbeds indicates that they commonly occur as couplets.

A typical couplet includes a lower conglomerate member with a sharp basal contact and an upper sandstone member with a poorly defined basal contact. Both members tend to be lenticular, generally less than a meter thick and no more than a few tens of meters in lateral extent. The conglomerate member typically is inversely graded and unstratified, whereas the sandstone member generally is normally graded and commonly shows stratification. Where seen in section parallel to paleotransport direction, the conglomerate coarsens in the down-transport direction until it terminates laterally as an accumulation of large clasts floating in the sandstone member. Clast long-axes that dip in an up-transport direction commonly define imbrication within the conglomerate. In contrast, clasts isolated in the sandstone member (and at the down-transport terminus of the conglomerate tend to lie with long axes normal to paleotransport.

The relation of the sandstone to the matrix of the conglomerate suggests that the couplet represents a single sedimentation unit, in which the pebbles moved as a traction carpet dispersed by intergranular collision. The textural gradations in the conglomerate and the general absence of small pebbles in the associated sandstone indicate nearly all of the gravel in the flow was deposited rapidly by frictional freezing. The associated sand flow had sufficient competence to roll a few of the larger clasts forward from the top and nose of the immobilized gravel bed. The size of the largest clasts (decimeters to meters) in the conglomerate suggests transport on relatively steep slopes. Forests within a few sandstone beds, however, that dip at near angle-of-repose in the general paleotransport direction, imply a very gentle gradient at the site of deposition. Comparison with other resedimented conglomerates indicates that the dominantly inversely graded conglomerate at Point Lobos represents a highly proximal facies.

Résumé

Les affleurements excellents de Point Lobos, Californie, fournissent une vue en trois dimensions de sédiments grossiers de remplissage d'un canyon sous-marin d'âge Paléocène. Quoique généralement, les matériaux de remplissage soient stratifiés à grande échelle (décimètres à plusieurs mètres), des unités sédimentaires individuelles peuvent être clairement délimitées lorsque les intercouches de grès isolent des couches individuelles de conglomérat. L'analyse des intercouches grès-conglomérat indiquent qu'elles se présentent communément comme des ensembles couplés.

Un ensemble couplé est formé d'un membre inférieur de conglomérat avec un contact très net à la base et un membre supérieur de grès avec un contact basal flou. Les deux membres tendent à être lenticulaires, généralement d'une épaisseur inférieure à un mètre, en n'ayant pas plus de quelques dizaines de mètres d'extension latérale. Le membre conglomératique est caractérisé par un granoclassement inverse et est non stratifié, tandis que le membre de grès est généralement à granoclassement normal et est fréquemment stratifié. Vue en section parallèle à la direction du paléocourant, le conglomérat devient plus grossier dans la direction avale de transport jusqu'à la limite où il forme latéralement une accumulation de fragments flottants dans le membre de grès. Les axes d'allongement des fragments inclinés dans la direction amont de transport définissent généralement l'imbrication dans le conglomérat. Contrairement, les fragments isolés au sein du membre du grès et à l'extrémité avale de transport du conglomérat tendent à s'aligner avec leurs axes d'allongement à angle droit au paléotransport.

La relation entre le grès et la matrice du conglomérat révèle que l'ensemble couplé représente un même unité sédimentaire, dans laquelle les galets ont été transportés dans une nappe par traction et furent dispersés par les chocs intergranulaires. Les gradations texturales dans le conglomérat et l'absence générale de petits cailloux dans le grès associé indiquent que presque tout le gravier de la coulée fut déposé rapidement par un blocage dû au frottement. La coulée de sable associée était suffisamment compétente pour rouler quelques uns des plus gros fragments de la partie supérieure et de l'extrémité de la couche de gravier immobilisée. La taille des plus gros fragments (décimètres à mètres) dans le conglomérat suggère un transport sur des pentes relativement abruptes. Les lits frontaux de quelques couches de grès, ces dernières cependant étant inclinées approximativement à l'angle de repos dans la direction générale du paléotransport, impliquent un gradient très faible à l'emplacement de l'accumulation des sédiments. La comparaison avec d'autres conglomérats resédimentés indique que le conglomérat avec prédominance d'un granoclassement inverse à Point Lobos représente un faciès hautement proximal.

Introduction

Numerous examples exist of conglomerate that has been carried by sediment gravity flows to depths below the range of effective wave-generated or tidal currents (see for example, summaries in Walker, 1975 and Stanley, 1980). Although such deep water conglomerates have been recognized nearly as long as have turbidites themselves (Natland and Kuenen, 1951), the mechanics of their transport and emplacement remain somewhat problematic. In part, this uncertainty derives from the difficulty of defining sedimentation units within the conglomerate. Many deep water conglomerates are visibly stratified into layers defined by variations in clast size; the bounding surfaces of these layers, however, are commonly obscure and can

---

[1] U.S. Geological Survey, 345 Middlefield Road, Menlo Park, California 94025, U.S.A.
  Many colleagues contributed useful observations, and raised important questions in the field; special acknowledgment is due here to Monty Hampton, Ralph Hunter, István Bérczi, and Roger Walker. Jennifer Bick and James Galloway assisted in the collection of field data and Ahmed Farah performed laboratory measurements of clast size and matrix proportion. The manuscript profited from valuable suggestions by Frances J. Hein and Ron J. Steel.

neither be precisely located nor traced laterally with confidence.

Sedimentation units of conglomerate can be better delineated where they are encased in sandstone. Such units are well exposed in a Paleocene submarine canyon fill at Point Lobos, California. This paper describes their geometry, contact relations, vertical and lateral textural gradations and patterns of clast orientation. The resulting model facilitates the recognition of less well-defined sedimentation units where they are amalgamated within the conglomerate and provides a basis for inferring the depositional processes involved. The paper also compares these deposits to other resedimented conglomerates.

The rocks in question crop out at Point Lobos, on the south side of Carmel Bay about 180 km south of San Francisco (Fig. 1). They are part of the Carmelo Formation of Bowen (1965), who assigned a Paleocene age to the unit on the basis of its fossil content. The rocks were initially attributed to deposition in shallow water, largely because of the abundance of coarse conglomerate and the presence of ripple marks on the surface of many of the sandstone beds (Herold, 1934). The first detailed sedimentologic study of the Carmelo Formation was made by Nili-Esfahani (1965) who on the basis of depositional structures inferred that sediment was deposited in deep water by a combination of gravity sliding, turbidity currents, traction currents and slumping.

In a further analysis of the deposit, I suggested that it formed within a submarine canyon (Clifton, 1981). The depositional contact with the underlying granodiorite is steep relative to stratification in the Paleocene deposits, broadly sinuous, and of substantial (at least tens of meters) relief. The walls are locally irregular and can be overhanging (Fig. 2). Paleocurrent indicators (pebble imbrication, ripple lamination, rare high-angle crossbedding) show a generally consistent pattern of transport parallel to the walls where they can be seen.

A poorly preserved gastropod in a mudstone clast was tentatively assigned by L.N. Marincovich to the genus *Tornatellaea*, which is known in Paleocene and possibly Eocene deposits elsewhere in California. In these occurrences it is associated with a fossil fauna that suggest water depths in the outer neritic zone (100-200 m) or deeper (L.N. Marincovich, writt. comm., 1980). This interpretation of paleobathymetry is supported by the sedimentary structures which show evidence only of unidirectional flow, thereby indicating deposition at depths below storm wave base, which is at 100 - 200 m on the present day Pacific Coast (Komar et al., 1972). On the other hand, the trace fossil assemblage suggests that deposition did not occur below mid-bathyal depths (Hill, 1981). Accordingly, the Carmelo Formation at Point Lobos appears to have been deposited in the upper reaches of a submarine valley, probably a canyon, at a depth of at least several hundred meters.

Depositional Features
Lithology

The Paleocene deposits at Point Lobos comprise a variety of lithologies. Conglomerate predominates at most of the outcrops (Fig. 3). Sandstone occurs as individual interbeds within the conglomerate, in amalgamated sets several meters thick, and as thin interbeds and laminae within mudstone. Mudstone forms thin interbeds separating sandstone layers or, more rarely, conglomerate and also occurs in intervals a few centimeters to more than 10 m thick, intercalating with thin sandstone layers. Mudstone intraclast breccia is a minor component of the sandstone and conglomerate. Pebbly mudstone, with deformed mudstone clasts and exotic pebbles scattered in a mudstone matrix, occurs locally.

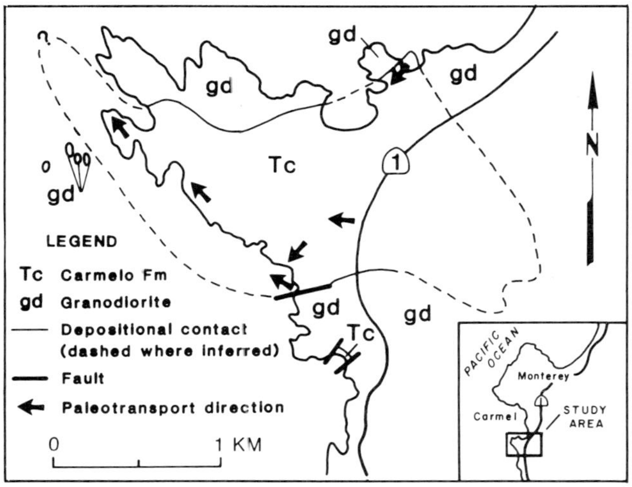

Fig. 1. Distribution of Paleocene deposits (the Carmelo Formation of Bowen, 1965) at Point Lobos. Arrows indicate paleotransport direction as determined from pebble imbrication, ripple lamination and rare high-angle crossbedding.

Fig. 2. Vertical-to-overhanging contact (above hammer) between conglomerate and granodiorite, northernmost exposure of the Paleocene at Point Lobos. Near-horizontal contact to left of hammer reflects actual attitude of stratification.

The conglomerate consists primarily of rounded to subrounded clasts of siliceous volcanic rock (Nili-Esfahani, 1965). These clasts range in size from less than a centimeter to several decimeters (Fig. 4). Granodiorite clasts are uncommon; where present, they generally are large (up to 2 m across) and subangular. The sandstone, which is mostly composed of angular grains of feldspar and quartz (Nili-Esfahani, 1965), appears to be more reflective of a local granodiorite source.

The conglomerate can be broadly categorized as disorganized or organized following Walker and Mutti (1973). Disorganized conglomerate shows neither internal textural grading nor stratification and its clasts show no preferred orientation (Fig. 5). The disorganized conglomerate commonly contains clasts of mudstone and some beds grade upward into pebbly mudstone.

Most of the conglomerate at Point Lobos is organized, in that it is crudely stratified into layers (Fig. 6) that may be inversely or, less commonly, normally graded. Most of the identifiable conglomerate beds are less than a meter thick. Channeling is prominent, particulary in sections transverse to paleotransport direction (Fig. 7). Although most clasts are flat-lying relative to stratification, imbrication is common (Figs. 4 and 8). Interbeds of sandstone, centimeters to decimeters thick, separate many of the beds or organized conglomerate. The sandstone beds typically lens out over a distance of a few tens of meters or less, and many contain isolated large clasts of conglomerate.

SEDIMENTATION UNITS AND SANDSTONE-CONGLOMERATE COUPLETS

The mechanics of transport and deposition of the organized conglomerates pose a generally unresolved problem, one that is exacerbated by the difficulty of defining sedimentation units within the conglomerate. Although contrasts in clast size define a stratification in the conglomerate, the differences generally are not striking enough to delineate bounding surfaces precisely (Figs. 4 and 6). Lateral or vertical textural change within conglomerate beds further

Fig. 3. Paleocene conglomerate at westernmost exposure at Point Lobos. Note contact relations with lighter colored granodiorite in the left background, whereby Paleocene strata on the right abut the steep depositional contact (arrow).

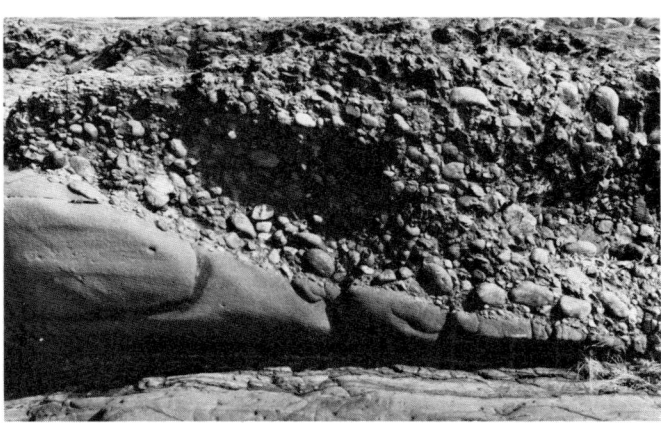

Fig. 5. Coarse disorganized conglomerate filling channel cut into sandstone. Pencil for scale near top of sandstone on left side of photograph.

Fig. 4. Coarse and fine conglomerate, Point Lobos. Note roundness of clasts (mostly siliceous volcanic rocks) and sub-horizontal orientation of largest cobbles.

Fig. 6. Organized conglomerate in exposure about 4 m high at westernmost exposure at Point Lobos. Note difficulty in defining sedimentation units, also scarcity and lenticularity of sandstone interbeds.

complicates their definition and the upper part of any conglomerate may be reworked into the basal part of an overlying bed. Such complications may be partly overcome by specifically examining conglomerate beds that are encased in sandstone, where the grain sizes differ sufficiently to define the units and indicate more clearly the effects of reworking.

In the examination of interbedded sandstone and conglomerate, a consistent association of lithologic characteristics emerged. Typically, conglomerate overlies sandstone with a well-defined basal contact, whereas the contact of sandstone over conglomerate is more poorly defined and commonly seems to grade into the matrix of the conglomerate (Fig. 9). For this reason, pairs of associated beds can be considered as couplets with an upper component of sandstone, typically centimeters to decimeters thick, and a lower component of conglomerate, typically decimeters thick.

Each component shows a different textural variation. Typically the conglomerate is inversely graded (Figs. 10-12) whereas the sandstone is normally graded. The normal grading within the sandstone is manifested in two ways. Not only does the sandstone bed itself commonly become progressively finer toward its top, but the sand matrix of the conglomerate also is graded, containing many small rock fragments 1-5 mm in diameter in the lower part that are absent higher in the bed (Fig. 13). Because samples cannot be collected in Point Lobos State Reserve, the texture of the sandstone was not analyzed. The position of this size change with the matrix differs from bed to bed. Although the proportion of matrix commonly appears to increase toward the top of the conglomerate bed (Fig. 11), measurement indicates that this is not always the case (Fig. 12A).

The internal structure and pebble orientation also differ within the two components of a couplet. The conglomerate lacks stratification, whereas the associated sandstone commonly shows planar lamination (Fig. 14). A few sandstone interbeds contain high-angle foresets, which dip in the direction of paleotransport as indicated by pebble

Fig. 8. Pebble imbrication in Paleocene conglomerate, Point Lobos. Compass points in direction of transport.

Fig. 9. Typical sandstone-conglomerate contact relations, Point Lobos. Note sharp contact of conglomerate with underlying sandstone, in contrast to ill-defined contact at base of sandstone.

Fig. 7. Section transverse to paleotransport direction, near westernmost Paleocene exposure at Point Lobos. Note large channels. Individual standing above cliff for scale.

Fig. 10. Inversely graded conglomerate encased in sandstone. Note dense packing of clasts. Note also gently inclined crossbeds in underlying sandstone that dip up-current (paleotransport right-to-left).

Fig. 11. Inversely graded conglomerate encased in sandstone. Note loose packing of clasts, particularly in upper part of bed. Paleotransport to left. Centimeter scale.

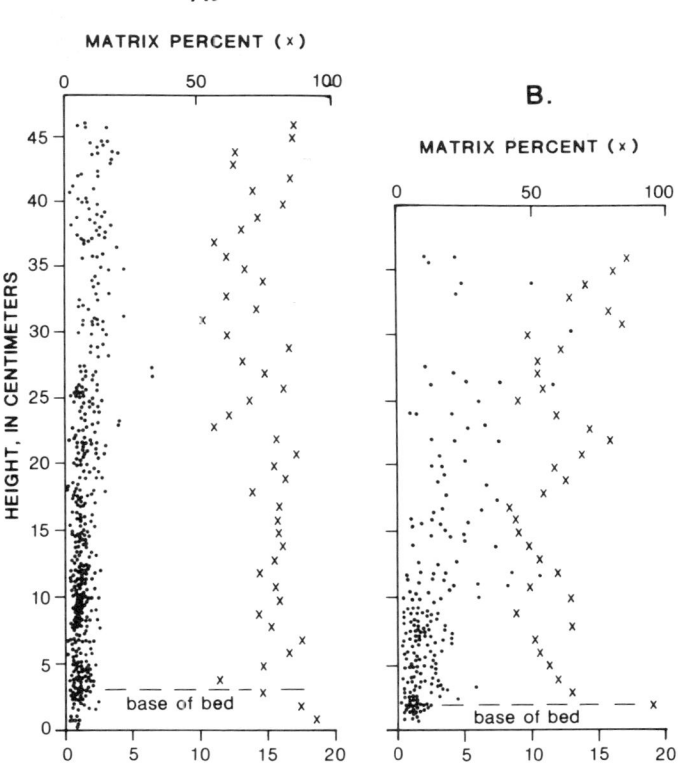

Fig. 12. **A.** Vertical change in clast size and matrix percent in conglomerate bed encased in sandstone. Data derived from measure of clast diameter as seen on photograph, and by measuring the percent of matrix along horizontal traverses spaced 1 cm apart. **B.** Vertical change in clast size and matrix percent in conglomerate bed shown in Figure 11. Note that, as in Figure 12A, inverse grading results from both an upward increase in clast size and an upward decline in the number of smaller pebbles. Note also that percent of matrix, unlike that shown in Figure 12A, increases rather consistently toward the top of the bed.

imbrication (Fig. 15). Low-angle foresets that dip in a direction opposite to paleotransport (Fig. 10) are not uncommon. Where pebbles or cobbles lie within a laminated sandstone, the laminations are deflected above and below the clast. Isolated cobbles within the sandstone bed commonly are larger than most of those in the associated conglomerate (Fig. 14). Within the conglomerate the pebbles are commonly aligned with long axes parallel to transport direction (Fig. 16), whereas the pebbles isolated within the sandstone are more likely to have long axes aligned transverse to flow direction (Fig. 17).

When viewed in sections parallel to transport direction, the conglomerate component of the couplet typically becomes progressively coarser laterally within a distance of several meters in a down-transport direction (Figs. 18 and 19). Such beds commonly terminate as a diminishing number of large clasts isolated in the overlying (and here laterally equivalent) sandstone bed (Fig. 18). Views completely transverse to transport direction are difficult to find in the sandstone-conglomerate couplets. The one example that could be clearly seen (Fig. 20) showed a lateral change over a distance of 3 m from inversely graded conglomerate to slightly thinner but coarser conglomerate in which grading, if present, was normal. The apparent attendant difference in clast orientation shown in Figure 20 may be due to exposure in transverse as opposed to longitudinal section relative to transport direction. Lateral textural variation was not noted in the associated sandstone.

The characteristics of the sandstone-conglomerate couplets suggest that the sedimentation units are fairly complex. The conglomerate component seems to occur as relatively small lobes, meters to tens of meters in length and width and seldom more than a meter thick. The conglomerate grades coarser from bottom to top, in a down-transport direction, and, possibly toward the margins in a direction transverse to transport direction. The sandstone component overlies the conglomerate and extends laterally from its downstream termination, typically for no more than a few tens of meters.

The character of the conglomerate provides a basis for perceiving sedimentation units within the organized conglomerate in the absence of sand interbeds. As noted, inverse grading is common, although the exact boundaries of the graded units are generally obscure. In sections parallel to transport, accumulations of unusually coarse conglomerate commonly grade progressively finer in an up-transport direction, but terminate abruptly against finer conglomerate at their down-transport end (Fig. 21). In sections parallel to bedding, many of the coarsest clasts at the down-transport termination are aligned with long axes transverse to flow direction (Fig. 22). The precise definition of sedimentation units remains a problem, however. With close scrutiny, one tends to subdivide the conglomerate into increasingly thinner units.

A fundamental question is whether the sandstone-conglomerate couplets themselves are sedimentation units,

Fig. 13. Textural changes in sandstone near base of sandstone interbed. Note increase in very coarse grains in matrix a short distance below top of conglomerate at the level of 1-4 cm on the scale.

Fig. 14. Graded and stratified sandstone between conglomerate beds. Note contact relations and presence of isolated 'outsized' cobbles within the sandstone.

Fig. 15. High-angle foreset bedding at top of sandstone interbed. Compass points in direction of paleotransport as indicated by abundant pebble imbrication elsewhere at this exposure.

Fig. 16. Long-axis orientation of clasts parallel to transport direction (indicated by compass). View is onto bedding plane surface of conglomerate seen in cross section in Figure 8.

Fig. 17. Long-axis orientation of clasts transverse to paleotransport direction (pencil) isolated in sandstone. Bedding plane exposure. Note lateral increase in clast size in the downtransport direction (to the right), also that a few of the clasts are sub-parallel to transport direction raising the possibility of a bimodal orientation.

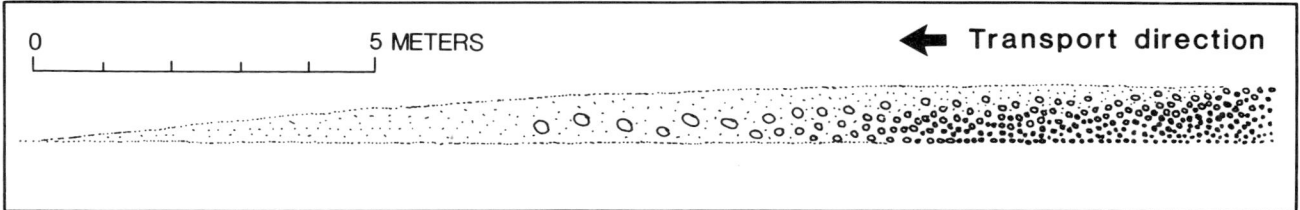

Fig. 18. Schematic representation of lateral change in a sandstone-conglomerate couplet relative to transport direction.

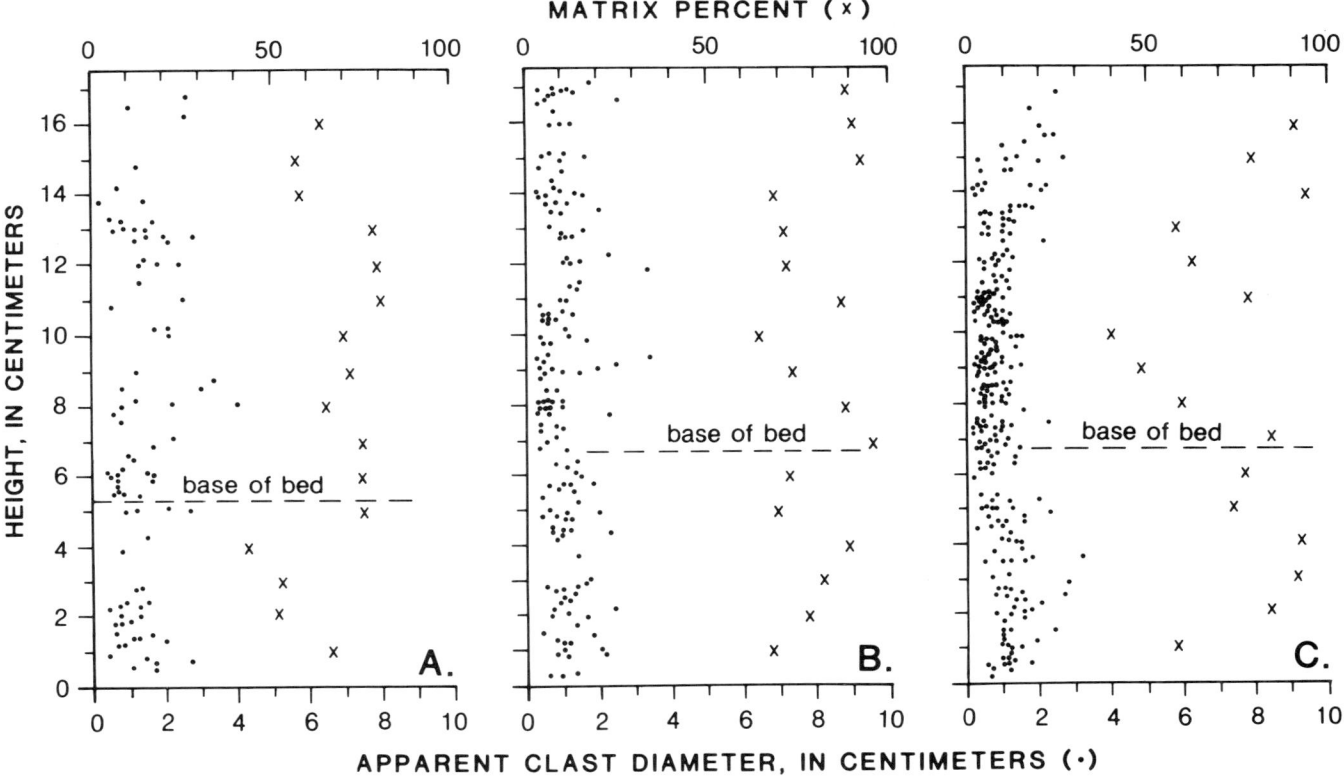

Fig. 19. Lateral change in clast size, matrix percent, and vertical variation of these parameters in a conglomerate bed exposed in section parallel to paleotransport direction. A. near down-transport terminus of conglomerate bed; B. 6 m up-transport and C. 9 m up-transport. Note down transport increase in clast size in bed for which the base is marked and down transport change from inverse grading to (if anything) normal grading. Note also that upward increase in matrix percent at location "C" disappears down transport.

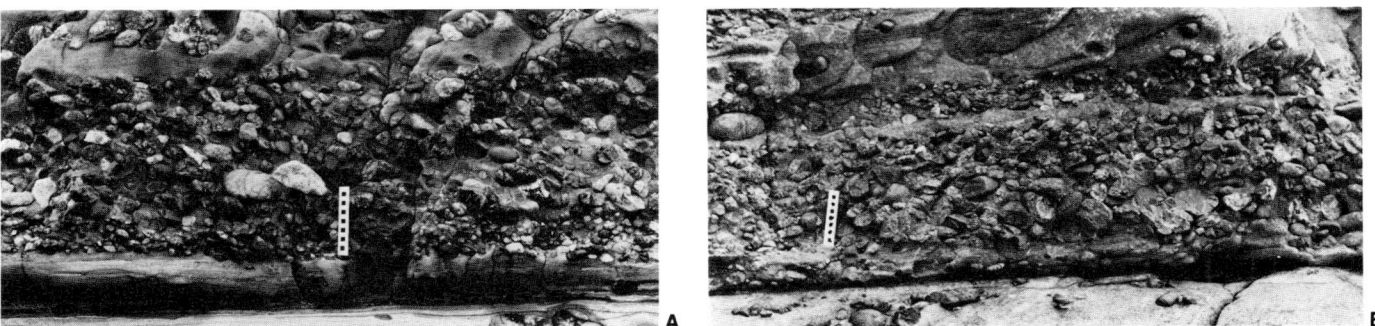

Fig. 20. Changes in conglomerate unit in section transverse to paleotransport direction. A. View of section parallel to transport direction through central part of unit. Note inverse grading and well-developed pebble imbrication; B. section transverse to paleotransport direction in same unit (marked by cm-divided ruler) 7 m in a direction normal to paleotransport from location of section figured in A. Note increased clast size and absence of inverse grading.

or whether the sandstone represents a subsequent (and possibly unrelated) event. The occurrence of numerous small 3-5 mm rock granules in the lower part of the conglomeratic matrix and their absence in the upper part of the matrix and in the overlying sandstone bed have several possible explanations. The granules may have been part of an admixture of sand and gravel that settled to the lower part of the flow as it moved. It is also possible that the granule-bearing matrix represents the sand that moved with the gravel, and that the overlying granule-free matrix filtered into place from a subsequent sand flow that lacked granules. Finally it is possible that the conglomerate was initially deposited with no matrix and that all of the matrix material was derived from a subsequent sand flow in which the initial deposition was granule-bearing sand that filtered to the bottom of the gravel bed.

Several lines of evidence suggest that all of the matrix material (and, accordingly, the overlying sandstone bed) was deposited simultaneously with the conglomerate. Nowhere does the conglomerate at Point Lobos lack matrix, even in a thick succession where sand interbeds are absent. Moreover, in such a succession the texture of the matrix material commonly differs perceptibly among the individual units of conglomerate, implying that each was accompanied by its own matrix material during emplacement or that each bed was followed by a possibly unrelated sand flow that emplaced the matrix as it otherwise bypassed an underlying conglomerate. The difference in texture between the matrix at the base of a conglomerate bed (Fig. 23) and the sandstone beneath the bed indicates that little of the matrix was derived by reworking of the sand from the underlying bed. Moreover the texture of the matrix commonly differs from bed to bed. These relations imply that every gravity flow of gravel had an accompanying volume of matrix-producing sand.

Mechanism of Emplacement

In terms of interpreting the processes of conglomeratic sedimentation in submarine canyons, the present is a very limited key to the past. Although conglomerate is considered a typical component of ancient submarine canyon fill (Nelson and Nilsen, 1974; Whitaker, 1974), the surface sediment on the floors of many modern canyons commonly consist predominantly of finer-grained sediment (c.f. Cacchione et al., 1978; Scott and Birdsall, 1978). Shepard and Dill (1966) report gravel as axial sediment in 18 of 38 canyons from which data were available, but the origin of the gravel remains problematic. In some canyons, such as Hudson Canyon, gravel in the upper reaches (to depths of about 150 m) is interpreted to have formed in shallow water during the late Pleistocene low stand (Stanley and Freeland, 1978) or to result from ice-rafting (Valentine et al., 1980). Localized patches of gravel in this canyon at greater depths (up to 450 m) is attributed to ice-rafting (Valentine et al., 1980). Rounded granules, pebbles, and cobbles of exotic composition have been collected from the axis of Monterey Canyon on the California coast at depths ranging from about 1400 to 3000 m (Martin, 1964). The absence of coarse sediment at shallower depths in the canyon axis led Martin to conclude that the gravel is derived from Pleistocene and younger sediment in the walls of the canyon.

The most common dynamic process in the axes of modern canyons seems to be currents that alternately flow up and down the canyon with velocities of several tens of centimeters per second (Keller and Shepard, 1978; Shepard and Marshall, 1978). The currents pulse in cycles that range in duration from 1 to 12 hours and seem to be at least partly induced by the tides. Evidence of such oscillatory flow is absent in the Paleocene deposits at Point Lobos, where all of the directional structures, from ripple lamination to pebble imbrication show a consistently unimodal direction of transport.

The other mechanism capable of transporting sand and coarser sediment within a canyon is by sediment gravity

Fig. 21. 'Nose' of coarse conglomerate sedimentation unit (under meter bar scale). Note abrupt lateral change to finer conglomerate on right. Paleotransport to right.

Fig. 22. Flow-transverse long-axis orientation of clasts near 'nose' of conglomerate sedimentation unit. View of bedding-plane section. Paleotransport direction (to right) indicated by pencil.

Fig. 23. Contrast in texture of sandy matrix at base of conglomerate and very coarse underlying sandstone. Note inverse-grading in conglomerate.

flow. Several descriptions exist of the episodic flow of sand at the heads of submarine canyons (Dill, 1964; Chamberlain, 1964; Marshall, 1978), but little is known about the processes that attend the movement of this material into deeper water. The best evidence for the mechanisms of transport and deposition of resedimented conglomerate continues to reside in the ancient analogues.

The nature of the sedimentation units provides a useful basis for interpreting the processes of emplacement of the gravel. The textural relations and clast orientation strongly suggest that the dispersive mechanism for the gravel was intergranular collision (Bagnold, 1954). This mechanism leads to the type of sediment gravity flow known as inertia flow (Sanders, 1965) or, subsequently, grain flow (Stauffer, 1967; Middleton and Hampton, 1973; Lowe, 1976a). In a grain flow, the larger clasts migrate to the top and front of a flow (Middleton and Hampton, 1976), producing a laterally and inversely graded deposit like that of the conglomerate sedimentation units at Point Lobos. The clasts in a grain flow deposit are aligned with long axes parallel to flow and imbricated up-flow (Rees, 1968) again like those in the body of the organized conglomerates at Point Lobos.

The sandstone component of the couplets appears to have been deposited by a mechanism other than grain flow. Specifically, the normal grading that is common to most of the sandstone beds is inconsistent with the textural sorting that accompanies grain flows. Accordingly, the sandstone appears to have been emplaced either by turbidity currents where fluid turbulence disperses the grains, or by fluidized flow, where upward movement of pore fluids supports the grains, or by some combination of the two processes (Middleton and Hampton, 1976; Lowe, 1976b).

If, as noted in the foregoing section, the couplets constitute sedimentation units it is possible that the gravel moved as a modified grain flow (Middleton, 1966) in the form of a 'traction carpet' beneath the sand flow as envisioned by Sanders (1965). Intergranular collisions between the pebbles and cobbles would create interstitial effects that would promote dispersal and transport and lead to the observed textural gradations. Such a situation exists on beach foreshores, where wave backwash causes the movement of dense clouds of sand above the bottom (Clifton, 1969). The grains in this cloud are sufficiently concentrated that the dispersive pressure from intergranular collision leads to inversely graded layers (Sallenger, 1979), but the impetus for transport is provided by the associated fluid gravity flow of the backwash. In the case of inversely graded conglomerate at Point Lobos, the actual impetus of flow may have been sediment gravity flow composed primarily of the sand fraction. The relatively high density that such a flow would impart to the fluid interstitial to the gravel would further promote its transport (Lowe, 1976a).

The pebbles isolated within the sand part of the couplet seem to have moved differently from those in the body of the conglomerate. The flow-parallel alignment of clasts in the conglomerate presumably reflects a mass flow process, whereas the flow-transverse orientation of those in the sand indicates emplacement by rolling (Johannson, 1963). The flow-transverse orientation of the coarsest pebbles at the front edge of a conglomerate unit also suggests that these clasts were rolling, at least just before deposition.

The disorganized conglomerates reflect a different mechanism of emplacement. The lack of organized fabric, internal textural gradation, or stratification, the upward transition of some disorganized conglomerates into pebbly mudstone, and the presence of outsized (up to 3 m) blocks of granodiorite in some beds indicate emplacement by debris flow whereby the shear strength of the matrix supports the clasts during transport (Middleton and Hampton, 1976). The debris flow deposits at Point Lobos typically are thicker and more irregular in shape than those emplaced by modified grain flow. In one place near the base of the Carmelo Formation on the northside of Point Lobos, gently dipping sandstone and conglomerate overlap onto a more steeply inclined upper surface of pebbly mudstone (Clifton, 1981). This relation suggests that the pebbly mudstone had surface relief of at least several meters.

Some beds seem to reflect an initial stage of transition from debris flow to another mechanism of transport. An example is a bed of pebbly sandstone, 50 cm thick, in which the pebbles a few centimeters in diameter are scattered through the sandstone. Vertical textural grading is absent from this bed, and stratification in the form of concentration of pebbles is faint at best. Nonetheless, the clasts clearly show a preferential orientation whereby long axes lie parallel to the bounding surfaces of the bed. A large (50 cm across) block of granodiorite floating in the upper part of this bed attests to a high level of competency that would normally be associated with a debris flow (Middleton and Hampton, 1976). The lack of grading and general absence of stratification also suggest a debris flow process; but the well-defined pebble orientation and faint hints of impersistent stratification locally within the bed indicate that the last phase, at least, of deposition involved a mechanism somewhat transitional between a debris flow and some other process, such as grain flow. One can speculate along the general line of Lowe (1982) that a fully disorganized debris flow might transform successively through phases that first show pebble orientation, then stratification and finally, if grain flow ensues, inverse grading. It is conceivable that many of the poorly defined sedimentation units in the organized conglomerate at Point Lobos in which inverse grading is not clearly developed reflect intermediate phases in the transformation between the 'end member' processes of debris flow and grain flow. It is also possible that some of these 'intermediate' units represent sections through the margins of a gravel grain flow unit, where the inverse grading may be less well defined.

Comparison with Other Resedimented Conglomerates

The organized conglomerates at Point Lobos bear both similarity and dissimilarity to resedimented conglomerates from other locations as described by other researchers. In developing generalized facies models for resedimented conglomerate, Walker (1975) postulated that the three dominant associations of lithologic features are inverse-to-normally graded, graded-stratified and disorganized-bed. At Point Lobos, disorganized conglomerates are common and graded-stratified beds relatively rare. Examples of apparently inverse-to-normally graded beds are present within the conglomerate, but in every case it is unclear whether the bed is truly an inverse-to-normally graded sedimentation unit or the result of two superimposed inversely graded units. On the other hand, the clearly defined conglomerate sedimentation units consistently show only inverse grading. Walker (1975) implies that inverse grading by itself is relatively uncommon in description of resedimented conglomerates. Surlyk (1978) describes no inversely graded beds (other than those showing inverse-to-normal grading) in a detailed analysis of Late Jurassic and Early Cretaceous deep-water conglomerate exposed on East Greenland. In contrast, I would infer here that it predominates in the organized conglomerates at Point Lobos. Similarly, Lowe (1982) shows numerous examples of well-defined beds in resedimented conglomerate that show only inverse grading.

The reason for this apparent difference in grading mode is uncertain. It may derive partly from the difficulty of delineating sedimentation units in interbedded conglomerate at Point Lobos and elsewhere. It is unlikely that conglomerate that is deposited with interbeds of sandstone (the basis for delineating most of the sedimentation units at Point Lobos) differs fundamentally in character of grading from that deposited without sandstone interbeds; at Point Lobos the few well-defined sedimentation units that lie within interbedded conglomerate are inversely graded. The prevalence of inversely graded beds in the conglomerate at Point Lobos might be due to the coarse size of the clasts. Kelling and Holroyd (1978) in a comparison of several deep-water conglomerates note that the type of grading depends in part on the grain size of the conglomerate; inverse grading is more prevalent in coarser, cobble-rich deposits. Davies and Walker (1974) note a similar relationship. The conglomerate, however, described by Surlyk (1978) as lacking inverse grading seems to be as coarse as that at Point Lobos, and many of the inversely graded beds at Point Lobos are relatively fine-grained. Finally, it is possible that predominantly inversely graded resedimented conglomerates originate in special paleogeographic/sedimentologic settings that somehow differ from those which produce inverse-to-normally graded and graded stratified beds.

It should be noted that the sandstone-conglomerate couplets at Point Lobos do display inverse-to-normal grading in the sense that the conglomerate is inversely graded and the overlying sandstone is normally graded. If one, however, considers the sandstone as an entity from matrix to overlying bed, it displays a normal grading that is superimposed on the inverse grading of the gravel. For the majority of conglomerate beds at Point Lobos, which lack a capping sandstone, this type of inverse-to-normal grading is absent.

Sandstone-conglomerate couplets in which the conglomerate is inversely graded and the sand normally graded are common in resedimented conglomerates in the Devonian Ksiaz Formation of southwestern Poland (Nemec *et al.*, 1980; Porębski, this volume). Where sandstone is interstratified with this conglomerate, the matrix percent typically increases upward within the conglomerate and the upper sandstone contacts are sharper than the lower boundaries, relations similar to those observed at Point Lobos. Rocheleau and Lajoie (1974) describe similar contact relations between sandstone and conglomerate in a Cambrian resedimented conglomerate at L'Islet, Québec, but note that normal grading dominates within the conglomerate beds.

High-angle foreset bedding in sandstone associated with the conglomerate has been reported from other resedimented water conglomerate (*e.g.*, Piper, 1970; Rocheleau and Lajoie, 1974; Hendry, 1976; Winn and Dott, 1977, 1978 and 1979; Kessler and McHargue, 1978; Hein and Walker, 1982). At Point Lobos the foreset units are rare and typically less than a few centimeters thick. Rocheleau and Lajoie (1974) found this structure to be more common (in 26% of the mapped beds) and thicker (10-40 cm). Large scale (up to 40 m) dune-like crossbedding in the conglomerate as described in upper Cretaceous deepsea fan deposits in southern Chile by Winn and Dott (1977) has not been observed at Point Lobos.

A consistent increase in clast size laterally toward the front of the resedimented conglomeratic units has not been reported previously. Johnson and Walker (1979) describe horizontal textural gradation in beds of Cambro-Ordovician resedimented conglomerate in Quebec, but could find no directional trends relative to paleoflow. Downstream coarsening in other deep water conglomerates may be overlooked owing to the requirement for good exposures in sections parallel to transport and because of its general subtlety. I examined the Point Lobos conglomerates at length before noting this lateral gradation; once observed it was fairly evident throughout the section.

The flow-parallel alignment and imbrication of clast long axes observed within the organized conglomerate at Point Lobos are characteristic of resedimented conglomerates (Walker, 1975). The flow-transverse alignment of long-axes displayed by clasts floating in sand or at the down-transport margin of conglomerate sedimentation units at Point Lobos is reported from few other deep water conglomerates (Walker, 1975). Hein (1982) describes bimodal orientations in the upper, finer parts of graded-stratified conglomerate of Cambro-Ordovician age in Quebec in which some clasts lie with long axes transverse to flow direction. Piper (1970) notes transverse orientation of clast long axes in a Silurian deep sea fan deposit in western Ireland. Some of the imbricated cobbles in this deposit are isolated in sandstone, but their long axis orientation is unspecified. These conglomerates differ from those at Point Lobos by being generally ungraded and having sharp upper contacts.

Some of the similarities and differences between the conglomerate at Point Lobos and other resedimented conglomerates can be explained by the model proposed on the basis of detailed study of deep-water conglomerates of Devonian age in southwestern Poland (Nemec et al., 1980; Porębski, this volume). In their model, the proximal-most deposits consist mostly of ungraded (disorganized) and inversely graded conglomerate and the distal-most deposits consist primarily of normally graded and graded-stratified conglomerates. In intermediate deposits, inversely to normally graded conglomerate may be particularly important. Concommitant with these changes, the maximum clast size and bed thickness decline in a down-transport direction and the abundance of sandstone 'cappings' (presumably the upper part of the couplets described in this report) increases. This model resembles that postulated by Walker (1975) in which disorganized conglomerates dominate within canyons or other channels into the basin. In these, the inversely graded part of inverse-to-normally graded beds becomes thinner and more poorly developed in a down-transport direction, in contrast to normal grading and stratification which become better developed in the upper part of the bed.

It is uncertain whether the transitions postulated above result from the evolution of individual flows during transport, or from the size of the available clasts and the competence of the flow (a function in part of slope) or from some combination of the two. Lowe (1982) proposes the following evolutionary continuum within individual flows: cohesive flow (producing disorganized conglomerate) to grain flow (producing inversely graded conglomerate) to high density turbidity current (successively producing inverse-to-normally graded, normally graded and finally graded-stratified conglomerate). However, the sedimentation units at Point Lobos with their well-defined coarse, downstream terminations appear to represent the freezing of the coarse component of a flow. It is likely that the associated sand flow continued down canyon, but the typical absence of small clasts in the upper part of the inversely graded conglomerate, or associated with the large clasts rolled forward from the front of the conglomerate, suggests that little if any gravel moved on with the sand flow.

The dispersive pressure in a flowing concentration of colliding grains differs with the square of the diameter of the clasts (Bagnold, 1954). The apparent relationship between coarse clast size and inverse grading (Kelling and Holroyd, 1978; Porębski, this volume) may reflect the increased dispersive pressured generated by the larger clasts. Conceivably, if the same flow carried only small pebbles, dispersive pressure might be subordinate to fluid turbulence as a means for dispersing the clasts, resulting in an inverse-to-normal or normally graded deposit. The size at which inertial collisional effects become dominant as a dispersive mechanism probably depends on the velocity, turbulence and density of the associated sand flow and the concentration of pebbles in that flow (Lowe, 1982). Inversely graded concentrations of granules and pebbles less than 1 cm in diameter at the base of some sandstone strata in the Annot sandstone (Stanley et al., 1978) attest to effective dispersive pressures in a fine traction carpet.

Discussion and Conclusions

The Paleocene deposits at Point Lobos consist primarily of inversely graded and disorganized conglomerates. As such they appear to be a highly proximal facies. The coarseness of the deposit and abundance of inverse grading suggest transport on relatively substantial slopes, although the high angle foresets in interbedded sandstone implies a nearly flat gradient at the site of deposition. It is likely that the conglomerate at Point Lobos accumulated at a break in slope along the canyon axis. The coarse traction carpet of a sediment gravity flow would freeze as the flow lost competence at this point (Lowe, 1982), producing the observed conglomerate sedimentation units. The associated sand-dominant flow continued down canyon to some depositional site in deeper water. Conceivably, flows with a finer gravel component might bypass the part of the canyon represented by Point Lobos and be deposited in graded or graded-stratified beds in more distal location.

One can only speculate on the origin of the sediment gravity flows that produced the Point Lobos conglomerates.

The roundness and polymict composition of the clasts indicate a complicated transport history prior to the initiation of the flows. Gravels similar in size, shape, and variability of composition presently accumulate on fan-deltas at the mouths of ephemeral streams on the southern coast of Spain. Deposition on these fan-deltas is episodic and rapid. As a consequence of floods during January and February, 1974, the fan-delta at the mouth of the Rio Adra advanced several tens of meters into water approaching 20 m deep. A few of the streams on the Mediterranean coast of Spain discharge directly into the throat of a submarine canyon. Diving with scuba at the mouth of the Rio Carboneras, we observed a sharp break in slope into a submarine canyon just seaward from a small submarine platform composed of boulder and cobble gravel. Bricks and other coarse artificial debris on the platform and embedded in sand on the slopes of the canyon attest to the present-day transport of coarse material into the canyon head. A major flood on the Rio Carboneras almost certainly would quickly deposit a large quantity of coarse gravel on the lip of this canyon, where it would be susceptible to mass failure and down-canyon transport by ensuing sediment gravity flows. A similar process is observed on the Yallahs fan-delta on southeastern Jamaica, where several submarine canyons that head onto the fan-delta are floored with coarse gravel (Wescott and Ethridge, 1980).

A number of descriptions exist of conglomerate in ancient submarine canyons (*cf.* Walker, 1975, Table 1) but relatively few provide details as to the degree and nature of the organization of the conglomerate. Disorganized conglomerates predominate in the canyon-fill facies of the Annot sandstone (Stanley, 1975) and graded beds are generally poorly developed, although both normal and inverse grading can be seen (Stanley, 1975; Stanley et al., 1978). The conglomerates in possible canyon fill sequences of Jurassic age in northern Tunisia are thick (up to 75 m), normally graded but otherwise disorganized debris flow deposits (Cossey and Erhlich, 1978). In submarine canyon or slope-channel deposits of Late Cretaceous age exposed on the central California coast, imbricated and inversely graded conglomerate was noted in association with disorganized conglomerate and pebbly mudstone (Lowe, 1972).

The available information supports Walker's (1975) premise that disorganized conglomerate predominates in the submarine canyon environment. The complexity, however, of the submarine canyon deposits at Point Lobos and the predominance of inversely graded beds there, demonstrate the potential danger in overgeneralizing about the nature of the deposits.

References

Bagnold, R.A. 1954. Experiments on a gravity-free dispersion of large solid spheres in a Newtonian fluid under shear. Proceedings of the Royal Society of London, Series A, v. 225, p. 49-63.

Bowen, O.E. 1965. Stratigraphy, structure and oil possibilities in Monterey and Salinas Quadrangles, California. *In:* Rennie, E.W., Jr. (Ed.), Symposium of Papers, Pacific Section, American Association of Petroleum Geologists, Bakersfield, p. 48-69.

Cacchione, D.A., Rowe, G.T. and Malahoff, A. 1978. Submersible investigation of outer Hudson Submarine Canyon: *In:* Stanley, D.J. and Kelling, G. (Eds.), Sedimentation in Submarine Canyons, Fans and Trenches. Stroudsburg: Dowden, Hutchinson and Ross, p. 42-50.

Chamberlain, T.K. 1964. Mass transport of sediment in the heads of Scripps Submarine Canyon, California. *In:* Miller, R.L. (Ed.), Papers in Marine Geology. New York: MacMillan, p. 42-64.

Clifton, H.E. 1969. Beach lamination: nature and origin. Marine Geology, v. 7, p, 553-559.

Clifton, H.E. 1981. Submarine canyon deposits, Point Lobos, California. *In:* Frizzel, V. (Ed.), Upper Cretaceous and Paleocene Turbidites, Central California Coast. Pacific Section, Society of Economic Paleontologists and Mineralogists, Guide Book to Field Trip No. 6, p. 79-92.

Cossey, S.P. and Ehrlich, R. 1978. Growth fault-controlled submarine carbonate debris flow and turbidite deposits from the Jurassic of northern Tunisia: possible submarine canyon fill sequences. *In:* Stanley, D.J. and Kelling, G. (Eds.), Sedimentation in Submarine Canyons, Fans and Trenches. Stroudsburg: Dowden, Hutchinson and Ross, p. 127-137.

Davies, I.C. and Walker, R.G. 1974. Transport and deposition of resedimented conglomerates: the Cap Enragé Formation, Cambro-Ordovician, Gaspé, Québec. Journal of Sedimentary Petrology, v. 44, p. 1200-1216.

Dill, R.F. 1964. Sedimentation and erosion in Scripps Submarine Canyon head. *In:* Miller, R.L. (Ed.), Papers in Marine Geology. New York: MacMillan, p. 23-41.

Hein, F.J. 1982. Depositional mechanisms of deep-sea coarse clastic sediments, Cap Enragé Formation, Québec. Canadian Journal of Earth Sciences, v. 19, p. 267-287.

Hein, F.J. and Walker, R.G. 1982, The Cambro-Ordovician Cap Enragé Formation, Québec, Canada: conglomeratic deposits of a braided submarine channel with terraces. Sedimentology, v. 29, p. 309-329.

Hendry, H.E. 1976. The orientation of discoidal clasts in resedimented conglomerates, Cambro-Ordovician, Gaspé, eastern Québec. Journal of Sedimentary Petrology, v. 46, p. 48-55.

Herold, C.L. 1934. Fossil markings in the Carmelo Series (Upper Cretaceous[?]), Point Lobos, California. Journal of Geology, v. 42, p. 630-640.

Hill, G.W. 1981. Ichnocoenoses of a Paleocene submarine-canyon floor, Point Lobos, California. *In:* Frizzell, V. (Ed.), Upper Cretaceous and Paleocene Turbidites, Central California Coast. Pacific Section, Society of Economic Paleontologists and Mineralogists, Guide Book to Field Trip No. 6, p. 93-104.

Johnson, B.A. and Walker, R.G. 1979. Paleocurrents and depositional environments of deep water conglomerates in the Cambro-Ordovician Cap Enragé Formation, Québec Appalachians. Canadian Journal of Earth Sciences, v. 16, p. 1375-1387.

Johannson C.E. 1963. Orientation of pebbles in running water. A laboratory study. Geografiska Annaler, v. 45, p. 85-112.

Keller, G.H. and Shepard, F.P. 1978. Currents and sedimentary processes in submarine canyons off the northeast United States. *In:* Stanley, D.J. and Kelling, G. (Eds.), Sedimentation in Submarine Canyons, Fans and Trenches. Stroudsburg: Dowden, Hutchinson and Ross, p. 15-32.

Kelling, G. and Holroyd, J. 1978. Clast size, shape, and composition in some ancient and modern fan gravels. *In:* Stanley, D.J. and Kelling, G. (Eds.), Sedimentation in Submarine Canyons, Fans and Trenches. Stroudsburg, Pennsylvania: Dowden, Hutchinson and Ross, p. 138-159.

Kessler, L.G. and McHargue, T.R. 1978. Depositional processes in submarine canyon and fan channels — some striking similarities to fluvial and delta plain environments. Geological Association of Canada/Geological Society of America 1978 Joint Annual Meeting, Toronto, Abstracts with Programs, v. 10, no. 7, p. 434.

Komar, P.D., Neudeck, R.H. and Kulm, L.D. 1972. Observations and significance of deep-water oscillatory ripple marks in the Oregon continental shelf. *In:* Swift, D.J.P., Duane, D.B. and Pilkey, O.H. (Eds.), Shelf Sediment Transport-Process and Pattern. Stroudsburg, Pennsylvania: Dowden, Hutchinson and Ross, p. 601-619.

Lowe, D.R. 1972. Submarine canyon and slope channel sedimentation model as inferred from Upper Cretaceous deposits, western California.

24th International Geological Congress, Montreal, Section 6, Stratigraphy and Sedimentology, p. 75-81.

Lowe, D.R. 1976a. Grain flow and grain flow deposits. Journal of Sedimentary Petrology, v. 36, p. 188-199.

Lowe, D.R. 1976b. Subaqueous liquefied and fluidized sediment flows and their deposits. Sedimentology, v. 23, p. 285-308.

Lowe, D.R. 1982. Sediment gravity flows: II. Depositional models with special reference to the deposits of high-density turbidity currents. Journal of Sedimentary Petrology, v. 52, p. 279-298.

Marshall, N.F. 1978. Large storm-induced slump reopens an unknown Scripps submarine canyon tributary. In: Stanley, D.J. and Kelling, G. (Eds.), Sedimentation in Submarine Canyons, Fans and Trenches. Stroudsburg, Pennsylvania: Dowden, Hutchinson and Ross, p. 73-84.

Martin, B.D. 1964. Monterey submarine canyon, California: genesis and relation to continental geology. Ph.D. Thesis, University of Southern California, Los Angeles, 249p.

Middleton, G.V. 1966. Experiments on density and turbidity currents: I. Motion of the head. Canadian Journal of Earth Sciences, v. 3, p. 523-546.

Middleton, G.V. and Hampton, M.A. 1973. Sediment gravity flows: mechanics of flow and deposition. In: Middleton, G.V. and Bouma, A.H. (Co-chairmen), Turbidites and Deep-Water Sedimentation. Pacific Section, Society of Economic Paleontologists and Mineralogists, Anaheim, Short Course Notes, p. 1-38.

Middleton, G.V. and Hampton, M.A. 1976. Subaqueous sediment transport and deposition by sediment gravity flows. In: Stanley, D.J. and Swift, D.J.P. (Eds.), Marine Sediment Transport and Environmental Management. New York: John Wiley and Sons, p. 197-218.

Natland, M.L. and Kuenen, Ph.H. 1951. Sedimentary history of the Ventura basin, California and the action of turbidity currents. In: Hough, J.L. (Ed.), Turbidity currents and the Transportation of Coarse Sediments to Deep Water. Society of Economic Paleontologists and Mineralogists, Special Publication 2, p. 76-107.

Nelson, C.H. and Nilsen, T.H. 1974. Depositional trends of modern and ancient deep-sea fans. In: Dott, R.H., Jr. and Shaver, R.H. (Eds.), Modern and Ancient Geosynclinal Sedimentation. Society of Economic Paleontologists and Mineralogists, Special Publication 19, p. 69-91.

Nemec, W., Porębski, S.J. and Steel, R.J. 1980. Texture and structure of resedimented conglomerates: examples from Książ Formation (Famennian-Tournaisian), southwest Poland. Sedimentology, v. 27, p. 519-538.

Nili-Esfahani, A. 1965. Investigation of Paleocene strata, Point Lobos, Monterey County, California. M.A. Thesis, University of California, Los Angeles, 228p.

Piper, D.J.W. 1970. A Silurian deep sea fan deposit in western Ireland and its bearing on the nature of turbidity currents. Journal of Geology, v. 78, p. 509-522.

Rees, A.I. 1968. The production of preferred orientation in a concentrated dispersion of elongated and flattened grains. Journal of Geology, v. 76, p. 457-465.

Rocheleau, M. and Lajoie, J. 1974. Sedimentary structures in resedimented conglomerate of the Cambrian flysch, L'Islet, Québec, Appalachians. Journal of Sedimentary Petrology, v. 44, p. 826-836.

Sallenger, A.H. 1979. Inverse grading and hydraulic equivalence in grain flow deposits. Journal of Sedimentary Petrology, v. 49, p. 553-562.

Sanders, J.E. 1965. Primary sedimentary structures formed by turbidity currents and related sedimentation mechanisms. In: Middleton, G.V. (Ed.), Primary Sedimentary Structures and their Hydrodynamic Interpretation. Society of Economic Paleontologists and Mineralogists, Special Publication 12, p. 192-219.

Scott, R.M. and Birdsall, B.C. 1978. Physical and biogenic characteristics of sediments from Hueneme Submarine Canyon, California coast. In: Stanley, D.J. and Kelling, G. (Eds.), Sedimentation in Submarine Canyons, Fans and Trenches. Stroudsburg, Pennsylvania: Dowden, Hutchinson and Ross, p. 51-64.

Shepard, F.P. and Dill, R.F. 1966. Submarine Canyons and Other Sea Valleys. Chicago: Rand McNally, 381p.

Shepard, F.P. and Marshall, N.F. 1978. Currents in submarine canyons and other sea valleys. In: Stanley, D.J. and Kelling, G. (Eds.), Sedimentation in Submarine Canyons, Fans and Trenches. Stroudsburg, Pennsylvania: Dowden, Hutchinson, and Ross, p. 3-14.

Stanley, D.J. 1975. Submarine canyon and slope sedimentation (Gres D'Annot) in the French Maritime Alps. 9th International Sedimentological Congress, Nice, France, Field Trip Guidebook, 129p.

Stanley, D.J. 1980. The Saint-Antonin conglomerate in the Maritime Alps: a model for coarse sedimentation on a submarine slope. Smithsonian Contributions to the Marine Sciences, no. 5, 25p.

Stanley, D.J. and Freeland, G.L. 1978. The erosion-deposition boundary in the head of Hudson submarine canyon defined on the basis of submarine observations. Marine Geology, v. 26, p. M37-M46.

Stanley, D.J., Palmer, H.D. and Dill, R.F. 1978. Coarse sediment transport by mass flow and turbidity current processes and downslope transformations in Annot Sandstone canyon-fan valley systems. In: Stanley, D.J. and Kelling, G. (Eds.), Sedimentation in Submarine Canyons, Fans and Trenches. Stroudsburg, Pennsylvania: Dowden, Hutchinson and Ross, p. 85-115.

Stauffer, P.H. 1967. Grain-flow deposits and their implications, Santa Ynez Mountains, California. Journal of Sedimentary Petrology, v. 37, p. 487-508.

Surlyk, F. 1978. Submarine fan sedimentation along fault scarps on tilted fault blocks, Jurassic-Cretaceous boundary, East Greenland. Grønlands Geologiske Undersøgelse Bulletin 128, 108p.

Valentine, P.L., Uzmann, J.R. and Cooper, R.A. 1980. Geology and biology of Oceanographer Submarine Canyon. Marine Geology, v. 38, p. 283-312.

Walker, R.G. 1975. Generalized facies models for resedimented and conglomerates of turbidite association. Geological Society of America Bulletin, v. 86, p. 737-748

Walker, R.G. and Mutti, E. 1973. Turbidite facies and facies association. In: Middleton, G.V. and Bouma, A.H. (Co-chairmen), Turbidites and Deep-Water Sedimentation. Pacific Section, Society of Economic Paleontologists and Mineralogists, Anaheim, Short Course Notes, p. 119-157.

Wescott, W.A. and Ethridge, F.A. 1980. Fan-delta sedimentology and tectonic setting - Yallahs fan delta, southeast Jamaica. American Association of Petroleum Geologists Bulletin, v. 64, p. 374-399.

Whitaker, J.H. McD. 1974. Ancient submarine canyons and fan valleys. In: Dott, R.H., Jr. and Shaver, R.H. (Eds.), Modern and Ancient Geosynclinal Sedimentation. Society of Economic Paleontologists and Mineralogists, Special Publication 19, p. 106-125.

Winn, R.D., Jr. and Dott, R.H., Jr. 1977. Large-scale traction-produced structures in deep-water fan-channel conglomerates in southern Chile. Geology, v. 5, p. 41-44.

Winn, R.D., Jr. and Dott, R.H., Jr. 1978. Submarine-fan turbidites and resedimented conglomerates in a Mesozoic arc-rear marginal basin in southern South America. In: Stanley, D.J. and Kelling, G. (Eds.), Sedimentation in Submarine Canyons, Fans and Trenches. Stroudsburg, Pennsylvania: Dowden, Hutchinson and Ross, p. 362-373.

Winn, R.D., Jr. and Dott, R.H. 1979. Deep-water fan-channel conglomerates of Late Cretaceous age, southern Chile. Sedimentology, v. 26, p. 203-228.